C0-AVX-617

INTRODUCTION TO CONVENTIONAL TRANSMISSION ELECTRON MICROSCOPY

Over the past 70 years transmission electron microscopy has developed into a sophisticated sub-nanometer-resolution imaging technique. This book covers the fundamentals of *conventional* transmission electron microscopy (CTEM) as applied to crystalline solids. Emphasis is on the experimental and computational methods used to quantify and analyze CTEM observations. A supplementary website containing interactive modules and free Fortran source code accompanies the text.

The book starts with the basics of crystallography and quantum mechanics providing a sound mathematical footing in both direct and reciprocal space. The next section deals with the microscope itself: lenses, correction coils, the electron gun, apertures, and electron detectors are all described in terms of the underlying theory. The second half of the book focuses on the dynamical theory of electron scattering in solids, including its applications to perfect and defective crystals, electron diffraction, and phase contrast techniques. Nearly all the electron micrographs in the book are taken from four study materials: Cu-15 at% Al, Ti, GaAs, and $BiTiO_3$, and detailed instructions for the preparation of thin foils are included. Detailed algorithm descriptions are included for a variety of computational problems, ranging from electron diffraction zone axis patterns to dynamical N-beam Bloch wave simulations.

Important features of this textbook include:

- highly illustrated and contains over 100 homework problems;
- supplementary website containing more than 30 000 lines of Fortran 90 source code;
- on-line interactive modules allowing the reader to try out real-time simulations.

The book is based on a lecture course given by the author in the Department of Materials Science and Engineering at Carnegie Mellon University, and is ideal for advanced undergraduate and graduate students as well as researchers new to the field.

MARC DE GRAEF was born in Antwerp (Belgium) on April 7, 1961. He studied physics at the University of Antwerp, and graduated with a Ph.D. in Physics from the Catholic University of Leuven in 1989. He was a post-doctoral researcher at the University of California at Santa Barbara before becoming a faculty member in the Department of Materials Science and Engineering at Carnegie Mellon University (Pittsburgh, PA). He is currently a Full Professor and co-director of the *J. Earle and*

Mary Roberts Materials Characterization Laboratory, which houses five TEMs and two SEMs. Professor De Graef has published more than 100 papers in reviewed journals, such as: *Ultramicroscopy, Micron, Journal of Microscopy, Philosophical Magazine, Physical Review, Journal of Applied Physics, Acta Materialia, Scripta Materialia*, and the *Journal of Solid State Chemistry*. In 1996 he received the George Tallman Ladd Research Award from the CMU Engineering College, and in 1998 he received the R.E. Peterson Award from the Society for Experimental Mechanics for Best Paper. He has co-edited a book on the characterization of magnetic materials in the series *Experimental Methods in the Physical Sciences* (Academic Press, 2000).

INTRODUCTION TO CONVENTIONAL TRANSMISSION ELECTRON MICROSCOPY

MARC DE GRAEF

PUBLISHED BY THE PRESS SYNDICATE OF THE UNIVERSITY OF CAMBRIDGE
The Pitt Building, Trumpington Street, Cambridge, United Kingdom

CAMBRIDGE UNIVERSITY PRESS
The Edinburgh Building, Cambridge CB2 2RU, UK
40 West 20th Street, New York, NY 10011-4211, USA
477 Williamstown Road, Port Melbourne, VIC 3207, Australia
Ruiz de Alarcón 13, 28014 Madrid, Spain
Dock House, The Waterfront, Cape Town 8001, South Africa

http://www.cambridge.org

First published 2003

Printed in the United Kingdom at the University Press, Cambridge

Typeface Times 11/14 pt *System* LaTeX 2_ε [TB]

A catalogue record for this book is available from the British Library

Library of Congress Cataloguing in Publication data

De Graef, Marc
Introduction to conventional transmission electron microscopy / Marc De Graef.
p. cm.
Includes bibliographical references.
ISBN 0 521 62006 6 (hb.) – ISBN 0 521 62995 0 (pb.)
1. Transmission electron microscopy. I. Title.
QH212.T7 D4 2003
620.1′1299–dc21 2002073926

ISBN 0 521 62006 6 hardback
ISBN 0 521 62995 0 paperback

for Pieter and Erika

Contents

Preface

About 70 years have passed since Ernst Ruska and Max Knoll constructed the first transmission electron microscope in Germany. During that time, the instrument has steadily developed from a microscope that barely matched the resolution of an optical microscope to a sophisticated and indispensable tool for materials research. Our theoretical understanding of the interaction of high-energy electrons with a solid thin foil and the subsequent propagation of those scattered electrons through a complex electron–optical system has also advanced significantly, to the point that, in principle at least, it is now possible to predict and simulate every possible image and imaging mode. The advent of affordable high-speed computing power has much to do with the transition from a qualitative characterization tool to a highly sensitive and quantitative research instrument.

This book has grown out of the lecture notes for the course "Introduction to Electron Optical Methods", taught every year or so in the Department of Materials Science and Engineering at Carnegie Mellon University (course 27-763, second year graduate level). While partial notes have existed since the Spring of 1993, work on the book itself started in the Summer of 1996. The subsequent publication of several new textbooks in the area of transmission electron microscopy (TEM)† has served to focus the topics treated in this text, and a significant effort was made to provide an approach that would be different from but complementary to the other new texts.

On a regular basis, the journal *Ultramicroscopy* publishes an article by Peter W. Hawkes, who reviews various publications in the field of TEM-based research and closely related fields. I would recommend that the microscopy student take a

† Among others: *Transmission Electron Microscopy, a textbook for Materials Science*, by D.B. Williams and C.B. Carter [WC96]; *Advanced Computing in Electron Microscopy*, by E.J. Kirkland [Kir98]; *Transmission Electron Microscopy and Diffractometry of Materials*, by B. Fultz and J.M. Howe [FH01].

look at one of those articles[†] to gain some appreciation of the breadth and depth of the work carried out in the international microscopy community. It is very difficult to cover a similar breadth in an introductory textbook. I have, therefore, limited the topics introduced in this book to a very small subset of all available topics. This book covers only *conventional* TEM techniques; in my interpretation, conventional is equivalent to "non-analytical", meaning that the book deals only with images and diffraction patterns formed by the primary electrons themselves. Analytical methods, such as energy dispersive x-ray spectroscopy and electron energy loss spectroscopy, are not described in this text. Since I have very little personal experience with scanning TEM (STEM) methods, I have also excluded those from the contents of this book.

I have attempted to write a practical and self-contained textbook; self-contained in the sense that nearly all the required background knowledge (crystallography, basic quantum mechanics, etc.) is included in dedicated chapters at the appropriate level; practical in the sense that the microscopy student should be able to take this book to the microscope (or computer) and follow its guidelines and descriptions. Nature provides us with a large variety of crystal structures and structural defects. In an introductory book, it would not be practical nor desirable to simply show image after image from many different material systems. I have intentionally limited the number of material systems from which electron micrographs are shown. The majority of images are taken from the following four study materials: Cu-15 at% Al, Ti, GaAs, and $BaTiO_3$. It should not be too difficult for the reader to acquire small amounts of these materials and prepare thin foils. The reader should be able (with some effort) to reproduce nearly every electron micrograph shown in this book. Similarly, extensive Fortran-90 source code is provided on a web site (`http://ctem.web.cmu.edu`); the computer-savvy reader should be able to compile the code and run many different programs related to TEM.

About the structure of this book

Chapters 1 and 2 provide the foundations of crystallography and quantum mechanics, at the level needed later on in the book. Many additional topics are covered in Chapter 2: wave interference, the theory of special relativity, the Bragg equation, the Ewald sphere construction, Fourier transforms, convolutions, the Dirac delta function, elastic scattering of electrons, shape functions, the phenomenological

[†] The series of reviews started in 1981; reviews from the past decade can be found in *Ultramicroscopy*, volumes **90**, p. 215 (2002); **87**, p. 213 (2001); **80**, p. 271 (1999); **76**, p. 139 (1999); **75**, p. 115 (1998); **72**, p. 83 (1998); **69**, p. 51 (1997); **65**, p. 239 (1996); **62**, p. 283 (1996); **60**, p. 325 (1995); **56**, p. 337 (1994); **52**, p. 205 (1993); **50**, p. 106 (1993); **49**, p. 436 (1993); **45**, p. 421 (1992).

description of absorption, and the numerical computation of the electrostatic lattice potential. All crystallographic topics in Chapter 1 are accompanied by extensively documented Fortran-90 source code on the web site.

In Chapter 3, the transmission electron microscope is dissected into its major components: lenses, correction coils, the electron gun, apertures, and electron detectors. All components are extensively described in terms of the underlying theory. The microscope is built up one step at a time, starting from a microscope without lenses. This chapter makes use of photographs of the cross-section of a real microscope column. The relation between reciprocal space and the objective lens back focal plane is introduced, along with higher-order Laue zones, and the simulation of electron diffraction zone axis patterns.

Standard operation procedures for the microscope are introduced in Chapter 4, along with the four study materials. For each material, the reader will find the crystallographic information, as well as a description of thin-foil preparation techniques. The second half of this chapter deals with microscope alignment, bright field and dark field imaging, diffraction patterns, microscope calibration, bend contours, stereographic projections as a tool to describe the orientation of the sample in the column, and the basics of convergent beam electron diffraction methods. The chapter concludes with an introduction to the Fresnel and Foucault modes of Lorentz microscopy.

The dynamical theory of electron scattering in solids is then treated extensively in Chapters 5–7. Chapter 5 introduces the standard methods to describe the scattering problem: the Darwin–Howie–Whelan equations, the scattering matrix formalism, the real-space formalism, and the Bloch wave approach. The symmetry of the diffraction process is also extensively investigated, and the 31 diffraction groups are introduced. In Chapter 6, the dynamical equations are solved for the two-beam case, which is the only case for which a general analytical solution exists. The kinematical model is derived, along with the explicit expressions for the two-beam transmitted and scattered amplitudes for the dynamical case. Numerical simulations are introduced, and the absorptive form factor is described, starting from a discussion of thermal diffuse scattering. Convergent beam electron diffraction (CBED) simulations and approximate thickness determination are introduced.

Chapter 7 deals with the full dynamical multi-beam problem, for both the systematic row case and the zone axis case. Solution methods include the scattering matrix approach, the differential equation approach, and the Bloch wave approach. We introduce the concept of thickness-averaged x-ray emission probability. Extensively documented source code is made available for all multi-beam simulations. The final sections of Chapter 7 deal with exit wave simulations (multi-slice, real-space, and Bloch wave), and with the quantum mechanical aspects of Lorentz microscopy.

In Chapter 8, we describe the general theory of defect image simulations, and then apply it to inclusions, dislocations, and various types of planar defects, including stacking faults, anti-phase boundaries, domain boundaries, inversion boundaries, and grain boundaries. In nearly all cases, the numerical images are compared with selected experimental observations. The emphasis of this chapter is on the simulation methodologies. Chapter 9 deals with a large variety of diffraction effects, including: how to index a zone axis pattern, twin patterns, orientation relations, overlapping crystals, double diffraction, Kikuchi and HOLZ lines, convergent beam patterns and crystal symmetry, modulated crystals, short-range order, and polyhedral inclusions.

Finally, in Chapter 10 we introduce phase contrast observation methods. This chapter focuses on the theory of image formation, and describes the microscope transfer function in detail. Point resolution and the information limit are defined, and the mathematical formalism of image formation and simulation is discussed in detail. Examples of high-resolution image simulations are provided, along with the source code to carry out these simulations. Image simulations for Lorentz microscopy are then explained in the context of the theory of phase contrast. The chapter ends with a brief introduction to exit wave reconstruction, as applied to magnetic materials through the transport-of-intensity formalism.

The electron micrographs in this text were all taken by the author, except where noted otherwise. Micrographs were taken on the following microscopes:

- At Carnegie Mellon University: Philips EM420, JEOL 2000EX, JEOL 4000EX;
- At the University of Antwerp: Philips CM20, Philips CM30, JEOL 4000EX.

About the software accompanying this book

It was not easy to decide which programming language to use for the source code accompanying this book. Modern object-oriented languages provide powerful programming environments but are often limited by the fact that relatively few numerical libraries are available. Fortran-77, the most commonly used scientific programming language, does not have dynamical memory allocation mechanisms and has a rather limited set of variable types. Extensive numerical libraries are available, optimized for many different hardware platforms. In the end, I decided to use Fortran-90 for all of the source code. Fortran-90 provides a good compromise between ease of use (the source code is easier to read than Fortran-77) and flexibility in type declarations and dynamical memory allocations. The code can also be linked to existing Fortran-77 libraries. Conversions to other programming languages should pose no major hurdles. The reader will find more than 30 000 lines of Fortran-90 source code on the web site. The code is grouped into two segments: a library of general routines, called *CTEMsoft*, and a series of programs that use the library routines. The source code is copyrighted using the GNU General Public License and is freely available. Copyright details are posted on the web site.

The book's web site can be found at the following Uniform Resource Locator or URL:

`http://ctem.web.cmu.edu/`

and is available 24 hours a day with occasional brief interruptions for site maintenance.

Many of the examples in the book are accompanied by interactive routines on the web site. Those routines typically require a few input parameters which can be set by the remote user and are then sent to the web server. The server runs the appropriate code, usually a routine written in the Interactive Data Language (IDL). The resulting output is sent back to the browser in the form of a JPEG or GIF document. About two dozen routines are currently available and more will be added in the future. The complete source code of all IDL routines is also available on-line, as well as the interface routines needed to run the entire web site. Those routines are written in the form of ION-scripts (ION = IDL ON the Web).

Software used in the preparation of this book

Some readers might find it interesting to know which software packages were used in the creation of this book. The following list provides the name of the software package, the vendor (for commercial packages), or author. Weblinks to all companies are provided through the book's web site.

- **Commercial packages**: Adobe Illustrator (Adobe); Adobe Photoshop (Adobe); Digital Micrograph 2.5 and 3.3 (Gatan); Interactive Data Language 5.0 (Research Systems); ION 1.1 (Research Systems); Pro Fortran 6.2 (Absoft); Textures (Blue Sky Research); Fortran-90 (Compaq);
- **Shareware packages**: Alpha editor (P. Keleher);
- **Free packages**: fftw (`www.fftw.org`); teTEX (`www.tug.org`); PERL 5.0 (`www.perl.com`); Apache web server (`www.apache.org`); gVIM editor (`www.vim.org`); ghostscript (`www.aladdin.com`).

The web site for this book runs on a dedicated 666 MHz Compaq TRU-64 UNIX workstation located in the author's office.

Acknowledgements

Many people have (knowingly or unknowingly) contributed to this book. I would like to thank all the graduate students who have taken my electron optical methods course at CMU since the Spring of 1993: Kuang-Ti Chang, Thomas Dupuy, Richard Hsu, Dakshinamurthy Kolluru, Yuichiro Nakamura, Balamuralikrishna Ramalingam, Ursula Sadiq, John Henry Scott, Mandyam Sriram, Levent Trabzon, Damien Gallet, Jin Yong Kim, Eric Ott, Patcharin Poosanaas, Min-Seub Shin, Candida Silva, Shiela Woodard, Chun-Te Wu, Xing Chu, Yu-Nu Hsu, Luana Iorio, Xiaoding Ma, Vijay Narayanan, Tao Pan, Eric Rehder, Edward Sanchez, Zhangmin Wang, Wei Yang, Nicholas Biery, Xuemei Gu, Hirofumi Iwanabe, Paul Lee, Ding-Chung Lu, Bryan Molloseau, Zafer Turgut, Andrew Westmeyer, Jie Zou, Amy Bayer, Juajie Chen, Bassem El-Dasher, Dorothy Farrell, Ali Gungor, Ayetekin Hitit, Amy Hsiao, Taehoon Jang, Sangki Jeong, Il-Seok Kim, Vidhya Ramachandran, Matthew Willard, Wayne Archibald, Kai-Chieh Chang, Yoon Suck Choi, Melik Demirel, Daiwon Choi, Hun Jae Chung, Yi Ding, Zhaohui Fan, Jose Garcia Gonzalez, Francis Johnson, Bala Kavaipatti, Srinivasan Kumar, Hwan Soo Lee, Feroz Mohammad, Robert Nooning, Barbara Osgood, Chad Parish, Ritesh Seal, Rajasekaran Swaminathan, Mihaela Tanase, and Shakul Tandon. Through a near-continuous stream of questions and comments they have improved the readability of the text, and on more than one occasion they have given me reason to rewrite an entire section and/or rework an illustration. Any errors that remain in the text are obviously my own, and the reader should feel free to report them through the book's web site.

There is a vast amount of literature on the subject of transmission electron microscopy. I have attempted to keep the number of citations down to a manageable number; a textbook should, in my opinion, not be an exhaustive source of all papers in the field. Consequently, I may have overlooked important work, or perhaps included topics without full citations of *all* seminal papers in that particular area. I would like to apologize to those readers who have contributed significantly to the knowledge in this field, but do not find their work cited. The omissions do not reflect on the quality of their work, but are a simple consequence of the page limitations of this book.

The final sections of this book were written during the Winter and Spring months of 2001/2002, while I was on a sabbatical stay at the University of Antwerp (EMAT laboratory). I would like to express my gratitude to Jef Van Landuyt, Gustaaf Van Tendeloo, Dirk Van Dyck, and Dominique Schrijvers for their hospitality and friendship. It was nice to return to the place where I started my undergraduate studies in physics in 1979, and to collaborate with my former teachers. I particularly enjoyed the help of and working with Severin Amelinckx, Ann De Belder, André De Munck, Jan Eysermans, Philippe Geuens, Els Jordaens, Jan Neethling, Pavel Potapov, Ludo Rossou, Freddy Schallenberg, Gertie Stoffelen, and Roger Van Ginderen.

I would like to acknowledge the Research Council of the University of Antwerp, the National Science Foundation (NSF), and Carnegie Mellon University for providing me with financial support during the writing of this book.

I would also like to thank several of my colleagues, currently or formerly at CMU, for their support during the many years it has taken me to complete the text: Greg Rohrer, Tresa Pollock, David Laughlin, and Alan Cramb. In particular, I would like to thank Tom Nuhfer, for his continuous support as supervisor of the electron optics facility at CMU. His knowledge of the practical and technological aspects of electron microscopy has had a tremendous influence on my understanding of the field and, therefore, on the contents of this book. Thanks also to Cathy Rohrer, for pointing out style and grammatical errors.

I would like to thank my editors at CUP, Tim Fishlock and Simon Capelin, for their patience. This book has taken me longer to complete than I had originally anticipated, and there was no pressure at all to hurry up and finish it off.

As any junior faculty member in the U.S. will know, it is not easy to combine a family life with a demanding job as a tenure track professor at a research school, and write a textbook during whatever spare time is left over at the end of the day. I would like to thank my wife, Marie, for her patience and understanding during several years of evening and weekend work; without her support I could not have been successful in my academic career.

Last but not least, I dedicate this book to my two young children, who were born after I started writing: Pieter, now five, and Erika, two. It has been a joy to watch their insatiable appetite for learning new things. It is my sincere hope that the reader of this book will experience a similar eagerness and enthusiasm to learn about transmission electron microscopy.

Antwerpen, Belgium *M.D.G.*
May 27th, 2002

Figure reproductions

I have made an effort to include mostly new figures in this book. A small number of figures were taken from other authors' published or unpublished work, and the following acknowledgements must be made: Dr. Duncan McKie (Fig. 1.17 taken from [MM86]); Dr M. Kersker, JEOL (Fig. 3.2), permission to publish photographs from the cross-section of a JEOL 120 CX (Figs. 3.2, 3.4, 3.19c, 3.20c and d, 3.23b, 3.32, 3.43b, c, and d); Dr B. Armbruster, Gatan, Inc. (Fig. 3.30); The Royal Society and Professor J.W. Steeds, FRS (Tables. 5.1 and 7.2 and Figs. 5.11 and 5.9, taken from [BESR76]); Academic Press, for permission to reproduce most of Chapter 2 of the book *Magnetic Microscopy and its Applications to Magnetic Materials*, edited by the author and Y. Zhu (Figs. 4.26–4.32, 7.32–7.36, and 10.38–10.39).

1

Basic crystallography

1.1 Introduction

In this chapter, we review the principles and basic tools of crystallography. A thorough understanding of crystallography is a prerequisite for anybody who wishes to learn transmission electron microscopy (TEM) and its applications to solid (mostly inorganic) materials. All diffraction techniques, whether they use x-rays, neutrons, or electrons, make extensive use of the concept of *reciprocal space* and, as we shall see repeatedly later on in this book, TEM is a unique tool for directly probing this space. Hence, it is important that the TEM user become as familiar with reciprocal space as with *direct* or *crystal space*.

This chapter will provide a sound mathematical footing for both direct and reciprocal space, mostly in the form of *non-Cartesian vector calculus*. Many textbooks on crystallography approach this type of vector calculus by explicitly stating the equations for, say, the length of a vector, in each of the seven crystal systems. While this is certainly correct, such tables of equations do not lend themselves to direct implementation in a computer program. In this book, we opt for a method which is independent of the crystal system and which can be implemented readily on a computer. We will introduce powerful tools for the computation of geometrical quantities (distances and angles) in both spaces and for a variety of coordinate transformations in and between those spaces. We will also discuss the *stereographic projection* (SP), an important tool for the analysis of electron diffraction patterns and crystal defects. The TEM user should be familiar with these basic tools.

Although many of these tools are available in commercial or public domain software packages, we will discuss them in sufficient detail so that the reader may also implement them in a new program. It is also useful to *understand* what the various menu-items in software programs really mean. We will minimize to the extent that it is possible the number of "black-box" routines used in this book. The reader may download ASCII files containing all of the routines discussed in this book from the website. All of the algorithms are written in standard Fortran-90, and can easily be translated into C, Pascal, or any of the object-oriented languages (C++, Java, etc.). The user interface is kept simple, without on-screen graphics. Graphics output, if any, is produced in PostScript or TIFF format and can be viewed on-screen with an appropriate viewer or sent to a printer. The source code can be accessed at the Uniform Resource Locator (URL) `http://ctem.web.cmu.edu/`.

1.2 Direct space and lattice geometry

From a purely mathematical point of view, crystallography can be described as *vector calculus in a rectilinear, but not necessarily orthonormal (or even orthogonal) reference frame*. A discussion of crystallographic tools thus requires that we define basic vector operations in a non-Cartesian reference frame. Such operations are the *vector dot product*, the *vector cross product*, the computation of the length of a vector or the angle between two vectors, and so on.

1.2.1 Basis vectors and unit cells

A *crystal structure* is defined as a regular arrangement of atoms decorating a periodic, three-dimensional *lattice*. The lattice is defined as the set of points which is created by all *integer* linear combinations of three *basis vectors* $\mathbf{a}$, $\mathbf{b}$, and $\mathbf{c}$. In other words, the lattice $\mathcal{T}$ is the set of all vectors $\mathbf{t}$ of the form:

$$\mathbf{t} = u\mathbf{a} + v\mathbf{b} + w\mathbf{c},$$

with (u, v, w) being an arbitrary triplet of integers. We will often denote the basis vectors by the single symbol $\mathbf{a}_i$, where the subscript i takes on the values 1, 2, and 3. We will restrict ourselves to *right-handed* reference frames; i.e. the mixed product $(\mathbf{a} \times \mathbf{b}) \cdot \mathbf{c} > 0$. The *lattice vector* $\mathbf{t}$ can then be rewritten as

$$\mathbf{t} = \sum_{i=1}^{3} u_i \mathbf{a}_i, \qquad (1.1)$$

with $u_1 = u$, $u_2 = v$, and $u_3 = w$. This expression can be shortened even further by introducing the following notation convention, known as the *Einstein summation convention*: *If a subscript occurs twice in the same term of an equation, then a summation is implied over all values of this subscript and the corresponding summation sign need not be written.* In other words, since the subscript i occurs twice on the right-hand side of equation (1.1), we can drop the summation sign and simply write

$$\mathbf{t} = u_i \mathbf{a}_i. \qquad (1.2)$$

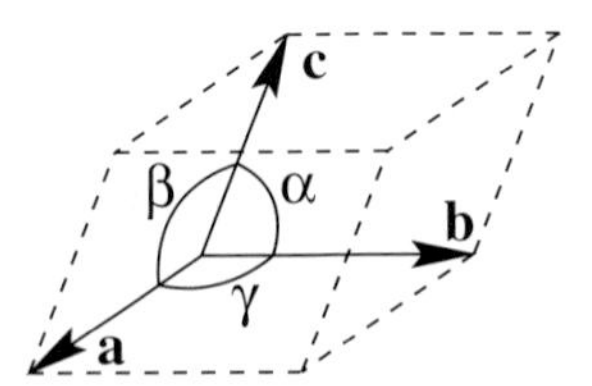

Fig. 1.1. Schematic representation of a general (triclinic or anorthic) unit cell.

The length of a vector is represented by the norm symbol | |; i.e. the length of the basis vector $\mathbf{a}_i$ is $|\mathbf{a}_i|$ with $|\mathbf{a}_1| \equiv a$, $|\mathbf{a}_2| \equiv b$, and $|\mathbf{a}_3| \equiv c$. The angles between the basis vectors are represented by the Greek letters α, β, and γ, as indicated in Fig. 1.1. The six numbers $\{a, b, c, \alpha, \beta, \gamma\}$ are known as the *lattice parameters* of the unit cell.

The lattice parameters can be used to distinguish between the seven *crystal systems*:

$\{a, b, c, \alpha, \beta, \gamma\}$	$a \neq b \neq c; \alpha \neq \beta \neq \gamma$	triclinic or anorthic (a);
$\{a, b, c, \frac{\pi}{2}, \beta, \frac{\pi}{2}\}$	$a \neq b \neq c; \beta \neq \frac{\pi}{2}$	monoclinic (m);
$\{a, a, c, \frac{\pi}{2}, \frac{\pi}{2}, \frac{2\pi}{3}\}$	$a = b \neq c$	hexagonal (h);
$\{a, a, a, \alpha, \alpha, \alpha\}$	$a = b = c; \alpha \neq \frac{\pi}{2}$	rhombohedral (R);
$\{a, b, c, \frac{\pi}{2}, \frac{\pi}{2}, \frac{\pi}{2}\}$	$a \neq b \neq c$	orthorhombic (o);
$\{a, a, c, \frac{\pi}{2}, \frac{\pi}{2}, \frac{\pi}{2}\}$	$a = b \neq c$	tetragonal (t);
$\{a, a, a, \frac{\pi}{2}, \frac{\pi}{2}, \frac{\pi}{2}\}$	$a = b = c$	cubic (c).

It is a basic property of a lattice that all lattice sites are equivalent. In other words, any site can be selected as the *origin*. The seven crystal systems give rise to seven *primitive lattices*, since there is only one lattice site per unit cell. We can place additional lattice sites at the endpoints of so-called *centering vectors*; the possible centering vectors are:

$$\mathbf{A} = \left(0, \frac{1}{2}, \frac{1}{2}\right);$$

$$\mathbf{B} = \left(\frac{1}{2}, 0, \frac{1}{2}\right);$$

$$\mathbf{C} = \left(\frac{1}{2}, \frac{1}{2}, 0\right);$$

$$\mathbf{I} = \left(\frac{1}{2}, \frac{1}{2}, \frac{1}{2}\right).$$

A lattice with an extra site at the **A** position is known as an *A-centered* lattice, and a site at the **I** position gives rise to a *body-centered* or *I-centered* lattice. When the

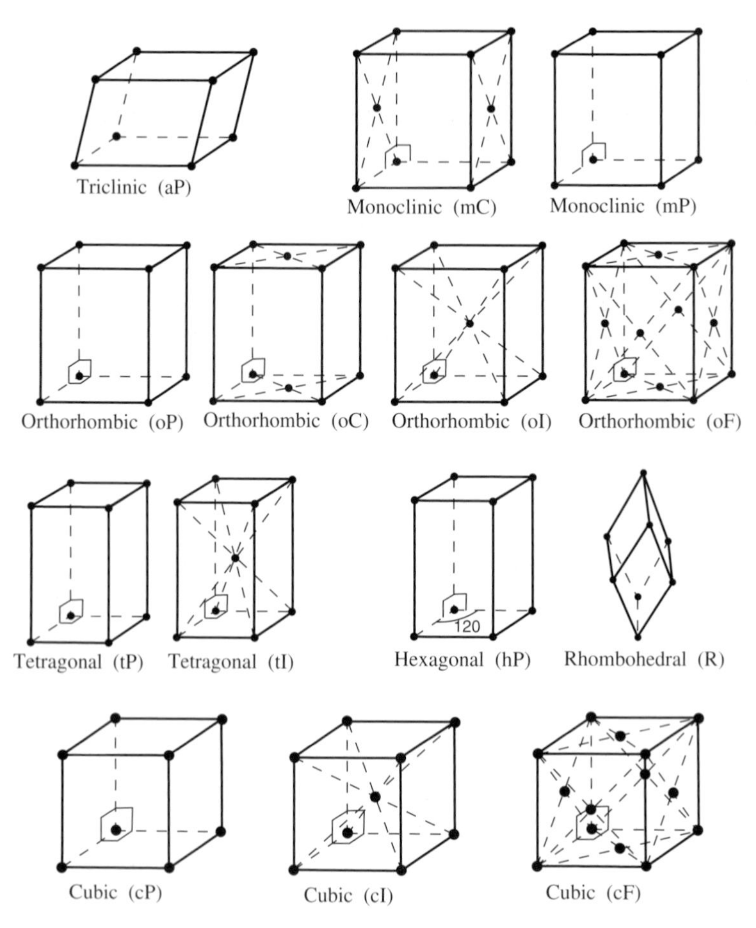

Fig. 1.2. The 14 Bravais lattices and their centering symbols.

three positions, **A**, **B**, and **C**, are simultaneously present as additional lattice sites, the lattice is *face centered* or *F-centered*.

When we combine these centering operations with each of the seven primitive unit cells, seven additional lattices are found. The 14 *Bravais lattices*, first derived by August Bravais in 1850 [Bra50], are shown in Fig. 1.2. They are commonly represented by two-letter symbols, the crystal system symbol followed by a centering symbol: *aP* (primitive anorthic), *mP* (primitive monoclinic), *mC* (C-centered monoclinic), *R* (primitive rhombohedral), *hP* (primitive hexagonal), *oP* (primitive orthorhombic), *oC* (C-centered orthorhombic), *oI* (body-centered orthorhombic), *oF* (face-centered orthorhombic), *tP* (primitive tetragonal), *tI* (body-centered tetragonal), *cP* (primitive cubic), *cI* (body-centered cubic), and *cF* (face-centered cubic). The choice of the lattice parameters of the Bravais lattices follows the conventions listed in the *International Tables for Crystallography, Volume A* [Hah96].

The vector **t** in equation (1.1) represents a *direction* in the crystal lattice. It is usually represented by the symbol $[uvw]$ (square brackets, no commas between the components). Negative components are denoted by a minus sign above the corresponding component(s), e.g. $[\bar{u}v\bar{w}]$ for the vector with components $(-u, v, -w)$. Note that there is no agreement in the literature on how to pronounce the symbol $[\bar{u}v\bar{w}]$; some researchers will pronounce the *bar* before the number (i.e. bar u, v, bar w), while others will pronounce it following the number (i.e. u bar, v, w bar). Since one is referring to the negative of a number, and usually this is pronounced as "negative u", or "minus u", it makes sense to pronounce the bar *before* the number to which it applies.†

The position of an atom inside the unit cell is described by the position vector **r**:

$$\mathbf{r} = x\mathbf{a} + y\mathbf{b} + z\mathbf{c} = \sum_{i=1}^{3} r_i \mathbf{a}_i = r_i \mathbf{a}_i,$$

where we have again made use of the summation convention. The numbers (x, y, z) are real numbers between 0 and 1, and are known as *fractional coordinates*.

1.2.2 The dot product and the direct metric tensor

It is important that we have a method of computing distances between atoms and angles between interatomic bonds in the unit cell. Distances in a Cartesian reference frame are typically computed by means of Pythagoras' Theorem: the distance D between two points P and Q with position vectors $\mathbf{p} = (p_1, p_2, p_3)$ and $\mathbf{q} = (q_1, q_2, q_3)$ is given by the length of the vector connecting the two points, or by the square root of the sum of the squares of the differences of the coordinates, i.e.

$$D = \sqrt{(p_1 - q_1)^2 + (p_2 - q_2)^2 + (p_3 - q_3)^2}.$$

In a non-Cartesian reference frame (and almost all crystallographic reference frames are non-Cartesian), this equation is no longer valid and it must be replaced by a more general expression that we shall now derive.

The dot product of two vectors **p** and **q** can be defined as the product of the lengths of **p** and **q** multiplied by the cosine of the angle θ between them, or

$$\boxed{\mathbf{p} \cdot \mathbf{q} \equiv |\mathbf{p}||\mathbf{q}| \cos\theta.} \quad (1.3)$$

† This is merely the author's personal preference. The choice is really up to the reader.

This definition does not depend on a particular choice of reference frame, so it can be taken as a general definition of the dot product. The dot product can be interpreted as the projection of one vector onto a second vector, multiplied by the length of the second vector. If the two vectors are identical, we find (since $\theta = 0$)

$$\mathbf{p} \cdot \mathbf{p} = |\mathbf{p}|^2,$$

from which we derive

$$|\mathbf{p}| = \sqrt{\mathbf{p} \cdot \mathbf{p}}.$$

If the vector $\mathbf{p}$ has components p_i with respect to the crystal basis vectors $\mathbf{a}_i$, we have[†]

$$|\mathbf{p}| = \sqrt{p_i \mathbf{a}_i \cdot p_j \mathbf{a}_j} = \sqrt{p_i (\mathbf{a}_i \cdot \mathbf{a}_j) p_j} \quad \left(= \sqrt{\sum_{i=1}^{3} \sum_{j=1}^{3} p_i (\mathbf{a}_i \cdot \mathbf{a}_j) p_j} \right).$$

We see that the *length* of a vector depends on all the dot products between the basis vectors (remember that there are two summations on the right-hand side of this equation!). The quantities $\mathbf{a}_i \cdot \mathbf{a}_j$ are of fundamental importance for crystallographic computations, and they are commonly denoted by the symbol

$$g_{ij} \equiv \mathbf{a}_i \cdot \mathbf{a}_j = |\mathbf{a}_i||\mathbf{a}_j| \cos \theta_{ij}. \tag{1.4}$$

The nine numbers g_{ij} form a 3×3 matrix which is known as the *direct metric tensor*. From Fig. 1.1, we find that this matrix is given explicitly by

$$g_{ij} = \begin{bmatrix} \mathbf{a} \cdot \mathbf{a} & \mathbf{a} \cdot \mathbf{b} & \mathbf{a} \cdot \mathbf{c} \\ \mathbf{b} \cdot \mathbf{a} & \mathbf{b} \cdot \mathbf{b} & \mathbf{b} \cdot \mathbf{c} \\ \mathbf{c} \cdot \mathbf{a} & \mathbf{c} \cdot \mathbf{b} & \mathbf{c} \cdot \mathbf{c} \end{bmatrix} = \begin{bmatrix} a^2 & ab \cos \gamma & ac \cos \beta \\ ab \cos \gamma & b^2 & bc \cos \alpha \\ ac \cos \beta & bc \cos \alpha & c^2 \end{bmatrix}. \tag{1.5}$$

The matrix g_{ij} is *symmetric*[‡] since $g_{ij} = g_{ji}$. It has only six independent components corresponding to the six lattice parameters $\{a, b, c, \alpha, \beta, \gamma\}$. In other words, the metric tensor contains the same information as the set of lattice parameters, but in a form that allows direct computation of the dot product between two vectors. Explicit expressions for all seven metric tensors are listed in Appendix A1 on page 661.

Example 1.1 *A tetragonal crystal has lattice parameters $a = \frac{1}{2}$ nm and $c = 1$ nm. Compute its metric tensor.*

[†] The indices i and j are known as *dummy indices*; it does not really matter which symbols we use for such summation indices, as long as we use them consistently throughout the computation.

[‡] In this textbook, the first subscript of g_{ij}, or any other matrix, will always refer to the rows and the second subscript to the columns of g.

Answer: *Substitution of these values into (1.5) results in*

$$g_{ij} = \begin{bmatrix} \frac{1}{4} & 0 & 0 \\ 0 & \frac{1}{4} & 0 \\ 0 & 0 & 1 \end{bmatrix}.$$

Note that the units of the metric tensor elements are (nanometer)^2, but for brevity we usually drop them until the end of a computation.

The length of the vector **p** can now be rewritten as

$$|\mathbf{p}| = \sqrt{p_i g_{ij} p_j}.$$

The argument of the square root contains a double summation over i and j. Since i is the row-index of the matrix g_{ij}, and since we can only multiply matrices that are *conformable*,[†] we find that the vector components p_i must be written in row form, while the components p_j must be written in column form, as follows:

$$|\mathbf{p}| = \sqrt{[p_1\ p_2\ p_3] \begin{bmatrix} a^2 & ab\cos\gamma & ac\cos\beta \\ ab\cos\gamma & b^2 & bc\cos\alpha \\ ac\cos\beta & bc\cos\alpha & c^2 \end{bmatrix} \begin{bmatrix} p_1 \\ p_2 \\ p_3 \end{bmatrix}}.$$

The dot product between two vectors **p** and **q** is given by

$$\boxed{\mathbf{p}\cdot\mathbf{q} = p_i\mathbf{a}_i \cdot q_j\mathbf{a}_j = p_i g_{ij} q_j,} \tag{1.6}$$

or explicitly

$$\mathbf{p}\cdot\mathbf{q} = [p_1\ p_2\ p_3] \begin{bmatrix} a^2 & ab\cos\gamma & ac\cos\beta \\ ab\cos\gamma & b^2 & bc\cos\alpha \\ ac\cos\beta & bc\cos\alpha & c^2 \end{bmatrix} \begin{bmatrix} q_1 \\ q_2 \\ q_3 \end{bmatrix}.$$

The angle θ between the two vectors is given by (from equation 1.3):

$$\boxed{\theta = \cos^{-1}\left(\frac{\mathbf{p}\cdot\mathbf{q}}{|\mathbf{p}||\mathbf{q}|}\right) = \cos^{-1}\left(\frac{p_i g_{ij} q_j}{\sqrt{p_i g_{ij} p_j}\sqrt{q_i g_{ij} q_j}}\right).} \tag{1.7}$$

Example 1.2 *For the tetragonal crystal in Example 1.1 on page 6, compute the distance between the points $(\frac{1}{2}, 0, \frac{1}{2})$ and $(\frac{1}{2}, \frac{1}{2}, 0)$.*

Answer: *The distance between two points is equal to the length of the vector connecting them, in this case $(\frac{1}{2} - \frac{1}{2}, 0 - \frac{1}{2}, \frac{1}{2} - 0) = (0, -\frac{1}{2}, \frac{1}{2})$. Using the tetragonal*

[†] A matrix $\mathcal{A}$ is said to be conformable with respect to $\mathcal{B}$ if the number of columns in $\mathcal{A}$ equals the number of rows in $\mathcal{B}$. Matrix multiplication is only defined for conformable matrices.

metric tensor derived previously, we find for the length of this vector:

$$|\mathbf{p}| = \sqrt{\begin{bmatrix} 0 & \frac{-1}{2} & \frac{1}{2} \end{bmatrix} \begin{bmatrix} \frac{1}{4} & 0 & 0 \\ 0 & \frac{1}{4} & 0 \\ 0 & 0 & 1 \end{bmatrix} \begin{bmatrix} 0 \\ \frac{-1}{2} \\ \frac{1}{2} \end{bmatrix}};$$

$$= \sqrt{\begin{bmatrix} 0 & \frac{-1}{2} & \frac{1}{2} \end{bmatrix} \begin{bmatrix} 0 \\ \frac{-1}{8} \\ \frac{1}{2} \end{bmatrix}} = \frac{\sqrt{5}}{4}\, nm.$$

Example 1.3 *For the tetragonal unit cell of Example 1.1 on page 6, compute the dot product and the angle between the vectors* [120] *and* [311].

Answer: *The dot product is found from the expression for the metric tensor, as follows:*

$$\mathbf{t}_{[120]} \cdot \mathbf{t}_{[311]} = [1\,2\,0] \begin{bmatrix} \frac{1}{4} & 0 & 0 \\ 0 & \frac{1}{4} & 0 \\ 0 & 0 & 1 \end{bmatrix} \begin{bmatrix} 3 \\ 1 \\ 1 \end{bmatrix} = [1\,2\,0] \begin{bmatrix} \frac{3}{4} \\ \frac{1}{4} \\ 1 \end{bmatrix} = \frac{5}{4}\, nm^2.$$

The angle is found by dividing the dot product by the lengths of the vectors, $|[120]|^2 = \frac{5}{4}\, nm^2$ *and* $|[311]|^2 = \frac{14}{4}\, nm^2$*, from which we find*

$$\cos\theta = \frac{\frac{5}{4}}{\sqrt{\frac{14}{4}}\sqrt{\frac{5}{4}}} = \frac{5}{\sqrt{70}} \rightarrow \theta = 53.30^\circ.$$

Example 1.4 *The angle between two direct space vectors can be computed in a single operation, instead of using the three individual dot products described in the previous example. Derive a procedure for computing the angle* θ *based on a* 2×3 *matrix containing the two vectors* $\mathbf{p}$ *and* $\mathbf{q}$.

Answer: *Consider the following formal relation:*

$$\begin{pmatrix} \mathbf{p} \\ \mathbf{q} \end{pmatrix} \cdot \begin{pmatrix} \mathbf{p} & \mathbf{q} \end{pmatrix} = \begin{pmatrix} \mathbf{p}\cdot\mathbf{p} & \mathbf{p}\cdot\mathbf{q} \\ \mathbf{q}\cdot\mathbf{p} & \mathbf{q}\cdot\mathbf{q} \end{pmatrix}.$$

The resulting 2×2 *matrix contains all three dot products needed for the computation of the angle* θ*, and only one set of matrix multiplications is needed. We can apply this short cut to the previous example:*

$$\begin{pmatrix} 1 & 2 & 0 \\ 3 & 1 & 1 \end{pmatrix} \begin{pmatrix} \frac{1}{4} & 0 & 0 \\ 0 & \frac{1}{4} & 0 \\ 0 & 0 & 1 \end{pmatrix} \begin{pmatrix} 1 & 3 \\ 2 & 1 \\ 0 & 1 \end{pmatrix} = \begin{pmatrix} \frac{5}{4} & \frac{5}{4} \\ \frac{5}{4} & \frac{14}{4} \end{pmatrix},$$

from which we find the same angle of $\theta = 53.30^\circ$.

Note that these equations are valid in every rectilinear coordinate frame[†] and, therefore, in every crystal system. Explicit expressions for distances and angles in the seven reference frames are listed in Appendix A1 on pages 663–664. For a Cartesian, orthonormal reference frame, the metric tensor reduces to the identity matrix. Indeed, the Cartesian basis vectors $\mathbf{e}_i$ have unit length and are orthogonal to each other; therefore, the metric tensor reduces to

$$g_{ij} = \begin{bmatrix} \mathbf{e}_1 \cdot \mathbf{e}_1 & \mathbf{e}_1 \cdot \mathbf{e}_2 & \mathbf{e}_1 \cdot \mathbf{e}_3 \\ \mathbf{e}_2 \cdot \mathbf{e}_1 & \mathbf{e}_2 \cdot \mathbf{e}_2 & \mathbf{e}_2 \cdot \mathbf{e}_3 \\ \mathbf{e}_3 \cdot \mathbf{e}_1 & \mathbf{e}_3 \cdot \mathbf{e}_2 & \mathbf{e}_3 \cdot \mathbf{e}_3 \end{bmatrix} = \begin{bmatrix} 1 & 0 & 0 \\ 0 & 1 & 0 \\ 0 & 0 & 1 \end{bmatrix} \equiv \delta_{ij}, \tag{1.8}$$

where we have introduced the *Kronecker delta* δ_{ij}, which is equal to 1 for $i = j$ and 0 for $i \neq j$. Substitution into equation (1.6) results in

$$\mathbf{p} \cdot \mathbf{q} = p_i \delta_{ij} q_j = p_i q_i = p_1 q_1 + p_2 q_2 + p_3 q_3,$$

which is the standard expression for the dot product between two vectors in a Cartesian reference frame. We will postpone until Section 1.9 a discussion of how to implement the metric tensor formalism on a computer.

1.3 Definition of reciprocal space

In the previous section, we have described how we can compute distances between atoms in a crystal and angles between the bonds connecting those atoms. In Chapter 2, we will see that *diffraction* of electrons is described by the *Bragg equation*, which relates the diffraction angle to the electron wavelength and the spacing between crystal planes. We must, therefore, devise a tool that will enable us to compute this spacing between successive lattice planes in an arbitrary crystal lattice. We would like to have a method similar to that described in the previous section, ideally one with equations identical in form to those for the distance between atoms or the dot product between direction vectors. It turns out that such a tool exists and we will introduce the *reciprocal metric tensor* in the following subsections.

1.3.1 Planes and Miller indices

The description of crystal planes has a long history going all the way back to *René-Juste Haüy* [Haü84] who formulated the *Second Law of Crystal Habit*, also known as the law of simple rational intercepts. This law prompted *Miller* to devise a system to label crystal planes, based on their intercepts with the crystallographic reference axes. Although the so-called *Miller indices* were used by several crystallographers before Miller, they are attributed to him because he used them extensively in his book and teachings [Mil39] and because he developed the familiar *hkl* notation.

[†] They are also valid for curvilinear coordinate frames, but we will not make much use of such reference frames in this book.

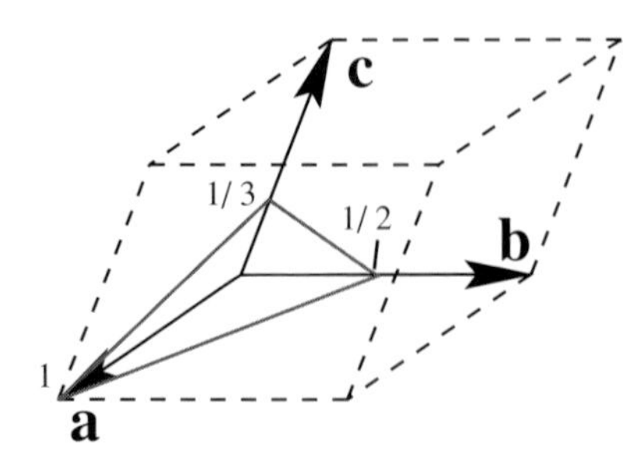

Fig. 1.3. Illustration of the determination of the Miller indices of a plane.

The Miller indices of a plane in an arbitrary crystal system are obtained in the following way.

(i) If the plane goes through the origin, then displace it so that it no longer contains the origin.

(ii) Determine the intercepts of the plane with the three basis vectors. Call those intercepts s_1, s_2, and s_3. The intercepts must be measured in units of the basis vector length. For the plane shown in Fig. 1.3, these values are $s_1 = 1$, $s_2 = \frac{1}{2}$, and $s_3 = \frac{1}{3}$. If a plane is parallel to one or more of the basis vectors, then the corresponding intercept value(s) must be taken as ∞.

(iii) Invert all three intercepts. For the plane in the figure we find $\frac{1}{s_1} = 1$, $\frac{1}{s_2} = 2$, and $\frac{1}{s_3} = 3$. If one of the intercepts is ∞, then the corresponding number is zero.

(iv) Reduce the three numbers to the smallest possible integers (relative primes). (This is not necessary for the example above.)

(v) Write the three numbers surrounded by round brackets, i.e. (123). This triplet of numbers forms the *Miller indices* of the plane.

In general, the Miller indices of a plane are denoted by the symbol (hkl). As for directions, negative indices are indicated by a *bar* or minus sign written above the corresponding index, e.g. $(\bar{1}2\bar{3})$. Although Miller indices were defined as relative primes, we will see later on that it is often necessary to consider planes of the type $(nh\, nk\, nl)$, where n is a non-zero integer. All planes of this type are parallel to the plane (hkl), but for diffraction purposes they are not the same as the plane (hkl). For instance, the plane (111) is parallel to the plane (222), but the two planes behave differently in a diffraction experiment.

1.3.2 The reciprocal basis vectors

It is tempting to interpret the triplet of Miller indices (hkl) as the components of a vector. A quick inspection of the orientation of the vector $\mathbf{n} = h\mathbf{a} + k\mathbf{b} + l\mathbf{c}$ with respect to the plane (hkl) in an arbitrary crystal system shows that, except

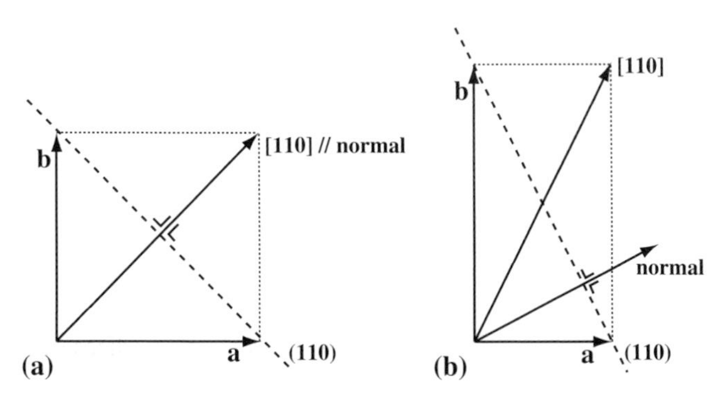

Fig. 1.4. (a) In a cubic (square) lattice, a direction vector [110] is normal to the plane with Miller indices (110); this is no longer the case for a non-cubic system, as shown in the rectangular cell (b).

in special cases, there is no fixed relation between the two (see Fig. 1.4). In other words, when the Miller indices are interpreted as the components of a vector with respect to the direct basis vectors $\mathbf{a}_i$, we do not find a useful relationship between this vector and the plane (hkl). We must then ask the question: can we find three new basis vectors $\mathbf{a}^*$, $\mathbf{b}^*$, and $\mathbf{c}^*$, related to the direct basis vectors $\mathbf{a}_i$, such that the vector $\mathbf{g} = h\mathbf{a}^* + k\mathbf{b}^* + l\mathbf{c}^*$ conveys meaningful information about the plane (hkl)? It turns out that such a triplet of basis vectors exists, and they are known as the *reciprocal basis vectors*. We will distinguish them from the direct basis vectors by means of an asterisk, $\mathbf{a}_j^*$.

The reciprocal basis vectors can be derived from the following definition:

$$\boxed{\mathbf{a}_i \cdot \mathbf{a}_j^* \equiv \delta_{ij},} \tag{1.9}$$

where δ_{ij} is the Kronecker delta introduced in equation (1.8). This expression fully defines the reciprocal basis vectors: it states that the vector $\mathbf{a}^*$ must be perpendicular to both $\mathbf{b}$ and $\mathbf{c}$ ($\mathbf{a}^* \cdot \mathbf{b} = \mathbf{a}^* \cdot \mathbf{c} = 0$), and that $\mathbf{a}^* \cdot \mathbf{a} = 1$. The first condition is satisfied if $\mathbf{a}^*$ is parallel to the cross product between $\mathbf{b}$ and $\mathbf{c}$:

$$\mathbf{a}^* = K(\mathbf{b} \times \mathbf{c}),$$

where K is a constant. The second condition leads to the value of K:

$$\mathbf{a} \cdot \mathbf{a}^* = K\mathbf{a} \cdot (\mathbf{b} \times \mathbf{c}) = 1,$$

from which we find

$$K = \frac{1}{\mathbf{a} \cdot (\mathbf{b} \times \mathbf{c})} \equiv \frac{1}{\Omega},$$

where Ω is the volume of the unit cell formed by the vectors $\mathbf{a}_i$.

A similar procedure for the remaining two reciprocal basis vectors then leads to the following expressions:

$$\left.\begin{aligned}\mathbf{a}^* &= \frac{\mathbf{b}\times\mathbf{c}}{\mathbf{a}\cdot(\mathbf{b}\times\mathbf{c})};\\ \mathbf{b}^* &= \frac{\mathbf{c}\times\mathbf{a}}{\mathbf{a}\cdot(\mathbf{b}\times\mathbf{c})};\\ \mathbf{c}^* &= \frac{\mathbf{a}\times\mathbf{b}}{\mathbf{a}\cdot(\mathbf{b}\times\mathbf{c})}.\end{aligned}\right\} \tag{1.10}$$

We define the *reciprocal lattice* $\mathcal{T}^*$ as the set of end-points of the vectors of the type

$$\mathbf{g} = h\mathbf{a}^* + k\mathbf{b}^* + l\mathbf{c}^* = \sum_{i=1}^{3} g_i\mathbf{a}_i^* = g_i\mathbf{a}_i^*,$$

where (h, k, l) are integer triplets. This new lattice is also known as the *dual lattice*, but in the diffraction world we prefer the name *reciprocal lattice*. We will now investigate the relation between the reciprocal lattice vectors $\mathbf{g}$ and the planes with Miller indices (hkl).

We will look for all the direct space vectors $\mathbf{r}$ with components $r_i = (x, y, z)$ that are perpendicular to the vector $\mathbf{g}$. We already know that two vectors are perpendicular to each other if their dot product vanishes. In this case we find:

$$0 = \mathbf{r}\cdot\mathbf{g} = (r_i\mathbf{a}_i)\cdot\left(g_j\mathbf{a}_j^*\right) = r_i\left(\mathbf{a}_i\cdot\mathbf{a}_j^*\right)g_j.$$

We also know from equation (1.9) that the last dot product is equal to δ_{ij}, so

$$\mathbf{r}\cdot\mathbf{g} = r_i\delta_{ij}g_j = r_ig_i = r_1g_1 + r_2g_2 + r_3g_3 = hx + ky + lz = 0. \tag{1.11}$$

The components of the vector $\mathbf{r}$ must satisfy the relation $hx + ky + lz = 0$ if $\mathbf{r}$ is to be perpendicular to $\mathbf{g}$. This relation represents the equation of a plane through the origin of the direct crystal lattice. If a plane intersects the basis vectors $\mathbf{a}_i$ at intercepts s_i, then the equation of that plane is given by [Spi68]

$$\frac{x}{s_1} + \frac{y}{s_2} + \frac{z}{s_3} = 1, \tag{1.12}$$

where (x, y, z) is an arbitrary point in the plane. The right-hand side of this equation takes on different values when we translate the plane along its normal, and, in particular, is equal to zero when the plane goes through the origin. Comparing

$$hx + ky + lz = 0$$

with

$$\frac{x}{s_1} + \frac{y}{s_2} + \frac{z}{s_3} = 0,$$

we find that the integers h, k, and l are reciprocals of the intercepts of a plane with the direct lattice basis vectors. This is exactly the definition of the *Miller indices* of a plane! We thus find the fundamental result:

> The reciprocal lattice vector $\mathbf{g}$, with components (h, k, l), is perpendicular to the plane with Miller indices (hkl).

For this reason, a reciprocal lattice vector is often denoted with the Miller indices as subscripts, e.g. $\mathbf{g}_{hkl}$.

Since the vector $\mathbf{g} = g_i \mathbf{a}_i^*$ is perpendicular to the plane with Miller indices $g_i = (hkl)$, the unit normal to this plane is given by

$$\mathbf{n} = \frac{\mathbf{g}_{hkl}}{|\mathbf{g}_{hkl}|}.$$

The perpendicular distance from the origin to the plane intersecting the direct basis vectors at the points $\frac{1}{h}$, $\frac{1}{k}$, and $\frac{1}{l}$ is given by the projection of any vector $\mathbf{t}$ ending in the plane onto the plane normal $\mathbf{n}$ (see Fig. 1.5). This distance is also, by definition, the *interplanar spacing* d_{hkl}. Thus,

$$\mathbf{t} \cdot \mathbf{n} = \mathbf{t} \cdot \frac{\mathbf{g}_{hkl}}{|\mathbf{g}_{hkl}|} \equiv d_{hkl}.$$

We can arbitrarily select $\mathbf{t} = \frac{\mathbf{a}}{h}$, which leads to

$$\mathbf{t} \cdot \mathbf{g}_{hkl} = \frac{\mathbf{a}}{h} \cdot \left(h\mathbf{a}^* + k\mathbf{b}^* + l\mathbf{c}^*\right) = \frac{\mathbf{a}}{h} \cdot h\mathbf{a}^* = 1 = d_{hkl}|\mathbf{g}_{hkl}|,$$

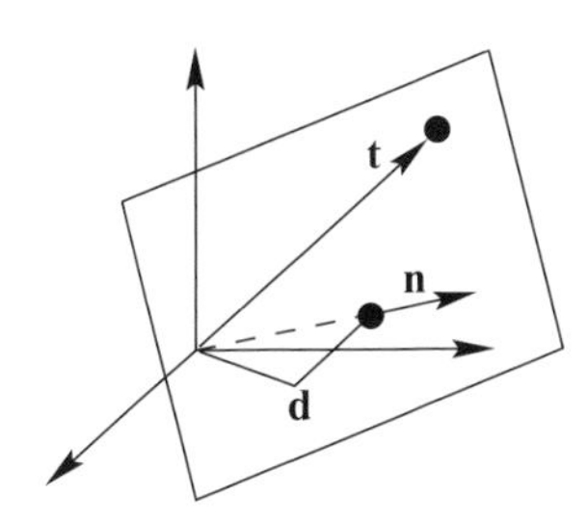

Fig. 1.5. The distance of a plane to the origin equals the projection of any vector **t** ending in this plane onto the unit plane normal **n**.

from which we find

$$\boxed{|\mathbf{g}_{hkl}| = \frac{1}{d_{hkl}}.} \tag{1.13}$$

The length of a reciprocal lattice vector is equal to the inverse of the spacing between the corresponding lattice planes.

We thus find that every vector $\mathbf{g}_{hkl}$ of the reciprocal lattice is parallel to the normal to the set of planes with Miller indices (hkl), and the length of $\mathbf{g}_{hkl}$ (i.e. the distance from the point (h, k, l) to the origin of the reciprocal lattice) is equal to the inverse of the spacing between consecutive lattice planes. At this point, it is useful to introduce methods for lattice calculations in the reciprocal lattice; we will see that the metric tensor formalism introduced in Section 1.2.2 can also be applied to the reciprocal lattice.

1.3.3 Lattice geometry in reciprocal space

We know that the length of a vector is given by the square root of the dot product of this vector with itself. Thus, the length of $\mathbf{g}$ is given by

$$\frac{1}{d_{hkl}} = |\mathbf{g}| = \sqrt{\mathbf{g}\cdot\mathbf{g}} = \sqrt{(g_i\mathbf{a}_i^*)\cdot(g_j\mathbf{a}_j^*)} = \sqrt{g_i(\mathbf{a}_i^*\cdot\mathbf{a}_j^*)g_j}.$$

Again we find that the general dot product involves knowledge of the dot products of the basis vectors, in this case the reciprocal basis vectors. We introduce the *reciprocal metric tensor*:

$$\boxed{g_{ij}^* \equiv \mathbf{a}_i^*\cdot\mathbf{a}_j^*.} \tag{1.14}$$

Explicitly, the reciprocal metric tensor is given by:

$$\begin{aligned} g^* &= \begin{bmatrix} \mathbf{a}^*\cdot\mathbf{a}^* & \mathbf{a}^*\cdot\mathbf{b}^* & \mathbf{a}^*\cdot\mathbf{c}^* \\ \mathbf{b}^*\cdot\mathbf{a}^* & \mathbf{b}^*\cdot\mathbf{b}^* & \mathbf{b}^*\cdot\mathbf{c}^* \\ \mathbf{c}^*\cdot\mathbf{a}^* & \mathbf{c}^*\cdot\mathbf{b}^* & \mathbf{c}^*\cdot\mathbf{c}^* \end{bmatrix}; \\ &= \begin{bmatrix} a^{*2} & a^*b^*\cos\gamma^* & a^*c^*\cos\beta^* \\ b^*a^*\cos\gamma^* & b^{*2} & b^*c^*\cos\alpha^* \\ c^*a^*\cos\beta^* & c^*b^*\cos\alpha^* & c^{*2} \end{bmatrix}, \end{aligned} \tag{1.15}$$

where $\{a^*, b^*, c^*, \alpha^*, \beta^*, \gamma^*\}$ are the reciprocal lattice parameters. Explicit expressions for all seven reciprocal metric tensors are given in Appendix A1 on page 662.

In Section 1.3.4, we will develop an easy way to compute the reciprocal basis vectors and lattice parameters; for now the explicit equations in the appendix are sufficient.

Example 1.5 *Compute the reciprocal metric tensor for a tetragonal crystal with lattice parameters* $a = \frac{1}{2}$ *and* $c = 1$.

Answer: *Substitution of the lattice parameters into the expression for the tetragonal reciprocal metric tensor in Appendix A1 yields*

$$g^*_{tetragonal} = \begin{bmatrix} 4 & 0 & 0 \\ 0 & 4 & 0 \\ 0 & 0 & 1 \end{bmatrix}.$$

We can now rewrite the length of the reciprocal lattice vector **g** as

$$\boxed{\frac{1}{d_{hkl}} = |\mathbf{g}| = \sqrt{\mathbf{g}\cdot\mathbf{g}} = \sqrt{g_i g^*_{ij} g_j}.} \tag{1.16}$$

The angle θ between two reciprocal lattice vectors **f** and **g** is given by the standard relation (equation 1.7):

$$\boxed{\theta = \cos^{-1}\left(\frac{f_i g^*_{ij} g_j}{\sqrt{f_i g^*_{ij} f_j}\sqrt{g_i g^*_{ij} g_j}}\right).} \tag{1.17}$$

Example 1.6 *Compute the angle between the* (120) *and* (311) *plane normals for the tetragonal crystal of Example 1.5.*

Answer: *Substitution of the vector components and the reciprocal metric tensor into the expression for the angle results in*

$$\cos\theta = \frac{[1\,2\,0]\begin{bmatrix} 4 & 0 & 0 \\ 0 & 4 & 0 \\ 0 & 0 & 1 \end{bmatrix}\begin{bmatrix} 3 \\ 1 \\ 1 \end{bmatrix}}{\sqrt{[1\,2\,0]\begin{bmatrix} 4 & 0 & 0 \\ 0 & 4 & 0 \\ 0 & 0 & 1 \end{bmatrix}\begin{bmatrix} 1 \\ 2 \\ 0 \end{bmatrix}}\sqrt{[3\,1\,1]\begin{bmatrix} 4 & 0 & 0 \\ 0 & 4 & 0 \\ 0 & 0 & 1 \end{bmatrix}\begin{bmatrix} 3 \\ 1 \\ 1 \end{bmatrix}}},$$

$$= \frac{20}{\sqrt{20\times 41}} = 0.698\,43,$$

$$\rightarrow \theta = 45.7^\circ.$$

Example 1.7 *Redo the computation of the previous example using the shorthand notation introduced in Example 1.4.*

Answer: *The matrix product is given by*

$$\begin{pmatrix} 1 & 2 & 0 \\ 3 & 1 & 1 \end{pmatrix} \begin{pmatrix} 4 & 0 & 0 \\ 0 & 4 & 0 \\ 0 & 0 & 1 \end{pmatrix} \begin{pmatrix} 1 & 3 \\ 2 & 1 \\ 0 & 1 \end{pmatrix} = \begin{pmatrix} 20 & 20 \\ 20 & 41 \end{pmatrix},$$

from which we find the same angle of $\theta = 45.7^\circ$.

Note that we have indeed managed to create a computational tool that is formally identical to that used for distances and angles in direct space. This should come as no surprise, since the reciprocal basis vectors are just another set of basis vectors, and the equations for direct space must be valid for *any* non-Cartesian reference frame. The particular choice for the reciprocal basis vectors (see equation 1.9) guarantees that they are useful for the description of lattice planes. In the next section, we will derive relations between the direct and reciprocal lattices.

1.3.4 Relations between direct space and reciprocal space

We know that a vector is a mathematical object that exists independently of the reference frame. This means that every vector defined in the direct lattice must also have components with respect to the reciprocal basis vectors and vice versa. In this section, we will devise a tool that will permit us to transform vector quantities back and forth between direct and reciprocal space.

Consider the vector $\mathbf{p}$:

$$\mathbf{p} = p_i \mathbf{a}_i = p_j^* \mathbf{a}_j^*,$$

where p_j^* are the reciprocal space components of $\mathbf{p}$. Multiplying both sides by the direct basis vector $\mathbf{a}_m$, we have

$$\left.\begin{aligned} p_i \mathbf{a}_i \cdot \mathbf{a}_m &= p_j^* \mathbf{a}_j^* \cdot \mathbf{a}_m, \\ p_i g_{im} &= p_j^* \delta_{jm} = p_m^*, \end{aligned}\right\} \tag{1.18}$$

or

$$p_m^* = p_i g_{im}. \tag{1.19}$$

It is easily shown that the inverse relation is given by

$$p_i = p_m^* g_{mi}^*. \tag{1.20}$$

We thus find that *post-multiplication by the metric tensor* transforms vector components from direct space to reciprocal space, and post-multiplication by the reciprocal metric tensor transforms vector components from reciprocal to direct space. These relations are useful because they permit us to determine the components of a direction vector $\mathbf{t}_{[uvw]}$ with respect to the reciprocal basis vectors, or the components of a plane normal $\mathbf{g}_{hkl}$ with respect to the direct basis vectors.

Example 1.8 *For the tetragonal unit cell of Example 1.1 on page 6, write down the reciprocal components of the lattice vector* [114].

Answer: *This transformation is accomplished by post-multiplication by the direct metric tensor:*

$$t^*_{[114]} = [1\,1\,4]\begin{bmatrix} \frac{1}{4} & 0 & 0 \\ 0 & \frac{1}{4} & 0 \\ 0 & 0 & 1 \end{bmatrix} = \begin{bmatrix} \frac{1}{4} & \frac{1}{4} & 4 \end{bmatrix}.$$

In other words, the [114] *direction is perpendicular to the* (1 1 16) *plane.*

Now we have all the tools we need to express the reciprocal basis vectors in terms of the direct basis vectors. Consider again the vector $\mathbf{p}$:

$$\mathbf{p} = p_i \mathbf{a}_i.$$

If we replace p_i by $p^*_m g^*_{mi}$, then we have

$$\mathbf{p} = p^*_m g^*_{mi} \mathbf{a}_i = p^*_m \mathbf{a}^*_m,$$

from which we find

$$\mathbf{a}^*_m = g^*_{mi} \mathbf{a}_i, \tag{1.21}$$

and the inverse relation

$$\mathbf{a}_m = g_{mi} \mathbf{a}^*_i. \tag{1.22}$$

In other words, *the rows of the metric tensor contain the components of the direct basis vectors in terms of the reciprocal basis vectors, whereas the rows of the reciprocal metric tensor contain the components of the reciprocal basis vectors with respect to the direct basis vectors.*

Finally, from equation (1.22) we find after multiplication by the vector $\mathbf{a}^*_k$:

$$\left.\begin{aligned} \mathbf{a}_m \cdot \mathbf{a}^*_k &= g_{mi} \mathbf{a}^*_i \cdot \mathbf{a}^*_k, \\ \delta_{mk} &= g_{mi} g^*_{ik}. \end{aligned}\right\} \tag{1.23}$$

In other words, the matrices representing the direct and reciprocal metric tensors are each other's inverse. This leads to a simple procedure to determine the reciprocal basis vectors of a crystal:

(i) compute the direct metric tensor;
(ii) invert it to find the reciprocal metric tensor;
(iii) apply equation (1.21) to find the reciprocal basis vectors.

Example 1.9 *For the tetragonal unit cell of the previous example, write down the explicit expressions for the reciprocal basis vectors. From these expressions, derive the reciprocal lattice parameters.*

Answer: *The components of the reciprocal basis vectors are given by the rows of the reciprocal metric tensor, and thus*

$$\mathbf{a}_1^* = 4\mathbf{a}_1;$$
$$\mathbf{a}_2^* = 4\mathbf{a}_2;$$
$$\mathbf{a}_3^* = \mathbf{a}_3.$$

The reciprocal lattice parameters are now easily found from the lengths of the basis vectors: $a^* = |\mathbf{a}_1^*| = |\mathbf{a}_2^*| = 4|\mathbf{a}_1| = 4 \times \frac{1}{2} = 2\ nm^{-1}$, *and* $c^* = |\mathbf{a}_3^*| = |\mathbf{a}_3| = 1\ nm^{-1}$. *The angles between the reciprocal basis vectors are all* 90°.

1.3.5 The non-Cartesian vector cross product

The attentive reader may have noticed that we have made use of the *vector cross product* in the definition of the reciprocal lattice vectors, without considering how the cross product is defined in a non-Cartesian reference frame. In this section, we will generalize the cross product to crystallographic reference frames.

Consider the two real-space vectors $\mathbf{p} = p_1\mathbf{a} + p_2\mathbf{b} + p_3\mathbf{c}$ and $\mathbf{q} = q_1\mathbf{a} + q_2\mathbf{b} + q_3\mathbf{c}$. The cross product between them is defined as

$$\boxed{\mathbf{p} \times \mathbf{q} \equiv \sin\theta\, |\mathbf{p}|\, |\mathbf{q}|\, \mathbf{z},} \tag{1.24}$$

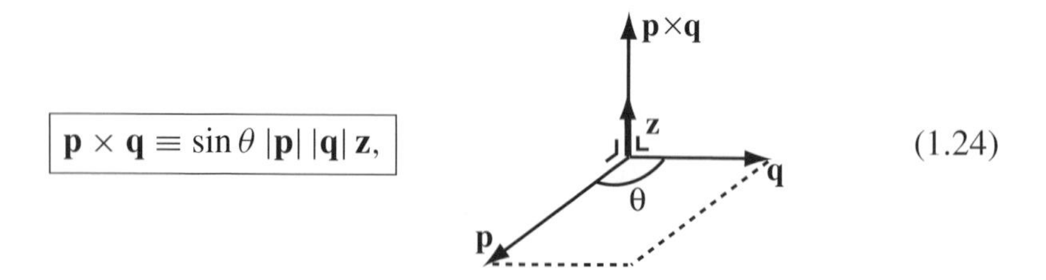

where θ is the angle between $\mathbf{p}$ and $\mathbf{q}$, and $\mathbf{z}$ is a unit vector perpendicular to both $\mathbf{p}$ and $\mathbf{q}$. The length of the cross product vector is equal to the area of the parallelogram enclosed by the vectors $\mathbf{p}$ and $\mathbf{q}$. It is straightforward to compute the components of the cross product:

$$\begin{aligned}\mathbf{p}\times\mathbf{q} = & \; p_1q_1\mathbf{a}\times\mathbf{a} + p_1q_2\mathbf{a}\times\mathbf{b} + p_1q_3\mathbf{a}\times\mathbf{c} \\ & + p_2q_1\mathbf{b}\times\mathbf{a} + p_2q_2\mathbf{b}\times\mathbf{b} + p_2q_3\mathbf{b}\times\mathbf{c} \\ & + p_3q_1\mathbf{c}\times\mathbf{a} + p_3q_2\mathbf{c}\times\mathbf{b} + p_3q_3\mathbf{c}\times\mathbf{c}.\end{aligned}$$

Since the cross product of a vector with itself vanishes, and $\mathbf{a}\times\mathbf{b} = -\mathbf{b}\times\mathbf{a}$, we can rewrite this equation as:

$$\begin{aligned}\mathbf{p}\times\mathbf{q} &= (p_1q_2 - p_2q_1)\,\mathbf{a}\times\mathbf{b} + (p_2q_3 - p_3q_2)\,\mathbf{b}\times\mathbf{c} + (p_3q_1 - p_1q_3)\,\mathbf{c}\times\mathbf{a} \\ &= \Omega[(p_2q_3 - p_3q_2)\,\mathbf{a}^* + (p_3q_1 - p_1q_3)\,\mathbf{b}^* + (p_1q_2 - p_2q_1)\,\mathbf{c}^*],\end{aligned} \tag{1.25}$$

where we have used the definition of the reciprocal basis vectors (equation 1.10). We thus find that the vector cross product between two vectors in direct space is described by a vector expressed in the reciprocal reference frame! This is to be expected since the vector cross product results in a vector perpendicular to the plane formed by the two initial vectors, and we know that the reciprocal reference frame deals with such normals to planes.

In a Cartesian reference frame, the reciprocal basis vectors are identical to the direct basis vectors $\mathbf{e}_i = \mathbf{e}_i^*$ (this follows from equation (1.21) and from the fact that the direct metric tensor is the identity matrix), and the unit cell volume is equal to 1, so the expression for the cross product reduces to the familiar expression:

$$\mathbf{p}\times\mathbf{q} = (p_2q_3 - p_3q_2)\,\mathbf{e}_1 + (p_3q_1 - p_1q_3)\,\mathbf{e}_2 + (p_1q_2 - p_2q_1)\,\mathbf{e}_3.$$

We introduce a new symbol, the *normalized permutation symbol* e_{ijk}. This symbol is defined as follows:

$$e_{ijk} = \begin{cases} +1 & \text{even permutations of 123,} \\ -1 & \text{odd permutations of 123,} \\ 0 & \text{all other cases.} \end{cases}$$

1
even
3 2
1
odd
3 2

The even permutations of the indices ijk are 123, 231, and 312; the odd permutations are 321, 213, and 132. For all other combinations, the permutation symbol vanishes. The sketch on the right shows an easy way to remember the combinations. We can now rewrite equation (1.25) as

$$\boxed{\mathbf{p}\times\mathbf{q} = \Omega e_{ijk} p_i q_j \mathbf{a}_k^*.} \tag{1.26}$$

Note that this is equivalent to the more conventional determinantal notation for the cross product:

$$\mathbf{p} \times \mathbf{q} = \Omega \begin{vmatrix} \mathbf{a}_1^* & \mathbf{a}_2^* & \mathbf{a}_3^* \\ p_1 & p_2 & p_3 \\ q_1 & q_2 & q_3 \end{vmatrix} \left(= \begin{vmatrix} \mathbf{e}_1 & \mathbf{e}_2 & \mathbf{e}_3 \\ p_1 & p_2 & p_3 \\ q_1 & q_2 & q_3 \end{vmatrix} \text{ Cartesian} \right).$$

Using equation (1.21), we also find

$$\boxed{\mathbf{p} \times \mathbf{q} = \Omega e_{ijk} p_i q_j g^*_{km} \mathbf{a}_m.} \tag{1.27}$$

The general definition of the cross product can be used in a variety of situations. A few examples are as follows.

(i) We can rewrite the definition of the reciprocal basis vectors (1.10) as a single equation, using the permutation symbol:

$$\mathbf{a}_i^* = \frac{1}{2\Omega} e_{ijk} \left(\mathbf{a}_j \times \mathbf{a}_k\right), \tag{1.28}$$

where a summation over j and k is implied. From this relation, we can also derive equation (1.21).

(ii) The volume Ω of the unit cell is given by the mixed product of the three basis vectors:[†]

$$\begin{aligned} \mathbf{a}_1 \cdot (\mathbf{a}_2 \times \mathbf{a}_3) &= \Omega \mathbf{a}_1 \cdot \left[e_{ijk} \mathbf{a}_{2,i} \mathbf{a}_{3,j} g^*_{km} \mathbf{a}_m\right], \\ &= \Omega e_{ijk} \delta_{2i} \delta_{3j} g^*_{km} \mathbf{a}_1 \cdot \mathbf{a}_m, \\ &= \Omega e_{23k} g^*_{km} g_{m1}, \\ &= \Omega e_{231} g^*_{1m} g_{m1}, \\ &= \Omega \delta_{11}, \\ &= \Omega. \end{aligned}$$

Example 1.10 *Determine the cross product of the vectors* [110] *and* [111] *in the tetragonal lattice of Example 1.1 on page 6.*

Answer: *From the general expression for the cross product we find*

$$\begin{aligned} \mathbf{t}_{[110]} \times \mathbf{t}_{[111]} &= \Omega e_{ijk} \mathbf{t}_{[110],i} \mathbf{t}_{[111],j} \mathbf{a}_k^*, \\ &= \frac{1}{4}\left[(1 \times 1 - 0 \times 1)\mathbf{a}_1^* + (0 \times 1 - 1 \times 1)\mathbf{a}_2^* + (1 \times 1 - 1 \times 1)\mathbf{a}_3^*\right], \\ &= \frac{1}{4}\left(\mathbf{a}_1^* - \mathbf{a}_2^*\right). \end{aligned}$$

Using the solution for Example 1.9 on page 18, this is also equal to $\mathbf{a}_1 - \mathbf{a}_2$ *or the direction vector* $[1\bar{1}0]$.

[†] $\mathbf{a}_{i,j}$ is the jth component of the basis vector $\mathbf{a}_i$, expressed in the direct reference frame $\mathbf{a}_i$.

It should be intuitively clear that if the cross product of two direct space vectors results in a reciprocal space vector, the reverse should also be true. This is indeed so, and it leads to an important tool for computing the direction indices of a *zone axis*. A zone axis is defined as a direction common to two or more planes. The concept of a zone axis will be crucial for standard electron diffraction techniques, as introduced in Chapter 4.

A direction common to two planes must be perpendicular to both plane normals, and, hence, the direction indices must be proportional to the components of the cross product of the two plane normals. Consider the two planes described by the normals $\mathbf{g}_1$ and $\mathbf{g}_2$. The cross product is given by (by analogy with equation 1.27)

$$\mathbf{g}_1 \times \mathbf{g}_2 = \Omega^* e_{ijk} g_{1,i} g_{2,j} g_{km} \mathbf{a}_m^* = \Omega^* e_{ijk} g_{1,i} g_{2,j} \mathbf{a}_k.$$

Explicitly working out the summations over i, j, and k we find

$$\mathbf{g}_1 \times \mathbf{g}_2 \parallel (k_1 l_2 - k_2 l_1)\,\mathbf{a}_1 + (l_1 h_2 - l_2 h_1)\,\mathbf{a}_2 + (h_1 k_2 - h_2 k_1)\,\mathbf{a}_3,$$

where we have dropped the reciprocal unit cell volume Ω^*, since direction indices are only defined in terms of relative prime integers.

This leads to a simple practical method for determining the direction indices of a zone axis. First, write down the Miller indices of the two vectors in horizontal rows, as follows:

$$\begin{array}{cccccc} h_1 & k_1 & l_1 & h_1 & k_1 & l_1 \\ h_2 & k_2 & l_2 & h_2 & k_2 & l_2. \end{array}$$

Then remove the first and the last column, i.e.

$$\begin{array}{cccccc} \not{h}_1 & k_1 & l_1 & h_1 & k_1 & \not{l}_1 \\ \not{h}_2 & k_2 & l_2 & h_2 & k_2 & \not{l}_2. \end{array}$$

Then compute the three 2×2 determinants formed by the eight remaining numbers above, as in

$$\begin{array}{ccccccc} k_1 & & l_1 & & h_1 & & k_1 \\ & \times & & \times & & \times & \\ k_2 & & l_2 & & h_2 & & k_2 \end{array}.$$

This leads to the following components for the direction vector $[uvw]$:

$$\left.\begin{aligned} u &= k_1 l_2 - k_2 l_1; \\ v &= l_1 h_2 - l_2 h_1; \\ w &= h_1 k_2 - h_2 k_1, \end{aligned}\right\} \qquad (1.29)$$

in agreement with the derivation above. This simple method for the computation of the zone axis indices is rather easy to remember, and it is sufficient for most computations involving cross products. One should keep in mind, however, that it can only be used if the actual length of the cross product vector is unimportant; i.e. if the resulting indices are rescaled to relative prime integers. If such a rescaling is not allowed, then the general expression (1.27) for the computation of the vector cross product must be used.

Example 1.11 *Determine the cross product of the lattice vectors* [110] *and* [111].

Answer: *Write the two vectors twice in rows, then remove the first and last columns, and work out the three* 2×2 *determinants:*

$$\begin{matrix} \not{1} & 1 & 0 & 1 & 1 & \not{0} \\ \not{1} & 1 & 1 & 1 & 1 & \not{1} \end{matrix} = (1\bar{1}0) \qquad (= \mathbf{g}_{1\bar{1}0}).$$

Example 1.12 *Determine the cross product of the reciprocal lattice vectors* $\mathbf{g}_{110}$ *and* $\mathbf{g}_{111}$.

Answer: *The answer is identical to that of the previous example, except that the resulting vector is a direct space vector:*

$$\begin{matrix} \not{1} & 1 & 0 & 1 & 1 & \not{0} \\ \not{1} & 1 & 1 & 1 & 1 & \not{1} \end{matrix} = [1\bar{1}0] \qquad (= \mathbf{t}_{1\bar{1}0}).$$

Sometimes we will need a general expression for the set of all planes containing a given direction $[uvw]$. Such a collection of planes is known as a *zone*. It is easy to see that a plane belongs to a zone only if its plane normal is perpendicular to the zone axis direction $[uvw]$. This means that the dot product $\mathbf{g}_{(hkl)} \cdot \mathbf{t}_{[uvw]}$ must vanish, or

$$\mathbf{g}_{(hkl)} \cdot \mathbf{t}_{[uvw]} = g_i t_j \mathbf{a}_i^* \cdot \mathbf{a}_j = g_i t_j \delta_{ij} = g_i t_i = 0,$$

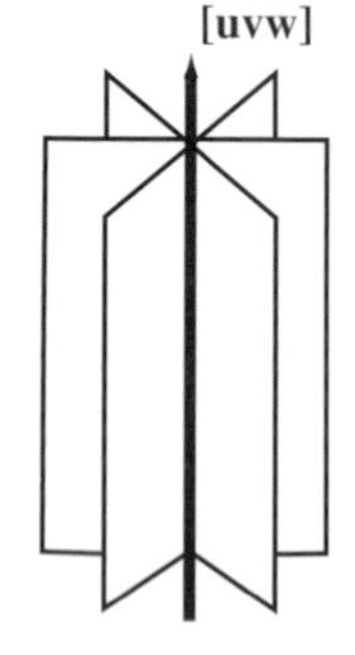

or in explicit component notation:

$$\boxed{hu + kv + lw = 0.} \tag{1.30}$$

This equation is known as the *zone equation* and it is valid for all crystal systems.

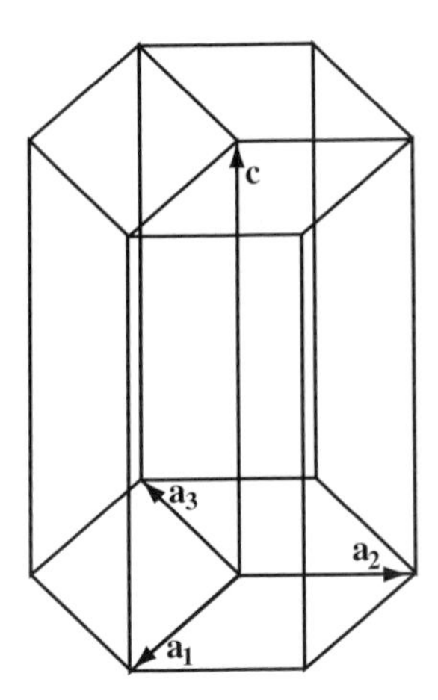

Fig. 1.6. The hexagonal unit cell is not uniquely defined and one can select any of the three cells shown above.

1.4 The hexagonal system

The hexagonal crystal system deserves a separate section because of some subtleties related to the indexing of directions and planes. The lattice parameters for the hexagonal system are given by $\{a, a, c, 90, 90, 120\}$. The choice of the unit cell is not unambiguous, since one can select any one of the three cells indicated in Fig. 1.6. The coplanar vectors $\mathbf{a}_i$, $i = 1, 2, 3$, and $\mathbf{c}$ form a *linearly dependent* set of four basis vectors. Three non-coplanar basis vectors always suffice for a three-dimensional (3D) crystal, but for the hexagonal system there are advantages to expressing vectors as linear combinations of the four basis vectors.

First of all, we can use the standard (hexagonal) basis vectors $\mathbf{a}_1$, $\mathbf{a}_2$, and $\mathbf{c}$ to express directions and Miller indices of planes. We will denote the three-index components of a lattice vector $\mathbf{t}$ by the symbol $[u'v'w']$. The Miller indices of a plane with intercepts $\frac{1}{h}, \frac{1}{k}, \frac{1}{l}$ are given by (hkl), and we can readily define the direct and reciprocal metric tensors for this 3-index system. All equations derived in the previous sections hold for this 3-index description of the hexagonal unit cell.

We can also express directions and plane normals with respect to the four basis vectors $\mathbf{a}_1$, $\mathbf{a}_2$, $\mathbf{a}_3$, and $\mathbf{c}$. A direction is then specified using the *Miller–Bravais* indices $[uvtw]$, with the third index t referring to the extra basis vector $\mathbf{a}_3$, i.e. $\mathbf{t} = u\mathbf{a}_1 + v\mathbf{a}_2 + t\mathbf{a}_3 + w\mathbf{c}$. Similarly, we can define the four-index symbol for a plane by $(hkil)$, with i being proportional to the reciprocal intercept of the plane with the basis vector $\mathbf{a}_3$. It is straightforward to show that $i = -(h + k)$, using elementary trigonometry based on Fig. 1.7(a). The 4-index Miller–Bravais system can be described by a 4×4 metric tensor. In the following two sections, we will discuss the subtleties and pitfalls of this 4-index notation.

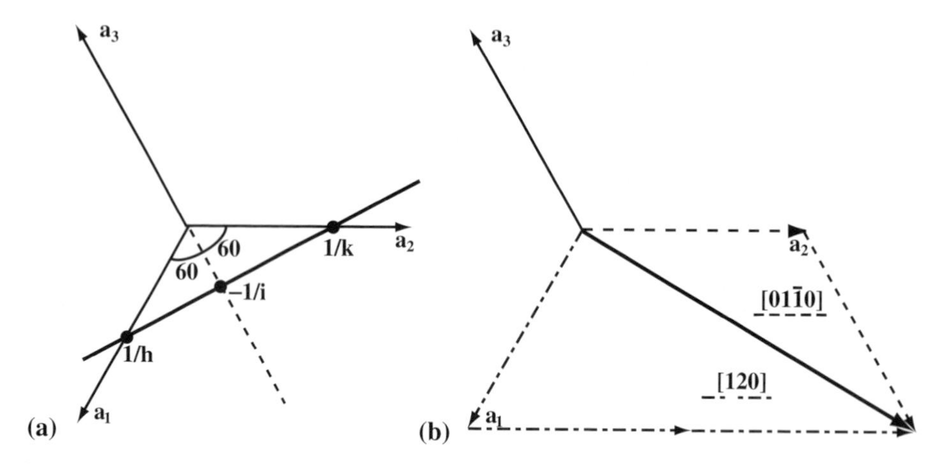

Fig. 1.7. (a) Schematic drawing used to show that $i = -(h + k)$; (b) the $[120] = [01\bar{1}0]$ direction, represented in the 3-index (dash-dotted lines) and 4-index (dashed lines) systems.

1.4.1 Directions in the hexagonal system

We note that the extra basis vector can be written as a linear combination of the other two vectors $\mathbf{a}_1$ and $\mathbf{a}_2$:

$$\mathbf{a}_3 = -(\mathbf{a}_1 + \mathbf{a}_2).$$

There are an infinite number of ways in which a 3D vector can be decomposed with respect to four basis vectors. For simplicity we select the particular decomposition for which $t = -(u + v)$, a relation similar to that for planes (see the next subsection). Let us now determine how this extra index relates to the 3-index notation $[u'v'w']$. The vector $\mathbf{t}$ is given by

$$\mathbf{t} = u'\mathbf{a}_1 + v'\mathbf{a}_2 + w'\mathbf{c} = u\mathbf{a}_1 + v\mathbf{a}_2 + t\mathbf{a}_3 + w\mathbf{c}.$$

Because the vector $\mathbf{t}$ is described by the *sum* of three or four vectors, one cannot simply leave out the third index t to convert from 4-index to 3-index notation. This is allowed for plane normals, as we will see in the next section, but *not* for lattice vectors. The correct transformation relations can easily be shown to be

$$\left.\begin{aligned} u' &= u - t = 2u + v; \\ v' &= v - t = 2v + u; \\ w' &= w. \end{aligned}\right\} \tag{1.31}$$

Table 1.1. *Equivalent indices in the Miller and Miller–Bravais indexing systems for hexagonal directions.*

Miller	Miller–Bravais	Miller	Miller–Bravais
[100]	$[2\bar{1}\bar{1}0]$	[010]	$[\bar{1}2\bar{1}0]$
[110]	$[11\bar{2}0]$	$[\bar{1}10]$	$[\bar{1}100]$
[001]	[0001]	[101]	$[2\bar{1}\bar{1}3]$
[011]	$[\bar{1}2\bar{1}3]$	[111]	$[11\bar{2}3]$
[210]	$[10\bar{1}0]$	[120]	$[01\bar{1}0]$
[211]	$[10\bar{1}1]$	[112]	$[11\bar{2}6]$

The inverse relations are given by

$$\left.\begin{aligned} u &= \frac{1}{3}\left(2u' - v'\right); \\ v &= \frac{1}{3}\left(2v' - u'\right); \\ t &= -\frac{1}{3}\left(u' + v'\right) = -(u + v); \\ w &= w'. \end{aligned}\right\} \qquad (1.32)$$

The difference between the two indexing systems is illustrated in Fig. 1.7(b): the lattice vector [120] is drawn (dash-dotted lines) as the sum $\mathbf{a}_1 + 2\mathbf{a}_2$, and also (dashed lines) as the sum $\mathbf{a}_2 - \mathbf{a}_3$, corresponding to the 4-index symbol $[01\bar{1}0]$. Table 1.1 lists a few common (low-index) directions in both 3- and 4-index notation.

The metric tensor relations derived in Section 1.2.2, and the equations presented in Tables A1.1–A1.4 are only valid for the 3-index system. To compute the length $|\mathbf{t}|$ of a lattice vector $\mathbf{t}$ in 4-index notation one can proceed in one of two ways:

(i) first convert the indices to the 3-index system using the relations (1.31) above, then use the standard relations described in this chapter and Table A1.1;
(ii) perform the calculation using the 4-index metric tensor formalism described in the following paragraphs.

Following [OT68], the 4-index metric tensor G is defined in the usual way as

$$G_{ij} \equiv \begin{bmatrix} \mathbf{a}_1\cdot\mathbf{a}_1 & \mathbf{a}_1\cdot\mathbf{a}_2 & \mathbf{a}_1\cdot\mathbf{a}_3 & \mathbf{a}_1\cdot\mathbf{c} \\ \mathbf{a}_2\cdot\mathbf{a}_1 & \mathbf{a}_2\cdot\mathbf{a}_2 & \mathbf{a}_2\cdot\mathbf{a}_3 & \mathbf{a}_2\cdot\mathbf{c} \\ \mathbf{a}_3\cdot\mathbf{a}_1 & \mathbf{a}_3\cdot\mathbf{a}_2 & \mathbf{a}_3\cdot\mathbf{a}_3 & \mathbf{a}_3\cdot\mathbf{c} \\ \mathbf{c}\cdot\mathbf{a}_1 & \mathbf{c}\cdot\mathbf{a}_2 & \mathbf{c}\cdot\mathbf{a}_3 & \mathbf{c}\cdot\mathbf{c} \end{bmatrix} = \frac{a^2}{2}\begin{bmatrix} 2 & -1 & -1 & 0 \\ -1 & 2 & -1 & 0 \\ -1 & -1 & 2 & 0 \\ 0 & 0 & 0 & 2c^2/a^2 \end{bmatrix}. \qquad (1.33)$$

The three vectors $\mathbf{a}_1$, $\mathbf{a}_2$, and $\mathbf{a}_3$ are coplanar, which means that the determinant of G vanishes. This has important implications for the definition of the reciprocal lattice vectors of the hexagonal system, as described in the next section.

The length of the lattice vector $\mathbf{t}$ in 4-index notation is then derived from the standard equations:

$$|\mathbf{t}|^2 = \mathbf{t}\cdot\mathbf{t} = t_i G_{ij} t_j = \frac{a^2}{2}[u\ v\ t\ w]\begin{bmatrix} 2 & -1 & -1 & 0 \\ -1 & 2 & -1 & 0 \\ -1 & -1 & 2 & 0 \\ 0 & 0 & 0 & 2c^2/a^2 \end{bmatrix}\begin{bmatrix} u \\ v \\ t \\ w \end{bmatrix},$$

from which we can easily derive

$$|\mathbf{t}| = \left[3a^2\left(u^2 + uv + v^2\right) + c^2w^2\right]^{1/2},$$

which is to be compared with the 3-index equation (Table A1.1)

$$|\mathbf{t}| = \left[a^2\left(u'^2 - u'v' + v'^2\right) + c^2w'^2\right]^{1/2}.$$

The angle θ between two lattice vectors in the 4-index notation can also be derived from the G tensor and the result is

$$\cos\theta = \frac{\frac{1}{2}a^2\left[6\left(u_1u_2 + v_1v_2\right) + 3\left(u_1v_2 + u_2v_1\right)\right] + c^2w_1w_2}{\left[3a^2\left(u_1^2 + u_1v_1 + v_1^2\right) + c^2w_1^2\right]^{1/2}\left[3a^2\left(u_2^2 + u_2v_2 + v_2^2\right) + c^2w_2^2\right]^{1/2}}.$$

1.4.2 The reciprocal hexagonal lattice

In Section 1.3.4, we have seen that the reciprocal basis vectors can be written as linear combinations of the direct basis vectors and the rows of the reciprocal metric tensor form the coefficients for this relation. To describe the 4-index reciprocal space, we must first obtain expressions for the reciprocal basis vectors. Since the direct 4-index metric tensor has a zero determinant, its inverse does not exist, so we cannot simply use the same relations to derive the reciprocal basis vectors. Instead we must explicitly derive the reciprocal 4-index basis vectors.

One of the primary motivations for constructing the reciprocal lattice was that triplets of Miller indices (hkl) could then be interpreted as the components of a vector $\mathbf{g}_{hkl}$. The 4-index notation for the Miller indices $(hkil)$, with $i = -(h+k)$, suggests that there are four reciprocal basis vectors, and that the coefficients h, k, i, and l are the components of the plane normal with respect to those basis vectors. Let us assume, again following [OT68], that there are indeed four basis vectors $\mathbf{A}_1^*$, $\mathbf{A}_2^*$, $\mathbf{A}_3^*$, and $\mathbf{C}^*$, chosen such that the following relation is valid:

$$\mathbf{g}_{hkl} = h\mathbf{a}_1^* + k\mathbf{a}_2^* + l\mathbf{c}^* = h\mathbf{A}_1^* + k\mathbf{A}_2^* + i\mathbf{A}_3^* + l\mathbf{C}^* = \mathbf{g}_{hkil}. \quad (1.34)$$

In other words, the vector $\mathbf{g}_{hkl}$ is identical to the vector $\mathbf{g}_{hkil}$, namely the normal to the plane $(hkl) = (hkil) \equiv (hk.l)$. The relation between the 4- and 3-index notation is rather simple: the 3-index notation can be obtained by dropping the third index i from the 4-index symbol. However, we must be careful when interpreting the indices $(hkil)$ as components of a vector! Indeed, from the 3-index reciprocal metric tensor for the hexagonal system follows:

$$\mathbf{a}_1^* = g_{1j}^* \mathbf{a}_j = \frac{4}{3a^2}\mathbf{a}_1 + \frac{2}{3a^2}\mathbf{a}_2;$$

$$\mathbf{a}_2^* = g_{2j}^* \mathbf{a}_j = \frac{2}{3a^2}\mathbf{a}_1 + \frac{4}{3a^2}\mathbf{a}_2;$$

$$\mathbf{c}^* = g_{3j}^* \mathbf{a}_j = \frac{\mathbf{c}}{c^2}.$$

After substitution of these relations into equation (1.34), we find for the 4-index reciprocal basis vectors:

$$\mathbf{A}_1^* = \frac{2}{3a^2}\mathbf{a}_1;$$

$$\mathbf{A}_2^* = \frac{2}{3a^2}\mathbf{a}_2;$$

$$\mathbf{A}_3^* = \frac{2}{3a^2}\mathbf{a}_3;$$

$$\mathbf{C}^* = \frac{\mathbf{c}}{c^2}.$$

It is easy to see that the 4-index reciprocal basis vectors *are not even parallel to the 3-index reciprocal basis vectors*! Instead, they are parallel to the original direct space basis vectors. The relation between the direct basis vectors and the two sets of reciprocal basis vectors is illustrated in Fig. 1.8(a). In Fig. 1.8(b), the $(11\bar{2}0)$ reciprocal lattice point is drawn as two linear combinations: $\mathbf{a}_1^* + \mathbf{a}_2^*$ and $\mathbf{A}_1^* + \mathbf{A}_2^* - 2\mathbf{A}_3^*$. Note that the reciprocal lattice vectors $\mathbf{A}_i^*$ give rise to a reciprocal lattice that is three times as dense as the actual reciprocal lattice; only those points for which $i = -(h+k)$, indicated by larger filled circles in Fig. 1.8(b), should be counted as real reciprocal lattice points.

The 4-index reciprocal metric tensor G^* is then given by

$$G_{ij}^* = \frac{2}{9a^2}\begin{bmatrix} 2 & -1 & -1 & 0 \\ -1 & 2 & -1 & 0 \\ -1 & -1 & 2 & 0 \\ 0 & 0 & 0 & 9a^2/2c^2 \end{bmatrix}. \qquad (1.35)$$

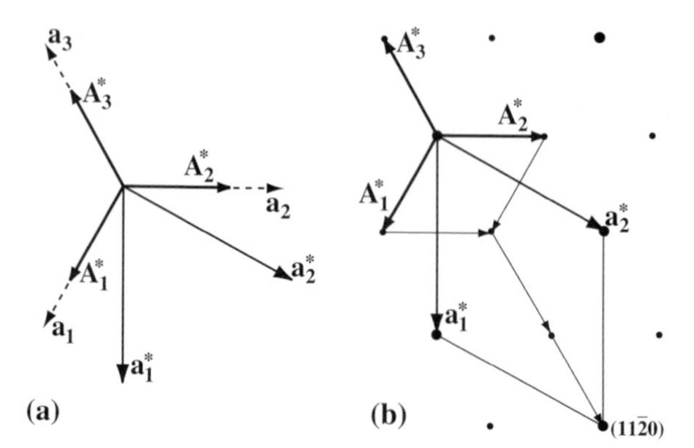

Fig. 1.8. (a) Geometrical representation of the direct and reciprocal basis vectors of the hexagonal lattice; (b) representation of the (11$\bar{2}$0) reciprocal lattice point.

It is straightforward to show that the length of a reciprocal lattice vector $\mathbf{g}_{hkil}$ is given by

$$|\mathbf{g}_{hkil}| = g_i G^*_{ij} g_j = \frac{4}{9a^2}\left[h^2 - hk + k^2 - i(h+k) + i^2\right] + \frac{l^2}{c^2},$$

and after substitution of $i = -(h+k)$, this expression becomes identical to that in Table A1.4. Since the 3- and 4-index equations for $|\mathbf{g}|$ are identical, there is no real need for the G^* reciprocal metric tensor to compute distances and angles in reciprocal space.

Summarizing, the Miller–Bravais and Miller symbols for planes in a hexagonal crystal lattice are related to each other by a simple additional index $i = -(h+k)$; when considered as components of the plane normal, they are expressed with respect to two *different* sets of reciprocal basis vectors. In contrast, the 3- and 4-index expressions for lattice vectors are expressed with respect to the *same* set of basis vectors, but the relation between the sets of indices is somewhat more complex (see equations 1.31 and 1.32). The simultaneous use of 3- and 4-index symbols may cause confusion, so we recommend choosing one of the systems rather than mixing the two systems, in particular when indexing hexagonal diffraction patterns.

As a final note, we point out that the 4-index system, when used as described above, is fully equivalent to the 3-index system and leads to the same type of relations between geometric quantities. As an important example, we mention the *zone equation*, which states when a plane (hkl) belongs to a zone $[uvw]$. In all crystal systems, the zone equation reads as

$$hu + kv + lw = 0.$$

In the 4-index description of the hexagonal system, this equation is still valid, provided the product of the additional indices is added. The dot product of a plane normal $\mathbf{g}$ with a lattice vector $\mathbf{t}$ in 4-index notation is given by

$$\begin{aligned}\mathbf{t}\cdot\mathbf{g} &= (u\mathbf{a}_1 + v\mathbf{a}_2 + t\mathbf{a}_3 + w\mathbf{c})\cdot\left(h\mathbf{A}_1^* + k\mathbf{A}_2^* + i\mathbf{A}_3^* + l\mathbf{C}^*\right)\\ &= (u\mathbf{a}_1 + v\mathbf{a}_2 + t\mathbf{a}_3 + w\mathbf{c})\cdot\left(\frac{2h}{3a^2}\mathbf{a}_1 + \frac{2k}{3a^2}\mathbf{a}_2 + \frac{2i}{3a^2}\mathbf{a}_3 + \frac{l}{c^2}\mathbf{c}\right)\\ &= hu + kv + it + lw,\end{aligned}$$

from which the hexagonal 4-index zone equation follows. Throughout this book, we will consistently use the 4-index notation for both directions and planes. The Fortran source code to be discussed in Section 1.9 uses the 3-index notation for all internal computations, but all input–output is performed in the 4-index format.

1.5 The stereographic projection

In this section, we introduce the concept of the *stereographic projection* and describe its importance for electron microscopy. As we shall see in later chapters, conventional TEM observations usually result in two-dimensional (2D) images or diffraction patterns. Since the object giving rise to those images is clearly three-dimensional (3D), we must find a way to represent the crystallographically relevant information in 2D drawings. This section deals with one of the most important 2D representations, the stereographic projection.

A stereographic projection is a 2D representation of a certain characteristic of a 3D object. This characteristic could be the set of normals to the bounding planes (or surfaces) of the object, or the set of directions parallel to all the edges of the object, or some other feature with a *directional* character. The stereographic projection was first proposed in 1839 by William H. Miller [Mil39] as a way to represent the normals to the faces of natural crystals in a 2D drawing, such that direct measurement of the angles between the face normals would be possible from the drawing.

Figure 1.9 shows a sphere of radius R. To obtain the stereographic projection (SP) of a point P on the sphere, we connect the point with the south pole (S) of the sphere and then determine the intersection of this connection line, PS, with the equatorial plane. The resulting point is the stereographic projection of the original point. The point on the sphere could represent the normal to a crystal plane, as shown in the figure, but it could also represent some other direction or directional feature. If the object is oriented such that a symmetry axis coincides with the north–south axis of the projection sphere, then the projection will show a corresponding 2D symmetry. We will discuss crystal symmetries in detail in the next section.

The location of the stereographic projection of a point is most easily computed using spherical coordinates. If the original point has coordinates (R, ϕ, θ), with ϕ

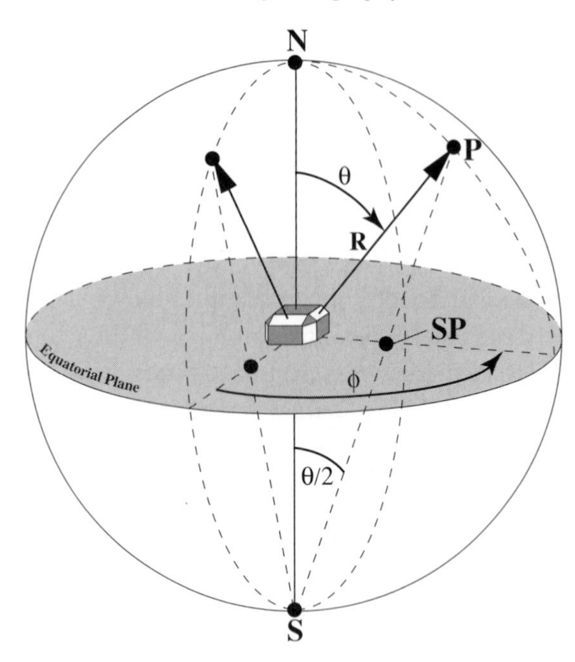

Fig. 1.9. Stereographic projection of the normals on crystal faces.

measured counterclockwise from a fixed axis in the equatorial plane and θ measured from the north pole (z-axis, see Fig. 1.9), then the stereographic coordinates are given by (ϕ, $R\tan\frac{\theta}{2}$). From this we see that a point on the equator circle, i.e. with coordinates ($R, \phi, \frac{\pi}{2}$), will have an SP on the equatorial circle with coordinates (ϕ, R). Points in the southern hemisphere will be projected outside the projection circle; to avoid having to make very large drawings, it is customary to project points in the southern hemisphere from the north pole, and represent the projections with open circles rather than filled circles for projections from the south pole.

The stereographic projection techniques make use of the so-called *Wulff net*, which is shown in Fig. 1.10. A standard Wulff net has a diameter of 20 cm (in the figure this is somewhat reduced to fit on the page). The net shows two sets of arcs: the first set intersects the points M' and M'' and represents the projections of *great circles*; i.e. circles with the same diameter as the projection sphere. If the line M'–M'' is taken as the origin for measurement of ϕ, then one can read for each of these great circles the value of θ from the line A–B. There is a great circle per degree, and every tenth circle is drawn with a slightly thicker line.

The second set of arcs on the net corresponds to the projection of a set of parallel planes, intersecting the projection sphere in circles. These planes are perpendicular to the equatorial plane and to the M'–M'' axis. If the projection sphere is rotated

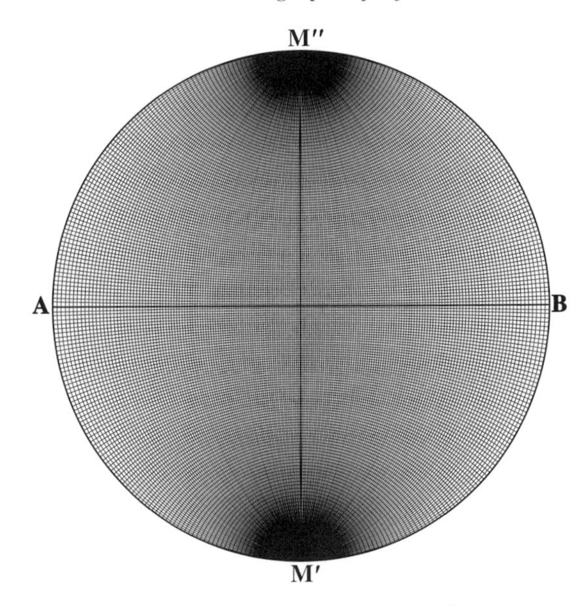

Fig. 1.10. Standard Wulff net for stereographic projections.

around this axis, then a point on the surface will trace a circular path; the projection of this path is given by the second set of arcs, which are again spaced by 1°.

Stereographic projections conserve angles; i.e. measurement of an angle on the projection will always correspond to the real 3D angle. It is this property that turns the SP into a rather useful technique for crystallography and TEM. In the following paragraphs, we will discuss several basic (manual) operations using the Wulff net. Some computer programs have all of these routines available, but it is often useful to know how to do them by hand.

The Wulff net is commercially available from various manufacturers (see website); one could also make a simple Wulff net using the PostScript file wulff.ps available from the website. The easiest way to use the net is to print or glue it on a thick piece of paper, and cut around the circle, leaving about 10 mm of extra space all around the edge. Figure 1.11 then shows how the net can be used to produce manual stereographic projections. A transparency is mounted with pins or tape onto a board. The Wulff net is then positioned underneath the transparency, such that the entire projection circle is covered by it. Then a pin is inserted through both the transparency and the Wulff net (through the center of the net), such that the whole assembly is firmly attached to the board. The Wulff net can then be turned around the pin, while the transparency remains fixed. All of the following sections assume that the reader has a similar setup available for manual stereographic projections.

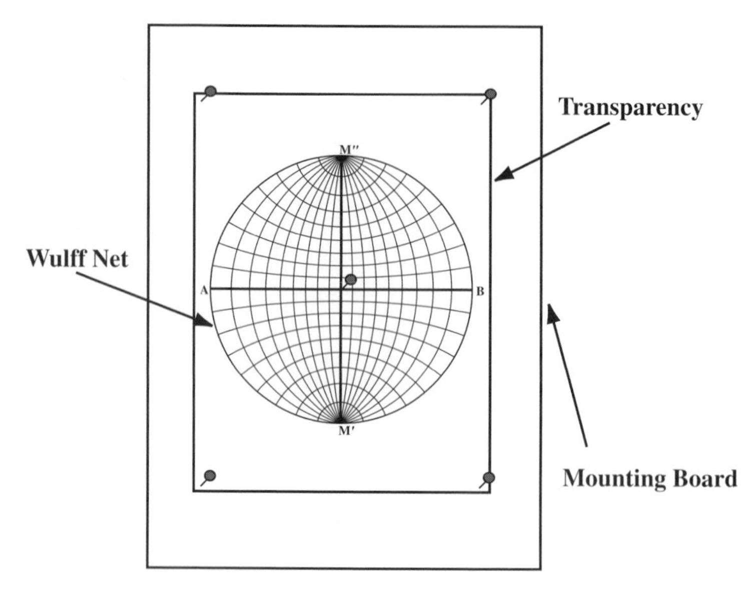

Fig. 1.11. Illustration of the manual use of a Wulff net for stereographic projections.

1.5.1 Drawing a point

Drawing a single point is the most elementary operation one can perform on a stereographic net. Let us assume that the standard net is being used, and that the 3D spherical coordinates of a point are known. They can be transferred onto the net in the following way: align the line A–B (by rotating the net) such that it is parallel to the bottom of the transparency. Then draw the line A–B onto the transparency; this will be your reference line and we will call it R–S (see Fig. 1.12a). If the coordinates of the point are (R, ϕ, θ), rotate the net over an angle ϕ (such that the line A–B on the net makes an angle ϕ with R–S). From the point M'', measure the angle θ along the outer circle;[†] connect the corresponding point with the point M'. The intersection of this line with the line A–B (the point p) is the SP of the original point.

1.5.2 Constructing a great circle through two poles

If two points p and q are known (Fig. 1.12b), then the great circle through those points can easily be found as follows: we know that the arcs connecting the points M' and M'' are projections of great circles. Therefore, we rotate the net until one of those arcs connects the two points. The resulting arc is then the projection of the great circle through the two given points.

[†] Alternatively one could measure the angle θ from the center of the circle along the line A–B.

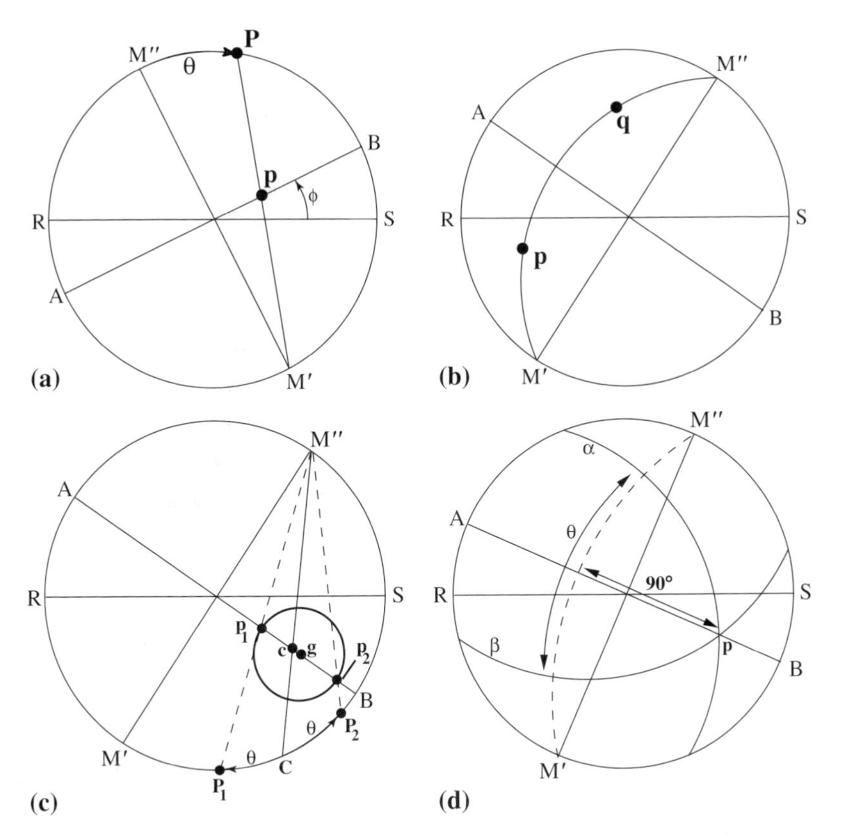

Fig. 1.12. (a) Determination of the SP of a single point, the spherical coordinates of which are known; (b) determination of the great circle through two points; (c) drawing a circle of radius θ around a point c; (d) determination of the angle between two great circles.

1.5.3 Constructing a small circle around a pole

We now wish to draw a circle with radius θ around a given pole. One general and two trivial solutions exist. If the center of the circle, i.e. the given pole, coincides with the center of the projection, then the circle can easily be drawn by reading the radius θ along the line A–B and using a compass set to this radius. If the circle has its pole on the equatorial plane, then all we need to do is rotate the net such that the point M' coincides with the pole; circles centered on M' are already drawn on the net and we only need to copy the one corresponding to θ.

The general case, where the pole does not lie in the center or on the equator, is shown in Fig. 1.12(c). Suppose the given pole is indicated by the point c. The center of the projection of a circle will, in general, not coincide with the projection of the center of the circle. The projected center can be found as follows: bring the line A–B onto the point c. From the point M'', draw a line through c which intersects the outer circle in C. Measure the angle θ on both sides of C, along the outer circle. Connect the resulting points P_1 and P_2 with M''. The intersections of these lines with the line A–B are called p_1 and p_2; both of these points lie on the projected circle. The center g of the projected circle lies midway between p_1 and p_2 and has radius gp_1. This completes the construction. The construction of circles is useful for error bars and for some aspects of trace analysis.

1.5.4 Finding the pole of a great circle

The pole of a great circle is defined as the north pole for that circle; i.e. it lies 90° away from each point of the circle. Since great circles on the Wulff net connect the points M' and M'', the corresponding pole can be found by measuring 90° along the line A–B from the intersection point of the great circle and this line. This construction is used to determine the location of a zone axis when two points of the zone are known.

1.5.5 Measuring the angle between two poles

Since the SP conserves angles, one can easily determine angles from the projection diagram. If two points, p and q, are known, one can draw the great circle through these points (see above) and then read the angle along that great circle from the Wulff net.

1.5.6 Measuring the angle between two great circles

The angle between two great circles is also easily determined: if the great circles α and β are given (Fig. 1.12d), bring the line A–B through their intersection point p, measure 90° along that line, and draw the great circle corresponding to the pole p. The angle between the intersections of this circle with the circles α and β is the required angle. We will use the various methods introduced in this section when we discuss trace analysis in Chapter 8.

1.6 Crystal symmetry

An object is said to have a certain symmetry if any operations exist which leave this object *invariant*. A symmetry operation is, thus, a mathematical operation that leaves the distances between all material points of an object unchanged (i.e. no stretching, twisting, shearing, or bending) or, equivalently, a symmetry operation is an *isometric* operation. The symmetry operations relevant for crystallography are translations, rotations, reflections, inversions, and their combinations. Only a finite number of combinations of such symmetry operations are compatible with the 14 Bravais lattices. The 230 allowed (and unique) combinations of symmetry elements are known as the three-dimensional (3D) *space groups*, and they are tabulated in [Hah96]. In this section, we will describe, with minimal derivations, those aspects of symmetry theory of importance for electron microscopy. In particular, we will describe the mathematical representation of symmetry operators, the 32 point groups, the 230 space groups, and how they can be used.

1.6.1 Symmetry operators

We distinguish between two basic kinds of symmetry operations: those that can be physically realized (rotations and translations), also known as operations of the first kind, or *proper operations*, and those that change the *handedness* of an object (reflection and inversion), operations of the second kind. All symmetry operations are represented by unique graphical symbols.

Operations of the first kind

(i) A pure rotation is characterized by a rotation axis $[uvw]$ and a rotation angle $\alpha = 2\pi/n$. The integer n is the *order* of the rotation and we say that a rotation is n-fold if its angle is given by $2\pi/n$. A pure rotation of order n is denoted by the symbol **n**. This is the so-called International or Hermann–Mauguin notation.[†] In drawings an n-fold rotation axis is represented by a filled regular polygon with n sides. A six-fold rotation axis perpendicular to the drawing plane is then indicated by the ⬢ symbol, a four-fold axis by ◆, a three-fold axis by ▲, and a two-fold axis by ⬮.

(ii) A pure translation is characterized by a translation vector **t**. We have already discussed translations earlier in this chapter. In drawings, translation vectors are indicated by arrowed lines.

Operations of the second kind

(i) A *pure reflection* is characterized by a plane (hkl), and the International symbol for a mirror plane is **m**. The Schœnflies symbol is the Greek letter σ. In a drawing a mirror plane is always indicated by a thick solid line, ——— .

(ii) An *inversion* is a point symmetry operation that takes all the points **r** of an object and projects them onto $-\mathbf{r}$. The operation is usually denoted by $\bar{\mathbf{1}}$ or sometimes by i. In drawings the inversion center is denoted by the symbol ∘ (i.e. a small open circle).

Symmetry operations are often represented by means of stereographic projections. Figure 1.13 shows the stereographic projections for a three-fold rotation, a

[†] There is a second notation system, the *Schœnflies* system, in which rotations are indicated by the symbol C_n. Even though the International system is the preferred system, we will also mention the Schœnflies notation because it is still frequently used.

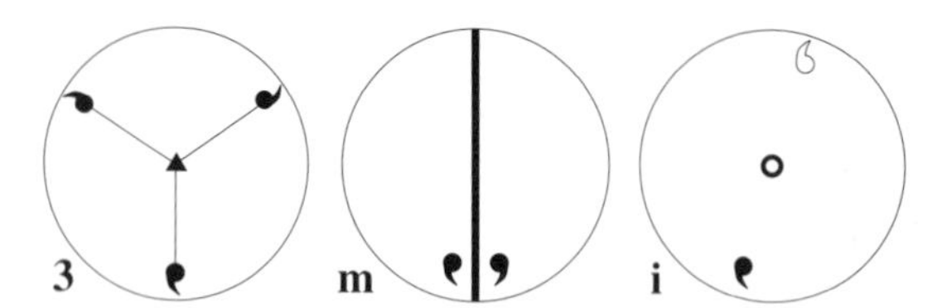

Fig. 1.13. Stereographic representation of a three-fold rotation, a mirror plane perpendicular to the projection plane, and an inversion.

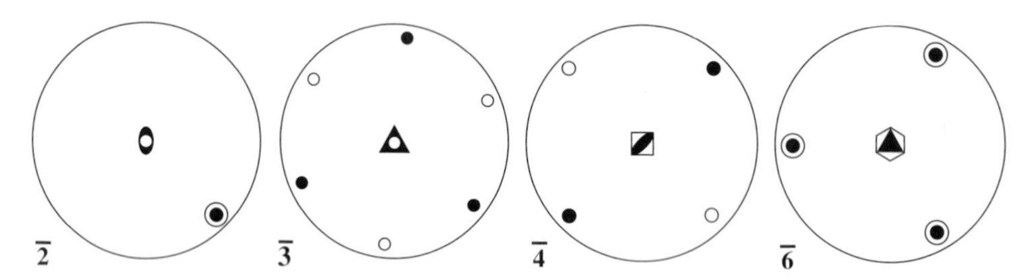

Fig. 1.14. Stereographic representation of the crystallographic rotoinversions.

mirror plane perpendicular to the projection plane, and an inversion. Note that the object does not have any symmetry itself (a circle with a curved tail), so that a change in handedness is readily observed.

The basic symmetry operations of the first and second kind can be combined with each other to create new symmetry operations. There are three combinations of interest to crystallography:

- combination of rotations with the inversion center;
- combination of rotations with translations;
- combination of mirrors with translations.

1.6.1.1 Combination of a rotation with the inversion center

The combination of a rotation axis with an inversion center located somewhere on that axis is called a *rotoinversion operation*. The rotoinversion rotates a point over an angle $2\pi/n$ **and** inverts the resulting point through the inversion center. A rotoinversion of order one is equivalent to the inversion operation i. Rotoinversions are represented by the symbol $\bar{\mathbf{n}}$ and the crystallographic rotoinversions are shown in the stereographic projections of Fig. 1.14. They are represented by special symbols: ⬮ for $\bar{\mathbf{2}}$, ▲ for $\bar{\mathbf{3}}$, ◆ for $\bar{\mathbf{4}}$, and ⬢ for $\bar{\mathbf{6}}$.

1.6.1.2 Combination of a rotation with a translation

For all symmetry operations except the translation, repeated operation eventually returns one to the initial point; e.g. three subsequent operations of the three-fold axis ▲ complete a 360° rotation, and two subsequent mirror operations reproduce the original object again. For combinations of symmetry elements involving the translation, this is no longer the case and one never returns to the initial point, no matter how many times the operation is repeated.

A *screw axis* $\mathbf{n_m}$ consists of a counterclockwise rotation through $2\pi/n$ followed by a translation $\mathbf{T} = \frac{m}{n}\mathbf{t}_{[uvw]}$ in the positive direction $[uvw]$ along the screw axis. The vector $\mathbf{T}$ is known as the *pitch* of the screw axis. Screw axes of the type $\mathbf{n_m}$ and $\mathbf{n_{n-m}}$ are mirror images of each other. A screw axis is called *right-handed* if $m < n/2$, *left-handed* if $m > n/2$, and *without hand* if $m = 0$ or $m = n/2$. Screw axes related to each other by a mirror operation are called *enantiomorphous*.

The crystallographic screw axes are (with their official graphical symbols): 2_1, 3_1, 3_2, 4_1, 4_2, 4_3, 6_1, 6_2, 6_3, 6_4, and 6_5. All screw axes are shown in Fig. 1.15: the number next to each circle refers to the height of the circle above the plane of the drawing. The axes are all perpendicular to the drawing.

1.6.1.3 Combination of a mirror plane with a translation

The last class of symmetry operators involves combinations of mirror planes and translations. We define the *glide plane* as the combined operation of a mirror with

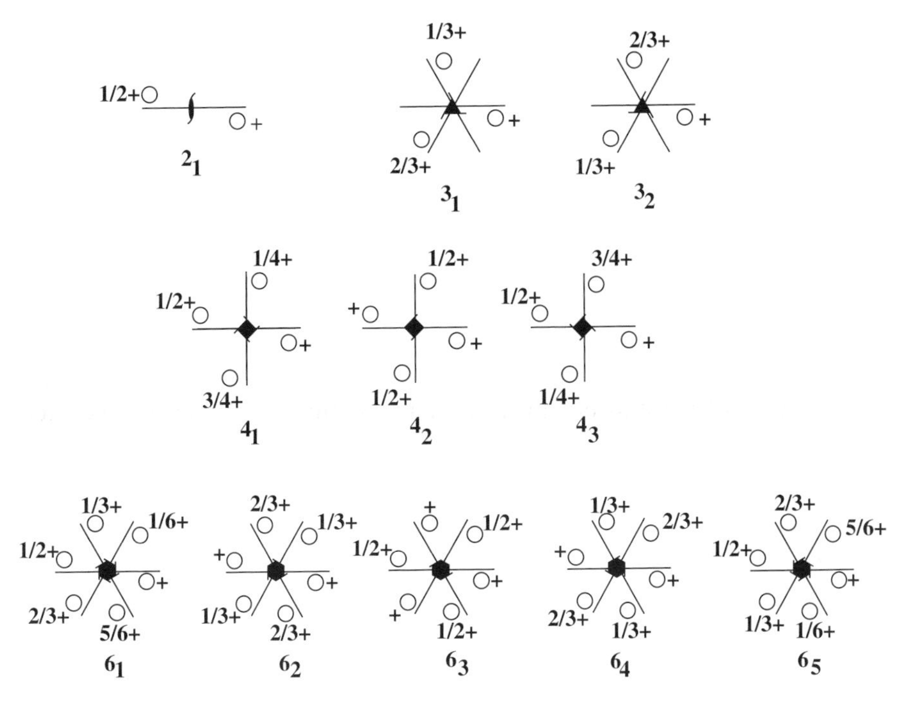

Fig. 1.15. Schematic representation of all crystallographic screw axes.

Table 1.2. *Different types of mirror and glide planes, their symbols and glide vectors.*

Name	Symbol	Glide vectors
Mirror	m	None
Axial glide	a	$\frac{\mathbf{a}}{2}$
	b	$\frac{\mathbf{b}}{2}$
	c	$\frac{\mathbf{c}}{2}$
Diagonal glide	n	$\frac{\mathbf{a}+\mathbf{b}}{2}, \frac{\mathbf{b}+\mathbf{c}}{2}, \frac{\mathbf{c}+\mathbf{a}}{2}, \frac{\mathbf{a}+\mathbf{b}+\mathbf{c}}{2}^{\dagger}$
Diamond glide	d	$\frac{\mathbf{a}\pm\mathbf{b}}{4}, \frac{\mathbf{b}\pm\mathbf{c}}{4}, \frac{\mathbf{c}\pm\mathbf{a}}{4}, \frac{\mathbf{a}\pm\mathbf{b}\pm\mathbf{c}}{4}^{\dagger}$

†These glide vectors are only possible in cubic and tetragonal systems.

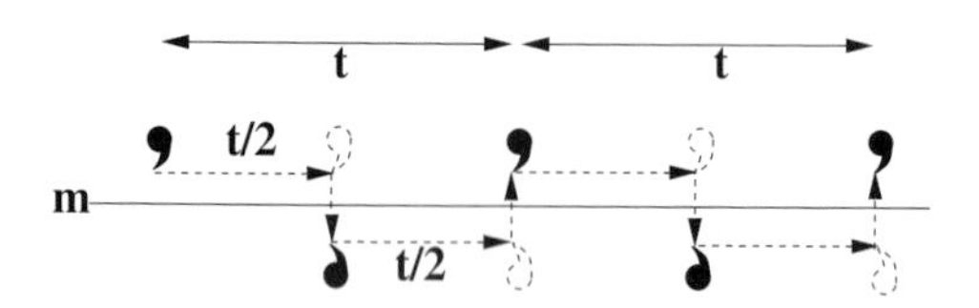

Fig. 1.16. Schematic representation of a glide plane.

a translation over *half* a lattice vector parallel to the mirror plane. An example is shown in Fig. 1.16.

The allowed glide vectors must be equal to one-half of the lattice vectors; in the case of centered Bravais lattices there are additional glide vectors, equal to one-half of the centering vectors $\frac{1}{2}(\mathbf{a}+\mathbf{b})$, $\frac{1}{2}(\mathbf{b}+\mathbf{c})$, $\frac{1}{2}(\mathbf{c}+\mathbf{a})$, and $\frac{1}{2}(\mathbf{a}+\mathbf{b}+\mathbf{c})$. Table 1.2 lists the various possibilities and names for all types of glide planes.

The official drawing symbols for glide planes depend on the orientation of the plane with respect to the drawing. If the plane is perpendicular to the drawing, then the plane is indicated by a solid bold line for m, a dashed bold line for a glide with translation in the plane of the drawing, a dotted bold line for a glide with translation perpendicular to the plane of the drawing, and a dash-dotted bold line for a diagonal glide (see Fig. 1.17). Note that a circle reflected in a mirror plane is represented by a circle with a comma in the center, indicating that an odd number of reflections relate that point to the original point. For glide planes parallel to the plane of the drawing, one uses a symbol based on ⌉, which represents a pure mirror. For an axial glide plane, an arrow is added to the symbol in the direction of the glide vector, i.e. ⌉ with an arrow. For a diagonal glide, the arrow points at an angle away from the corner, as in ⌉ with a diagonal arrow. Examples are shown in the right-hand column of Fig. 1.17.

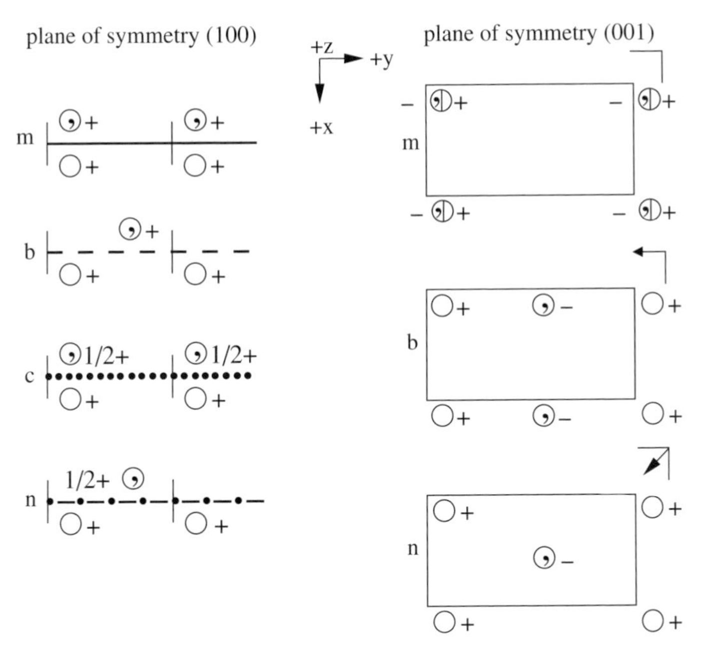

Fig. 1.17. Graphical representation of glide planes perpendicular (left) and parallel (right) to the plane of the drawing. (Figure based on Fig. 4.16 in [MM86].)

1.6.2 Mathematical representation of symmetry operators

A symmetry operator $\mathcal{O}$ operating on a material point $\mathbf{r}$ transforms its coordinates into a new vector $\mathbf{r}'$ with components (x'_1, x'_2, x'_3) as follows:

$$\mathbf{r}' = \mathcal{O}[\mathbf{r}]. \tag{1.36}$$

In the previous section we saw that the most general crystallographic symmetry operator $\mathcal{O}$ consists of a rotation/inversion/reflection and, sometimes, a translation. This means that in mathematical terms the coordinates of the new point can be derived from those of the old point by first applying the rotation/inversion/reflection and then adding the translation vector. This is explicitly described by

$$\begin{aligned} x'_1 &= D_{11}x_1 + D_{12}x_2 + D_{13}x_3 + u_1; \\ x'_2 &= D_{21}x_1 + D_{22}x_2 + D_{23}x_3 + u_2; \\ x'_3 &= D_{31}x_1 + D_{32}x_2 + D_{33}x_3 + u_3. \end{aligned}$$

Using the *Einstein summation convention* this set of equations can be rewritten as

$$x'_i = D_{ij}x_j + u_i, \tag{1.37}$$

or, in operator form,

$$\mathbf{r}' = \mathbf{D}\mathbf{r} + \mathbf{t} = (\mathbf{D}|\mathbf{t})[\mathbf{r}\,] = \mathcal{O}[\mathbf{r}\,]. \tag{1.38}$$

The matrix **D** describes the *point symmetry transformation* and the vector **t** describes the translational component of the operator $\mathcal{O}$. In the crystallographic literature, the point symmetry part is usually called the *linear part* and the whole operator is called a *motion*. The compact notation $(\mathbf{D}|\mathbf{t})$ is called the *Seitz symbol.*

The symmetry matrices can be defined with respect to a generic Cartesian reference frame, or with respect to the Bravais lattice vectors $\mathbf{a}_i$. The latter choice is the most convenient one, and it is straightforward to determine the entries of the matrix D_{ij} for any given symmetry operation. All that is needed are the transformations of the basis vectors [100], [010], and [001]. The columns of the matrix D_{ij} then contain the components of the new points; i.e. the components of the point $\mathcal{O}([100])$ are given by the column D_{i1}, $\mathcal{O}([010])$ is given by D_{i2} and so on. An example will illustrate the procedure.

Example 1.13 *Determine the transformation matrix D_{ij} for the operator* **4**, *parallel to the* $\mathbf{a}_3$ *direction of a tetragonal unit cell.*

Answer: *From the drawing below one can determine readily that the operator rotates the* [100] *vector onto the vector* [010], [010] *is rotated onto the vector* $[\bar{1}00]$, *and the vector* [001] *is invariant since it lies on the rotation axis. The columns of the matrix are then formed by the vectors* [010], $[\bar{1}00]$, *and* [001], *or*†

$$D_{ij} = \begin{pmatrix} 0 & -1 & 0 \\ 1 & 0 & 0 \\ 0 & 0 & 1 \end{pmatrix}.$$

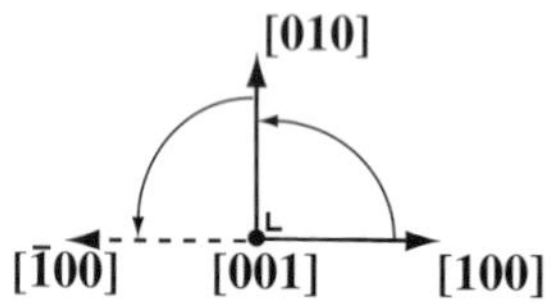

For computational purposes, it is convenient to work with four-dimensional vectors instead of the regular three-dimensional vectors. This sounds complicated, but it is really very simple: to each triplet of vector components (x_1, x_2, x_3) we simply add the number 1 as the fourth component, i.e. $(x_1, x_2, x_3, 1)$. This type of coordinate is called a *normal coordinate*. The fourth component must always be equal to 1, and is denoted by x_4.

† We will use the convention that a regular matrix is surrounded by parentheses, whereas a matrix representing a tensor will be denoted by square brackets.

The reason for the introduction of normal coordinates is the following: let us define a 4×4-matrix $\mathcal{W}$ of the following form:

$$\mathcal{W} = \begin{pmatrix} D_{11} & D_{12} & D_{13} & u_1 \\ D_{21} & D_{22} & D_{23} & u_2 \\ D_{31} & D_{32} & D_{33} & u_3 \\ 0 & 0 & 0 & 1 \end{pmatrix}. \tag{1.39}$$

In other words, $\mathcal{W}$ contains the matrix $\mathbf{D}$ and the translation vector $\mathbf{t}$. If we denote a four-dimensional vector by the symbol $\bar{\mathbf{r}} = (x_1, x_2, x_3, 1)$, then the equation

$$\bar{\mathbf{r}}' = \mathcal{W}\bar{\mathbf{r}}$$

fully describes the symmetry operation, as can be verified easily by writing out the component equations.

Example 1.14 *Determine the matrix $\mathcal{W}$ for a screw axis parallel to the $\mathbf{a}_3$ axis of a hexagonal unit cell.*

Answer: *This screw axis of order six has a 60° rotation combined with a pitch vector $\mathbf{T} = \frac{1}{6}\mathbf{a}_3$. As can be verified from the drawing below we have the following transformations of the basis vectors: $\mathcal{O}([100]) = [110]$, $\mathcal{O}([010]) = [\bar{1}00]$, and $[001]$ is invariant. The matrix $\mathcal{W}$ is therefore given by*

$$\mathcal{W} = \begin{pmatrix} 1 & \bar{1} & 0 & 0 \\ 1 & 0 & 0 & 0 \\ 0 & 0 & 1 & \frac{1}{6} \\ 0 & 0 & 0 & 1 \end{pmatrix}.$$

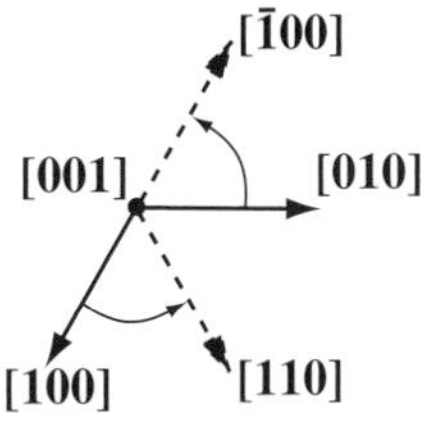

[$\bar{1}$00]
[001]
[010]
[100]
[110]

The main advantage of expressing the transformation matrices in terms of the basis vector of the Bravais lattice is that the components of the matrix D_{ij} are always simple integers (-1, 0, or 1). The 60° rotation of Example 1.14 would have the numbers $\frac{1}{2}$ and $\pm\sqrt{3}/2$ as entries in the transformation matrix if the operator were expressed with respect to a standard Cartesian reference frame. The translational part of the matrix $\mathcal{W}$ may, of course, contain fractions or integer numbers. This will become important when we implement space groups in a computer program (see Section 1.9.3).

Table 1.3. *Multiplication table for the three-fold rotation group **3**.*

	$\mathcal{E}$	**3**	$\mathbf{3}^2$
$\mathcal{E}$	$\mathcal{E}$	**3**	$\mathbf{3}^2$
3	**3**	$\mathbf{3}^2$	$\mathcal{E}$
$\mathbf{3}^2$	$\mathbf{3}^2$	$\mathcal{E}$	**3**

1.6.3 Point groups

Group theory plays an important role in crystallography because it allows for a systematic description of all possible crystal symmetries, and consequently, all crystal structures. The central concept in group theory is that of a *group*. A group is a set $\mathcal{G}$ of elements $\mathcal{O}_i$, $i = 1, \ldots, N$, where N may be finite or infinite. If a multiplication operation is defined by writing two elements of the set next to each other, i.e. $\mathcal{O}_i\mathcal{O}_j$ (meaning that the operation $\mathcal{O}_j$ is performed first, followed by $\mathcal{O}_i$), then the set $\mathcal{G}$ is a group if the following four conditions are satisfied:

(i) every product $\mathcal{O}_i\mathcal{O}_j$ belongs to the set $\mathcal{G}$ (the set is *closed* under the multiplication);
(ii) for every three elements we have $\mathcal{O}_i(\mathcal{O}_j\mathcal{O}_k) = (\mathcal{O}_i\mathcal{O}_j)\mathcal{O}_k$ (the multiplication is *associative*);
(iii) there exists an *identity element* $\mathcal{E}$, such that $\mathcal{E}\mathcal{O}_i = \mathcal{O}_i\mathcal{E} = \mathcal{O}_i$ for every operator $\mathcal{O}_i$;
(iv) and every element $\mathcal{O}_i$ has an *inverse element* $\mathcal{O}_k$, such that $\mathcal{O}_i\mathcal{O}_k = \mathcal{O}_k\mathcal{O}_i = \mathcal{E}$.

If, in addition, the product of *any* pair of elements is independent of the order of multiplication, then the group is said to be *Abelian* or *commutative*.

A simple example of a finite group is the set consisting of the identity element and a three-fold rotation axis **3**; there are three elements in the set, $\mathcal{E}$, **3**, and $\mathbf{3}^2$ (i.e. a rotation over $240° = -120°$). It is straightforward to verify that this set satisfies all five conditions above, so that the set $\{\mathcal{E}, \mathbf{3}, \mathbf{3}^2\}$ is an Abelian group. A standard tool for verifying whether or not a set is a group is the *multiplication table* (also known as *Cayley's square*). The multiplication table for the three-fold rotation group is shown in Table 1.3. If every element of the set occurs once in each row and once in each column, then the set is a group. If, in addition, the table is symmetric with respect to the main diagonal, then the group is an Abelian group. There are many examples of infinite groups. The simplest one is perhaps the set of integer numbers, which, as is easily checked, forms an infinite Abelian group under the addition operation.

Before considering the application of group theory to crystallography we must determine which symmetry operations are possible in crystals. One can easily

Table 1.4. *Allowed angles between rotation axes of orders A, B, and C, compatible with the Bravais lattices.*

A	B	C	$\hat{BC}$	$\hat{AC}$	$\hat{AB}$
2	2	2	90°	90°	90°
2	2	3	90°	90°	60°
2	2	4	90°	90°	45°
2	2	6	90°	90°	30°
2	3	3	70°32′	54°44′	54°44′
2	3	4	54°44′	45°	35°16′

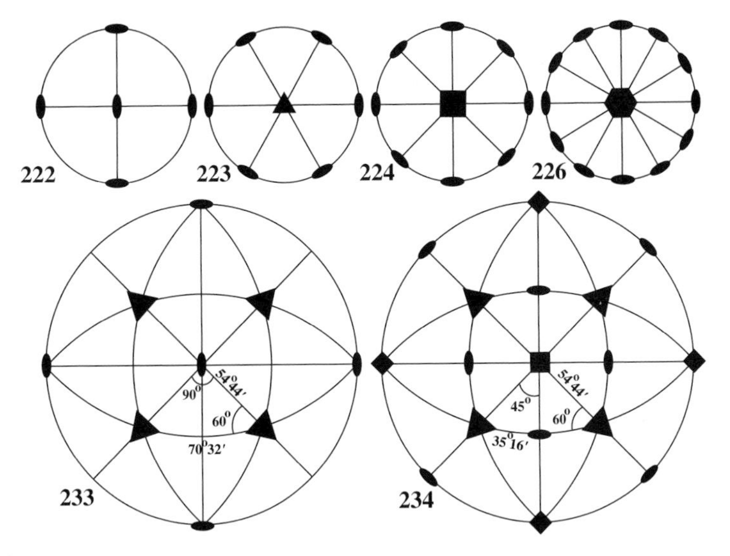

Fig. 1.18. Stereographic projections of the six crystallographically allowed combinations of rotational axes.

show that the only rotational symmetry operations compatible with the translational periodicity of the Bravais lattices are the rotations of order 1, 2, 3, 4 and 6 (e.g. [MM86]). Furthermore, only six combinations of multiple rotation axes are possible in any given Bravais lattice†: 222, 223, 224, 226, 233, and 234. The angles between those rotation axes are fixed and given in Table 1.4. Stereographic projections of these combinations of rotation elements are shown in Fig. 1.18.

When the simple rotational symmetries 1, 2, 3, 4, and 6, and the six combinations in Fig. 1.18 are combined with all possible combinations of mirror planes and the inversion symmetry, 32 distinct crystallographic point symmetries are found (they

† For an explicit derivation of these statements we refer the interested reader to [MM86] or [BG90].

Table 1.5. ***International***–[*Schœnflies*] *notation, group order N and crystal system S (a, m, h, R, o, t or c) for the 32 crystallographic point groups.*

Point group	N	S	Point group	N	S
1–[C_1]	1	a	$\bar{\mathbf{3}}$–[S_6]	6	R
$\bar{\mathbf{1}}$–[C_i]	2	a	**32**–[D_3]	6	R
2–[C_2]	2	m	**3m**–[C_{3v}]	6	R
m–[C_s]	2	m	$\bar{\mathbf{3}}$**m**–[D_{3d}]	12	R
2/m–[C_{2h}]	4	m	**6**–[C_6]	6	h
222–[D_2]	4	o	$\bar{\mathbf{6}}$–[C_{3h}]	6	h
mm2–[C_{2v}]	4	o	**6/m**–[C_{6h}]	12	h
mmm–[D_{2h}]	8	o	**622**–[D_6]	12	h
4–[C_4]	4	t	**6mm**–[C_{6v}]	12	h
$\bar{\mathbf{4}}$–[S_4]	4	t	$\bar{\mathbf{6}}$**m2**–[D_{3h}]	12	h
4/m–[C_{4h}]	8	t	**6/mmm**–[D_{6h}]	24	h
422–[D_4]	8	t	**23**–[T]	12	c
4mm–[C_{4v}]	8	t	**m3**–[T_h]	24	c
$\bar{\mathbf{4}}$**2m**–[D_{2d}]	8	t	**432**–[O]	24	c
4/mmm–[D_{4h}]	16	t	$\bar{\mathbf{4}}$**3m**–[T_d]	24	c
3–[C_3]	3	R	**m**$\bar{\mathbf{3}}$**m**–[O_h]	48	c

are called *point symmetries* because all symmetry elements intersect each other at one point). These 32 combinations are known as the 32 *point groups*. Their names in both International (Hermann–Mauguin) notation and Schœnflies notation, number of elements (order), and corresponding crystal systems are listed in Table 1.5. Stereographic projections can be found in Appendix A4, together with rendered drawings of the point group symmetry elements. Detailed drawings for all point groups, as well as Quicktime, AVI, and animated GIF movies illustrating the 3D structure of each point group are available from the website. Examples of rendered drawings for the point groups **4/m** and **m**$\bar{\mathbf{3}}$**m** are shown in Fig. 1.19; the general position in the point group is represented by a right-handed helix and it can be seen that the rotation axis preserves the handedness, whereas the mirror plane reverses the handedness of the helices. The handedness of the general point is not usually represented in stereographic projections of the point groups (such as those shown in Appendix A4).

1.6.4 Families of planes and directions

In a given crystal system the set of directions equivalent to $\mathbf{t} = [uvw]$ is given by

$$t_i\left(D^{(k)}\right) = D^{(k)}_{ij} t_j \qquad k = 1, \ldots, N, \tag{1.40}$$

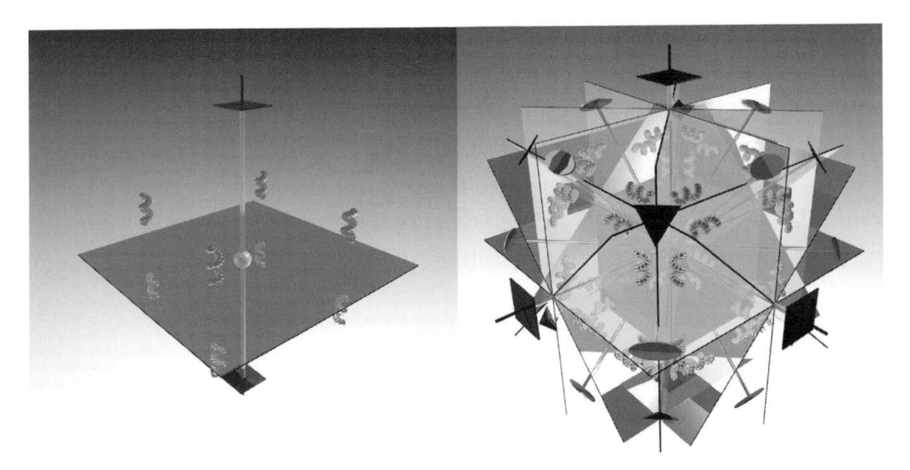

Fig. 1.19. Rendered images of the point groups **4/m** and **m$\bar{3}$m**. The general point is represented by a short helix, so that the action of the operators of the second kind is properly shown. Details about how this image and those in Appendix A4 were created can be found in [DG98] and on the website.

where the N matrices $D^{(k)}$ are the point symmetry matrices of the corresponding *point group*. The set of equivalent directions is known as a *family*, and is denoted by the symbol $\langle uvw \rangle$. Note that the number of members in a family depends on the point symmetry of the crystal.

Since the symmetry matrices $D^{(k)}$ are defined with respect to the direct basis vectors, we cannot always use the same matrices to determine the set of reciprocal lattice vectors equivalent to $\mathbf{g}_{hkl}$. Both the symmetry matrix and the reciprocal lattice vector must be expressed *in the same reference frame* before the equivalent vectors can be determined. This means that we must transform the reciprocal lattice vector $\mathbf{g}_{hkl}$ to direct space, compute the equivalent vectors using the set of N matrices $D^{(k)}$, and then transform all equivalent vectors back to reciprocal space. This is expressed mathematically by

$$g_m\left(D^{(k)}\right) = g_{ml} D^{(k)}_{lj} g^*_{ji} g_i. \qquad (1.41)$$

Alternatively, we can first compute the components of the symmetry matrices in reciprocal space, resulting in a set of N matrices $D^{*(k)}$, and then operate with these matrices on the reciprocal lattice vector $\mathbf{g}_{hkl}$. The set of reciprocal symmetry matrices is given by

$$D^{*(k)}_{mi} = g_{ml} D^{(k)}_{lj} g^*_{ji}. \qquad (1.42)$$

The set of plane normals equivalent to (hkl) is denoted by the family $\{hkl\}$; the number of members again depends on the point symmetry of the crystal.

The concept of *family* is used whenever the components of the vectors are all integers (as in $[uvw]$ or (hkl)). For non-integer components one can define similar concepts: the set of points $\mathbf{r}^{(k)}$ equivalent to a given general point $\mathbf{r}$ is known as the *orbit* of $\mathbf{r}$. In reciprocal space the set of points $\mathbf{q}^{(k)}$ equivalent to $\mathbf{q}$ is known as the *star* of $\mathbf{q}$. The number of members in a family, orbit, or star is generally known as its *multiplicity*. In Section 1.9 we will discuss a method to determine numerically the point symmetry matrices of an arbitrary point group, based on the space group symmetry.

1.6.5 Space groups

The point groups enumerate all combinations of symmetry elements consistent with the translational symmetry of the Bravais lattices. They describe the symmetry of an object with respect to a single fixed point. When the point group symmetries are combined with the translational symmetry operators of the Bravais lattices, one can show that there are precisely 230 unique combinations. This was shown independently by Fedorov in Russia [Fed90], Schœnflies in Germany [Sch91], and Barlow in England [Bar94] during the last decade of the 19th century. This is a rather important finding because it means that every crystal structure must belong to one of those 230 symmetries. Since a combination of a point group and a Bravais lattice creates an infinite lattice, the resulting symmetry groups (of infinite order) are known as *space groups*.

The simplest space groups can be constructed by selecting one of the point groups and one of the Bravais lattices for the corresponding crystal system, and copying the symmetry elements of the point group onto every lattice site of that Bravais lattice. In this way, one can construct 73 different combinations, known as the *symmorphic space groups*. A table with all 73 symmorphic space groups is reproduced in Appendix A4 on page 674 (Table A4.2). The International (or Hermann–Mauguin) symbol for a symmorphic space group consists of the centering symbol of the Bravais lattice (P, A, B, C, R, I, or F), followed by the point group symbol. It is not necessary to include the crystal system in the symbol, since this should be obvious from the point group (see Table 1.5).

The remaining $230 - 73 = 157$ space groups are obtained by systematically replacing every rotation axis in the point groups by all possible screw axes, and every mirror plane by all possible glide planes consistent with the Bravais lattice. These groups all have either a glide plane or a screw axis or both in their symbols, and they are known as the *non-symmorphic space groups*. They are ranked by point group in Appendix A4, Table A4.3 (page 675). Several space groups are defined with respect to two different origins, known as *first setting* and *second setting*; they are indicated in the tables by an asterisk after the space group symbol. In addition some of the trigonal space groups can be defined with respect to either a hexagonal or a rhombohedral reference frame (see Section 1.7.3) and, hence, also have two settings.

For a rather extensive derivation of the space groups and a description of some of their uses in materials science we refer to Burns and Glazer [BG90]. The complete tables of all point and space groups can be found in the *International Tables for Crystallography*, Volume A [Hah96]. This book, and its companion volumes on reciprocal space [Shm96], and mathematical, physical, and chemical tables [WP99], are amongst the most important reference works for electron microscopists and crystallographers. When working on a crystallographic problem one should always keep them at hand.

It is now a straightforward matter to compute the coordinates of all points $\mathbf{r}^{(k)}$, $k = 2, \ldots, N$ equivalent to a given point $\mathbf{r}^{(1)}$, i.e. the members of the *orbit*. All that is required is a list of N 4×4 transformation matrices corresponding to all the symmetry elements of the space group.† Multiplication of each matrix with the normal coordinates of the first point then results in a list of equivalent points. If the original point lies on one or more of the symmetry elements of the space group, then there will be duplicate entries in this list, and they will need to be removed. The largest possible multiplicity $N = 192$ is encountered in the face-centered cubic space group $\mathbf{Fm\bar{3}m}$. Although all symmetry operators for all space groups are explicitly listed in the *International Tables for Crystallography*, Vol. A, it is not advisable to enter all these matrices into a computer program. Through the clever use of group theory one can store all the necessary space group information in a surprisingly small number of bytes, as is discussed in detail in Section 1.9.3.

1.7 Coordinate transformations

The mathematical relations derived in this chapter allow us to compute any geometrical quantity in any of the seven crystal systems. One may now ask the question: how do these relations change when we change the reference frame? The need to change from one reference frame to another arises frequently in the study of solid state phase transformations, when the crystal structure changes with temperature or applied field (electric or magnetic). In addition, in the presence of planar defects, such as twins, it is frequently necessary to express crystallographic quantities on one side of the defect in the reference frame of the crystal on the other side. In this section we will describe in detail how one can convert vectors and the metric tensors from one reference frame to another one.

† That is, all the elements within the fundamental unit cell; neighboring unit cells can be reached by adding Bravais translation vectors to all equivalent points.

1.7.1 Transformation rules

Let us consider two crystallographic reference frames, $\{\mathbf{a}_1, \mathbf{a}_2, \mathbf{a}_3\}$ and $\{\mathbf{a}'_1, \mathbf{a}'_2, \mathbf{a}'_3\}$. In general, the relation between the two sets of basis vectors can be written as

$$\left.\begin{aligned} \mathbf{a}'_1 &= \alpha_{11}\mathbf{a}_1 + \alpha_{12}\mathbf{a}_2 + \alpha_{13}\mathbf{a}_3; \\ \mathbf{a}'_2 &= \alpha_{21}\mathbf{a}_1 + \alpha_{22}\mathbf{a}_2 + \alpha_{23}\mathbf{a}_3; \\ \mathbf{a}'_3 &= \alpha_{31}\mathbf{a}_1 + \alpha_{32}\mathbf{a}_2 + \alpha_{33}\mathbf{a}_3. \end{aligned}\right\} \tag{1.43}$$

This is a *linear* relation, known as a *coordinate transformation*. We will assume throughout this book that the origin itself does not change during the coordinate transformation. If there were a change in the position of the origin as well, then we would need to add a translation vector to the equations above.

The nine numbers α_{ij} can be grouped as a square matrix:

$$\alpha_{ij} = \begin{pmatrix} \alpha_{11} & \alpha_{12} & \alpha_{13} \\ \alpha_{21} & \alpha_{22} & \alpha_{23} \\ \alpha_{31} & \alpha_{32} & \alpha_{33} \end{pmatrix}, \tag{1.44}$$

and the transformation equations can be rewritten in short form as

$$\mathbf{a}'_i = \alpha_{ij}\mathbf{a}_j. \tag{1.45}$$

The *inverse* transformation must also exist and is described by the inverse of the matrix α_{ij}:

$$\mathbf{a}_i = \alpha^{-1}_{ij}\mathbf{a}'_j. \tag{1.46}$$

Consider the position vector $\mathbf{p}$. This vector is independent of the reference frame, and has components in both the unprimed and primed reference frames. We must have the following relation:

$$\mathbf{p} = p_i\mathbf{a}_i = p'_j\mathbf{a}'_j. \tag{1.47}$$

Using the inverse coordinate transformation we can rewrite the first equality as

$$p_i\mathbf{a}_i = p_i\alpha^{-1}_{ij}\mathbf{a}'_j,$$

and after comparison with the last equality of equation (1.47) we find

$$p'_j = p_i\alpha^{-1}_{ij}. \tag{1.48}$$

Note the order of the indices of the matrix α; the summation index is the index i, which means that we must *premultiply* the matrix by the *row vector* p_i. Similarly, one can readily show that

$$p_i = p'_j\alpha_{ji}. \tag{1.49}$$

We interpret equations (1.48) and (1.49) as follows: the vector $\mathbf{p}$ is independent of the chosen reference frame if its components with respect to two different reference frames are related to each other by equations (1.48) and (1.49). This relation obviously also holds for direction vectors $[uvw]$, since they are a special case of position vectors $\mathbf{p}$ (integer components instead of rational ones).

It is now straightforward to derive the transformation relation for the direct metric tensor:

$$\begin{aligned} g'_{ij} &= \mathbf{a}'_i \cdot \mathbf{a}'_j; \\ &= \alpha_{ik}\mathbf{a}_k \cdot \alpha_{jl}\mathbf{a}_l; \\ &= \alpha_{ik}\alpha_{jl}\mathbf{a}_k \cdot \mathbf{a}_l, \end{aligned}$$

and hence

$$g'_{ij} = \alpha_{ik}\alpha_{jl}g_{kl}. \tag{1.50}$$

The inverse relation is given by

$$g_{ij} = \alpha^{-1}_{ik}\alpha^{-1}_{jl}g'_{kl}. \tag{1.51}$$

One can use these relations to *define* a second-rank tensor: any mathematical quantity h_{ij} which satisfies the above transformation rules is a second-rank tensor. Similarly, any mathematical quantity p_i, satisfying the transformation rules (1.48) and (1.49), is a vector. It is straightforward to extend these definitions to higher-order tensors, but we will not need them in this book. For a detailed overview of tensor calculus we refer to [Wre72].

Next, we will derive the transformation relations for quantities in reciprocal space. We have already seen that if the components of a vector $\mathbf{p}$ are known in direct space, then its components in the reciprocal reference frame are given by

$$p^*_i = g_{ij}p_j.$$

Using equation (1.49) we have

$$p^*_i = g_{ij}\alpha_{kj}p'_k. \tag{1.52}$$

From equation (1.51) we find, after multiplying both sides of the equation by α_{kj},

$$g_{ij}\alpha_{kj} = \alpha^{-1}_{il}g'_{lk},$$

and substitution in equation (1.52) leads to

$$\begin{aligned} p^*_i &= \alpha^{-1}_{il}g'_{lk}p'_k, \\ &= \alpha^{-1}_{il}p'^*_l, \end{aligned}$$

where we have once again used the properties of the direct metric tensor. The components of a vector in reciprocal space thus transform as follows:

$$p_i^* = \alpha_{il}^{-1} p_l'^*, \tag{1.53}$$

and the inverse relation is given by

$$p_l'^* = \alpha_{li} p_i^*. \tag{1.54}$$

In particular, these equations are valid for the reciprocal lattice vectors $\mathbf{g}$.

The reciprocal basis vectors satisfy similar transformation relations, which are derived as follows:

$$\begin{aligned}\mathbf{g} = g_i^*\mathbf{a}_i^* &= g_l'^*\mathbf{a}_l'^*,\\ &= \alpha_{li} g_i^* \mathbf{a}_l'^*,\end{aligned}$$

from which we find:

$$\mathbf{a}_i^* = \mathbf{a}_l'^* \alpha_{li}. \tag{1.55}$$

The corresponding inverse relation is given by

$$\mathbf{a}_i'^* = \mathbf{a}_j^* \alpha_{ji}^{-1}. \tag{1.56}$$

Finally, it is again easy to show that the reciprocal metric tensor transforms according to the rules

$$g_{ij}^* = \alpha_{ki}\alpha_{lj} g_{kl}'^*, \tag{1.57}$$

and

$$g_{ij}'^* = \alpha_{ki}^{-1}\alpha_{lj}^{-1} g_{kl}^*. \tag{1.58}$$

The transformation rules derived in this section are summarized in Table 1.6. All that is required to carry out *any* coordinate transformation is the matrix α_{ij}, expressing the new basis vectors in terms of the old ones. The relations in Table 1.6 require, in addition to α_{ij}, the inverse α_{ij}^{-1} and transpose α_{ij}^T matrices, and the transpose of the inverse matrix $(\alpha_{ij}^{-1})^T$. These transformation rules seem easy enough, but one must actually pay close attention to the indices in order to avoid mistakes. In the following subsection we will illustrate coordinate transformations by means of a few examples.

1.7.2 Examples of coordinate transformations

Example 1.15 *Consider the two sets of basis vectors shown in Fig. 1.20; write down the transformation matrix α_{ij}. Show by explicit computation using the relations in*

Table 1.6. *Overview of all transformation relations for vectors and the metric tensor in direct and reciprocal space. Pay close attention to the order of the indices!*

Quantity	Old to new	New to old
Direct basis vectors	$\mathbf{a}'_i = \alpha_{ij}\mathbf{a}_j$	$\mathbf{a}_i = \alpha^{-1}_{ij}\mathbf{a}'_j$
Direct metric tensor	$g'_{ij} = \alpha_{ik}\alpha_{jl}g_{kl}$	$g_{ij} = \alpha^{-1}_{ik}\alpha^{-1}_{jl}g'_{kl}$
Direct space vectors	$p'_i = p_j\alpha^{-1}_{ji}$	$p_i = p'_j\alpha_{ji}$
Reciprocal basis vectors	$\mathbf{a}'^*_i = \mathbf{a}^*_j\alpha^{-1}_{ji}$	$\mathbf{a}^*_i = \mathbf{a}'^*_j\alpha_{ji}$
Reciprocal metric tensor	$g'^*_{ij} = g^*_{kl}\alpha^{-1}_{ki}\alpha^{-1}_{lj}$	$g^*_{ij} = g'^*_{kl}\alpha_{ki}\alpha_{lj}$
Reciprocal space vectors	$k'^*_i = \alpha_{ij}k^*_j$	$k^*_i = \alpha^{-1}_{ij}k'^*_j$

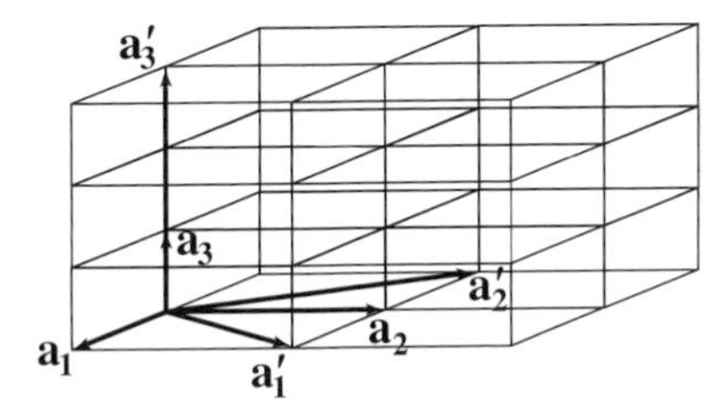

Fig. 1.20. Unit cell drawing for Example 1.15.

Table 1.6 how the $[uvw]$ direction and the (hkl) plane normal change to their new values in the primed reference frame.

Answer: *The transformation matrix α_{ij} is easily derived from a visual inspection of Fig. 1.20:*

$$\alpha_{ij} = \begin{pmatrix} 1 & 1 & 0 \\ -1 & 1 & 0 \\ 0 & 0 & 3 \end{pmatrix}.$$

Direction indices transform according to the inverse of this matrix, or

$$\begin{aligned} [u'v'w'] &= [u\,v\,w]\,\alpha^{-1}_{ji}; \\ &= \frac{1}{2}[uvw]\begin{pmatrix} 1 & -1 & 0 \\ 1 & 1 & 0 \\ 0 & 0 & \frac{2}{3} \end{pmatrix}; \\ &= \left[\frac{u+v}{2}\,\frac{v-u}{2}\,\frac{w}{3}\right]. \end{aligned}$$

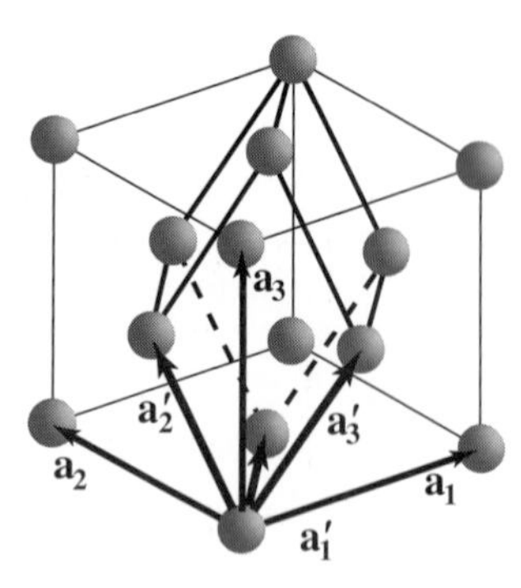

Fig. 1.21. Unit cell drawing for Example 1.16.

Reciprocal lattice vectors transform according to the matrix α_{ij} itself, or

$$\begin{bmatrix} h' \\ k' \\ l' \end{bmatrix} = \alpha_{ij} \begin{bmatrix} h \\ k \\ l \end{bmatrix} = \begin{pmatrix} 1 & 1 & 0 \\ -1 & 1 & 0 \\ 0 & 0 & 3 \end{pmatrix} \begin{bmatrix} h \\ k \\ l \end{bmatrix} = \begin{bmatrix} h+k \\ k-h \\ 3l \end{bmatrix}.$$

Example 1.16 *Consider the face-centered cubic lattice shown in Fig. 1.21; we can define a primitive rhombohedral unit cell for this structure, as indicated by the primed basis vectors. Determine the transformation matrix α_{ij}, and express the reciprocal basis vectors of the new reference frame in terms of those of the old reference frame. Then compute the direct metric tensor for the primitive cell using the transformation equations and show that the result is identical to the direct computation of the metric tensor using the equations in Appendix A1.*

Answer: *The transformation matrix α_{ij} is easily derived from a visual inspection of Fig. 1.21:*

$$\alpha_{ij} = \frac{1}{2}\begin{pmatrix} 1 & 1 & 0 \\ 0 & 1 & 1 \\ 1 & 0 & 1 \end{pmatrix}.$$

The reciprocal basis vectors transform according to the inverse of this matrix, or

$$\begin{aligned} \left(\mathbf{a}_1'^* \, \mathbf{a}_2'^* \, \mathbf{a}_3'^*\right) &= \left(\mathbf{a}_1^* \, \mathbf{a}_2^* \, \mathbf{a}_3^*\right) \alpha_{ji}^{-1}; \\ &= \left(\mathbf{a}_1^* \, \mathbf{a}_2^* \, \mathbf{a}_3^*\right) \begin{pmatrix} 1 & -1 & 1 \\ 1 & 1 & -1 \\ -1 & 1 & 1 \end{pmatrix}; \\ &= \left(\mathbf{a}_1^* + \mathbf{a}_2^* - \mathbf{a}_3^* \,|\, -\mathbf{a}_1^* + \mathbf{a}_2^* + \mathbf{a}_3^* \,|\, \mathbf{a}_1^* - \mathbf{a}_2^* + \mathbf{a}_3^*\right). \end{aligned}$$

The direct metric tensor transforms according to $g'_{ij} = \alpha_{ik} g_{kl} \alpha_{jl}$, or (note that the matrix α_{jl} must be transposed before multiplication since the summation index *l must be the row index!)*

$$g'_{ij} = \frac{1}{4}\begin{pmatrix} 1 & 1 & 0 \\ 0 & 1 & 1 \\ 1 & 0 & 1 \end{pmatrix}\begin{bmatrix} a^2 & 0 & 0 \\ 0 & a^2 & 0 \\ 0 & 0 & a^2 \end{bmatrix}\begin{pmatrix} 1 & 0 & 1 \\ 1 & 1 & 0 \\ 0 & 1 & 1 \end{pmatrix};$$

$$= \frac{a^2}{4}\begin{pmatrix} 1 & 1 & 0 \\ 0 & 1 & 1 \\ 1 & 0 & 1 \end{pmatrix}\begin{bmatrix} 1 & 0 & 1 \\ 1 & 1 & 0 \\ 0 & 1 & 1 \end{bmatrix};$$

$$= \frac{a^2}{4}\begin{bmatrix} 2 & 1 & 1 \\ 1 & 2 & 1 \\ 1 & 1 & 2 \end{bmatrix}.$$

The rhombohedral metric tensor is given by

$$g_{ij} = b^2\begin{bmatrix} 1 & \cos\alpha & \cos\alpha \\ \cos\alpha & 1 & \cos\alpha \\ \cos\alpha & \cos\alpha & 1 \end{bmatrix},$$

where b and α are the lattice parameters of the primitive unit cell. From the drawing one can easily show that $b = a\sqrt{2}$ and $\cos\alpha = \frac{1}{2}$, which leads to the same expression for g_{ij}.

1.7.3 Rhombohedral and hexagonal settings of the trigonal system

The derivation of the 14 Bravais lattices in Section 1.2.1 used the fact that all lattice points must be equivalent; i.e. they must have the same environment. The centered lattices can be obtained from the primitive lattices by placing additional sites at positions given by the centering vectors **A**, **B**, **C**, or **I**. No other centering vectors are allowed, except for the hexagonal crystal system. A number of space groups describe symmetries that can be indexed in terms of a hexagonal reference frame or a rhombohedral reference frame. This ambiguity has given rise to considerable confusion in the literature. We will follow Burns and Glazer [BG90] (but with a simplified notation) and derive the transformation relations between the two descriptions.

Consider the hexagonal lattice shown in Fig. 1.6 on page 23. The lattice parameters are given by $\{a, a, c, 90, 90, 120\}$ or $\{a, c\}$ for short. When the standard centering vectors are considered, no new lattices are obtained. However, when the

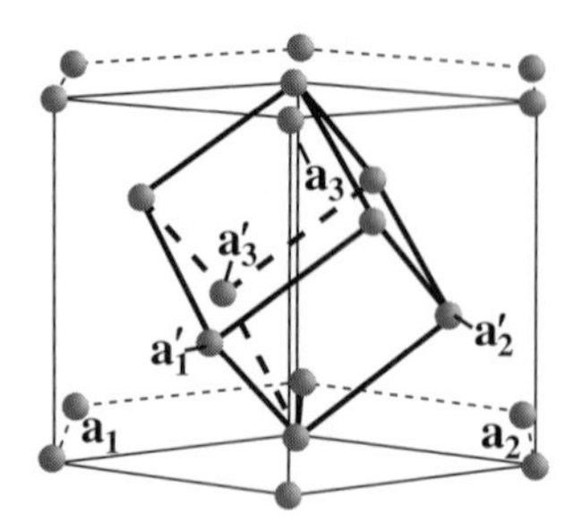

Fig. 1.22. Obverse setting of the rhombohedral lattice, with both hexagonal and rhombohedral basis vectors indicated.

following centering vectors are used

$$\mathbf{R}_1 = \left(\frac{2}{3}, \frac{1}{3}, \frac{1}{3}\right);$$
$$\mathbf{R}_2 = \left(\frac{1}{3}, \frac{2}{3}, \frac{2}{3}\right),$$

a new lattice is obtained, since now all lattice points have identical surroundings. The resulting lattice is shown in Fig. 1.22, with both hexagonal and rhombohedral basis vectors indicated. The centering vectors above refer to the so-called *obverse setting*; the *reverse setting* has the third components of the centering vectors exchanged. The obverse setting is the preferred setting, according to the *International Tables for Crystallography* [Hah96, page 79].

The transformation relation between the rhombohedral (primed) and hexagonal (unprimed) lattices can be derived from Fig. 1.22:

$$\begin{pmatrix} \mathbf{a}'_1 \\ \mathbf{a}'_2 \\ \mathbf{a}'_3 \end{pmatrix} = \frac{1}{3}\begin{pmatrix} 2 & 1 & 1 \\ -1 & 1 & 1 \\ -1 & -2 & 1 \end{pmatrix}\begin{pmatrix} \mathbf{a}_1 \\ \mathbf{a}_2 \\ \mathbf{a}_3 \end{pmatrix}.$$

The inverse of this transformation matrix is given by

$$\alpha^{-1} = \begin{pmatrix} 1 & -1 & 0 \\ 0 & 1 & -1 \\ 1 & 1 & 1 \end{pmatrix}.$$

We can once again use the standard transformation relations in Table 1.6 to convert quantities from one reference frame to the other.

The relations between the hexagonal lattice parameters $\{a, c\}$ and $\{a_r, \alpha\}$ of the rhombohedral system can be derived as follows: the parameter a_r is given by the length of the vector $\mathbf{a}'_1$, or, using the hexagonal metric tensor,

$$a_r^2 = |\mathbf{a}'_1|^2 = \frac{1}{18}\begin{bmatrix} 2 & 1 & 1 \end{bmatrix} \begin{bmatrix} 2a^2 & -a^2 & 0 \\ -a^2 & 2a^2 & 0 \\ 0 & 0 & 2c^2 \end{bmatrix} \begin{bmatrix} 2 \\ 1 \\ 1 \end{bmatrix} = \frac{a^2}{3} + \frac{c^2}{9},$$

or

$$a_r = \frac{a}{3}\sqrt{3 + \left(\frac{c}{a}\right)^2}. \tag{1.59}$$

The rhombohedral angle α follows from

$$\mathbf{a}'_1 \cdot \mathbf{a}'_2 = a_r^2 \cos\alpha.$$

We have

$$\mathbf{a}'_1 \cdot \mathbf{a}'_2 = \frac{1}{18}\begin{bmatrix} 2 & 1 & 1 \end{bmatrix} \begin{bmatrix} 2a^2 & -a^2 & 0 \\ -a^2 & 2a^2 & 0 \\ 0 & 0 & 2c^2 \end{bmatrix} \begin{bmatrix} -1 \\ 1 \\ 1 \end{bmatrix} = \frac{c^2}{9} - \frac{a^2}{6},$$

from which we find, after some simple rearrangements:

$$\cos\alpha = 1 - \frac{9}{2(c/a)^2 + 6}. \tag{1.60}$$

It is left as an exercise for the reader to invert these equations and to show that

$$a = a_r\sqrt{2 - 2\cos\alpha}; \tag{1.61}$$

$$c = a_r\sqrt{3 + 6\cos\alpha}. \tag{1.62}$$

1.8 Converting vector components into Cartesian coordinates

We have seen in previous sections that there is a distinct advantage to working in crystal coordinates (i.e. in a non-Cartesian reference frame) in both direct and reciprocal space. However, at the end of a simulation or calculation, the results are almost invariably represented on a computer screen or on a piece of paper, both of which are 2D media with essentially Cartesian reference frames. We must, therefore, provide a way to transform direct and reciprocal crystal coordinates into Cartesian coordinates. It is not difficult to carry out such a conversion for the crystal systems of high symmetry (cubic, tetragonal, and orthorhombic) since their coordinate axes are already at right angles to each other. However, for a monoclinic or triclinic system the conversion to Cartesian coordinates is a bit more difficult and it becomes important to have an algorithm that will do the conversion independent of the crystal system. Such a conversion exists and is derived below. The derivation

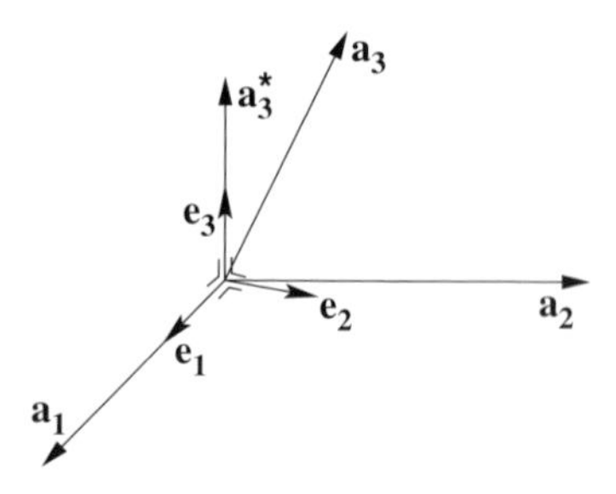

Fig. 1.23. Definition of the Cartesian reference frame from the direct and reciprocal reference frames.

is somewhat tedious, but the resulting transformation is quite general and can be used for both direct and reciprocal space quantities.

The transformation can be carried out by means of the so-called *direct and reciprocal structure matrices*.† Let us assume a crystal reference frame $\mathbf{a}_i$, and the corresponding reciprocal reference frame $\mathbf{a}^*_j$. From these two reference frames we can construct a Cartesian reference frame $\mathbf{e}_i$ as follows: $\mathbf{e}_1$ is the unit vector along $\mathbf{a}_1$, $\mathbf{e}_3$ is the unit vector along the *reciprocal* basis vector $\mathbf{a}^*_3$ (and is, therefore, by construction normal to $\mathbf{e}_1$), and $\mathbf{e}_2$ completes the right-handed Cartesian reference frame (see Fig. 1.23):

$$\left.\begin{aligned} \mathbf{e}_1 &= \frac{\mathbf{a}_1}{|\mathbf{a}_1|}; \\ \mathbf{e}_2 &= \mathbf{e}_3 \times \mathbf{e}_1; \\ \mathbf{e}_3 &= \frac{\mathbf{a}^*_3}{|\mathbf{a}^*_3|} = \frac{\mathbf{a}_1 \times \mathbf{a}_2}{\Omega|\mathbf{a}^*_3|}. \end{aligned}\right\} \qquad (1.63)$$

We will refer to this reference frame as the *standard Cartesian frame*. Now consider a vector $\mathbf{r}$ with components r_i with respect to the basis vectors $\mathbf{a}_i$. The components of $\mathbf{r}$ in the Cartesian reference frame are given by x_j, or

$$\mathbf{r} = r_i\mathbf{a}_i = x_j\mathbf{e}_j.$$

The components r_i and x_j are related to one another by a linear coordinate transformation represented by the matrix a_{ij}:

$$x_i = a_{ij}r_j.$$

The elements of the transformation matrix can be determined as follows: equations (1.63) are rewritten in terms of the direct and reciprocal metric

† This section follows the same conventions as those used in the *EMS* software package [Sta87].

tensors as

$$\mathbf{e}_1 = \frac{\mathbf{a}_1}{\sqrt{g_{11}}};$$
$$\mathbf{e}_2 = \frac{\mathbf{a}_3^* \times \mathbf{a}_1}{\sqrt{g_{11}g_{33}^*}};$$
$$\mathbf{e}_3 = \frac{\mathbf{a}_3^*}{\sqrt{g_{33}^*}} = \frac{g_{3m}^*\mathbf{a}_m}{\sqrt{g_{33}^*}}.$$

From the definition of $\mathbf{a}_3^*$ we derive

$$\Omega\left(\mathbf{a}_3^* \times \mathbf{a}_1\right) = (\mathbf{a}_1 \times \mathbf{a}_2) \times \mathbf{a}_1 = -\mathbf{a}_1 \times (\mathbf{a}_1 \times \mathbf{a}_2).$$

The triple vector product can be simplified using the vector identity

$$\mathbf{u} \times (\mathbf{v} \times \mathbf{w}) = (\mathbf{u}\cdot\mathbf{w})\mathbf{v} - (\mathbf{u}\cdot\mathbf{v})\mathbf{w},$$

which leads to

$$\Omega\left(\mathbf{a}_3^* \times \mathbf{a}_1\right) = g_{11}\mathbf{a}_2 - g_{12}\mathbf{a}_1,$$

and finally

$$\mathbf{e}_2 = \frac{g_{11}\mathbf{a}_2 - g_{12}\mathbf{a}_1}{\Omega\sqrt{g_{11}g_{33}^*}}.$$

The vector $\mathbf{r}$ can now be written as follows:

$$\mathbf{r} = x_j\mathbf{e}_j = \left[\frac{x_1}{\sqrt{g_{11}}} - \frac{g_{21}x_2}{\Omega\sqrt{g_{11}g_{33}^*}} + \frac{g_{31}^*x_3}{\sqrt{g_{33}^*}}\right]\mathbf{a}_1 + \left[\frac{g_{11}x_2}{\Omega\sqrt{g_{11}g_{33}^*}} + \frac{g_{32}^*x_3}{\sqrt{g_{33}^*}}\right]\mathbf{a}_2 + \frac{g_{33}^*x_3}{\sqrt{g_{33}^*}}\mathbf{a}_3.$$

Using the fact that the direct and reciprocal metric tensors are each other's inverse, we can explicitly write the matrix a_{ij} as

$$a_{ij} = \begin{pmatrix} \sqrt{g_{11}} & \frac{g_{21}}{\sqrt{g_{11}}} & \frac{g_{31}}{\sqrt{g_{11}}} \\ 0 & \Omega\sqrt{\frac{g_{33}^*}{g_{11}}} & -\frac{\Omega g_{32}^*}{\sqrt{g_{33}^*g_{11}}} \\ 0 & 0 & \frac{1}{\sqrt{g_{33}^*}} \end{pmatrix};$$
$$= \begin{pmatrix} a & b\cos\gamma & c\cos\beta \\ 0 & b\sin\gamma & -\frac{c\mathcal{F}(\beta,\gamma,\alpha)}{\sin\gamma} \\ 0 & 0 & \frac{\Omega}{ab\sin\gamma} \end{pmatrix}, \qquad (1.64)$$

where (see Appendix A1)

$$\mathcal{F}(\alpha, \beta, \gamma) = \cos\alpha\cos\beta - \cos\gamma.$$

The matrix a_{ij} is known as the *direct structure matrix* and it transforms crystal coordinates to Cartesian coordinates. Note that its elements depend both on the direct and reciprocal metric tensors and, thus, on the lattice parameters $\{a, b, c, \alpha, \beta, \gamma\}$, as shown by the second equality in (1.64). The inverse transformation is given by the inverse matrix:

$$r_i = a_{ij}^{-1} x_j \qquad \text{with} \quad a_{ij}^{-1} = \begin{pmatrix} \frac{1}{a} & \frac{-1}{a\tan\gamma} & \frac{bc\mathcal{F}(\gamma,\alpha,\beta)}{\Omega\sin\gamma} \\ 0 & \frac{1}{b\sin\gamma} & \frac{ac\mathcal{F}(\beta,\gamma,\alpha)}{\Omega\sin\gamma} \\ 0 & 0 & \frac{ab\sin\gamma}{\Omega} \end{pmatrix}. \tag{1.65}$$

The direct structure matrix is particularly useful if one wants to create a drawing of a crystal structure, and an outline of a computer program for crystal structure drawings is presented in the next section.

Example 1.17 *Compute the Cartesian coordinates of the lattice point* (2, 3, 1) *in the tetragonal lattice of Example 1.1 on page 6.*

Answer: *From the lattice parameters* $a = \frac{1}{2}$ *and* $c = 1$ *we find for the direct structure matrix:*

$$a_{ij} = \begin{pmatrix} \frac{1}{2} & 0 & 0 \\ 0 & \frac{1}{2} & 0 \\ 0 & 0 & 1 \end{pmatrix}.$$

Hence the Cartesian components of the vector (2, 3, 1) *are* $(1, \frac{3}{2}, 1)$.

One can use the same formalism to determine the Cartesian coordinates of a reciprocal lattice point; such coordinates would be used to draw a diffraction pattern or any other representation of reciprocal space. To preserve the relative orientation of crystal and reciprocal space, we look for a second structure matrix b_{ij}, which represents the transformation from the reciprocal reference frame to *the same* Cartesian reference frame. Consider the reciprocal space vector **k**, with components k_j with respect to the reciprocal basis vectors $\mathbf{a}_j^*$ and Cartesian components q_i,

$$\mathbf{k} = q_i\mathbf{e}_i = k_j\mathbf{a}_j^*.$$

This can be rewritten in terms of the direct basis vectors $\mathbf{a}_l$ as

$$\mathbf{k} = q_i\mathbf{e}_i = k_j g_{jl}^*\mathbf{a}_l = r_l\mathbf{a}_l \quad \text{with} \quad r_l = k_j g_{jl}^*.$$

We can now use the direct structure matrix a_{il} to relate q_i to r_l:

$$q_i = a_{il}r_l = a_{il}g^*_{jl}k_j = a_{il}g^*_{lj}k_j = b_{ij}k_j.$$

The *reciprocal structure matrix* b_{ij} is thus defined by

$$b_{ij} = a_{il}g^*_{lj}. \tag{1.66}$$

This matrix converts reciprocal space coordinates into Cartesian coordinates. The inverse relation is given by

$$k_i = b^{-1}_{ij}q_j.$$

One can use the fact that the length of a vector must be independent of the reference frame to show that the transpose of the reciprocal structure matrix b is equal to the inverse of the direct structure matrix a, or

$$b^T = a^{-1}.$$

The reciprocal structure matrix is thus given by the transpose of the matrix in equation (1.65). Note that only the lattice parameters are used to compute the structure matrices. In addition, the lattice parameters are used to compute the direct and reciprocal metric tensors. In a computer implementation it is thus convenient to have a single routine which computes all four matrices.

Example 1.18 *For the tetragonal crystal in the previous example, compute the Cartesian components of the reciprocal lattice point* (221).

Answer: *The Cartesian components q_i require the reciprocal structure matrix b_{ij}, which is the transpose of the inverse of a_{ij}:*

$$b_{ij} = \begin{pmatrix} 2 & 0 & 0 \\ 0 & 2 & 0 \\ 0 & 0 & 1 \end{pmatrix},$$

from which the Cartesian components of (221) *follow as* (4, 4, 1).

1.9 Crystallographic calculations on the computer

1.9.1 Preliminary remarks

In this section we will briefly discuss how one can implement the crystallographic calculations of the preceding sections in a computer program. The subroutines discussed below are grouped in a library that can be used for most computations throughout this textbook, and can be downloaded in ASCII format from the website. All routines were written in Fortran-90. Fortran compilers are widely available for most computer platforms so that the reader should have no major problems implementing the routines. While there are many other, more advanced programming languages, Fortran was selected because it is easy to use, and because it is very well suited for the type of computations presented in this textbook.

The website provides the source code for all Fortran routines used in this book. In addition to individual programs, the website also provides a number of *libraries* containing routines that are frequently used. The following library files are available:

- local.f90 — This module defines parameters that are specific to the location where the programs are executed; the parameters in this file may be customized by the user.
- io.f90 — This module contains most of the *input* and *output* routines used by other routines. All user interactions occur via routines defined in this module, so that ideally this is the only module that should have to be modified if the programs are implemented on a different platform.
- error.f90 — This module contains some simple error reporting routines.
- constants.f90 — This module contains the definitions of various physical constants and parameters used by many different programs.
- math.f90 — This module groups a number of basic mathematical routines, such as matrix multiplication and inversion.
- crystalvars.f90 — This module defines the variable types for lattice parameter information, direct and reciprocal metric tensors and structure matrices.
- symmetryvars.f90 — This module defines all variables needed to use the 3D space groups and related group theoretical objects.
- crystal.f90 — This module contains all routines that are related to geometrical crystallographic computations, such as dot and cross products and coordinate transformations, and also basic routines for creating crystal data files.
- symmetry.f90 — This module implements the 3D space groups and related group theoretical objects.
- files.f90 — This module contains all file operations, such as opening and closing of files, including crystal data files.
- postscript.f90 — Crystallographic computations generally produce graphical output, and, unfortunately, there is no standard graphics language or library that will work on every computer platform. For this reason graphics output from the programs in this book is provided in PostScript format, so that it can be printed on paper or displayed on the screen using a PostScript viewer. PostScript is platform independent, so it should work on UNIX, Windows, and Macintosh platforms. The file postscript.f90 groups all routines that produce PostScript output. Additional routines can easily be created as needed, based on the templates in this file.
- tiff.f90 — This module provides a simple routine to create a TIFF image file containing a grayscale image.

diffraction.f90 All routines related to elastic scattering are grouped in this library. This includes computation of the relativistic electron wavelength, Fourier coefficients of the electrostatic lattice potential, extinction distances, excitation errors, and so on. In addition this module provides many of the support routines for multi-beam and systematic row computations.

others.f90 In this module all routines provided by other researchers are grouped together.

graphics.f90 This module contains a number of routines for the creation of graphics output in the form of 2D plots, shaded surfaces and contour plots. The routines make extensive use of the postscript.f90 routines.

All library files are combined into a single object code library during the compilation process. This object library libtem.a is then linked with the individual program as needed. Detailed instructions for compilation of all routines are available from the website.

In the interest of brevity we will not reproduce any Fortran source code in this book. Instead, we will use a simple form of *pseudo code* to describe the outline of the most important computations. All pseudo code will be labeled by symbols of the type PC-0. Pseudo code is a textual description of the logic of a program, and it should be rather straightforward to follow the various steps in the actual code. In the example PC-0 a simple pseudo code section is shown. The top line, beginning with **Input:**, states which other sections of code must be completed before executing the current section. The next line briefly describes the type of output produced by the code, if any. The remaining lines are the pseudo code statements (on the left) with the corresponding subroutines or functions indicated on the right. When appropriate the pseudo code will also state which equation from the text is used in the Fortran code. The program flow statements (if, repeat until, while, ...) are printed in bold-face characters.

Pseudo Code PC-0 Example of pseudo code.

Input: –
Output: no output
compute something
if this something is positive **then**
use equation 1-15 {subroutine CalcMatrices}
else
use equation 2-11
end if
display result on screen

1.9.2 Implementing the metric tensor formalism

Nearly all crystallographic computations require that the user enter the lattice parameters, the space group, and the atom positions for the atoms in the *asymmetric unit*. From the lattice parameters we can compute the direct and reciprocal metric tensors and structure matrices, which contain all geometrical information about crystal space and reciprocal space in a form useful for numerical computations. The following routines are provided for user input of the crystallographic information (their use is described in pseudo codes PC-1 and PC-2 and the source code can be found in the crystal.f90 and symmetry.f90 modules).

- GetLatParm: This routine asks the user to enter the crystal system, followed by the independent lattice parameters in nanometers. If the crystal system is trigonal, the user is asked to specify hexagonal or rhombohedral lattice parameters. GetLatParm then calls CalcMatrices.
- CalcMatrices: This routine computes the direct and reciprocal metric tensors, and the direct and reciprocal structure matrices. Direct and reciprocal lattice parameters and all matrices are stored in a user-defined variable type (the equivalent of the Fortran-77 "common block") for use by other routines.
- GetSpaceGroup: This routine lists the space groups for the selected crystal system, and asks the user to select by number. For space groups with two origin settings, the user is also asked which setting should be used. GetSpaceGroup then calls GenerateSymmetry.
- GenerateSymmetry: From the space group number and setting a recursive algorithm creates all 4×4 symmetry matrices, as described in Section 1.9.3. The matrices are stored in a special variable type for use by other routines.
- GetAsymPos: The user is prompted for the coordinates of the atoms in the asymmetric unit, as well as the site occupation parameters and the Debye–Waller factors.

The vector dot and cross products are then readily implemented using the metric tensor formalism. The following general purpose routines are available in the

---

Pseudo Code PC-1 Define a new crystal structure.

Input: –
Output: crystal data file [*.xtal]

prompt for crystal system and lattice parameters	{GetLatParm}
prompt for space group number and setting	{GetSpaceGroup}
prompt for contents of asymmetric unit	{GetAsymPos}
save data in file	{SaveData}

Pseudo Code PC-2 Load crystal structure data in memory.

Input: –
Output: crystal data stored in common blocks
  if this is a new crystal **then**
    do PC-1
  else
    read data from file [*.xtal] {LoadData}
    compute metric tensors and structure matrices {CalcMatrices}
    generate all symmetry matrices {GenerateSymmetry, PC-3}
  end if

---

crystal.f90 module (the variables between square brackets are the arguments of the function or subroutine calls).

- CalcDot [**p**, **q**, s_1] Computes the dot product between two vectors. The last argument of the function call, s_1, is a switch; when the switch is "d", the dot product is computed using the direct metric tensor, "r" uses the reciprocal metric tensor, and "c" is a dot product between a direct space vector and a reciprocal space vector, which implies using the Kronecker delta as a metric tensor.
- CalcCross [**p**, **q**, **r**, s_1, s_2, v] Computes the cross product between two vectors. There are two switches, one to indicate the space in which the input vectors are defined and the other to select the space in which the output vector will be expressed. The last parameter indicates whether or not the cross product should be scaled by the unit cell volume.
- CalcLength [**p**, s_1] Computes the length of a vector. This routine calls CalcDot and uses the same switch values.
- CalcAngle [**p**, **q**, s_1] Computes the angle between two vectors. This routine calls CalcDot three times, using the same value for the switch.
- TransSpace [**p**, **q**, s_1, s_2] Transforms the components of a vector **p** from space s_1 to space s_2, and returns the result in the variable **q**.
- TransCoor [**p**, **q**, α, s_1, d] Transforms a vector in space s_1 from the old to the new reference frame (d = "on") or from new to old (d = "no"), using the transformation matrix α. This implements all vector coordinate transformations listed in Table 1.6.
- MilBrav [**p**, **q**, d] Transforms the components of a direction vector in a hexagonal system from the 3-index (**p**) to the 4-index (**q**) version when the switch $d = 34$, and the reverse when $d = 43$.

The Fortran program latgeom.f90 illustrates how these routines can be used to implement basic lattice geometry computations. We leave it as an exercise for the reader to add an option to convert from Miller to Miller–Bravais indices in a hexagonal system, using the MilBrav subroutine. The **ION**-routine latgeom.pro on the website implements some of the crystallographic computations described in this chapter in an interactive module. The user defines the lattice parameters and selects two vectors. The routine returns the metric tensors, structure matrices, and the lengths of and angles between the two vectors in both direct and reciprocal space.

1.9.3 Using space groups on the computer

The section describing the 230 space groups in *International Tables for Crystallography*, Volume A [Hah96], is 606 pages long and contains a wealth of information. To implement the space groups into a computer program appears to be a formidable task, considering the amount of information available. It is therefore quite surprising that it is possible to compile the relevant information on all 230 space groups into one single ASCII file which is only 4104 bytes long! Here again, we find an indication of the powerful tools group theory provides.

In this section we describe a simple space group encoding scheme.† From the definition of a group on page 42 we know that every combination of two symmetry operations must again be a symmetry operation. A group can then be characterized by a limited number of operators, known as the *generators*, from which all other operations can be derived by repeated multiplication. The Bravais lattice translation vectors are obviously generators, otherwise one could never generate the infinite lattice. For the highest symmetry space group ($\mathbf{Fm\bar{3}m}$), there are only seven generators from which all 192 symmetry elements in the fundamental unit cell can be derived.

Since each generator can be freely chosen, as long as the set of generators generates the entire group, a judicious choice of the generators for the 230 space groups results in a list of only 14 different point symmetry operators. In other words, if we label and store the 3×3 matrices of those 14 operations, then for each group we only need to list the labels of the matrices to be used as generators and the components of the translation vectors associated with the generators, and this results in only one line of data per space group. It is then up to the algorithm to take the generators and compute all possible products between them until no more new operators are found.

The algorithm creates all the 4×4 matrices for the operations inside the fundamental unit cell (the cell enclosed by $\mathbf{a}_i$). Adding a translation vector $\mathbf{t} = t_i \mathbf{a}_i$ to the positions generated by the matrices then translates the position to a neighboring

† The author would like to acknowledge G. Ceder (MIT) for permission to make his source code available.

unit cell. For the purposes of this book, it is sufficient to only use the operators in the fundamental unit cell.

To generate all atom positions inside a unit cell when the positions in the *asymmetric unit* are given, we compute all the symmetry matrices first, and then apply all matrices to the coordinates of each input atom. Then we remove identical positions (which can occur if the input coordinates coincide with one of the special positions in the space group) to end up with the orbit of the input atom. The coordinates thus generated are fractional coordinates with respect to the unit cell basis vectors. They can be used in subsequent computations of, for example, the structure factor or Fourier coefficients of the electrostatic lattice potential (see later chapters). In Section 1.8 we described a general method to convert the fractional coordinates to Cartesian coordinates, so that crystal structure drawings or diffraction patterns can be generated.

The following lines list the ASCII codes for a few selected space groups, together with the space group symbol and number:

P1	1	*01aOOO0*
C222	21	*03aDDObOOOcOOO0*
Cmcm	63	*13aDDObOODcOOD0*
P4/nbm	125	*04bOOOgOOOcOOOhDDO1YYO*
Fd$\bar{3}$m	227	*07aODDaDODbODDcDDOdOOOeFBFhBBB1ZZZ.*

The encoding scheme works as follows: consider a crystal structure described by the orthorhombic space group **Cmcm** (space group # 63). This space group is completely encoded by the character string *13aDDObOODcOOD0*. This string should be read as follows: the first digit takes on the value 1 or 0, to indicate whether or not the space group contains the inversion operator as a symmetry element. In the case of **Cmcm** the inversion operator is indeed one of the generators. The next digit gives the number of additional generator elements, in this case 3. Each generator is then described by four characters, and there are three such quadruplets in the string: *aDDO*, *bOOD*, and *cOOD*. Their meaning will be discussed in the next paragraph. The next character in the string (*0*) is only different from zero if the space group has more than one origin setting in the *International Tables*. For the space group **Cmcm** this character is equal to zero, indicating that there is only one origin setting. For the space group #125 on the other hand, the character following the four generator quadruplets is equal to one, indicating that a second origin setting is available. Since a change in origin implies a translation, there are three additional uppercase characters in the string, which represent the components of the translation vector that transforms coordinates in the first setting to coordinates in the second setting (using the conventions of the *International Tables*). For space group #125 the translation vector for the second setting is represented by the character triplet

YYO which, according to the conversion Table A3.2 in Appendix A3, corresponds to the vector with components $[-\frac{1}{4}, -\frac{1}{4}, 0]$.

Each generator matrix is a 4×4 matrix; the 3×3 submatrix corresponding to the point symmetry operation is labeled by one of the 14 lowercase letters *a* through *n*. The actual matrix elements are listed in Table A3.1 in Appendix A3. Since a space group operator may also contain a translational part, the four-character string *aDDO* contains three additional uppercase characters, representing the three components of the translational part of the symmetry operator. From the table in Appendix A3 one can see that the character *D* corresponds to the value $\frac{1}{2}$, and the character *O* to zero. The string *aDDO* thus represents the 4×4 matrix

$$aDDO \equiv \begin{pmatrix} 1 & 0 & 0 & \frac{1}{2} \\ 0 & 1 & 0 & \frac{1}{2} \\ 0 & 0 & 1 & 0 \\ 0 & 0 & 0 & 1 \end{pmatrix},$$

and the string *bOOD* represents the matrix

$$bOOD \equiv \begin{pmatrix} -1 & 0 & 0 & 0 \\ 0 & -1 & 0 & 0 \\ 0 & 0 & 1 & \frac{1}{2} \\ 0 & 0 & 0 & 1 \end{pmatrix},$$

and so on. The remaining elements of the space group **Cmcm** are then generated from the three matrices *aDDO*, *bOOD*, and *cOOD*, and the inversion operator by first computing all the powers of the generators, and then all products between generators until no further unique symmetry elements can be found (see PC-3). This computation is performed by the subroutine GenerateSymmetry, which then stores all symmetry operator matrices in a special user-defined variable type for use by other routines. The Fortran program listSG.f90 can be used to list all the equivalent point positions for a general point (x, y, z) for any given space group.

Once all symmetry matrices are known it is straightforward to compute the members of families, orbits, and stars. Pseudo code PC-4 shows how one can compute the equivalent points for an atom in the asymmetric unit. The routines CalcOrbit and CalcStar implement the pseudo code. Note that these routines can also be used to generate equivalent point positions for the point groups. Both routines have a switch to turn on or off the reduction of the coordinates to the fundamental unit cell. If reduction is turned off, then the coordinates of the equivalent points will represent the point group symmetry with respect to the origin of the unit cell. The Fortran program orbit.f90 implements the CalcOrbit routine to create all the equivalent points for an arbitrary space group; the program prompts the user for the

Pseudo Code PC-3 Generate space group symmetry matrices.

Input: space group number {GetSpaceGroup}
Output: array of all non-equivalent symmetry matrices
load and decode the space group code string
determine the generator matrices
repeat
multiply generators amongst themselves and resulting matrices
until no more new matrices found
reduce all translation components to the fundamental unit cell
identify the point operations $D^{(k)}$ and store their 3×3 matrices
compute the reciprocal point symmetry matrices $D^{*(k)}$ {eqn. 1.42}
store all independent symmetry matrices

Pseudo Code PC-4 Compute the orbit of a given point.

Input: crystal data file [*.xtal]
Output: equivalent atom positions
load crystal data {PC-2}
for each symmetry matrix **do**
multiply matrix with atom coordinates
if this is a new position **then**
keep position and increment counter
if reduction required **then**
reduce position to fundamental unit cell
end if
else
discard position
end if
end for
return list and number of equivalent positions

space group number and an atom position, and reduces all generated coordinates to the basic unit cell. Members of families of directions or planes can be computed in the same way, simply by passing the indices of a general direction or plane to the CalcFamily routine. The CalcFamily routine is particularly useful to generate stereographic projections, as described in Section 1.9.5.

1.9.4 Graphical representation of direct and reciprocal space

There are many different ways to represent crystal structures: amongst the more commonly used ones are the ball-and-stick representation, the space-filling representation, and polyhedral models to emphasize local coordinations. All of these representation methods can be used with both orthogonal and perspective projections. All of these methods consist of two basic steps:

(i) computation of the Cartesian coordinates of each atom in the unit cell(s);
(ii) computation of the projected (or screen) coordinates of each atom.

The latter step will not be discussed in detail in this book. We refer to the extensive literature on computer graphics libraries (e.g. [NDW93, SS87] and many others). This computation essentially involves perspective or parallel projections, for a given location and orientation of the object (the crystal) and the observer. Note that the 4×4 matrices used to represent symmetry operations are also used for perspective projections; it turns out that the elements on the fourth row of a 4×4 matrix can be used to describe various geometric projections in a very compact way.

The first step above can easily be completed with the tools derived earlier in this chapter. After generating the *orbit* of each atom in the asymmetric unit (CalcOrbit),

---

Pseudo Code PC-5 Draw a crystal structure or reciprocal space.

```
Input: crystal data file [*.xtal]
Output: PostScript file(s)
  Load crystal data                                                        {PC-2}
  if Real Space Drawing then
    compute atom coordinates for each atom in asymmetric unit         {CalcOrbit}
    how many unit cells in the drawing?
    complete unit cell faces, edges and corners
    convert all coordinates to a Cartesian reference frame           {TransSpace}
  else
    how many reciprocal unit cells in the drawing?
    for each reciprocal lattice point do
      compute structure factor                                             {PC-9}
      convert all coordinates to a Cartesian reference frame         {TransSpace}
    end for
  end if
  repeat
    construct Cartesian projection matrix for given viewing direction
    create PostScript drawing (using atom radii or |structure factor|^2)
  until no more drawings needed
```

the coordinates are converted from crystal space to the standard Cartesian reference frame (TransSpace). The resulting Cartesian coordinates are then entered into the graphics system pipeline, where they are combined with the projection information to produce a 2D drawing. The viewing or projection direction can usually be defined in two ways: either with respect to a Cartesian reference frame, or, more commonly, by means of the direction indices $[uvw]$. If the latter mechanism is used, care must be taken to also transform the direction indices to the standard Cartesian reference frame, using the direct structure matrix a_{ij} (TransSpace). The details of drawing the atom (radius, color, texture, etc.) are left to the reader; using the program drawcell.f90 a simple PostScript drawing can be generated for an arbitrary unit cell. The pseudo code for this program is shown in PC-5. An example drawing for the structure of rutile (TiO_2, space group **P4₂/mnm** (#136), $a = 0.4594$ nm, $c = 0.2958$ nm, Ti at 2a $(0, 0, 0)$, O at 4f $(x, x, 0)$ with $x = 0.3$) is shown in Fig. 1.24.

Drawings of the reciprocal space lattice can be made in the same way: first we determine the set of reciprocal lattice points to be included in the drawing, then the reciprocal structure matrix b_{ij} is used to convert the (hkl) triplets to Cartesian coordinates (TransSpace), and finally these coordinates are passed on to the graphics

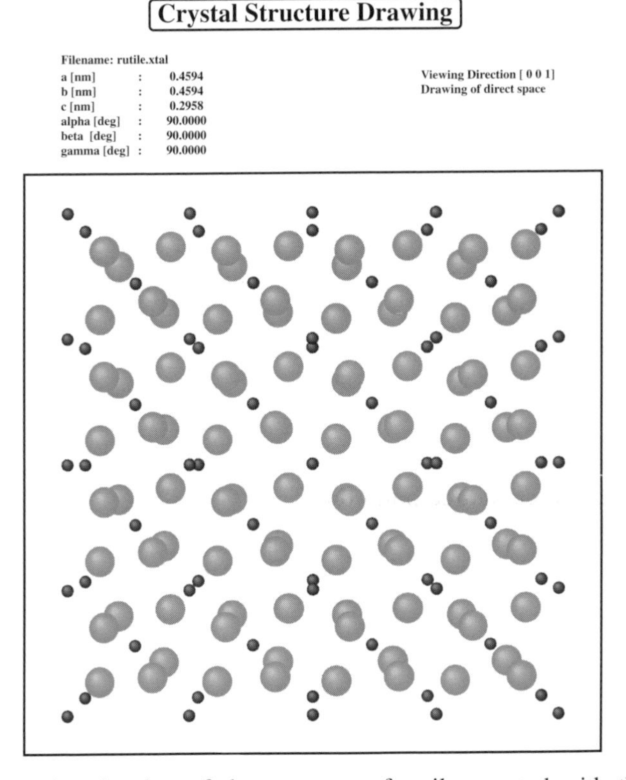

Fig. 1.24. Perspective drawing of the structure of rutile, created with the drawcell.f90 program.

pipeline for display. The main advantage of using the direct and reciprocal structure matrices defined in Section 1.8 is that the relative orientation of direct and reciprocal space is properly represented and, in addition, the algorithm is independent of the particular choice of Bravais lattice. We will postpone discussion of a program for drawing electron diffraction patterns until Chapter 4. The program drawcell.f90 can also be used to create a perspective or parallel projection drawing of the reciprocal lattice; the "size" of each reciprocal lattice point is determined by the value of the corresponding structure factor, as discussed in Chapter 2.

1.9.5 Stereographic projections on the computer

There are several computer programs available that allow the user to draw stereographic projections of direct and reciprocal space. While these programs are often sufficient for regular use, there may be situations when the TEM user will need to perform a certain operation that is not a feature of any of the existing programs. In such a situation it may be necessary to write a dedicated program to generate certain patterns. In this subsection, we will describe a straightforward method to compute stereographic projections for arbitrary crystals, in both direct and reciprocal space.

We already know from Section 1.8 how to convert crystal or reciprocal coordinates into Cartesian coordinates. Since the SP represents directions, the next step is to *normalize* the Cartesian components so that the direction corresponds to a point on the unit sphere. Let us assign the coordinates (x, y, z) to this point P, as illustrated in Fig. 1.25. If we connect the point P with the south pole S, then the

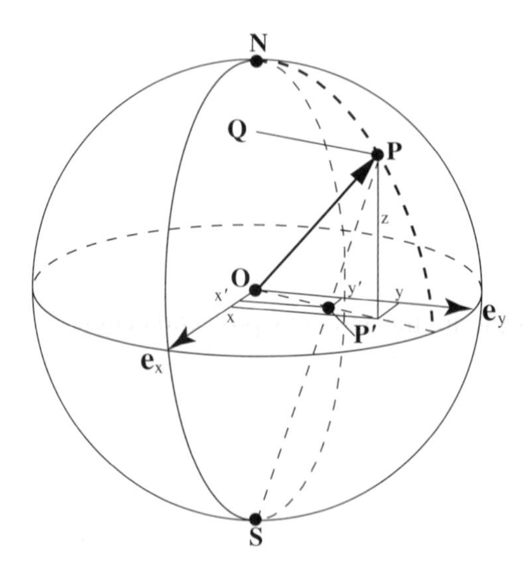

Fig. 1.25. Schematic representation of the relation between 3D Cartesian coordinates and stereographic projection coordinates.

triangle PQS is congruent with $P'OS$, from which we derive:

$$x' = \frac{x}{1+z};$$
$$y' = \frac{y}{1+z}.$$

If the projection circle has radius R, then the above equations are simply multiplied by R to arrive at the proper 2D coordinates.

A projection of the reciprocal space directions (or plane normals) is easily generated by first transforming the coordinates of each reciprocal lattice point to Cartesian coordinates, and then proceeding in the same way as described above. We will now illustrate the use of stereographic projections in three programs: the first program draws the projection of all plane normals or directions (with indices in a given range) for an arbitrary crystal in an arbitrary orientation. The second program does the same but only for selected families of planes or directions, and the third program allows for superposition of the projections from two crystals, with a given orientation relation between them.

The pseudo code for the first program is shown in PC-6. The program can be found on the website under the name stereo.f90. Multiple projections can be made for a given crystal structure, and a [111] projection of a cubic crystal is shown in Fig. 1.26a. The second program, family.f90, is described in PC-7, and is only slightly different from stereo.f90. One member of each family must be specified and the point symmetry operators are then used to compute the entire family

---

Pseudo Code PC-6 Stereographic projection of a range of directions/plane normals.

Input: PC-2, PC-3
Output: stereographic projection in PostScript format
  repeat
    determine projection direction
    real space or reciprocal space projection?
    open a new PostScript file and draw projection circle
    repeat
      which range of directions/plane normals should be drawn?
      transform all members to Cartesian space and project
      draw (and label) the points
    until drawing complete
    close PostScript file
  until no further patterns required

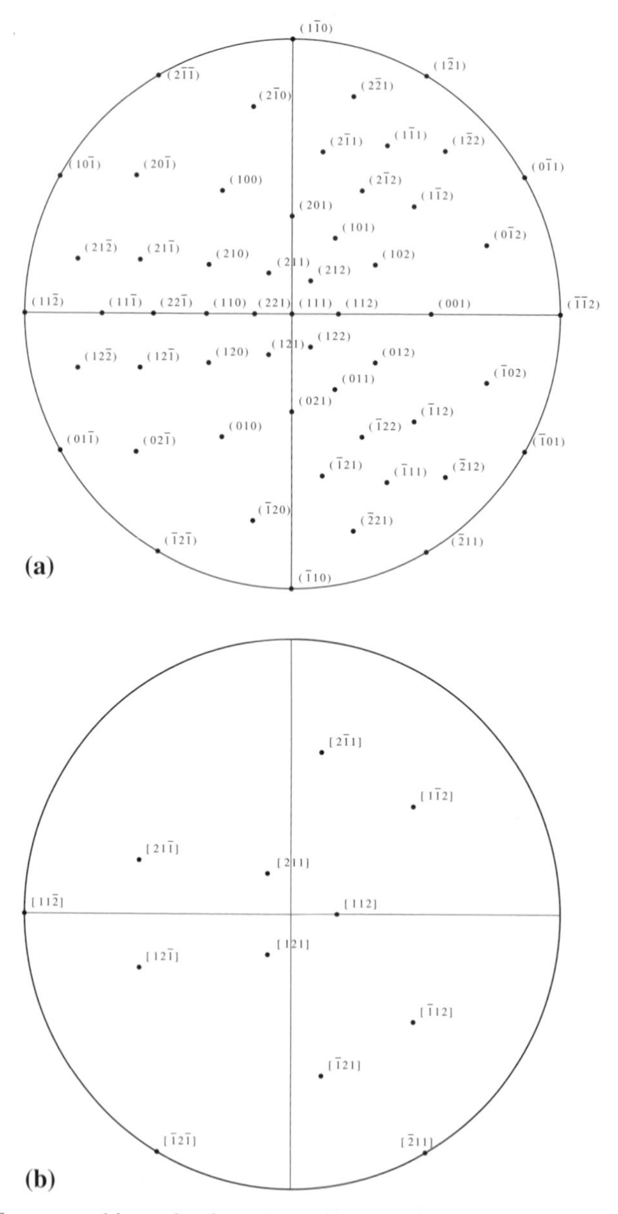

Fig. 1.26. (a) Stereographic projection of the plane normals of a cubic crystal viewed along the [111] direction (created with stereo.f90); (b) stereographic projection of the ⟨112⟩ family in a cubic crystal (created with family.f90).

(CalcFamily). Example program output is shown in Fig. 1.26b for the {112} family in a cubic crystal.

The third program, orient.f90, is described in PC-8. The program asks for two crystal structure files, and a complete orientation relation (two parallel planes and

Pseudo Code PC-7 Stereographic projection of a family.

```
Input: PC-2, PC-3
Output: stereographic projection in PostScript format
  repeat
    determine projection direction
    real space or reciprocal space projection?
    open a new file and draw projection circle
    repeat
      which direction/plane normal should be drawn?
      compute members of its family                          {CalcFamily}
      transform all members to Cartesian space and project
      draw (and label) the points
    until drawing complete
    close PostScript file
  until no further patterns required
```

Pseudo Code PC-8 Stereographic projection of orientation relation for two crystals.

```
Input: PC-2, PC-3                                            {for two crystals}
Output: stereographic projection in PostScript format
  obtain orientation relation
  real space or reciprocal space projection?
  compute transformation matrix α_ij                         {1.67}
  determine range of directions/plane normals
  transform all quantities to reference frame of crystal A
  transform all quantities to the standard Cartesian frame
  repeat
    determine projection direction and projection matrix
    proceed as in PC-6 for both crystals
  until no further patterns required
```

two parallel directions in those planes), and then draws the two patterns superimposed on the same drawing. This program is a bit more complex than the other two because it combines nearly all aspects of crystallography discussed in this chapter. The orientation relation (OR) between the two crystals consists of the following information:

$$\mathbf{g}_A = (hkl)_A \parallel (hkl)_B = \mathbf{g}_B;$$
$$\mathbf{t}_A = [uvw]_A \parallel [uvw]_B = \mathbf{t}_B.$$

The direction **t** must lie in the plane **g**, and hence the zone equation $\mathbf{t} \cdot \mathbf{g} = 0$ must be satisfied for each crystal. All crystallographic information must now be converted to one single reference frame, say that of crystal A. We must therefore determine the transformation matrix α_{ij}, which expresses the orientation relation above. This can be done as follows.

(i) Construct a Cartesian reference frame in each crystal; the z-direction of this reference frame is taken parallel to the direction vector $\mathbf{t}_{A,B}$, the x-direction is parallel to the plane normal $\mathbf{g}_{A,B}$, and the y-direction completes the right-handed orthonormal frame:

$$\mathbf{e}_x^{A,B} = \frac{\mathbf{g}_{A,B}}{|\mathbf{g}_{A,B}|};$$
$$\mathbf{e}_y^{A,B} = \mathbf{e}_z^{A,B} \times \mathbf{e}_x^{A,B};$$
$$\mathbf{e}_z^{A,B} = \frac{\mathbf{t}_{A,B}}{|\mathbf{t}_{A,B}|}.$$

(ii) Rewrite the equations above in matrix form:

$$\mathbf{e}_i^A = \mathbf{E}_{ij}\mathbf{a}_j^A;$$
$$\mathbf{e}_i^B = \mathbf{E}'_{ij}\mathbf{a}_j^B.$$

(iii) Equate the Cartesian reference frames of both crystals, so that the proper OR is obtained:

$$\mathbf{E}_{ik}\mathbf{a}_k^A = \mathbf{E}'_{ij}\mathbf{a}_j^B.$$

(iv) Solve this equation for $\mathbf{a}_j^A$:

$$\mathbf{a}_l^A = \mathbf{E}_{li}^{-1}\mathbf{E}'_{ik}\mathbf{a}_k^B = \alpha_{lk}\mathbf{a}_k^B, \tag{1.67}$$

and this defines the transformation matrix α. We can now apply the transformation relations in Table 1.6 to convert quantities between the properly oriented crystals A and B.

Now that the transformation matrix for the OR is known, all remaining computations proceed as before and the stereographic projection of the two crystals can be produced for an arbitrary viewing direction. The same method can be used to compute diffraction patterns for two crystals with a given OR. An example of the output of the orient.f90 program is shown in Fig. 1.27 for two cubic crystals with a standard twin orientation relationship $(111)_A \parallel (111)_B$ and $[11\bar{2}]_A \parallel [\bar{1}\bar{1}2]_B$.

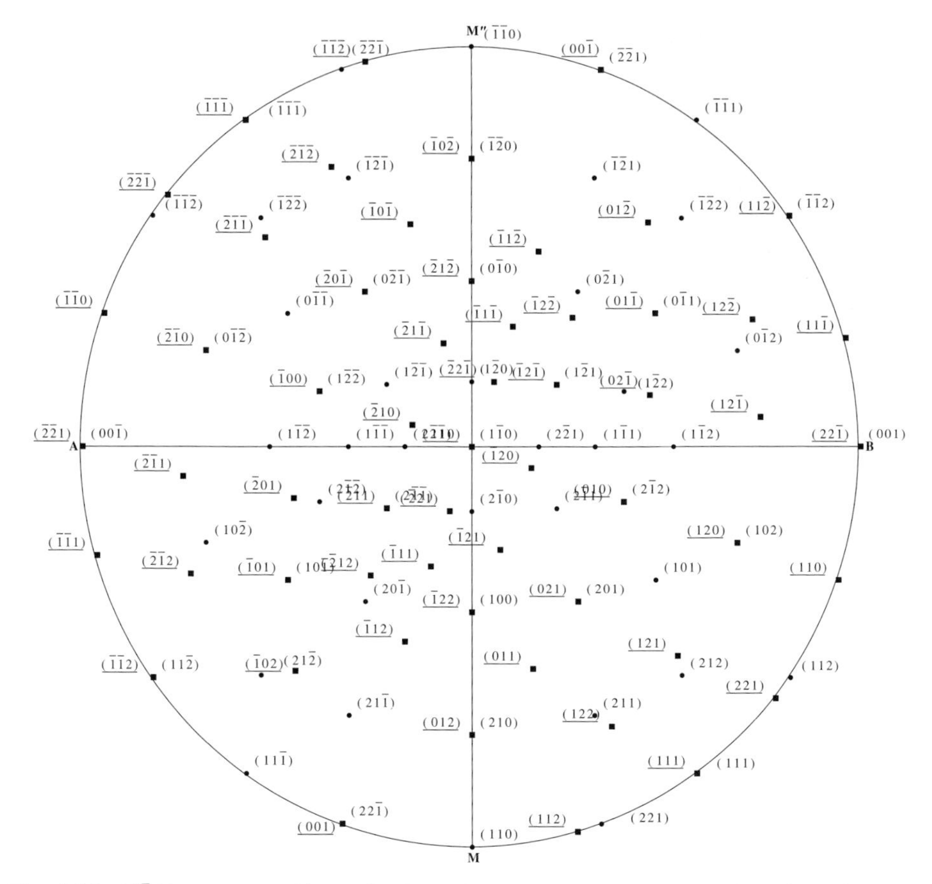

Fig. 1.27. $[1\bar{1}0]$ stereographic projection of two cubic crystals in twin orientation (created using orient.f90); the size of the projection circle was reduced to fit on the page, and consequently some of the indices overlap.

Example 1.19 *Consider the orientation relation* $(111)_A \parallel (111)_B$ *and* $[11\bar{2}]_A \parallel [\bar{1}\bar{1}2]_B$ *in a cubic crystal. Compute the transformation matrix* α_{ij}.

Answer: *For the first basis vector* $\mathbf{e}_x$ *we have*

$$\mathbf{g} = g_i \mathbf{a}_i^* = g_i g_{im}^* \mathbf{a}_m = \frac{1}{a^2} g_m \mathbf{a}_m,$$

which after normalization becomes

$$\mathbf{e}_x = \frac{g_m}{a\sqrt{\sum_j g_j^2}} \mathbf{a}_m.$$

This relation is valid for each crystal A and B. For $\mathbf{g} = (111)$ *we find*

$$\mathbf{e}_x^{A,B} = \frac{1}{a\sqrt{3}}\left(\mathbf{a}_1^{A,B} + \mathbf{a}_2^{A,B} + \mathbf{a}_3^{A,B}\right).$$

The third basis vector $\mathbf{e}_z$ *is given by*

$$\mathbf{e}_z = \frac{t_i}{a\sqrt{\sum_j t_j^2}}\mathbf{a}_i,$$

and therefore

$$\mathbf{e}_z^A = \frac{1}{a\sqrt{6}}\left(\mathbf{a}_1^{A,B} + \mathbf{a}_2^{A,B} - 2\mathbf{a}_3^{A,B}\right); \quad \mathbf{e}_z^B = -\mathbf{e}_z^A.$$

The vector $\mathbf{e}_y$ *is then determined from the cross product:*

$$\mathbf{e}_y = \frac{\epsilon_{ijk}t_i g_j}{a\sqrt{\sum_m t_m^2 \sum_n g_n^2}}\mathbf{a}_k,$$

from which we find

$$\mathbf{e}_y^A = \frac{1}{a\sqrt{2}}\left(\mathbf{a}_1^A - \mathbf{a}_2^A\right); \quad \mathbf{e}_y^B = -\mathbf{e}_y^A.$$

The matrices E and E′ are given by

$$E_{ij} = \frac{1}{a\sqrt{6}}\begin{pmatrix} \sqrt{2} & \sqrt{2} & \sqrt{2} \\ \sqrt{3} & -\sqrt{3} & 0 \\ 1 & 1 & -2 \end{pmatrix}; \qquad E'_{ij} = \frac{1}{a\sqrt{6}}\begin{pmatrix} \sqrt{2} & \sqrt{2} & \sqrt{2} \\ -\sqrt{3} & \sqrt{3} & 0 \\ -1 & -1 & 2 \end{pmatrix}.$$

The inverse of E is readily shown to be equal to

$$E_{ij}^{-1} = a\sqrt{6}\begin{pmatrix} \frac{1}{3\sqrt{2}} & \frac{1}{2\sqrt{3}} & \frac{1}{6} \\ \frac{1}{3\sqrt{2}} & \frac{-1}{2\sqrt{3}} & \frac{1}{6} \\ \frac{1}{3\sqrt{2}} & 0 & \frac{-2}{6} \end{pmatrix}.$$

The matrix $\alpha = E^{-1}E'$ *is then given by*

$$\alpha_{ij} = \frac{1}{3}\begin{pmatrix} -1 & 2 & 2 \\ 2 & -1 & 2 \\ 2 & 2 & -1 \end{pmatrix}.$$

Note that this is a symmetric unitary matrix, and therefore it is equal to its own inverse and to its transpose. The same matrix can therefore be used for all the transformations listed in Table 1.6.

1.10 Recommended additional reading

The reader may find useful additional information on crystallography and symmetry in the following books:

- *Crystal Structures: A Working Approach*, H.D. Megaw, W.B. Saunders Company (Philadelphia, 1973).
- *Essentials of Crystallography*, D. McKie and C. McKie, Blackwell Scientific Publications (Oxford, 1986).
- *Manual of Mineralogy*, C. Klein and C.S. Hurlbut, Jr., 20th edition, John Wiley & Sons (New York, 1985).
- *Space Groups for Solid State Scientists*, G. Burns and A.M. Glazer, 2nd edition, Academic Press (Boston, 1990).
- *Crystallography*, W. Borchardt-Ott, 2nd edition, Springer (Berlin, 1995).
- *Point Groups, Space Groups, Crystals, Molecules*, R. Mirman, World Scientific (Singapore, 1999).

A good introduction to vector calculus in non-Cartesian coordinate frames can be found in:

- *Vector and Tensor Analysis with Applications*, A.I. Borisenko and I.E. Tarapov, Dover Publications (New York, 1979).
- *Introduction to Vector and Tensor Analysis*, R.C. Wrede, Dover Publications (New York, 1972).

A basic introduction to programming in Fortran-77 and Fortran-90 can be found in:

- *Numerical Recipes in FORTRAN: the Art of Scientific Computing*, W.H. Press, B.P. Flannery, S.A. Teukolsky, and W.T. Vetterling, 2nd edition, Cambridge University Press (New York, 1992).
- *FORTRAN 77 for Engineers and Scientists: with an Introduction to FORTRAN 90*, L. Nyhoff and S. Leestma, 4th edition, Prentice Hall (New Jersey, 1996).
- *Fundamentals of Engineering Programming with C and Fortran*, H.R. Myler, Cambridge University Press (New York, 1998).
- *Fortran 90/95 for Scientists and Engineers*, S.J. Chapman, McGraw-Hill (Boston, 1998).

Exercises

1.1 A monoclinic crystal has lattice parameters $\{1, \frac{1}{2}, 2, 90, 60, 90\}$ (distances in nm).
   (a) Write down the explicit expression for its metric tensor.
   (b) Compute the distance between the origin and the point $\left(\frac{1}{2}, \frac{1}{2}, \frac{1}{2}\right)$.
   (c) Compute the angle between the directions [100] and [111].
   (d) Compute the reciprocal metric tensor.

(e) Compute the angle between the (001) and (111) plane normals.
(f) Compute the direct space components of the plane normal $\mathbf{g}_{113}$.
(g) Write down the explicit expressions for the reciprocal basis vectors. From these expressions, derive the reciprocal lattice parameters.
(h) Compute the cross product between the direction vectors [100] and $[01\bar{1}]$, and express it in both the direct and reciprocal reference frames.
(i) Compute the Cartesian components of the point $(1, -1, 1)$ (standard Cartesian reference frame).
(j) Compute the Cartesian coordinates of the reciprocal lattice point (111).

1.2 Consider a monoclinic unit cell, with lattice parameters $\{2, 1, 1, 90, 45, 90\}$. Use a unit distance of 4 cm (i.e. a lattice parameter of 1 should correspond to 4 cm on your drawing). The **b** axis points down into the plane of the drawing, normal to both **a** and **c** which lie in the plane of the drawing.
(a) Draw the basis vectors and the following planes (in different colors or linestyles): (101), (201), $(10\bar{1})$, and (102).
(b) Draw the reciprocal basis vectors $\mathbf{a}^*$ and $\mathbf{c}^*$ to scale and in the proper orientation with respect to the basis vectors $\mathbf{a}_i$.
(c) Draw the **g**-vectors for the four planes given above, based on the reciprocal basis vectors. All **g**-vectors should start in the origin and use the same linestyle/color as the corresponding planes.
(d) *Measure* the d-spacings on the drawing (in units of 4 cm), and *calculate* $|\mathbf{g}|$ for each plane; verify that $d = 1/|\mathbf{g}|$ (to within the accuracy of your drawing).

1.3 Show that for a cubic crystal system, the direction $[uvw]$ is always parallel to the plane normal on the plane with Miller indices (uvw).

1.4 Use equation (1.28) to derive relation (1.21).

1.5 Show that the inverse of the Seitz operator $(\mathbf{D}|\mathbf{t})$ is given by $(\mathbf{D}^{-1}| - \mathbf{D}^{-1}\mathbf{t})$.

1.6 Consider the six-fold axis ⬢ located at the point $(\frac{1}{2}, \frac{1}{2}, 0)$ and parallel to the $\mathbf{e}_3$-direction. What is the corresponding matrix $\mathcal{W}$? (Hint: Decompose the operator into two translations and a rotation, in the proper order.)

1.7 Write down all the 3×3 symmetry matrices for the elements of the point group $\bar{\mathbf{6}}\mathbf{m2}$ with respect to the hexagonal basis vectors.

1.8 Consider the body-centered cubic Bravais lattice. Define a primitive unit cell for this structure and determine the transformation equations for directions and planes with respect to the conventional cubic basis vectors.

1.9 Derive equations (1.61) and (1.62) on page 55.

1.10 Work out the transformation matrices $\mathbf{E}_{ij}$, $\mathbf{E}'_{ij}$, and α_{ij} for the following orientation relation in a tetragonal crystal with lattice parameters $\{a, c\}$:
$(011)_A \parallel (011)_B$, $[0\bar{1}1]_A \parallel [01\bar{1}]_B$.

2

Basic quantum mechanics, Bragg's Law, and other tools

2.1 Introduction

One of the principal goals of transmission electron microscopy (TEM) is to observe and determine the crystallographic nature of a variety of features in materials, from the actual crystal structure to the detailed atomic configurations and chemistry around defects. In the previous chapter, we introduced the mathematical tools needed to describe crystal structures and to perform geometric computations in an arbitrary Bravais lattice. In the present chapter, we introduce another set of tools that are used to describe the interactions between electrons and the specimen.

Since observation of a phenomenon or an object always implies an interaction between the observer and that object, it is important that such observations be carried out under the proper conditions. For instance, to determine the color of an object, we would typically illuminate the object with white light, and analyze the frequencies of the reflected light, while to determine the fluorescent properties of that same object we would illuminate it with ultraviolet light, and so on. Establishing the proper observation conditions is thus crucial to the subsequent success of the observation. Since the object *modifies* the incident illumination, we can extract valuable information about the object by carefully analyzing this modified signal.

To determine the crystallographic characteristics of a material, we *illuminate* it with electrons. It is thus intuitively clear that the TEM must consist of three major stages:

(i) the *illumination stage*, which creates a beam of electrons in a well-defined reference state;
(ii) the *interaction stage*, where the sample modifies the reference state of the electron beam; and
(iii) the *observation stage*, where the modified electron beam is converted into a signal that can be detected by the human eye.

In the early part of the twentieth century, the importance of the observer as part of an experiment was recognized, and it became clear that a system of interacting particles cannot be observed without actually modifying that system by the very act of observing it. In *classical mechanics*, which deals with the motion of macroscopic objects, the observer is rarely an important part of the equation, but in the atomistic world things are quite different. The development of *quantum mechanics* (QM) in the first quarter of the twentieth century led to the incorporation of the observer into the physical system, and, more importantly, led to a new understanding of what a *physical observable* is. Since the electron microscope uses atomistic particles to probe the internal structure of materials, it is clear that the theory describing image formation in TEM must by necessity be of a quantum mechanical nature.

In addition to behaving by the rules of quantum mechanics, the electrons in a typical TEM move at a velocity that is a substantial fraction of the velocity of light. Motion at such high velocities must follow the rules of the *special theory of relativity*. We thus conclude that an observation in a TEM is essentially an experiment in *relativistic quantum mechanics*. Fortunately it is not necessary to have an advanced physics degree to understand how a TEM works, or to interpret the images obtained in various observation modes.

In this chapter, we shall discuss basic quantum mechanics and the special theory of relativity at the level required for the later chapters in this book. We will see that, once the proper starting equation (or *governing* equation) is obtained, we need no longer explicitly worry about the relativistic nature of a TEM observation. Then we introduce Bragg's Law in direct and reciprocal space and, starting from quantum mechanics, we define the Fourier transform, which essentially transforms quantities between the two spaces. We derive expressions for diffraction from infinite and finite lattices, and conclude the chapter with a description of a numerical procedure to compute the electrostatic lattice potential. Along the way, we introduce an expression for the wavelength of a relativistic electron.

2.2 Basic quantum mechanics

Classical physics describes how a macroscopic object behaves when it is subjected to external forces. When the properties of the object are known (mass, mass distribution, size, shape, etc.), together with the forces acting on that object, then the equations of classical physics allow us to compute how this object will behave as a function of time. We can calculate the position and the orientation of the object at any given moment in time. More importantly, we can measure the position, momentum, angular momentum, and energy of the object *simultaneously* (with proper measuring tools). Classical physics is therefore *deterministic*, since knowledge of the observables at one point in time allows integration of the equations of motion to any later (or earlier) moment in time. The mathematical framework of classical physics can be formulated at various levels, using Lagrangian or Hamiltonian equations of motion, and it can be shown that all of classical physics can be derived from the *principle of least action* [Gol78].

With the discovery of quantum mechanics in the early part of the twentieth century, it became clear that the laws of classical physics are no longer valid in the atomistic realm. The fundamental difference between classical physics and quantum mechanics is that a physical observable cannot always be measured with absolute certainty and, instead, one can only compute *probabilities* that a particle will be at a certain location, with a certain momentum or angular momentum, etc. In quantum mechanics, *physical observables* no longer satisfy deterministic equations, and, therefore, the classical concept of a *particle trajectory* is no longer valid. The probabilities or expectation values of physical observables are computed from a function which is the solution to the *Schrödinger equation*. This function is known as the *wave function* of the system.

2.2.1 Scalar product between functions

The basis of the mathematical formalism of quantum mechanics lies in the proposition that the state of a system of N particles can be described by a definite, in general complex-valued, function Ψ of the particle coordinates $\mathbf{r}_i$ $(i = 1, \ldots, N)$.[†] The square of the modulus of this function, $|\Psi|^2$, describes the *probability distribution* of the value of the coordinates. The function Ψ is called the *wave function* of the system, and it expresses the *probability amplitude*. All physical observables are expressions bilinear in Ψ and Ψ^*, where the asterisk denotes complex conjugation. If an observable is represented by the symbol f, then the *expectation value* of f is given by

$$\langle f \rangle = \iiint \Psi^*(\mathbf{r}) f \Psi(\mathbf{r}) \, \mathrm{d}\mathbf{r},$$

where the integration extends over all of space. The primary goal of quantum mechanics is thus to compute the wave function Ψ for a given system of particles. Once the wave function is known, the expectation values of all physical observables can be computed using expressions of the type above. Note that a wave function Ψ need only be determined to within a constant phase factor of the form $\mathrm{e}^{\mathrm{i}\alpha}$, with α a real number, since the phase factor cancels out when substituted in the expression for the expectation value of an observable.

[†] In this section we closely follow Chapter 2 of the book *Quantum Mechanics* by L.D. Landau and E.M. Lifschitz [LL74].

Since many computations in QM involve integrals of the above type, it is convenient to introduce a shorthand notation, first suggested by Dirac [Dir47, pp. 14–22]. For a given observable f, and two wave functions Ψ and Φ, we define the following symbol:

$$\langle \Phi | f | \Psi \rangle \equiv \iiint \Phi^* f \Psi. \tag{2.1}$$

The symbol $\langle \, | | \, \rangle$ is known as the *bra-ket*. Note that there is no mention of the coordinate variable at all; the definition (2.1) does not depend on the particular *representation* or basis in which the wave functions are expressed.

The bra-ket notation is rather similar to that of the dot product between two vectors. The symbol $\langle \Phi | f | \Psi \rangle$ means: take the function $|\Psi\rangle$, operate on it with the operator or function f, and project the result onto the function $\langle \Phi |$. In the absence of an operator f, we can define the *scalar product* of two functions Ψ and Φ as (taking $f = 1$):

$$\langle \Phi | \Psi \rangle = \iiint \Phi^* \Psi. \tag{2.2}$$

If $\langle \Phi | \Psi \rangle = 0$, then we say that the functions Ψ and Φ are *orthogonal functions*, in the same way that two vectors $\mathbf{p}$ and $\mathbf{q}$ are orthogonal when $\mathbf{p} \cdot \mathbf{q} = 0$. If, in addition, $\langle \Psi | \Psi \rangle = 1$ and $\langle \Phi | \Phi \rangle = 1$, then the functions Ψ and Φ are *normalized orthogonal* or *orthonormal* functions.

2.2.2 Operators and physical observables

In QM, every physical quantity is described by an *operator* $\hat{f}$. An operator can be represented by a simple multiplication by a number, or a combination of partial derivatives, or an integral, etc. The allowed values of this physical quantity are the *eigenvalues* of the operator. The spectrum of eigenvalues can be discrete, e.g. the energy levels of the bound states of the hydrogen atom, or continuous, e.g. the momentum of a free particle. To compute the value of a certain physical observable, we must determine which wave functions are *eigenfunctions* of the corresponding operator; i.e. we must solve the equations

$$\hat{f} | \Psi_n \rangle = f_n | \Psi_n \rangle, \qquad \text{with } n = 1, 2, \ldots. \tag{2.3}$$

This results in a (possibly infinite) set of eigenfunctions $|\Psi_n\rangle$ with (possibly complex) eigenvalues f_n. The real physical state of the system is then described by a *linear superposition* of all these *eigenstates*, i.e.

$$|\Psi\rangle = \sum_n c_n | \Psi_n \rangle, \tag{2.4}$$

where c_n are complex numbers. The expectation value of the operator $\hat{f}$ for a system in the state $|\Psi\rangle$ is then given by

$$\langle \hat{f} \rangle = \langle \Psi | \hat{f} | \Psi \rangle = \left(\sum_m c_m^* \langle \Psi_m | \right) \hat{f} \left(\sum_n c_n | \Psi_n \rangle \right). \tag{2.5}$$

Using equation (2.3) this can be rewritten as

$$\langle \hat{f} \rangle = \sum_m \sum_n c_m^* c_n \langle \Psi_m | \hat{f} | \Psi_n \rangle = \sum_{m,n} f_n c_m^* c_n \langle \Psi_m | \Psi_n \rangle. \tag{2.6}$$

If the set of eigenfunctions Ψ_n is *complete* [Ode79] and *orthonormal*, then the final bra-ket can be rewritten as

$$\langle \Psi_m | \Psi_n \rangle = \delta_{mn}, \tag{2.7}$$

from which

$$\langle \hat{f} \rangle = \sum_{m,n} f_m c_m^* c_n \delta_{mn} = \sum_n f_n |c_n|^2. \tag{2.8}$$

The expectation value of the physical observable f is hence equal to the sum of the squares of the moduli of the coefficients c_n, weighted by the eigenvalues of the operator $\hat{f}$. This is a very general result, valid for all operators.

An example may clarify the use of these equations. In classical mechanics, the kinetic energy T of a free particle is described by

$$T = \frac{1}{2} m_0 v^2 = \frac{p^2}{2m_0}, \tag{2.9}$$

with m_0 the rest mass of the particle. The momentum of classical mechanics is translated into a momentum operator $\hat{p} = -\mathrm{i}\hbar\nabla$, where $\hbar$ is Planck's constant $h = 6.626\,075 \times 10^{-34}$ J s, divided by 2π, and ∇ is the gradient differential operator.†
The eigenvalues $\mathbf{p}$ and eigenfunctions $|\Psi\rangle$ of this operator are determined from the equation:

$$\hat{p}|\Psi\rangle = -\mathrm{i}\hbar\nabla|\Psi\rangle = \mathbf{p}|\Psi\rangle, \tag{2.10}$$

or, in the coordinate representation:

$$-\mathrm{i}\hbar\nabla\Psi(\mathbf{r}) = \mathbf{p}\Psi(\mathbf{r}).$$

This equation has solutions of the form

$$\Psi(\mathbf{r}) = C\mathrm{e}^{\frac{\mathrm{i}}{\hbar}\mathbf{p}\cdot\mathbf{r}}. \tag{2.11}$$

† The differential form of the momentum operator is a consequence of the invariance of a closed system of particles with respect to infinitesimal translations. For a derivation of the operator we refer to Section 12 in [LL74], or Section 25 in [Dir47].

The constant C follows from the normalization condition $\langle\Psi|\Psi\rangle = 1$ and we leave it as an exercise for the reader to show that $C = h^{-3/2}$.

The allowed values of the momentum eigenvalue $\mathbf{p}$ form a continuous spectrum. The kinetic energy operator becomes

$$\hat{T} = \frac{\hat{p}^2}{2m_0} = -\frac{\hbar^2\nabla^2}{2m_0}. \tag{2.12}$$

The eigenvalue equation for the kinetic energy operator has the same eigenfunctions as the momentum operator (for a free particle), but with different eigenvalues:

$$-\frac{\hbar^2}{2m_0}\nabla^2 \mathrm{e}^{\frac{\mathrm{i}}{\hbar}\mathbf{p}\cdot\mathbf{r}} = \frac{p^2}{2m_0}\mathrm{e}^{\frac{\mathrm{i}}{\hbar}\mathbf{p}\cdot\mathbf{r}}. \tag{2.13}$$

The average kinetic energy of a free particle can hence be computed as follows:

$$\langle\hat{T}\rangle = \langle\Psi|\hat{T}|\Psi\rangle = \frac{p^2}{2m_0}\langle\Psi|\Psi\rangle = \frac{p^2}{2m_0}, \tag{2.14}$$

and we find the classical expression for the kinetic energy.

2.2.3 The Schrödinger equation

In classical mechanics, the equations of motion can be derived from either the Lagrangian or the Hamiltonian functions, depending on the approach [Gol78]. The *Hamiltonian* of a classical system of interacting particles can be written as

$$H = T + V, \tag{2.15}$$

where V is the potential energy of the particles. Using variational principles, one can derive the Hamiltonian equations of motion, or, equivalently, the Euler–Lagrange equations of motion, from which all of classical mechanics can be derived, including Newton's equations.

In quantum mechanics, the Hamiltonian function is converted into a Hamiltonian operator:

$$\hat{H} = \hat{T} + \hat{V}. \tag{2.16}$$

We can again formulate an eigenvalue equation for these operators:

$$\hat{H}|\Psi\rangle = (\hat{T} + \hat{V})|\Psi\rangle = E_t|\Psi\rangle, \tag{2.17}$$

where E_t is the total energy of the system. The Hamiltonian operator describes the time evolution of the system and can be rewritten in differential form as

$$\hat{H}|\Psi\rangle = \mathrm{i}\hbar\frac{\partial}{\partial t}|\Psi\rangle. \tag{2.18}$$

This expression follows from the fact that the time derivative of the wave function at any given instant is determined by the value of the function $|\Psi\rangle$ itself at that instant, which leads to a linear (first-order) relation [LL74]. Combining the above equations results in the *time-dependent Schrödinger equation*:

$$\mathrm{i}\hbar\frac{\partial}{\partial t}|\Psi\rangle = (\hat{T}+\hat{V})|\Psi\rangle = E_t|\Psi\rangle. \tag{2.19}$$

For a free particle, $\hat{V}=0$, the solution to this equation is given by $\Psi(\mathbf{r},t) = C\mathrm{e}^{-\frac{\mathrm{i}}{\hbar}(E_t t-\mathbf{p}\cdot\mathbf{r})}$. This solution represents a *plane wave* with energy $E_t = p^2/2m_0$ and momentum $\mathbf{p}$. The wave has a frequency $\nu = E/\hbar = 2\pi\omega$.

The *stationary* (or time-independent) *Schrödinger equation* takes the form:

$$\frac{\hbar^2}{2m_0}\Delta|\Psi\rangle + [E_t - \hat{V}]|\Psi\rangle = 0, \tag{2.20}$$

where we have replaced the kinetic energy operator by its differential form. These equations were obtained by Schrödinger in 1926 [Sch26] and from them most of QM can be derived. We refer to standard QM textbooks for detailed discussions of the Schrödinger equation and other aspects of quantum mechanics, such as angular momentum and the Heisenberg *uncertainty principle* (e.g. [Hei25]).

The examples above show that the wave function is the central object in quantum mechanics. When the wave function is known, the expectation values of all physical observables can be calculated. The momentum eigenfunctions are very important for electron scattering, since a scattering process generally involves *momentum transfer*. As we will see in later chapters, all of electron diffraction theory can be expressed with the tools introduced in this and the following sections.

2.2.4 The de Broglie relation

In a seven-page doctoral thesis, Louis de Broglie [de 23] postulated in 1924 that all particles have some wave-like properties (wave–particle duality); he associated a wavelength λ with the momentum, p, of a particle:

$$\lambda = \frac{h}{p}. \tag{2.21}$$

Defining the wave number k as $k = 1/\lambda$, this relation can be rewritten as $p = hk$, or, in vector form, $\mathbf{p} = h\mathbf{k}$. This establishes a linear relation between momentum space and reciprocal space, since $|\mathbf{k}|$ has the dimensions of a reciprocal length. As a consequence of the de Broglie relation, we find that *reciprocal space is identical to momentum space*, apart from a scaling factor (Planck's constant).

We can now consider the wave function of a particle in coordinate space and write it as a linear combination of the momentum eigenfunctions:

$$\Psi(\mathbf{r}) = C\sum_n c_n e^{\frac{i}{\hbar}\mathbf{p}_n\cdot\mathbf{r}};$$

$$= \sum_{\mathbf{k}} c_{\mathbf{k}} e^{2\pi i\mathbf{k}\cdot\mathbf{r}}. \tag{2.22}$$

The constant C has been absorbed in the coefficients $c_{\mathbf{k}}$. If the eigenspectrum of the momentum operator is continuous, then the summation must be replaced by an integral. Expansion of a wave function in terms of plane waves is very important for the description of diffraction phenomena, as we will see in later chapters and in Section 2.5.

Since plane waves form the basis for much of electron diffraction theory, it is perhaps useful to take a closer look at the properties of a plane wave. By definition, a plane wave is an eigenfunction of the momentum operator $\hat{p}$. It is easy to see why such a function is called a *plane* wave: if the argument of the exponential is constant, i.e. $\mathbf{k}\cdot\mathbf{r} = d$, then the value of the wave function is constant:

$$|\Psi\rangle = e^{2\pi i d} = \cos(2\pi d) + i\sin(2\pi d).$$

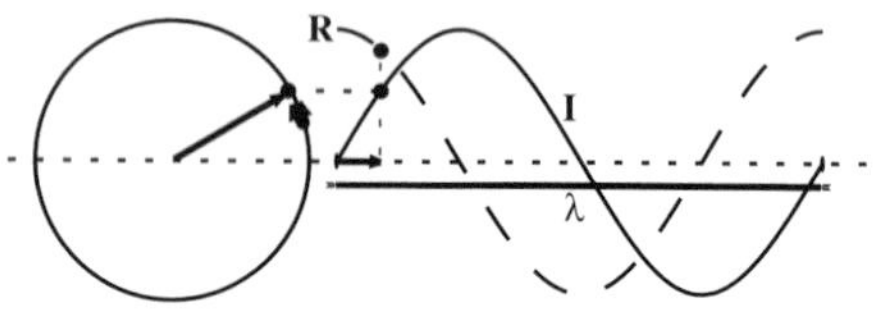

It is not difficult to see that the geometric locus of all the vectors $\mathbf{r}$ satisfying the equation $\mathbf{k}\cdot\mathbf{r} = d$ is a plane perpendicular to the vector $\mathbf{k}$. The wave function takes on all values on the complex circle with unit radius when d goes to $d+1$ or, equivalently, when the wave advances one wavelength λ. The sketch above shows the real (R) and imaginary (I) components of one wavelength of the plane wave.

2.2.5 The electron wavelength (non-relativistic)

Assume that an electron is accelerated by an electrostatic potential drop E. The potential energy of the electron is then equal to eE and when the electron leaves the field region, it will have acquired a kinetic energy given by

$$T = \frac{p^2}{2m_0} = eE, \tag{2.23}$$

Table 2.1. *Non-relativistic electron wavelengths (in pm) for selected acceleration voltages E (in volts). The relativistic values are given in the last column (using equation 2.33).*

E (volt)	λ_{nr} (pm)	λ (pm)
100	122.64	122.63
500	54.84	54.83
1 000	38.78	38.76
5 000	17.34	17.30
10 000	12.26	12.20
20 000	8.67	8.59

where m_0 is the rest mass of the electron. As we have seen above, the de Broglie equation relates the momentum p to the (non-relativistic) electron wavelength λ_{nr} by

$$\lambda_{nr} = \frac{h}{p},$$

and thus

$$eE = \frac{h^2}{2m_0\lambda_{nr}^2},$$

which can be rewritten as

$$\lambda_{nr} = \frac{h}{\sqrt{2m_0eE}} = \frac{1226.39}{\sqrt{E}}. \tag{2.24}$$

When the voltage is expressed in volts, the wavelength is given in picometers. This equation is valid for low acceleration potentials E (up to about 10 000 V), and a selection of wavelengths is shown in Table 2.1. For comparison the table also shows the relativistic values, computed from equation (2.33) on page 92. It can be seen that the corrections are small, which means that for the acceleration voltages typically used in a scanning electron microscope (SEM) it is a reasonable approximation to use the non-relativistic expression (2.24).

2.2.6 Wave interference phenomena

Interference between particle waves is a purely quantum mechanical effect, and it lies at the heart of transmission electron microscopy. In classical mechanics, the superposition of two waves, e.g. surface waves on a liquid, can give rise to interference patterns: at positions where the waves are in-phase, constructive interference will give rise to a displacement equal to the sum of the two wave amplitudes, whereas partial or complete destructive interference occurs when the waves are out-of-phase. A crucial requirement for this type of classical wave interference to occur is that both waves must be present *simultaneously*. In quantum mechanics it is possible for a single particle (or wave) to give rise to an interference pattern.

The *double-slit experiment* provides a standard illustration of this phenomenon [FLS63]. A screen with two identical slits A and B is placed between a particle source S and a detection screen D. If one of the slits, say A, is covered, then the particles which pass through the other slit, B, will arrive at the detector with a certain spatial distribution, described by a probability function $|\psi_B|^2$ (indicated in Fig. 2.1 by a thin solid line). The same happens when slit B is covered. When both slits are uncovered, the detector cannot determine through which of the slits the particle has traveled, and, hence, the intensity distribution on the detector is given by the function $|\psi_A + \psi_B|^2$ (the oscillating thicker line in Fig. 2.1), and not by the sum $|\psi_A|^2 + |\psi_B|^2$. If, on the other hand, the experiment is modified such that it can be determined through which of the two slits the particle traveled, then the intensity distribution is given by the sum of the individual distributions $|\psi_A|^2 + |\psi_B|^2$ (the dashed line in Fig. 2.1). The difference between the two patterns is the interference term $\psi_A\psi_B^* + \psi_A^*\psi_B$.

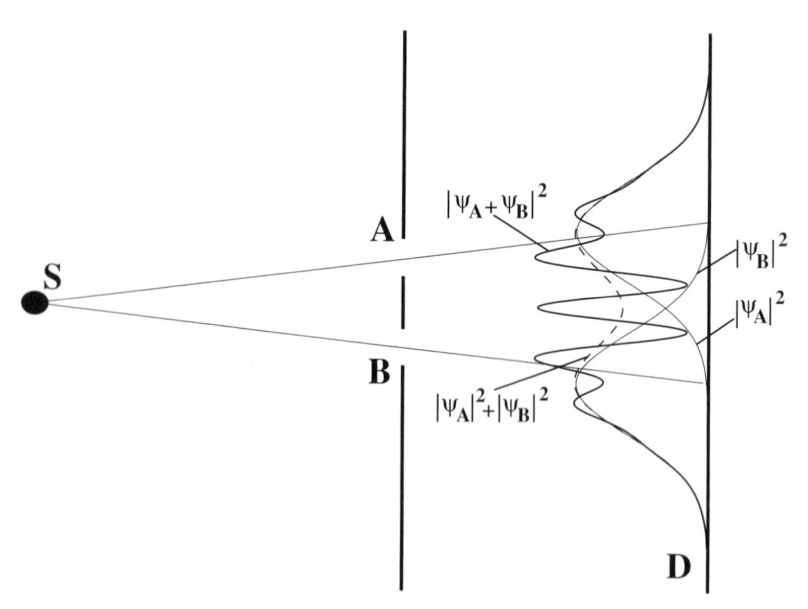

Fig. 2.1. Schematic illustration of the double-slit interference experiment (details given in text).

It is important to emphasize that this type of quantum mechanical interference is entirely due to the way the experiment is carried out. This is an example of how the observer is intricately linked to the experiment. In the standard double-slit experiment one cannot distinguish electrons that passed through slit A from those which passed through slit B, and therefore interference will occur. This interference will occur even when the time between subsequent particles is large, say one particle per minute. Even then interference will occur because we still cannot distinguish between the two slits. We can say that the particle interferes with itself, in a way determined by the layout of the experiment. This is precisely what happens in a TEM. At any given moment in time, only one single beam electron is present in the sample (see Section 3.7.6). The crystalline specimen acts as a "screen" with multiple slits. Since we do not attempt to determine where exactly the electron traveled through the sample (i.e. by which atom(s) it was scattered), the wave function of the electron will interfere with itself, and form an interference pattern on the detector. We will see later on that the microscope operator plays an important role in determining the type of interference pattern that will be formed, and which component of the pattern is used to obtain an image.

2.3 Elements of the special theory of relativity

2.3.1 Introduction

In modern transmission electron microscopes, the electrons travel at substantial fractions of c, the velocity of light. Thanks to the efforts of Lorentz, Einstein, and many others, we now know that such fast motion obeys the laws of the *special theory of relativity*; the classical Newtonian laws of motion must be modified to take into account the fact that c is the maximum velocity for *any* type of motion. In our daily lives this is not an important fact, since we "know" from personal (i.e. classical) experience that velocities are *additive* quantities. When observer B moves with respect to observer A with a velocity v along the x-direction, and observer C moves with respect to B with velocity u in the same direction, then the relative velocity between C and A is simply the sum $w = u + v$. This is the *principle of relativity*, first formulated by Newton. The coordinate transformation associated with such relative velocities is known as a *Galilean transformation*, and is mathematically described by

$$\begin{cases} x' = x - vt; \\ y' = y; \\ z' = z; \\ t' = t. \end{cases}$$

All laws of classical physics are invariant under a Galilean transformation. It was noted by various researchers around the end of the nineteenth century, that the Maxwell equations of electrodynamics are not invariant under such a simple transformation. In 1887 Voigt [Voi87] and subsequently Lorentz in 1904 [Lor04] discovered that the Maxwell equations are invariant under the following remarkable transformation:

$$\begin{cases} x' = \gamma(x - vt); \\ y' = y; \\ z' = z; \\ t' = \gamma(t - vx/c^2), \end{cases} \tag{2.25}$$

where

$$\gamma = \frac{1}{\sqrt{1-\beta^2}} \quad \text{and} \quad \beta = \frac{v}{c}. \tag{2.26}$$

It is easy to see that for small velocities v, these equations reduce to those for the Galilean transformation, since $\lim_{c\to\infty} \gamma = 1$. Equations (2.25) are now known as a *Lorentz transformation*, sometimes referred to as a *boost*, and the modified principle of relativity states that all laws of physics must be invariant under such a transformation.

From these equations, we can derive the Einstein addition law for velocities (referring to the same reference frames A, B and C as before):

$$w = \frac{u+v}{1+uv/c^2}. \tag{2.27}$$

Einstein was the first to fully understand the implications of the Lorentz transformation on the concepts of space and time. He showed that Newton's equations of motion could be made invariant by making the mass of a particle dependent on its velocity with respect to a reference frame. In particular, Newton's equation of motion can be made invariant under a Lorentz transformation by the substitution:

$$m = \gamma m_0 = \frac{m_0}{\sqrt{1-v^2/c^2}}. \tag{2.28}$$

Newton's equation of motion under an applied force $\mathbf{F}$ then reads as

$$\mathbf{F} = \frac{\mathrm{d}}{\mathrm{d}t}(m\mathbf{v}) = m_0 \frac{\mathrm{d}}{\mathrm{d}t}\frac{\mathbf{v}}{\sqrt{1-v^2/c^2}} = m_0 \frac{\mathrm{d}}{\mathrm{d}t}(\gamma \mathbf{v}). \tag{2.29}$$

This will be the starting equation for the analysis of electron trajectories in an electromagnetic field; the force $\mathbf{F}$ is then the *Lorentz force*, as we will see in Chapter 3.

For an extensive in-depth discussions of the special theory of relativity we refer the reader to the following books: [Wey22], [FLS63], and [MTW73]. For the purposes of this book it is sufficient to list a few consequences of the theory that are of importance for electron microscopy.

(i) *Equivalence of mass and energy*: the kinetic energy T of a particle traveling with a velocity v is given by

$$T = c^2(m - m_0) = m_0c^2(\gamma - 1).$$

In the non-relativistic limit when $\beta \rightarrow 0$, one can readily show that

$$\gamma \rightarrow 1 + \frac{1}{2}\frac{v^2}{c^2} + \cdots$$

and substitution in the above expression for the kinetic energy T results in the standard expression $T = \frac{1}{2}m_0v^2$. The energy relation is often rewritten as

$$mc^2 = m_0c^2 + T, \tag{2.30}$$

which states that the total energy equals the sum of the rest energy and the kinetic energy. This relation expresses the equivalence of mass and energy.

(ii) *Lorentz–Fitzgerald contraction of length scales*: The Lorentz–Fitzgerald contraction refers to the fact that an object of length $l = x_2 - x_1$ to one observer acquires a length $x_2' - x_1' = l/\gamma$ to an observer in the primed coordinate system. The contraction occurs along the direction of motion. In the TEM this means that when an electron traveling at a velocity $v = \beta c$ encounters a crystal of thickness z_0 (measured in the reference frame of the microscope), then the electron will "see" a crystal of thickness z_0/γ in its own reference frame. As we will see in the next section, a 400 kV electron travels at 83% of the velocity of light, which means that a 100 nm thick crystal appears only 55.7 nm thick to the electron.

(iii) *Einstein dilatation of time scales*: the duration of an interval in the unprimed reference system, $t_2 - t_1$ (for a given location in space) becomes $t_2' - t_1' = \gamma(t_2 - t_1)$ in the primed reference system. This is often restated as "stationary clocks run slow".

In the following subsection, we will rederive the expression for the wavelength of the electron, this time making use of the relativistic relations for mass and energy.

2.3.2 The electron wavelength (relativistic)

The total energy of the relativistic electron is given by equation (2.30):

$$E_t = mc^2.$$

When the electron is accelerated by a potential drop E, it acquires a potential energy eE, which is completely converted into kinetic energy T when the particle leaves the electrostatic field. The potential energy is then given by

$$eE = (m - m_0)c^2. \tag{2.31}$$

The wavelength according to de Broglie is given by

$$\lambda = \frac{h}{p} = \frac{h}{mv}. \tag{2.32}$$

The mass of a relativistic particle increases with increasing velocity according to relation (2.28) which, using (2.31), can be rewritten as (following [HK89a])

$$\gamma = 1 + \frac{eE}{m_0c^2} = 1 + \omega,$$

where ω is introduced as a shorthand notation. From the definition of γ we can derive the electron velocity:

$$v = c\sqrt{1 - \frac{1}{\gamma^2}} = \frac{c}{1+\omega}\sqrt{(1+\omega)^2 - 1}.$$

The momentum p is then given by

$$p = mv = m_0(1+\omega)v = m_0c\sqrt{2\omega + \omega^2} = \sqrt{2m_0eE\left(1 + \frac{eE}{2m_0c^2}\right)}.$$

Substitution into the de Broglie expression for the wavelength yields:

$$\lambda = \frac{h}{\sqrt{2m_0eE\left(1 + \frac{e}{2m_0c^2}E\right)}}.$$

Substitution of the physical constants listed in Appendix A2 then leads to

$$\boxed{\lambda = \frac{1\,226.39}{\sqrt{E + 0.978\,45 \times 10^{-6}E^2}},} \tag{2.33}$$

where again the wavelength is given in picometers when the voltage E is expressed in volts. Table 2.2 lists the relativistic wavelengths for a few commonly used voltages. Note that the relativistic correction increases with increasing voltage; at $E = 400$ kV, the electrons travel at 83% of the speed of light and have a mass 78% heavier than the rest mass, while at $E = 3$ MeV the electron mass increases by nearly 700%.

Table 2.2. *Relativistic acceleration potential $\hat{\Psi}$, electron wavelength λ, wavenumber $K_0 = 1/\lambda$, mass ratio $\gamma = m/m_0$, relative velocity $\beta = v/c$, and interaction constant σ for various acceleration voltages E.*

E (kV)	$\hat{\Psi}$ (V)	λ (pm)	K_0 (nm^{-1})	m/m_0	$\beta = v/c$	σ (V^{-1} nm^{-1})
100	109 784	3.701	270.165	1.196	0.548	0.009 244
120	134 090	3.349	298.577	1.235	0.587	0.008 638
200	239 139	2.508	398.734	1.391	0.695	0.007 288
300	388 062	1.969	507.937	1.587	0.777	0.006 526
400	556 556	1.644	608.293	1.783	0.828	0.006 121
800	1 426 224	1.027	973.761	2.566	0.921	0.005 503
1000	1 978 475	0.872	1146.895	2.957	0.941	0.005 385
1250	2 778 867	0.736	1359.228	3.446	0.957	0.005 296
2000	5 913 900	0.504	1982.876	4.914	0.979	0.005 176
3000	11 806 277	0.357	2801.657	6.871	0.989	0.005 122

It is convenient to introduce a new notation for the relativistically corrected acceleration potential [HK89a]:

$$\hat{\Psi} \equiv E\left(1 + \frac{e}{2m_0c^2}E\right) = E + \epsilon E^2, \tag{2.34}$$

where $\epsilon = 0.978\,45 \times 10^{-6}$ V^{-1}. The quantity $\hat{\Psi}$ is known as the *relativistic acceleration potential*. At an acceleration voltage of 200 kV, the electron experiences a relativistic acceleration potential of 239.1 kV.

We also introduce a new parameter σ, defined by

$$\sigma \equiv \frac{2\pi\gamma m_0 e\lambda}{h^2}, \tag{2.35}$$

which is known as the *interaction constant*. It has dimensions of V^{-1} nm^{-1} when lengths are expressed in nanometers. The interaction constant will become useful later on in this book when we introduce the dynamical diffraction equations. One can surmise from the last column in Table 2.2 that σ approaches a constant value with increasing acceleration voltage E. Formally this limit is computed as follows:

$$\lim_{E\to\infty} \sigma = \frac{2\pi m_0 e}{h^2} \lim_{E\to\infty} \gamma\lambda = \frac{2\pi m_0 e}{h^2}\frac{h}{m_0 c} \underbrace{\lim_{\omega\to\infty} \sqrt{\frac{(1+\omega)^2}{2\omega+\omega^2}}}_{\lim=1} = \frac{e}{\hbar c}.$$

The numerical value of the limit is given by $0.005\,067\,73$ V^{-1} nm^{-1}. The ratio h/m_0c is known as the *Compton wavelength* $\lambda_C = 2.4263$ pm. The Compton wavelength is the non-relativistic wavelength of an electron traveling at the velocity of light (from the de Broglie equation 2.21), i.e. an electron with momentum $p = m_0c$.

Starting from the acceleration voltage E, the Fortran function CalcWaveLength in the module diffraction.f90, available from the website, computes all the entries on a single line of Table 2.2.

2.3.3 Relativistic correction to the governing equation

Dirac was among the first to note that the Schrödinger equation is not invariant under a Lorentz transformation, and he then proceeded to derive the equation which now bears his name: the *Dirac equation* [Dir28a, Dir28b]. While relativistic quantum theory is far beyond what we need for this textbook, it is interesting to note that both the spin of the electron and the existence of the positron (and anti-matter in general) are direct consequences of this theory. We refer to [Tha92] and [Gre90] for an in-depth introduction to relativistic quantum theory.

One can explicitly work out the Dirac equation for a relativistic plane wave incident upon a crystal, but the mathematics is fairly involved and beyond the scope of this book. Fujiwara [Fuj61] has shown in a fundamental paper that the Dirac equation can be simplified for scattering of high-energy electrons, and that spin can be neglected entirely. He concluded that the stationary Schrödinger equation (2.20) can be used to describe high-energy electron scattering, *provided the electron wavelength and mass are replaced by their relativistically corrected values.* For a critical analysis of this derivation we refer to [FHS86].

In this section, we will follow an alternative derivation, which starts from the general expression for a time-independent wave equation of a wave with wavenumber K_0 (known as the *Helmholtz* equation):

$$\Delta\Psi + 4\pi^2 K_0^2 \Psi = 0. \tag{2.36}$$

This equation is valid for electromagnetic radiation with wavenumber K_0, but it can also be used to derive the relativistically corrected Schrödinger equation for the electron scattering process [Rei93].

We define the electron-optical refractive index n as the ratio of the wavelength of the electron outside and inside the crystal. If the crystal is represented by a positive potential energy term V, then the energy of the electron inside the crystal is given by $E + V$. The wavelength of the electron can be written in terms of the relativistic acceleration potential as

$$\lambda = \frac{h}{\sqrt{2m_0e\hat{\Psi}}} \quad \text{and} \quad \lambda_c = \frac{h}{\sqrt{2m_0e\hat{\Psi}_c}},$$

where $\hat{\Psi}_c$ is the relativistic acceleration potential inside the crystal. Explicit computation of $\hat{\Psi}_c$ yields

$$\begin{aligned}\hat{\Psi}_c &= E + V + \epsilon(E+V)^2;\\ &= E + \epsilon E^2 + V(1+2\epsilon E) + \epsilon V^2;\\ &\cong \hat{\Psi} + \gamma V,\end{aligned}$$

where we have used the fact that ϵV^2 is much smaller than the other terms in the equation. Since V depends on position in the crystal, $\hat{\Psi}_c$ does so as well. The *electron-optical refractive index* is then computed as follows:

$$n = \frac{\lambda}{\lambda_c} = \sqrt{\frac{\hat{\Psi}_c}{\hat{\Psi}}} = \sqrt{1 + \frac{\gamma V}{\hat{\Psi}}}.$$

In the presence of *refraction* (i.e. when n is not equal to 1) we must replace the factor K_0^2 in equation (2.36) by $n^2K_0^2$, which yields

$$\Delta\Psi + 4\pi^2 n^2 K_0^2 \Psi = \Delta\Psi + 4\pi^2\left(1 + \frac{\gamma V}{\hat{\Psi}}\right)\frac{2m_0 e}{h^2}\hat{\Psi}\Psi = 0,$$

from which we find, after rearranging terms and substituting m for γm_0:

$$\boxed{\Delta\Psi + 4\pi^2 K_0^2 \Psi = -4\pi^2 U(\mathbf{r})\Psi.} \tag{2.37}$$

$K_0 = 1/\lambda$ is the relativistic wavenumber and

$$U(\mathbf{r}) \equiv \frac{2me}{h^2}V(\mathbf{r}) = \frac{\sigma}{\pi\lambda}V(\mathbf{r}), \tag{2.38}$$

where we have used the definition of the interaction constant σ in equation (2.35). Equation (2.37) will form the starting point for all theoretical derivations in the remainder of this book. For a much more detailed and mathematically more rigorous approach to the derivation of the relativistic wave equation we refer to [Fuj61], [FHS86] and to Chapters 55 and 56 in [HK94]. Note that some authors (e.g. [Rei93]) define the electrostatic potential to be a negative potential, so that the minus sign in equation (2.37) changes to a plus sign. The differences between the standard non-relativistic Schrödinger equation (2.20) and equation (2.37) are the relativistic values for both the electron mass ($m = m_0\gamma$) and wavelength (λ from equation 2.33). Technically, equation (2.37) is *not* the Schrödinger equation, but a scalar form of the Dirac equation (known as the *Klein–Gordon equation*), which looks just like the standard Schrödinger equation. Equation (2.37) correctly includes all relativistic effects relevant to the scattering of high-energy electrons, and we will return to it in Chapter 5.

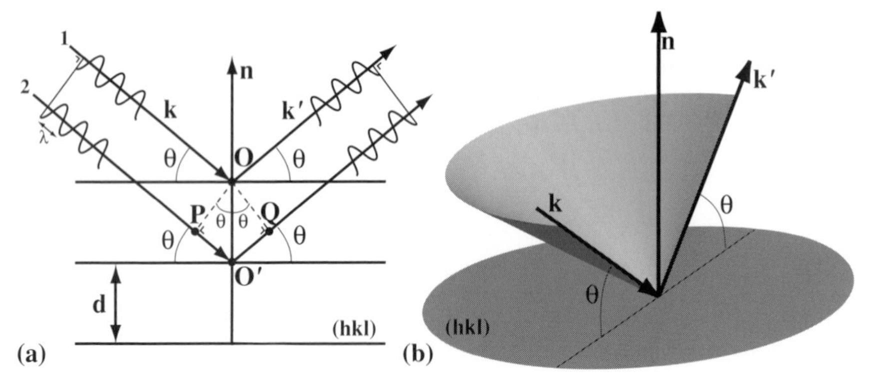

Fig. 2.2. (a) Geometrical construction leading to the direct space Bragg equation; (b) the incident and diffracted directions and the plane normal must lie in a planar section through a conical surface with its top in the plane.

2.4 The Bragg equation in direct and reciprocal space

2.4.1 The Bragg equation in direct space

Consider the drawing in Fig. 2.2(a). A plane wave with wavelength λ is incident upon a set of parallel planes with Miller indices (hkl). The incidence angle is θ. If we regard each plane as a semi-transparent mirror, then part of the incident intensity will pass through the first plane, and part will be reflected. For an ideal mirror we can apply Snell's law, which states that the incident and reflected angles must be equal, and both directions must be coplanar with the normal **n** to the (hkl) plane.

The same process occurs on the second plane. This wave has traveled a bit further than the first wave, so for an observer far away from the crystal the two reflected waves may arrive somewhat out of phase. The condition for in-phase arrival is derived easily as follows: the path length difference Δ between waves 1 and 2 is given by the sum of the distances PO' and $O'Q$. These distances can be expressed in terms of the interplanar spacing d_{hkl} and the angle θ as follows:

$$\begin{aligned}\Delta &= PO' + O'Q,\\ &= d_{hkl}\sin\theta + d_{hkl}\sin\theta,\\ &= 2d_{hkl}\sin\theta.\end{aligned}$$

For in-phase arrival at the observation point, also known as *constructive interference*, this path length difference must be equal to an integral number of wavelengths $n\lambda$, or

$$2d_{hkl}\sin\theta = n\lambda. \tag{2.39}$$

This is the direct space *Bragg equation*, first derived in 1915 by W.H. Bragg [Bra15]. The integer n labels the various *diffraction orders* for a given set of planes.

In the electron microscopy world it is common practice to move the integer n to the left-hand side of equation (2.39),

$$2\frac{d_{hkl}}{n}\sin\theta = \lambda.$$

We recall from the discussion of Miller indices in the previous chapter that the planes with Miller indices $(nh\,nk\,nl)$ are parallel to the planes (hkl), but with an interplanar spacing equal to

$$d_{nh\,nk\,nl} = \frac{d_{hkl}}{n}.$$

So instead of talking about nth-order diffraction from the planes (hkl) we may equivalently talk about first-order diffraction from the planes $(nh\,nk\,nl)$, even when those planes do not physically exist in the crystal lattice (i.e. they do not have any atoms located on them). We thus arrive at the working version of the Bragg equation:

$$\boxed{2d_{hkl}\sin\theta = \lambda.} \tag{2.40}$$

The angle θ is known as the *Bragg angle*. This equation should be used with the understanding that, from a diffraction point of view, the planes $(nh\,nk\,nl)$ are not equivalent to the planes (hkl), and that only first-order diffraction should be considered.

The Bragg equation describes the geometric condition for diffraction to occur. It does not say anything about the *intensity* of the diffracted wave. The Bragg equation is derived with respect to a specific plane, and states that the incident and diffracted waves must travel in directions which lie on a conical surface with top in the diffracting plane and opening angle $\pi/2-\theta$, as illustrated in Fig. 2.2(b). In other words, for each plane (hkl) in a crystal there is a conical surface centered around the plane normal with opening angle determined by the interplanar spacing d_{hkl} and the radiation wavelength λ. While the Bragg equation in direct space is rather simple and elegant, it is not very useful when one wishes to determine the absolute direction of a diffracted wave for a given crystal structure, since both the incident wave direction and the crystal orientation must be specified with respect to some reference frame. The direct space version of the Bragg equation is mostly used to compute the diffraction angle 2θ for a given wavelength and crystal structure; this has been implemented in the function CalcDiffAngle in the module diffraction.f90, available from the website. To incorporate absolute wave directions it is common practice to convert the Bragg equation to the *reciprocal reference frame*, which leads to the Ewald sphere construction, as described in the next section.

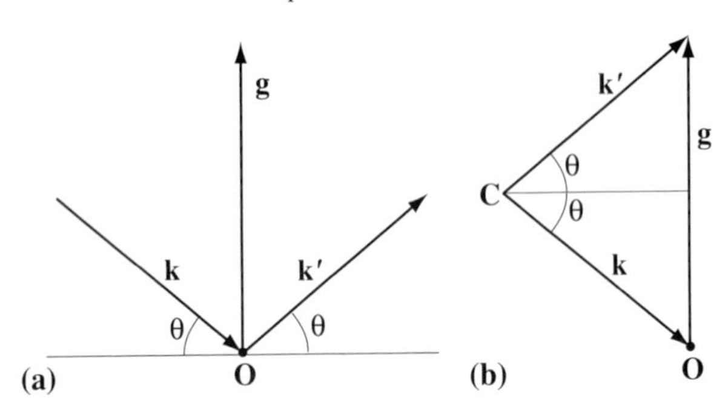

Fig. 2.3. (a) Geometrical construction leading to the reciprocal space Bragg equation; (b) the diffracted wave vector is translated to complete the vector sum $\mathbf{k}' = \mathbf{k} + \mathbf{g}$.

2.4.2 The Bragg equation in reciprocal space

We have already seen in Section 2.2.4 that a plane wave can be represented by its momentum vector **p**, or, using the de Broglie relation, by its wave vector **k**. **k** is a vector in reciprocal space which fully characterizes the direction and wavelength of a plane wave with respect to the crystal reference frame. Since the plane (hkl) is also represented by a vector $\mathbf{g}_{hkl}$ in reciprocal space, we can derive the reciprocal space equivalent of the Bragg equation (2.40).

Consider the drawing in Fig. 2.3(a). It shows the wave vectors corresponding to the incident and diffracted wave directions of Fig. 2.2(a). The reciprocal lattice vector $\mathbf{g}_{hkl}$ is also drawn in the proper orientation. Since vectors can always be translated parallel to themselves, we can redraw the diffracted vector $\mathbf{k}'$ so that its starting point coincides with the starting point **C** of **k**, as shown in Fig. 2.3(b). The relation between the three vectors can now be expressed as

$$\boxed{\mathbf{k}' = \mathbf{k} + \mathbf{g}.} \tag{2.41}$$

Note that the length of the wave vectors is constant for elastic scattering processes. It is easy to see that this vector relation is equivalent to the direct space Bragg equation (2.40). Projecting the above equation onto the vector **g** leads to

$$\begin{aligned}
\mathbf{k}' \cdot \mathbf{g} &= \mathbf{k} \cdot \mathbf{g} + \mathbf{g} \cdot \mathbf{g}; \\
|\mathbf{k}|\,|\mathbf{g}| \sin\theta &= -|\mathbf{k}|\,|\mathbf{g}| \sin\theta + |\mathbf{g}|^2; \\
\frac{\sin\theta}{\lambda} &= -\frac{\sin\theta}{\lambda} + \frac{1}{d_{hkl}},
\end{aligned}$$

from which the Bragg equation follows. The Bragg relation is thus satisfied whenever the point **C** in Fig. 2.3(b) falls on the perpendicular bisector plane of the vector **g**.

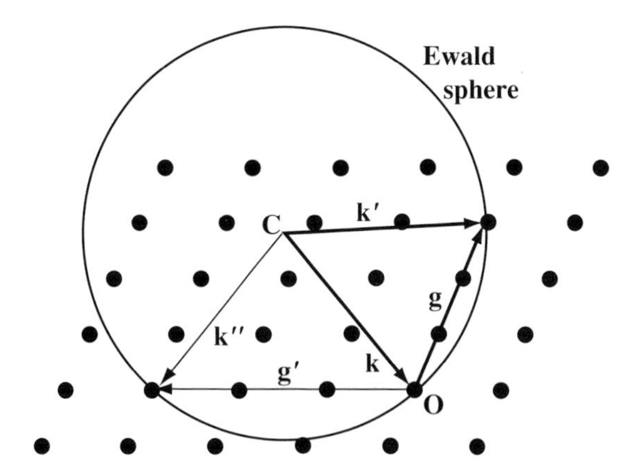

Fig. 2.4. Ewald sphere construction.

The perpendicular bisector plane to the vector $\mathbf{g}_{hkl}$ consists of all the points that are at the same distance from the origin of reciprocal space and the reciprocal lattice point hkl. This leads to a simple geometric construction for the direction of a diffracted wave, when the incident wave vector and the crystal orientation are given. The construction is known as the *Ewald sphere construction* and is shown in Fig. 2.4. First draw the reciprocal lattice with origin **O**. Then draw the incident wave vector **k** such that its end point coincides with **O**. The starting point **C** is then taken to be the center of a sphere (the *Ewald sphere*) with radius $|\mathbf{k}| = 1/\lambda$. Whenever a reciprocal lattice point falls on this sphere, the Bragg condition is satisfied and a diffracted wave with wave vector $\mathbf{k} + \mathbf{g}$ may occur.† Note that there are two reciprocal lattice points on the Ewald sphere in Fig. 2.4; this means that there are two diffracted beams in the directions $\mathbf{k}'$ and $\mathbf{k}''$.

Recall from the de Broglie relation that the momentum vector and the wave vector are identical, apart from a scaling factor h. The Bragg equation in reciprocal space can thus be rewritten as

$$\mathbf{p}' = \mathbf{p} + h\mathbf{g} = \mathbf{p} + \Delta\mathbf{p}. \tag{2.42}$$

This equation reflects the fact that the incident electron with momentum **p** undergoes a change of direction corresponding to $h\mathbf{g}$. Reciprocal lattice vectors thus correspond to the allowed elastic momentum changes (in units of Planck's constant) for a given crystal structure.‡ It is somewhat surprising that the allowed momentum changes for elastic scattering do not depend on the energy or type of

† Note that we say "may occur", because the Bragg equation only states the geometrical condition for diffraction. As we shall see later on, this does not necessarily mean that there will be diffracted intensity in that direction.

‡ It is easy to verify that $h\mathbf{g}$ indeed has the dimensions of momentum.

incident radiation, but are determined entirely by the crystal structure. We find that the reciprocal lattice not only provides us with a way to describe crystal planes and interplanar spacings, it also determines the allowed momentum changes for elastic scattering events.

Example 2.1 *Compute the momentum transfer for elastic scattering of an electron from the* (220) *planes in aluminum.*

Answer: *The lattice parameter of face-centered cubic aluminum is* $a = 0.4049$ *nm [VC85]. From the relations derived in Chapter 1 we find for the length of the vector* $\mathbf{g}_{220}$: $|\mathbf{g}_{220}| = \sqrt{8}/a = 6.9855$ nm^{-1}*. Multiplying by Planck's constant then results in the elastic momentum transfer* $\Delta p = 4.628 \times 10^{-24}$ $kg\,m\,s^{-1}$.

It can be seen from the Ewald sphere construction that diffraction is a rather improbable process. It requires that a mathematical point (with zero volume) should fall on to a spherical shell of zero thickness. For real crystals and for realistic types of radiation we will see that the reciprocal lattice point occupies a finite volume with a well-defined shape, and that the Ewald sphere has a finite thickness related to the wavelength spread of the radiation. These two factors combine to make diffraction a much more probable process. In almost all diffraction experiments one has the ability to either move the incident beam direction (which corresponds to moving the Ewald sphere through the reciprocal lattice), or move the crystal with a stationary incident beam direction (which corresponds to moving the reciprocal lattice through the Ewald sphere). In a transmission electron microscope both options are present: we can change the incident beam direction within a small range of directions using beam tilt controls and the sample can be tilted around one or more axes, so that the orientation of the crystal lattice with respect to the incident beam direction can be changed.

2.4.3 The geometry of electron diffraction

It will be useful to have a mathematical expression for the Ewald sphere. The general equation of a sphere can be written down in vector form as follows:

$$(\mathbf{r} - \mathbf{r}_0) \cdot (\mathbf{r} - \mathbf{r}_0) = R^2,$$

where R is the radius and $\mathbf{r}_0$ is the vector locating the center of the sphere. This equation is valid in every reference frame, and in the simple case of the Cartesian frame we recover the standard equation

$$(x - x_0)^2 + (y - y_0)^2 + (z - z_0)^2 = R^2.$$

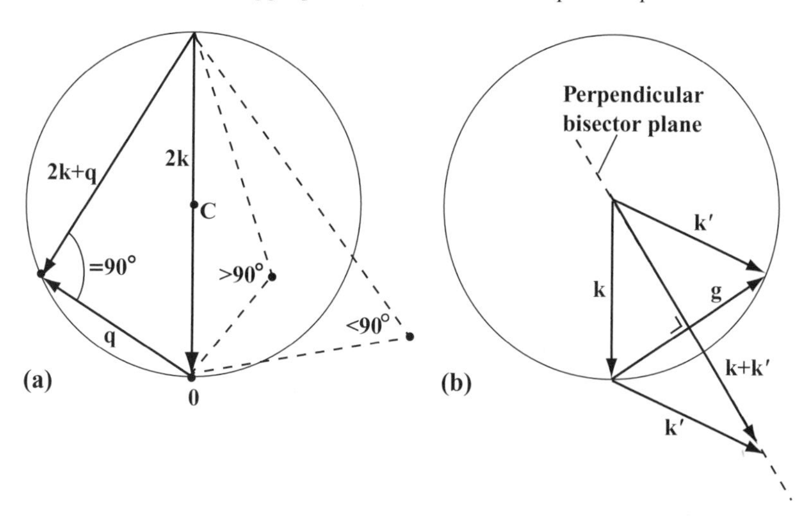

Fig. 2.5. (a) Illustration of the vector equation (2.43) for the Ewald sphere; (b) Bragg orientation for the planes **g**.

In reciprocal space we replace the position vector **r** by a reciprocal space vector **q**; the center of the Ewald sphere is located at the position given by the vector −**k**, and the radius is equal to the inverse of the wavelength $R = 1/\lambda$. This leads to

$$(\mathbf{q} + \mathbf{k}) \cdot (\mathbf{q} + \mathbf{k}) = \frac{1}{\lambda^2}.$$

Since $|\mathbf{k}| = 1/\lambda$ we can rewrite this equation as

$$\boxed{\mathbf{q} \cdot (2\mathbf{k} + \mathbf{q}) = 0.} \tag{2.43}$$

This is the equation for the Ewald sphere: every vector **q** that is perpendicular to the vector $2\mathbf{k} + \mathbf{q}$ must end on the Ewald sphere. The dot product is negative when **q** ends inside the Ewald sphere, and positive when it ends outside the sphere, as can be seen from Fig. 2.5(a). The Bragg equation in reciprocal space can therefore be rewritten for a given reciprocal lattice vector **g** as

$$\mathbf{g} \cdot (2\mathbf{k} + \mathbf{g}) = \mathbf{g} \cdot (\mathbf{k} + \mathbf{k}') = 0.$$

This relation expresses the fact that the bisector plane of **g** must contain the vector $\mathbf{k} + \mathbf{k}'$ for the exact Bragg orientation, as shown in Fig. 2.5(b).

Fig. 2.6 shows the diffraction angle 2θ (in mrad)† for the (200), (400), and (600) lattice planes in aluminum ($a = 0.4049$ nm) as a function of the acceleration voltage E (log–linear plot). The vertical lines indicate the commonly used voltages

† Recall that 1 mrad equals $0.057\,2958^\circ = 3'26''$.

Table 2.3. *Diffraction angles 2θ for the* (200), (400), *and* (600) *lattice planes in aluminum for $E = 200$ kV and $E = 1$ MV in mrad (degrees). The last column shows the corresponding angles for x-ray diffraction using Cu-K_α radiation with $\lambda = 0.154\,2838$ nm; the* (600) *planes do not give rise to a diffracted beam for this wavelength.*

Plane	$2\theta_{200\,\mathrm{kV}}$	$2\theta_{1\,\mathrm{MeV}}$	Cu-K_α x-rays
(200)	12.38 (0.71)	4.31 (0.25)	781.31 (44.76)
(400)	24.77 (1.42)	8.61 (0.49)	1731.52 (99.21)
(600)	37.16 (2.13)	12.92 (0.74)	—

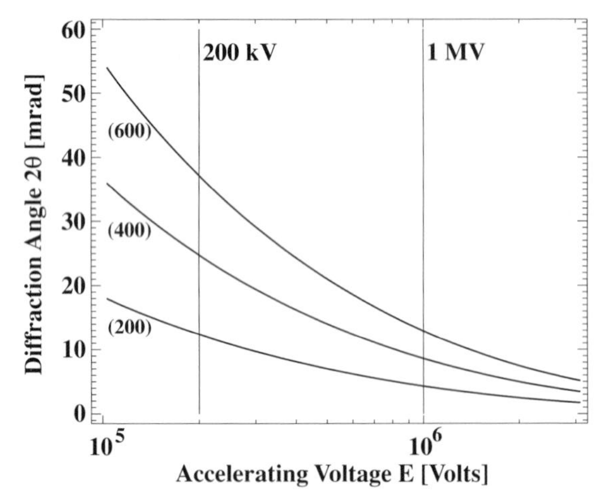

Fig. 2.6. Log–linear plot of the diffraction angle 2θ (mrad) versus acceleration voltage E (volts) for the (200), (400), and (600) lattice planes in aluminum.

$E = 200$ kV and $E = 1$ MV. Table 2.3 lists the corresponding diffraction angles 2θ in mrad and degrees, computed using

$$2\theta_{hkl} = 2\sin^{-1}\left(\frac{\lambda}{2d_{hkl}}\right).$$

For comparison, the table also shows the diffraction angle for diffraction using Cu-K_α x-rays. It can be seen from the table that it is a good approximation to replace $\sin\theta$ by θ for electron diffraction, since the diffraction angles increase nearly linearly with $|\mathbf{g}_{hkl}|$.

Table 2.2 on page 93 lists the length of the wave vector $\mathbf{K}_0$ (or, equivalently, the radius of the Ewald sphere) for the commonly used acceleration voltages. Note that these numbers are two to three orders of magnitude larger than the typical length of a reciprocal lattice vector $\mathbf{g}_{hkl}$. This means that the Ewald sphere has a large radius

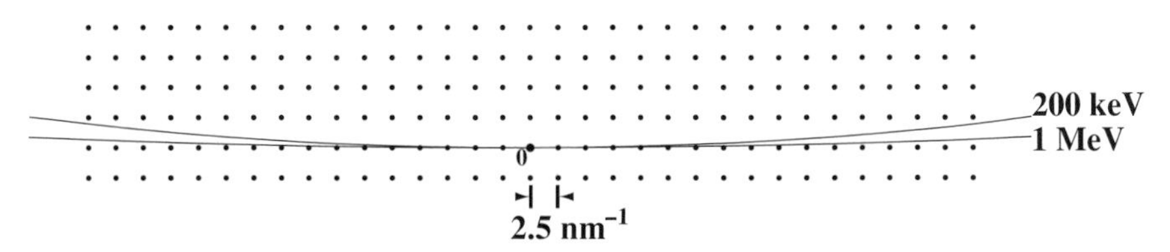

Fig. 2.7. Ewald sphere drawn to scale for the reciprocal lattice of a square crystal with lattice parameter 0.4 nm, and a 200 keV and 1 MeV incident electron beam.

compared to the dimensions of the reciprocal lattice. Near the origin of reciprocal space the Ewald sphere is nearly planar, as shown in Fig. 2.7, which represents a to-scale drawing of a square reciprocal lattice with lattice parameter $a^* = 2.5\ \text{nm}^{-1}$. The Ewald spheres corresponding to acceleration voltages of 200 kV and 1 MV are superimposed on the drawing. The radii of the Ewald spheres (on the scale of the drawing) are 340 and 977 mm, respectively, for 200 and 1000 kV electrons. Diffracted beams are therefore nearly parallel to the incident beam, with diffraction angles 2θ of the order of a degree or so. Diffraction of high-energy electrons is then essentially a forward-scattering process, and most electron trajectories are oriented at small angles with respect to the incident beam.

2.5 Fourier transforms and convolutions

2.5.1 Definition

Consider a particle described by a wave function $\Psi(\mathbf{r})$. In general, this function will be a superposition of different momentum eigenstates, each with its own momentum eigenvalue $\mathbf{p}$. Since we have established the relation between $\mathbf{p}$ and $\mathbf{k}$, we will continue to use the wave vector description from here on. Suppose we ask the question: how much does a certain wave vector $\mathbf{k}$ contribute to this function $\Psi(\mathbf{r})$? In other words, if Ψ is written as a superposition of momentum eigenfunctions, then what is the weight $\Psi(\mathbf{k})$ of the eigenfunction with wave vector $\mathbf{k}$?

This question is answered readily by considering the analogy with vectors: given a vector $\mathbf{t}$, what is the contribution of the basis vector $\mathbf{e}_1$ to $\mathbf{t}$? We know that the components of a vector can be determined by projecting this vector onto the three basis vectors, so we find:

$$\begin{aligned}\mathbf{t}\cdot\mathbf{e}_1 &= (u\mathbf{e}_1 + v\mathbf{e}_2 + w\mathbf{e}_3)\cdot\mathbf{e}_1,\\ &= u.\end{aligned}$$

The dot product thus allows us to determine the components (or projection) of a vector with respect to a set of basis vectors. All we need to determine those components is a set of independent (basis) vectors $\mathbf{e}_i$.

The dot product can also be defined for continuous functions, as we have seen in Section 2.2.1, and if we apply equation (2.2) to the unit amplitude plane wave we find

$$\langle e^{2\pi i\mathbf{k}\cdot\mathbf{r}} | e^{2\pi i\mathbf{k}'\cdot\mathbf{r}} \rangle = \iiint e^{2\pi i(\mathbf{k}'-\mathbf{k})\cdot\mathbf{r}} d\mathbf{r} = \delta(\mathbf{k}' - \mathbf{k}).$$

Now we are ready to answer the question posed at the beginning of this section: the contribution of the wave vector $\mathbf{k}$ to the wave function $\Psi(\mathbf{r})$ is determined from the projection of $\Psi(\mathbf{r})$ onto the momentum eigenfunction corresponding to $\mathbf{k}$:

$$\Psi(\mathbf{k}) = \langle e^{2\pi i\mathbf{k}\cdot\mathbf{r}} | \Psi(\mathbf{r}) \rangle. \tag{2.44}$$

We thus have an integral relation between the functions $\Psi(\mathbf{r})$ and $\Psi(\mathbf{k})$, which is given explicitly by

$$\Psi(\mathbf{k}) = \iiint e^{-2\pi i\mathbf{k}\cdot\mathbf{r}} \Psi(\mathbf{r})\, d\mathbf{r}.$$

We define the *direct Fourier transform* (DFT) of the function $\Psi(\mathbf{r})$ as

$$\boxed{\Psi(\mathbf{k}) = \mathcal{F}[\Psi(\mathbf{r})] \equiv \iiint \Psi(\mathbf{r}) e^{-2\pi i\mathbf{k}\cdot\mathbf{r}}\, d\mathbf{r},} \tag{2.45}$$

where the integral extends over all of 3D space. The operator $\mathcal{F}$ represents the DFT. The *inverse Fourier transform* (IFT) is then defined by

$$\boxed{\Psi(\mathbf{r}) = \mathcal{F}^{-1}[\Psi(\mathbf{k})] \equiv \iiint \Psi(\mathbf{k}) e^{2\pi i\mathbf{k}\cdot\mathbf{r}}\, d\mathbf{k}.} \tag{2.46}$$

The Fourier transform thus establishes a mechanism for converting a direct space wave function into its reciprocal space equivalent, which is essentially a momentum space representation of that function (apart from a constant scaling factor). We will represent the DFT of a function by a function with the same symbol, but with a reciprocal vector as its argument; in other words, the functions $\Psi(\mathbf{r})$ and $\Psi(\mathbf{k})$ form a *Fourier transform pair*.

At this point we should warn the reader that there are a few different notational systems around in the literature (see [Spe88, pp. 155–156] and [VAJ99, pp. 333–338] for a comparative list of notations, and a discussion of the history of the two conventions). First of all, it is possible to change the order of the functions in equation (2.44); this means that the sign of the exponent of the DFT will change to positive instead of negative, and the IFT will now have the minus sign (this is known as the *crystallographic sign convention*). Both conventions are in use, and the reader should be careful in comparing expressions from various textbooks. In addition, solid-state physicists often incorporate the factor of 2π in the definition of the wave vector, i.e. $|\mathbf{k}| = 2\pi/\lambda$; in other words, they *scale* reciprocal space by a factor of 2π. The reciprocal basis vectors, defined in equation (1.9) on page 11, must then also be multiplied by 2π, and the de Broglie relation then contains the factor $\hbar$ instead of h. Finally, the factor of 2π is sometimes put in front of both Fourier integrals, in the form $1/(2\pi)^{D/2}$, where D is the dimensionality of the space in which the transform is computed. Hence, when comparing formulas in the literature, be aware of the fact that the Fourier transform can be defined in several different ways, all of which are correct, *provided subsequent calculations are performed in a consistent way*. For an in-depth study of Fourier transforms and related mathematical operations as applied to diffraction phenomena we refer the reader to [Don69], [BW75], [Cow81], and [Bra86]. In this book, we will adhere to the *quantum mechanical sign convention*, which has a minus sign in the exponential of the direct Fourier transform.

2.5.2 The Dirac delta-function

The Dirac delta-function is a rather strange mathematical object which seems to defy normal properties of continuity and differentiability and still remains useful in various types of integral calculations. Although there is a rigorous mathematical definition of this and related functions in the theory of *distributions* [Don69, pp. 91–96], we will restrict ourselves to some simple definitions and properties (see, e.g., [Bra86]).

The Dirac δ-function is defined by

$$\delta(\mathbf{r}-\mathbf{a}) = \begin{cases} 0 & \mathbf{r} \neq \mathbf{a}; \\ \infty & \mathbf{r} = \mathbf{a}. \end{cases} \tag{2.47}$$

It is represented by an infinitely narrow peak of infinite height at the position $\mathbf{r} = \mathbf{a}$, with the additional property that the area under the peak is equal to one:

$$\iiint \delta(\mathbf{r}-\mathbf{a})\,\mathrm{d}\mathbf{r} = 1. \tag{2.48}$$

If we multiply the δ-function by a constant c, then the function $c\delta(\mathbf{r}-\mathbf{a})$ is a delta-function of *weight* c.

We can regard the δ-function as the limiting case of the Gaussian distribution, when the width of the distribution goes to zero and the amplitude to infinity, keeping the area under the curve normalized; in the 1D case this is described by

$$\delta(x) = \lim_{a\to\infty}\left[\frac{a}{\sqrt{\pi}}\mathrm{e}^{-a^2x^2}\right]. \tag{2.49}$$

Another useful property is the complex representation of the delta-function:

$$\delta(\mathbf{k}) = \mathcal{F}[1] = \iiint e^{-2\pi i \mathbf{k}\cdot\mathbf{r}} \, d\mathbf{r}. \tag{2.50}$$

Comparing this relation to the definition of the DFT, we find that the Fourier transform of the constant 1 is equal to $\delta(\mathbf{k})$. The Fourier transform of a constant c is a δ-function of weight c ($\mathcal{F}[c] = c\mathcal{F}[1] = c\delta(\mathbf{k})$).

2.5.3 The convolution product

The convolution product $C(\mathbf{r})$ (also known as *folding* or *faltung*) of two functions $f(\mathbf{r})$ and $g(\mathbf{r})$ is defined by

$$C(\mathbf{r}) \equiv f(\mathbf{r}) \otimes g(\mathbf{r}) = \iiint f(\mathbf{R})g(\mathbf{r} - \mathbf{R}) \, d\mathbf{R}. \tag{2.51}$$

The Dirac δ-function is the identity operator of the convolution product:

$$f(\mathbf{r}) \otimes \delta(\mathbf{r}) = \iiint f(\mathbf{R})\delta(\mathbf{r} - \mathbf{R}) \, d\mathbf{R} = f(\mathbf{r}). \tag{2.52}$$

This is known as the *sifting* property of the δ-function. The following two theorems are useful:

$$\text{the multiplication theorem:} \qquad \mathcal{F}[f(\mathbf{r})g(\mathbf{r})] = F(\mathbf{k}) \otimes G(\mathbf{k}); \tag{2.53}$$

$$\text{the convolution theorem:} \qquad \mathcal{F}[f(\mathbf{r}) \otimes g(\mathbf{r})] = F(\mathbf{k})G(\mathbf{k}). \tag{2.54}$$

The proof for these theorems is quite straightforward and is left to the reader (see, e.g., [Cow81]).

Convolutions are fundamental to many different types of experiments. In a rather general sense, we can describe an experiment as follows: a reference signal $\mathcal{R}$ is created and directed at the material to be investigated. This could be a beam of electrons, or monochromatic light, or a heat pulse, or anything else. The sample then modifies this reference signal, and ideally we would like to measure the modified signal $\mathcal{R}'$. In practice, however, the detector system, which converts the modified signal into an observable signal (e.g. visible light, or an electrical current or voltage) introduces its own *fingerprint* onto the signal $\mathcal{R}'$. In the simplest case this would be a linear scaling of $\mathcal{R}'$, but in many cases the relationship between detected signal $\mathcal{R}''$ and modified signal $\mathcal{R}'$ is non-linear and can be described by a convolution of $\mathcal{R}'$ with a function $\mathcal{T}$, known as the *instrument point spread function*:

$$\mathcal{R}'' = \mathcal{T} \otimes \mathcal{R}'. \tag{2.55}$$

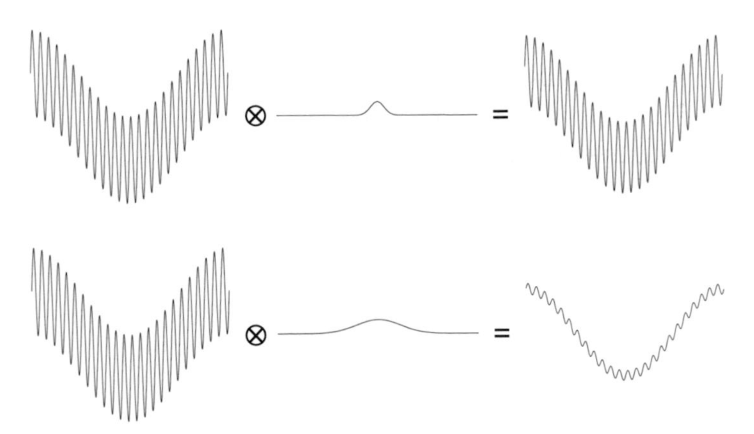

Fig. 2.8. Illustration of the convolution product of two functions. The function on the second row is broader than on the first row, and the resulting convolution product significantly reduces (blurs) the high-frequency detail of the original function.

Almost invariably, this means that the instrument imposes a *resolution limit*, which results in blurring of the smallest details in the signal $\mathcal{R}'$. An example of such blurring is shown in Fig. 2.8, which shows the convolution of a high-frequency signal with two point spread functions, one narrow and the other wide. The high-frequency details are suppressed by the convolution with the wider point spread function.

We have seen above that the Dirac delta-function is the identity operator for the convolution product, so it follows that for an ideal instrument, i.e. one that does not leave its fingerprint on the modified signal, the point spread function should be equal to the Dirac delta-function. In the case of electron microscopy, it is the task of the microscope manufacturer to design a detector system (which includes all microscope lenses and the viewing screen or camera) that has a minimal impact on the modified signal $\mathcal{R}'$. We will see in later chapters that this is only partially possible and as microscope users we need to be aware of how the microscope affects the wave function of the electrons after they have passed through the sample.

One might think that if the function $\mathcal{T}$ were fully known, then a measurement of $\mathcal{R}''$, followed by a mathematical operation known as *deconvolution*, would yield the function $\mathcal{R}'$ directly. Using the convolution theorem we can reconstruct the signal function as:

$$\mathcal{R}' = \mathcal{F}^{-1}\left[\frac{\mathcal{F}[\mathcal{R}'']}{\mathcal{F}[\mathcal{T}]}\right]. \tag{2.56}$$

It is obvious that this deconvolution procedure only works when $\mathcal{F}[\mathcal{T}]$ does not contain any zeros. As we will see in the chapter on phase contrast microscopy

(Chapter 10), we are in the unfortunate situation that the Fourier transformed point spread function for the TEM usually contains multiple zero crossings.

2.5.4 Numerical computation of Fourier transforms and convolutions

The Fourier transform, as defined in equations (2.45) and (2.46), is a continuous integral transform, suitable for analytical computations. Numerical computations operate on *discrete* functions rather than continuous functions. The *discrete direct Fourier transform* (DDFT) of a one-dimensional function $f(x_j)$ defined on a discrete grid of N equidistant points $x_j = j\delta$, with δ the grid stepsize, is defined as

$$f(q_i) = \frac{1}{N}\sum_{j=0}^{N-1} f(x_j)\mathrm{e}^{-2\pi \mathrm{i} ij/N}. \tag{2.57}$$

The "frequencies" q_j are defined by $q_j = j/N\delta$. A straightforward (but naive) numerical implementation of this equation would require the computation of $2N$ trigonometric functions (cos and sin). In terms of processor cycles, this would be a very expensive computation.

It is customary to introduce a shorthand notation for the exponential factor:

$$\omega_N \equiv \mathrm{e}^{\frac{2\pi}{N}\mathrm{i}}, \tag{2.58}$$

and hence

$$f(q_i) = \frac{1}{N}\sum_{j=0}^{N-1} f(x_j)\omega_N^{-ij}. \tag{2.59}$$

The *discrete inverse Fourier transform* (DIFT) is obtained by changing the sign of the exponent of ω_N:

$$f(x_j) = \sum_{i=0}^{N-1} f(q_i)\omega_N^{ij}. \tag{2.60}$$

Higher-dimensional discrete Fourier transforms are defined in a similar way. The two-dimensional DDFT is defined by

$$f(q_i, r_j) = \frac{1}{NM}\sum_{k=0}^{N-1}\sum_{l=0}^{M-1} f(x_k, y_l)\omega_N^{-ik}\omega_M^{-jl}. \tag{2.61}$$

The DDFT is essentially a matrix multiplication. If we denote the function values $f(x_i)$ by X_i, and $f(q_j)$ by Q_j, then it is easy to see that for $N = 4$, equation (2.59) is explicitly given by

$$\begin{pmatrix} Q_0 \\ Q_1 \\ Q_2 \\ Q_3 \end{pmatrix} = \frac{1}{4} \begin{pmatrix} 1 & 1 & 1 & 1 \\ 1 & \omega_4^{-1} & \omega_4^{-2} & \omega_4^{-3} \\ 1 & \omega_4^{-2} & \omega_4^{-4} & \omega_4^{-6} \\ 1 & \omega_4^{-3} & \omega_4^{-6} & \omega_4^{-9} \end{pmatrix} \begin{pmatrix} X_0 \\ X_1 \\ X_2 \\ X_3 \end{pmatrix}, \tag{2.62}$$

or, in matrix notation:

$$\mathbf{Q} = \mathcal{M}\mathbf{X}. \tag{2.63}$$

The matrix $\mathcal{M}$ is a symmetric matrix, and only four different numbers need be computed in the case $N = 4$; since $\omega_N^{-k} = \omega_N^{-k \pm N}$, we have $\omega_4^{-6} = \omega_4^{-2}, \omega_4^{-9} = \omega_4^{-1}$, and $\omega_4^{-4} = 1$, and the matrix reduces to

$$\mathcal{M} = \frac{1}{4} \begin{pmatrix} 1 & 1 & 1 & 1 \\ 1 & \omega_4^{-1} & \omega_4^{-2} & \omega_4^{-3} \\ 1 & \omega_4^{-2} & 1 & \omega_4^{-2} \\ 1 & \omega_4^{-3} & \omega_4^{-2} & \omega_4^{-1} \end{pmatrix}. \tag{2.64}$$

In general, the $N \times N$ matrix $\mathcal{M}$ contains only N different numbers, which are all powers of a single complex number ω_N. Note that ω_N requires two evaluations of trigonometric functions, using Euler's formula $\mathrm{e}^{\mathrm{i}x} = \cos x + \mathrm{i} \sin x$. The number ω_N and its powers are known as *twiddle factors*, and their computation can be performed in a very efficient way using the so-called *Singleton algorithm* [Sin67], which minimizes the total number of trigonometric function calls. For a description of the Singleton algorithm we refer the reader to Chu and George [CG00, pp. 14–15].

After the computation of the twiddle factors, we could in principle construct the matrix $\mathcal{M}$ and compute the product $\mathcal{M}\mathbf{X}$. This would again be inefficient from the computational point of view, and it would also require significant computer memory for large values of N, and for multi-dimensional arrays. In 1965 Cooley and Tukey [CT65] published an algorithm for the fast computation of summations of the type (2.59). Their algorithm became known as the *fast Fourier transform* (FFT) algorithm. The FFT algorithm makes use of the fact that if N is a power of 2, then the original summation (2.59) can be split into two summations:

$$Q_i = \frac{1}{2N} \sum_{j=0}^{N/2-1} X_{2j} \omega_N^{-i(2j)} + \frac{\omega_N^{-i}}{2N} \sum_{j=0}^{N/2-1} X_{2j+1} \omega_N^{-i(2j)}. \tag{2.65}$$

Since $\omega_N^2 = \omega_{N/2}$ and also $\omega_{N/2}^{ij} = \omega_{N/2}^{(i+N/2)j}$, we find

$$Q_i = \frac{1}{2N} \sum_{j=0}^{N/2-1} X_{2j} \omega_{N/2}^{-ij} + \frac{\omega_N^{-i}}{2N} \sum_{j=0}^{N/2-1} X_{2j+1} \omega_{N/2}^{-ij}, \tag{2.66}$$

in which each summation is now a DDFT of size $N/2$ [CG00, pp. 22–23]. Each of the new summations can again be split into two summations of size $N/4$, and we repeat this k times until $N/k = 2$, which is a trivial sum. The computational gains obtained from using the FFT algorithm are significant. A conventional "brute force" DDFT of a 1-D array with $N = 2^n$ entries takes of the order of N^2 floating point operations. The FFT algorithm takes only $N \log_2 N = nN$ operations, which represents a significant time saving factor: for $N = 1024$ we have $n = 10$ and hence the FFT takes of the order of 10 240 operations compared with 1048 576 for the naively implemented DDFT. For multi-dimensional discrete Fourier transforms, the time saving factors are even larger. We will make extensive use of the FFT in Chapters 7 and 10, when we talk about the *multi-slice* image simulation method.

There are many different numerical implementations of this recursive algorithm. The algorithm can be extended to arrays of arbitrary size N; in such cases the transform is decomposed into summations corresponding to $N = 2^a 3^b 5^c 7^d 11^e 13^f \ldots$, in other words, N is decomposed into powers of prime numbers. While the FFT algorithm is intuitively easy to understand, the reader is advised *against* attempting to implement the algorithm him/herself; there are many commercial and public domain implementations available, and most of them have been optimized for speed. In fact, in a 1995 review article, Sorensen, Burrus, and Heideman [SBH95] list more than 30 different implementations, and a total of about 3500 papers devoted to FFTs, since their inception in 1965.

For a detailed description of the basic algorithm and its implementation we refer to *Numerical Recipes: The Art of Scientific Computing (FORTRAN Version)* [PFTV89], Section 12.2. There are several public domain implementations available, among others the fftpack library (`www.netlib.org`). The source code in this book makes use of the FFTW public domain library of FFT routines;† the source code can be downloaded from `http://www.fftw.org`. The library is easy to install and test. Although all of the routines are written in C, the package provides a set of *wrapper*-routines that convert the internal data to a column-based ordering, so that Fortran programs can call all of the C routines. Instructions for installing the FFTW library can be found on the website. The main reasons for using the FFTW library are its speed, and the fact that the algorithm automatically selects the best decomposition of N into prime numbers, so that the restriction $N = 2^n$, common to many implementations, is not present. This will become useful in later chapters, where we will need to compute FFTs of arrays for which the dimensions are not simple powers of 2.

† FFTW was written by Matteo Frigo and Steven G. Johnson. The source code is included on the website under the GNU General Public Licence.

Convolution products can be computed readily using the direct and inverse FFT routines and the *convolution theorem* (equation 2.54):

$$f(\mathbf{r}) \otimes g(\mathbf{r}) = \mathcal{F}^{-1}[\mathcal{F}[f(\mathbf{r})]\mathcal{F}[g(\mathbf{r})]], \tag{2.67}$$

and therefore do not require a specialized algorithm. The use of this equation assumes that both functions have been defined on the same discrete grid.

2.6 The electrostatic lattice potential

The Schrödinger equation (2.37) contains two important contributions: the kinetic energy term, which is experimentally determined by the acceleration voltage of the electron microscope, and the electrostatic lattice potential $V(\mathbf{r})$, which describes the interaction of the beam electrons with the sample. Finding the correct potential for a given crystal structure is one of the more fundamental problems of solid-state physics, and most so-called *first-principles* methods produce this potential, often as the result of a self-consistent iteration. Fortunately for electron microscopists, it turns out that, to a very good approximation, we can consider a solid to consist of *spherical point scatterers*, represented by atomic scattering factors $f^e(\mathbf{s})$. The true lattice potential is then the potential due to spherical scatterers plus contributions from interatomic bonds. For most TEM applications, we can safely ignore bonding contributions and assume that the solid consists of isolated spherical point scatterers. In the following sections, we will first define the atomic scattering factors, and then derive expressions for the Fourier coefficients of the electrostatic lattice potential for an infinite and a finite crystal. We will only consider *elastic scattering events*; i.e. scattering events for which only the direction of the wave vector changes. The electron wavelength, or, equivalently, its energy, remains unchanged.

2.6.1 Elastic scattering of electrons by an individual atom

The scattering of electrons by a single atom is determined by the total charge distribution of that atom:

$$\rho(\mathbf{r}) = |e|[\rho_n(\mathbf{r}) - \rho_e(\mathbf{r})],$$

where ρ_n is the nuclear charge distribution and is ρ_e the electron charge distribution. The electrostatic atom potential $V^a(\mathbf{r})$ is related to the total charge distribution through Poisson's equation:

$$\Delta V^a(\mathbf{r}) = -\frac{|e|}{\epsilon_0}[\rho_n(\mathbf{r}) - \rho_e(\mathbf{r})], \tag{2.68}$$

where Δ is the Laplacian differential operator. The probability amplitude for an electron to be scattered in a certain direction can now be computed using the *bra-ket* concept introduced earlier in this chapter. The probability amplitude P that an incident plane wave with wave vector $\mathbf{k}$ will be scattered by the atomic potential $V^a(\mathbf{r})$ into the direction $\mathbf{k}'$ is given by

$$P = \left\langle \mathrm{e}^{2\pi \mathrm{i}\mathbf{k}'\cdot\mathbf{r}} \middle| V^a(\mathbf{r}) \middle| \mathrm{e}^{2\pi \mathrm{i}\mathbf{k}\cdot\mathbf{r}} \right\rangle. \tag{2.69}$$

We *define* the atomic scattering amplitude, $f^e(\Delta\mathbf{k})$, to be equal to this probability amplitude, or, equivalently, to the Fourier transform of the atomic potential distribution $V^a(\mathbf{r})$:

$$f^e(\Delta\mathbf{k}) \equiv \iiint V^a(\mathbf{r})\mathrm{e}^{-2\pi \mathrm{i}\Delta\mathbf{k}\cdot\mathbf{r}}\,\mathrm{d}\mathbf{r}, \tag{2.70}$$

where $\Delta\mathbf{k} = \mathbf{k}' - \mathbf{k}$ is the momentum transfer vector. The atomic scattering amplitude thus describes how the momentum of an incident electron changes upon elastic scattering by the atom. The atomic scattering factor expresses the probability amplitude that a certain elastic momentum transfer will occur.

Using the fact that the x-ray scattering amplitude $f^X(\Delta\mathbf{k})$ is defined as the Fourier transform of the *electron* charge density $\rho_e(\mathbf{r})$,† [Shm96], and approximating the nuclear charge density by a delta-function of weight Z (the atomic number) located in the origin, we can introduce the inverse Fourier transforms of all quantities in Poisson's equation:

$$\Delta\left[\iiint f^e(\Delta\mathbf{k})\mathrm{e}^{2\pi \mathrm{i}\Delta\mathbf{k}\cdot\mathbf{r}}\,\mathrm{d}\Delta\mathbf{k}\right] = \frac{|e|}{\epsilon_0}\iiint \left[f^X(\Delta\mathbf{k}) - Z\right]\mathrm{e}^{2\pi \mathrm{i}\Delta\mathbf{k}\cdot\mathbf{r}}\,\mathrm{d}\Delta\mathbf{k}. \tag{2.71}$$

Bringing the Laplacian operator (operating on $\mathbf{r}$) inside the integral and dropping all integrals we find:

$$f^e(\Delta\mathbf{k}) = \frac{|e|}{4\pi^2\epsilon_0|\Delta\mathbf{k}|^2}\left[Z - f^X(\Delta\mathbf{k})\right]. \tag{2.72}$$

This relation is valid for an isolated atom. Since we will be studying crystals, with regular arrangements of atoms, we will often rewrite the scattering vector $\Delta\mathbf{k}$ as $\Delta\mathbf{k} = 2\mathbf{s}$; for the Bragg condition, $\mathbf{k}' = \mathbf{k} + \mathbf{g}$ or, equivalently, $\mathbf{s} = \mathbf{g}/2$. Using the

† The contribution of the nuclear charge to x-ray photon scattering is several orders of magnitude smaller than that of the orbital electrons because of the mass difference between electrons and nucleons.

Bragg equation we find for the magnitude of **s**:

$$|\mathbf{s}| = \frac{1}{2d_{hkl}} = \frac{\sin\theta}{\lambda} = \frac{|\mathbf{g}_{hkl}|}{2}. \tag{2.73}$$

This means that for a given crystal structure the atomic scattering factor is "sampled" only at scattering vectors **s** corresponding to half the reciprocal lattice vectors. Since **s** is independent of the wavelength of the electrons ($1/2d$ is constant for a given crystal structure!), the following expression for the atomic scattering factor is independent of the experimental conditions:

$$f^e(s) = \frac{|e|}{16\pi^2\epsilon_0|s|^2}\left[Z - f^X(s)\right] \qquad \text{(Mott–Bethe formula).} \tag{2.74}$$

Fig. 2.9 shows the electron scattering factor for a copper atom; the vertical lines indicate the reciprocal lattice spacings for the face-centered cubic Cu structure ($a = 0.361\,47$ nm). Scattering functions for other atoms can be displayed using the **ION**-routine scatfac.pro on the website.

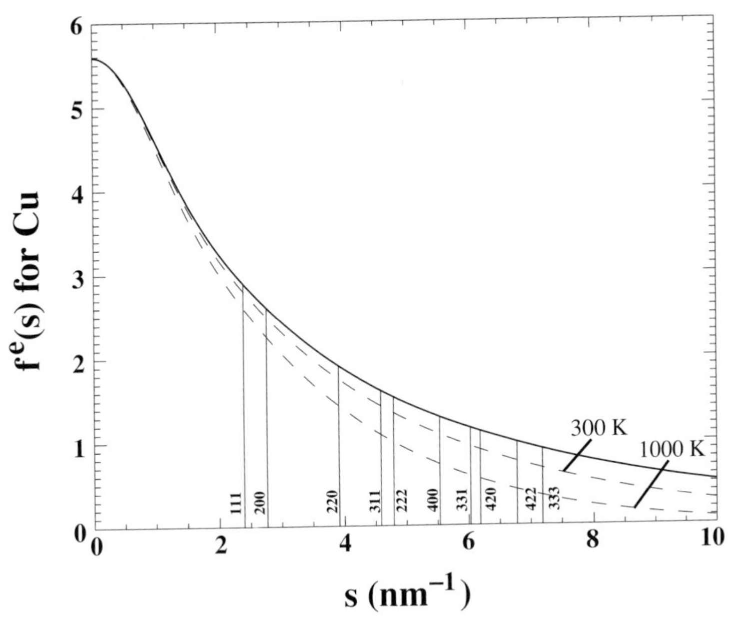

Fig. 2.9. Atomic scattering factor for Cu, with superimposed lines indicating the first 10 reciprocal lattice spacings $|\mathbf{g}|/2$ for face-centered cubic (fcc) Cu. The scattering factor is "sampled" at those values of $s = \sin\theta/\lambda$. The values in this figure must be multiplied by $0.047\,878\,01$ to yield scattering factors in units of V nm^3. The dashed lines correspond to the scattering factor for fcc copper at $T = 300$ K ($B(300) = 0.005\,75$ nm^2) and $T = 1000$ K ($B(1000) = 0.018\,69$ nm^2) [GP99].

The electron scattering factors have been calculated by many researchers and are often parameterized in a power series or an exponential expansion. For the low atomic number atoms, this computation is actually not very difficult and one can use programs such as the Herman–Skillman code (based on non-relativistic Hartree–Fock–Slater equations) [HS80] to compute the atomic scattering factors. For heavier atoms, relativistic corrections become important and the computation of $f^e(s)$ becomes far more involved. There are several parameterizations of the electron and/or x-ray scattering factors available in the literature:

- Smith and Burge [SB62];
- Doyle and Turner [DT68];
- Weickenmeier and Kohl [WK91];
- Rez, Rez and Grant [RRG94].

A more extensive list and discussion can be found in Appendix D of E.J. Kirkland's book on *Advanced Computing in Electron Microscopy* [Kir98, pp. 199–219]. The source code available from the website uses two sets of scattering factors: a combination of Doyle–Turner and Smith–Burge parameterizations, identical to that used in the EMS package [Sta87], and the Weickenmeier–Kohl FSCATT routine.[†]

Both Doyle–Turner and Smith–Burge parameterizations are exponential expansions of the form:

$$f^e(s) = \frac{|e|}{16\pi^2\epsilon_0}\left(\frac{Z - f^X(s)}{s^2}\right) = 0.047\,878\,01 \sum_{j=1}^{N} a^i_j \mathrm{e}^{-b^i_j s^2}, \qquad (2.75)$$

where $N = 4$ for Doyle–Turner parameters and $N = 3$ for Smith–Burge parameters; s must be expressed in nm^{-1}. The units of $f^e(s)$ are then V nm^3. This expansion is accurate for values of s up to about 20 nm^{-1}. The parameterization can be found in the module diffraction.f90 on the website.

Example 2.2 *Compute the electron scattering factor f^e_{Cu} for the (111) planes of copper, with lattice parameter $a = 0.361\,47$ nm.*

Answer: *First, we must evaluate the scattering parameter $s = \sin\theta/\lambda = 1/2d$ for this particular situation. The value for d_{111} in a cubic crystal is easily found using*

$$d_{hkl} = \frac{a}{\sqrt{h^2 + k^2 + l^2}}$$

from which we find $d_{111} = 0.361\,47/\sqrt{3} = 0.208\,695$ nm. The scattering parameter s is then equal to 2.3958 nm^{-1}. Substitution of this value, and the parameters for a_j

[†] The original source code of FSCATT was developed by Andreas Wieckenmeier and Helmut Kohl, then at the Technische Hochschule in Darmstadt, Germany, and is included on the website with the permission of the authors.

and b_j for copper (see diffraction.f90 *file), into the atomic scattering factor equation results in*

$$f^e_{Cu}(2.3958) = 0.047\,878\,01 \times \left\{1.579\mathrm{e}^{-1.820\times(2.3958)^2} + 1.820\mathrm{e}^{-0.124\,53\times(2.3958)^2} + 1.658\mathrm{e}^{-0.025\,04\times(2.3958)^2} + 0.532\mathrm{e}^{-0.003\,33\times(2.3958)^2}\right\};$$

$$= 0.047\,878\,01 \times 2.891\,086 = 0.138\,42\ V\,nm^3.$$

Note that this number is independent of the wavelength of the electrons being used since the number $1/2d_{hkl}$ is independent of the wavelength.

The *International Tables for Crystallography* provide tables of the electron scattering factors for all elements and their most common ions [WP99, pp. 226–244].† The values in these tables must be multiplied by the factor of 0.047 878 01 to give the numerical values above. The scattering factor for copper, shown in Fig. 2.9, must also be multiplied by this constant to have units of V nm^3. When comparing scattering factors from various sources, the reader should study carefully how the factors have been defined in each source.

The Weickenmeier–Kohl expansion uses seven fitting parameters for the elastic scattering amplitudes of the neutral atoms. The fitting function is of the form

$$f^e(s) = \frac{1}{s^2}\sum_{i=1}^{6} A_i\left(1 - \mathrm{e}^{-B_i s^2}\right), \tag{2.76}$$

where

$$A_1 = A_2 = A_3 = 0.023\,95\frac{Z}{3(1+V)}\ \text{Å}^{-1}; \quad A_4 = A_5 = A_6 = V A_1.$$

The specific definition of A_1 guarantees that the asymptotic behavior of the scattering amplitude for large s has the correct shape. The values of the six parameters B_i and V are listed in Table 1 in [WK91], and are tabulated in the FSCATT function in the others.f90 module. Since this function also returns a second parameter, the *absorptive form factor*, we will defer a more complete description until Section 6.5.1.

† The reader should be warned that the older edition of the *International Tables for Crystallography* [HL52] uses a different definition of the electron scattering factors; they are commonly denoted by f^B (B for Born approximation) and are related to the current definition by

$$f^B(s) = \frac{2\pi\gamma m_0|e|}{h^2} f^e(s).$$

The disadvantage of this formulation is that these scattering factors are dependent upon the acceleration voltage (through γ). The factors f^B are expressed in nm when s is in nm^{-1}. For a more detailed discussion of the relation between the two sets of factors see [SZ92, pp. 30–35].

2.6.2 Elastic scattering by an infinite crystal

An infinite point lattice can be described by a set of unit-weight delta-functions, located at the lattice positions:

$$\mathcal{T}(\mathbf{r}) = \sum_{u,v,w=-\infty}^{+\infty} \delta(\mathbf{r} - \mathbf{t}_{uvw}). \qquad (2.77)$$

Each delta-function corresponds to a single unit cell of the lattice. Every unit cell contains N atoms located at positions $\mathbf{r}_j$ and the potential of a single unit cell is then described by

$$V_{\text{cell}}(\mathbf{r}) = \sum_{j=1}^{N} V_j^a(\mathbf{r} - \mathbf{r}_j).$$

The complete lattice potential is determined by "copying" the single unit cell potential at every lattice site; this operation is described by a convolution product:

$$V(\mathbf{r}) = V_{cell}(\mathbf{r}) \otimes \mathcal{T}(\mathbf{r}).$$

The potential $V(\mathbf{r})$ has, by construction, the periodicity of the underlying Bravais lattice. This allows us to expand the potential as a discrete Fourier series; this is essentially a plane wave expansion, where the "planes" are the actual lattice planes described by the reciprocal lattice vectors $\mathbf{g}$:

$$V(\mathbf{r}) = \sum_{\mathbf{g}} V_{\mathbf{g}} \mathrm{e}^{2\pi \mathrm{i}\mathbf{g}\cdot\mathbf{r}}; \qquad (2.78)$$

the coefficients $V_{\mathbf{g}}$ are the Fourier coefficients of the electrostatic lattice potential. There is one coefficient (in general, a complex number) for each set of planes in the crystal. Since the potential $V(\mathbf{r})$ is real, we must have $V_{-\mathbf{g}} = V_{\mathbf{g}}^*$. If, in addition, the origin of the reference frame is located at an inversion center, then the Fourier coefficients are all real numbers and we have $V_{-\mathbf{g}} = V_{\mathbf{g}}$.

We anticipate that, for an infinite crystal, the Fourier transform of the lattice potential will consist of discrete peaks at the reciprocal lattice positions. As we will show below, for a finite crystal this is no longer true, and the Fourier coefficients become continuous functions of a reciprocal space vector $\mathbf{q}$, or

$$V(\mathbf{q}) = \frac{1}{\Omega} \iiint V(\mathbf{r}) \mathrm{e}^{-2\pi \mathrm{i}\mathbf{q}\cdot\mathbf{r}}\, \mathrm{d}\mathbf{r}. \qquad (2.79)$$

This can be rewritten as

$$V(\mathbf{q}) = \frac{1}{\Omega} \mathcal{F}[V(\mathbf{r})],$$

and using the convolution theorem we have

$$V(\mathbf{q}) = \frac{1}{\Omega} \underbrace{\mathcal{F}[V_{cell}(\mathbf{r})]}_{I} \underbrace{\mathcal{F}[\mathcal{T}(\mathbf{r})]}_{II}. \tag{2.80}$$

The first factor on the right-hand side of equation (2.80) describes the contribution of a single unit cell. Using the definition of the atomic scattering factors, we arrive at

$$\mathcal{F}[V_{cell}(\mathbf{r})] = \sum_{j=1}^{N} f_j^e \mathrm{e}^{-2\pi i \mathbf{q}\cdot\mathbf{r}_j}. \tag{2.81}$$

It is easy to show that the second factor in equation (2.80) can be rewritten as a sum of delta-functions at the reciprocal lattice positions (e.g. [Cow81, p. 43]):

$$\mathcal{F}[\mathcal{T}(\mathbf{r})] = \sum_{\mathbf{g}} \delta(\mathbf{q} - \mathbf{g}). \tag{2.82}$$

We define the Fourier coefficients of the electrostatic lattice potential as

$$V_{\mathbf{g}} \equiv \frac{1}{\Omega} \sum_{j=1}^{N} f_j^e \mathrm{e}^{-2\pi i \mathbf{g}\cdot\mathbf{r}_j}, \tag{2.83}$$

and we find for the Fourier transform of the potential

$$V(\mathbf{q}) = \sum_{\mathbf{g}} V_{\mathbf{g}} \delta(\mathbf{q} - \mathbf{g}). \tag{2.84}$$

As we anticipated earlier, the Fourier transform (for an infinite crystal) is a discrete function which is non-zero only at the reciprocal lattice points. The numerical computation of the Fourier coefficients $V_{\mathbf{g}}$ will be discussed in Section 2.6.7.

Example 2.3 *Compute the probability $\mathcal{P}$ that an electron will be scattered from a state with wave vector $\mathbf{k}$ into a state with wave vector $\mathbf{k}'$ by a periodic electrostatic potential $V(\mathbf{r})$.*

Answer: *A periodic potential $V(\mathbf{r})$ can be written as a Fourier series:*

$$V(\mathbf{r}) = \sum_{\mathbf{g}} V_{\mathbf{g}} \mathrm{e}^{2\pi i \mathbf{g}\cdot\mathbf{r}}.$$

The probability $\mathcal{P}$ is given by the bra-ket:

$$\begin{aligned}
\mathcal{P} &= \left\langle e^{2\pi i \mathbf{k}'\cdot\mathbf{r}} \middle| V(\mathbf{r}) \middle| e^{2\pi i \mathbf{k}\cdot\mathbf{r}} \right\rangle, \\
&= \sum_{\mathbf{g}} V_{\mathbf{g}} \left\langle e^{2\pi i \mathbf{k}'\cdot\mathbf{r}} \middle| e^{2\pi i \mathbf{g}\cdot\mathbf{r}} \middle| e^{2\pi i \mathbf{k}\cdot\mathbf{r}} \right\rangle, \\
&= \sum_{\mathbf{g}} V_{\mathbf{g}} \iiint e^{2\pi i (\mathbf{k}+\mathbf{g}-\mathbf{k}')\cdot\mathbf{r}}\, d\mathbf{r}, \\
&= \sum_{\mathbf{g}} V_{\mathbf{g}} \delta(\mathbf{k}+\mathbf{g}-\mathbf{k}').
\end{aligned}$$

This expression is only different from zero whenever $\mathbf{k}+\mathbf{g}=\mathbf{k}'$, which is the Bragg equation in reciprocal space. In such a situation, the probability amplitude $\mathcal{P}$ is equal to the Fourier coefficient $V_{\mathbf{g}}$.

Example 2.4 *Determine for which reciprocal lattice vectors* $\mathbf{g}$ *the Fourier coefficient $V_{\mathbf{g}}$ (equation 2.83) vanishes in a structure with a face-centered cubic lattice.*

Answer: *The summation over all the atoms in the unit cell can be written as a summation over one-quarter of the atoms and a summation over all atoms equivalent to each other by the Bravais lattice N_c centering vectors $\mathbf{R}_k$:*

$$\begin{aligned}
V_{\mathbf{g}} &= \frac{1}{\Omega} \sum_{j=1}^{N} f_j^e e^{-2\pi i \mathbf{g}\cdot\mathbf{r}_j}; \\
&= \frac{1}{\Omega} \sum_{j=1}^{N/4} f_j^e e^{-2\pi i \mathbf{g}\cdot\mathbf{r}_j} \sum_{k=1}^{N_c} e^{-2\pi i \mathbf{g}\cdot\mathbf{R}_k}; \\
&= \frac{1}{\Omega} \sum_{j=1}^{N/4} f_j^e e^{-2\pi i \mathbf{g}\cdot\mathbf{r}_j} \left[1 + e^{-\pi i \mathbf{g}\cdot(\mathbf{a}+\mathbf{b})} + e^{-\pi i \mathbf{g}\cdot(\mathbf{a}+\mathbf{c})} + e^{-\pi i \mathbf{g}\cdot(\mathbf{b}+\mathbf{c})}\right]; \\
&= \left[1 + (-1)^{h+k} + (-1)^{h+l} + (-1)^{k+l}\right] \frac{1}{\Omega} \sum_{j=1}^{N/4} f_j^e e^{-2\pi i \mathbf{g}\cdot\mathbf{r}_j}.
\end{aligned}$$

We have used the fact that $\mathbf{g}\cdot(\mathbf{a}+\mathbf{b}) = h+k$, and also $e^{-\pi i n} = (-1)^{-n} = (-1)^n$. All reciprocal lattice points for which the Miller indices have mixed parity have a vanishing Fourier coefficient $V_{\mathbf{g}}$, and will not give rise to a diffracted beam. Reflections which are absent because of the Bravais lattice centering symmetry are known as systematic absences or systematic extinctions.

2.6.3 Finite crystal size effects

We define the shape function $D(\mathbf{r})$ of a crystal as a function which equals 1 inside the crystal and 0 outside:

$$D(\mathbf{r}) = \begin{cases} 1 & \text{inside;} \\ 0 & \text{outside.} \end{cases} \tag{2.85}$$

The finite crystal potential $V_f(\mathbf{r})$ then becomes the product of the potential for the infinite crystal and the shape function:

$$V_f(\mathbf{r}) = V(\mathbf{r})D(\mathbf{r}). \tag{2.86}$$

We can now apply the multiplication theorem to determine how the finite character of the crystal affects the Fourier transform of the lattice potential:

$$V_f(\mathbf{q}) = \mathcal{F}[V(\mathbf{r})] \otimes \mathcal{F}[D(\mathbf{r})].$$

Since the δ-function is the identity operator for the convolution product, we find, after substitution of equation (2.84):

$$V_f(\mathbf{q}) = \sum_{\mathbf{g}} V_{\mathbf{g}} D(\mathbf{q} - \mathbf{g}). \tag{2.87}$$

The Fourier transform of the lattice potential of a finite crystal is, hence, described by a similar formula as for the infinite crystal, except that the reciprocal lattice points now have a certain shape in reciprocal space. The exact shape of this volume is determined by the external shape of the crystal. Let us consider two examples of importance to electron microscopy: a planar thin foil, and a finite rectangular crystal.

Example 2.5 *Compute the reciprocal shape function $D(\mathbf{q})$, also known as the shape amplitude, for a thin foil of thickness z_0 with parallel top and bottom surfaces.*

Answer: *Taking the origin of the reference frame in the center of the foil, and the z-direction normal to the foil, the reciprocal shape function is given by*

$$\begin{aligned} D(\mathbf{q}) &= \int_{-\infty}^{+\infty} \mathrm{e}^{-2\pi \mathrm{i} q_x x}\, \mathrm{d}x \int_{-\infty}^{+\infty} \mathrm{e}^{-2\pi \mathrm{i} q_y y}\, \mathrm{d}y \int_{-z_0/2}^{+z_0/2} \mathrm{e}^{-2\pi \mathrm{i} q_z z}\, \mathrm{d}z; \\ &= -\delta(q_x)\delta(q_y) \left[\frac{\mathrm{e}^{-2\pi \mathrm{i} q_z z}}{2\pi \mathrm{i} q_z} \right]_{-z_0/2}^{z_0/2}, \end{aligned}$$

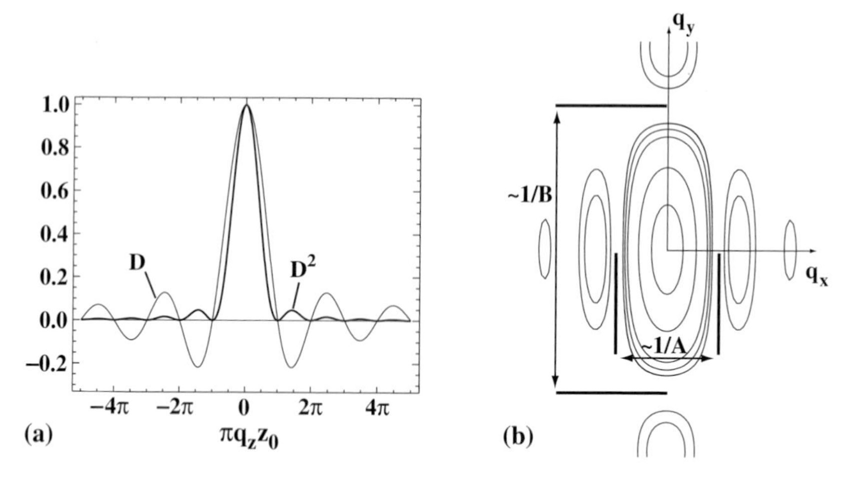

Fig. 2.10. (a) Reciprocal shape functions D and D^2 for a thin foil of thickness z_0 along the direction normal to the foil surface; (b) contour plot of a planar cut through the reciprocal shape function of a rectangular crystal $D^2(q_x, q_y, 0)$.

which results in

$$D(\mathbf{q}) = z_0\delta(q_x)\delta(q_y)\frac{\sin(\pi q_z z_0)}{\pi q_z z_0}.$$

Fig. 2.10(a) shows the function $D(0, 0, q_z)/z_0$ *and its square.*

Example 2.6 *Compute the reciprocal shape function* $D(\mathbf{q})$ *for a rectangular crystal with dimensions* A, B, *and* C.

Answer: *Using the same reference frame as in the previous example, and carrying out three identical integrations, we have*

$$\begin{aligned} D(\mathbf{q}) &= \int_{-A/2}^{+A/2} \mathrm{e}^{-2\pi \mathrm{i} q_x x}\,\mathrm{d}x \int_{-B/2}^{+B/2} \mathrm{e}^{-2\pi \mathrm{i} q_y y}\,\mathrm{d}y \int_{-C/2}^{+C/2} \mathrm{e}^{-2\pi \mathrm{i} q_z z}\,\mathrm{d}z; \\ &= ABC\frac{\sin(\pi A q_x)}{\pi A q_x}\frac{\sin(\pi B q_y)}{\pi B q_y}\frac{\sin(\pi C q_z)}{\pi C q_z}. \end{aligned}$$

The square of this function is shown in Fig. 2.10(b) as a contour plot $D^2(q_x, q_y, 0)$.

As we will see in the next chapter, nearly all TEM observations are carried out with a circular electron beam; in other words, the illuminated volume in the specimen is to a good approximation cylindrical with top and bottom surfaces not necessarily flat or parallel. Since the illuminated volume is always finite, but with a diameter that depends on the beam diameter, the volume and shape of reciprocal lattice points will vary with the incident beam "shape" and the shape function $D(\mathbf{r})$ is determined by the intersection of the incident beam and the specimen. For typical thin foils used in TEM the third dimension (usually the z-axis) is much smaller than the lateral dimensions (see the first example above) and therefore the reciprocal lattice points will be shaped as rods with their length axes perpendicular to the foil. Since these rods reside in reciprocal space, they are called *reciprocal lattice rods* or *relrods*.

We conclude from the examples above that, for a finite crystal, every reciprocal lattice point is extended infinitely in reciprocal space, but, because of the rapidly decreasing character of these functions, each point can be approximated by a finite volume in reciprocal space. The shape of this volume can become quite complicated for samples which have a non-rectangular shape. In Section 9.8 on page 575 we will discuss in detail how the *shape amplitude* $D(\mathbf{q})$ for a polyhedral particle can be calculated analytically.

Another important consequence of the finiteness of a crystal is that the Bragg condition for diffraction can be relaxed; whereas diffraction by an infinite crystal requires a mathematical point to coincide with the surface of the reflecting sphere (a rather improbable event), diffraction by a finite crystal becomes more probable because of the extent of the reciprocal lattice points.

2.6.4 The excitation error or deviation parameter $s_{\mathbf{g}}$

Since diffraction is possible even when the Bragg relation is not precisely satisfied, it is useful to have a parameter that measures the deviation from the exact Bragg orientation. For a thin foil, the Fourier transform of the shape function can be approximated by a relrod, as we have seen in the previous section. The deviation from the exact Bragg orientation can then be measured as the distance between the reciprocal lattice point and the Ewald sphere along the major axis of the relrod, or, equivalently, along the foil normal. We represent this distance by a vector $\mathbf{s} = s\mathbf{e}_z$, with $\mathbf{e}_z$ the unit normal to the foil, pointing down towards the viewing screen of the microscope such that $\mathbf{k} \cdot \mathbf{e}_z$ is positive.

The length of the vector $\mathbf{s}$ can be computed as follows: from the equation for the Ewald sphere (2.43) we have

$$(\mathbf{g} + \mathbf{s}) \cdot (2\mathbf{k} + \mathbf{g} + \mathbf{s}) = 0,$$

since the vector $\mathbf{g} + \mathbf{s}$ ends on the Ewald sphere, by construction. This relation can be rewritten as

$$2(\mathbf{k} + \mathbf{g}) \cdot \mathbf{s} + s^2 = -\mathbf{g} \cdot (2\mathbf{k} + \mathbf{g}); \tag{2.88}$$

we recognize the equation of the Ewald sphere on the right-hand side of this equation. If the angle between $\mathbf{k}+\mathbf{g}$ and the foil normal is denoted by α, then we also have

$$s + \frac{s^2}{2|\mathbf{k}+\mathbf{g}|\cos\alpha} = \boxed{\frac{-\mathbf{g}\cdot(2\mathbf{k}+\mathbf{g})}{2|\mathbf{k}+\mathbf{g}|\cos\alpha} \equiv s_{\mathbf{g}}.} \qquad (2.89)$$

We introduce† the new variable $s_{\mathbf{g}}$ which is equal to the right-hand side of the equation above; $s_{\mathbf{g}}$ is known as the *excitation error* or the *deviation parameter*. It has units of reciprocal distance, since it is measured in reciprocal space. For relativistic electrons we have $|\mathbf{k}+\mathbf{g}| \approx |\mathbf{k}|$, which is of the order of several hundred nm^{-1} (see Table 2.2); the second term on the left-hand side of equation (2.89) is thus very small and can be safely ignored. The deviation parameter $s_{\mathbf{g}}$ is then to a very good approximation equal to the distance s between the reciprocal lattice point $\mathbf{g}$ and the Ewald sphere, measured along the relrod. The difference between $s_{\mathbf{g}}$ and s increases with decreasing acceleration voltage and only becomes important for low-energy electron diffraction.

According to Fig. 2.5(a), $\mathbf{g}\cdot(2\mathbf{k}+\mathbf{g})$ is negative when $\mathbf{g}$ is inside the Ewald sphere, which means that the excitation error $s_{\mathbf{g}}$ defined in equation (2.89) is *positive* when the reciprocal lattice point $\mathbf{g}$ lies *inside* the Ewald sphere. $s_{\mathbf{g}}$ is *negative* when $\mathbf{g}$ lies *outside* the Ewald sphere, and is equal to zero when the Bragg orientation is exactly satisfied. The deviation parameter is entirely determined by the orientation of the crystal lattice with respect to the Ewald sphere, and is therefore an experimental parameter. The microscope operator has full control over the magnitude and sign of the deviation parameter. The degree to which $s_{\mathbf{g}}$ can be varied is determined by the type of specimen holder used for the observations; we will discuss the major types of specimen holders in Section 3.9.2. The deviation parameter is one of the most important parameters for conventional TEM observations, and, as we shall see in Chapters 6 and 8, the image contrast for both perfect and defective crystals is to a large extent determined by the value of $s_{\mathbf{g}}$.

The excitation error can be computed numerically using the function Calcsg in the module diffraction.f90. The function takes a $\mathbf{g}$-vector, the incident wave vector $\mathbf{k}$, and the foil normal $\mathbf{F}$ as input, and returns the excitation error in nm^{-1}.

2.6.5 Phenomenological treatment of absorption

In the preceding sections, we have only discussed *elastic* scattering processes, during which the electron energy does not change. This is an approximation of what

† The reason for selecting this particular expression will become clear in Chapter 5 where we will derive the dynamical diffraction equations. The deviation parameter as defined above appears directly in those equations.

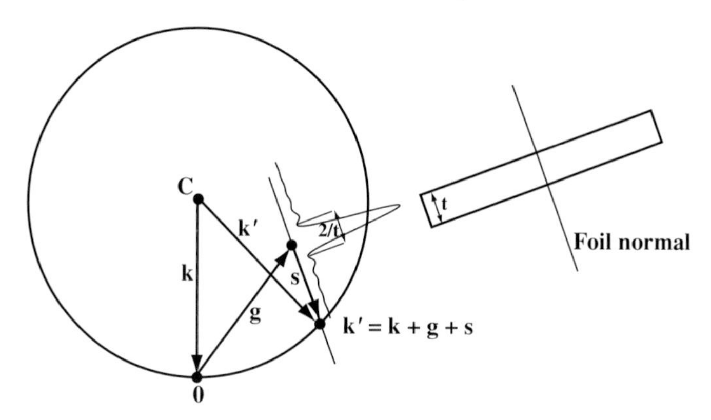

Fig. 2.11. Definition of the excitation error in terms of the incident and diffracted wave vectors and the reciprocal lattice vector **g**. The excitation error is measured along the normal to the foil surface, or, equivalently, along the relrod. The angles between the vectors are greatly exaggerated; in reality **g** is nearly perpendicular to **k** (for electron diffraction).

really happens when a beam electron interacts with the crystal. An electron can exchange energy with the crystal in a variety of processes, which are commonly grouped under the name *inelastic scattering processes*. Electrons that are inelastically scattered no longer travel in the directions predicted by the Bragg equation, and give rise to a background image intensity, which reduces the contrast of the elastic image.

The interaction between the beam electron and the crystal can be taken into account by including the ground state and excited states of the crystal in the Schrödinger equation. Following Yoshioka [Yos57], we write the Schrödinger equation as

$$\left(\hat{T} + \hat{H}_{cr} + \hat{H}_{int}\right)|\Psi\rangle = E_t|\Psi\rangle, \tag{2.90}$$

where $\hat{H}_{cr}$ is the crystal Hamiltonian, and $\hat{H}_{int}$ is the electrostatic (Coulomb) interaction term between the beam electron and the crystal. The wave function $|\Psi\rangle$ represents the state of the crystal *and* the beam electron, and can therefore be written as the product of two wave functions:

$$|\Psi\rangle = \sum_n |\psi_n\rangle|a_n\rangle, \tag{2.91}$$

where the functions $|a_n\rangle$ represent the nth excited state of the crystal ($|a_0\rangle$ is the ground state). The beam electron wave function is given by $|\psi_n\rangle$, for an electron that leaves the crystal in the excited state $|a_n\rangle$. The crystal wave functions $|a_n\rangle$ are the eigenfunctions of the crystal Hamiltonian

$$\hat{H}_{\mathrm{cr}}|a_n\rangle = \epsilon_n|a_n\rangle,$$

where ϵ_n is the energy of the excited state. The crystal wave functions contain all the information about lattice excitations, including phonons, plasmons, excitons, polarons, and so on. The computation of the $|a_n\rangle$ is therefore a non-trivial problem, which lies at the heart of solid-state physics.

If we assume that the crystal wave functions and energies are known, then we can substitute equation (2.91) into (2.90) and multiply the resulting equation by the state $\langle a_m|$. Since $\langle a_m|a_n\rangle = \delta_{mn}$ for a complete set of functions, we find

$$\left(E_t - \epsilon_m - \hat{T}\right)|\psi_m\rangle = \sum_n \langle a_m|\hat{H}_{int}|a_n\rangle|\psi_n\rangle \equiv \sum_n H_{mn}|\psi_n\rangle, \tag{2.92}$$

where we have introduced the matrix elements H_{mn} of the electrostatic interaction between beam electron and crystal. These elements describe the probability amplitude that a beam electron will change the excitation state of the crystal from the state $|a_n\rangle$ to the state $|a_m\rangle$. For elastic scattering, the crystal state does not change, and, hence, the diagonal matrix elements H_{nn} describe elastic scattering processes. Equation (2.92) is usually rewritten as a set of two equations, commonly known as the *Yoshioka equations* [Yos57], by separating out the term $m = 0$:

$$\left(E_t - \epsilon_0 - \hat{T} - H_{00}\right)|\psi_0\rangle = \sum_{n\neq 0} H_{0n}|\psi_n\rangle; \tag{2.93}$$

$$\left(E_t - \epsilon_m - \hat{T} - H_{mm}\right)|\psi_m\rangle = \sum_{n\neq m} H_{mn}|\psi_n\rangle. \tag{2.94}$$

In the absence of inelastic scattering, the off-diagonal matrix elements H_{mn} $(m \neq n)$ vanish, and the right-hand sides of both Yoshioka equations vanish. The diagonal term on the left-hand side represents the interaction of the electron with the crystal, and is commonly known as the *potential energy* of the beam electron in the crystal. We have already seen that in the ground state of the crystal this potential energy is given by $H_{00} = eV(\mathbf{r})$.

The computation of the matrix elements H_{mn} is a difficult problem in general and one usually approaches the computation of inelastic scattering probabilities by only considering one effect at a time; e.g. the contribution of phonons, or plasmons, and so on. Such computations are beyond the scope of this book. We refer to Z.L. Wang's book *Elastic and Inelastic Scattering in Electron Diffraction and Imaging* for a recent review of inelastic scattering [Wan95]. We will instead use the phenomenological approach of Molière [Mol39] for x-ray diffraction, which was first applied to electron diffraction by Honjo and Mihama [HM54]; Yoshioka [Yos57] showed formally that inelastic scattering in electron diffraction can be taken into account by adding a complex potential $W(\mathbf{r})$ to the electrostatic lattice potential. The imaginary part of this potential describes the inelastic scattering processes in the form of an attenuation of the wave amplitude (*absorption*); the real part of $W(\mathbf{r})$ is associated with virtual inelastic scattering and has been shown by P. Rez [Rez78] to be small for high-energy electrons. We will hence assume that the so-called *optical potential* $W(\mathbf{r})$ is purely imaginary and of the form $\mathrm{i}W(\mathbf{r})$.

Next, we show that such an imaginary term can describe absorption: following [GBA66] we have

$$V(\mathbf{r}) \to V_c(\mathbf{r}) = V(\mathbf{r}) + \mathrm{i}W(\mathbf{r}). \tag{2.95}$$

Now consider the time-dependent Schrödinger equation:

$$\mathrm{i}\hbar\frac{\partial\Psi}{\partial t} = -\frac{\hbar^2}{2m}\Delta\Psi - eV_c\Psi, \tag{2.96}$$

where V_c is the complex lattice potential defined in equation (2.95). The complex conjugate of this equation is given by

$$-\mathrm{i}\hbar\frac{\partial\Psi^*}{\partial t} = -\frac{\hbar^2}{2m}\Delta\Psi^* - eV_c^*\Psi^*. \tag{2.97}$$

Multiplying the first equation by Ψ^*, the second by Ψ, and subtracting the second from the first results in

$$\mathrm{i}\hbar\frac{\partial\Psi\Psi^*}{\partial t} = -\frac{\hbar^2}{2m}(\Psi^*\Delta\Psi - \Psi\Delta\Psi^*) - e\big(V_c - V_c^*\big)\Psi\Psi^*. \tag{2.98}$$

This in turn can be rewritten as

$$\frac{\partial\Psi\Psi^*}{\partial t} - \frac{\mathrm{i}\hbar}{2m}\nabla\cdot(\Psi^*\nabla\Psi - \Psi\nabla\Psi^*) = \frac{\mathrm{i}e}{\hbar}\big(V_c - V_c^*\big)\Psi\Psi^*. \tag{2.99}$$

Since $\Psi\Psi^* = \rho$ equals the probability density and the argument of the divergence operator is the *electron flux* $\mathbf{J}$,[†] we can rewrite this equation one more time (using equation 2.95):

$$\frac{\partial\rho}{\partial t} + \nabla\cdot\mathbf{J} = -\frac{2e}{\hbar}W\rho. \tag{2.100}$$

We thus find that, if $W > 0$, then the total number of electrons will decrease with time, i.e. *absorption* will occur. If $W = 0$, then equation (2.100) reduces to a standard continuity equation for the probability density. Note that this description does not contain any details concerning the exact nature of the inelastic scattering process; all that is required to describe absorption is the positive imaginary part of the lattice potential.

For the remainder of this book we will use the phenomenological description of absorption and inelastic scattering contributions.

[†] See, for instance, [LL74].

2.6.6 Atomic vibrations and the electrostatic lattice potential

The derivation of the atomic scattering factor $f^e(s)$ assumed that the atom was at rest at the origin of the reference frame. In reality, atoms vibrate around their equilibrium lattice positions, and the amplitude of this vibration depends on the absolute temperature. The theory of lattice vibrations (or *phonons*) is well developed, and we refer the reader to [Bir84] for a detailed discussion of lattice dynamics.

Lattice vibration frequencies for most solids are in the range 10^{10}–10^{12} Hz. A 200 keV electron travels at a velocity of $v = 0.695c$, which corresponds to a distance of about 208 μm in a 10^{-12} s time interval. This is very large compared to the atomic dimensions, which means that subsequent electrons will "see" this atom located at different positions, but essentially at rest for each individual electron. Each atom, therefore, appears to be spread out over a larger volume and this affects the electron scattering factor $f^e(s)$. The correction factor is commonly known as the *Debye–Waller factor*, and we will next derive an explicit expression based on the definition of the atomic scattering factor.

Consider an isolated atom at rest at the origin, with atomic potential $V^a(\mathbf{r})$. The atomic scattering factor $f^e(s)$ expresses the probability amplitude that an incident electron with wave vector $\mathbf{k}$ will be scattered by the atom in the direction $\mathbf{k}'$; this is expressed by the *bra-ket*:

$$f^e(\Delta\mathbf{k}) = \left\langle \mathrm{e}^{2\pi \mathrm{i}\mathbf{k}'\cdot\mathbf{r}} \middle| V^a(\mathbf{r}) \middle| \mathrm{e}^{2\pi \mathrm{i}\mathbf{k}\cdot\mathbf{r}} \right\rangle.$$

Assume next that the atom vibrates around this equilibrium position with a vibration amplitude $\mathbf{u}(t)$. For simplicity and without loss of generality we will assume that the vibration occurs along the x-direction. The instantaneous atomic potential at a time t is given by

$$V^a(\mathbf{r}, t) = V^a(\mathbf{r} + \mathbf{u}(t)) = V^a(\mathbf{r} + u(t)\mathbf{e}_x).$$

For small vibration amplitudes the instantaneous potential can be expanded in a Taylor series in u, as follows (dropping the argument $\mathbf{r}$):

$$V^a(t) = V^a + \frac{u(t)}{1!}\left.\frac{\mathrm{d}V^a}{\mathrm{d}x}\right|_{u=0} + \frac{u(t)^2}{2!}\left.\frac{\mathrm{d}^2V^a}{\mathrm{d}x^2}\right|_{u=0} + \cdots.$$

If we average this expression over time, then all odd powers of $u(t)$ will vanish, since positive and negative excursions are equally likely over a sufficiently long time interval. Denoting the time average by $\langle\,\rangle$, we have

$$\left\langle V^a(t)\right\rangle = V^a + \frac{\langle u^2\rangle}{2!}\left.\frac{\mathrm{d}^2V^a}{\mathrm{d}x^2}\right|_{u=0} + \cdots.$$

Next, we can calculate how the atomic scattering factor is modified by this average potential:

$$\begin{aligned}
\langle f^e(\Delta\mathbf{k})\rangle &= \langle \mathrm{e}^{2\pi\mathrm{i}\mathbf{k}'\cdot\mathbf{r}} | \langle V^a(t)\rangle | \mathrm{e}^{2\pi\mathrm{i}\mathbf{k}\cdot\mathbf{r}}\rangle; \\
&= \iiint \mathrm{e}^{-2\pi\mathrm{i}\Delta\mathbf{k}\cdot\mathbf{r}} \left(V^a + \frac{\langle u^2\rangle}{2!} \left.\frac{\mathrm{d}^2 V^a}{\mathrm{d}x^2}\right|_{u=0} + \cdots \right) \mathbf{dr}; \\
&= \left[1 + (2\pi\mathrm{i}\Delta\mathbf{k})^2 \frac{\langle u^2\rangle}{2!} + \cdots \right] \iiint \mathrm{e}^{-2\pi\mathrm{i}\Delta\mathbf{k}\cdot\mathbf{r}} V^a \,\mathbf{dr}; \\
&\approx \mathrm{e}^{-2\pi^2\langle u^2\rangle(\Delta\mathbf{k})^2} f^e(\Delta\mathbf{k}),
\end{aligned}$$

where we have used the relation $\mathcal{F}_u[\mathrm{d}^n f(x)/\mathrm{d}x^n] = (2\pi\mathrm{i}u)^n f(u)$ (e.g. see [Cow81, p. 32]). Introducing $|\Delta\mathbf{k}| = 2s$ as before we arrive at the time-averaged atomic scattering factor:

$$\langle f^e(s)\rangle = f^e(s)\mathrm{e}^{-8\pi^2\langle u^2\rangle s^2} = f^e(s)\mathrm{e}^{-Bs^2}; \tag{2.101}$$

the symbol B is proportional to the atomic mean-square vibration amplitude $\langle u^2\rangle$, which is expressed in units of nm^2 and depends on absolute temperature. For realistic simulations, the Debye–Waller factor must be taken into account. We have assumed in this simplified derivation that the vibrations are isotropic. For anisotropic vibrations B becomes a second-rank tensor, which can be represented by a "vibrational ellipsoid". The mean-square vibration amplitude then depends on the particular reciprocal lattice vector $\mathbf{g}$ selected for the computation of $f^e(s)$.

The Debye–Waller factor B is a function of temperature $B = B(T)$; exact expressions for B can be derived for a given vibration model, and for details we refer to the *International Tables for Crystallography*, volume C [Shm96], and references therein. Note that B does not vanish at 0 K because of the zero-point motion of the atoms.

For elemental solids, the Debye–Waller factor can be calculated from the phonon density of states $g(\omega)$, using the following relation:

$$B(T) = \frac{4\pi^2\hbar}{m} \int_0^{\omega_m} \coth\left(\frac{\hbar\omega}{2k_B T}\right) \left[\frac{g(\omega)}{\omega}\right] \mathrm{d}\omega,$$

where m is the atomic mass, T is the temperature (in kelvin), k_B is the Boltzmann constant, and ω_m is the maximum phonon frequency. The phonon density of states can be measured experimentally using neutron diffraction techniques. This relation has been used by Gao and Peng [GP99] to parameterize the Debye–Waller factors for 68 elemental crystals and a number of compounds in the temperature range from $T = 0$ to 1000 K. The dashed lines in Fig. 2.9 represent the atomic scattering factor for copper, at temperatures of $T = 300$ and 1000 K; the scattering factor is significantly reduced for large scattering angles.

2.6.7 Numerical computation of the Fourier coefficients of the lattice potential

The quantity $V_{\mathbf{g}}$ (2.83) is known as the *electron structure factor* [SZ92]. Sometimes it is convenient to rewrite the summation over all N atoms in the unit cell as a double summation over all N_a atoms in the asymmetric unit and over all symmetry operators of the space group. If we use the Seitz symbol to denote a symmetry element (see equation 1.38 on page 40), then the electron structure factor becomes

$$V_{\mathbf{g}} = \frac{1}{\Omega} \sum_{j=1}^{N_a} f_j^e(s) \sum_{(\mathbf{D}|\mathbf{t})} \mathrm{e}^{-2\pi \mathrm{i}\mathbf{g}\cdot(\mathbf{D}|\mathbf{t})[\mathbf{r}_j]}.$$

The second summation is then essentially a sum over the orbit of each of the atoms in the asymmetric unit. This expression is particularly useful for implementation in a computer program, since one would typically ask the user to enter the space group and the coordinates of the atoms in the asymmetric unit. One must take care to remove double positions generated by the symmetry operators before computing the summation.

We can now use equation (2.83) combined with the parameterization of the electron scattering factors discussed in Section 2.6.1 to compute numerically the Fourier coefficients of the electrostatic lattice potential for an arbitrary crystal structure. This computation is implemented in the subroutine CalcUcg in the module diffraction.f90 and described in PC-9; the routine returns the modulus and phase of the complex Fourier coefficient (and optionally also the Fourier coefficient of the absorption potential $W_{\mathbf{g}}$):

$$V_{\mathbf{g}} = |V_{\mathbf{g}}|\mathrm{e}^{\mathrm{i}\theta_{\mathbf{g}}}; \qquad W_{\mathbf{g}} = |W_{\mathbf{g}}|\mathrm{e}^{\mathrm{i}\theta'_{\mathbf{g}}},$$

from which the real and imaginary part can be computed through the use of Euler's formula $\mathrm{e}^{\mathrm{i}x} = \cos x + \mathrm{i}\sin x$. The routine uses either the Doyle–Turner/Smith–Burge or the Weickenmeier–Kohl atomic scattering and absorptive form factors. A more complete discussion of the absorptive form factor can be found in Section 6.5.1 on page 371.

The source code for this book incorporates the standard isotropic Debye–Waller factor in all scattering factor computations. The Debye–Waller factor must be entered in nm^2 for each atom in the asymmetric unit. The Fourier coefficients $V_{\mathbf{g}}$

Pseudo Code PC-9 Compute the Fourier coefficient $V_\mathbf{g}$ (and $W_\mathbf{g}$).

Input: crystal data file [*.xtal], Miller indices h, k, l
Output: $|V_\mathbf{g}|, \theta_\mathbf{g}$

```
load crystal data                                              {PC-2, page 63}
generate all atoms in the unit cell                            {CalcPositions}
compute d_hkl                                                  {CalcLength}
convert to scattering parameter s = 1/(2 d_hkl)
for each atom type do
    compute f^e(s) (including Debye–Waller factor)             {2.75 or 2.76}
    for each atom in orbit do
        compute χ_j = 2π g · r_j
        compute cos χ_j and sin χ_j
    end for
    add all cos χ_j and sin χ_j together for the orbit
    multiply sum with f_j^e(s)
    add to total sum for V_g                                   {2.102}
end for
convert V_g to modulus and phase
```

of the electrostatic lattice potential are then given by (using the Doyle–Turner parameterization):

$$V_\mathbf{g} = \frac{0.047\,878\,01}{\Omega} \sum_{j=1}^{N_a} \left(\mathrm{e}^{-B_j s^2} \sum_{i=1}^{N} a_i^j \mathrm{e}^{-b_i^j s^2} \right) \sum_{(\mathbf{D}|\mathbf{t})} \mathrm{e}^{-2\pi \mathrm{i}\mathbf{g}\cdot(\mathbf{D}|\mathbf{t})[\mathbf{r}_j]}. \tag{2.102}$$

A similar expression is obtained for the Weickenmeier–Kohl version of the elastic scattering factor fitting function; it is left to the reader to write down the explicit expression for this case. When Ω is given in nm^3 and s in nm^{-1}, then $V_\mathbf{g}$ is expressed in volts. The factor between large parentheses must be computed only once for each atom in the asymmetric unit, and for every value of the scattering parameter s; the orbit $(\mathbf{D}|\mathbf{t})[\mathbf{r}_j]$ of each atom in the asymmetric unit can be precomputed before the potential Fourier coefficients are determined. This has been implemented in the routine CalcPositions in the module symmetry.f90.

As an example of the use of the CalcUcg routine we will next discuss the computation of the direct space electrostatic lattice potential for an arbitrary crystal structure. The direct space potential is given by the 3D inverse Fourier transform:

$$V(\mathbf{r}) = \sum_\mathbf{g} V_\mathbf{g} \mathrm{e}^{2\pi \mathrm{i}\mathbf{g}\cdot\mathbf{r}}.$$

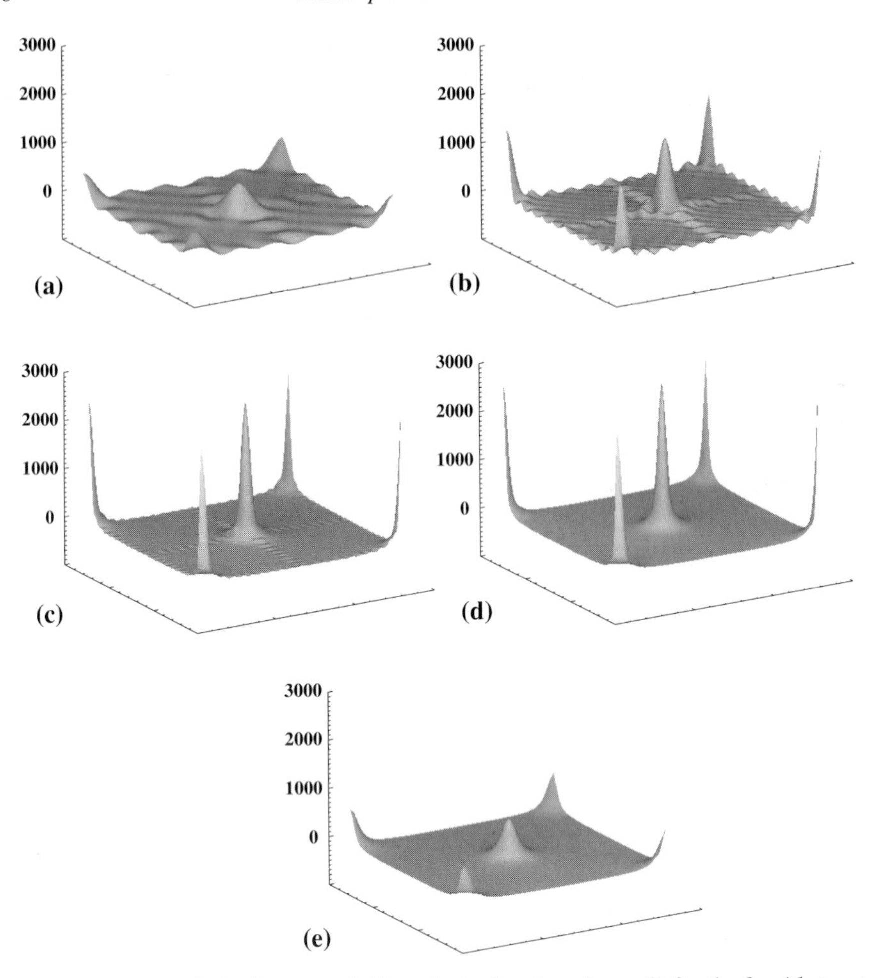

Fig. 2.12. Electrostatic lattice potential in volts in the plane $(x, y, 0)$ for the fcc Al structure. The number of contributing Fourier coefficients increases from (a) to (d), according to the entries in Table 2.4. Part (e) includes the room-temperature Debye–Waller factor for aluminum $B(300) = 0.008\,33\ \text{nm}^2$.

The potential at any given point $\mathbf{r}$ is thus determined by *all* reciprocal lattice points $\mathbf{g}$. Since the potential is strongly peaked at the atom positions we expect to need a large number of Fourier coefficients to represent the potential accurately.

Figure 2.12 represents the electrostatic potential $V(x, y, 0)$ for a single unit cell of the face-centered structure of aluminum. The atoms are located at the corners and at the center of the unit cell face. The computation used the Doyle–Turner parameterization of the elastic scattering factors. The potential is represented as a shaded surface and was calculated on an array of 128×128 points. The "pixel size" for this computation is thus $404.9/128 = 3.15$ pm. The only difference between the four shaded surfaces in Fig. 2.12 is the number of Fourier coefficients entering

Table 2.4. *Total number of families contributing to the electrostatic lattice potential of aluminum as a function of the radius $|\mathbf{g}_{max}|$ (in nm^{-1}). The numbers in this table correspond to the computed potentials in Fig. 2.12.*

	$\vert\mathbf{g}_{max}\vert = 10$	$\vert\mathbf{g}_{max}\vert = 20$	$\vert\mathbf{g}_{max}\vert = 40$	$\vert\mathbf{g}_{max}\vert = 80$
No. of independent families	7	27	145	559
No. of allowed reflections	58	510	4140	19 236

the summation. The details of the computation are tabulated in Table 2.4: the total number of Fourier coefficients is determined by the truncation radius $|\mathbf{g}_{max}|$. The larger the number of Fourier coefficients, the more accurate the electrostatic lattice potential. Since $V(\mathbf{r})$ was defined to be a positive potential (see page 94), the occurrence of negative values is often an indication that an insufficient number of Fourier coefficients have been used. Oscillations in the potential, in particular in regions where no atoms are present, are also caused by premature truncation of the Fourier summation.

In the first chapter we defined the concept of a *family* of lattice planes. All lattice planes (hkl) related to each other by the point group symmetry operations form a family $\{hkl\}$. It would be tempting, but incorrect, to assume that all members of a family have the same structure factor F_{hkl} (or V_{hkl}). The lattice potential has the symmetry of the complete space group, so that the symmetry of the potential Fourier coefficients will depend on all the space group symmetry elements and not just on the point group symmetry. There is no a priori reason why the final summation in equation (2.102),

$$\sum_{(\mathbf{D}|\mathbf{t})} \mathrm{e}^{-2\pi i \mathbf{g}\cdot(\mathbf{D}|\mathbf{t})[\mathbf{r}_j]},$$

should be the same for all members of a family $\{hkl\}$. The magnitude $|V_{\mathbf{g}}|$ is the same for all members of a given family, but the phase factor $\mathrm{e}^{i\theta_{\mathbf{g}}}$ is not necessarily the same. For centrosymmetric space groups, the phase factor amounts to a simple plus or minus sign; for non-centrosymmetric space groups, the relations between the phases $\theta_{\mathbf{g}}$ in a given family can be more complicated. Ultimately the atom positions in the asymmetric unit cell determine what the phase factor will be.

The computation of $V(\mathbf{r})$ for an arbitrary rectangular section through the unit cell of an arbitrary crystal structure has been implemented in the Fortran program vr.f90. The algorithm is explained in PC-10. To determine the members of each family, it is convenient to use a 3D array of Booleans to make sure no reciprocal

Pseudo Code PC-10 Compute the electrostatic lattice potential $V(\mathbf{r})$.

```
Input: crystal data file [*.xtal]
Output: Nx × Ny array of potential values
  load crystal data                                              {PC-2}
  generate all atoms in the unit cell                            {CalcPositions}
  ask user for |g_max|
  for each triplet for which |g| < |g_max| do
    if not member of previously considered family then
      compute family {hkl}                                       {CalcFamily}
      if reflection allowed by Bravais lattice centering then
        compute and store V_hkl for each family member           {PC-9}
      end if
      mark entire family in Boolean array
    end if
  end for
  ask for section plane (three corner points)
  ask for resolution Nx × Ny
  for every allowed family {hkl} do
    for every member (hkl) of the family do
      for each location in section plane do
        compute χ = 2πi g · r
        compute cos χ + i sin χ
        multiply by V_hkl                                        {complex number!}
        add to total potential                                   {only real part needed}
      end for
    end for
  end for
  store pixel coordinates and potential values
```

lattice points are either missed or double-counted. The program generates an ASCII output file with pixel coordinates and potential values (in volts), which can then be processed by an appropriate graphics program.† The implementation of the program vr.f90 does not make use of fast Fourier transforms; while one would normally use 3D FFTs to speed up the transformation from reciprocal to direct

† Most rendered drawings and contour plots in this book were generated with the *Interactive Data Language* [Res98].

space, the implementation in vr.f90 illustrates in more direct terms how the direct space potential $V(\mathbf{r})$ is determined by *all* Fourier coefficients $V_{\mathbf{g}}$. A second program, vrfft.f90, is provided which makes use of the FFTW library to compute the 3D Fourier transform.

Figure 2.12(e) represents the same section of the electrostatic lattice potential of Al, but includes the Debye–Waller factor $B(300) = 0.008\,33$ nm^2 [GP99]. Since vibrations cause the atom to appear larger, i.e. to occupy more space with a resulting smaller electron density than the atom at rest, the elastic scattering factor for large scattering parameters s is affected more strongly than the small s region. The peaks in Fig. 2.12(e) are hence broadened and have a lower peak value.

2.7 Recommended additional reading

In addition to the references cited in this chapter, the following books may be of interest to the reader. Basic texts on quantum mechanics:

- *Quantum Mechanics: a Modern Introduction*, A. Das and A.C. Melissinos, Gordon and Breach Science Publishers (New York, 1986).
- *Introduction to Quantum Mechanics*, B.H. Bransden and C.J. Joachain, Longman Scientific & Technical (New York, 1989).
- *An Introduction to Theory and Applications of Quantum Mechanics*, A. Yariv, Wiley (New York, 1982).

Introductory books on the special theory of relativity:

- *Introduction to Special Relativity*, R. Resnick, Wiley (New York, 1968).
- *Introduction to Special Relativity*, W. Rindler, Oxford University Press (New York, 1982).
- *Space and Time in Special Relativity*, N.D. Mermin, McGraw-Hill (New York, 1968).

A basic introduction to diffraction can be found in:

- *The Basics of Crystallography and Diffraction*, C. Hammond, Oxford University Press (New York, 1997).

Fourier transforms, convolutions, and their numerical implementation are discussed in:

- *The Fourier Transform and its Applications*, R.N. Bracewell, 3rd edition, McGraw Hill (Boston, 2000).
- *Inside the FFT Black Box: Serial and Parallel Fast Fourier Transform Algorithms*, E. Chu and A. George, CRC Press (Boca Raton, 2000).
- *Algorithms for Discrete Fourier Transform and Convolution*, R. Tolimieri, M. An, and C. Lu, Springer-Verlag (New York, 1989).

Additional information on inelastic electron scattering may be found in:

- *Fundamentals of Inelastic Electron Scattering*, P. Schattschneider, Springer-Verlag (Wien, 1986).
- *Elastic and Inelastic Scattering in Electron Diffraction and Imaging*, Z.L. Wang, Plenum Press (New York, 1995).

Exercises

2.1 Show that the normalization constant C for the momentum plane wave eigenfunctions (page 84) is given by $h^{-3/2}$.

2.2 Derive a proof for the multiplication theorem (2.53) and the convolution theorem (2.54) on page 106.

2.3 Derive an expression, similar to equation (2.102), for the Weickenmeier–Kohl parameterization of the elastic scattering factor.

2.4 Consider the drawing you have made for Exercise 1.2; determine graphically the direction of an incident beam (with wave vector in the plane of the drawing) for which the (201) planes and the $(10\bar{1})$ planes will simultaneously be in Bragg orientation. What is the angle between the two diffracted beams (both the computed angle and the angle measured from the drawing)? What is the wavelength of this incident beam?

2.5 Use the Fortran libraries on the website to write a program that creates a list of scattering angles 2θ for an arbitrary crystal structure and a given acceleration voltage. Each family of reflections $\{hkl\}$ should only occur once in this list.

2.6 Show that the DFT of a normalized Gaussian function

$$\frac{a}{\sqrt{\pi}}\mathrm{e}^{-a^2x^2}$$

is again a Gaussian function. What is the relation between the widths of this Fourier transform pair?

2.7 Use the Fortran routines on the website, in particular the fftw library, to write a program to compute the convolution product of two one-dimensional functions of arbitrary length.

2.8 Use the Fortran libraries on the website to write a program that lists the real and imaginary Fourier coefficients of the electrostatic lattice potential for an arbitrary crystal structure, using the Weickenmeier–Kohl atomic scattering factor parameterization.

2.9 Consider a row of reciprocal lattice points $n\mathbf{g}$ (Fig. 2.13). If the angle between the vectors $\mathbf{g}$ and $\mathbf{k}_0 = \frac{1}{\lambda}\mathbf{e}_z$ is given by ω, show that, to a good approximation, the excitation error of the reflection $n\mathbf{g}$ is given by

$$s_{n\mathbf{g}} \approx -n\frac{|\mathbf{g}|}{2}\left(\frac{2\cos\omega + n\lambda|\mathbf{g}|}{1 + n\lambda|\mathbf{g}|\cos\omega}\right).$$

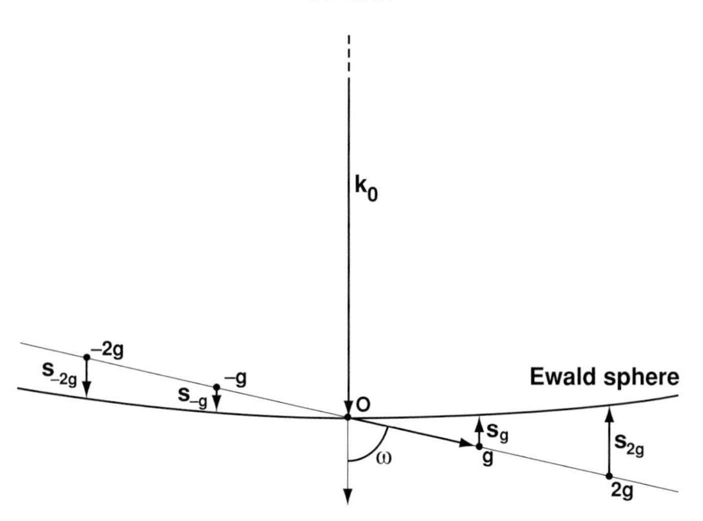

Fig. 2.13. Excitation errors for a row of reflections in a random orientation with respect to the Ewald sphere.

In particular, for $\omega = \pi/2$ show that the following parabolic approximation is valid:

$$s_{n\mathbf{g}} \approx -\frac{n^2}{2}\lambda|\mathbf{g}|^2 \quad \left(\approx -\frac{1}{\lambda}\left(1 - \sqrt{1 - n^2\lambda^2|\mathbf{g}|^2}\right)\right), \tag{E2.1}$$

where the second expression is the exact solution which will be derived in Section 3.9.5.

2.10 Determine the conditions for *systematic absences* for all different Bravais lattice centering types.

2.11 Building on the previous exercise, write the structure factor for all seven non-primitive Bravais lattices as the product of a structure factor due to lattice centering and a term containing the atoms in the asymmetric unit. As an example, here is the expression for the face-centered cubic lattice:

$$F_{hkl}^{fcc} = \left[1 + (-1)^{h+k} + (-1)^{h+l} + (-1)^{k+l}\right] \sum_{j=1}^{N/4} f_j \mathrm{e}^{2\pi \mathrm{i}\mathbf{g}\cdot\mathbf{r}_j}. \tag{E2.2}$$

Under which conditions can the second term become equal to zero?

3

The transmission electron microscope

3.1 Introduction

In this chapter, we introduce the components of the transmission electron microscope. After a brief review of the history of the electron and the electron microscope, we will consider a cross-section of a real microscope (JEOL 120CX) and identify the various stages and components inside the column. We will discuss peripheral support systems (vacuum, compressed air, water cooling, etc.) without which the microscope could not possibly function. Then we focus our attention on the general properties of round magnetic lenses and discuss basic charged particle dynamics in magnetic fields. We will illustrate the behavior of electrons in a magnetic field by means of simplified paraxial trajectory simulations, and introduce the concept of cardinal elements of a lens. Then we will consider deviations from the paraxial behavior and introduce various lens aberrations of importance in electron microscopy. We conclude the section on magnetic lenses with a brief overview of multipole lenses, used as deflectors and correctors.

The electron gun is next, and we will introduce the basic gun types and their operational parameters, along with a discussion of beam properties (current, brightness, coherence, etc.). In the remainder of the chapter, we break the microscope column down into four regions: the illumination system, the goniometer stage and objective lens, the imaging stage, and the detector stage. For each stage, we will describe the common components, why they are present, and how the microscope operator can affect their behavior. In particular, we will relate the geometry of the goniometer stage to reciprocal space and Bragg's Law. We will conclude this chapter with a brief review of electron detection systems.

3.2 A brief historical overview

When Max Knoll and Ernst Ruska built the first transmission electron microscope† in Berlin in 1931–1934 [KR32a, KR32b, Rus34a, Rus34b], it took only a few years before the first commercial machines became available (Metropolitan Vickers, 1936; Siemens, 1939; RCA, 1941; Hitachi, 1941; Philips, 1949; JEOL, 1949; etc.) [Hal53, Fuj86, Haw85]. Ruska's 1933 microscope had three lenses, a magnification of 12 000×, and a resolution of 50 nm, significantly better than the best optical microscope of that time.

The basic components of the TEM have not changed substantially since the late 1940s. The electronic circuitry, however, has changed significantly from the pre-transistor electronics of the first microscopes to the sophisticated modern machines that permit remote operation of nearly all microscope functions. There has been a steady improvement of the quality of the main lenses in the microscope, the stability of high voltage and lens currents, and an increased theoretical understanding of lens aberrations and how they can be corrected. This evolution continues to date, and in this chapter we will on occasion cite current research in various areas of electron optics.

It is fascinating to consider the history of the discovery of the electron, which preceded the electron microscope by about 35 years. Indirect evidence for the existence of a fundamental charged particle was only slowly obtained, starting with the extensive electrolysis experiments by Faraday in the 1830s [Far39]. The name *electron* was first used by Stoney in 1891, when he named the fundamental unit of charge [Sto91]. It was only after the actual discovery of the electron itself in 1897‡ that the name was also used to denote the particle. For an overview of selected important events around the turn of the nineteenth century leading up to the construction of the first electron microscope we refer to Table 3.1.

The currently accepted value for the unit of electric charge is $e = 1.602\,177 \times 10^{-19}$ C, and the rest mass of the electron is $m_0 = 9.109\,389 \times 10^{-31}$ kg [CT95]. Additional physical constants associated with the electron may be found in Table A2.1 on page 665 (Appendix A2).

For a detailed account of the early days of the electron we refer to the wonderful book by A. Pais: *Inward Bound: of Matter and Forces in the Physical World* [Pai86], which provides a wealth of historical information and anecdotes about the main players in the field. The first chapter of the book *Electron: a centenary volume* discusses the role of J.J. Thomson in the discovery of the electron [Spr97]. For a detailed account of the history of electron microscopy in various countries worldwide

† An earlier two-lens version of an electron microscope with a magnification of 17× was built in 1931, but its resolution was worse than that of an optical microscope.

‡ One could debate about the actual date of the discovery. Some authors prefer the year 1899 as the year of discovery, since J.J. Thomson determined e in that year.

Table 3.1. *Partial chronology of the history of the electron and the electron microscope, and other events which have had a significant impact on the microscopy field, with references to fundamental papers.*

Year	Event
1871	Cromwell Fleetwood Varley suggests that the carriers of electricity are corpuscular, with a negative charge [Var71]
1876	Eugene Goldstein studies discharges in gases, and coins the name *cathode rays*, starting a long debate about their nature [Gol76]
1891	George Johnstone Stoney coins the word *electron* for the unit of charge [Sto91]
1897	Emil Wiechert is the first to obtain reasonable bounds on the magnitude of e/m (January) [Wie97]
1897	Walter Kaufmann and J.J. Thomson independently measure e/m (April) [Kau97, Tho97]
1899	J.J. Thomson determines the value of e, which makes him the discoverer of the electron [Tho99]
1905	Albert Einstein publishes the *special theory of relativity* and establishes the equivalence of mass and energy [Ein05]
1913	Niels Bohr introduces a model for the structure of the hydrogen atom [Boh13a, Boh13b, Boh13c]
1923	Maurice de Broglie establishes the principle of wave–particle duality [de 23]
1925	Wolfgang Pauli discovers the exclusion principle; Werner Heisenberg develops matrix quantum mechanics [Pau25, Hei25]
1926	Erwin Schrödinger develops quantum mechanics based on differential equations; Hans Busch develops the theory of magnetic lenses [Sch26, Bus26]
1927	Clinton Davisson and Lester Germer discover electron diffraction [DG27]
1928	Paul Dirac formulates the relativistic theory of the electron; Hans Bethe develops the first dynamical theory of electron diffraction [Dir28a, Bet28]
1931–4	Ernst Ruska and Max Knoll build the first electron microscope [KR32a, KR32b, Rus34a, Rus34b]

we refer the interested reader to the book *The Growth of Electron Microscopy*, edited by T. Mulvey [Mul96].

3.3 Overview of the instrument

Since the first electron microscope was constructed by Ruska and Knoll, TEMs have evolved into complex computer-controlled machines, and many of the underlying physical principles are hidden from the user in modern machines. Students who learn to operate the latest (digital) model as their first machine may find it hard to operate some of the earlier models, especially the ones where lens and gun alignment are mostly mechanical. An inexperienced user should be aware of the fact that different microscope models may have a different layout of the most

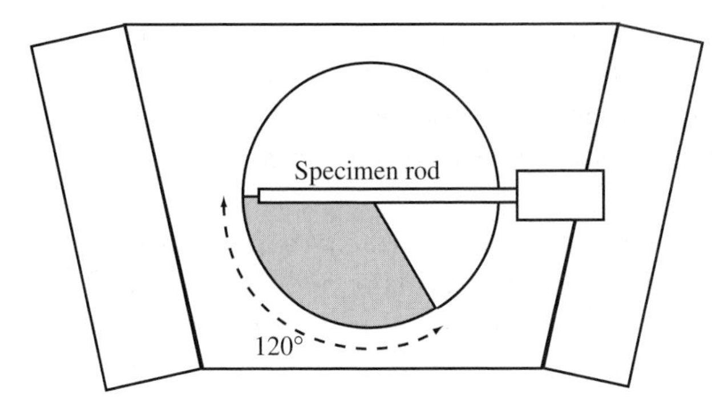

Fig. 3.1. Schematic top view of the section removed from the JEOL 120CX microscope column; the specimen holder location is also indicated.

commonly used controls and that some "mental adjustments" may be needed when moving from one instrument to another. The contents of this and the following chapter should provide some common ground for the operation of all microscope models.

An *optical* microscope, used in transmission mode, typically consists of five parts: the light source; the condensor lens, which focuses the light beam onto the sample; a transparent sample, usually a thin section of rock or tissue; the objective lens, sometimes in contact with the sample through a contact medium (oil); and the magnifying lens(es), often combined with the *ocular* or eyepiece. The number of lenses may vary but most optical microscopes have all five components. A TEM can be divided into similar sections. The five sections of a TEM are (from top to bottom): the electron gun, the illumination stage, the objective lens, with an electron transparent sample immersed into its magnetic field, the magnification and projection system, often with three or more lenses, and the detector (a viewing screen, photographic camera, charge-coupled device (CCD) camera, etc.

A cross-section of a JEOL 120CX microscope column is shown on the right in Fig. 3.2.[†] A corresponding schematic drawing, taken from the microscope manual, is shown on the left, along with an outline of the five main components.

[†] Several of the photographs in this chapter were obtained from a cross-section of a real TEM, a JEOL model 120CX, built in 1978 (serial number EM 156049-17). Throughout the second half of 1999 the author, with the help of Mr N.T. Nuhfer and Mr G. Biddle, removed a 120° lengthwise wedge from the column, using a high-capacity band saw. One edge of the wedge cut is along the leftmost (when facing the microscope) side of the goniometer stage, the other cut is at 60° from the opposite side, as illustrated in the schematic top view in Fig. 3.1. The cut extends from the top of the electron gun down to the top of the viewing chamber, over a length of approximately 1060 mm; the viewing chamber itself was left intact, to provide sufficient mechanical support to the column. The surfaces of the cuts were sanded to a smooth finish and covered with a protective layer of polyurethane, to minimize corrosion. After completion of all the cuts, the microscope column was reassembled at the bottom of the stairwell on the ground floor of Roberts Engineering Hall, Carnegie Mellon University campus. Additional photographs of the cross-section are available from the website.

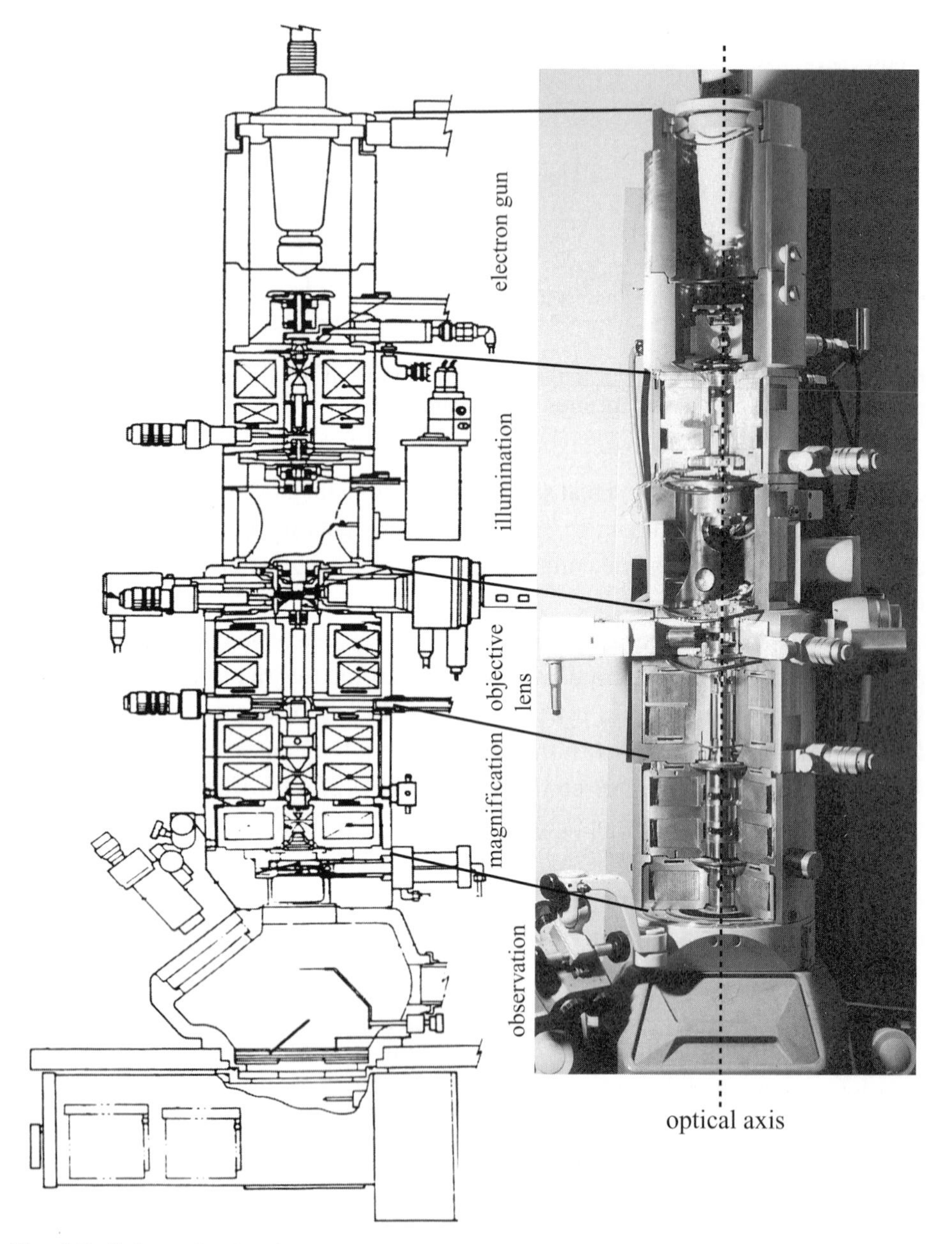

Fig. 3.2. Schematic drawing and photograph of a real cross-section of a JEOL 120CX transmission electron microscope, with the main stages outlined. (Drawing reproduced with permission.)

In addition to the five main stages listed above, a TEM contains a number of subsystems that are vital for proper operation, but are of secondary importance to the operator during routine use of the microscope.†

- *The high voltage system.* This system is usually located in a separate unit and is connected to the electron gun by a thick cable. For acceleration voltages of 400 kV and higher, this type of cable connection is no longer possible and the accelerator is placed just behind the microscope column or on top of the column.
- *The vacuum system.* Electrons need a good vacuum (in the range of 10^{-7}–10^{-10} Pa (or 10^{-5}–10^{-8} torr, where 1 torr = 133.322 Pa) to travel the typically 1–2 m length of the column, hence every TEM is equipped with a *high vacuum* system. The system generally consist of one or more roughing pumps, possibly with a vacuum buffer tank, one or more oil diffusion pumps and/or ion getter pumps and/or turbomolecular pumps, and a number of independent vacuum circuits with manual or automatic (pneumatic or electronic) valves. Some microscopes are modified for *ultra-high vacuum* conditions, which means that the partial pressures of *all* gases in the column are better than 10^{-8} torr. In older microscopes, such as the Siemens Elmiskope 102, the operator had to be familiar with the pumping diagrams and cycles and often manually open or close valves. Nowadays, this task is fully automated and interlocks on the high voltage system and the filament emission system will prevent the operator from using the machine if the vacuum level is inadequate.

  The operator should always take note of the proper vacuum level at which the microscope must be operated. This should be part of the prework checklist. The vacuum is usually best in the electron gun and slightly worse at the specimen stage. Some microscopes allow observations at near ambient pressures in a variety of gas mixtures through the use of an *environmental specimen cell*; in those cases the sample region is separated from the rest of the microscope column by *differential pumping apertures*.

  In older microscopes, such as the JEOL 120CX, the vacuum region of the microscope is separated from the rest by about two-dozen O-rings. In modern instruments the electrons travel through so-called *liner tubes*, long cylindrical sections with a small inner diameter (a few millimeters), which extend across several lenses and reduce the total number of vacuum seals needed.
- *The cooling system.* Large electric currents can generate a lot of heat (Ohm's law), hence the lens coils, which can carry several amperes of current, must be water-cooled. This system is fully automatic and the operator should never have to intervene at any point. The cooling system is interlocked with other systems, so that the microscope cannot be used if insufficient cooling water is available.
- *Radiation shields.* All electrons emitted by the filament must eventually find their way back into the electrical circuit, and must get rid of their kinetic energy. All surfaces

† A TEM operator should be aware of the meaning of the warning signs and indicator lights should any of the subsystems described in this section malfunction or cause problems. We refer to the microscope manual for a detailed description of all subsystems.

inside the column that may be exposed to electrons are therefore made of a conductive material. A more serious problem is posed by the conversion of kinetic energy into x-ray photons wherever the electrons enter solid matter. All areas of the microscope where x-rays may be generated are therefore surrounded by a double shield: the first layer of the shield is a low atomic number material, typically aluminum, which converts most of the kinetic energy into low-energy x-ray photons. The second layer is a high atomic number material, typically lead, which efficiently absorbs the low-energy x-ray photons. The viewing window is made of a special lead containing glass and can be rather expensive. The radiation shields contribute significantly to the weight of the column, and are thicker for higher-energy microscopes.

- *The photographic camera system.* A photographic camera, consisting of two removable boxes, is mounted below or behind the viewing screen: one box contains unexposed plates and the other one receives the exposed plates (depending on the microscope type the number of available plates can vary anywhere from 20 to 50). The exposure time and film sensitivity can be set from the microscope console, although one should not change the latter unless the type of film is changed. All microscopes indicate the magnification (or camera length for diffraction) and the plate serial number near the edge of the negative; some models also allow for the inclusion of 30 or so characters of text, to identify the operator and sample.

This concludes the enumeration of the more common microscope subsystems. Possible additional subsystems include energy dispersive analyzers for x-rays or electron energy loss spectroscopy (EDS and EELS), cameras (TV rate, CCD, or imaging plate), scanning attachments, various detectors before and after the sample, imaging energy filters, etc. With the exception of electron detectors, we will not address these additional subsystems in this book.

3.4 Basic electron optics: round magnetic lenses

3.4.1 Cross-section of a round magnetic lens

In a light optical microscope, glass lenses are used to first focus a light beam onto the sample, and then magnify the image to the desired magnification. Each lens is a carefully shaped piece of glass. The radius of curvature of each surface along with the refractive index of the glass determine the optical properties and quality of the lens. Glass lenses have been used since the early compound (multi-lens) microscopes built by Zacharias Jansen in 1595 and the high-quality single-lens microscopes built by Anton Leeuwenhoek in the second half of the seventeenth century [Jon97]. Light optics has become a highly developed field, with numerous applications that affect us in our daily lives. Cameras, CD players, lens and mirror-based telescopes are only a few examples of the widespread availability of high-quality optics; light

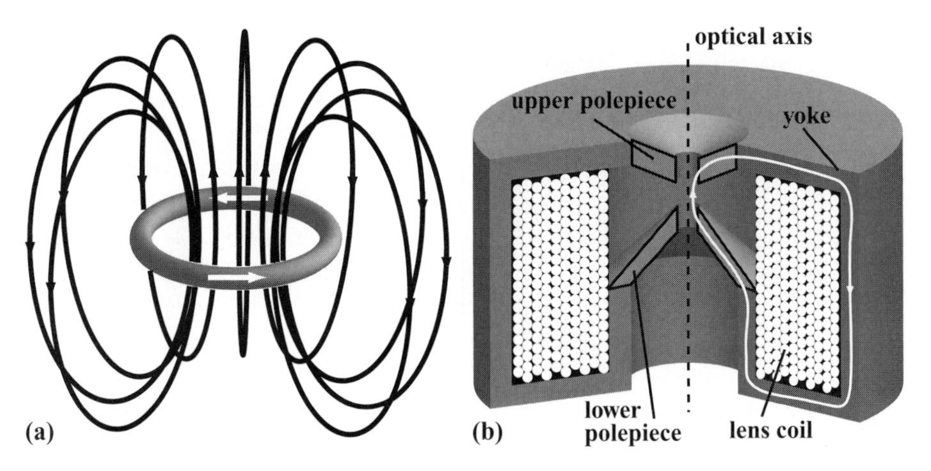

Fig. 3.3. (a) Schematic representation of the magnetic field lines around a circular current-carrying coil; (b) a magnetic yoke concentrates the field lines near the center of the coil and provides a particular geometry to the field.

optical systems can be designed and built to any specifications, with virtually no aberrations at all.

Deflection of electrons requires magnetic or electrostatic lenses. All modern transmission electron microscopes contain at least six and often more than six *round magnetic lenses*. The main purpose of a round magnetic lens is to focus the electron beam towards the optical axis. The basic principle behind such a lens is shown in Fig. 3.3(a): a circular current running through a coil produces a magnetic field that is nearly *axial* at the center of the coil. We know from electromagnetism that the strength of the field obtained at the center of the coil is proportional to the magnetic permeability of vacuum (or air) [Jac75]. We can significantly increase the attainable field strength by using a solid *yoke* with a large magnetic permeability to guide the field lines and increase the field strength near the center of the coil (Fig. 3.3b). The details of the shape of the pole pieces near the center will determine the precise geometry of the magnetic field.

A cross-section of a real magnetic lens (the first condensor lens of a JEOL120CX) is shown in Fig. 3.4; the large rectangular area to the right is the tightly packed copper winding, as shown in the inset. The upper and lower portions of the yoke are separated by a brass spacer. The lens *pole pieces*, typically made of a high-permeability Fe–Co alloy, are visible to the left of the winding. The magnetic field strength is highest near the optical axis in between the upper and lower pole pieces. The distance between the upper and lower pole pieces is known as the *lens gap*. While the geometrical details may vary from one lens to the next, all round magnetic lenses are basically identical in design to the lens shown in Fig. 3.4. Next, we will

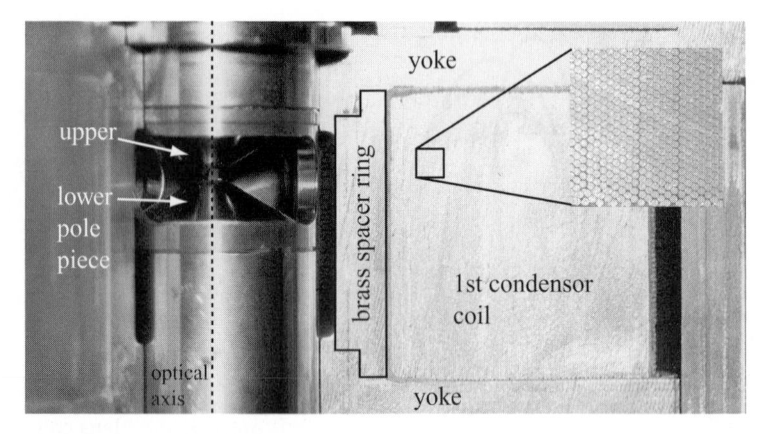

Fig. 3.4. Cross-section of the first condensor lens of a JEOL 120CX microscope. The coil winding has an areal density of nearly 300 copper wires per cm^2. The pole pieces, visible through the slot on the left, are made of a high-permeability Fe–Co alloy.

take a closer look at the magnetic field components inside a round magnetic lens.

3.4.2 Magnetic field components for a round lens

Let us consider a magnetic field with circular symmetry around its central axis. It is convenient to employ cylindrical coordinates (r, φ, z), related to the standard Cartesian coordinates (x, y, z) by the relations:

$$\begin{cases} x = r\cos\varphi; \\ y = r\sin\varphi; \\ z = z. \end{cases} \tag{3.1}$$

The z-axis is taken to be the symmetry axis of the field. In the absence of field-producing currents, we can use a scalar magnetic potential $\omega(r, z)$, satisfying the homogeneous Laplace equation $\Delta\omega = 0$, to derive the magnetic field components $\mathbf{B} = -\mu\nabla\omega$ [Jac75] (μ is the permeability). This potential does not depend on the azimuthal angle φ because of the rotational symmetry.

The Laplace equation in cylindrical coordinates is given by (e.g. [Jac75])

$$\frac{1}{r}\frac{\partial}{\partial r}\left(r\frac{\partial\omega}{\partial r}\right) + \frac{1}{r^2}\frac{\partial^2\omega}{\partial\varphi^2} + \frac{\partial^2\omega}{\partial z^2} = 0,$$

and the magnetic field components are given by

$$B_r = -\mu\frac{\partial\omega}{\partial r} \qquad \text{and} \qquad B_z = -\mu\frac{\partial\omega}{\partial z}.$$

Following Szilagyi [Szi88], we will express the potential $\omega(r, z)$ as a radial series, as follows:

$$\omega(r, z) = \sum_{i=0}^{\infty} a_i(z) r^i. \tag{3.2}$$

After substitution into the Laplace equation we find

$$\sum_{i=-2}^{\infty} (i + 2)^2 a_{i+2}(z) r^i + \sum_{i=0}^{\infty} a_i^{(2)}(z) r^i = 0,$$

where the superscript (2) denotes the second-order derivative with respect to z. For $i = -1$ we find $a_1(z) = 0$ and therefore all odd functions $a_{2i+1}(z)$ vanish. Denoting $a_0(z)$ by $U(z)$ we also find

$$a_2(z) = -\frac{U^{(2)}(z)}{4};$$
$$a_4(z) = -\frac{a_2^{(2)}(z)}{16} = \frac{U^{(4)}(z)}{64};$$
$$a_6(z) = \cdots$$

from which we find the general relation

$$a_{2i}(z) = \frac{(-1)^i}{4^i (i!)^2} U^{(2i)}(z). \tag{3.3}$$

This leads to the following expression for the scalar magnetic potential:

$$\omega(r, z) = \sum_{i=0}^{\infty} \frac{(-1)^i U^{(2i)}(z)}{(i!)^2} \left(\frac{r}{2}\right)^{2i}. \tag{3.4}$$

If we define the *axial flux density* $B(z)$ by

$$B(z) = -\mu \left.\frac{\partial \omega}{\partial z}\right|_{r=0},$$

then we find for the field components

$$B_r = \sum_{i=1}^{\infty} \frac{(-1)^i B^{(2i-1)}(z)}{i!(i-1)!} \left(\frac{r}{2}\right)^{2i-1} = -B'(z)\frac{r}{2} + B^{(3)}(z)\frac{r^3}{16} - \cdots; \tag{3.5}$$

$$B_z = \sum_{i=0}^{\infty} \frac{(-1)^i B^{(2i)}(z)}{(i!)^2} \left(\frac{r}{2}\right)^{2i} = B(z) - B''(z)\frac{r^2}{4} + B^{(4)}(z)\frac{r^4}{64} + \cdots. \tag{3.6}$$

In other words, the magnetic field components for a round lens at any point in space are completely determined by the flux density $B(z)$ along the optical axis and all of its derivatives. We will use the series expansions above to derive the paraxial ray equation for a round magnetic lens in the next sections.

There are many different techniques for computing electrostatic and magnetic fields, many of them relying on some form of Laplace's equation for a domain with a boundary of complicated shape. Finite element methods are among the more popular methods for the calculation of fields. For an extensive review of these methods and experimental measurements of field configurations we refer to [HK89a, Chapters 6–13] and [Szi88, Chapter 3].

3.4.3 The equation of motion for a charged particle in a magnetic field

We have seen in Chapter 2 that electrons in a TEM must be treated as relativistic charged particles. Newton's equation of motion,

$$\frac{\mathrm{d}\mathbf{p}}{\mathrm{d}t} = \mathbf{F},$$

with $\mathbf{p} = m\mathbf{v}$ the so-called *kinetic momentum* and $\mathbf{F}$ the total external force, can be rewritten for a particle with charge $-|e|$ in an arbitrary, time-dependent electromagnetic field:

$$\frac{\mathrm{d}\mathbf{p}}{\mathrm{d}t} = \frac{\mathrm{d}\,(m\mathbf{v})}{\mathrm{d}t} = -|e|[\mathbf{E}(\mathbf{r},t) + \mathbf{v} \times \mathbf{B}(\mathbf{r},t)]. \tag{3.7}$$

The right-hand side is the *Lorentz force*, a velocity-dependent force. If the mass m is written in the relativistic form γm_0, then equation (3.7) describes the relativistic motion of an electron. We recall that the magnetic field $\mathbf{B}$ can be written as the curl of the magnetic vector potential $\mathbf{A}$ ($\mathbf{B} = \nabla \times \mathbf{A}$) or as the gradient of the magnetic scalar potential ω in the absence of currents in the field region, and the electric field $\mathbf{E}$ is equal to $-\nabla\Phi$ with Φ the electrostatic potential [Jac75]. For a conventional transmission electron microscope, the electrostatic potential is zero along most of the electron trajectory, except for the initial acceleration by the voltage E, and the electrostatic lattice potential $V(\mathbf{r})$ inside the crystal.

Equation (3.7), with $\mathbf{B} = 0$, is used to describe the electron optical characteristics of electron guns and electrostatic deflectors. For $\mathbf{E} = 0$, the equation describes a purely magnetic field, and it is the starting point of *trajectory analysis* procedures for magnetic lenses. The design of an electromagnetic device thus consists of two separate problems:

(i) for a given electrode or lens geometry, the field must be computed;
(ii) then, with this field, the trajectory equation must be solved for particles traveling at various distances from the electrode surfaces or the optical axis of the lens.

For a very general derivation of the relativistic trajectory equations for a combined electrostatic and magnetic field distribution in arbitrary curvilinear coordinates we refer to Szilagyi [Szi88, Chapter 2], whose derivation starts from the Euler–Lagrange equations of motion. The equations of motion derived in this way are valid for a general time-dependent magnetic field, and are thus applicable to a wide range of field geometries, including synchrotrons, particle accelerators, electron microscopes, cathode ray tubes, and so on. It should come as no surprise that the literature on the subject of charged particle dynamics is vast, going back to the second half of the nineteenth century. We refer the interested reader to the volumes on *Principles of Electron Optics* [HK89a, HK89b] for an extensive list of papers relevant to electron microscopy.

For the purposes of this textbook, it will be sufficient to derive the relevant equations from the Lorentz equation of motion (3.7) in cylindrical coordinates. The basis vectors of the cylindrical coordinate system are dependent upon the particular point were they are considered:

$$\left.\begin{aligned} \mathbf{e}_r &= \cos\varphi\,\mathbf{e}_x + \sin\varphi\,\mathbf{e}_y; \\ \mathbf{e}_\varphi &= -\sin\varphi\,\mathbf{e}_x + \cos\varphi\,\mathbf{e}_y; \\ \mathbf{e}_z &= \mathbf{e}_z. \end{aligned}\right\} \tag{3.8}$$

The time derivatives are given by (denoting a derivative with respect to time by a dot above the variable):

$$\left.\begin{aligned} \dot{\mathbf{e}}_r &= \dot{\varphi}\,\mathbf{e}_\varphi; \\ \dot{\mathbf{e}}_\varphi &= -\dot{\varphi}\,\mathbf{e}_r; \\ \dot{\mathbf{e}}_z &= 0. \end{aligned}\right\} \tag{3.9}$$

The velocity vector is given by

$$\mathbf{v} = \dot{r}\,\mathbf{e}_r + r\dot{\varphi}\,\mathbf{e}_\varphi + \dot{z}\,\mathbf{e}_z. \tag{3.10}$$

We now have all the tools we need to express the Lorentz equation in cylindrical coordinates, and after a simple computation (taking $\mathbf{E} = 0$) we find the following three component equations:

$$\frac{\mathrm{d}(m\dot{r})}{\mathrm{d}t} - mr\dot{\varphi}^2 = -|e|r\dot{\varphi}B_z; \tag{3.11}$$

$$\frac{\mathrm{d}(mr\dot{\varphi})}{\mathrm{d}t} + m\dot{r}\dot{\varphi} = -|e|[\dot{z}B_r - \dot{r}B_z]; \tag{3.12}$$

$$\frac{\mathrm{d}(m\dot{z})}{\mathrm{d}t} = |e|r\dot{\varphi}B_r;$$

$$\text{where}\quad m = \frac{m_0}{\sqrt{1 - \frac{1}{c^2}(\dot{r}^2 + r^2\dot{\varphi}^2 + \dot{z}^2)}}. \tag{3.13}$$

In this derivation we have used the fact that $B_\varphi = 0$, as it must be for an axially symmetric field distribution. These equations are non-linear, coupled partial differential equations and they are rather difficult to solve because of the explicit dependence of the relativistic mass on the velocity components. The non-relativistic version of these equations is obtained by setting $m = m_0$.

The second equation can be rewritten by noting that [Rei93]

$$\frac{1}{r}\frac{\mathrm{d}(mr^2\dot{\varphi})}{\mathrm{d}t} = \frac{\mathrm{d}(mr\dot{\varphi})}{\mathrm{d}t} + m\dot{r}\dot{\varphi}.$$

This can be verified easily by explicitly writing out the two derivatives. If we truncate the expressions for the field components B_r and B_z (equations 3.5 and 3.6) after the first term, then we find for equation (3.12):

$$\frac{1}{r}\frac{\mathrm{d}(mr^2\dot{\varphi})}{\mathrm{d}t} = \frac{|e|}{2}\left[r\dot{B}(z) - 2\dot{r}B(z)\right],$$

which reduces to

$$\frac{\mathrm{d}(mr^2\dot{\varphi})}{\mathrm{d}t} = \frac{|e|}{2}\frac{\mathrm{d}\left[r^2 B(z)\right]}{\mathrm{d}t}.$$

This can be integrated readily to give

$$r^2\left[m\dot{\varphi} - \frac{|e|}{2}B(z)\right] = C, \tag{3.14}$$

where C is an integration constant.[†] C must be zero for any trajectory that intersects the optical axis at some point (for such a point $r = 0$). Such trajectories are known as *meridional trajectories*. Trajectories that are not meridional are *skew trajectories*. C can be computed for an arbitrary initial point with coordinates (r_0, φ_0, z_0).

3.4.4 The paraxial approximation

It is common practice to transform the time dependence of the equations of motion into a dependence on one of the coordinates. In most cases the z-coordinate is taken to be the independent variable, and the goal of the calculation is then to determine $r(z)$ and $\varphi(z)$ for a charged particle traveling through the field described by $B(z)$. The azimuthal equation can be rewritten using the following relation:

$$\frac{\mathrm{d}}{\mathrm{d}t} = \dot{z}\frac{\mathrm{d}}{\mathrm{d}z} = v\frac{\mathrm{d}}{\mathrm{d}z},$$

where v is the axial velocity of the electron. It is a good approximation to assume that v is very nearly constant, so that $\ddot{z} = \dot{v} = 0$. In what follows we will

[†] If we had chosen to derive the equations of motion from the variational principle, then this equation would have followed directly from the fact that the coordinate variable φ does not explicitly appear in the Lagrangian of the system, and therefore $\dot{\varphi}$ must be a constant of motion [Szi88].

denote a derivative with respect to z by a prime, i.e. $\mathrm{d}f/\mathrm{d}z \equiv f'$. The azimuthal equation (3.14) for a meridional trajectory ($C = 0$) becomes

$$\dot{\varphi} - \frac{|e|}{2mv}B(z) = \dot{\varphi} - \frac{|e|}{2\sqrt{2em_0\hat{\Psi}}}B(z) = 0.$$

Using the notation of the previous chapter we find

$$\boxed{\dot{\varphi} - \frac{\eta B(z)}{2\sqrt{\hat{\Psi}}} = 0.} \tag{3.15}$$

The total rotation angle of the electron trajectory around the optical axis is derived by integrating this equation:

$$\varphi(z) = \varphi_0 + \frac{\eta}{2\sqrt{\hat{\Psi}}}\int_{z_0}^{z} B(z)\,\mathrm{d}z. \tag{3.16}$$

The rotational frequency for a paraxial electron intersecting the optical axis can then be derived from equation (3.14),

$$\dot{\varphi} = \frac{e}{2m}B(z),$$

and this quantity is known as the *Larmor frequency*.

The radial equation can be simplified using the so-called *paraxial* approximation: we will only consider electrons traveling close to the optical axis (r remains small), with motion predominantly in the z-direction (both radial and azimuthal velocities are small compared to the forward velocity, or $r' \ll 1; r\varphi' \ll 1$). The radial trajectory equation (3.11) can then be rewritten using the following relation:

$$\ddot{r} = v^2 r'' + r'\ddot{z} \approx v^2 r''.$$

Since $\dot{v} = 0$, we also have $\dot{m} = 0$, and we leave it as an exercise for the reader to show that this leads to the following radial paraxial equation:

$$\boxed{r'' + \frac{\eta^2 B^2(z)}{4\hat{\Psi}}r = 0.} \tag{3.17}$$

It is important to note that the prefactor

$$\frac{\eta^2 B^2(z)}{4\hat{\Psi}}$$

in the radial paraxial equation is always positive, and becomes equal to zero only for a vanishing magnetic field strength. This implies that an electron trajectory is always bent *towards* the optical axis (for a round lens),[†] regardless of the electron energy and the details of the magnetic field configuration.

3.4.5 Numerical trajectory computation

The radial equation (3.17) can be solved analytically for a few particular choices of the axial magnetic flux distribution $B(z)$, but in general requires a numerical approach. We refer to [HK89a, HK89b] for a detailed discussion of the analytical solutions for *Glaser's bell-shaped* field, given by

$$B(z) = \frac{B_0}{1 + (z/a)^2}.$$

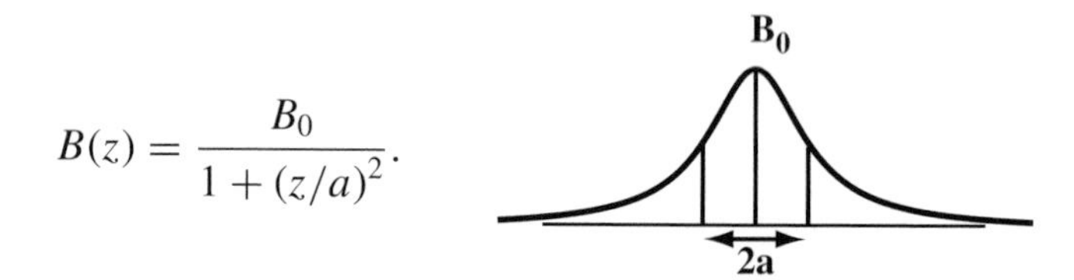

a is a parameter with the dimension of length, which indicates the extent of the field: small values of a correspond to sharply peaked fields. Although this field does not represent a realistic magnetic field distribution (the field strength does not drop off sufficiently fast at large distances), it is often employed in theoretical calculations because it allows closed-form solutions for almost all relevant parameters, including the lens aberrations.

In this section we opt for a simple numerical approach to determine the paraxial electron trajectory for an arbitrary axial field distribution $B(z)$. From the website the reader may download the program lens.f90, which implements a simple paraxial meridional trajectory computation. The program is written for Glaser's bell-shaped field, but can be modified easily for other axial field distributions. The reader is encouraged to modify the field parameters to study how a magnetic field can be used to create lenses with varying focal length. It is particularly interesting to use lens.f90 to simulate how a two-condensor lens system works. The following paragraphs describe the program and the PostScript output.

The program lens.f90 integrates both the azimuthal and radial equations. The azimuthal equation is straightforward to solve by replacing the integral (3.16) by a summation. There are many different algorithms which can be used in this case: trapezoidal integration, Simpson's rule, Romberg integration and Gaussian quadrature are amongst the more commonly used methods [Szi88]. Since we will be using fourth-order Runge–Kutta method to solve the radial equation, we will use Simpson's rule for the azimuthal equation.[‡] If the integration interval (z_0, z_N) is

[†] A positive second term in the radial equation can only be canceled by a negative first term, which means that the curvature of the trajectory must be negative, i.e. the electron trajectory bends towards the optical axis.

[‡] It can be shown that Simpson's rule is equivalent to a fourth-order Runge–Kutta method.

divided into N segments of length Δz, then the total rotation angle $\varphi(z_N)$ is given by [AS77]

$$\varphi(z_N) = \varphi_0 + \frac{\eta \Delta z}{6\sqrt{\hat{\Psi}}} \left[B(z_0) + B(z_N) + 2 \sum_{j=1}^{N/2-1} B(z_0 + 2j\Delta z) + 4 \sum_{j=1}^{N/2} B(z_0 + (2j-1)\Delta z) \right] + \mathcal{O}\left[(\Delta z)^4\right] \quad (3.18)$$

with φ_0 the initial rotation angle at the point z_0. The error in the computation is of order $(\Delta z)^4$, as indicated by the last term. This equation is rather simple to implement in a Fortran routine. In the same computational loop, we can also calculate an approximate value for the focal length, using Busch's formula [Bus26]:

$$\frac{1}{f} = \frac{\eta^2}{4\hat{\Psi}} \int_{z_0}^{z_N} B^2(z)\,\mathrm{d}z; \quad (3.19)$$

$$= \frac{\eta^2 \Delta z}{12\hat{\Psi}} \left[B^2(z_0) + B^2(z_N) + 2 \sum_{j=1}^{N/2-1} B^2(z_0 + 2j\Delta z) + 4 \sum_{j=1}^{N/2} B^2(z_0 + (2j-1)\Delta z) \right] + \mathcal{O}\left[(\Delta z)^4\right].$$

Both numerical values can then be compared† with the theoretical values for Glaser's bell-shaped field [HK89a]:

$$\varphi = \frac{\eta B_0}{2\sqrt{\hat{\Psi}}} \pi a;$$

$$\frac{1}{f} = \frac{\eta^2 B_0^2}{8\hat{\Psi}} \pi a.$$

The program lens.f90 employs fourth-order Runge–Kutta integration with fixed step size to integrate the radial paraxial equation (the public domain rksuite90 software package is available from the `www.netlib.org` website; for a general description of Runge–Kutta procedures we refer the interested reader to the *Numerical Recipes* book [PFTV89]). The Runge–Kutta procedure operates on a

† In comparing these equations with others quoted in the literature for Glaser's bell-shaped field one should bear in mind that there is a difference between the real focal length and the asymptotic focal length. The equations above all refer to asymptotic quantities. Since Glaser's field has a slow drop off at large distances from the maximum field strength, the numerical procedure will in general underestimate the total rotation angle φ and overestimate the focal length.

system of two coupled first-order differential equations which are equivalent to the radial equations:

$$\frac{\mathrm{d}y_1}{\mathrm{d}z} = y_2; \tag{3.20}$$

$$\frac{\mathrm{d}y_2}{\mathrm{d}z} = -\frac{\eta^2 B^2(z)}{4\hat{\Psi}} y_1, \tag{3.21}$$

from which the solution $r(z) = y_1$ follows. The function y_1 provides the distance from the optical axis, and the function y_2 represents the slope of the electron trajectory with respect to the optical axis. The initial values to be inserted into this routine are then the initial distance $r(z_0)$ and slope $r'(z_0)$.

Pseudo Code **PC-11** Simulation of paraxial meridional electron trajectories.

Input: axial flux density $B(z)$, acceleration voltage
Output: PostScript drawing of trajectories

```
integrate rotation angle φ(z) and focal length f      {equation 3.18}
use a fourth-order Runge–Kutta procedure              {equations 3.20 and 3.21}
(rksuite.f) to obtain r(z)
generate PostScript output
```

The source code of the lens.f90 program is rather simple and is represented in pseudo code PC-11; the radial equation is solved by a driver subroutine rksol2. The initial values are taken to be $r(z_0) = 1.0$ and $r'(z_0) = 0.0$, i.e. an electron entering the field parallel to the optical axis at unit distance. The initial value for the azimuthal angle is taken as $\varphi(z_0) = \pi/2$, i.e. the electron enters the field traveling in the plane (y, z).

The output of the program is a PostScript file (default name: lens.ps) which can be displayed with a PostScript viewer or sent directly to a printer. Typical program output is shown in Fig. 3.5. The list near the top of the page shows the acceleration voltage (both the input value and the relativistic value), the electron wavelength, the maximum axial field strength along the trajectory, the total azimuthal rotation angle $\varphi(z_N)$, and the focal length computed using Busch's formula. The three rectangular boxes show:

(i) top left: profile of the axial field distribution $B(z)$ (arbitrary scale, thinnest line), and the radial trajectory $r(z)$ (thickest line). When the radial distance becomes larger than a preset truncation value, the trajectory is clipped;

(ii) top right: projection of the trajectory along the positive z-direction (i.e. looking away from the electron source). The position of each point is given by $(x(z), y(z)) = (r(z)\cos\varphi(z), r(z)\sin\varphi(z))$;

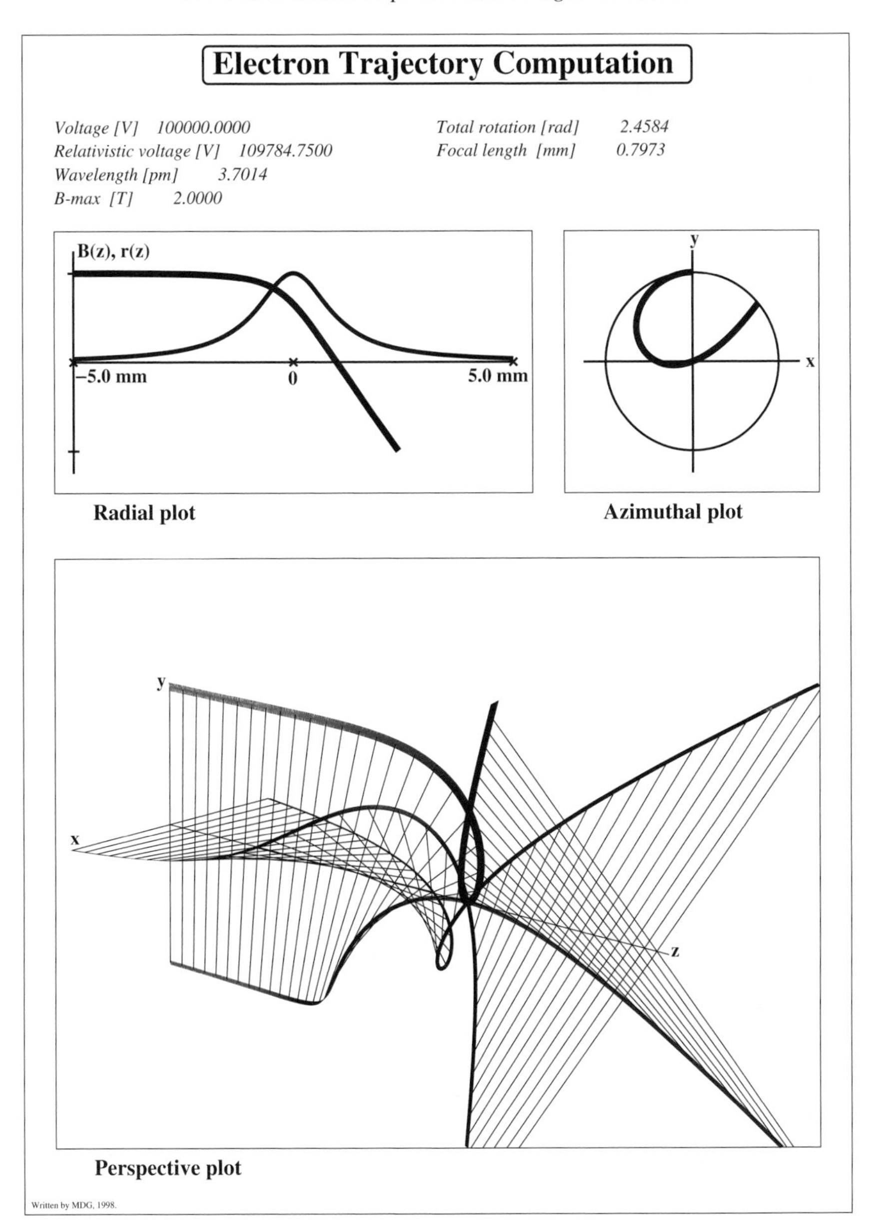

Fig. 3.5. Representative output of the lens.f program, for Glaser's bell-shaped field. The parameters of the field are $B_{max} = 2.0$ T, $a = 1.0$ mm, and the acceleration voltage is $V = 100$ kV.

(iii) bottom: perspective representation of four identical trajectories, obtained from the computed trajectory by rotation by $[0, 1, 2, 3] \times \pi/2$ in the (x, y) plane. The trajectories are clipped by the frame of the drawing, to prevent them from overlapping other parts of the output.

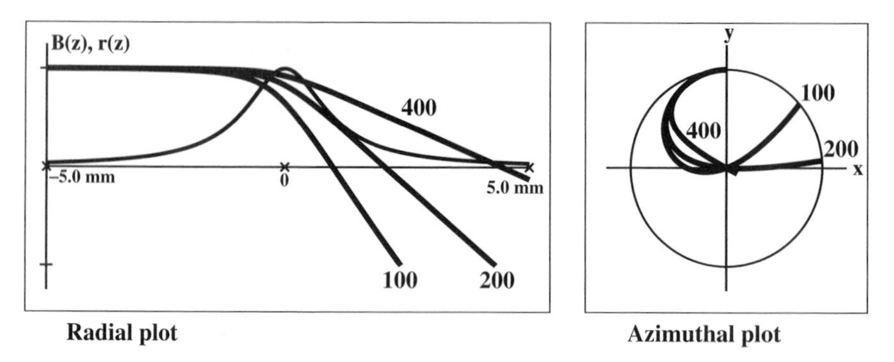

Fig. 3.6. Radial and azimuthal trajectories for the field of Fig. 3.5, for acceleration voltages $V = 100$, 200 and 400 kV.

This program can be used to study how various parameters affect the focal length of the lens and the electron trajectory. For instance, by entering a number of different acceleration voltages for a given lens field, we find (Fig. 3.6) that a faster electron is less affected by the field or, equivalently, that a higher acceleration voltage requires more powerful lenses, or lenses that act over a longer distance along the electron trajectory.

Stationary (i.e. time-independent) magnetic fields cannot accelerate electrons, they can only change the travel direction of the electrons. Accordingly, in the absence of electrostatic fields, magnetic fields leave the energy of the electron unchanged. The action of an axially symmetric magnetic field on the trajectory of an electron can then be understood as follows: for an electron with an initial axial velocity component v_z, the radial component of the magnetic flux density B_r creates an azimuthal force F_φ. The particle thus acquires an azimuthal velocity component $v_\varphi = r\dot{\varphi}$, which will make the trajectory curve around the optical axis. This azimuthal component will then interact with the z component of the flux density to create a radial force component F_r, which brings the particle closer to the optical axis.

If we place two lenses along the optical axis the electron trajectory becomes more complex. Figure 3.7 shows a configuration of two lenses ($B_{1,max} = 1.2$ T, $B_{2,max} = -1.8$ T, $a = 1.0$ mm), spaced by 20 mm. The acceleration voltage is $V = 400$ kV. This configuration mimics a double-condensor system, where the first lens has a fixed cross-over and the second a variable cross-over; by changing the current through the second lens the focal length of the system can be changed. Note that the negative magnitude of the second field does not affect the focal length, but does change the overall rotation angle.

The programs discussed in this section are greatly simplified versions of the types of computations a lens designer would have to carry out to create an entire

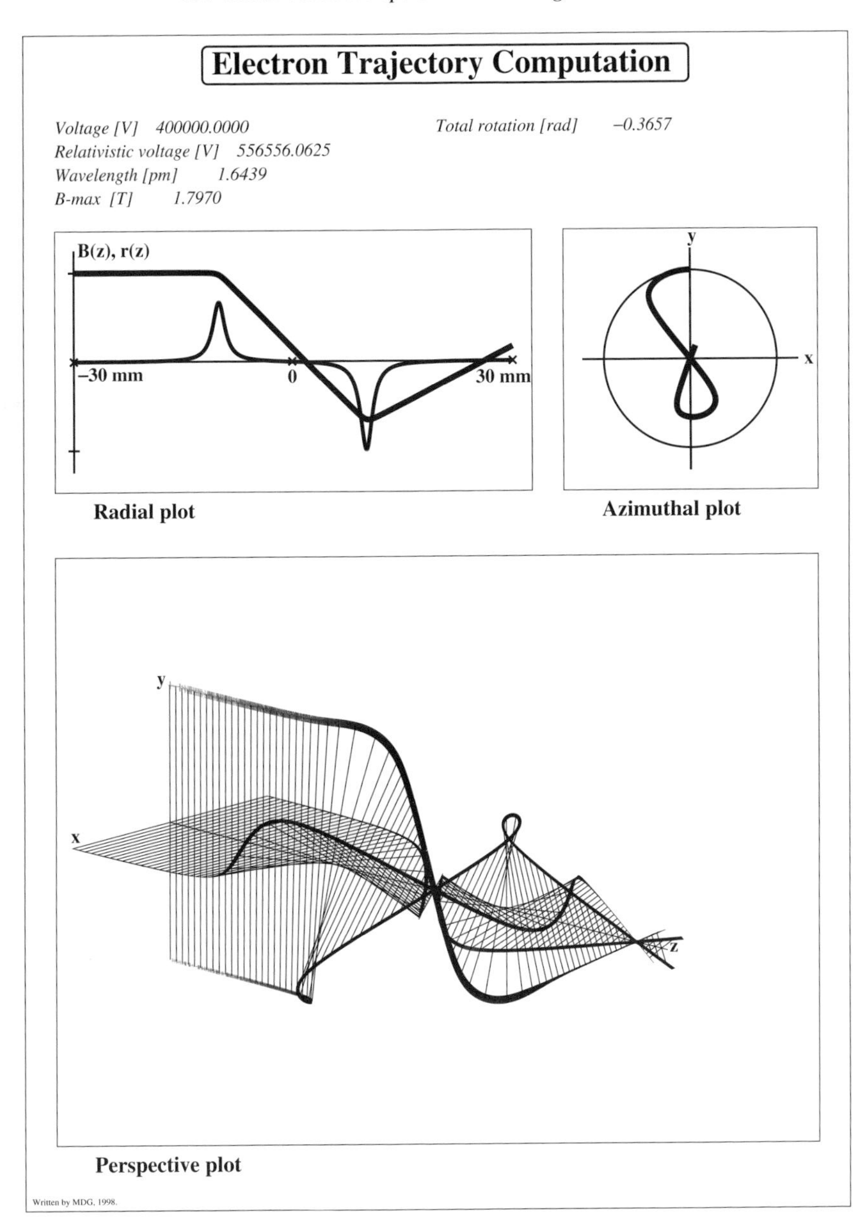

Fig. 3.7. Double-lens configuration; the centers of the lenses are 20 mm apart. The parameters of the field are $B_{max} = 1.2, -1.8$ T, $a = 1.0$ mm, and the acceleration voltage is $V = 400$ kV.

microscope column, or any other lens configuration. Professional designers would typically employ variable step-size Runge–Kutta integration algorithms or the Burlisch–Stoer algorithm [PFTV89] to integrate the equations of motion accurately for realistic field distributions. Although most microscopists will rarely need to worry about the physics of a magnetic lens, it is useful to play with simple programs such as lens.f90 to acquire some understanding of how a magnetic lens works.

3.4.6 General properties of round magnetic lenses

The general solution to any second-order differential equation (and therefore also equation 3.17) can be written as a linear combination of two independent solutions. We select two solutions with special properties:† the first solution, $r_1(z)$, satisfies the condition

$$\lim_{z\to-\infty} r_1(z) = 1, \tag{3.22}$$

or, the *incident* ray is *asymptotically* parallel to the optical axis, at unit distance from the axis. The second solution, $r_2(z)$, is asymptotically parallel to the *outgoing* optical axis, i.e.

$$\lim_{z\to+\infty} r_2(z) = 1. \tag{3.23}$$

A general solution of equation (3.17) thus has the form:

$$r(z) = Ar_1(z) + Br_2(z). \tag{3.24}$$

We will refer to each solution as a *ray*. Since magnetic fields have infinite range, a ray can only become asymptotically identical to a straight line.

The rays $r_1(z)$ and $r_2(z)$ become asymptotically straight lines far from the lens center, and hence we have (using the equation for a straight line intersecting the optical axis):

$$\lim_{z\to+\infty} r_1(z) = (z - z_{Fi})r'_{1i};$$
$$\lim_{z\to-\infty} r_2(z) = (z - z_{Fo})r'_{2o}.$$

We will call z_{Fi} the *asymptotic image focus* (see Fig. 3.8) and z_{Fo} the *asymptotic object focus*. As before, a prime denotes differentiation with respect to z.

A ray parallel to $r_1(z)$, but at a different distance from the optical axis, can be written as $\lambda r_1(z)$, where λ is a real number; the asymptotic outgoing ray belonging to $\lambda r_1(z)$ is $\lambda(z - z_{Fi})r'_{1i}$, according to the relations above. This ray *intersects the optical axis in the same point* z_{Fi}, hence the name *image focus*.

We can now determine the intersection of the incident asymptote to the ray $r_1(z)$ and its emergent asymptote; the intersection point lies in the plane $z = z_{Pi}$

† We will follow closely Section 16.2 in [HK89a] and Chapter 4 in [Szi88], but with a slightly different notation.

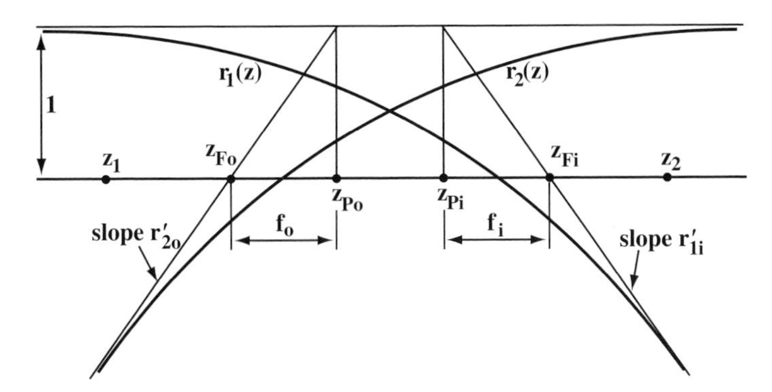

Fig. 3.8. Definition of the asymptotic ray parameters and location of the general points z_1 and z_2.

(see Fig. 3.8), which is called the *asymptotic principal image plane*:

$$\text{incident asymptote} \rightarrow 1 = (z_{Pi} - z_{Fi})r'_{1i} \leftarrow \text{emerging asymptote}, \tag{3.25}$$

from which

$$z_{Pi} = z_{Fi} + \frac{1}{r'_{1i}} = z_{Fi} - f_i, \tag{3.26}$$

where f_i is the *asymptotic focal length*. Similarly, the asymptotes to $r_2(z)$ intersect in the *asymptotic principal object plane* $z = z_{Po}$, with associated asymptotic focal length $f_o = 1/r'_{2o}$. The equality $f_i = f_o$ is valid, provided the electron does not experience any change in its kinetic energy while traveling through the lens. Inserting the limit values for $r_1(z)$ and $r_2(z)$ into equation (3.24), we obtain the following equations for the asymptotic general ray:

$$\lim_{z\to-\infty} r(z) = A + B\frac{z - z_{Fo}}{f_o};$$

$$\lim_{z\to+\infty} r(z) = -A\frac{z - z_{Fi}}{f_i} + B.$$

There are two components to each ray: the distance to the optical axis, and the angle of the trajectory with respect to the optical axis. It is then convenient to introduce a two-component *ray-vector*:

$$\mathbf{U}(z) = \begin{pmatrix} r(z) \\ r'(z) \end{pmatrix}. \tag{3.27}$$

We leave it up to the reader to eliminate A and B from equation (3.27), for two positions z_1 and z_2 (see Fig. 3.8), z_1 somewhere on the incident asymptote and z_2 somewhere on the emergent asymptote of the general solution. We obtain the matrix equation

$$\mathbf{U}_2 = \mathcal{T}\,\mathbf{U}_1, \tag{3.28}$$

where $\mathcal{T}$ is the *lens transfer matrix*, which describes the paraxial behavior of the lens. The matrix is given explicitly by

$$\mathcal{T} = \begin{pmatrix} \frac{-(z_2 - z_{Fi})}{f_i} & f_o + \frac{(z_1 - z_{Fo})(z_2 - z_{Fi})}{f_i} \\ -\frac{1}{f_i} & \frac{(z_1 - z_{Fo})}{f_i} \end{pmatrix}. \tag{3.29}$$

The transfer matrix can be used as a compact tool to describe the relevant imaging features of the lens. For a system with multiple lenses sharing the same optical axis, the image plane of one lens becomes the object plane of the next lens, and we can describe the behavior of the entire system by multiplication of the individual transfer matrices in the proper order. The paraxial (asymptotic) imaging properties of an entire microscope column can thus be reduced to a set of transfer matrices.

Let us now consider a few important special cases. The points z_1 and z_2 used to determine the transfer matrix $\mathcal{T}$ are randomly chosen on the incident and emerging asymptotes. Consider a point P_o in the plane $z_1 = z_o$, such that all rays leaving P_o converge to a single point P_i in the plane $z_2 = z_i$. The two planes z_o and z_i are then said to be *conjugate planes*. Mathematically, this means that the component $r(z_i)$ of the ray-vector $\mathbf{U}_2(z_i)$ must be independent of the initial slope $r'(z_o)$ of the trajectory, or, equivalently

$$(z_o - z_{Fo})(z_i - z_{Fi}) = -f_i f_o. \tag{3.30}$$

This relation is known as *Newton's lens equation*, and it plays a central role in both light and electron optics [BW75]. In the absence of electrostatic fields (i.e. when $f_o/f_i = 1$) the transfer matrix relation for conjugate planes can be rewritten as

$$\begin{pmatrix} r_i \\ r_i' \end{pmatrix} = \begin{pmatrix} M & 0 \\ -1/f_i & M_\alpha \end{pmatrix} \begin{pmatrix} r_o \\ r_o' \end{pmatrix}. \tag{3.31}$$

These equations are interpreted as follows: the radial image distance from the optical axis is related to the object distance by the relation

$$r_i = M r_o.$$

M is hence known as the *transverse magnification*. It is easy to see that the transverse magnification depends on the location of the image plane $z = z_i$ according to $M = -(z_i - z_{Fi})/f_i$. In particular, the transverse magnification can be smaller than 1 when the distance between image and principal image planes is shorter than the focal length. In such a case M is a *de*magnification factor.

The slopes of the trajectories at the object and image locations are related by

$$r_i' = -\frac{r_o}{f_i} + M_\alpha r_o',$$

where M_α is the *angular magnification*. The angular magnification depends on the location of the object plane $z = z_o$ according to $M_\alpha = (z_o - z_{Fo})/f_i$.

Since the determinant of the transfer matrix is equal to $f_o/f_i = 1$, we can easily show (using Newton's lens equation 3.30) that

$$MM_\alpha = \frac{f_o}{f_i} = 1.$$

We thus find that a lens with a large transverse magnification must have a small angular magnification and vice versa. We will see later on in this chapter that the objective lens of a TEM is a strongly excited lens with a large transverse magnification.

In addition to the two focal planes $z = z_{Fo}$ and $z = z_{Fi}$, we can define the following pairs of planes:

- *principal planes*: the two conjugate planes related to each other by a *unit transverse magnification* are known as the *principal planes*. They are denoted in Fig. 3.8 by the symbols $z = z_{Po}$ and $z = z_{Pi}$;
- *nodal planes*: the nodal planes are located at the intersection points of asymptotic rays for which the slopes of the incident and emergent sections are equal. The points are related to each other by *unit angular magnification*, and are indicated by the symbols $z = z_{No}$ and $z = z_{Ni}$. For round magnetic lenses the nodal points coincide with the principal points [HK89a].

Because of the positive character of the factor $\eta^2 B^2(z)/4\hat{\Psi}$ in the radial trajectory equation, the image nodal point is located closer to the object than the object nodal point, from which it follows that both nodal and principal planes for a round magnetic lens are crossed (i.e. the image principal plane is closer to the object plane than to the image plane).

The various points (focal, principal, and nodal) are collectively called the *asymptotic cardinal elements* of the lens. Using the lens transfer matrix it is straightforward to compute the magnification properties of a set of coaxial lenses, by direct matrix multiplication. The theory derived in this section is valid when the details of the field inside the lens are unimportant, i.e. when one is only interested in the input and output of a lens. This is the case for most lenses in the TEM, except for the objective lens.

The objective lens cannot be treated in the same way as the other lenses, since the specimen is inserted into the magnetic field, thereby dividing the lens into two parts. For this case, we can use the so-called *osculating cardinal elements*; this is beyond

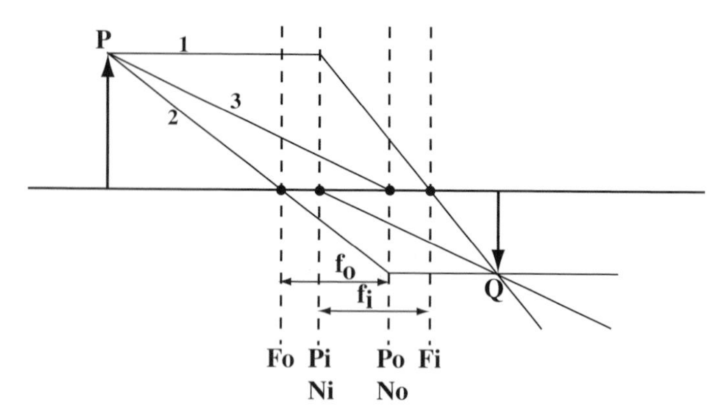

Fig. 3.9. Graphical construction of the image **Q** of an object point **P**, using the asymptotic cardinal elements. Any two of the three rays shown can be used to determine the location **Q**.

the scope of this textbook and we refer to [Szi88], [Spe88] and [HK89a, Chapter 17] for more detailed information.

The relations between the cardinal elements and the focal lengths can be used to determine graphically the image formed by a lens for an arbitrary object location. Consider the drawing in Fig. 3.9: an object (arrow ending in **P**) is located to the left of the lens. The six cardinal elements of the lens are indicated on the drawing (with crossed principal planes). The image point **Q** corresponding to **P** can then be determined by constructing any two out of the three rays shown in the figure. Ray **1** is drawn parallel to the optical axis until it reaches the image principal plane **Pi**; then a straight line is drawn from this intersection point through the image focal point **Fi**. Ray **2** is a straight line through the object focal point **Fo** until it reaches the object principal plane **Po**, from where the line is continued parallel to the optical axis into image space. The third ray **3** goes from the point **P** to the object nodal point **No**; a parallel line is then drawn starting in the image nodal point **Ni** and continues into image space. The three lines intersect each other at the image point **Q** conjugate to **P**. For the particular location of **P** in this drawing, the transverse magnification is less than one, and the image is a demagnified (and inverted) version of the object. By varying the current through a magnetic lens, one can change the location of the focal planes continuously, and therefore the position of the image plane conjugate to the object. In particular, if the distance between the object and the object focal plane is less than the object focal length, the image will be magnified rather than demagnified. The image space focal plane is commonly known as the *back focal plane*.

The graphical construction of Fig. 3.9 can be used to illustrate the concepts of *depth of field* and *depth of focus*. Consider three points in an object, shown in Fig. 3.10. The middle point (o_2) lies in the object plane conjugate to the image plane z_2. The point o_1 to the left is at a larger distance from the lens plane, and

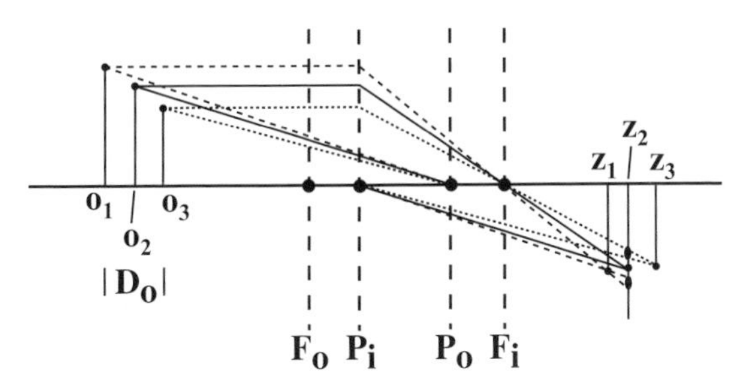

Fig. 3.10. Graphical illustration of depth of field.

its image plane is therefore to the left of the plane $z = z_2$. The point o_3 has its conjugate image plane at $z = z_3$. Since we observe the image in the plane $z = z_2$, the rays originating from the points o_1 and o_3 intersect the image plane $z = z_2$ in circles, commonly known as *circles of confusion*. The radius of a circle of confusion depends on the range of angles α_o between the optical axis and the electron rays that give rise to the image; a smaller angular range produces a smaller circle of confusion. In optical photography this is typically obtained by reducing the size of the aperture, i.e. "stopping down" the aperture. It can be shown that the *depth of field* D_o, the distance in object space over which the rays will give rise to an in-focus image with a circle of confusion smaller than δ, is given by $D_o = 2\delta/\alpha_o$, where α_o is the semi-aperture angle [WC96,HHN$^+$77]. In other words, if we wish to resolve features of the order of $\delta = 1$ nm, and the largest angle between the electron trajectories and the optical axis is $\alpha_o = 10$ mrad, then the depth of field is given by $D_o = 200$ nm. All the object details larger than $\delta = 1$ nm are simultaneously in focus if the sample thickness is less than $D_o = 200$ nm, when the apertures limit the maximum angle to 10 mrad. The *depth of focus* D_i is the distance in image space corresponding to D_o in object space. It is easy to show that $D_i = D_o M^2$, where M is the lateral magnification. For a magnification of $M = 10^5$, the depth of focus for the parameters above is $D_i = 2 \times 10^{12}$ nm $= 2$ km, which means that the precise location of the image plane is not important since the image will be in-focus over a distance of 2 km! The large depth of focus makes it possible to observe an in-focus image on the fluorescent screen of the microscope, and then expose a negative located several centimeters below the image screen *without having to refocus the image*. We will return to the circle of confusion concept when we define the *resolution* of a microscope in Chapter 10.

3.4.7 Lenses and Fourier transforms

In the preceding sections, we have described how magnetic lenses work, using a geometrical description. In view of the particle–wave duality of quantum mechanics we must also consider the action of a lens in terms of the wave function of the

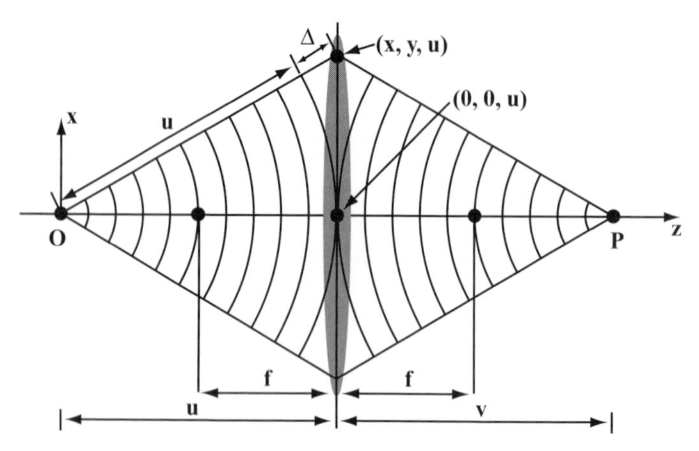

Fig. 3.11. Schematic illustration of the action of a perfect lens. The y-direction comes out of the plane of the drawing.

electrons passing through the lens. This leads to the field of *wave optics*. At this point, we will introduce only briefly the most important concepts of wave optics; we will return to wave optics in Chapter 10, when we will discuss phase contrast techniques.

Consider a perfect point source located at the origin on the optical axis of a round lens, as shown schematically in Fig. 3.11. The point source emits spherical waves. The distance between the source and the lens plane is the *object distance* u. A perfect lens would *invert the curvature of the spherical waves* and create a point image P at a distance v from the lens plane. Since the phase of the spherical wave is constant along the spherical surfaces, the phase cannot be constant in the lens plane (i.e. at $z = u$). The phase difference between the points $(0, 0, u)$ and (x, y, u) is determined by the path length difference Δ:

$$\Delta = u\sqrt{1 + \frac{x^2 + y^2}{u^2}} - u.$$

Since we are mostly interested in the paraxial behavior of the lens we can assume that $x^2 + y^2$ will be much smaller than u^2, and after expansion of the square root (using $\sqrt{1+x} \approx 1 + x/2 + \cdots$) we find:

$$\Delta \approx \frac{x^2 + y^2}{2u}.$$

The quadratic approximation to the phase distribution in the lens plane is then given by

$$\exp\left(\mathrm{i}\frac{2\pi}{\lambda}\frac{x^2 + y^2}{2u}\right) = \exp\left(\pi \mathrm{i}k\frac{x^2 + y^2}{u}\right),$$

where $k = 1/\lambda$ is the wave number. This function is commonly known as the *Fresnel propagator* and is represented by the symbol

$$\boxed{\mathcal{P}_u(x, y) \equiv \exp\left(\pi \mathrm{i}k\frac{x^2 + y^2}{u}\right).} \tag{3.32}$$

The propagator $\mathcal{P}_u$ describes how a spherical point source affects the phase distribution in a plane normal to the propagation direction at a distance u from the source.†

If instead of a single point source we have a distribution $\psi(x, y)$ of point sources in the object plane, then we can invoke *Huygens' principle* which states that the propagation of a wave $\psi(x, y)$ through space from the plane $z = 0$ to the plane $z = u$ can be viewed as a superposition of spherical scatterers in the plane $z = 0$, each with an amplitude and phase given by $\psi(x, y)$. The combination of all the spherical wave fronts at the plane $z = u$ becomes the new wave front. In mathematical terms this is represented by the convolution product:

$$\psi(x, y, u) = \psi(x, y, 0) \otimes \mathcal{P}_u(x, y). \tag{3.33}$$

This is the fundamental propagation equation, valid under paraxial conditions. A more rigorous derivation of the Fresnel propagator would start from the Kirchoff formula and is beyond the scope of this book; we refer the interested reader to the many textbooks on electrodynamics and wave optics (e.g. [BW75, Jac75, Wan79]). The exact expression for the Fresnel propagator is then given by

$$\mathcal{P}_u(x, y) \equiv \frac{\mathrm{i}}{\lambda u}\exp\left(\pi \mathrm{i}k\frac{x^2 + y^2}{u}\right). \tag{3.34}$$

Next, we consider the lens itself. Newton's lens equation (3.30) can be rewritten using the notation of Fig. 3.11:

$$\begin{aligned} f^2 &= -(z_o - z_{Fo})(z_i - z_{Fi}); \\ &= (u - f)(v - f), \end{aligned}$$

from which we derive

$$\frac{1}{f} = \frac{1}{u} + \frac{1}{v}. \tag{3.35}$$

A perfect lens inverts the curvature of the wave field that enters it, and this reversal of curvature is described by a multiplication by a pure phase factor. If we again use the quadratic approximation (also known as the *small-angle approximation*)

† Note that many textbooks use the opposite sign convention for the Fresnel propagator (i.e. a negative exponent). The convention used in this textbook is consistent with the sign convention for the Fourier transform.

we anticipate that the lens phase shift will be described by a function of the form

$$\mathcal{L}(x) = \mathrm{e}^{\pi \mathrm{i} k q x^2},$$

where q is to be determined, and we restrict ourselves to the one-dimensional case. For an ideal lens, the image amplitude of a perfect point source (i.e. a delta-function $\delta(x)$) should be proportional to a delta-function (apart from unimportant phase factors). In mathematical terms this statement reads as (referring to Fig. 3.11)

$$\begin{aligned}\delta(x) &= [[\delta(x) \otimes \mathcal{P}_u(x)]\,\mathcal{L}(x)] \otimes \mathcal{P}_v(x) \qquad \text{(ideal lens)}, \\ &= [\mathcal{P}_u(x)\mathcal{L}(x)] \otimes \mathcal{P}_v(x).\end{aligned} \tag{3.36}$$

The second equality follows because the delta-function is the identity function with respect to the convolution product. Using the definition of the convolution product (equation 2.51 on page 106) we have (e.g. [Cow81]):

$$\begin{aligned}[\mathcal{P}_u(x)\mathcal{L}(x)] \otimes \mathcal{P}_v(x) &= \int \mathcal{P}_u(X)\mathcal{L}(X)\mathcal{P}_v(x - X)\,\mathrm{d}X; \\ &= \int \mathrm{e}^{\pi \mathrm{i} k X^2/u} \mathrm{e}^{\pi \mathrm{i} k q X^2} \mathrm{e}^{\pi \mathrm{i} k (x-X)^2/v}\,\mathrm{d}X; \\ &= \int \mathrm{e}^{\pi \mathrm{i} k X^2(\frac{1}{u}+\frac{1}{v}+q)} \mathrm{e}^{\pi \mathrm{i} k x^2/v} \mathrm{e}^{-2\pi \mathrm{i} k x X/v}\,\mathrm{d}X.\end{aligned}$$

We can use the lens formula and substitute $1/f$ for $1/u + 1/v$. The first exponential can then be removed provided we set $q = -1/f$. The remaining integral is

$$\begin{aligned}[\mathcal{P}_u(x)\mathcal{L}(x)] \otimes \mathcal{P}_v(x) &= \mathrm{e}^{\pi \mathrm{i} k x^2/v} \int \mathrm{e}^{-2\pi \mathrm{i} k x X/v}\,\mathrm{d}X; \\ &= \mathrm{e}^{\pi \mathrm{i} k x^2/v}\,\delta\left(\frac{kx}{v}\right).\end{aligned}$$

The image amplitude is hence proportional to a delta-function, so that the image of a point source is again a point. The perfect lens is therefore described by the phase factor:

$$\boxed{\mathcal{L}_f(x) = \mathrm{e}^{-\pi \mathrm{i} k x^2/f}.} \tag{3.37}$$

Next, we can describe how a lens produces an image or diffraction pattern for an arbitrary wave $\psi(x)$ (one dimensional for simplicity). We are interested in the wave in the image plane, and also the wave in the lens back focal plane, at $v = f$. The wave in the back focal plane is given by

$$\phi_{bfp}(x) = [[\psi(x) \otimes \mathcal{P}_u(x)]\mathcal{L}_f(x)] \otimes \mathcal{P}_f(x). \tag{3.38}$$

Working out the integrals as before we find

$$\phi_{bfp}(x) = \mathrm{e}^{\pi \mathrm{i}kx^2/f} \int \psi(Y)\mathrm{e}^{\pi \mathrm{i}kY^2/u} \left(\int \mathrm{e}^{\pi \mathrm{i}kX^2/u} \mathrm{e}^{-2\pi \mathrm{i}kX\left(\frac{Y}{u}+\frac{x}{f}\right)} \,\mathrm{d}X \right) \mathrm{d}Y. \quad (3.39)$$

The integral over X can be carried out by means of the substitution $Q = k(Y/u + x/f)$, which results in

$$\phi_{bfp}(x) = \mathrm{e}^{\pi \mathrm{i}kx^2\left(\frac{1}{f}-\frac{u}{f^2}\right)} \int \psi(Y)\mathrm{e}^{-2\pi \mathrm{i}kxY/f} \,\mathrm{d}Y. \quad (3.40)$$

The integral over Y is a direct Fourier transform with the reciprocal space variable $kx/f = x/\lambda f$. We find that the wave function in the back focal plane of the lens is given by the direct Fourier transform of the object wave function (assuming paraxial conditions hold):

$$\phi_{bfp}(x) = \mathrm{e}^{\pi \mathrm{i}kx^2(\frac{1}{f}-\frac{u}{f^2})} \psi\left(\frac{kx}{f}\right). \quad (3.41)$$

It is now straightforward to compute the wave in the image plane, by application of the propagation equation between the back focal plane and the image plane:

$$\begin{aligned} \phi_{ip}(x) &= \phi_{bfp}(x) \otimes \mathcal{P}_{v-f}(x); \\ &= \mathrm{e}^{\pi \mathrm{i}kx^2/(v-f)} \int \mathrm{e}^{\pi \mathrm{i}kX^2(\frac{1}{f}-\frac{u}{f^2}+\frac{1}{v-f})} \psi\left(\frac{kx}{f}\right) \mathrm{e}^{-2\pi \mathrm{i}kxX/(v-f)} \,\mathrm{d}X. \end{aligned}$$

Using the lens equation we can show that the first exponential inside the integral is equal to 1. After the substitution $Q = kX/f$ and using $f/(v-f) = u/v$ we find

$$\begin{aligned} \phi_{ip}(x) &\sim \mathrm{e}^{\pi \mathrm{i}kx^2/(v-f)} \int \psi(Q)\mathrm{e}^{-2\pi \mathrm{i}kuxX/v} \,\mathrm{d}Q; \\ &\sim \psi\left(-\frac{ux}{v}\right). \end{aligned}$$

Despite the negative exponent, the integral over Q is an inverse Fourier transform, which accounts for the minus sign in the last line above. The image amplitude is therefore proportional to the inverse Fourier transform of the amplitude in the back focal plane. The image is inverted and (de)magnified by a factor of $\frac{v}{u}$.

When the distances from the electron source to the crystal and from the crystal to the observer are much larger than the dimensions of the crystal, then we say that the *Fraunhofer diffraction conditions* apply. For an incident plane wave the scattered wave at a large distance from the crystal is described by the Fourier transform of the object function [Cow81]. The diffraction pattern is thus effectively located at infinity. One could imagine a lens with vanishingly small field strength $B(z)$, which, according to Busch's formula (3.19), would have an infinite focal length. When we turn on the lens current, the focal length will become finite and the back focal plane will be located closer to the lens plane. The main purpose of a lens is then to bring the diffraction pattern from infinity to a finite location (one focal length from the lens center), and consequently there will also be an image plane, conjugate to the object plane. The relation between an object and its diffraction pattern is described by the Fourier transform, regardless of the position of the back focal plane (apart from scaling factors).

It is this simple fact that is employed in the transmission electron microscope to create both diffraction patterns and images of the specimen. It is important to remember from this discussion that the diffraction pattern exists regardless of the presence or absence of a lens; the lens only determines where along the optical axis of the system the diffraction pattern and image can be found. For a detailed review of Fraunhofer diffraction we refer to Born and Wolf [BW75], Cowley [Cow81], and Hawkes and Kasper [HK89a]. We will return to the subject of wave optics when we talk about phase contrast observations in Chapter 10.

3.5 Basic electron optics: lens aberrations

3.5.1 Introduction

Magnetic lenses are not perfect, i.e. the paraxial ray equation fails for rays at increasing distances from the optical axis. For such rays one must include the higher-order terms in the expansions of the magnetic field components with respect to the derivatives of the axial flux density $B(z)$ (equations 3.5 and 3.6). While the mathematics involved in the derivation of aberration coefficients is not very difficult, "a series of elementary operations is needed that requires quite a lot of patience" and there is little point in immersing ourselves "into the mind-boggling maze of elementary mathematical manipulations" [Szi88, page 221]. Instead, we refer the patient and mathematically skilled reader to Chapter 5 in [Szi88] and Chapters 21–31 in [HK89a].

For the purposes of this book, it will suffice to define the five primary aberrations for a round magnetic lens and illustrate their influence on image formation. The full mathematical treatment of lens aberrations requires a perturbation analysis of the paraxial trajectory equation including higher-order terms in the expansions for the field components.

3.5.2 Aberration coefficients for a round magnetic lens

Although an explicit derivation of round lens aberrations will not be attempted here, it is instructive to describe briefly the general procedure. Consider an object plane, a lens, and an image plane. A point P_o in the object plane (Fig. 3.12) is

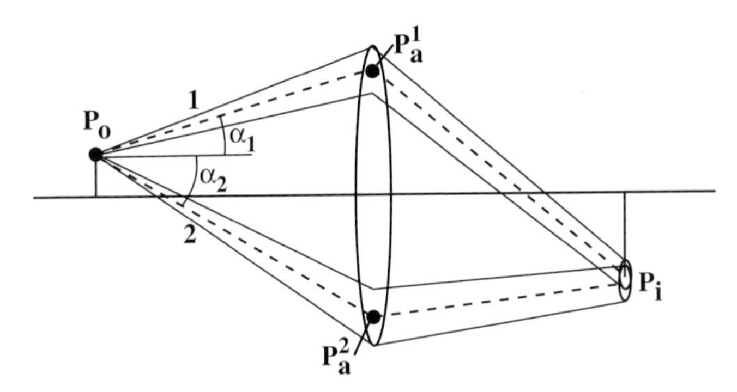

Fig. 3.12. Schematic illustration of image formation for an imperfect lens.

conjugate to the point P_i in the image plane. For a perfect lens, all electrons that leave the object point reach the image point, and the corresponding wave front is known as the *Gaussian wave front*. Apart from a magnification and possibly an image rotation and/or inversion, the distance between object points is conserved in the image, regardless of the trajectory followed by the electrons between the object and image plane.

For an imperfect lens, the wave surface exiting the lens deviates from a Gaussian (spherical) surface. The magnitude of the deviation depends on the coordinates of the points P_o and P_i, and also on the coordinates of the point where the electrons passed the lens plane. This is illustrated schematically in Fig. 3.12: electrons which travel within a certain angular range from the trajectory at an angle α_1, indicated by the thin solid lines on both sides of the dashed line, will reach the image plane in a disk (not necessarily round). Electrons which travel along the direction α_2 will also reach the image plane, but not necessarily on the same disk. The overall image conjugate to the object point P_o is obtained by considering all possible angles α_i in object space. The union of all disks in the image plane is then the *aberrated image* of the point P_o.

The shape of the real wavefront which exits the lens can be compared with that of the spherical Gaussian wave front. It is then common practice to expand the difference between the two wave fronts in terms of the coordinates of P_o, P_i, and the point P_a at which the electrons cross the lens plane. Since the Gaussian surface is quadratic in the coordinates, the lowest-order terms in the difference will be of third order in the coordinates. They are the so-called *third-order aberrations*, also known as the *primary Seidel aberrations*. Higher-order aberrations can also be computed, and we refer to the sources mentioned in the *Introduction* for the details of the derivation.

There are many different ways in which the aberrations can be described mathematically. We will follow Hawkes and Kasper [HK89a, equation 24.38b] and use complex variables to denote the location of a point in a plane normal to the optical axis. The point P_o has coordinates (x_o, y_o) in the object plane, and is represented by the complex number $u_o = x_o + \mathrm{i}y_o$. The distance between P_o and the optical axis is then given by $r_o^2 = u_o u_o^*$. The point P_a is similarly described by the complex number $u_a = x_a + \mathrm{i}y_a$. We will assume that the image rotation along the optical axis has been absorbed into the definition of the respective coordinate frames. It can then be shown that the image point P_i is located at the position $u_i + \Delta u_i$, where u_i is the Gaussian position, and

$$\begin{aligned}
\Delta u_i = {} & C r_a^2 u_a && \leftarrow \text{spherical aberration} \\
& + 2(K + ik) r_a^2 u_o + (K - ik) u_a^2 u_o^* && \leftarrow \text{coma} \\
& + (A + ia) u_o^2 u_a^* && \leftarrow \text{astigmatism} \\
& + F r_o^2 u_a && \leftarrow \text{field curvature} \\
& + (D + id) r_o^2 u_o && \leftarrow \text{distortion.}
\end{aligned} \tag{3.42}$$

The eight constants C, K, k, A, a, F, D, and d are the real† aberration coefficients; they describe the five primary or Seidel aberrations. The magnification factor M has been omitted for clarity. The primary aberrations can be grouped into three categories: the aperture aberrations (terms independent of the object position u_0 in equation 3.42), the chromatic aberrations (terms independent of u_a, discussed in Section 3.7.4) and the parasitic aberrations (in general these are linear in u_o or linear in u_a). Geometric aberrations are caused by the properties of the magnetic field. Chromatic aberrations are caused by the instabilities in the lens current and/or instabilities in the acceleration potential; they give rise to slightly different electron trajectories in the lens and hence to a blurring of the image. Finally, parasitic aberrations are caused by inhomogeneities or imperfections in the lens pole pieces, deviations from perfect circular symmetry, etc.

The first term in equation (3.42), *spherical aberration*, is of third order in the coordinates u_a, and hence also of third order in the angle α between the trajectory and the optical axis. The next distortion, *coma*, is of second order in α, and is described by two numbers K and k, which can be converted into an amplitude and angle, i.e. a single complex number. The next two aberrations, *astigmatism and field curvature*, are linear in α; astigmatism is also represented by two numbers A and a. Finally, the fifth aberration, known as *distortion*, only depends on the position in the object plane, and is also described by two numbers D and d. Each of the aberrations describes a particular way in which the wave front can depart from the spherical shape. The true wave front shape is a linear superposition of the five aberrated wave fronts. Next, we will describe briefly the nature of the five Seidel aberrations.

† One can also define asymptotic aberration coefficients for projector lenses.

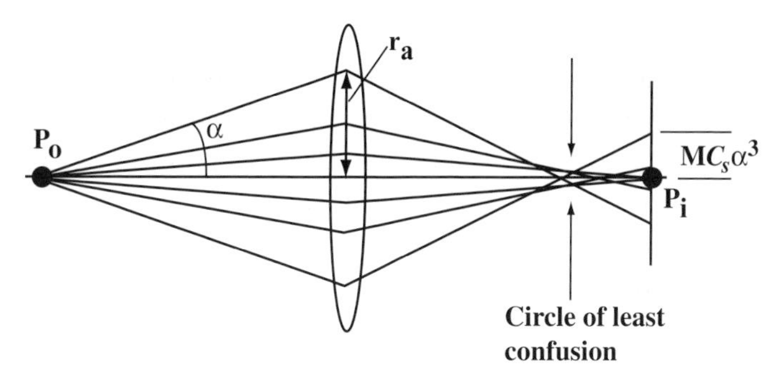

Fig. 3.13. Spherical aberration of rays far from the optical axis.

3.5.2.1 Spherical aberration

The nature of spherical aberration can be understood by considering the first term in equation (3.42) and rewriting the position vector u_a in polar coordinates for a circle of radius r_a in the lens plane: $u_a = r_a(\cos\theta + \mathrm{i}\sin\theta)$, where θ ranges from 0 to 2π. The correction term then becomes

$$\Delta x_i + \mathrm{i}\Delta y_i = MCr_a^3(\cos\theta + \mathrm{i}\sin\theta),$$

which is a circle with radius proportional to the cube of the distance r_a. The circle does not depend on the location u_o of the object point. Figure 3.13 shows schematically how electrons leaving an axial object point are affected by spherical aberration: rays passing through the outer zone of the lens (i.e. through a circle with a large radius r_a) are focused more strongly than the paraxial rays, and will hence come to a focus before the Gaussian image plane.† The spherical aberration coefficient C_s (which is proportional to the coefficient C defined above) can be computed in terms of the axial magnetic field distribution $B(z)$ by:‡

$$C_s = \frac{\eta^2}{8\hat{\Psi}} \int_{z_0}^{z_1} \left\{ B'(z)^2 + \frac{3e}{8m\hat{\Psi}} B^4(z) - B^2(z) \left[\frac{h'(z)}{h(z)} \right]^2 \right\} h^4(z)\,\mathrm{d}z, \tag{3.43}$$

where $h(z)$ is a paraxial trajectory, which leaves an axial object point with unit slope, i.e. $h'(z_o) = 1$ [HK89b]. The presence of powers of the magnetic field in this expression makes the spherical aberration particularly sensitive to small variations in the field. In the Gaussian image plane, a point object is imaged as a disk, with radius

$$r_i = MC_s\alpha^3. \tag{3.44}$$

† The Gaussian image plane is the plane conjugate to the object plane in the absence of aberrations; it can be computed from Newton's lens equation, if the object position is known.

‡ There are many different ways to write down the spherical aberration coefficient; the expression above is special in that it clearly shows that the spherical aberration of a round lens is always a positive quantity. This is known as *Scherzer's theorem*.

The narrowest diameter of the electron beam corresponds to the so-called *circle of least confusion*. The spherical aberration coefficient C_s is typically expressed in millimeters; modern microscopes have an objective lens with C_s in the range 0.5–1.5 mm, while for older models the objective C_s can be several millimeters.

From the expression above, we can see that the spherical aberration coefficient of a round magnetic lens cannot become a negative quantity, i.e. rays affected by spherical aberration intersect the optical axis *before* reaching the image plane. This also means that it is impossible to create a round magnetic lens with zero spherical aberration, since one can show that a lens with $C_s = 0$ does not have a real image [Cre77]. Spherical aberration correction has been a hot research topic ever since Scherzer's discovery of the theorem that now bears his name. Only recently has technology reached the point where C_s-correctors are becoming available commercially, and perhaps the microscopes of the future will have such built-in correctors as a standard component. The interested reader is encouraged to consult the following sources: [HK89b, Chapter 41], [CK80], [Ros71], [HU97], [KDL99]. C_s reduction and/or correction remains an area of very active research and development.

3.5.2.2 Coma

The second term in equation (3.42) can be rewritten without loss of generality by considering the following special case: the object point is fixed at the location $u_o = x_o$, the aberration coefficient has only one non-zero component ($k = 0$), and the electrons pass the lens plane in concentric circles of radius r_a centered on the optical axis:

$$
\begin{aligned}
\Delta x_i + \mathrm{i}\Delta y_i &= MKx_o\left(2r_a^2 + u_a^2\right);\\
&= MKr_a^2x_o\left(2 + \mathrm{e}^{\mathrm{i}2\theta}\right);\\
&= MKr_a^2x_o\left[(2 + \cos 2\theta) + \mathrm{i}\sin 2\theta\right].
\end{aligned}
$$

Figure 3.14(a) shows the aberration figure for the conditions: $x_o = 1$, $M = 1$, $K = 0.1$, and $u_i = 1 + \mathrm{i}0$. The images of the five concentric circles with increasing radii r_a are displaced along the positive x-direction in such a way that the entire aberration figure is located between two straight lines. This aberration only affects object points that are not located on the optical axis. The coma coefficients can also be expressed in terms of special electron trajectories and the axial field $B(z)$; for explicit expressions we refer the reader to [HK89a].

A more general example of coma is shown in Fig. 3.14(b); eight points are selected in the object plane at the locations $(\pm 1, 0)$, $(0, \pm 1)$, and $(\pm 1, \pm 1)$. The

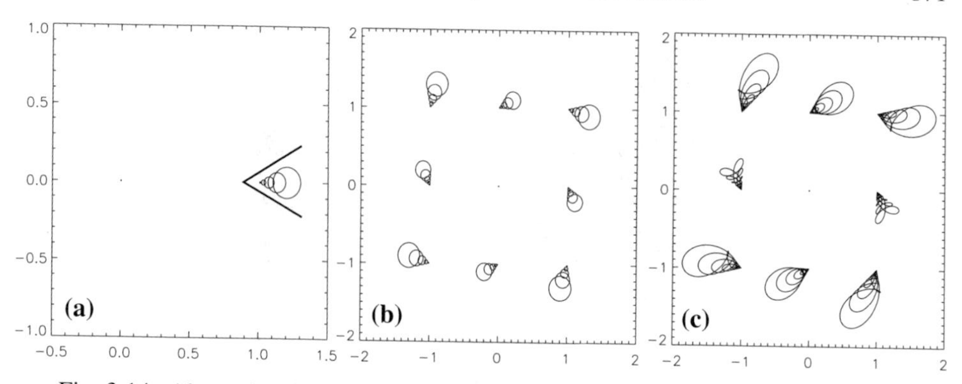

Fig. 3.14. Aberration figures due to coma for the parameters described in the text.

coma coefficient is taken to be $0.05 - i0.1$, and the concentric circles are centered on the optical axis. The resulting aberration figures are computed for the same eight points in the image plane, ignoring the magnification factor. If the concentric circles are centered on an off-axis point in the lens plane, say the point $u_a = 0.4 + i0.2$, then the aberration figures become more complex and are different for different object points, as shown in Fig. 3.14(c). Note that the aberration figures remain located between the same two straight lines, regardless of the position of the point u_a. The orientation of the lines is determined by the value of $K + ik$.

It is clear from the examples in Fig. 3.14 that coma is an undesirable lens aberration, since it generates extended, asymmetric images of point objects. It is possible to perform a *coma-free alignment* of an electron microscope, and we will return to this in Chapter 10 on phase contrast microscopy.

3.5.2.3 Astigmatism and field curvature

Both astigmatism and field curvature are linear in u_a, so it is convenient to treat them together. In the image plane the combined effect of astigmatism (represented in polar notation as $A + ia = \mathcal{A}e^{i\theta_a}$) and field curvature is given by

$$\Delta u_i = \mathcal{A}r_a e^{i(\theta_a - \theta)}\left(x_o^2 - y_o^2 + 2ix_o y_o\right) + F r_o^2 r_a e^{i\theta}, \tag{3.45}$$

where the angle θ once again describes a circle of radius r_a in the lens plane. The effect of the field curvature term is illustrated in Figs 3.15(a) and (b) for $F = 0.05$ and 0.15, respectively. The aberration coefficients are shown in the form $[\mathcal{A}, \theta_a, F]$. Note that the angle θ is only varied over three-quarters of a circle; this will be useful to illustrate the effect of the astigmatism coefficients. The further the object point is located from the optical axis, the larger its image will be. This is to be contrasted with the effect of spherical aberration, which results in disks of the same size, regardless of the position in the object plane.

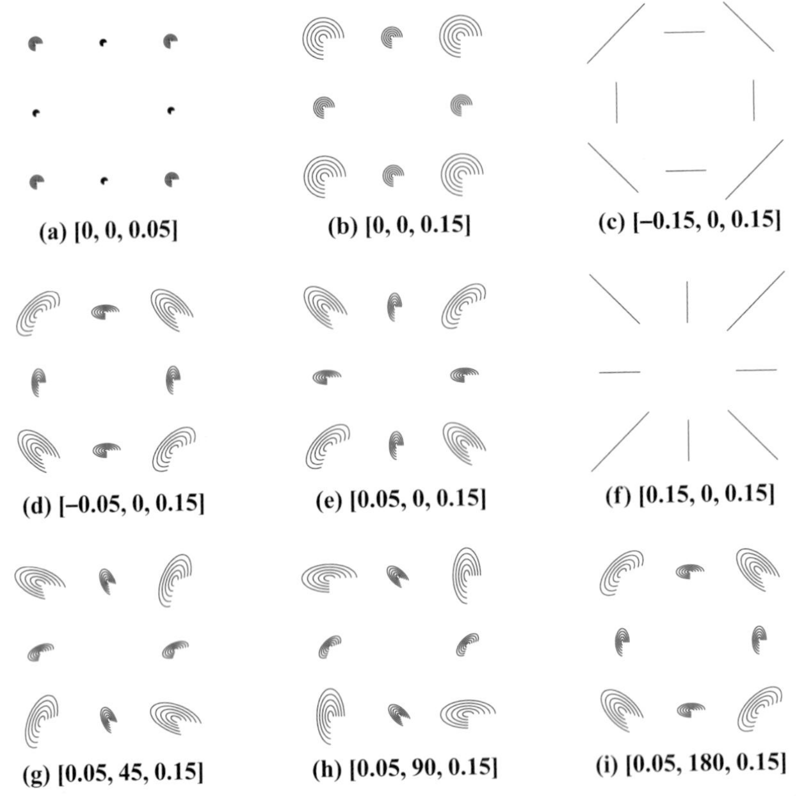

Fig. 3.15. Illustration of the effect of astigmatism and field curvature (represented in the form $[\mathcal{A}, \theta_a, F]$) on the image of eight object points aligned on a square. The circle segments correspond to five circles of increasing diameter in the lens plane.

Consider the point $u_o = 1 + \mathrm{i}0$, and take $\theta_a = 0$; the expression above then simplifies to

$$\begin{aligned}\Delta u_i &= r_a\left(\mathcal{A}\mathrm{e}^{-\mathrm{i}\theta} + F\mathrm{e}^{\mathrm{i}\theta}\right);\\ &= r_a\left[(F + \mathcal{A})\cos\theta + \mathrm{i}(F - \mathcal{A})\sin\theta\right].\end{aligned}$$

It is clear from this expression that for $F = \pm\mathcal{A}$ the aberration figure collapses into a straight line; this is illustrated graphically in Fig. 3.15(c)–(f).

Another way to represent the effect of astigmatism is through a trajectory diagram. Figure 3.16 shows that the focal length of an astigmatic lens becomes dependent on the azimuthal angle, and varies continuously between two extreme values, indicated by arrows in the figure. This lens aberration can be completely canceled by means of *quadrupole stigmators*. Most microscopes have two sets of stigmators: one for the condensor lens, to remove astigmatism from the incident beam, and one for the objective lens. Manual astigmatism correction requires considerable

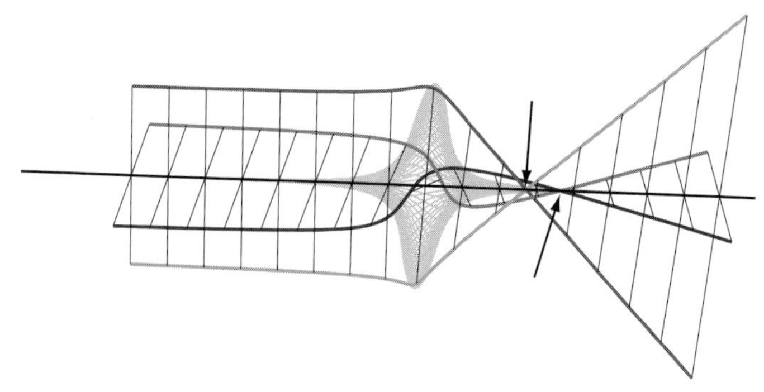

Fig. 3.16. Effect of astigmatism on paraxial rays for a 400 kV electron in Glazer's bell-shaped model with $B_0 = 0.5$ T and $a = 3.0$ mm; the magnitude of the astigmatism corresponds to a 25% azimuthal change of the magnetic field strength.

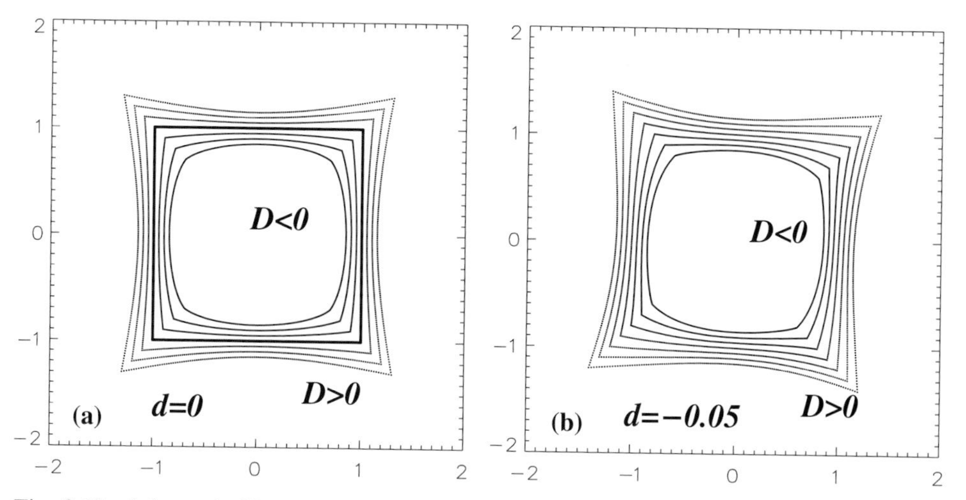

Fig. 3.17. Schematic illustration of the distortion aberration for $D = [-0.3 \ldots 0.3]$ with (a) $d = 0$ and (b) $d = -0.05$. The original square is outlined in (a).

practice and is generally regarded as one of the more difficult aspects of transmission electron microscopy. With modern computer-controlled microscopes, astigmatism correction can be performed entirely and very accurately by a computer algorithm.

3.5.2.4 Distortion

The last Seidel aberration does not depend on the position u_a in the lens plane and is illustrated in Fig. 3.17. The distortions for $d = 0$ are known as *pin-cushion distortion* (for $D > 0$) and *barrel distortion* (for $D < 0$). For $d \neq 0$ an additional torsional distortion is superimposed on the barrel or pin-cushion shapes, as illustrated in Fig. 3.17(b). The distortion aberration is only important in the low-magnification

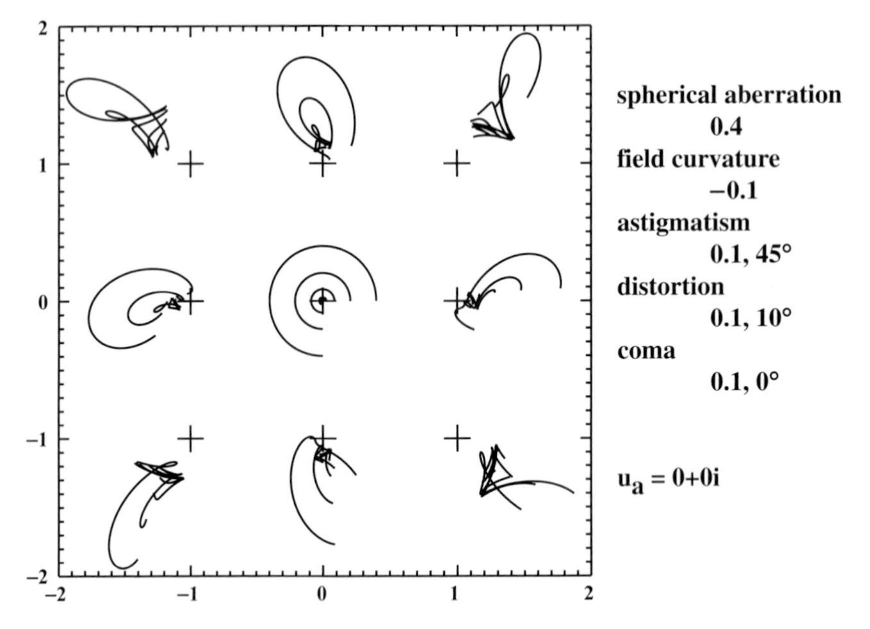

Fig. 3.18. Illustration of the output of the **ION** routine seidel.pro for the lens aberrations stated.

range of a microscope and can be ignored safely for most higher-magnification work.

3.5.2.5 Combination of the five Seidel aberrations

It is now straightforward to combine all five Seidel aberrations into a single illustration (Fig. 3.18), so that the total effect of the aberrations can be studied. The reader may access the **ION** routine seidel.pro from the website to enter various combinations of primary aberration coefficients. The routine will return the aberration figures for nine points on a square in the object plane, with five circles of linearly increasing radii around an arbitrary point in the lens plane.

3.6 Basic electron optics: magnetic multipole lenses

The example trajectory computations of Section 3.4 illustrate a rather simple way to deal with electron optical elements. In the absence of currents inside the field region, one starts from expressions for the magnetic scalar potential ω obtained as solutions of Laplace's equation. By making judicious use of the symmetry of the problem, significant simplifications can often be introduced. For instance, the general expression for the magnetic scalar potential in cylindrical coordinates is given by

$$\omega(r, \varphi, z) = \sum_{m=0}^{\infty} [a_m(r, z) \cos m\varphi + b_m(r, z) \sin m\varphi].$$

For a round magnetic lens only the term in $m = 0$ survives, and this leads to the radial series expansion in equation (3.2). There are many other magnetic or electrostatic elements which do not have rotational symmetry, but instead display a number N of symmetry elements; such elements are known as *multipole lenses*. In the presence of N symmetry planes, the magnetic scalar potential must satisfy the relation [Szi88]

$$\omega(r, \varphi, z) = \omega\left(r, \varphi \pm \frac{2\pi l}{N}, z\right),$$

from which we can derive that

$$\omega(r, \varphi, z) = \sum_{n=0}^{\infty} a_{nN}(r, z) r^{nN} \cos(nN\varphi).$$

For a detailed discussion of this expression and relations derived from it we refer to [Szi88, Chapter 3] and references therein. For the purposes of this textbook it will suffice to discuss briefly the standard magnetic quadrupole, and electrostatic deflectors, because they are common components in a microscope column.

A round magnetic lens does not exert a direct radial force on an electron traveling nearly parallel to the optical axis. The small radial force which brings the electron closer to the optical axis is entirely due to the azimuthal velocity component acquired as a consequence of the cross product nature of the Lorentz force. Multipole elements can be used to exert mostly radial forces on the electron beam, and the simplest multipole field is that present in a beam deflector. The main purpose of a beam deflector is to provide a lateral shift of the entire beam, with as little disturbance of the beam structure as possible [HK89a]. In other words, the ideal deflector simply moves the beam around, without changing its shape. The currents required to run magnetic deflectors and quadrupoles are substantially less than those required for round magnetic lenses. While round lenses require anywhere between 1 and 15 A, deflection coils typically use 10–500 mA and are reversible, i.e. the direction of the current can be reversed. Correction coils use even less current, typically in the range of 1–200 mA [Szi88].

There are several beam deflection systems in a standard TEM. On the microscope console they are commonly labeled as *beam shift*, *beam tilt*, *image shift*, and so on. In addition, many microscopes have a *scanning attachment* as part of the illumination system; the incident beam can then be scanned periodically along a line or a rectangular area on the sample, and various detectors measure a scattered signal (e.g. secondary or back-scattered electrons), fully synchronized with the incident beam. The *scan coils* used for this purpose are essentially high-speed beam deflectors, somewhat similar in nature to the deflection system in a television tube.

While a good deflector does not appreciably change the shape of the electron beam, a quadrupole lens (or rather a pair of quadrupole lenses) is used to adjust the beam shape, in particular to correct astigmatism. In the next two sections we will discuss briefly the main characteristics of deflectors and quadrupole lenses. There is also an extensive literature on the aberrations of multipole elements. This is far beyond the scope of this book, and we refer the interested reader to [HK89a, Chapters 29 and 32] and references therein.

3.6.1 Beam deflection

Beam deflectors can consist of electrostatic or magnetic components. The main purpose of a deflector is to change the direction of the electron beam with respect to the optical axis. Electrostatic deflectors usually consist of two pairs of electrodes oriented perpendicular to each other, one providing deflection in the x-direction, and the other in the y-direction. More complex designs with multiple electrodes are also commonly used. Magnetic deflectors are usually used for scanning applications.

While the electron optical characteristics of deflectors are interesting in their own right, it is not necessary to discuss them in detail. For the "average microscopist" it suffices to point out that there are two different operation modes for deflectors: *beam tilt* and *beam shift*. They are shown schematically in Fig. 3.19 for a converged beam. In the beam shift mode, the first set of deflectors changes the trajectory without changing the beam shape; the second deflector changes the

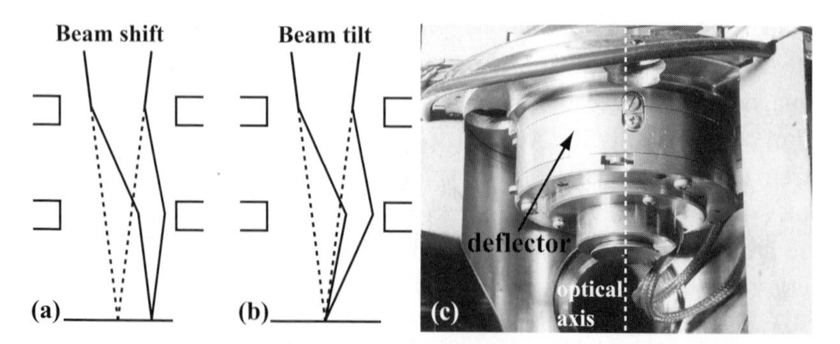

Fig. 3.19. Schematic illustration of (a) beam shift mode and (b) beam tilt mode. Part (c) is a photograph of the beam deflector housing of the JEOL 120CX microscope.

trajectory by precisely the opposite angle, so that the electron beam is translated or shifted across the sample. The field of the individual deflector elements is rather short along the beam path, so that the deflection can be represented as an abrupt change of direction. In beam tilt mode the second deflector provides a deflection *twice as large* as that of the first deflector, so that the illumination spot does not move. Only the angle between the beam and the optical axis at the illumination spot changes. The same pair of deflectors can be used for both beam tilt and beam shift by changing the relative excitations of the various deflector surfaces or coils. It goes without saying that the deflection system must be properly aligned so that beam tilt and beam shift can operate independently. Figure 3.19(c) shows the beam deflector of the JEOL 120CX microscope; the round housing contains the deflector plates and is located below the condensor stigmator which is shown in Fig. 3.20(c).

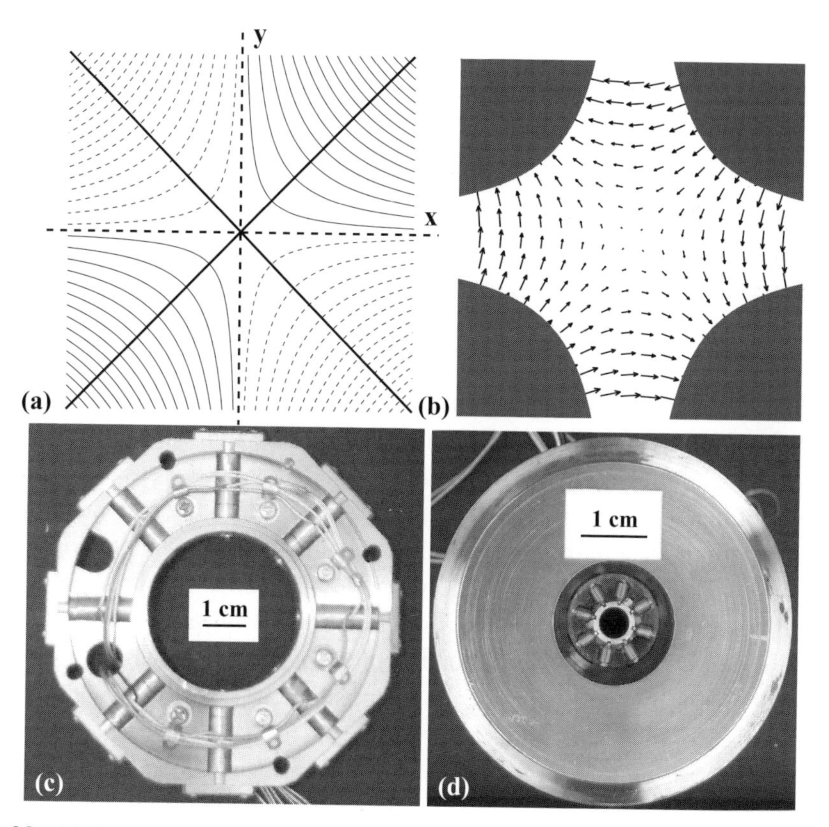

Fig. 3.20. (a) Equipotential contours for an ideal quadrupole field with a potential proportional to xy. The symmetry planes are inclined at 45° with respect to the coordinate axes; (b) magnetic field configuration corresponding to the potential in (a). (c) and (d) are photographs of the condensor and objective stigmators of the JEOL 120CX microscope, respectively.

3.6.2 Quadrupole elements

A *quadrupole lens* is a lens with two orthogonal planes of symmetry, and two additional planes of anti-symmetry oriented at 45° with respect to the symmetry planes (Fig. 3.20a). In the absence of currents in the field region, the magnetic induction inside such a lens can be derived from the scalar magnetic potential $\omega(x, y, z)$. One can show [Szi88] that this potential is to a good approximation described by

$$\omega(x, y, z) = F(z)xy,$$

where $F(z)$ is a function which describes the axial variation of the flux density. Equipotential contours for this potential are shown in Fig.3.20(a). The magnetic field components are determined from the gradient of the scalar potential:

$$\begin{aligned} B_x &= -\mu F(z)y \\ B_y &= -\mu F(z)x \\ B_z &= -\mu F'(z)xy. \end{aligned}$$

Figure 3.20(b) shows such a magnetic field configuration in between the pole pieces. Note that if $F(z)$ is constant inside the quadrupole lens, then there is no axial component to the field and the field components lie entirely in a plane perpendicular to the optical axis. For electrons traveling nearly parallel to the optical axis the resulting Lorentz force will also be normal to the axis. Because of the cross product nature of the Lorentz force we find that the force vectors are tangent to the equipotential lines for an electron traveling parallel to the optical axis. Such a field geometry may hence be used to change the shape of the electron beam. Conversely, a beam affected by astigmatism can be restored to a circular shape. The main use of quadrupole lenses in the transmission electron microscope is therefore to correct astigmatism in the beam. This is typically done at two locations: the first set of quadrupole lenses is located between the last condensor lens and the beam deflectors. A photograph of such a correction coil is shown in Fig. 3.20(c). This correction coil is actually a pair of quadrupoles oriented at 45° with respect to each other. This permits adjustments of the beam shape in two orthogonal directions. The second set of astigmatism correctors is located immediately below the objective lens lower pole piece, and is typically much smaller than the condensor stigmator – a photograph is shown in Fig. 3.20(d). These correction coils are among the most important components of a modern electron microscope. We will see in Chapter 10 that without astigmatism correction coils, the resolution of a TEM would be around 1 nm, not good enough for observations at the length scale of interatomic distances.

3.7 Basic electron optics: electron guns

3.7.1 Introduction

In this section, we will discuss one of the most essential components of a TEM: the electron gun. The main purpose of the electron gun is to provide a steady stream of electrons with a constant kinetic energy, i.e. a *monochromatic* electron beam. Ideally, the electrons should appear to emanate from a single point on the filament surface, so that the gun can be described as a *point source*. In addition, since we have seen in the preceding sections that several lens aberrations depend on the angle between the trajectory and the optical axis, the electrons should travel close to the optical axis, at very small angles to the axis. Finally, after passing through the entire column (including the sample) there must be sufficient electrons left over to form an image on the detector (screen, micrograph, etc.) to acquire the data in a reasonable amount of time (exposure time). This means that the *image current density*, the number of electrons per second that hit a unit area of the detector, must be reasonably high.

These qualitative statements can be cast into more precise mathematical terms, and the design of electron guns is a highly specialized field, similar to the design of magnetic lens elements. The basic design process goes as follows: for a given geometry of the filament and all potential surfaces around the tip, the electric equipotential surfaces are computed, typically by finite element analysis. The trajectory equation is solved for a large number of trajectory starting points, and the properties of the assembly are computed, including aberrations. The design parameters are then varied until the desired functionality is obtained.

We will discuss the various types of electron guns in the following subsections. First, we will describe the physics behind the process of *electron emission*.

3.7.2 Electron emission

In this section, we will closely follow a paper by Fransen et al. [FCRT$^+$99] and Chapter 44 in [HK89b]. For simplicity we will represent the surface of the filament tip as a flat surface of unspecified lateral dimensions. The z-axis is taken to be normal to the surface. At absolute zero, the energy distribution of the electrons inside the metal is rather simple: all energy levels up to a maximum energy are occupied. This maximum energy is known as the *Fermi energy* E_F. When the temperature is increased, some electrons will gain sufficient thermal energy to occupy higher energy states. The distribution of occupied states is described by the *Fermi–Dirac distribution*:

$$f(E,T) \equiv \frac{1}{1+\mathrm{e}^{[E-(E_F-V_0)]/k_BT}}, \tag{3.46}$$

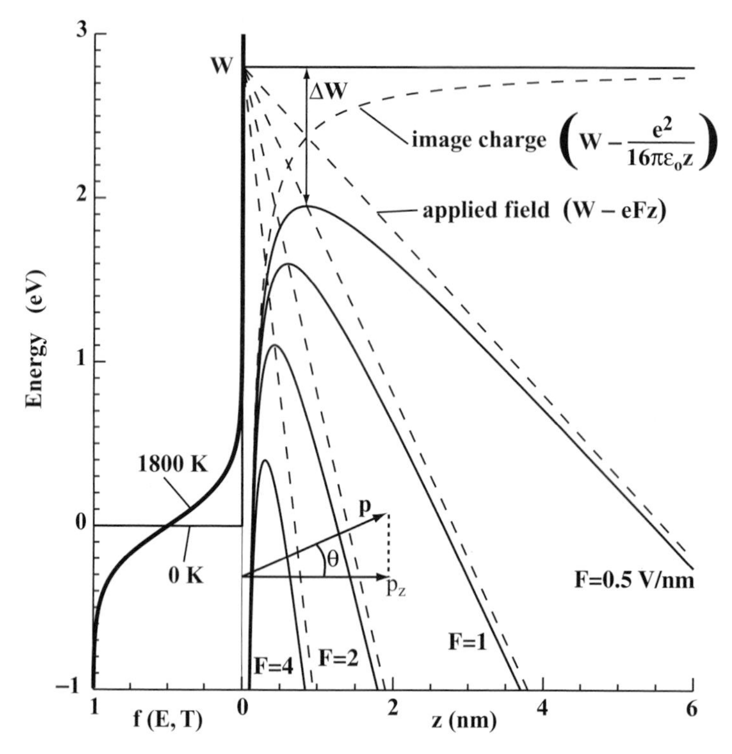

Fig. 3.21. Schematic illustration of the Fermi–Dirac function at 0 and 1800 K (left-hand side). The barrier potential $V(z)$ (in eV) is shown as a solid line for applied field strengths of $F = 0.5$, 1, 2, and 4 V nm^{-1}, and a work function $W = 2.8$ eV. The components of $V(z)$ are shown as dashed lines. (This figure is based on Fig. 2 in [FCRT$^+$99].)

where k_B is the Boltzmann constant ($8.618\,16 \times 10^{-5}$ eV K^{-1}), and V_0 is a constant used to adjust the energy origin. As the temperature increases, the Fermi–Dirac function develops a high-energy tail, as shown on the left in Fig. 3.21 for a temperature of $T = 1800$ K and $V_0 = E_F$.

The energy needed for an electron to escape the metal is given by the *work function*, W. The work function is equal to the difference between the vacuum energy level and the Fermi level, and depends on both the atom type and the crystal plane forming the metal surface. For pure tungsten the work function is 4.5 eV, while for ZrO-coated (001) tungsten, W is reduced to 2.8 eV. If the filament temperature is sufficiently high so that energy states above the vacuum level are occupied, then electrons will be emitted into the vacuum. This process is known as *thermal emission*. If an electric field is applied between the metal surface and an external electrode, then the vacuum level becomes a function of the distance to the surface and is represented by the *barrier potential* $V(z)$. It is customary in the electron emission literature to represent this electric field by the symbol F instead of E, since E is already taken to represent the microscope acceleration voltage.

The energy $V(z)$ depends on the workfunction W, the applied field F, and also on the Coulomb interaction between the emitted electron and its mirror image in the metal surface. An explicit expression is given by

$$V(z) = W - eFz - \frac{e^2}{16\pi\epsilon_0 z} \qquad (z > 0); \tag{3.47}$$

$$= W - Fz - \frac{0.359\,95}{z}\ [\text{eV}] \qquad (\text{with } F \text{ in V nm}^{-1},\ z \text{ in nm}).$$

The zero of potential energy in this case is taken at the Fermi level. Figure 3.21 shows the function $V(z)$ for several field strengths F and a work function of 2.8 eV [FCRT$^+$99]. The dashed lines indicate the Coulomb term and the electric field term. It is clear that with increased applied field strength the barrier energy becomes lower by an amount ΔW. This lowering is known as the *Schottky effect*. It is left to the reader to show that

$$\Delta W = \sqrt{\frac{e^3 F}{4\pi\epsilon_0}} = 1.1998\sqrt{F}\ [\text{eV}] \quad (\text{with } F \text{ in V nm}^{-1}). \tag{3.48}$$

The main question to be answered is then: what is the emitted current density in the z-direction at a given temperature and applied field? While the general solution to this problem is still unknown, there are several special cases for which an approximate analytical solution is possible. For an electron that has left the metal with momentum $\mathbf{p}$ we have a total energy of

$$E = \frac{\mathbf{p}\cdot\mathbf{p}}{2m} + V(z).$$

This energy can be decomposed into components normal and tangential to the surface. We define (following [HK89a])

$$U_n \equiv \frac{1}{2m}p_z^2 + V(z); \qquad U_t \equiv \frac{1}{2m}\left(p_x^2 + p_y^2\right).$$

Not every electron with the correct energy will actually escape from the metal. A fraction of the electrons will be reflected from the surface and return to the bulk of the metal. The fraction of electrons that is actually emitted is described by $D(U_n)$, the *transmission coefficient*.

Next, we consider an elementary volume in momentum space, $\mathrm{d}^3 p$. If we use cylindrical coordinates, as shown in Fig. 3.21, then we can rewrite this volume element as

$$\begin{aligned} d^3 p &= \sin\theta\ p^2\, dp\, d\theta\, d\phi, \\ &= \sin\theta\ p(p\, dp)\, d\theta\, d\phi, \\ &= \sin\theta \sqrt{2m(E - V(z))}(m\, dE)\, d\theta\, d\phi, \end{aligned}$$

where we have used $p^2 = 2m[E - V(z)]$ and $p\,dp = \frac{1}{2}d(p^2) = m\,d(p^2/2m) = m\,dE$. The total number $N(p)$ of electrons with momentum between p and $p + dp$ that have enough energy to escape the metal is given by the product of the number of quantum states per elementary volume in momentum space ($2/h^3$), the probability that the electron will have the correct energy (the Fermi–Dirac function), and the probability that the electron will not be reflected ($D(U_n)$), or

$$N(p) = \frac{2}{h^3} \frac{D(U_n)}{1 + e^{(E-E_F)/k_B T}}.$$

The current density element $d^3\mathbf{j}$ is then given by

$$\begin{aligned} d^3\mathbf{j} &= e\frac{\mathbf{p}}{m} N(p)\, d^3 p \\ &= \frac{2e}{mh^3}\mathbf{p} f(E, T) D(U_n)\, d^3 p. \end{aligned} \tag{3.49}$$

Since we are mostly interested in the current density component along the z-direction we can use $p_z = \sqrt{2m[E - V(z)]}\cos\theta$ to obtain [FCRT+99]:

$$d^3 j_z = \frac{2me}{h^3} \frac{[E - V(z)]}{1 + e^{(E-E_F)/k_B T}} D(U_n) \sin 2\theta\, dE\, d\theta\, d\phi. \tag{3.50}$$

For a given model for $D(U_n)$ and $V(z)$, this equation can be integrated over all angles ($0 < \phi < 2\pi; 0 < \theta < \pi/2$), and over all appropriate energies to obtain the total emitted current density. We will now consider some special cases for which these integrations can be carried out analytically.

3.7.2.1 Thermionic emission

In the case of *thermionic emission*, there is no external applied field and the origin of the energy scale can conveniently be chosen to coincide with the vacuum level. The potential barrier in the absence of a field vanishes outside the crystal, and the transmission coefficient $D(U_n) = 1$ for positive U_n. At high temperature, the Fermi–Dirac function can be simplified by dropping the 1 in the denominator and we find for the current density (3.50):

$$j_z = \frac{2me}{h^3} \int_0^{2\pi} \mathrm{d}\phi \int_0^{\pi/2} \sin 2\theta \,\mathrm{d}\theta \int_0^{+\infty} E\mathrm{e}^{-(E+W)/k_BT} \,\mathrm{d}E,$$
$$= AT^2\mathrm{e}^{-W/k_BT} \equiv j_T, \tag{3.51}$$

with

$$A = \frac{4\pi mek_B^2}{h^3} = 1.201\,75 \times 10^6 \frac{\mathrm{A}}{\mathrm{m^2\,K^2}}.$$

This equation is known as the *Richardson–Dushman equation*, and can be used to determine the work function W from measurements of the current density $j_z(T)$ at different temperatures. For a tungsten filament with $W = 4.5$ eV, operated at $T = 2700$ K, we have $j_T \approx 3.5\ \mathrm{A\,cm^{-2}}$. A lanthanum-hexaboride filament (LaB_6) with a work function of $W = 2.7$ eV is typically operated between $T = 1800$ K ($j_T = 10.7\ \mathrm{A\,cm^{-2}}$) and $T = 1900$ K ($j_T = 29.9\ \mathrm{A\,cm^{-2}}$).

3.7.2.2 Schottky emission

We have seen previously that in the presence of an electric field F of the order of $1\ \mathrm{V\,nm^{-1}}$ the barrier energy is reduced by an amount ΔW. The emission current density of a Schottky emitter is then given by equation (3.51) with W replaced by $W - \Delta W$, or

$$j_S = \mathrm{e}^{\Delta W/k_BT} j_T. \tag{3.52}$$

The emission current is hence enhanced with respect to the thermionic emission current. For a ZrO-coated tungsten single crystal filament with a work function of $W = 2.8$ eV and an applied field of $1\ \mathrm{V\,nm^{-1}}$ ($\Delta W = 1.2$ eV), the current density is $j_S = 1.3 \times 10^4\ \mathrm{A\,cm^{-2}}$, about three orders of magnitude larger than for thermionic emission.

Equation (3.52) is valid as long as *tunneling* through the potential barrier can be ignored. If the applied field is sufficiently strong, so that the potential barrier becomes narrow (Fig. 3.21), then the transmission coefficient $D(U_n)$ must be computed explicitly. This is beyond the scope of this textbook and we refer to [FCRT+99] for an explicit derivation of the so-called *extended Schottky emission*.

3.7.2.3 Thermal field emission

In thermal field emission the temperature of the filament is kept low (room temperature) so that the high-energy tail of the Fermi–Dirac function nearly vanishes. Electrons can only escape from near the Fermi level by tunneling through the narrow potential barrier, as shown in Fig. 3.21 for $F = 4\ \mathrm{V\,nm^{-1}}$. The applied field is typically several volts per nanometer and the current density computation becomes rather lengthy; we refer to Fransen et al. [FCRT+99] for a brief description and Kasper [Kas82] for a detailed account. The current density for thermal field emission is given by

$$j_{TF} = \frac{4\pi me}{h^3} d^2 \mathrm{e}^{-bW/d} \frac{\pi p}{\sin \pi p}. \tag{3.53}$$

The factor d is a function of the applied field and the work function (see [Kas82]):

$$d = \frac{ehF}{4\pi t(\Delta W/W)\sqrt{2mW}} = \frac{9.7598 \times 10^{-2}}{t(\Delta W/W)} \frac{F}{\sqrt{W}} \text{ [eV]} \quad (F \text{ in V nm}^{-1}, W \text{ in eV}),$$

with

$$t(y) = 1 + 0.1107 y^{1.33},$$

$b \approx 0.6$, and $p = k_B T/d$. Equation (3.53) at $T = 0$ K is known as the *Fowler–Nordheim equation*. For non-zero temperatures the equation is only accurate for $p < 0.7$. Comparison with the Richardson–Dushman equation shows that d takes on the role of $k_B T$. For a tungsten field emission tip operated at room temperature ($T = 300$ K) and an applied field of 7 V nm^{-1}, we have $d = 0.3$ eV, $p = 0.085$, and $j_{TF} = 1.9 \times 10^5$ A cm^{-2}.

3.7.2.4 Beam brightness

Let us assume that all electrons emanate from a circular region of radius r on the tip of a filament. An observer at P, a distance q from the filament (Fig. 3.22b), will then "see" an electron source with a surface area of πr^2. The size of this source is typically expressed as a *solid angle* rather than a surface area, since the apparent

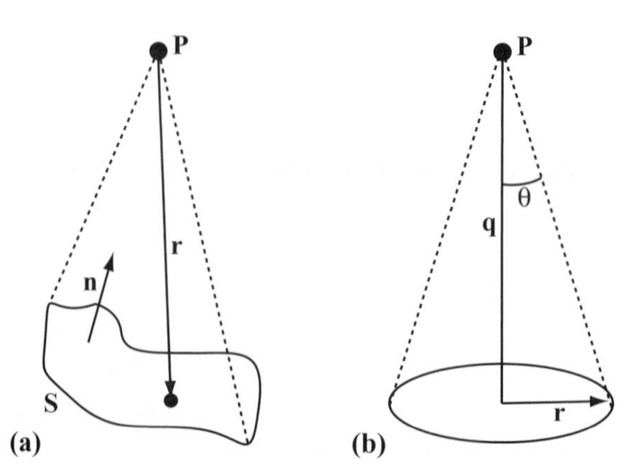

Fig. 3.22. Definition of the solid angle and the aperture semi-angle θ; (a) general case and (b) for a cone.

surface area depends on the distance q. Solid angle is defined as *the area subtended by a surface S when looking at that surface from a reference point, divided by the square of the distance to that surface*, or

$$\Omega(S) = \iint_S \frac{\mathbf{r} \cdot \mathbf{n}}{r^3} \, \mathrm{d}S$$

with $\mathbf{n}$ the outward normal to the surface (Fig. 3.22a). For a sphere, the solid angle, when looking from the center of the sphere, is equal to

$$\Omega = \int_0^{\pi} \int_0^{2\pi} \sin\theta \, \mathrm{d}\theta \, \mathrm{d}\phi = 4\pi.$$

The unit of solid angle is the *steradian*. The solid angle subtended by the base of a cone with base radius r when viewed from the top is

$$\Omega = \frac{\pi r^2}{q^2} = \pi\theta^2$$

(see Fig. 3.22b) with $q = r/\theta$ the height of the cone.

The total current density j passing through the point P is proportional to the solid angle subtended by the source at the point P, and the proportionality factor β is known as the *beam brightness*:

$$\beta = \frac{j}{\pi\theta^2}. \tag{3.54}$$

Brightness is measured in amperes per unit area per steradian. If we represent the current density j by the current I divided by the area A, then we can determine how a magnetic lens affects the beam brightness. A magnetic lens with lateral magnification M reduces the current density I/A to I/M^2A^2; the angular aperture θ is increased by the angular magnification M_α by a factor $M_\alpha^2 = 1/M^2$, so that the total brightness remains unchanged. Brightness is hence a property of an electron beam and is left unchanged by (aberration-free) magnetic lenses.

From the definition of brightness we find that a higher current density will increase the brightness; in other words, if we can get more electrons to emanate from the filament emission area with a momentum *inside* the cone centered at P, then the current density and hence the brightness would increase. In Section 3.7.2 we have seen that the energy of the emitted electron could be separated into two components U_n (corresponding to the momentum p_z and the barrier potential) and U_t (corresponding to the momentum components p_x and p_y). It is then easy to see that a rapid decrease of current density with increasing tangential momentum components (and therefore with U_t) will produce a higher brightness. The (axial) brightness can hence be defined as [FCRT+99, SM84]:

$$\beta \equiv \frac{e\hat{\Psi}}{\pi}\left(\frac{\mathrm{d}j_z}{\mathrm{d}U_t}\right)_{U_t=0}, \tag{3.55}$$

with $\hat{\Psi}$ the relativistic acceleration potential introduced in Chapter 2. We can compute the brightness for the various emission mechanisms discussed in the previous section. The resulting brightnesses are given by

$$\beta_T = \frac{e\hat{\Psi}j_T}{\pi k_B T} \quad \text{(thermionic emission);} \tag{3.56}$$

$$\beta_S = \frac{e\hat{\Psi}j_S}{\pi k_B T} \quad \text{(Schottky emission);} \tag{3.57}$$

$$\beta_{TF} = \frac{e\hat{\Psi}j_{TF}}{\pi d} \quad \text{(thermal field emission).} \tag{3.58}$$

The **ION** routine emission.pro on the website can be used to display the barrier potential $V(z)$ and various quantities (current density, brightness, work function reduction, etc.) for all three emission mechanisms described in this section. The drawings produced by this routine are similar to that shown in Fig. 3.21.

Using the gun brightness we can compute the current density at any point along the optical axis of the microscope, using equation (3.54). The current density in the illumination spot after an aperture of diameter r, at a distance D from the aperture plane, is given by

$$j_0 = \pi\beta\left(\frac{r}{D}\right)^2.$$

For a tungsten hair-pin filament with $\beta = 5 \times 10^5$ A cm^{-2} sr^{-1} [Spe88] the current density for a 400 μm diameter condensor aperture ($r = 0.2$ mm) is then approximately 1 A cm^{-2}, using an approximate value for $D = 250$ mm, measured directly on the cross-section of the microscope column discussed earlier in this chapter. If a smaller condensor aperture is used, the current density goes down but the brightness remains constant. The image current density at the phosphor screen for a magnification of $M = 500\,000$ is then 4×10^{-12} A cm^{-2}, i.e. current densities at the viewing screen are of the order of a few pA cm^{-2}. This determines the exposure time for electron micrographs, when it is combined with the *sensitivity* of the emulsion (see Section 3.11 on electron detectors). The maximum brightness for a given electron gun is proportional to the acceleration potential and increases exponentially with temperature; the lifetime of a filament decreases with increased temperature, so there is an optimum operating temperature for the filament.

Table 3.2. *Approximate parameters for different electron guns (sources: [Spe88, Rei93, WC96, FCRT+99]).*

	Thermionic emission		Field emission	
	Tungsten	LaB_6	Schottky	Field emission
Work function [eV]	4.5	2.7	2.8	4.5
Operating temperature [K]	2700	1700	1800	300
Current density [A cm^{-2}]	1–3	25–100	10^4–10^6	10^4–10^6
Virtual source diameter [nm]	50 000	10 000	10–15	3–5
Brightness [A cm^{-2} sr^{-1}]	10^4	10^6	10^8–10^9	10^8–10^9
Energy spread [eV]	1–3	0.5–2	0.3–1.0	0.2–0.3
Required gun vacuum [Pa]	10^{-3}	10^{-5}	10^{-7}	10^{-8}
Lifetime [h]	25–100	500–1000	>1000	>1000
Cost [U.S.$]	10–100	400–1200	1500–3500[a]	1500–3500

[a]Does not include installation costs.

3.7.3 Electron guns

In the previous section, we have described how many electrons are emitted from the filament tip for a given work function, temperature, and applied field. The shape of the filament tip and the geometry of the applied field did not enter the discussion at any point. In reality, the geometry becomes very important because it determines the trajectory of the electron once it has been emitted from the tip. Solving the trajectory equations for a given gun geometry is far beyond the scope of this book and we refer the interested reader to Chapters 43–50 in [HK89b] and references therein.

In this section, we will describe the structure of the electron gun. Several different gun types are available based on thermionic emission, (cold) field emission, or Schottky field emission. Table 3.2 shows some of the relevant parameters for the different electron guns. Descriptions of measurement techniques for gun characteristics can be found in [Spe88, Section 7.2] and [WC96, Section 5.5].

3.7.3.1 Thermionic electron guns

Figure 3.23(a) shows a schematic drawing of the components of a thermionic electron gun, also known as a *triode gun*, along with a photograph of the cross-section of the gun of a JEOL 120CX TEM (Fig. 3.23b). The gun consists of three basic components: the *filament* (F), the *grid* or *Wehnelt cap*, and the *anode*. The filament is typically a hairpin-shaped tungsten wire or a pointed tungsten wire. In the latter case the pointed wire is spot-welded to a tungsten support wire. A voltage V produces a resistive heating current I_F through the filament wire. Tungsten

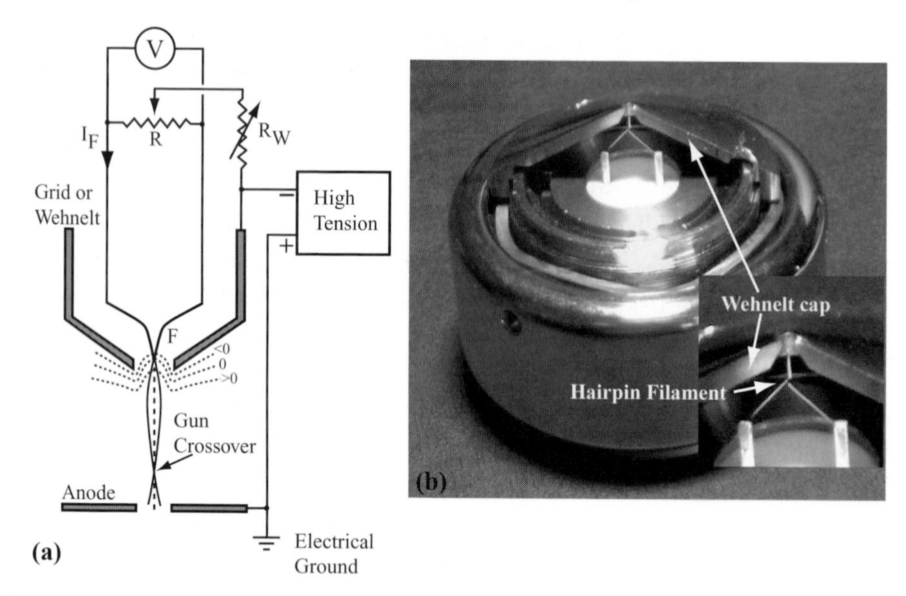

Fig. 3.23. (a) Schematic of a thermionic electron gun; (b) photograph of the cross-sectioned Wehnelt assembly of the JEOL 120CX microscope.

filaments, with a work function of $W = 4.5$ eV, are operated at 2600–2800 K. Single-crystal LaB_6 emitters have a work function of 2.7 eV, and are operated in the range 1700–1900 K.

The anode surface is kept at ground potential, and the high-tension supply generates a voltage (the acceleration voltage E) between the filament tip and the anode plate. The filament is kept at a negative potential of $-E$ V with respect to the anode, so that electrons emitted from the tip are accelerated towards the anode. Thermal emission occurs not only from the tip but also from the flanks of the tip, and this would lead to electrons traveling at large angles with respect to the optical axis. To prevent emission from the filament flanks, a conical surface is introduced around the filament. This surface is known as the *grid* or, more commonly, the *Wehnelt cap*. The grid is kept at a small negative voltage with respect to the tip by means of a variable resistor R_W. On the microscope console the resistance R_W can be changed by means of the *gun bias* control. The bias resistor is tied into the filament heating circuit, so that the energy of the accelerated electron will always be E eV, regardless of the bias setting.

The negative Wehnelt cap produces a zero-potential contour, indicated with dotted lines in Fig. 3.23(a). The bias voltage is set so that this contour intersects the filament close to the tip. Regions along the flanks of the filament experience a negative potential and hence thermionic emission will be suppressed. Changing the gun bias amounts to changing the location of this zero-potential surface, and hence the size of the emitting region. If the bias voltage is too large then there will be no intersection between the zero-potential contour and the tip, which means that there will be no emission. If the bias voltage is too low, then the size of the emitting region will be large, and therefore the gun brightness will decrease. There is an optimal gun bias setting for which the brightness is maximized. This voltage setting depends sensitively on the precise location of the tip within the Wehnelt cap and on the shape of the Wehnelt cap. Typical filament currents at maximum brightness are about 100 μA for tungsten filaments and this is indicated by the *beam current* on the microscope console.

3.7.3.2 Schottky and field emission electron guns

Schottky and field emission electron guns are quite similar in construction, so we will describe them both in this section. A Schottky electron gun employs a tungsten single-crystal filament with a flat circular (100) facet normal to the optical axis; the diameter of this facet is typically around 1 μm. The filament is coated with a thin ZrO layer, which reduces the work function from 4.5 to about 2.8 eV. The operating temperature can hence also be lowered to about 1800 K. A *suppressor cap* (see Fig. 3.24a) surrounds the filament so that only the very tip protrudes through the cap. The suppressor cap takes on the role of the Wehnelt cap in a thermionic gun: a voltage of −2 kV prevents electron emission from the flanks of the filament crystal.

An *extractor plate*, kept at a positive potential in the range 2–7 kV, provides the electric field needed to extract electrons from the tip. The field is inversely proportional to the tip radius so that fields of several volts per nanometer can be obtained. The electrons are then accelerated by the *anode plate* and emerge through a small hole along the optical axis. The electrons appear to emanate from a point

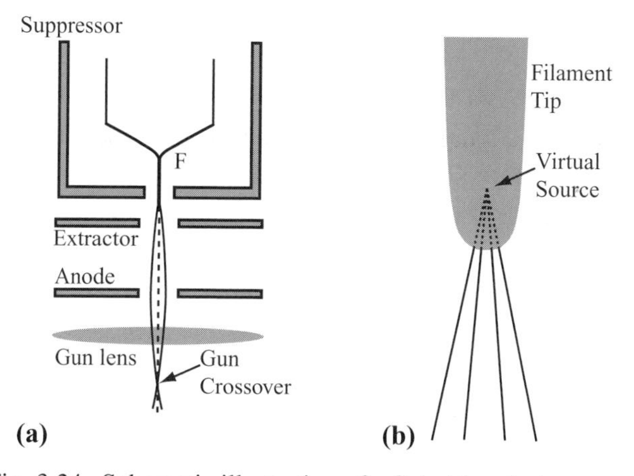

Fig. 3.24. Schematic illustration of a Schottky electron gun.

inside the filament (Fig. 3.24b), the *virtual source* with a size of about 10 nm, and do not form a gun cross-over, so that a *gun lens* is needed to intersect the electron trajectories with the optical axis. The gun lens is an electrostatic lens with a variable potential in the range 0.5–2 kV. A strongly excited gun lens will bring the gun cross-over close to the gun, whereas a weakly excited gun lens will have a lower cross-over (closer to the first condensor lens). The emission current is typically about 150 μA.

A cold field emission gun is very similar in construction to the Schottky gun. It has the same components: suppressor, extractor, anode, and gun lens. The filament tip (typically tungsten) has a radius of about 100 nm and the extractor voltage provides a field strength sufficient to cause electrons near the Fermi level to tunnel through the barrier potential (which is narrow because of the applied field). The gun vacuum for a cold field emitter must be around 10^{-8} Pa, whereas a Schottky emitter requires about 10^{-6} Pa. The Schottky emitter also provides a more stable emission current over a long period of time. Cold field emitter tips must be *flashed* regularly, to remove contaminants from the emitter surface. Parameter ranges for both Schottky and cold field emission guns can be found in Table 3.2.

3.7.4 Beam energy spread and chromatic aberration

Electrons which leave a thermionic emission tip have initial energies of at least the work function W. In this energy range the Fermi–Dirac distribution function (3.46) reduces to a Maxwell–Boltzmann distribution (taking the zero of the energy scale at the Fermi level):

$$f(E,T) = \frac{1}{1+\mathrm{e}^{-E/k_BT}} \approx \mathrm{e}^{-E/k_BT}. \tag{3.59}$$

We can then use equation (3.51) to define the *normalized energy distribution* $g(E)$ [HK89b]:

$$g(E)\,\mathrm{d}E \equiv \frac{\mathrm{d}j_z}{j_z} = \frac{E}{(k_BT)^2} f(E,T)\,\mathrm{d}E. \tag{3.60}$$

The distribution $k_BTg(E)$ is shown in Fig. 3.25 as a function of the variable E/k_BT. The mean electron energy is given by $\langle E\rangle = 2k_BT$, the *most probable energy* corresponds to the maximum of the curve at $E_m = k_BT$, and the energy spread (full width at half maximum, or FWHM) is given by $\Delta E = 2.446k_BT$. Table 3.3 shows the relevant energies for a tungsten emitter and a LaB_6 emitter, using a typical operating temperature. The last column shows the corresponding expressions for a thermal field emission gun. The energy spread for field emission is in the range 0.2–0.4 eV. The energy distribution should be superimposed on the final energy of the electrons after acceleration.

Table 3.3. *Energy distribution parameters for a tungsten emitter and a LaB_6 emitter; energies in eV, temperature in K. The last column shows the explicit expressions for a thermal field emitter [HK89b].*

	W	LaB_6	Thermal field emission
Work function	4.5	2.7	
Temperature	2800	1900	
$f(W, T)$	7.95×10^{-9}	6.89×10^{-8}	
$\langle E \rangle$	0.482	0.327	$\pi pd \cot(\pi pd)$
E_m	0.241	0.164	$d\pi p / \sin(\pi p)$
ΔE	0.590	0.400	$k_B T[\ln p - \ln(1 - p)]$

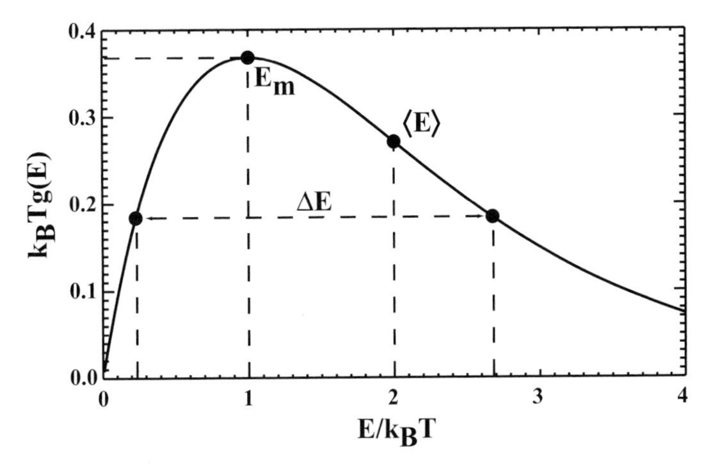

Fig. 3.25. Normalized energy distribution $k_B T g(E)$ as a function of $E/k_B T$ for thermionic emission. The mean energy, most probable energy, and energy spread are indicated.

It was discovered experimentally by Boersch in 1954 [Boe54] that there is an anomalous broadening of the energy distribution for thermionic emission, that the energy distribution tends to become symmetric around its maximum E_m (which is not the case for the Maxwell–Boltzmann distribution shown in Fig. 3.25), and that the mean energy $\langle E \rangle$ is shifted. These effects increase with increasing beam current. Although there is at present no single theory explaining all of these effects, it has been shown that they are due to *stochastic Coulomb interactions* between the electrons, whenever they come close together (as in a cross-over). This has led to modifications of the shape of the Wehnelt cap; the net result of these modifications is that the cross-over moves further away from the cathode (telescopic effect) and can sometimes be completely avoided. The Boersch effect occurs whenever electrons are forced into a narrow beam, e.g. in electron lithography where high probe currents and small probe diameters are required. The energy spread of a thermionic gun is increased to about 1 eV by the Boersch effect.

Recent developments in electron optical design of *monochromators* make it possible to reduce the energy spread of the beam down to the range 50–200 meV. Such *energy filters* are becoming available on dedicated microscopes and may revolutionize the study of electron energy loss mechanisms.

Chromatic aberration is caused by the fact that the focal length of a lens depends on the wavelength and hence on the electron energy. Small fluctuations in the high-voltage supply or in the lens power supply will cause small variations in λ (and thus in $\hat{\Psi}$) and in $B(z)$, respectively. There are four important sources of fluctuations and/or energy changes in a microscope:

(i) intrinsic energy spread of the electrons leaving the filament, typically $\Delta E/E \approx 10^{-5}$ (see Tables 3.1 and 3.2);
(ii) high-voltage instabilities, $\Delta V/V \approx 10^{-6}\,\mathrm{min}^{-1}$;
(iii) fluctuations $\Delta I/I$ of the lens currents;
(iv) energy losses in the specimen.

As a consequence of these fluctuations the focal length f of a magnetic lens will no longer be precisely defined and instead we can talk about the *defocus spread* Δ [Spe88]:

$$\Delta = C_c\left[\frac{\sigma^2(V)}{V^2}+\frac{4\sigma^2(I)}{I^2}+\frac{\sigma^2(E)}{E^2}\right]^{1/2}, \tag{3.61}$$

where σ indicates the standard deviation. The first term represents fluctuations of the acceleration voltage, the second represents fluctuations of the focal length due to fluctuations of the lens current I (recall that the focal length depends on the square of the field $B(z)$, hence the factor of 4 above), and the last term corresponds to the intrinsic energy spread of the electron gun, including the Boersch effect. The *chromatic aberration coefficient* C_c, typically expressed in mm, can be written in terms of the axial field distribution $B(z)$ as [HK89a]:

$$C_c = \frac{e}{8m\hat{\Psi}}\int_{z_0}^{z_1} B^2(z)h^2(z)\,\mathrm{d}z, \tag{3.62}$$

where $h(z)$ is a paraxial trajectory that leaves an axial object point with unit slope. The net effect of chromatic aberration is that the image of a point will be blurred into a disk. Chromatic aberration can be removed in principle by combining the lens with an electrostatic mirror with negative C_c. We will return to chromatic aberration in Chapter 10.

3.7.5 Beam coherence

The primary purpose of the electron gun and the illumination system as a whole is to create a stream of electrons with a well-known reference state. We will see in Chapter 5 that the equations describing the interaction between beam electrons and the sample can be solved when the incident electron can be represented by a *plane wave*. We must, therefore, introduce a measure for how well a beam electron can be described by a plane wave.

The *Heisenberg uncertainty principle* states that the product of the uncertainty or spread in energy and the uncertainty in time must be larger than Planck's constant, or

$$\Delta E \Delta t > h. \tag{3.63}$$

For a thermionic gun with an energy spread ΔE of 1 eV we find that $\Delta t \approx 4 \times 10^{-15}$ s. During this time interval the electron travels a distance λ_{tc}, known as the *temporal coherence length*, given by

$$\lambda_{tc} = c\Delta t\sqrt{1 - \frac{1}{\gamma^2}}. \tag{3.64}$$

At 200 kV, we have $\lambda_{tc} = 861$ nm. The temporal coherence length of optical lasers varies from a few centimeters to many meters. The temporal coherence length would be infinite for an electron gun with zero energy spread, i.e. a perfectly monochromatic electron gun. The longer the temporal coherence length, the more "identical" the beam electrons are.

Ideally, all beam electrons would emanate from a single point on the filament. In reality, the electrons that reach the sample appear to leave a point which coincides with the gun cross-over for the thermionic gun and the virtual source for the field emission gun. This virtual source has a finite lateral extent (lateral means normal to the optical axis) of a few nanometers for field emission guns to 10–50 μm for thermionic guns. We will see in Section 3.8 that in most cases the electron beam passes through an aperture (the second condensor aperture) before reaching the sample. The beam divergence angle θ_c subtended by this aperture at the sample can be used to define the *transverse coherence width* or *spatial coherence width* λ_{sc} of the illumination:

$$\lambda_{sc} = \frac{\lambda}{2\pi\theta_c}. \tag{3.65}$$

If two object points A and B are separated from each other by a distance $r < \lambda_{sc}$, then the *amplitudes* of the waves scattered by these two points must be added before converting to intensities:

$$I = |\Psi_A + \Psi_B|^2; \tag{3.66}$$
$$= |\psi_A|^2 + |\psi_B|^2 + \left(\Psi_A\Psi_B^* + \Psi_A^*\Psi_B\right); \tag{3.67}$$
$$= I_A + I_B + I_{AB} \qquad \text{(coherent)}. \tag{3.68}$$

If the separation between the two object points is much larger than the coherence width, then the final scattered intensity is given by

$$I = I_A + I_B = |\psi_A|^2 + |\psi_B|^2 \qquad \text{(incoherent)}. \tag{3.69}$$

The difference between the coherent and incoherent cases is the interference term I_{AB}. The intermediate range, i.e. distances comparable to the coherence width, is described by the theory of partial coherence.

Under normal observation conditions with a thermionic electron gun, the illuminating aperture is *incoherently* filled with statistically independent electron sources. For field emission systems the entire illuminating aperture is *coherently* filled and the transverse coherence width at the aperture plane [SZ92] is defined by

$$\lambda_{sc,a} = \frac{\lambda}{2\pi\theta_s}, \tag{3.70}$$

where θ_s is the angle subtended by the source at the aperture plane. A smaller aperture radius will increase the transverse coherence width, as will a decrease of the acceleration voltage.

It is intuitively clear that an infinitely small aperture will only allow one single wave vector to contribute to the incident electron beam, hence a pure plane wave is obtained. Experimentally this cannot be established without severe losses in current density. For high-resolution observations, where interference between the different diffracted beams is employed to obtain structural images, the coherence width is important and an appropriate choice of the condensor aperture radius must be made by the microscope operator. A smaller aperture will increase the coherence width but at the same time the current density will be reduced, thus requiring longer exposure times. Image contrast caused by interference from beams emanating from regions of the sample within the coherence width is called *phase contrast*. In Chapter 10 we will deal extensively with various types of phase contrast.

For in-depth discussions of coherence in optical and electron optical systems we refer the reader to [Goo68, Chapter 6], [Spe88, Chapter 4], [SZ92, Chapter 8], and [BW75, Section 7.5.8].

3.7.6 How many electrons are there in the microscope column?

It is a simple exercise to calculate how many beam electrons are present in the microscope column at any given moment in time. A beam current I of 1 nA corresponds to 6.25×10^{18} electrons $C^{-1} \times 10^{-9}$ C s^{-1}, or 6.25×10^{9} electrons s^{-1}. Electrons accelerated by a potential of E volts travel at a velocity $v = \beta c$, and in one second travel a distance of $\Delta = v$ m. If electrons are emitted from the gun at regular intervals in time (a reasonable approximation), then we can compute the *linear electron density* n_e per nA of beam current in the column:

$$n_e = \frac{6.25 \times 10^9}{v}.$$

At $E = 200$ kV, electrons travel at a velocity of $v = 0.695c$, which results in a linear electron density of about 30 electrons m^{-1} nA^{-1}. The average distance between consecutive electrons is then about 3 cm, which is much larger than the thickness of the sample. As far as the sample is concerned, it is thus a very good approximation to consider each beam electron separately, and this is precisely what the equations derived in Chapter 2 accomplish.

This illustrates once again that an electron diffraction experiment is, in fact, the quintessential quantum mechanics experiment; individual electrons are sent through a sample with a very large number of "slits", and since we do not attempt to determine precisely which "slit" the electron traveled through, an interference pattern will result at the detector.

It takes an electron a time $\Delta t = z_0/v$ to traverse a sample of thickness z_0; for a 100 nm thick sample, and a 200 kV electron, $\Delta t = 4.8 \times 10^{-16}$ s. This time interval is about four orders of magnitude shorter than the typical duration of a single cycle of a lattice vibration, and therefore it is a very good approximation to consider the atoms to be stationary throughout the interaction with the beam electron. The time interval between two consecutive electrons is 1.6×10^{-10} s for a 1 nA beam of 200 kV electrons; during this time interval atoms have gone through about 100 vibration periods, which means that there is no correlation between the position of an atom for one beam electron and its position for the next beam electron. This justifies the approximations made in the derivation of the Debye–Waller factor in Chapter 2.

3.8 The illumination stage: prespecimen lenses

The reader might be tempted to ask the question: why are there so many lenses in an electron microscope? This is a relevant question, and when we compare the early microscopes with the computer-controlled machines currently on the market we cannot avoid noticing that the current number of lenses and correction coils is far greater than that of the older machines. The easiest way to understand *why* there are so many lenses is to analyze what each lens does in relation to its neighbors. It is instructive to start with a microscope without lenses and to add lenses, one at a time. We will divide the microscope column into prespecimen, specimen, and post-specimen regions, and describe the functionality of the main lenses in each of those regions. While the actual number of lenses for any particular microscope may vary, we will focus in the next sections on those lenses that are present in nearly every recent microscope model.

The illumination system of a TEM serves one important purpose: to create a beam of electrons in a well-defined reference state (either a plane wave or a converged fine probe). Only if we know what the reference state is can we hope to extract information concerning the sample by analyzing the modified signal $\mathcal{R}''$ (recall the discussion on page 106). It is thus important for the microscope operator to understand the purpose of each of the lenses in the illumination system. This situation is somewhat similar to that encountered when one tries to solve a differential equation: the equation can only be completely solved if the initial conditions are specified, and solutions may depend in a rather sensitive way on those initial conditions. We have already seen that, in the case of the TEM, the relevant differential equation is the stationary Schrödinger equation. If we wish to compare theoretical solutions to this equation with experimental observations, we must from the beginning ensure that both have the same initial conditions. The illumination system is, hence, of crucial importance for the interpretation of experimental images based on a theoretical description of the image formation mechanism.

Let us consider the configuration shown in Fig. 3.26(a): an electron gun directly illuminates a specimen, without any magnetic lenses present. This configuration offers only limited control over the electron beam: the Wehnelt voltage or *gun bias* combined with the geometry of the entire assembly determine the location and size of the cross-over. The total number of electrons that reach any specific area on the specimen is rather small and can be changed somewhat by adjusting the gun bias. Since this changes the area on the filament from which electrons are emitted, the beam brightness would also change. This is clearly an undesirable configuration since we would like to be able to form a very fine probe on the sample surface or use a parallel incident beam. So we decide to add a single magnetic lens, the condensor lens C_1, positioned close to the anode, as shown in Fig. 3.26(b).

If this lens C_1 is excited such that the image of the gun cross-over (with a diameter of $s = 10$–100 μm) is formed on the specimen surface, then the size of the smallest illuminated area (the *spot size*) equals sv/u, generally of the same order of magnitude as the cross-over diameter itself. This is again undesirable because we may wish to investigate details on a length scale of nanometers, not microns, and this will require nanometer-sized probes.

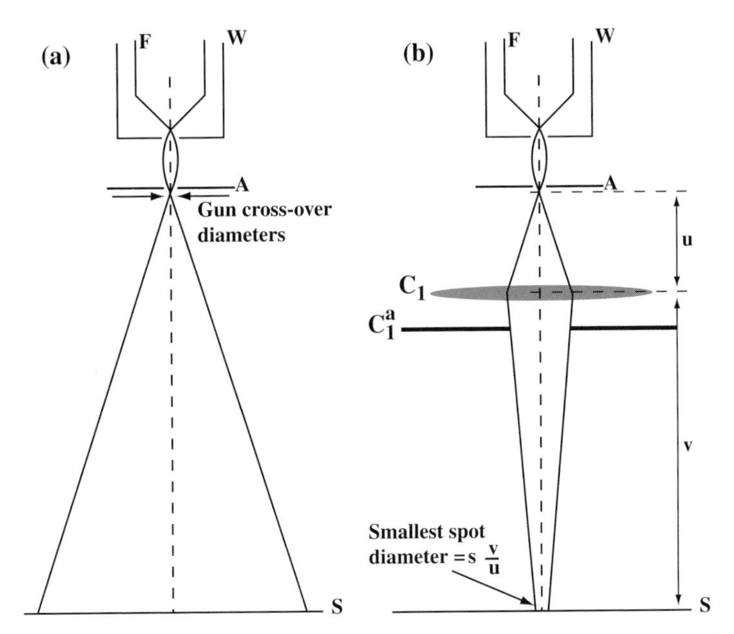

Fig. 3.26. (a) Prespecimen ray diagram without lenses. The gun cross-over diameter s is 10–100 μm. (b) Ray diagram with a single condensor lens C_1. The image of the gun cross-over is magnified by a factor v/u. The fixed aperture C_1^a limits the angular range of electron trajectories.

Most microscopes have a fixed aperture, the *first condensor aperture* C_1^a, located just below the lens C_1. The purpose of this aperture is to capture electrons which travel at large angles with respect to the optical axis and to prevent them from hitting the inner surfaces of the subsequent lens pole pieces. This cuts down on the number of x-rays that are generated in the column.

If we want the spot size on the specimen to be much smaller than the diameter of the cross-over, then we must make v much smaller than u; this can be done by strongly exciting the lens C_1, as is shown in the ray diagram of Fig 3.27(a). The stronger the excitation, the shorter the focal length will be and the smaller the magnification factor v/u (which is then known as a *demagnification* factor). When v is much smaller that u, as indicated in Fig. 3.27(a), most electrons will travel at rather large angles with respect to the optical axis, even after passing through C_1^a, and we obviously need another lens to focus the beam into a small area on the specimen. This second lens is known as C_2 (condensor 2) and it has a variable current control and hence variable focal length.

The second condensor lens has the C_1 cross-over as its object plane, and by varying the C_2 current we can position the conjugate image on any plane above, at, or below the sample. If the C_2 lens current is such that the focal length is longer than the distance from the lens center to the sample, then the lens is said to be

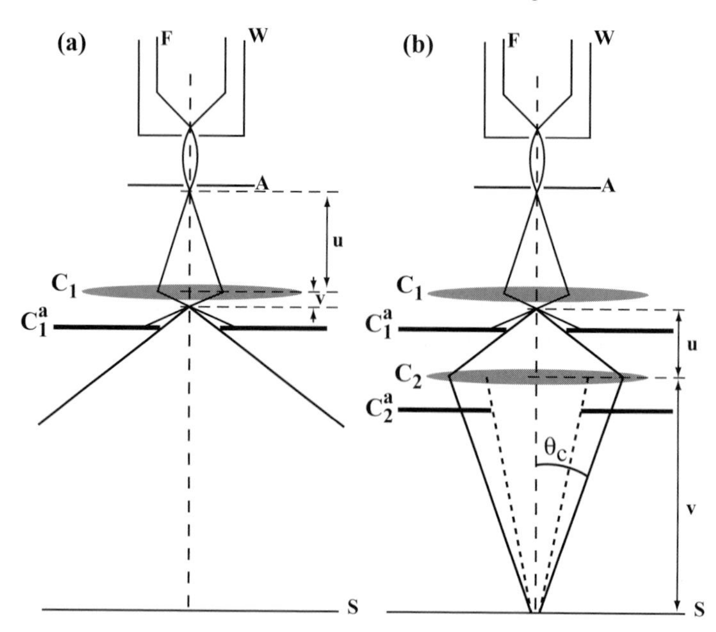

Fig. 3.27. (a) Ray diagram for a strongly excited C_1 lens. Electrons leave the C_1^a aperture at large angles with respect to the optical axis. (b) Ray diagram with a second condensor lens C_2. The lens has the first condensor cross-over as the object plane, and the specimen as the image plane. The diameter of the C_2 cross-over on the specimen is known as the *spot size*. The variable diameter aperture C_2^a limits the angular range of electron trajectories at the expense of beam current.

under-focused; if the image cross-over falls before the sample, then the lens is said to be *over-focused*. The lens is *in-focus* when the smallest possible spot diameter is observed on the sample. Although over- and under-focused conditions may give rise to a spot with as identical diameter on the sample, one must be aware of the fact that the *beam divergence angle* θ_c is different in the two cases. Observation modes that require a nearly parallel incident beam (such as high-resolution observations, see Chapter 10) must hence be carried out with as underf-ocused C_2 lens (unless additional condensor lenses are present). On microscopes with rotary lens control knobs, the under-focus condition usually corresponds to a counterclockwise rotation (i.e. less current through the lens).

Using the two-lens condensor system, we can routinely obtain 100 nm spot sizes. The first condensor lens control is often known on the microscope console as the *spot size*; the spot size can only be varied in discrete steps. The C_2 current is often labeled as *intensity* or *brightness* on the microscope console; it can be varied continuously. Since the C_1 cross-over changes location whenever the spot size is changed it is necessary to change the C_2 current to obtain the smallest beam diameter on the sample. The *second condensor aperture* C_2^a can be set to one of three or four different diameters. It can be seen from Fig. 3.27(b) that a change in C_2^a diameter does not change the size of the illuminated area on the specimen, only the beam intensity changes (less beam current for a smaller aperture diameter). A smaller aperture will also change the maximum beam divergence angle by limiting the solid angle subtended by the C_1 cross-over at the sample. This improves the coherence of the incident beam at the expense of lower image intensity. Modern microscopes often have a third so-called *mini-condensor* lens, located right above the objective lens upper pole piece. Since this lens is an integral part of the objective lens stage, we will discuss it in the next section.

3.9 The specimen stage

3.9.1 Types of objective lenses

We have seen that the objective lens of a TEM is a type of immersion lens, since the sample is located well within the magnetic field of the lens. There are two major types of objective lenses: the high-contrast lenses, also known as asymmetric lenses, and the twin or symmetric lenses. In an asymmetric objective lens the sample is located at one-third of the distance between the upper and lower pole pieces, as shown in Fig. 3.28. The objective or diffraction aperture is located two-thirds of the way down and coincides with the lens back focal plane. The upper pole piece usually has a narrower bore diameter than the lower pole piece, hence the name "asymmetric". In a symmetric lens design, the sample is located in the center between the two pole pieces, as indicated in Fig. 3.29. Both bore diameters are the same in the symmetric lens.

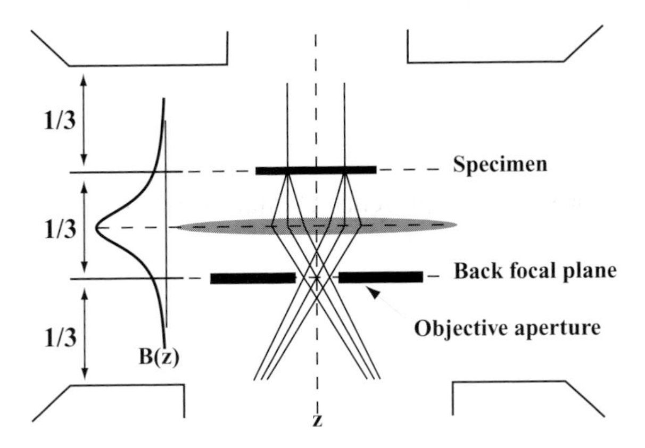

Fig. 3.28. Schematic diagram of a high-contrast objective lens. The specimen and back focal plane are located at one-third and two-thirds the distance along the axis of the lens. The angles between the rays and the optical axis are highly exaggerated and are actually only a few tens of milliradians. The lens forms a *first intermediate image*.

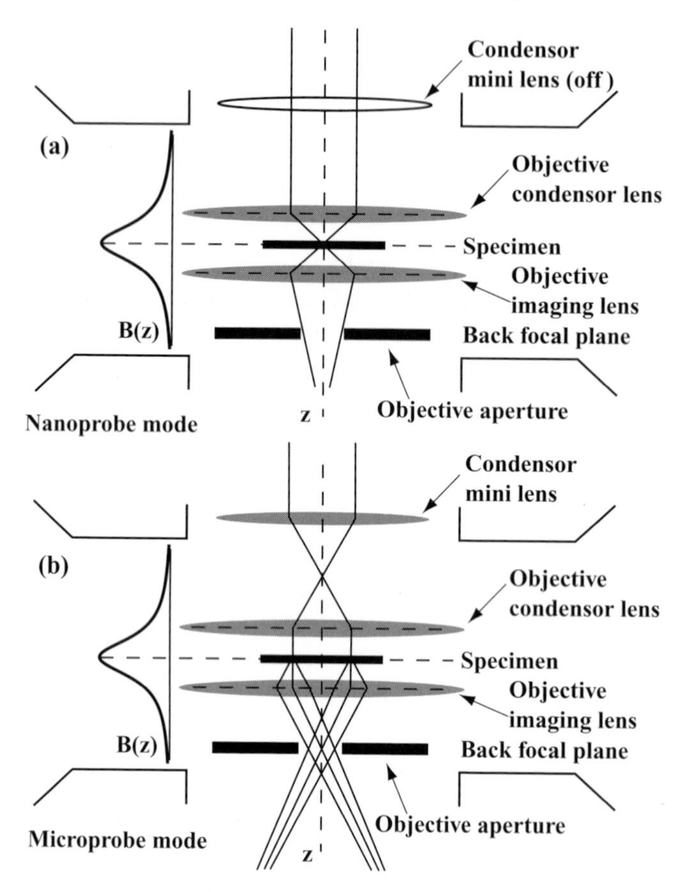

Fig. 3.29. Schematic diagram of a symmetric objective lens. With the condensor mini-lens switched off (a) a fine probe is obtained (nanoprobe mode). When the condensor mini-lens is switched on (b) a parallel beam is obtained (microprobe mode).

Asymmetric lenses have a much longer focal length than symmetric lenses (often in the range 5–10 mm), and hence a smaller lateral magnification M or, equivalently, a larger angular magnification M_α, which makes them well suited for large camera length diffraction observations. Since the incident electrons experience only a fraction of the objective lens magnetic field before reaching the sample (as indicated by the $B(z)$ profile in Fig. 3.28), this prefield acts only as a weak third condensor lens. Recalling that the primary task of the condensor lenses is to *demagnify* the gun cross-over, it is clear that the asymmetric objective lens cannot be used to obtain nanometer probe sizes.

In the symmetric lens design (the so-called *Riecke–Ruska lens*) the electron experiences nearly half of the objective lens magnetic field before reaching the sample, and this upper half of the field is known as the *objective condensor lens*. It can be used to further demagnify the incident probe, down to the single nanometer size range. This observation mode is thus known as *nanoprobe mode* (Fig. 3.29a). When an additional mini-condensor lens is present, often as part of the objective lens circuit, then this lens can be used to create another cross-over, this time right above the objective condensor lens (Fig. 3.29b). The electrons leaving this cross-over are then focused in a parallel but very narrow beam and after passing through the sample and the *objective imaging lens*, they go through a symmetric cross-over. This mode is known as *microprobe mode*. In microprobe mode, the electrons travel in a parallel beam through the sample, whereas in nanoprobe mode they are focused into a small probe. Microprobe mode is therefore useful as a diffraction technique, since the parallel beam can be focused readily into diffraction spots, while nanoprobe mode is more useful for analytical observation techniques. The focal length of the symmetric lens is much smaller than that of the asymmetric lens, and is usually in the range 1–3 mm. This means that the objective aperture is located very close to the sample, which imposes limitations on the tilt range of the specimen holder. The *lens gap* of a symmetric lens (the distance between the upper and the lower pole pieces) is of the order of a few millimetres; the asymmetric lens has a much larger lens gap, in the range of 10–20 mm.

While the primary function of the illumination stage is to create an electron beam in a well-defined reference state, the primary function of the objective lens is to bring the various diffracted electron beams to a cross-over while introducing minimal lens aberrations (see also Section 3.5). When we discussed the Fraunhofer diffraction conditions we saw that we do not need to have a lens in order to have a diffraction pattern; indeed, simply projecting a laser beam through a fine optical grating onto a wall on the other side of a room will produce a diffraction pattern without using a lens. The main purpose of the objective lens is to reduce the infinite "focal length" of the lensless system to a finite length, so that both diffraction pattern and image can then be magnified by other lenses. It is, in principle, possible to create microscopes without lenses, since all of the relevant physics occurs in the interaction of the electron beam with the specimen. One such microscope is the point projection microscope [SQM93], which uses no lenses at all, and hence does not suffer from lens aberrations.

3.9.2 Side-entry, top-entry and special purpose stages

There are two different types of specimen stages: those where the sample enters the objective lens from the side (*side-entry*) and those where the sample is lowered into the lens from the top (*top-entry*).

The side-entry stage is by far the most common stage and is used for both conventional and analytical electron microscopy and for most *in situ* experiments. As shown in the schematic drawing in Fig. 3.30, the sample is mounted at the end

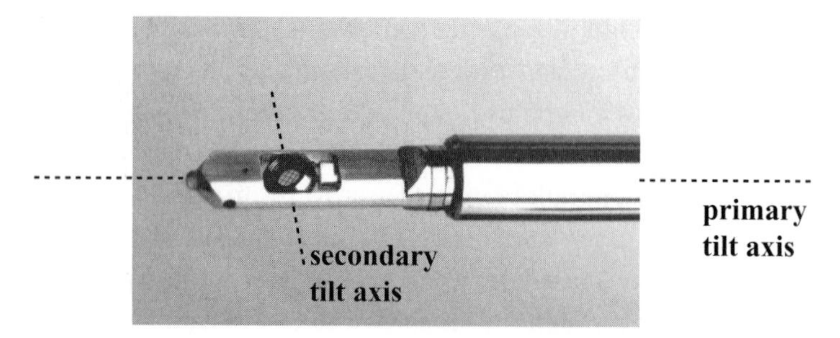

Fig. 3.30. Side-entry double-tilt holder; the primary and secondary tilt axes are indicated. (Photograph reproduced with permission of B. Armbruster, Gatan, Inc.)

of a rod-like specimen holder. The sample fits into a 3.05 mm diameter cup, which can either pivot around one axis in the plane of the sample, or rotate around an axis perpendicular to the sample. The rod-axis represents the primary rotation axis. These holders are referred to as double-tilt and rotation-tilt, respectively. Depending on the objective lens gap the maximum tilt angle may be as high as 60° or as low as 15°. Note that most specimen holders are made out of a soft aluminum alloy (or pure Be for some analytical holders), so utmost care is necessary whenever a holder is manipulated. Depending on the diameter of the microscope column, the specimen rod is anywhere from about 25–60 cm long. This rod is entered into the column through a vacuum *airlock*. Refer to the microscope manuals for the correct procedure to insert a sample holder into the column.

For most microscopes the sample has five degrees of freedom as illustrated in Fig. 3.31: two lateral movements in the horizontal plane (x and y, controlled by the specimen translation wheels or a joystick/trackball system), one vertical movement (z, specimen height) and two rotation angles α and β (either double-tilt or rotation-tilt). The five numbers $\{x, y, z, \alpha, \beta\}$ describe the specimen *attitude*. The rotation angles are defined as positive for counterclockwise rotations, when looking from the positive side of the x, y, or z axes towards the specimen. The reader should carefully study the geometry of the holder(s) he/she will be using, in particular how the holder is oriented inside the microscope column; for several microscopes, the holder is actually placed upside down in the column.† For top-entry stages, the reader should consult the microscope manual to determine the relative orientation of the specimen holder and rotation axes with respect to the microscope column.

In modern microscopes, some or all of the five coordinates may be computer controlled, while in older instruments x, y, and z motion are manually controlled. The primary tilt angle α can be changed manually or via a motor, and the secondary tilt/rotation angle β is always controlled by a motor. Both motors can be actuated

† Upside down with respect to the orientation of the specimen cup during loading.

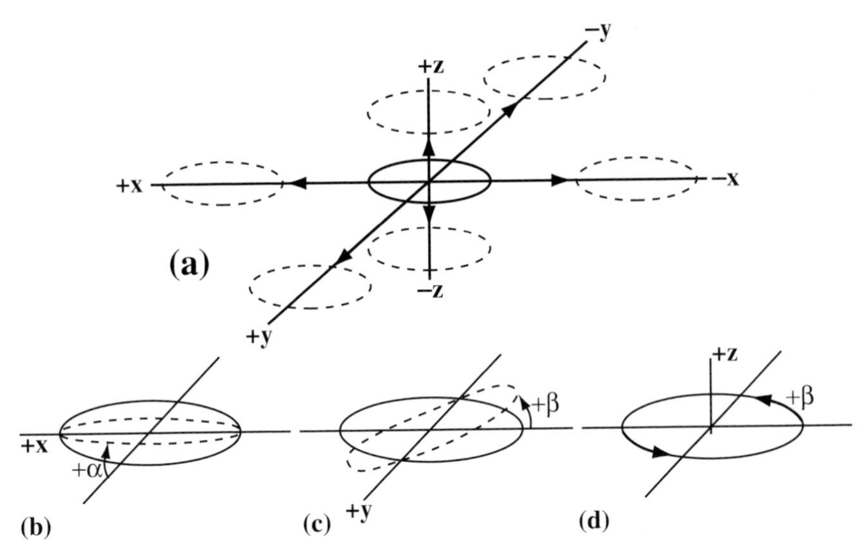

Fig. 3.31. Degrees of freedom for a typical specimen holder: (a) translation axes (positive x points along the length axis of the holder, positive y points towards the front of the microscope column); (b) primary tilt angle α; (c) secondary tilt angle β for a double-tilt holder; (d) secondary angle β for a rotation-tilt holder.

by floor-mounted pedals, or, in some microscope models, by rotary or push buttons on the microscope console near the column. More recent computer-controlled microscopes are equipped with a joystick-like mechanism to control the specimen attitude.

We have seen in Section 3.5 that lens aberrations increase rapidly for electrons traveling at large angles with respect to the optical axis of the microscope. This axis is defined as the line connecting the filament tip to the center of the viewing screen, and for a well-aligned microscope column the center line of each lens should coincide with the optical axis. The size of the electron transparent region in a thin foil can range from a few to several hundred micrometers, and the specimen translation controls x and y are used to bring a region of interest in the sample onto the optical axis of the column and therefore in the electron beam.

For both top- and side-entry stages, the specimen is located inside the objective lens magnetic field. Surrounding the specimen holder is the so-called *cold-finger* or *anti-contamination device* (ACD). It is a piece of metal, typically copper, linked to a liquid nitrogen dewar. By keeping the cold finger at liquid nitrogen temperature, residual gases in the microscope vacuum (or petroleum derivatives from an oil diffusion pump) are trapped onto this cold surface instead of on the sample. The ACD keeps the sample clean so that prolonged observations are possible. On some microscope models, one can insert an ACD-heater into the dewar at the end of a microscope session so that the remaining nitrogen is boiled off.

A transmission electron microscope is an ideal instrument for *in situ* experiments. A variety of special purpose stages have been designed:

- heating and cooling;
- magnetic field (one must use special low-field pole pieces in order to apply a controlled magnetic field on the sample);
- electric field, e.g. to study ferroelectric domain walls;
- straining stages, mostly tensile, some torsion or mode I deformation.

Combinations of two or more externally applied fields (e.g. electric field and deformation) are also possible. We refer to [Val79] for an extensive overview of different stages. Each stage has its own peculiarities and one should always be aware of the fact that performing an experiment on a *thin-foil* sample may not give rise to the same sample response as that expected for a bulk sample. The recent advances in spherical aberration correctors will undoubtedly rekindle interest in *in situ* observations, since the lens gap was kept small to reduce the spherical aberration. If C_s can be varied at will, then there is no longer a need for a small lens gap.

3.9.3 The objective lens and electron diffraction geometry

The objective lens is used with a high excitation current which means that a high lateral magnification is obtained. From our discussion on electron optics we know that the angular magnification M_α is the inverse of the lateral magnification M, which means that rays leaving the objective lens will travel at very small angles to the optical axis in all subsequent lenses – this is why most aberrations of the subsequent lenses are unimportant and can be ignored. The aberrations of the objective lens cannot be ignored since in this lens electrons travel at the largest angles with respect to the optical axis. The spherical aberration C_s and the chromatic aberration C_c determine the final image quality for conventional TEM work (assuming correctable aberrations, such as astigmatism, have been taken care of).

A cross-sectional view of the goniometer area of the JEOL 120CX microscope is shown in Fig. 3.32(a). The specimen rod enters from the right and is firmly seated in a metal cup which is linked to one of the specimen translation controls. The specimen holder sits right above the bottom plate of the anti-contamination device (cold finger) which, in turn, sits above the objective aperture. The objective stigmator follows a few millimeters below the lower objective pole piece. The OL pole pieces are located *above* the lens coil which is visible in the lower left-hand corner. Figure 3.32(b) shows the entire imaging stage of the microscope, from the objective lens area down to the projector lens; the bottom three lenses will be discussed in more detail in the following section.

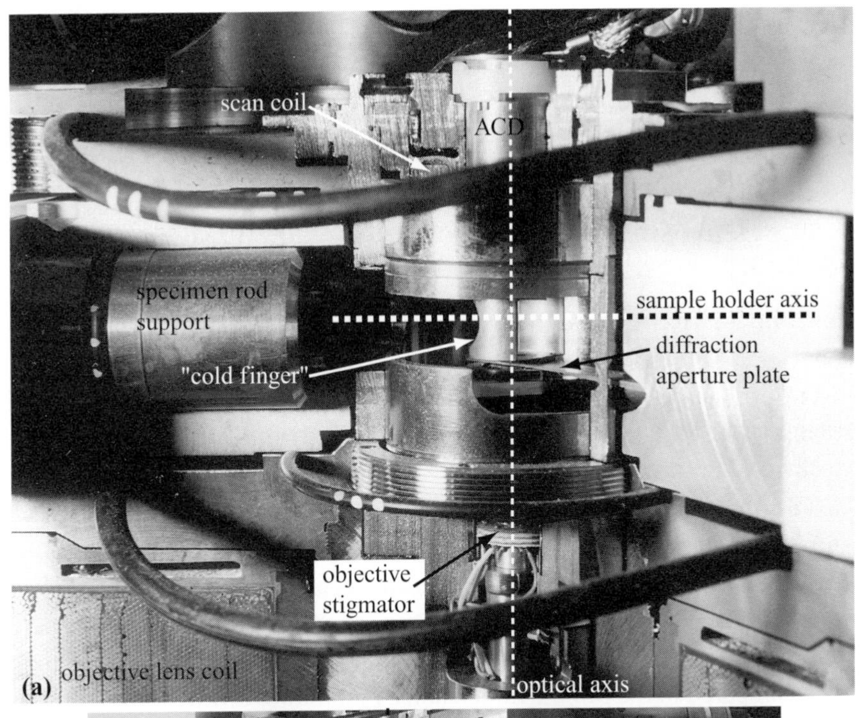

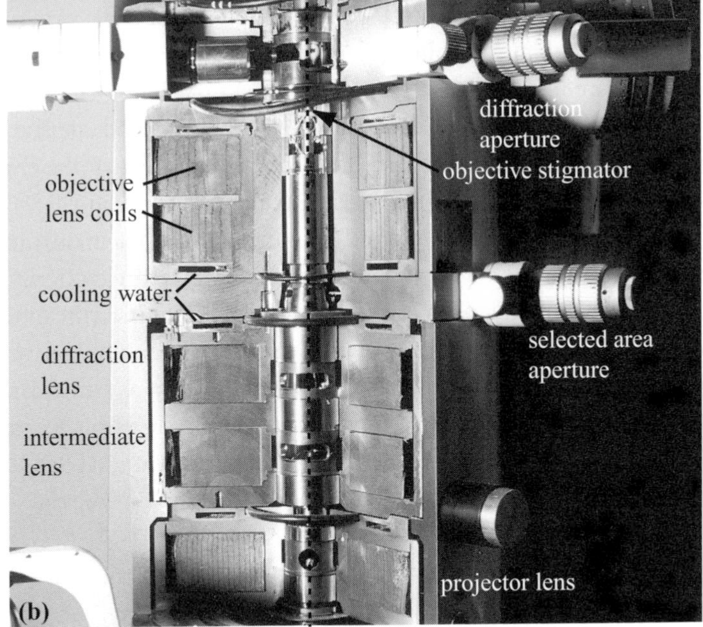

Fig. 3.32. (a) Cross-sectional view of the goniometer area of the JEOL 120CX microscope; (b) cross-sectional view of the objective lens and magnifying lenses.

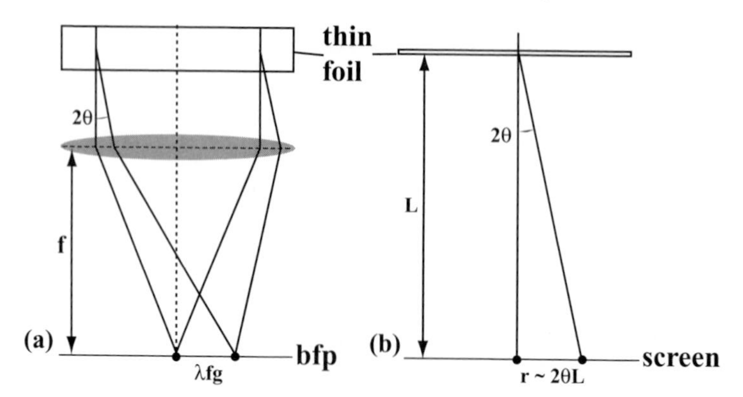

Fig. 3.33. (a) Schematic representation of the diffraction geometry; (b) definition of the diffraction camera length L.

The OL is characterized by two different planes: the back focal plane, in which the diffraction pattern is formed, and the image or *selected area* plane. In the back focal plane we can introduce an aperture, typically known as the *diffraction aperture*. Similarly, in the image plane we can introduce the so-called *selected area* aperture (see Fig. 3.32b). Since electrons leaving the OL form an image at a large distance compared to the sample dimensions, Fraunhofer diffraction conditions are satisfied and hence the back focal plane contains the Fourier transform of the object function. The reciprocal lattice points in the back focal plane are at a distance $\lambda f|\mathbf{g}|$ from the origin of reciprocal space, as shown in Fig. 3.33(a) (this follows from equation 3.41). The BFP hence contains an "image" of reciprocal space, magnified by a factor of λf. Only those reciprocal lattice points that lie near the Ewald sphere and satisfy the Bragg condition will give rise to a diffracted beam.

Figure 3.33(b) defines the *camera length L* of the microscope in diffraction mode. The camera length is a measure for the magnification of the diffraction pattern. The relation between the camera length L, the electron wavelength λ, the lattice spacing d and the measured spacing r on an electron micrograph can be derived easily as follows: from the Bragg equation we know that $2d\sin\theta = \lambda$. Since Bragg angles for high-energy electrons are small, it is a good approximation to replace $\sin\theta$ by θ, and we have $2d\theta = \lambda$. From Fig. 3.33(b) we have $r = L\tan(2\theta) \approx 2L\theta$. Combining the two relations we find

$$r = \frac{\lambda L}{d} = \lambda L g, \tag{3.71}$$

where $g = 1/d$. The constant L is known as the camera length, and the product λL as the *camera constant*; it has dimensions of [nm mm]. For a given lattice spacing d, an increase in the acceleration voltage will cause a corresponding decrease in the electron wavelength and therefore a decrease of the distance r in the diffraction pattern. The camera length L converts the diffraction angle 2θ into the distance r.

For the aluminum example in Table 2.3 on page 102, $2\theta = 12.38$ mrad for the (200) planes (at 200 kV), and for a typical camera length (the distance between the sample and the observation screen) of $L = 1000$ mm we find $r = 12.38$ mm. Rather than changing the magnitude of L by physically changing the location of the observation screen (a bit impractical), we can use the imaging lenses to change the *effective* value of L, while keeping the object and image planes constant. All microscopes have a predefined set of camera lengths, and the microscope operator can select the value of L best suited for the observation at hand. This value would typically be shown on the microscope console and on the diffraction micrograph. Care must be taken to accurately calibrate the camera length; this will be discussed in Section 4.5 in the following chapter.

From the preceding paragraphs we can deduce the following geometrical interpretation of electron diffraction (see Fig. 3.34a): the incident electron beam is represented by the wave vector **k**, which is parallel to the optical axis of the objective lens. According to the Ewald sphere construction, this wave vector ends in the origin **O** of reciprocal space, which is located at the intersection of the optical

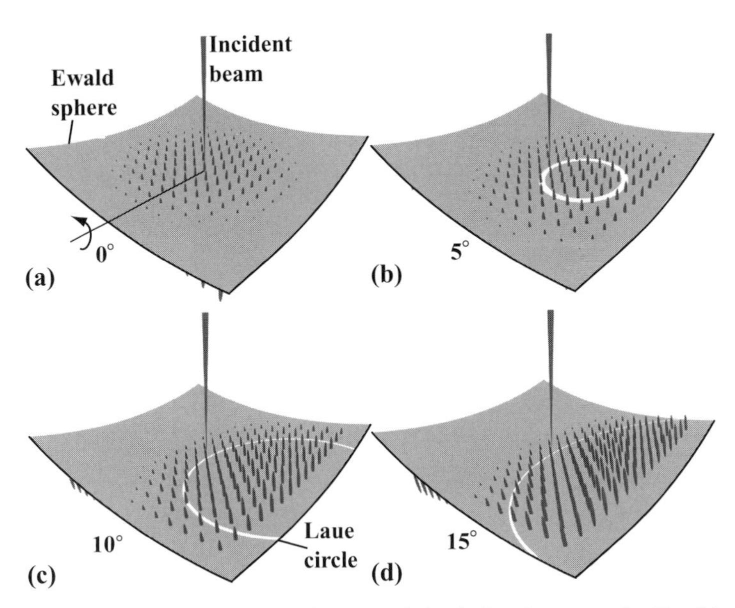

Fig. 3.34. Perspective representation of the spatial relation between the Ewald sphere and a plane in reciprocal space. In (a) the plane is tangential to the sphere, and the relrods corresponding to individual reciprocal lattice points intersect the sphere. When the reciprocal lattice plane is tilted around an axis in the plane, the intersection of this plane and the Ewald sphere becomes a circle through the origin, the Laue circle, indicated by a white arc. With increasing tilt angle (b, c, and d) the circle becomes larger.

axis and the BFP. By construction the Ewald sphere is then at a *tangent to the BFP*. If the crystal is oriented such that a zone axis $\mathbf{t}_{[uvw]}$ is parallel to the optical axis, then all reciprocal lattice vectors $\mathbf{g}_{hkl}$ satisfying the zone equation $\mathbf{t}_{[uvw]} \cdot \mathbf{g}_{hkl} = 0$ will lie in the BFP. The distance s between a reciprocal lattice point in the BFP and the Ewald sphere, measured parallel to the optical axis, is approximately given by $s \approx \lambda g^2/2$; if the crystal surface normal is parallel to the electron beam, then this distance s is identical to the *excitation error* $s_{\mathbf{g}}$.

By using the *beam tilt* controls on the microscope console, the operator can change the angle between the incident wave vector and the optical axis. The Ewald sphere will then intersect the objective lens back focal plane along a circle known as the *Laue circle*; the diameter of this circle increases with increasing beam tilt. Alternatively, we could keep the beam along the optical axis, and use the specimen tilt controls to change the orientation of the sample with respect to the incident beam. This also changes the orientation of reciprocal space with respect to the back focal plane, and again we can observe a ring of diffracted beams along the intersection of the Ewald sphere and the BFP (see Figs 3.34b–d). Since electronic control of the beam tilt is usually much more sensitive than the mechanical control of the specimen orientation, the beam tilt can be used to precisely align the beam along a crystal zone axis, so that the BFP contains all reflections corresponding to the planes of the zone. The subsequent imaging lenses are then used to magnify the diffraction pattern by a factor of λL with respect to reciprocal space.

3.9.4 Numerical computation of electron diffraction patterns

Numerical simulation of a kinematical electron diffraction pattern for a given zone axis direction $[uvw]$ consists of two steps.

(i) Computation of the intensity I_{hkl} for all reciprocal lattice points that belong to the zone $[uvw]$; the intensity can be taken to be proportional to $|V_{\mathbf{g}}|^2$, which is known as the *kinematical approximation*. The Fortran subroutine CalcUcg described in PC-9 can be used along with the routine CalcFamily, which is based on the CalcOrbit routine described in PC-4.

(ii) Computation of the geometry of the diffraction pattern. From two low-index reciprocal lattice points the entire pattern can be generated by simple vector additions. The Cartesian 2D coordinates of each reflection can then be computed using the reciprocal structure matrix b_{ij}.

When the correct value of the camera length L is used, the computed and experimental diffraction patterns should have the same magnification and can be overlaid on one another. In practice, it is useful to have a program that can generate a set of independent zone axis patterns, say, for all the zone axis for which $-2 < u, v, w < 2$.

Such a program would use symmetry arguments to compute the intensity for only one member of each family of reciprocal lattice points, and also one member from every family of directions. Pseudo code for the diffraction pattern algorithm is shown in PC-12. The source code for a fully functional program zap.f90, with PostScript output, can be downloaded from the website.

Pseudo Code PC-12 Zone axis diffraction patterns.

Input: PC-2, PC-3
Output: zone axis patterns in PostScript format
 ask for microscope voltage and camera length
 for all reciprocal lattice points in a given range **do**
  select a reciprocal lattice vector $\mathbf{g}_{hkl}$
  compute the family $\{hkl\}$ {PC-4}
  compute $I_\mathbf{g} = |V_\mathbf{g}|^2$ for one member of family
  skip computation of $I_\mathbf{g}$ for all other family members
 end for
 for every direction $[uvw]$ in a given range **do**
  compute the entire family $\langle uvw \rangle$ {PC-4}
 end for
 rank all directions $[uvw]$ according to increasing $|\mathbf{t}_{uvw}|$
 list all independent directions
 for each direction selected by user **do**
  compute the geometry of the zone axis pattern
  determine the diameter of the reflections {equation 3.72 or 3.73}
  produce PostScript code
 end for

Example output from the zap.f90 program is shown in Fig. 3.35; the program draws six patterns per page, in this case the lowest-order zone axis patterns for rutile (TiO_2, 200 kV, $L = 1600$ mm). The computed patterns show reflections as solid circles with a diameter related to the calculated intensity $I_\mathbf{g}$. The program implements two different intensity scales.

- **Logarithmic intensity scale**. This scale computes the radius $r_\mathbf{g}$ to be used for a given diffraction spot using the following relation [GW87]:

$$r_\mathbf{g} = c_1 \log(1 + c_2 I_\mathbf{g}), \tag{3.72}$$

  with c_1 and c_2 constants. The zap.f90 program uses $c_1 = 1.27$ mm $(= \frac{1}{20}$ in) and $c_2 = 0.1$.
- **Exponential intensity scale**. An alternative intensity scale compresses the entire range of intensities by suppressing the higher intensities and increasing the lowest intensities.

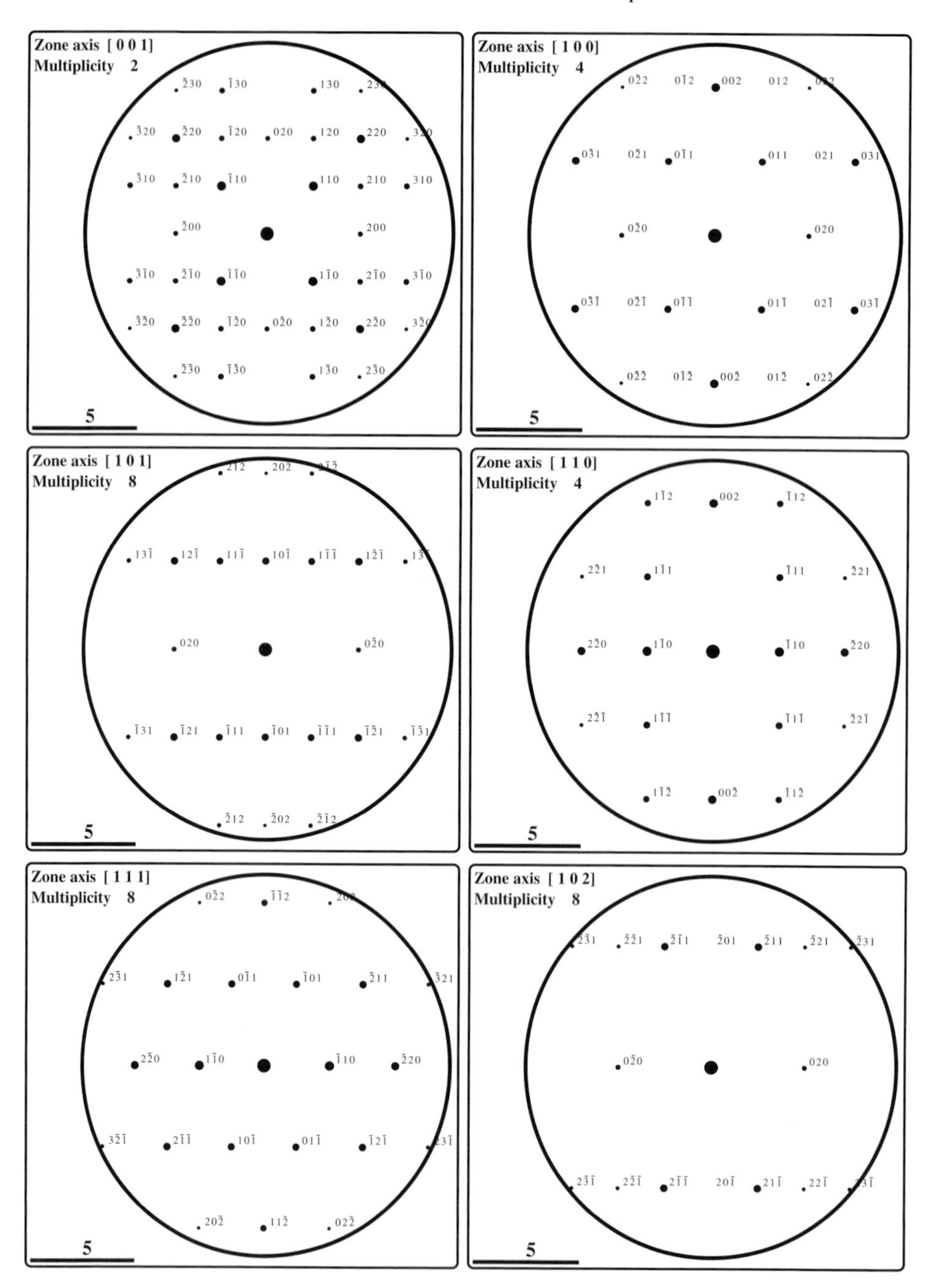

Fig. 3.35. Example output of the zap.f90 program for the lowest-order zone axis patterns of rutile (TiO_2, 200 kV, $L = 1600$ mm). The figure has been reduced to 70% to fit on the page.

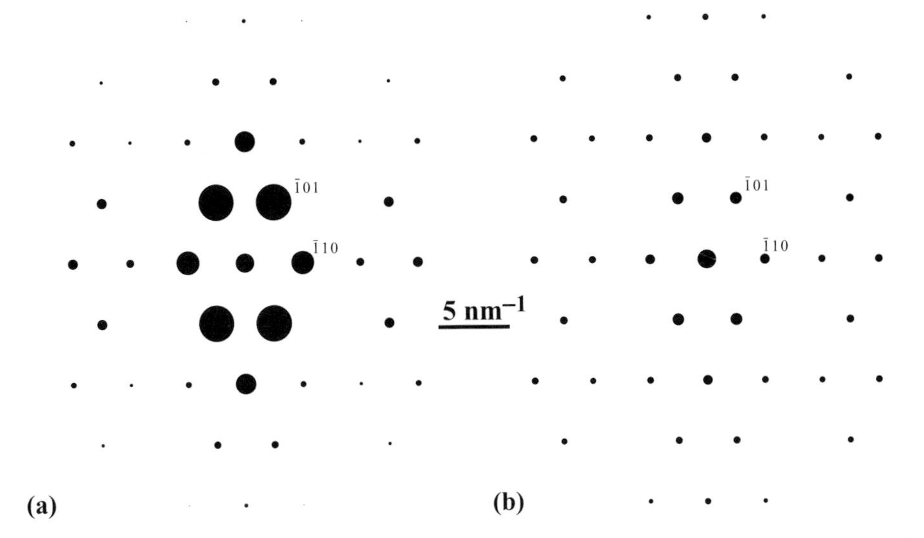

Fig. 3.36. [11.3] zone axis pattern for hexagonal Ti (200 kV, $L = 800$ mm): (a) logarithmic and (b) exponential intensity scaling.

This can be accomplished by a scaling relation of the following type:

$$r_{\mathbf{g}} = c_1 \left(\frac{I_{\mathbf{g}}}{I_{max}} \right)^{c_2}, \tag{3.73}$$

where I_{max} is the maximum intensity. The zap.f90 program uses $c_1 = 1.27$ mm and $c_2 = 0.2$.

A comparison of the two intensity scales is shown in Fig. 3.36 for the [11.3] zone axis pattern of hexagonal Ti, for a microscope acceleration voltage of 200 kV and camera length $L = 800$ mm. Figure 3.36(a) uses logarithmic intensity scaling and Fig. 3.36(b) shows the exponential scaling. For comparison with experimental diffraction patterns the exponential scaling often provides better visual agreement.

3.9.5 Higher-order Laue zones

3.9.5.1 Definition and examples

The reciprocal lattice is a three-dimensional lattice and in any given zone axis pattern we observe a nearly planar section through this lattice. The reciprocal lattice planes parallel to the plane imaged in a zone axis pattern (Fig. 3.37) also intersect the Ewald sphere, but at a rather large distance from the origin of reciprocal space. We recall from Chapter 1 that the set of reciprocal lattice points belonging to a zone axis pattern can be described by means of the *zone equation*:

$$\mathbf{g} \cdot \mathbf{t} = hu + kv + (it) + lw = 0, \tag{3.74}$$

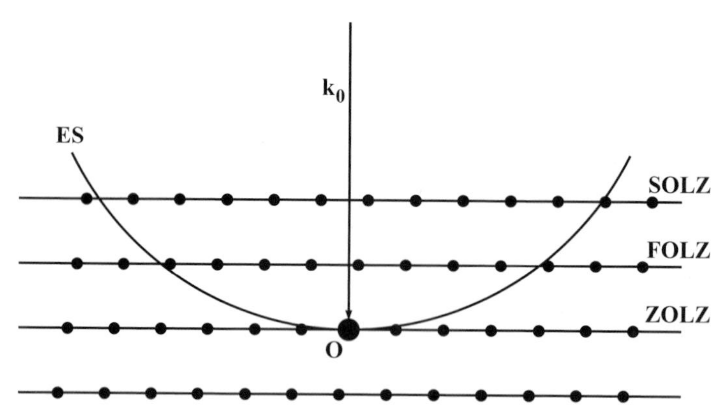

Fig. 3.37. Schematic illustration of the geometry of higher-order Laue zones or HOLZ.

where the term *it* only occurs for the 4-index hexagonal case. This set of reciprocal lattice points is also known as the *zero-order Laue zone*, or *ZOLZ*.

Reciprocal lattice planes parallel to the ZOLZ satisfy the equation

$$hu + kv + (it) + lw = N, \tag{3.75}$$

where N is a non-zero integer. The exact values of N depend on the zone axis direction $\mathbf{t}$. The integer N effectively numbers the reciprocal lattice planes, which are then known as the *higher-order Laue zones*, or *HOLZ*. The lowest non-zero value for N for which reflections are allowed corresponds to the *first-order Laue zone* (FOLZ), the next value to the *second-order Laue zone* (SOLZ), and so on.[†]

3.9.5.2 Simulation of HOLZ diffraction patterns

The simulation of electron diffraction patterns that contain reflections from HOLZ layers is not very difficult and only requires a careful definition of an appropriate reference frame. Consider a zone axis diffraction pattern, recorded with the beam along the $\mathbf{t} = [uvw]$ direction. We can always select two short reciprocal lattice vectors as basis vectors of the ZOLZ, say $\mathbf{g}_1$ and $\mathbf{g}_2$ (see Fig. 3.38). The vectors are selected so that the reference frame is right-handed, i.e. $\mathbf{g}_1 \times \mathbf{g}_2 \parallel \mathbf{t}$. The third basis vector $\mathbf{g}_3$ is selected along $\mathbf{t}$ so that its length is equal to the distance between the plane formed by $\mathbf{g}_1$ and $\mathbf{g}_2$ and the next plane (i.e. the plane for which $N = 1$ in equation 3.75). Because of the duality between real space and reciprocal space we know that this distance is given by the inverse of the length of $\mathbf{t}$ or $1/|\mathbf{t}| \equiv H$. We will use the vectors $\mathbf{g}_i (i = 1, \ldots, 3)$ as basis vectors for all computations involving HOLZ reflections, including Kikuchi lines and HOLZ line computations discussed

[†] It is possible for entire planes in reciprocal space to consist of forbidden reflections. In such a case there may be no reflections for $N = 1$, while the first ring of visible reflections has $N = 2$. This ring is still known as the first-order Laue zone. In other words, the value of N is not necessarily equal to the order of the HOLZ layer [Ead90].

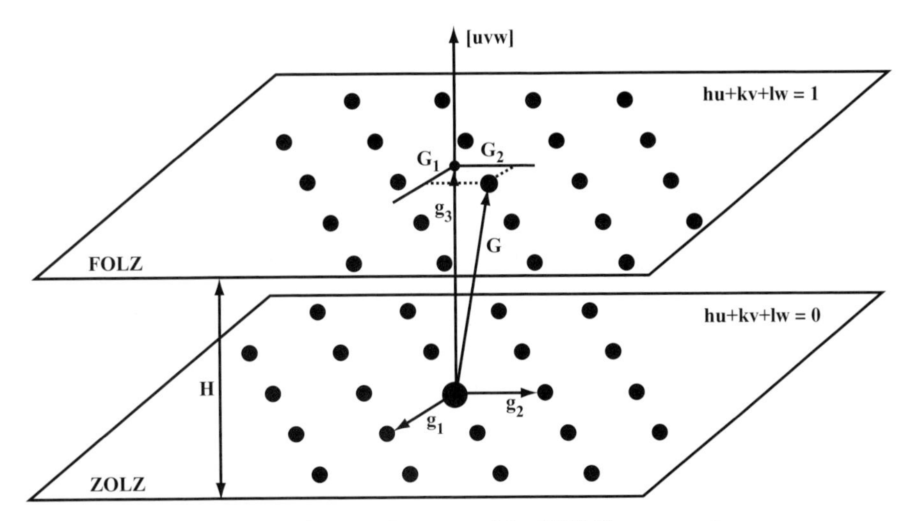

Fig. 3.38. Reference frame used for HOLZ computations.

in Chapter 9. A numerical procedure for determining the new reference frame will be introduced in Chapter 7.

In some crystal structures, the reciprocal vector $\mathbf{g}_3$ will be a reciprocal lattice vector (meaning that it can be written as an integer linear combination of the vectors $\mathbf{a}_j^*$). In other structures, or for other zone axes, the vector $\mathbf{g}_3$ will not be a reciprocal lattice vector. In such cases the reflections on the plane $N = 1$ will be shifted with respect to those of the ZOLZ. We will take the vector $\mathbf{G}$ to be the shortest reciprocal lattice vector of the $N = 1$ plane (Fig. 3.38). Any vector in the FOLZ can then be written as the sum of a vector in the HOLZ and $\mathbf{G}$. More generally, any reciprocal lattice vector $\mathbf{h}$ can be written as

$$\mathbf{h} = (n_1 + NG_1)\mathbf{g}_1 + (n_2 + NG_2)\mathbf{g}_2 + N\mathbf{g}_3, \tag{3.76}$$

where n_1 and n_2 are integers, and G_i are the components of $\mathbf{G}$ along the vectors $\mathbf{g}_i$. It is not difficult (exercise) to show that

$$G_1 = \frac{\mathbf{G}\cdot\mathbf{g}_1\gamma_{22} - \mathbf{G}\cdot\mathbf{g}_2\gamma_{12}}{\gamma_{11}\gamma_{22} - \gamma_{12}^2};$$
$$G_2 = \frac{\mathbf{G}\cdot\mathbf{g}_2\gamma_{12} - \mathbf{G}\cdot\mathbf{g}_1\gamma_{11}}{\gamma_{11}\gamma_{22} - \gamma_{12}^2},$$

with $\gamma_{ij} = \mathbf{g}_i \cdot \mathbf{g}_j$.

The projection of the vector $\mathbf{G}$ onto the ZOLZ plane can also be found using a method proposed by Jackson [Jac87]. If $\mathbf{t}^*$ represents the reciprocal space vector corresponding to $\mathbf{t}$, i.e. $t_i^* = t_j g_{ji}$ (see Chapter 1), then the vector connecting the

ZOLZ to the Nth HOLZ layer is given by

$$\mathbf{G}_{\perp} = \frac{\mathbf{t}^*}{|\mathbf{t}^*|} HN.$$

Subtracting this vector from the vector $N\mathbf{G}$ results in the projection $N\mathbf{g}_P$, i.e.

$$N\mathbf{g}_P = N\mathbf{G} - \mathbf{G}_{\perp}.$$

This vector clearly lies in the ZOLZ plane. It is then simply a matter of expressing the vector $\mathbf{g}_P$ with respect to the basis vectors $\mathbf{g}_1$ and $\mathbf{g}_2$ to recover the components G_i.

Next, we must determine where a diffracted beam corresponding to a HOLZ reflection will intersect the viewing screen. Consider Fig. 3.39: the reciprocal lattice points are represented by *relrods* parallel to the foil normal $\mathbf{F}$. A diffracted beam corresponding to the reflection $\mathbf{g}$ will intersect the ZOLZ in a point on the line $\mathbf{k}' = CQ$. The reader should note that the foil normal $\mathbf{F}$ may have a component out of the plane of the drawing, so that the scattered vector $\mathbf{k}'$ need not lie in the plane formed

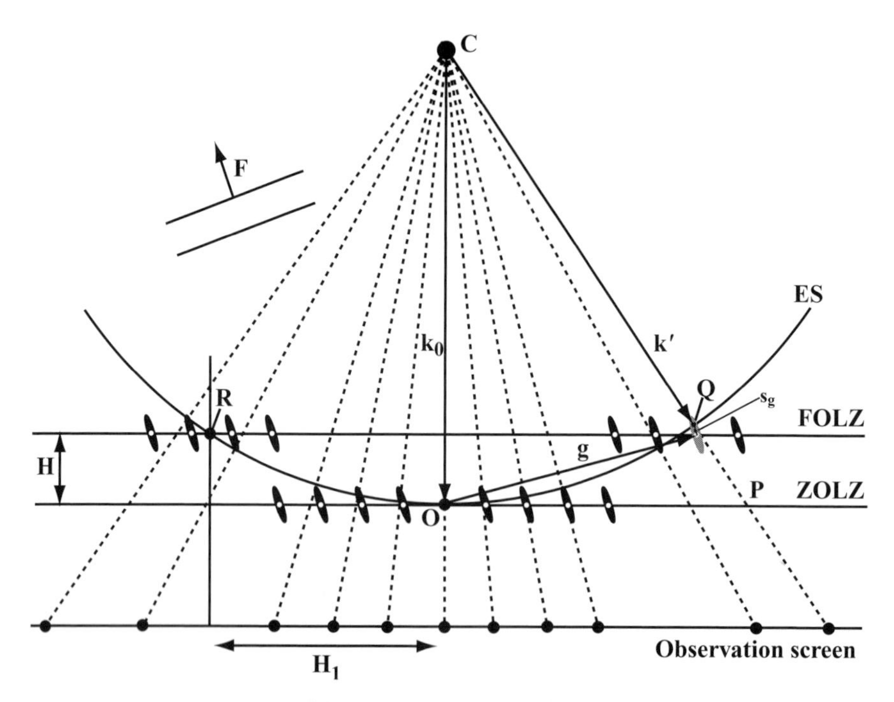

Fig. 3.39. Intersection of FOLZ relrods with the Ewald sphere result in a diffracted beam in the direction $\mathbf{k}' = \mathbf{k}_0 + \mathbf{g} + \mathbf{s}_\mathbf{g}$.

by $\mathbf{k}_0$ and $\mathbf{g}$. The coordinates of the point Q can be found from the vector sum:

$$\mathbf{k}' = \mathbf{k}_0 + \mathbf{g} + \mathbf{s}.$$

These three vectors can be expressed in the $\mathbf{g}_i$ reference frame:

$$\begin{aligned}
\mathbf{g} &= (n_1 + NG_1)\mathbf{g}_1 + (n_2 + NG_2)\mathbf{g}_2 + N\mathbf{g}_3;\\
\mathbf{k}_0 &= -\frac{1}{\lambda H}\mathbf{g}_3;\\
\mathbf{s} &= s_\mathbf{g}(F_1\mathbf{g}_1 + F_2\mathbf{g}_2 + F_3\mathbf{g}_3),
\end{aligned}$$

where F_i are the components of the unit vector $\hat{\mathbf{F}}$ with respect to $\mathbf{g}_i$. Combining these equations we find for $\mathbf{k}'$:

$$\mathbf{k}' = (n_1 + NG_1 + s_\mathbf{g}F_1)\mathbf{g}_1 + (n_2 + NG_2 + s_\mathbf{g}F_2)\mathbf{g}_2 + \left(N - \frac{1}{\lambda H} + s_\mathbf{g}F_3\right)\mathbf{g}_3.$$

If we scale this vector so that the point Q moves to P along the line CQ, then the ZOLZ coordinates of the point P can be found by making the $\mathbf{g}_3$ component of $\mathbf{k}'$ equal to $-1/\lambda H$. The resulting ZOLZ position vector is then given by

$$\mathbf{g}_P = \frac{1}{1 - \lambda H(N + s_\mathbf{g}F_3)}\sum_{i=1}^{2}(n_i + NG_i + s_\mathbf{g}F_i)\mathbf{g}_i. \qquad (3.77)$$

This vector can then be scaled by the camera constant λL to find the position of the HOLZ reflection $\mathbf{g}$ on the viewing screen. Equation (3.77) is valid for all HOLZ reflections, including the ZOLZ reflections, and takes into account the foil normal and the curvature of the Ewald sphere. We leave it to the reader to derive an expression for $\mathbf{g}_P$ when the incident beam direction is tilted away from the zone axis $\mathbf{t}$, so that $\mathbf{k}_0$ acquires components along $\mathbf{g}_1$ and $\mathbf{g}_2$.

The point R in Fig. 3.39 indicates the intersection of the Ewald sphere with the HOLZ layer. The intersection forms a ring, known as the HOLZ ring. The radius R_n of the nth HOLZ ring can be computed from the equation of the Ewald sphere as follows:

$$R_n = \sqrt{\frac{2H_n}{\lambda} - H_n^2}, \qquad (3.78)$$

where $H_n = nH$. The term H_n^2 is often small and can be ignored. This radius R_n can be measured from an experimental electron diffraction pattern. The procedure will be illustrated in Chapter 9.

In Chapters 5–7, we will describe how the intensity of a diffracted beam depends on the distance of the reciprocal lattice point to the Ewald sphere, i.e. the excitation error $s_\mathbf{g}$. In the simplest approximation, the intensity $I_\mathbf{g}$ of a reflection $\mathbf{g}$ has a sinc dependence on the excitation error, i.e.

$$I_{\mathbf{g}} \rightarrow I_{\mathbf{g}} \left[\frac{\sin(\pi z_0 s_{\mathbf{g}})}{\pi z_0 s_{\mathbf{g}}} \right]^2 = I_{\mathbf{g}} \mathrm{sinc}^2(\pi z_0 s_{\mathbf{g}}),$$

where z_0 is the foil thickness. This follows from the so-called *kinematic theory* introduced in Chapter 6; it can also be derived from the expression for the relrod intensity in Chapter 2 (Section 2.6.3). For each reflection closer to the Ewald sphere than a preset threshold distance, s_{max}, we compute the scaled intensity and use it to determine the radius of the circle representing the reflection in a drawing of the diffraction pattern.

Pseudo code PC-13 outlines the major computational steps of the program HOLZDP.f90; example output from this program is shown in Fig. 3.40 for the [00.1] zone axis pattern of Ti and the [112] pattern of GaAs. We will return to HOLZ-related computations and observations in Chapters 7 and 9.

Pseudo Code PC-13 Zone axis diffraction patterns with HOLZ reflections.

Input: PC-2, PC-3
Output: HOLZ zone axis patterns in PostScript format

```
ask for microscope voltage and camera length
for each incident beam direction do
   determine g_i basis vectors and G
   get foil normal F and Laue center
   compute radii R_n of HOLZ circles
   determine contributing reflections for ZOLZ
   plot ZOLZ pattern
   for every HOLZ layer do
      determine contributing reflections
      plot HOLZ pattern
   end for
   label selected reflections
end for
```

3.10 The magnification stage: post-specimen lenses

The post-specimen part of the objective lens field is used to magnify the image, typically by a factor of 30–50 times. Most microscopes have at least three other lenses following the objective lens. They are usually known as (from top to bottom) the *diffraction lens* (DL), the *intermediate lens* (IL), and the *projector lens* (PL). A

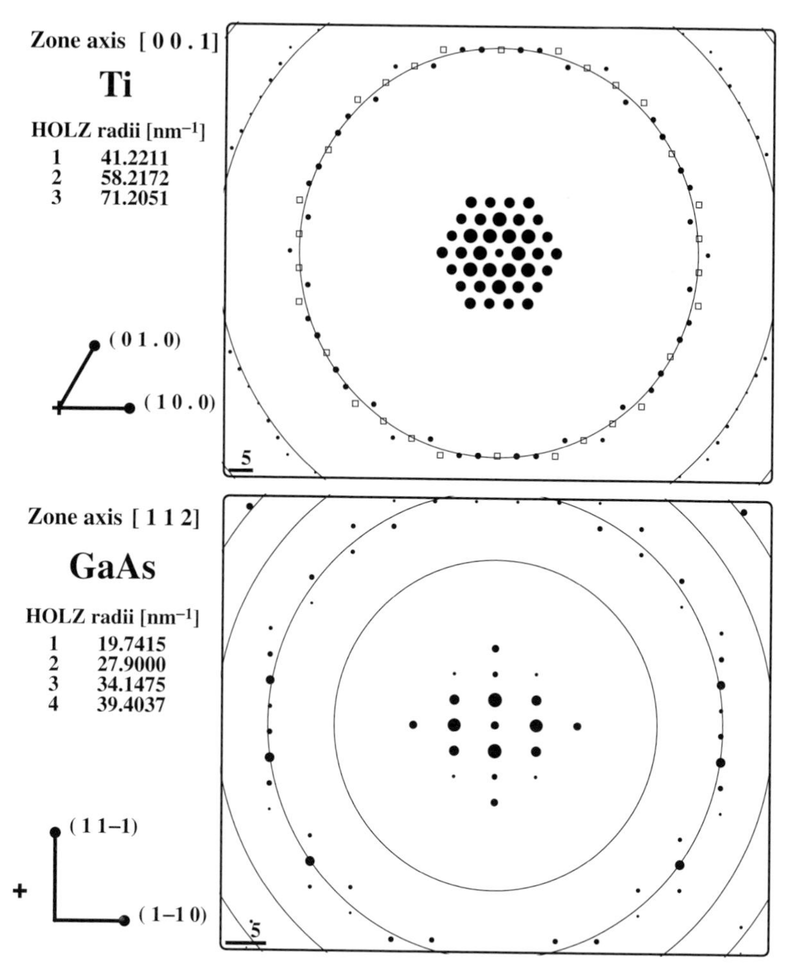

Fig. 3.40. Simulated zone axis patterns with HOLZ reflections for Ti and GaAs. The HOLZ ring radii are indicated. The small + sign indicates the shift vector **G** of the FOLZ with respect to the HOLZ layer.

cross-sectional view of the imaging stage of the JEOL 120CX microscope is shown in Fig. 3.32(b). In the following paragraphs we will add these lenses one at a time, to make clear why they are all needed.

If the objective lens were the only imaging lens present, then the situation would be rather simple. Since both the specimen and the observation screen have a fixed location, the object and image distances are fixed, and this in turn determines the focal length (and magnification) of the objective lens (see Section 3.4.6). The objective lens current can be varied to bring the image plane into exact coincidence with the screen (this is known as *focusing*), at which point everything is fixed. To obtain a system with a *variable* magnification we must add a second lens, the *projector lens* PL.

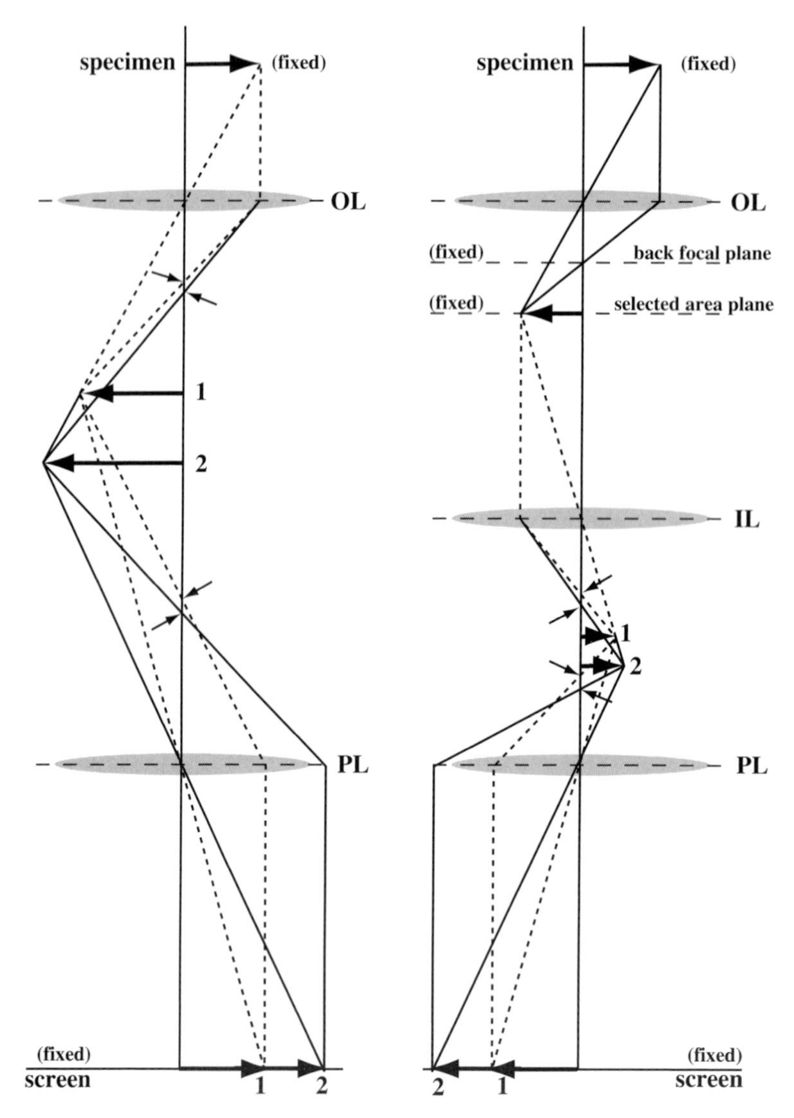

Fig. 3.41. (a) Two-lens system consisting of objective OL and projector PL lenses. Ray diagrams for two different overall magnifications are shown, and the locations of the focal points are arrowed; (b) three-lens system with fixed OL current. Again two different overall magnifications are shown.

The ray diagram in Fig. 3.41(a) shows how two images **1** and **2** with different magnification can be generated by the two lenses OL and PL. The ray diagram is constructed based on Fig. 3.9, and for simplicity we have made the two principal planes **Po** and **Pi** coincide (this is known as the *thin-lens approximation*); while this is not necessarily a realistic approximation, it does simplify the ray diagrams substantially and it is sufficient to illustrate how sets of lenses work.

We will use ray pairs **1–3** or **2–3** (in the notation of Fig. 3.9) to obtain the location of image and object planes in Fig. 3.41. Two sets of rays are shown (solid and dashed lines) for two different excitations of the lenses. When the OL current is decreased, the image focal length increases (the focal points are arrowed) and the image plane moves down the optical axis. This plane then becomes the object plane for the PL and two images with different overall magnifications are obtained. The PL current must be increased to bring the object focal point (arrowed) closer to the center of the lens. The overall magnification is then the product of the magnifications of each lens, as can be verified easily by direct multiplication of the two transfer matrices (Section 3.4.6).

Although we have reached our goal of a variable magnification system, there are several drawbacks to this setup. First of all, it is actually desirable to have an objective lens with a fixed current: for high-magnification observations small variations in the lens temperature can cause large specimen drifts, so large current changes in the objective lens must be avoided. In most observation modes, we use the objective lens at a given current, and focusing is carried out by small adjustments of this current (changes of a few tenths of a percent). There is another advantage to having a fixed OL current: the image plane conjugate to the specimen has a fixed location along the optical axis. This plane is rather important for conventional TEM work, and it is known as the *selected area* or SA plane. From here on, we will assume that the OL current is such that the SA plane, a plane fixed by the location of the SA aperture, is conjugate to the specimen plane. This also fixes the back focal plane and the diffraction aperture.

Since the fixed OL current also fixes the PL current for which the object plane is conjugate to the screen, we must add a third lens (intermediate lens, IL) to regain control over the magnification of the system. Adding a third lens to a system with a fixed OL current has the same effect as adding the projector lens to the variable OL: a variable magnification is again obtained, as shown in the ray diagram of Fig. 3.41(b). The SA plane is the object plane for the *intermediate lens*, while the observation screen is the image plane for the projector lens. Changing the current in both IL and PL creates a variable magnification system.

There is still one significant drawback to a three-lens system: a given magnification can only be reached in one way. To reduce the combined effects of the lens aberrations it is desirable to have more than one way to create a given magnification, so that the method with the lowest overall aberration can be used. This can be obtained by adding yet another lens, the *diffraction lens* DL. As shown in Fig. 3.42(a), several combinations of IL and PL lens currents can be used to obtain the same overall magnification. We can also vary the current in the DL to have the OL back focal plane as the object plane, and hence a magnified image of the electron diffraction pattern can be obtained, as shown in Fig. 3.42(b).

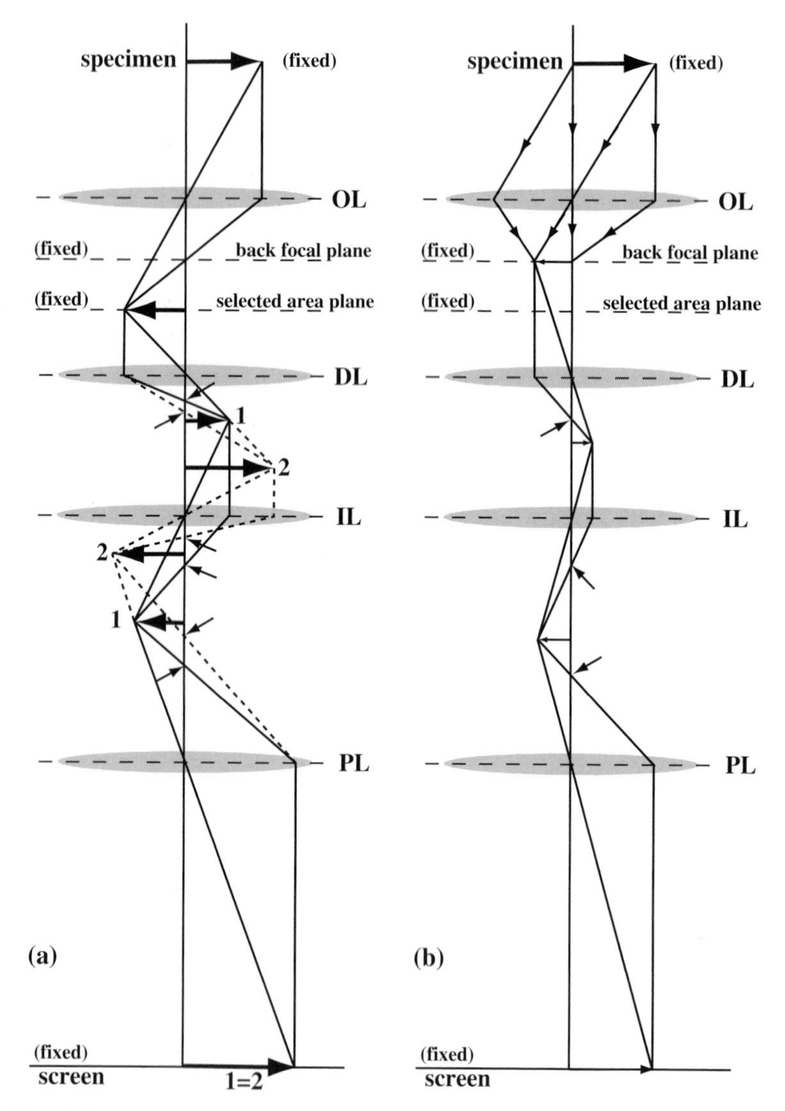

Fig. 3.42. (a) Four-lens system. Two ray diagrams with the same overall magnification are shown, illustrating that the same magnification can be obtained in more than one way. (b) Four-lens system with the OL back focal plane as object plane for the diffraction lens DL; this produces a magnified diffraction pattern on the viewing screen.

Since the focal length of a magnetic lens, and, hence, its magnification, depend on the electron energy, and the SA plane is a fixed plane, it is clear that a change in the microscope acceleration voltage must be accompanied by a change in lens currents, to maintain the appropriate relations between the fixed conjugate planes. Electron microscopes, therefore, have several preset lens currents for a number of acceleration voltages.

Not all lenses are excited for all observation conditions. Whenever a lens switches on or off, the microscope operator will observe an inversion of the image, since a cross-over is added to or removed from the beam path. Some microscopes have an objective minilens, a weak lens located between the OL and DL; this lens is used for low-magnification observations, for which the main OL is switched off completely. In other microscopes there is a second projector lens to compensate for the fact that the electron beam must go through a narrow differential pumping aperture just below the IL. We can discern a general rule of thumb: for every new *fixed* cross-over an additional lens must be added to keep the flexibility of the system the same. Finally, by carefully matching the lens currents for all imaging lenses in the column one can compensate for the image rotation caused by the Lorentz force and obtain a column with nearly zero overall image rotation over the entire magnification range.

3.11 Electron detectors

3.11.1 General detector characteristics

A good electron detector should have the following characteristics:

- sensitivity down to single electrons;
- a linear relation between the incident intensity and the detector signal for a wide range of intensities;
- high spatial resolution;
- low noise level;
- high detector readout speed or frequency;
- reproducibility.

Additional characteristics, such as ease of use, lifetime, and cost, may also be important. In the following subsections, we will discuss the discreteness of digital images and factors affecting the intensity measured by a detector. We will conclude this chapter with a brief description of the viewing screen, photographic emulsions, and charge-coupled device (CCD) cameras.

3.11.1.1 Detector pixel size

We will assume that we can define a detector *pixel size*, i.e. the size of the smallest region (usually a square) that contains a readable signal. For a charge-coupled device camera the pixel size is set by the physical dimensions of the sensitive area, typically around $25 \times 25 = 625\ \mu m^2$. For a photographic negative the pixel size is of the order of the silver halide grain size. After digitization of a photographic negative the pixel size would be determined by the scanning resolution. Digitization at 1200 "dots-per-inch" or dpi is equivalent to a pixel size of 21 μm, similar to that of a CCD camera [Ead96].

Since electronic images are always discrete images, sampled on a square grid of pixels, we must take a closer look at the details of *sampling*; in particular, we must ask the question: when is a sampled image a reliable representation of the actual image?

Consider a continuous function $f(x)$ with Fourier transform $f(q)$, i.e.

$$f(q) = \mathcal{F}[f(x)] = \int_{-\infty}^{+\infty} f(x)\mathrm{e}^{-2\pi \mathrm{i} q x}\,\mathrm{d}x.$$

The function $f(x)$ can be sampled with a *sampling interval* δ, so that we obtain a series of discrete values

$$f_i \equiv f(i\delta), \qquad i = -\infty, \ldots, -1, 0, 1, \ldots, +\infty.$$

The inverse of the sampling interval is known as the *sampling rate*. We define the *Nyquist critical frequency* as half of the sampling rate:

$$q_c \equiv \frac{1}{2\delta}. \tag{3.79}$$

The *sampling theorem* then states the following [PFTV89]: *if a continuous function $f(x)$, sampled at an interval δ, is band-width limited to frequencies smaller than q_c (i.e. $f(q) = 0$ for all frequencies $q > q_c$), then that function $f(x)$ is completely determined by the samples f_i*. The value $f(x)$ at any point in between two sampling points may be determined using the following interpolation relation:

$$f(x) = \sum_{i=-\infty}^{+\infty} f_i \mathrm{sinc}[2\pi q_c(x - i\delta)]. \tag{3.80}$$

If the sampling interval δ is too large, so that $f(q)$ is non-zero for $q > q_c$, then the sampling theorem fails, and the sampled function f_i will not be a faithful representation of the actual function $f(x)$. If the function $f(x)$ varies on a length scale smaller than δ, then the spatial frequencies corresponding to this length scale will not be properly represented in the Fourier transform, and the low-frequency Fourier coefficients will contain information on the high-frequency behavior of the function. This phenomenon is known as *aliasing*.

The main conclusion from this brief discussion of sampling is then that a digital representation of a continuous signal should have a sampling rate that is at least twice that needed to represent the smallest details of interest in the signal. Since the detector pixel size often effectively sets the sampling interval δ, this means that *images must be acquired at the proper microscope magnification*. In numerical computations with discrete images, it is often convenient to explicitly set all function values in the direct discrete Fourier transform (DDFT) of the image to zero for spatial frequencies beyond the Nyquist limit q_c.

3.11.1.2 Intensity levels for individual pixels

An ideal detector would be capable of detecting single electrons. Each pixel (i, j) in the image generated by such a detector would therefore contain an *intensity* $I(i, j)$ equal to the number of electrons that reached the pixel (i, j). For "real" detectors, the situation is somewhat more complicated because: (a) the signal from an individual electron may be spread over multiple pixels; (b) not all electrons may be detected; and (c) the inherent noise level in the detector electronics may place a lower limit on the number of electrons detected.

The quality of a detector can be quantified by the so-called *detective quantum efficiency* or DQE. Rose [Ros46] has defined the DQE as the squared ratio of the signal-to-noise ratio of the detector output to the signal-to-noise ratio of the input signal, or

$$\mathrm{DQE} \equiv \frac{(S/N)^2_{output}}{(S/N)^2_{input}}. \tag{3.81}$$

The DQE varies from 0 (only noise as the detector output) to 1 (the detector adds no noise to the signal). The *signal-to-noise ratio* can be defined as the mean image intensity $\bar{I}$ divided by the *standard deviation* of the intensity $\sigma(I)$:

$$S/N = \frac{\bar{I}}{\sigma(I)} \tag{3.82}$$

with

$$\bar{I} = \frac{1}{MN}\sum_{i=1}^{N}\sum_{j=1}^{M} I(i, j);$$

$$\sigma(I) = \left\{\frac{1}{MN-1}\sum_{i=1}^{N}\sum_{j=1}^{M}\left[I(i, j) - \bar{I}\right]^2\right\}^{1/2}.$$

The square of the standard deviation is known as the *variance* $\mathrm{var}(I)$.

The DQE defined above implicitly assumes that a single electron can only cause a signal in a single pixel. In reality, a single electron may cause a signal in a small neighborhood of pixels. This "spreading" of the signal is mathematically described by the *point spread function* T. The signal I_d measured by the detector is then the convolution of the actual incident intensity, $I(i, j)$, and the point spread function $T(i, j)$, or

$$I_d(i, j) = I(i, j) \otimes T(i, j). \tag{3.83}$$

It is often convenient to express this relation in Fourier space as

$$\mathcal{F}[I_d(i, j)] \equiv I_d(\mu, \nu) = I(\mu, \nu)T(\mu, \nu),$$

where $T(\mu, \nu) = \mathcal{F}[T(i, j)]$ is the *modulation transfer function* or MTF, and the components (μ, ν) indicate the spatial frequency. The point spread function concept is valid for both photographic and electronic detectors, and we will make extensive use of this concept in Chapter 10. Note that if $T(\mu, \nu)$ is known *and* non-zero for all spatial frequencies, then the incident intensity I can be retrieved from the detected intensity I_d by a simple division in Fourier space:

$$I(\mu, \nu) = \frac{I_d(\mu, \nu)}{T(\mu, \nu)}.$$

Fortunately, the MTF for photographic film and for CCD cameras does not contain any zero-crossings for frequencies up to q_c, and therefore this type of deconvolution is possible, assuming $I_d(\mu, \nu)$ has been zeroed for spatial frequencies beyond q_c. As a consequence of the frequency-dependent response of the detector system, the DQE also becomes frequency dependent, i.e. DQE(μ, ν) [dRMK93].

In the absence of an incident signal, a detector would ideally produce a zero-intensity signal. In practice, there will be a small signal due to noise in the detector electronics, noise generated by cosmic radiation hitting the active areas of the detector, and so on. An image acquired without any incident intensity is known as a *dark image*, $D(i, j)$, or a *background image*. The dark image should be subtracted from each acquired image, since it is common to all images. A dark image should be acquired with the same exposure time and other recording settings as the actual image.

The sensitivity of pixels may vary across the detector. In the case of a charge-coupled device camera the incident electrons are typically converted to light by a scintillator layer, and subsequently this light signal travels through a fiber-optic connector to the CCD surface. The internal structure of this fiber-optic connection creates spatial variations of the detected signal, even for a homogeneous incident illumination. This sensitivity must be removed from the image before any image processing may be carried out. This is accomplished by the acquisition of a *gain reference image*, $G(i, j)$, using uniform illumination. Correction of the sensitivity variations of individual pixels is known as *gain normalization*.

The true incident intensity $I(i, j)$ can be derived from the detector signal $I_d(i, j)$ by means of the following operations:

$$I = \mathcal{F}^{-1}\left[\frac{\mathcal{F}\left[\frac{I_d - D}{G - D}\overline{(G - D)}\right]}{\mathcal{F}[T]}\right], \tag{3.84}$$

where the overbar denotes the average over the entire image. The first operation (background subtraction and gain normalization) can be performed in real time, since the inverse of the gain reference image can be stored in memory. Deconvolution of the point spread function is slightly more time consuming, and is usually not performed in real time. Note that this procedure is quite general and applies to all detectors with a non-zero modulation transfer function $T(\mu, \nu)$.

The intensity $I_d(i, j)$ has a range from 0 to a maximum intensity I_{max}, which depends on how many electrons a pixel can detect before saturation occurs. For electronic detectors, this is given by the number of bits per pixel; if a detector uses 12 bits per pixel, then the range of detectable intensities is from 0 to $2^{12} - 1$ or 4096 gray levels. This interval is known as the *dynamic range* of the detector. It would appear that a 12-bit detector would be able to distinguish 4096 incident intensity levels. However, since the arrival of electrons at the detector is itself a statistical process, there is no statistically significant difference between, say, 4000 counts and 4001 counts. In fact, the standard deviation of the number N of electrons which reaches a detector element obeys Poisson statistics, and is given by $\sqrt{N}$. This means that for $N = 4000$ the standard deviation is about 63, so that all pixels with counts between $N = 3937$ and $N = 4063$ are statistically equivalent. The random arrival of electrons is known as *shot noise* or *Poisson noise*. For a detailed discussion of shot noise and related topics we refer the interested reader to [Spe88, Section 6.8].

The human eye itself has a non-linear response to incident intensity. Physiological measurements indicate that the "electrical output" of the human eye is proportional to *the logarithm of the stimulus* (Weber–Fechner's law). This is reflected in the magnitude scale used for the apparent brightness of stars; the original magnitude scale covers the range from 1 (brightest) to 6 (faintest), which corresponds to a factor of 100 in absolute intensity change. Therefore the linear magnitude scale actually corresponds to an exponential scale with base 2.512, since $(2.512)^5 = 100$. As a direct consequence of its logarithmic response, the human eye can only distinguish between a limited number of gray values in an image, typically on the order of 30 different gray levels. The eye is, therefore, not a very good instrument to compare quantitatively the intensity distributions of experimental and simulated images.

After the "true" image intensity has been determined by means of equation (3.84), other image processing operations may be performed on $I(i, j)$. Typical operations are high-pass and low-pass filtering, edge enhancement, Fourier analysis, smoothing, and so on. The literature on image processing is extensive, and the reader is encouraged to consult *The Image Processing Handbook*, by J.C. Russ [Rus92], for further information. The field of *image algebra* formalizes the ways in which images can be manipulated. For an introduction to image algebra, the reader is referred to Chapters 70–76 in [HK94].

The last step in any image processing procedure is always the visualization of the final image on a computer screen or in printed form. Grayscale images are typically represented on a linear 8-bit intensity scale, so that there are 256 output gray levels available. Image intensities need not be mapped linearly into the available gray levels. A commonly used mapping is based on equation (3.73) on page 211, with

$c_1 = 255$ and $c_2 = \gamma$ (in some image processing systems, c_2 is set equal to $1/\gamma$). A gamma-factor of $\gamma = 1$ produces a linear intensity mapping between 0 and 255. For values of $\gamma > 1$ the lowest intensities are compressed into fewer gray levels, while for $\gamma < 1$ the higher intensity values are compressed. This type of non-linear scaling enhances the *image contrast*; the percentage contrast between two pixels is defined by

$$\%C = \frac{|I_1 - I_2|}{I_1 + I_2} \times 100. \tag{3.85}$$

The shot noise in an image along with the saturation value of each pixel determine the smallest percentage contrast that can be obtained. If a pixel has I counts and the neighboring pixel has $I + \mathrm{d}I$ counts, then the percentage contrast is given by [Ead96]:

$$\begin{aligned} \%C &= \frac{100}{2I/\mathrm{d}I + 1}; \\ &> \frac{100}{\sqrt{I} + 1} \approx \frac{100}{\sqrt{I}}. \end{aligned}$$

We have used the fact that two pixels have a statistically different intensity only when $\mathrm{d}I > 2\sqrt{I}$. For large values of I the last expression provides a good approximation. For $I = 10\,000$ the lowest percentage contrast that is statistically meaningful would be around 1.0%.

The average intensity of the image, $\bar{I}$, is the image brightness. Image brightness can be changed by a mapping of the form

$$I_{display} = c_{min} + (c_{max} - c_{min}) \left(\frac{I}{I_{max}} \right)^{\gamma}.$$

If c_{max} is smaller than 255, the overall image brightness decreases; if c_{min} is larger than 0, then the image brightness increases. It is obvious that brightness is changed at the expense of the number of available display gray levels. Figure 3.43(a) shows two different contrast scaling functions for $\gamma = 0.3$ and 2.5 for the display intensity range [0, 255]. The effect of both scalings on a reference image is shown in Figs 3.43(c) and (d), which represent gamma-scaled versions of the image in Fig. 3.43(b).

The reader should be warned that with modern image processing packages it is all too easy and often quite tempting to *enhance* experimental images, by changing the contrast or by sharpening the details. If images derived from an experiment (i.e. after application of equation 3.84 to the detected images $I_d(i, j)$) are to be used for comparison with numerically simulated images, *no image altering operations should be performed before the comparison has been completed!* If image

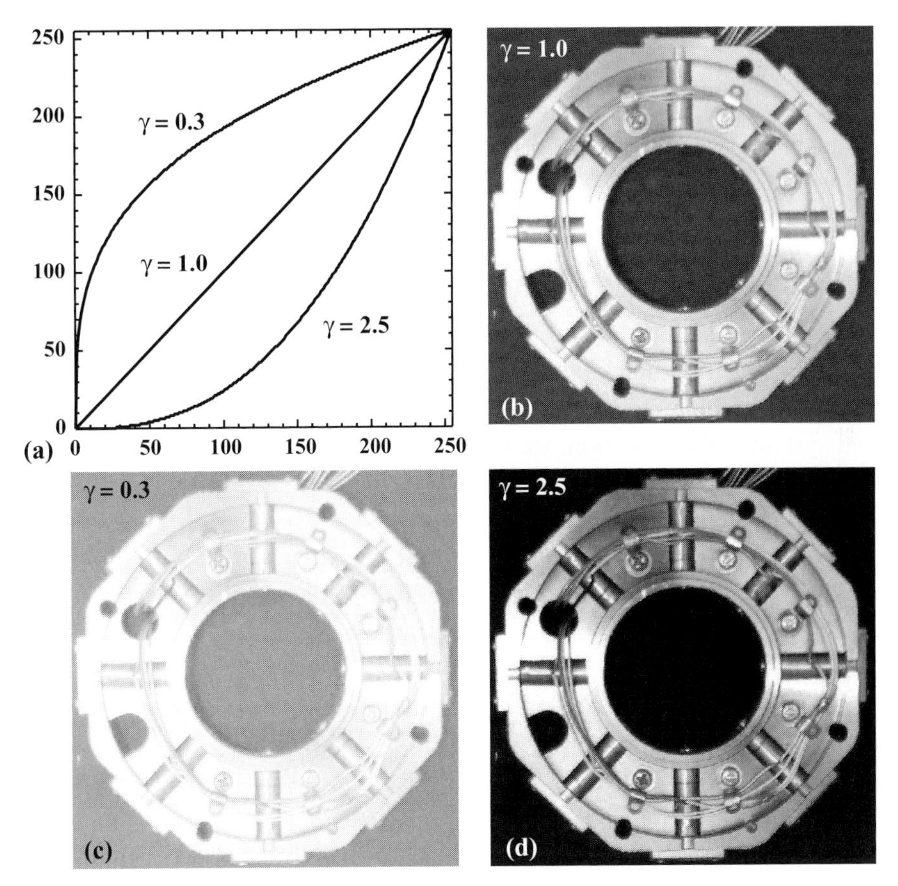

Fig. 3.43. (a) Gamma correction curves; (b) reference image (condensor stigmator of JEOL 120CX); (c) gamma-scaled image for $\gamma = 0.3$; (d) gamma-scaled image for $\gamma = 2.5$.

modifications or enhancements are made (smoothing, filtering of noise, etc.), then this should be stated explicitly in the figure caption or text accompanying the figure.

3.11.2 Viewing screen

The final stage in the microscope detects the electrons and forms an observable image. This is routinely accomplished with a phosphor (or ZnS) *viewing screen*, where the incident electrons excite outer-shell electrons and "green" photons† are emitted when these fall back to their low-energy states. With high-intensity electron beams, some "burn-in" can occur if the beam is focused on one point for too long. Viewing screens are therefore replaced every couple of years. The resolution of a

† The human eye is most sensitive in the green portion of the visible electromagnetic spectrum, and therefore the screen phosphor is designed to emit most strongly in this frequency range.

phosphor screen is about 50 μm (the smallest detail that can be resolved on the screen depends on the phosphor grain size). The apparent image brightness on the viewing screen depends on the microscope acceleration voltage (a higher voltage corresponds to a dimmer image) and on the thickness of the lead glass viewing window.

3.11.3 Photographic emulsions

Photographic emulsions are perhaps the most commonly used electron detectors. Although digital detection systems are becoming increasingly important, it is doubtful that photographic film will be replaced completely in the near future by purely digital systems. In this section, we will describe briefly how a photographic emulsion reacts to an incident electron and how one may extract quantitative information from the exposed and developed negative.

When an incident electron beam with current density j_0 hits a photographic emulsion, then the total *exposure* E is defined as the product of current density (also known as *illuminance*) and exposure time τ:

$$E = j_0\tau.$$

The energy of the absorbed electrons locally converts at least one and often more than one silver halide grains into metallic silver. Those regions of the emulsion that have been converted to metallic silver are then *developed*, which essentially means that the entire halide grain is converted to metallic silver by a suitable reducing agent (i.e. the developer). Grains that have not been exposed are then removed by the *fixing* or *fixation* process, and the exposed regions appear as dark patches on the transparent acetate support film. For extensive details on the theory of the photographic process we refer the interested reader to the book *Fundamentals of Photographic Theory* by James and Higgins [JH68].

When the developed negative is illuminated by a light source of intensity I_0, a fraction $I(x, y)$ is transmitted, depending on the density of metallic silver at the location (x, y). We define the photographic *density* D by

$$D(x, y) = \log_{10} \frac{I_0}{I(x, y)}.$$

The ratio I/I_0 is known as the *intensity transmittance* of the negative. The photographic density can be measured by a *densitometer* or a scanner equipped with a transparency adapter. Instead of using the actual intensity of the light source as I_0, it is often easier to use the scanned intensity from a region near the edge of the negative (a region which was not exposed to electrons, but underwent the same development and fixation process as the exposed region).

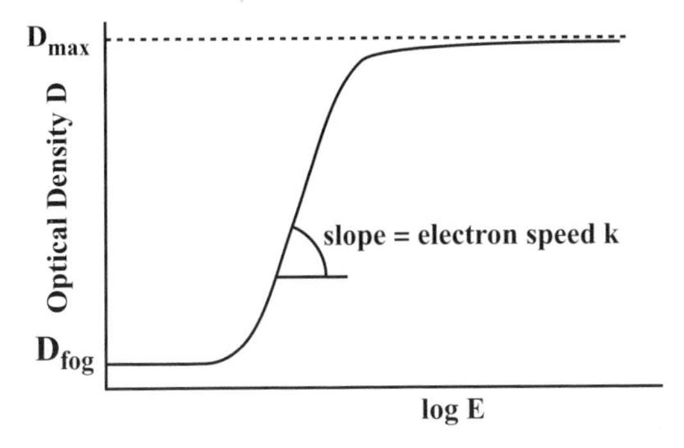

Fig. 3.44. Schematic representation of the Hurter–Driffield curve, relating optical density D to the logarithm of the exposure E. The fog level represents the density of an unexposed but developed region of the negative. The maximum density level is typically around $D \approx 3$.

The relation between the density D and the exposure E is commonly represented by a logarithmic plot of D versus $\log E$, known as the *Hurter–Driffield curve* or the *characteristic curve*. A schematic example of such a curve is shown in Fig. 3.44. For exposure to light, such a curve properly describes the relation between E and D. Up to a minimum exposure level, the film density is constant and equal to a small number, known as the *fog level*. Then the curve becomes linear with slope k, the *film speed*, sometimes known as the *photographic contrast*. Electron speeds for modern emulsions are in the range 0.3–1.1 μm^2 per electron. For large exposures, all halide grains are converted into metallic silver and the film saturates at a limiting density D_{max}. The practical value of the saturation density is between $D_{max} = 2$ and 3 for most applications, $D = 3$ indicating that only $\frac{1}{1000}$ of the illuminated intensity is transmitted through the negative. The precise shape of the characteristic curve depends in a sensitive way on the experimental conditions during development and fixation (temperature, time, etc.), and it is not a trivial matter to extract reliable intensity data from photographic negatives. For an extensive discussion of this topic the reader is referred to the early work by Frieser, Klein, and Zeitler [FKZ59] and the more recent approach to quantitative analysis of electron negatives by Campbell, Cohen and King [CCK97]. Zeitler [Zei92] has shown that for electron exposures at densities of less than 1.5 the relation between D and E is essentially linear.

It is possible to define and measure the point spread function for an electron negative. It can be shown (e.g. [Zei92]) that the radially symmetric PSF $T(r)$ is given by

$$T(r) = \frac{1}{2\pi x_0^2} K_0\left(\frac{r}{x_0}\right), \tag{3.86}$$

where K_0 is a modified Bessel function and x_0 is a decay length, which depends on the electron energy. The corresponding modulation transfer function $T(q)$ is given by the Fourier transform:

$$T(q) = \frac{1}{1 + q^2 x_0^2}. \tag{3.87}$$

This function has no zero-crossings and can, therefore, be deconvolved from the recorded transmittance. In practice, such a deconvolution will be performed numerically, which means that the negative must be digitized, either by means of a densitometer, or by regular scanning using a transparency adapter. Whichever digitization method is used, the reader should be aware that *the scanning device itself may impose its own modulation transfer function* on the digitized image. For densitometers such an instrumental MTF is well understood and can be determined easily. For flat-bed scanners on the other hand, the MTF depends on the details of the illumination and also on the discreteness of the detector array, typically a diode array.

3.11.4 Digital detectors

Charge-coupled device detectors, commonly known as CCD detectors, have been around since 1970, when Boyle and Smith proposed the first charge-coupled semiconductor device [BS70]. While there have been many advances in the performance of CCD detectors since their invention, the basic operational principles have not changed significantly. A silicon substrate (Fig. 3.45a) is lightly doped with about 10^{15} acceptor atoms cm^{-3} so that it acts as a p-type semiconductor. A thin (100 nm) silicon dioxide layer is deposited on top of the substrate, followed by a metal gate. When a positive voltage is applied to the gate the majority carriers, holes in the case of a p-type semiconductor, are repelled from the area underneath the gate

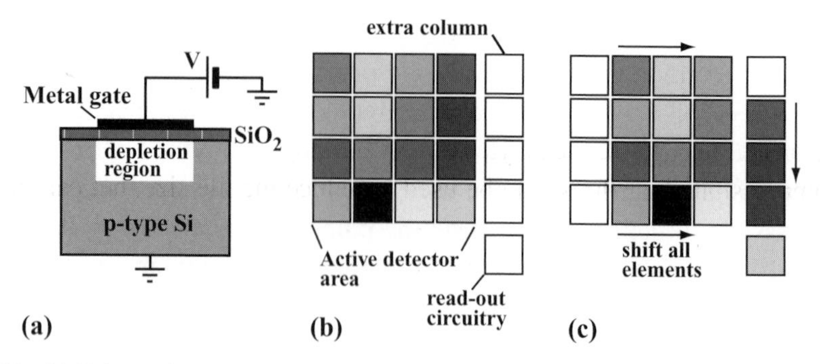

Fig. 3.45. (a) Schematic representation of the structure of a single element of a CCD device; (b) all elements of a 4 × 4 detector array are exposed to incident light; (c) the array is first shifted one element to the right, and then the elements of the vertical column are shifted into the read-out circuitry. Then the array is shifted again, until all elements have passed through the read-out circuitry.

and a depletion layer is formed. The bias voltage therefore creates a potential well, the depth of which is determined by the bias voltage, the thickness of the oxide layer and the size of the metal contact [BL80]. This potential well can subsequently be filled with minority carriers (electrons), which will form a thin layer close to the oxide. As the number of electrons increases, the depletion region shrinks, until a maximum number of electrons has been accumulated. At this point the potential well is saturated and can accept no additional electrons.

When an array of CCD elements is exposed to light, the potential wells will be filled with electrons, and the number of electrons in each well will be proportional to the incident light intensity for that particular well. Once the exposure has been completed, the charge in each CCD element must be determined. This is accomplished by a clever combination of voltage changes applied to neighboring detector elements. For a detailed discussion of this mechanism as well as a description of other CCD geometries we refer to [BL80] or [HM79]. All the charges accumulated along a line of CCD elements are simultaneously shifted one element sideways (Fig. 3.45b), and the final column is shifted down towards the read-out circuitry which measures the total charge in each individual element (Fig. 3.45c). When all elements have been read, the CCD device is ready for the next exposure.

Scientific CCD detectors are typically square detectors, with element sizes in the range of 10–25 μm on the side. For TEM imaging, the beam electrons are first converted into photons in a phosphor or YAG (yttrium aluminum garnet) scintillator. Subsequently, the photons hit the active area of the CCD elements and generate electrons that fill the potential wells in the depleted regions of the p-type semiconductor. The scintillator is coupled to the CCD detector array through a fiber-optic coupling, or is directly cemented onto the array. To reduce electronic noise, CCD detectors are cooled down to the range -30 to $-50\,^\circ$C using Peltier cooling elements. The dominant sources of noise are in the scintillator electron–photon conversion and the detector photon–electron conversion processes.

Modern CCD detectors can have as many as 4096×4096 detector elements, and each element has a dynamic range of 12, 14, or 16 bits. A single image from such a detector will occupy 32 megabytes (33 554 432 bytes) of memory or disk space. File compression algorithms may be used to reduce the file size, but care must be taken to use only lossless compression algorithms when raw image data is stored.[†] Image analysis should only be performed on raw image data, not on data that has been stored using a lossy compression algorithm.

The imaging characteristics of a CCD detector are well described using the MTF formalism introduced in Section 3.11.1.2. For a rather complete description of the

[†] Lossless image compression is typically available in the tagged image file format, or *tiff*-format. Lossy image compression algorithms are used for the Joint Photographic Experts Group format, or *jpeg*-format.

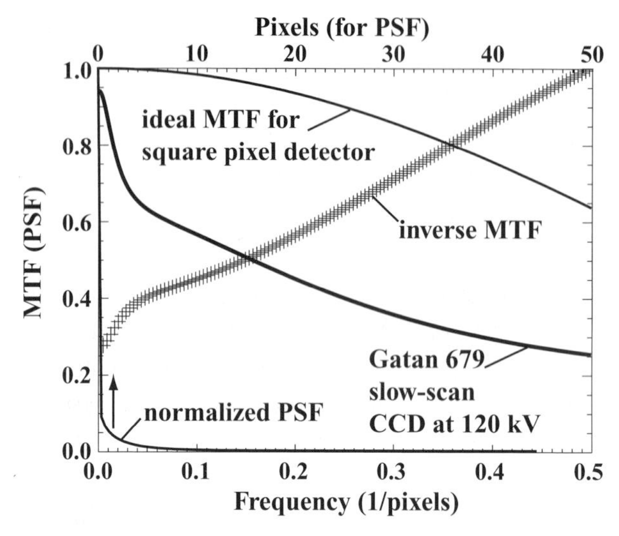

Fig. 3.46. Modulation transfer function for a 1 K × 1 K Gatan 679 slow-scan CCD camera with a 120 kV incident electron beam [Zuo96]. The inverse of the MTF (MTF_{max}/MTF) is also shown, along with the ideal MTF for a signal lined up with one of the rows or columns of a square detector array. The bottom curve is the normalized point spread function $T(r)$ versus distance measured in pixels (scale on top).

imaging characteristics of CCD cameras we refer the interested reader to de Ruijter [dR95] and Zuo [Zuo96]. In this section, we will only state the most important facts. The MTF of a CCD camera is similar in form to that of a photographic negative. Zuo [Zuo96] has modeled slow-scan CCD cameras with different scintillators at different microscope acceleration voltages using the following MTF:

$$T(q) = \frac{a}{1+\alpha q^2} + \frac{b}{1+\beta q^2} + c, \tag{3.88}$$

where the first two terms model the head and tail part of the point spread function and c is a constant. Figure 3.46 shows an example MTF for a Gatan model 679 slow-scan CCD camera operated at 120 kV. The camera has 1024 × 1024 pixels with a dynamic range of 12 bits, a YAG scintillator, and Peltier cooling to −30 °C. The fitting coefficients are $a = 0.3293$, $\alpha = 3328.8$, $b = 0.46423$, $\beta = 13.511$, and $c = 0.147\,94$[†] [Zuo96]. The frequency range goes from zero to the Nyquist frequency q_c; in other words, the frequency axis is labeled in units of the sampling rate.

[†] The reader should note that these are approximate parameters; at zero frequency, the MTF should be equal to unity, or $a + b + c = 1$, and for the case shown we have $a + b + c = 0.941\,47$.

The corresponding point spread function consists of two modified Bessel functions K_0 plus a delta-function (the Fourier transform of the constant c):

$$T(r) = \frac{1}{2\pi\alpha} K_0\left(\frac{r}{\sqrt{\alpha}}\right) + \frac{1}{2\pi\beta} K_0\left(\frac{r}{\sqrt{\beta}}\right) + c\delta(r), \tag{3.89}$$

which is a sharply peaked function with a slowly decaying tail. For the fitting parameters introduced above, 85% of the area under the PSF falls within the first six pixels, while 99% falls within the first 30 pixels. This indicates that an incident electron has an 85% probability to be detected inside a circle with radius six pixels centered on the point of impact on the scintillator. The main consequence of the point spread function is that the highest spatial frequencies are suppressed with respect to the low spatial frequencies, and this in turn causes sharp features in the image to be smoothed. The inverse MTF, also shown in Fig. 3.46, can be used to deconvolve the PSF from the image. Care must be taken that the various operations in equation (3.84) are performed in the proper order.

The detective quantum efficiency, DQE, is in the range 0.5–0.7 for electron doses between 100 and 1000 counts pixel^{-1}. For explicit expressions for the frequency-dependent DQE (DQE(μ, ν)) we refer to [dR95] and [Zuo96]. It can also be shown from experimental measurements that the number of CCD well electrons is between 10 and 30 per incident beam electron, depending on the details of the scintillator.

Exercises

3.1 Derive the paraxial radial trajectory equation (3.17).

3.2 Use the lens.f90 program to determine an empirical relation between a and B_0 (for Glaser's bell-shaped field) for which the lens has the so-called "telescopic property", i.e. the incident and exciting trajectories are both parallel to the optical axis, but at a different distance from the axis. Such a lens would change the beam diameter of a cylindrical beam.

3.3 Use the lens.f90 program to create a configuration consisting of two unequal lenses which, for a given acceleration voltage, will produce an exit trajectory at half the distance from the optical axis (i.e. $r(z_i) \approx 0.5r(z_0)$ and $r'(z_i) = 0.0$) and which has zero total image rotation.

3.4 Derive the components of the lens transfer matrix (3.29).

3.5 Verify graphically that the image **Q** of the object point **P** will be magnified when the object is located at a distance shorter than the object focal length from the object focal plane (see Fig. 3.9 and the discussion on page 160).

3.6 Draw ray diagrams similar to the one in Fig. 3.27(b) for the under-focus and over-focus conditions of the second condensor lens.

3.7 Show that the barrier reduction due to the Schottky effect is given by equation (3.48) on page 181.

3.8 Equation (3.77) was derived for HOLZ reflections with the incident beam along a zone axis. If the incident beam is tilted slightly away from the zone axis, so that the wave vector $\mathbf{k}_0$ acquires the components L_1 and L_2 with respect to the vectors $\mathbf{g}_1$ and $\mathbf{g}_2$, what will be the resulting changes to equation (3.77)?

4

Getting started

4.1 Introduction

This chapter covers a wide range of topics, all of which are related to standard, routine microscope operation. Microscope operation can be divided into three levels: (1) operations common to all microscope models, regardless of manufacturer; (2) operations specific to a particular model; and (3) operations specific to a particular laboratory or location. In this chapter, we will only deal with the first kind of operations.

For operations specific to a given microscope model, we refer the reader to the microscope manual, which comes with every installation. Most manuals are very detailed, with photographs of microscope components and the layout of the console. In addition, routine procedures such as specimen exchange, photographic plate exchange, switching between imaging modes (which buttons to press), and so forth, are detailed in the microscope manual. For the beginning microscopist, it is a good idea to keep this manual at hand throughout the first series of microscope sessions.

In addition, there may be special (local) rules for such things as:

- specimen and specimen holder handling (where the gloves are, what kind of tweezers should be used, how the specimen holder should be stored, etc.);
- how to turn up the filament current (i.e. how much time should be left between subsequent filament current increments, should the acceleration voltage be left on between users, etc.);
- what type of log should one keep? Is there a general log for all micrographs, or does every user keep a personal log? How does one sign up for microscope time?
- in what state should the microscope be left for the next user (e.g. the minimum number of unexposed plates to be left in the camera, the position of all apertures, the magnification at which the microscope should be left, etc.).

Such rules are often written up in a laboratory manual, or verbally passed on to new microscope users. Rather than attempt to cover every possible laboratory situation, for the remainder of this book we will assume that: (1) the reader has access to a transmission electron microscope; (2) there is an experienced microscopist or microscope technician in the laboratory who can be consulted when problems arise; (3) the microscope manual is at hand; and (4) ample microscope time is available. The microscope manual and additional local procedures make up what we shall call the *standard procedures manual*. The reader is encouraged to become familiar with all such procedures, and to follow them at all times. Whenever we use statements such as "switch to diffraction mode" we will assume that the reader can find out from the standard procedures manual how to do that.

What this chapter does deal with can then be summarized as follows:

- startup and alignment procedures;
- basic observation modes;
- how to take "good" micrographs;
- microscope calibration;
- basic preparation to begin observations on a "new" material.

The last item includes thin-foil preparation as well as gathering crystallographic information concerning the material and producing a *crystal information package*.

This chapter does not provide the reader with a detailed overview of all possible thin-foil preparation methods. Specimen preparation is simply too important to be discussed in only one short chapter. Specimen preparation may take well over 80% of the overall experimental effort; it is not unusual for a microscopist to spend five or more times as much time on the specimen preparation part than on the actual microscopy observations! With the advent of atomic resolution microscopes and advanced analytical instruments it has become clear that without a clean, well-prepared foil there is little or no point in reserving time on a microscope. Conventional TEM techniques are perhaps a bit more forgiving about the quality of the foil (e.g. foils which are useless for analytical measurements may still be useful for conventional imaging), but it is always better to invest time in preparing good foils, regardless of the observations subsequently carried out on those foils. However, because of overall page limitations, this textbook does not devote a proportional number of pages to specimen preparation. We refer the reader to one of the following books for a detailed account of various specimen preparation techniques: *Specimen Preparation for Transmission Electron Microscopy of Materials* (P. Goodhew, [Goo84]); the four volumes in the MRS conference series *Specimen Preparation for Transmission Electron Microscopy of Material* [BAM88, And90, ATB92, AW97]; and volume 5 of the original Edington monographs on *Practical Electron Microscopy in Materials Science* (difficult to find but an excellent source, [Edi76]).

As stated in the Preface on page xiv, one of the main goals of this text is to provide a good starting point for the microscopy student. The author has selected four materials from which nearly all micrographs in this book have been obtained. The materials will be discussed in Section 4.3 and are relatively easy to obtain. Detailed specimen preparation instructions are included for these four materials, so that the reader may prepare his/her own foils and then reproduce most of the micrographs in the book. These four materials will also be used in later chapters for a variety of theoretical and numerical computations of image contrast and diffraction features. While this approach limits the range of microstructural features and defects that are considered, the advantage is that the reader can practice and compare his/her observations with the images in this book. After a discussion of specimen preparation for the four study materials, Cu-15 at% Al, Ti, GaAs, and $BaTiO_3$, we conclude the chapter with a series of typical observations, both in image and diffraction modes. It will then be the primary goal of the theoretical chapters to explain and model the various contrast features observed, and to create a framework for the routine simulation of conventional TEM image contrast.

4.2 The xtalinfo.f90 program

Whenever one starts to work on a new material, it is useful to gather all the necessary crystallographic information about this material. In addition, it is often useful to calculate several relevant, usually low-index, zone axis diffraction patterns. We will see in Chapter 5 that there are also a few quantities, such as the *extinction distance*, which are useful for the interpretation of conventional TEM micrographs.

The program xtalinfo.f90, available from the website, creates several pages of PostScript output, listing the most relevant information about the crystal structure in a form useful to the TEM operator. All of the entries on these pages can be obtained readily from various public domain and commercial software packages, but it is convenient to have everything available in one single program, formatted in a consistent way. The input to the program is the basic crystallographic information (lattice parameters, space group, and asymmetric unit), the microscope acceleration voltage and camera length; the output consists of crystallographic information, structure factor amplitudes, intensities, and extinction distances ranked by family, stereographic projections for the three basic crystallographic directions (both direct and reciprocal space projections), and computed electron diffraction patterns for a series of low-index zone axes (drawn to the correct scale). The program combines nearly all of the routines described in Chapters 1 and 2, and for information on the individual modules we refer to the relevant sections in those chapters.

Table 4.1. *Material parameters for Cu-15 at% Al (taken from [Vil97]); [copper–aluminum Bronze; Pearson Symbol cF4; Strukturbericht Notation A1].*

Space group	$Fm\bar{3}m$	
Lattice parameters	$a = 0.365\,26$ nm	# 225
Asymmetric unit	Cu,Al at 4a	(0, 0, 0)

There are many crystallographic resources available, both in electronic and paper form. The following list provides some of the most important reference works and journals for crystal structure data:

- *Pearson's Handbook (Desk Edition)* [Vil97]; contains 27 686 different compounds.
- *Atlas of Crystal Structure Types* [DVvV91]; contains complete geometric descriptions of all structure types referenced in *Pearson's Handbook*.
- *Crystal Structures* [Wyc63]; a six-volume detailed description of many structure prototypes.
- *Chemical Abstracts*, published by the American Chemical Society, contains a large amount of crystallographic information on both organic and inorganic compounds.
- *Structure Reports* (first started as *Strukturbericht*), published by the International Union of Crystallography.
- *Acta Crystallographica* also published by the International Union of Crystallography.
- *Powder Diffraction File*; a large collection of (x-ray) powder diffraction patterns, available from the International Center for Diffraction Data.†
- *Collaborative Computational Project Number 14 for Single Crystal and Powder Diffraction*; a large collection of crystallography related links.‡

You should be able to find crystallographic information for nearly all known compounds in at least one of the above sources.

4.3 The study materials

4.3.1 Material I: Cu-15 at% Al

4.3.1.1 Crystallography

Cu–Al alloys are commonly known as *bronzes*, and are solid solutions of up to 18.8 at% Al (at room temperature) in a face-centered cubic matrix of Cu atoms. Space group and lattice parameter information can be found in Table 4.1, and

† `www.icdd.com` (Since the web is a dynamic medium and links can change, the reader is referred to the website for a reasonably up-to-date listing of crystallography related links.)
‡ `www.ccp14.ac.uk`

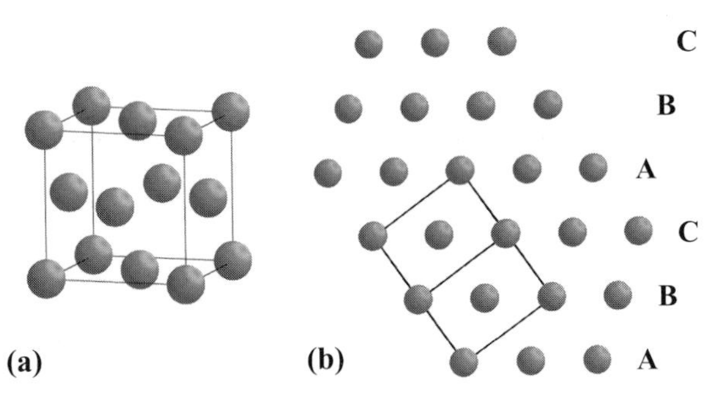

Fig. 4.1. (a) Parallel projection of the close-packed cubic structure; (b) [110] projection illustrating the *ABC* stacking sequence of the fcc lattice.

structure drawings are shown in Fig. 4.1. There are several reasons for selecting this material as one of the four study materials: first of all, the crystal structure is cubic, which is one of the easier structures to deal with for the beginning microscopist. Secondly, Cu–Al bronzes have a relatively low stacking fault energy, so the probability of finding a stacking fault is reasonably high. Lastly, after a proper heat treatment, the Al atoms are no longer randomly substituted on the Cu lattice, but instead will form a *short-range ordered* state.

It is easy to obtain a Cu–Al bronze:† simply melting pure Cu and Al together in a sealed quartz tube will do the trick. The melting temperature of a Cu-15 at% Al alloy is around 1050 °C [Com73, page 259]. Alternatively, Cu–Al bronzes can be obtained easily from metals supplies companies.

Any Cu–Al bronze will do, but if you are going to make your own material the Al content should be between 10 and 15 at%, so that the short-range ordering effect can be made more pronounced. Higher Al contents should be avoided, unless you want to deal with hexagonal and monoclinic martensitic phases! If you cannot obtain a Cu–Al bronze, then the next best thing would be pure copper; with the exception of SRO, you will be able to reproduce most of the Cu–Al-based figures in this text. We have also prepared pure copper foils with a small grain size (around 2 μm) and will use them for some of the observations in this chapter.

4.3.1.2 Thin-foil preparation

If you made your own alloy and slowly cooled it from the melt, then you should have a reasonably coarse-grained microstructure. If you purchased a rod of Cu–Al bronze it might be a good idea to subject it to a solutioning heat treatment before you begin to prepare a TEM foil; quite often the microstructure of extruded or swaged

† In some countries bronze coins are still in use. You could simply take one of these coins, check its composition, and use it to prepare your samples!

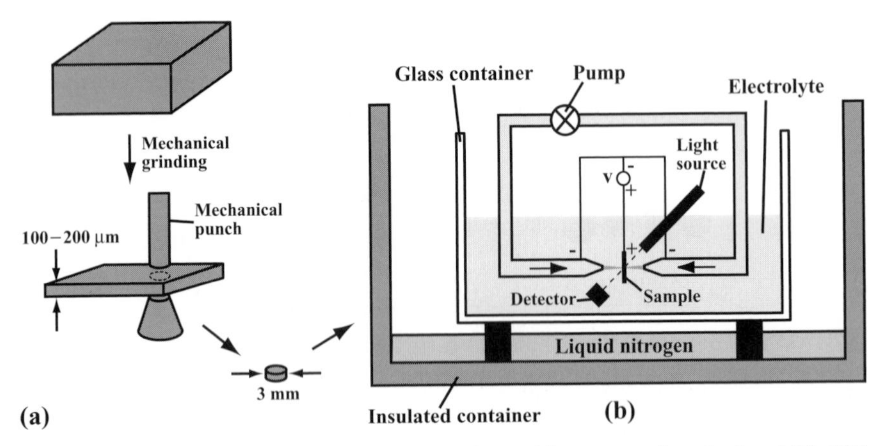

Fig. 4.2. (a) Mechanical grinding and mechanical punching are used to obtain a 100–200 μm thick 3 mm diameter thin foil; (b) schematic illustration of the double-jet electro-polishing process.

rods will have a very high defect density, something we wish to avoid in this case. A heat treatment of 1 h at 700 °C will drastically reduce the defect density.

Use a diamond saw or wire electro-discharge machining (EDM) to cut a thin slice of material, about 300–500 μm thick. Using coarse sandpaper, reduce the thickness of the slice to around 100–200 μm, by removing material from both sides (Fig. 4.2a). Then use fine sandpaper to get a smooth surface. Use a mechanical punch to punch out 3 mm diameter disks from the thin sheet. If you want to introduce a significant dislocation density in your samples, then you can bend the sheet a bit (making sure you deform the material plastically) and bend it back to a flat sheet before you punch out the disks.

For many metals and alloys the best thin foils are obtained using *double-jet electro-polishing*. The basic electrolyte used for many different copper alloys consists of 70 vol% methanol and 30 vol% nitric acid (HNO_3).† Other mixtures are also used and often a mixture used for a material similar to the one you intend to study can be used as a starting point.

The basic electro-polisher is shown schematically in Fig. 4.2(b): a glass container with the electrolyte is placed on supports in a cooling vessel. Typically this vessel contains a small amount of liquid nitrogen at the bottom; evaporation of the nitrogen then cools down the glass container and its contents. The sample is mounted in between two teflon support plates and positioned in between two nozzles. A rotary

† A word of caution: chemicals can be dangerous to your health in many different ways. Always read the safety instructions that come with the chemical, usually in the form of a *Material Safety Data Sheet* or *MSDS*. Those sheets can be obtained from the manufacturer. Since every microscopy laboratory will have its own safety rules and guidelines, we urge the reader to familiarize himself/herself with all applicable rules and to follow them at all times.

Table 4.2. *Material parameters for Ti (taken from [Vil97]); [titanium; Pearson Symbol hP2; Strukturbericht Notation A3].*

Space group	$P6_3/mmc$	
Lattice parameters	$a = 0.2950$ nm $c = 0.4681$ nm	# 194
Asymmetric unit	Ti at 2c	$(\frac{1}{3}, \frac{2}{3}, \frac{1}{4})$

pump is used to produce two fine jets that are directed at the center of the disk. Typically, the jets are about 0.5 mm in diameter and the pump velocity should be controlled to give straight jets at the sample location. A voltage is then applied between the sample (positive) and the two nozzles (negative). As the voltage is increased, material is removed from the sample and a current will begin to flow between the sample and the nozzles. At the lowest voltages, only a small amount of material is removed and the sample is "etched". At a slightly higher voltage, the sample is "polished", and at the highest voltage "pitting" (i.e. a rapid, uncontrolled removal of material) occurs. It takes some experience to find this polishing plateau in the current–voltage curve; the type of electrolyte, the applied voltage, the electrolyte temperature, and a host of others parameters (including jet alignment, age and purity of the electrolyte, and "good fortune") will determine the quality of the thin foil.

It may take a bit of experimenting to find the right conditions for your material. For the Cu-15 at% Al foils we used the following conditions: electrolyte methanol – 20 vol% HNO_3, $T = -30\,°C$, voltage $V = 5$–8 V. Polishing took 30–60 s per foil, for an initial foil thickness of 100 μm. The applied voltage will depend on the age of the electrolyte; the more Cu-ions are present in the electrolyte, the larger the voltage must be to obtain well-polished foils.

4.3.2 Material II: Ti

4.3.2.1 Crystallography

Pure Ti has a hexagonal crystal structure with lattice parameters $a = 0.2950$ nm and $c = 0.4679$ nm. The space group information and atom positions for the asymmetric unit are given in Table 4.2. There are two atoms per unit cell, and the structure can be described as a two-layer stacking of close-packed planes, usually represented by the stacking symbol **AB**, the Ramsdell symbol **2H**, or in triangle notation $(\Delta\nabla)^n$ [HB84]. The stacking symbol indicates that the close-packed planes alternate between two different positions. Each close-packed plane has two kinds of triangular dimples in between groups of three atoms. Those dimples are represented

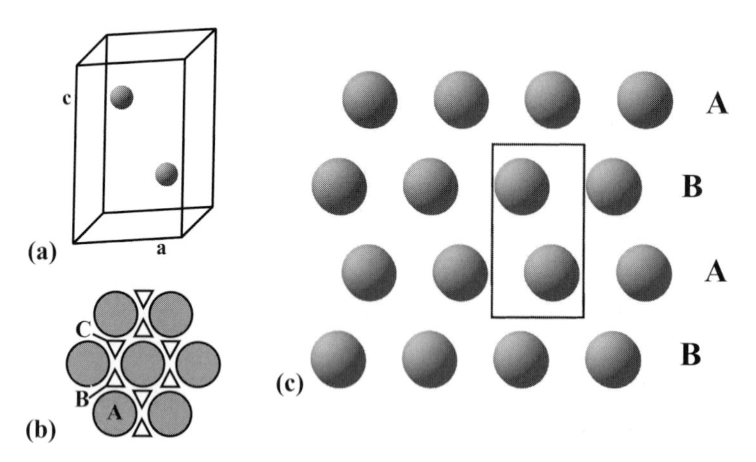

Fig. 4.3. (a) Parallel projection of the Ti unit cell; (b) single close-packed layer showing two sets of triangular dimples; (c) projection along the $[2\bar{1}.0]$ direction showing the *ABAB* stacking sequence.

by the symbols Δ and ∇, as shown in Fig. 4.3(b). If we represent the first plane by the letter **A**, then the next plane can be located on the Δ dimples, producing a **B** plane, or on the ∇ dimples, producing a **C** plane. If we alternate the stacking of close-packed planes such that dimples of opposite orientation are filled, then we obtain the hexagonal close-packed structure. If we always fill only one kind of dimple, say $(\nabla)^n$, then we obtain the face-centered cubic structure shown in Fig. 4.1(b), with a stacking sequence of **ABCABC** or **3R** (**R** stands for rhombohedral, which is the primitive unit cell corresponding to the fcc cell). Several defects, such as *stacking faults* and *twins*, can be described in terms of such stacking symbols.

The reasons for including titanium as a study material are twofold: first of all, titanium has a hexagonal structure, which will permit us to illustrate the use of the 4-index system along with the conventional 3-index system for the other three study materials. Secondly, it is rather straightforward to produce *inclusions* in titanium; as described in more detail in the next section, if we use an electrolyte containing hydrogen ions (and many commonly used electrolytes do) then we can cause precipitation of TiH_2 hydride particles in the thin foil. This is an artifact of specimen preparation, but we will use it to our advantage to study image formation and diffraction for a material containing precipitates.

Pure titanium can be obtained from many different sources. We ordered a $15 \times 15\ cm^2$ thin sheet (0.9 mm thick) of grade 2 titanium from McMaster-Carr Supply Company for about U.S. $80. Other sources are obviously also acceptable. Ti manufacturing is technologically demanding because of the high reactivity of Ti in an oxygen atmosphere; this makes it unlikely that you will have easy access to Ti manufacturing equipment.

4.3.2.2 Thin-foil preparation

Thin-foil preparation for titanium proceeds along the same lines as for the first study material. We used the following conditions: electrolyte 95 vol% methanol, 5 vol% sulfuric acid (H_2SO_4), $T = -35\,^\circ\mathrm{C}$, voltage $V = 35$–40 V. Polishing took 4–6 min per foil, for an initial foil thickness of 190 μm. At lower temperatures, polishing was significantly slower and upon perforation resulted in foils with rapidly increasing thickness away from the hole. At higher temperature a significant number of hydride particles were produced in the foil. This is something you will need to avoid if you are preparing Ti-based thin foils. It is possible to use electrolytes that do not contain hydrogen ions, such as a solution of 40 g of CaCl in 500 ml of methanol at $-45\,^\circ\mathrm{C}$ and 35 V [Pie95]. For the purposes of this text we will actually make use of the hydride particles so do not worry too much about them as you prepare the thin foils. You could also use the dimple grinding and ion milling method, described in the next section, if you want to avoid completely the introduction of hydrides in your foil.

4.3.3 Material III: GaAs

4.3.3.1 Crystallography

We have selected GaAs as the third study material because it is easy to obtain in single-crystal form, and because it has a non-centrosymmetric space group. As we will describe in detail in the next chapter, there are a few subtle differences both in image formation and diffraction between centrosymmetric and non-centrosymmetric crystals. GaAs will be used to illustrate those differences. The crystal structure of GaAs, shown in Fig. 4.4, is that of zinc-blende (also known as sphalerite); it is a face-centered cubic structure with two atoms in the asymmetric unit, and four formula units per unit cell. GaAs can be regarded as the prototype of an important class of materials, the binary semi-conductors. A significant part

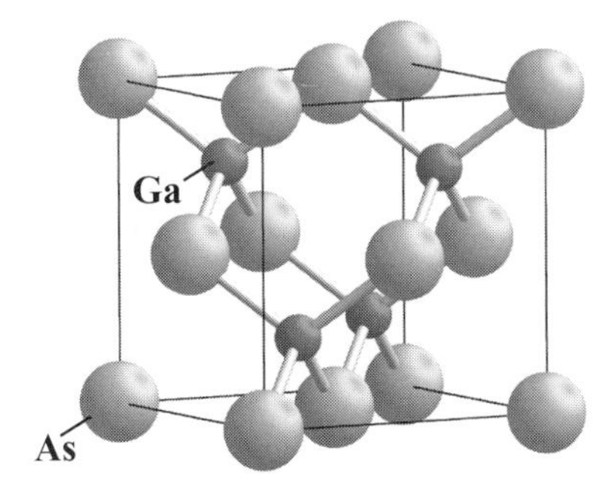

Fig. 4.4. Parallel projection of a single unit cell of the GaAs structure.

Table 4.3. *Material parameters for GaAs (taken from [Vil97]); [zinc-blende or sphalerite; Pearson Symbol cF8; Strukturbericht Notation B3].*

Space group	$F\bar{4}3m$	
Lattice parameters	$a = 0.5652$ nm	# 216
Asymmetric unit	As at 4a	(0, 0, 0)
	Ga at 4c	$\left(\frac{1}{4}, \frac{1}{4}, \frac{1}{4}\right)$

of all TEM-based research worldwide is centered on materials with this particular crystal structure, or on ternary structures derived from it.

As an illustration of the use of the xtalinfo.f90 program we list a portion of the output of this program for the GaAs structure in Fig. 4.5 on the following pages. The output has been reduced in size to fit within the margins of this book. The first portion of this figure shows standard crystallographic output, along with various matrices useful for crystallographic computations. The second part lists information for all independent families of planes with Miller indices less than 5. Some of the entries in this table will be defined in Chapter 5. The third portion of Fig. 4.5 shows a number of representative zone axis diffraction patterns. These patterns can be drawn to the correct scale for a given microscope camera length. Crosses indicate systematic absences. The multiplicity mentioned alongside each pattern is actually a useful piece of information because it indicates how easy it will be to find that particular zone axis. A small multiplicity means that it will be more difficult to obtain this zone axis orientation, whereas a multiplicity of 48 indicates that you will not have to look for very long to find it. The probability of finding a particular zone axis orientation will be modified if the (polycrystalline) sample has a preferential texture.

4.3.3.2 Thin-foil preparation

GaAs is typically manufactured as a [001]-oriented wafer (Fig. 4.6a). The {110} planes in the [001] zone are good cleavage planes. We prepared two different samples: a [001]-oriented foil and a [110]-oriented foil. The [001]-oriented foil is rather easy to prepare. Cleave the wafer until you have a square sample of approximately 2×2 mm^2. GaAs can be cleaved in the same way that you would cut glass: use a sharp razor blade and a ruler to score one surface along a [110] direction. Then support the wafer and use the ruler to press down on the part that you wish to cleave (see Fig. 4.6b). You should have no trouble obtaining perfectly straight cleavage planes.

gaas.xtal

Direct Space

a :	0.5653	*Space Group* :	**F -4 3 m [#216]**
b :	0.5653	*Bravais Lattice* :	**cF**
c :	0.5653		
α :	90.0000	*Volume V [nm^3]* :	0.1806
β :	90.0000		
γ :	90.0000		

Reciprocal Space

*a** :	1.7690	*Volume V* [nm^–3]* :	5.5356
*b** :	1.7690		
*c** :	1.7690		
$\alpha*$:	90.0000		
$\beta*$:	90.0000		
$\gamma*$:	90.0000		

Important Matrices

$$g = \begin{bmatrix} 0.31956 & 0.00000 & 0.00000 \\ 0.00000 & 0.31956 & 0.00000 \\ 0.00000 & 0.00000 & 0.31956 \end{bmatrix} \qquad g^* = \begin{bmatrix} 3.12926 & 0.00000 & 0.00000 \\ 0.00000 & 3.12926 & 0.00000 \\ 0.00000 & 0.00000 & 3.12926 \end{bmatrix}$$

$$a = \begin{bmatrix} 0.56530 & 0.00000 & 0.00000 \\ 0.00000 & 0.56530 & 0.00000 \\ 0.00000 & 0.00000 & 0.56530 \end{bmatrix} \qquad b = \begin{bmatrix} 1.76897 & 0.00000 & 0.00000 \\ 0.00000 & 1.76897 & 0.00000 \\ 0.00000 & 0.00000 & 1.76897 \end{bmatrix}$$

Asymmetric Unit

atom	x	y	z	occ.	B
Ga [31]	0.0000	0.0000	0.0000	1.0000	0.0058
As [33]	0.2500	0.2500	0.2500	1.0000	0.0063

Distances [nm], angles [degrees]

Fig. 4.5. Partial output of the xtalinfo.f90 program for the GaAs structure.

Structure Factor Information

h	k	l	p	d	g	2θ	|V|	Vphase	|V'|	V'phase	ξ_g	ξ'_g
0	0	0	1	0.00000	0.00000	0.0000	21.08094	0.00000	1.11911	0.00000	0.00	535.93
1	1	-1	4	0.32638	3.06395	7.6839	9.71481	0.84450	0.47377	0.84165	61.74	1265.94
1	1	1	4	0.32638	3.06395	7.6839	9.71481	-0.84450	0.47377	-0.84165	61.74	1265.94
2	0	0	6	0.28265	3.53794	8.8726	0.76727	-3.14159	0.03662	-3.14159	781.70	16375.89
2	2	0	12	0.19986	5.00341	12.5478	8.89642	0.00000	0.56329	0.00000	67.42	1064.74
3	1	-1	12	0.17044	5.86702	14.7136	5.30499	-0.82069	0.37275	-0.84480	113.06	1609.02
3	1	1	12	0.17044	5.86702	14.7136	5.30499	0.82069	0.37275	0.84480	113.06	1609.02
2	2	-2	4	0.16319	6.12790	15.3678	0.22302	-3.14159	0.03069	-3.14159	2689.36	19541.95
2	2	2	4	0.16319	6.12790	15.3678	0.22302	-3.14159	0.03069	-3.14159	2689.36	19541.22
4	0	0	6	0.14132	7.07589	17.7453	6.05159	0.00000	0.48008	0.00000	99.11	1249.30
3	3	-1	12	0.12969	7.71077	19.3376	3.85773	0.79954	0.32425	0.84446	155.47	1849.69
3	3	1	12	0.12969	7.71077	19.3376	3.85773	-0.79954	0.32425	-0.84446	155.47	1849.68
4	2	0	24	0.12640	7.91109	19.8399	0.06867	-3.14159	0.02662	-3.14159	8734.28	22531.40
4	2	2	12	0.11539	8.66616	21.7336	4.70112	0.00000	0.42597	0.00000	127.58	1407.99
4	2	-2	12	0.11539	8.66616	21.7336	4.70112	0.00000	0.42597	0.00000	127.58	1407.99
3	3	3	4	0.10879	9.19185	23.0521	3.07112	0.79504	0.28985	0.84346	195.29	2069.21
3	3	-3	4	0.10879	9.19185	23.0521	3.07112	-0.79504	0.28985	-0.84345	195.29	2069.20
4	4	0	12	0.09993	10.00682	25.0960	3.85023	0.00000	0.38411	0.00000	155.78	1561.45
4	4	-2	12	0.09422	10.61383	26.6184	0.03757	-3.14159	0.02074	-3.14159	15965.04	28921.36
4	4	2	12	0.09422	10.61383	26.6184	0.03757	-3.14159	0.02074	-3.14159	15965.04	28921.36
4	4	-4	4	0.08159	12.25580	30.7366	2.78370	0.00000	0.31926	0.00000	215.46	1878.59
4	4	4	4	0.08159	12.25580	30.7366	2.78370	0.00000	0.31926	0.00000	215.46	1878.59

Distances [nm], angles [mrad], potential [V,rad]

Fig. 4.5 (*cont.*).

Measure the thickness of the sample using a micrometer. In our case we measured 630 μm. Mount the small sample on a dimple grinder support stub (Fig. 4.6c) using *a small amount* of low melting point mounting wax. Let the wax cool down and center the stub on the dimpler. Gently lower the dimpler arm with a wheel in place onto the sample, and set the amount of material that you wish to remove.† Then place a small amount of 6 μm diamond polishing compound and a drop of a water-soluble lubricating oil on the sample. Lower the wheel onto the sample and remove a preset amount of material.

We used a 20 mm diameter dimpling wheel on a Gatan Model 656 Dimple Grinder to obtain a shallow dimple, and set the dimple weight at 15–20 g. The wheel speed was kept rather low, around 2–3 rev s^{-1}. Remove 80–90% of the thickness using a 6 μm diamond compound (550 μm in our case, with three steps of 150 μm each, followed by one step of 100 μm). In between steps, gently wash away the diamond compound using a methanol squirt bottle and check the surface with an optical microscope. The nominal thickness of the sample at the center of the dimple should

† There are several models of dimple grinders or "dimplers" available and we refer the reader to the instructions that come with each machine for more details.

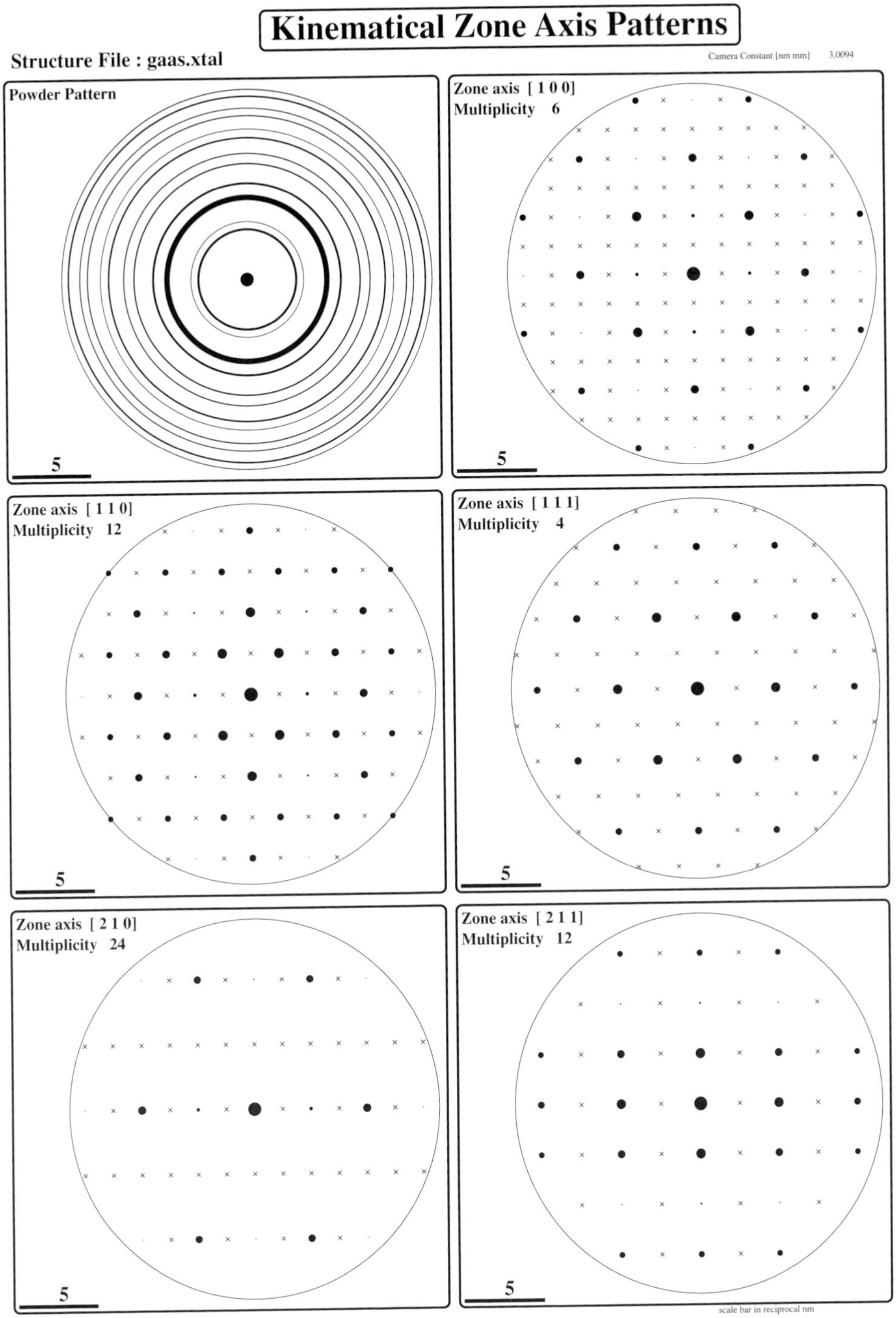

Kinematical Zone Axis Patterns

Structure File : gaas.xtal

Camera Constant [nm mm] 3.0094

Powder Pattern

5

Zone axis [1 0 0]
Multiplicity 6

5

Zone axis [1 1 0]
Multiplicity 12

5

Zone axis [1 1 1]
Multiplicity 4

5

Zone axis [2 1 0]
Multiplicity 24

5

Zone axis [2 1 1]
Multiplicity 12

5

scale bar in reciprocal nm

Fig. 4.5 (*cont.*).

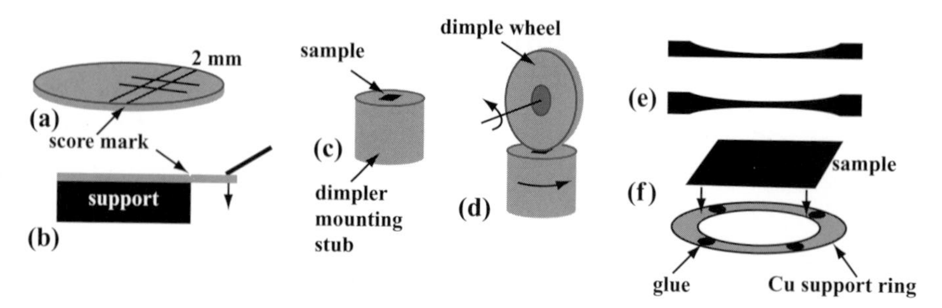

Fig. 4.6. Illustration of the steps involved in the preparation of a GaAs [001] thin foil.

be around 80–100 μm. This is of course an approximate number, since the accuracy of the amount of material removed is probably not much better than about 5 microns per step.

Remove 60 μm using a 1 μm diamond compound (two steps of 30 μm each), and then continue with three more steps of about 10 μm each, leaving (theoretically) about 10 μm of material at the center of the dimple. We were rather conservative on each of the last five steps, so the true thickness at the dimple center was probably a little larger than 10 μm. It is possible to estimate the thickness by means of an optical microscope with a calibrated focusing knob; simply measure the difference between the in-focus condition at the dimple center and the in-focus condition of the mounting stub surface, and subtract the thickness of the mounting wax layer (this assumes that you measured the total thickness, stub + wax + sample, before you started dimpling). With some experience you should be able to determine the thickness to within a few microns. After completion of the dimpling steps you should soak both stub and sample in acetone to dissolve the mounting wax. After a little while the sample will simply fall off of the stub.

Since the thin foil you have just produced is mechanically not very strong you should mount it on a copper support ring (see Fig. 4.6f). Many different types of mounting rings and disks are available from microscopy supply companies. We used a copper ring with an inner diameter of 1.5 mm. Use super-glue or two-component glue to attach the foil to the support ring, taking care to keep the glue away from the center of the sample. Only a very small amount of glue is needed to attach the sample to the support ring; typically three or four "dabs" of glue spread out around the ring will suffice.[†] The dimpled side should face away from the support ring. Some researchers prefer to attach the sample to a support ring before any dimpling

[†] You should use your own judgment on how much glue to apply. If the glue squirts out around the edge when you place the ring on the sample, then you probably used too much of it. Make sure the sample and support ring do not get stuck on the bench top or glass slide!

is carried out. This may be necessary for materials with high porosity or very low strength.

It is possible to use double-sided dimpling, i.e. to dimple both surfaces of the disk. This requires a bit more care in terms of thickness measurements and alignment of the two dimples with respect to each other. The reader should experiment with various methods; after a while, you will find a procedure that works for many different materials.

The dimpled foil is now ready for *ion milling*. The surface is typically exposed to argon ions (Ar^+), which have been accelerated by a potential of ~1–6 kV. The ions may reach the sample in a wide beam or in a focused beam, depending on the type of ion miller you have. We used a Gatan Precision Ion Polishing System Model 691 (PIPS). Both guns were first set at an incidence angle of 10° above the sample. After about 1 h of milling at a potential of 3 kV, one of the guns was set at an angle of 10° below the sample surface, and an additional 10 min of milling was carried out. Finally both angles were reduced to 8° and the potential was dropped slightly to 2.8 kV for a final 7 min polishing step.

The second GaAs sample was oriented along the [110] zone axis, at 90° from the first sample orientation. A bit more work is required to prepare a good thin foil in this orientation. The method outlined below can also be used for other materials, in particular for the study of thin films.

We cleaved a strip from the GaAs wafer along one of the {110} planes. We cut the strip into three pieces of roughly equal size, and glued the three strips together to obtain a 1.8 mm thick stack, which is a bit easier to handle. It is important to keep track of the orientation of each of the pieces, so that all three are in the same orientation. Since one of the sides of the stack is parallel to the (110) plane, and we wish to obtain a thin foil with a [110] foil normal, we can use a diamond saw to cut a 300 μm thin strip normal to the [110] direction. The final sample was then cut from this strip and had dimensions of 2.0×1.8 mm^2. We mechanically ground down the sample to a smooth polished finish, and glued it onto a copper disk with a central hole (polished side facing the support ring). The remainder of the sample preparation was identical to that of the [001]-oriented foil.

We should point out to the reader that the ease with which GaAs cleaves is both "a curse and a blessing"; it is a blessing because it is straightforward to obtain straight cleavage facets in the early stages of sample preparation. On the downside it is also easy to lose the entire thin area of your foil when the foil is mounted in the TEM sample holder. The air pressure exerted on the sample when you move the holder across the room from the mounting table or desk-top to the microscope column may be sufficient to cause a cleavage fracture of the entire thin area, leaving a large facetted hole in your carefully prepared thin foil!

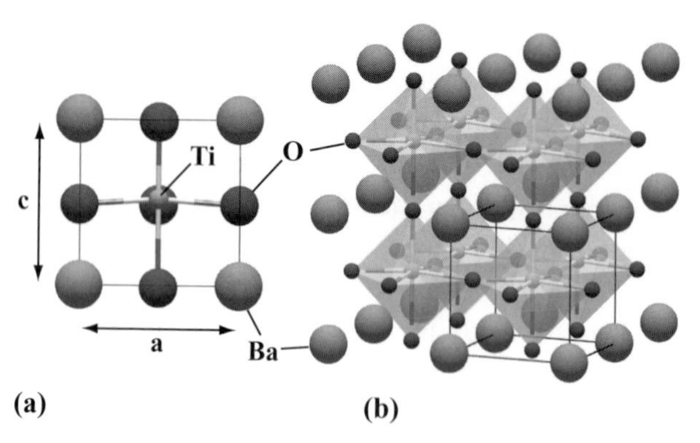

Fig. 4.7. (a) [100] projection of the structure of $BaTiO_3$; the front oxygen atom has been removed so that the small [001] shift of Ti can be observed. (b) Parallel projection of $2 \times 2 \times 2$ unit cells with oxygen coordination octahedra outlined.

The reader is encouraged to experiment with different specimen preparation methods to gain experience in sample preparation.† One easy experiment is to glue the pieces of the wafer together with a particular orientation relation between the pieces. For instance, if the central piece in the stack of three is rotated by 45° with repect to the top and bottom pieces, and the hole in the thin foil is grown sufficiently large so that it overlaps one or both of the glued interfaces, then you will be able to obtain both [110] and [100] zone axis orientations in the same sample. Since sample preparation is a time consuming portion of microscopy research, it usually pays to think carefully about your sample geometry *before* you start sample preparation.

4.3.4 Material IV: $BaTiO_3$

4.3.4.1 Crystallography

The fourth and last study material, $BaTiO_3$, is a representative of an important class of structures, the *perovskites*. The commercial significance and application potential of perovskite-based materials is mind-boggling. Perovskites are used for their dielectric, piezoelectric, electro-optical, magneto-resistive, catalytic, superconducting, or protonic conductivity properties [Roh01]. We have selected $BaTiO_3$ because it is easy to obtain in powder form, or you can make it yourself, starting from BaO and TiO_2 powders. The room-temperature structure is non-centrosymmetric, as shown in Fig. 4.7 and Table 4.4. We will use this material to illustrate the image contrast of defects in a non-centrosymmetric structure.

† The website will contain a repository for specimen preparation methods for the study materials of this book. Any reader who has an alternative method of preparing the thin foils can add his/her procedure to the repository.

Table 4.4. *Material parameters for $BaTiO_3$ (taken from [Wyc63]); [barium titanate; Pearson Symbol tP5 Strukturbericht Notation $E2_1$].*

Space group	P4mm	# 99
Lattice parameters	$a = 0.39947$ nm $c = 0.40336$ nm	
Asymmetric unit	Ba at 1a	(0, 0, 0)
	Ti at 1b	$(\frac{1}{2}, \frac{1}{2}, 0.512)$
	O at 2c	$(\frac{1}{2}, 0, 0.486)$
	O at 1b	$(\frac{1}{2}, \frac{1}{2}, 0.023)$

The basic perovskite structure is primitive cubic, with $CaTiO_3$ as the prototype structure. The space group is **Pm$\bar{3}$m** (# 221) and the Strukturbericht notation is $E2_1$. The Ti atom is octahedrally coordinated by six O atoms, and is located at the center of the octahedron. The room-temperature structure of $BaTiO_3$ differs from the cubic perovskite only in the small displacement of the positive and negative centers of charge, which results in a small tetragonal distortion with respect to the cubic cell. The Ti atom is displaced from the center of the cell by about 7.2 pm along the [001] direction. This small displacement is sufficient to cause the positive and negative centers of charge to no longer coincide, which in turn creates a permanent electrostatic dipole moment in the unit cell. Regions with identical orientation of this dipole moment form *ferroelectric domains*, and in Chapter 8 we will take a close look at the boundary between two such domains.

4.3.4.2 Thin-foil preparation

We obtained $BaTiO_3$ in powder form from Alfa Aesar (99.7% purity), mixed the powder with a polyvinyl alcohol binder, and pressed a pellet (11 mm diameter, 3 mm thick) using a 150 MPa uniaxial pressure. The pellet was then heated to 900 °C to burn off the binder, sintered for 12 h at 1230 °C, and then heated to 1430 °C for 3 h to grow the grains. The resulting pellet was light gray in color, and was about 99% dense. You could also start from the basic oxides BaO and TiO_2, mix them in the proper proportions, and then subject them to the same sequence of processing steps; you may need to repeat this procedure twice to get a single-phase microstructure (i.e. after the first cycle you crush the pellet to a fine powder and then repeat the cycle).

A thin foil can be prepared as follows: use a low-speed diamond saw to cut strips of about 350 μm thickness from the pellet. Use a sharp razor blade to cut small squares (2×2 mm^2) from the strips. Mount one such square on a dimple support stub, using low melting point mounting wax, and remove about 100 μm of material, using 6 μm diamond paste. Then use a 1 μm compound to remove an additional 50 μm. After removing the sample from the dimple support, attach a copper support ring (1.5 mm inner diameter) to the dimpled side of the sample, to provide mechanical strength.

Dimple the second side of the disk using 6 μm diamond paste (remove about 100 μm), followed by removal of 75 μm using 1 μm paste. At that point the sample should become translucent and with an optical microscope (in transmitted light mode) you may be able to observe a grainy substructure. The sample is now ready for ion milling.

Place the supported sample in an ion mill holder and mill at an angle of 10° and an acceleration voltage of around 3 kV until a small hole is formed. Once a hole appears, remove the sample from the mill and place it in a carbon coater. Most ceramics are poor conductors of electricity, so a thin conducting layer must be deposited onto the foil to reduce charging due to the electron beam. The exact evaporation conditions will depend on the type of coater you have. After deposition of a thin carbon layer your sample is ready to be placed in the microscope.

4.4 A typical microscope session

4.4.1 Startup and alignment

As the reader may have noticed by now, the learning curve for the TEM is quite a bit steeper than that for, for example, the optical microscope; it may take many hours of (sometimes frustrating) work to become familiar with the operation of the TEM. It may take an even longer time to learn how to interpret electron micrographs, whether they be images or diffraction patterns. This is quite normal and the reader should be patient and persistent. While this book deals with the basics of TEM operation and image interpretation and leaves out a great deal of advanced techniques (in particular, all analytical observation modes), the techniques covered in this and the following chapters are fundamental to all observation modes, including analytical methods. A thorough study of this book combined with many hours on the microscope will provide the reader with a firm background from which more advanced techniques can be learned. That this requires a fair amount of mathematics is unavoidable and the reader should divide his/her time between the experimental work described in this chapter and the study of the theory of image formation in the following chapters. The reader is encouraged to try out all examples in this chapter and, since the main illustrations are taken from the four study materials, it should not be too difficult to reproduce nearly every micrograph shown.

Every microscope session starts with the following basic steps (assuming a thin foil is available).

(i) *Load the sample in the sample holder.* Take extreme care not to touch the part of the holder that will be in the vacuum of the microscope. If you do need to touch the holder, wear lint-free gloves. Specimen holders are delicate and very expensive. You should treat the sample and the sample holder as if they were precious gems, and you should always take your time and use a steady hand when handling them.

(ii) *Clean the sample holder (optional).* For modern field emission microscopes it is recommended that the sample and sample holder be subjected to a short plasma cleaning step. This procedure removes most hydrocarbons from the sample surface and improves the overall cleanliness of the sample. This is particularly important for analytical microscopy. More information on plasma cleaning can be found in [IFO+99].

(iii) *Insert the holder into the column.* Since top-entry and side-entry machines have different standard operating procedures, we refer the reader to those instructions. In general, the specimen holder is first partially inserted to activate the pumping system. Then, after a fixed pumping delay, the holder is further inserted through an airlock mechanism until it is firmly seated in position. The pumpdown time can be as short as a few minutes, or much longer for some top-entry microscopes. One can use this time to dim the lights in the microscope room, to give the eyes ample time to adjust to the lower light level.

(iv) *Turn on the acceleration voltage.* This again depends on the particular machine you are using. On older machines you would typically use push-buttons or a rotary knob to increase the voltage in steps of 20 kV or so up to the maximum acceleration voltage. On many modern machines the acceleration voltage is never turned off, so that this step can be skipped. On higher-voltage machines, such as the JEOL 4000 series, the voltage is slowly ramped up to the maximum, a process which may take an hour or longer; for such machines it is a good idea to start them up a few hours before your session begins.

(v) *Turn on the filament current.* To extend the lifetime of the filament you should always take the time to turn the filament current up from zero to near saturation. For thermal emission guns, the filament has to heat up to a temperature where thermal emission starts (usually in the range 2000–2700 K), and since heating is always associated with thermal expansion and stresses, it should come as no surprise that slow is better than fast. As a rule of thumb one could say that the time it takes to heat up the filament to near saturation should be proportional to the cost of the filament: a tungsten hairpin filament can be turned on in a matter of a few seconds, whereas LaB_6 single-crystal filaments should be turned on over a period of a few minutes. For hot field emission tips the extraction voltage is often left on and the beam is simply deflected away from the optical axis of the microscope. Opening the gun airlock will bring the beam back onto the optical axis, and no changes in filament current are needed.

This brings us to the most important part of the startup procedure: the alignment step. Nothing will be more frustrating to the microscopy student (or for that matter to an experienced user) than an incorrectly aligned microscope. It is important that a routine alignment be carried out at the beginning of every session, even when others have used and aligned the instrument before you begin your session. The details of this routine alignment will depend on the particular microscope being used, but there are a few general steps which should be part of any alignment. While the more modern machines often have the built-in computer prompt the user for the various alignment steps, many of the older machines do not have an on-board computer and the user has to execute the steps from the user manual. Alignments can vary from purely electronic or software-driven steps to partially mechanical alignment, and this depends mostly on the age of the instrument. The following list itemizes the most common alignment steps.

(i) *Gun alignment*. Use the gun tilt controls to obtain a symmetric filament image, as shown in Fig. 4.8(a) for a tungsten source and a LaB_6 source. If no undersaturated image can be obtained, then the gun tilt controls should be adjusted so that maximum brightness is observed on the viewing screen.
(ii) *Condensor aperture alignment*. The electron beam should expand and contract concentrically when the second condensor lens current is changed. If the center of the

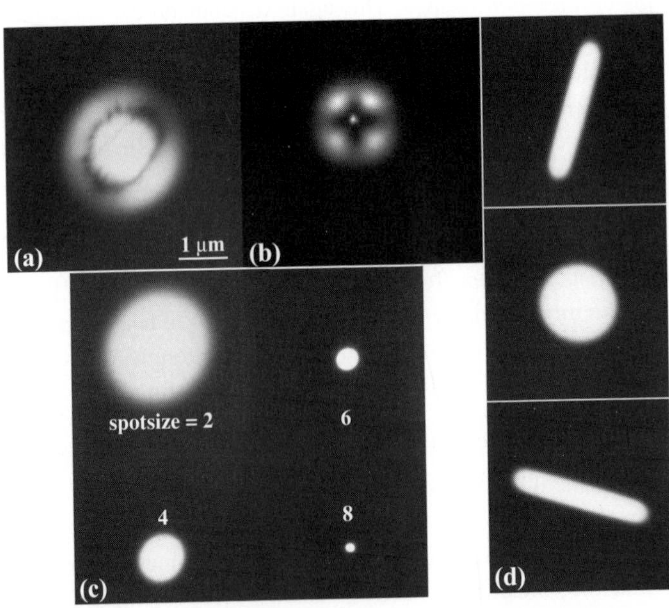

Fig. 4.8. Undersaturated filament tip images for (a) tungsten and (b) LaB_6 filaments; (c) shows four different spot sizes for the JEOL 2000EX microscope; (d) shows the effect of severe astigmatism on the beam shape: top and bottom are over-focus and under-focus images, respectively, while the center image is in-focus (all images are taken at the same magnification).

beam moves upon changing this current, then the condensor aperture should be translated until no further movement occurs for a wide range of beam diameters. An application which needs a high beam current should use a large-diameter condensor aperture, whereas observations which need a high beam coherence should sacrifice some current by selecting a smaller diameter aperture.

(iii) *Illumination stage alignment.* This step ensures that the electron trajectory is aligned with the optical axes of both first and second condensor lenses. The alignment procedure is usually an iterative procedure whereby the spot size (first condensor lens) is alternated between two settings, and the beam displacement on the screen is minimized by means of the beam and gun shift deflection coils. The beam alignment is satisfactory when no substantial beam shifts are observed when the spot size is changed.[†] Figure 4.8(c) shows a typical series of spot sizes for a JEOL 2000EX microscope.

(iv) *Condensor astigmatism correction.* The incident electron beam should have a circular cross-section for all settings of the second condensor lens. Condensor lens astigmatism can be recognized readily as an elliptical distortion of the beam, which changes major axis when the second condensor lens current goes from an under-focus to an over-focus condition, as shown in Fig. 4.8(d). The condensor stigmator coils must be used to correct the beam shape. Condensor astigmatism can also be corrected in the in-focus condition (smallest beam diameter) by ensuring that the filament tip produces the sharpest possible image (under-saturated condition).

(v) *Filament saturation.* When all four previous steps have been carried out properly, the filament current (in the case of a thermal emission gun) can be increased to the saturation level, often indicated on the microscope console by the position of a so-called *beam-stop*, a metal rod which prevents the user from turning the current up any further.

(vi) *Sample eucentricity.* The height of the sample inside the microscope column is an important variable. The primary tilt axis is a fixed axis with respect to the microscope column; when the primary tilt is changed, the entire goniometer stage rotates around this axis. The sample holder axis, i.e. the center line along the holder rod, does not necessarily coincide with the primary tilt axis, as illustrated in Fig. 4.9(a). When the goniometer is tilted, a feature on the sample, indicated by a black circle, will also rotate around the primary tilt axis, and its projection on the viewing screen will move sideways.[‡] The *eucentric position* is then obtained by adjusting the height of the sample, the z-control, until there is no further lateral movement during tilting (Fig. 4.9b). This adjustment could be manual or electronic, depending on the microscope model. Eucentric alignment is important because it facilitates sample tilting while in diffraction mode (the region of interest will stay on or near the optical axis). The eucentric

[†] Note that the beam diameter *will* change when changing the spot size; this is due to the fact that the object plane of the second condensor lens changes with spot size, as discussed in Chapter 3. One must then modify the second condensor lens current to obtain the same beam diameter on the screen as before.

[‡] This is actually useful, because it provides a simple way to identify the orientation of the primary tilt axis on the viewing screen.

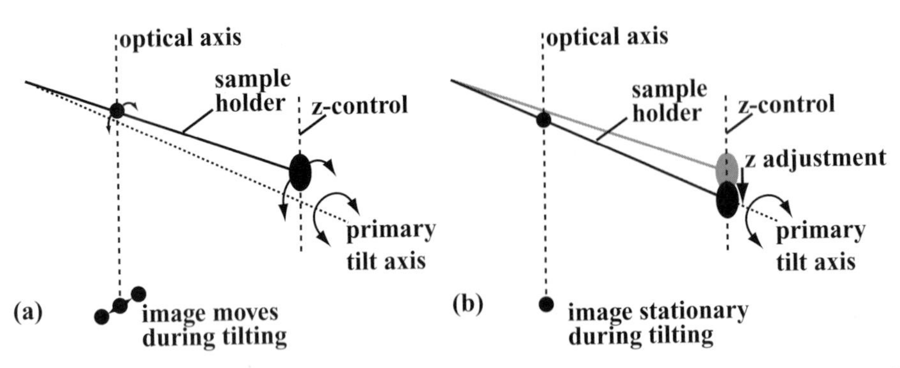

Fig. 4.9. (a) A sample point (black circle) that does not lie on the primary tilt axis will move laterally during tilting. (b) By adjusting the height of the sample in the column, the lateral movement can be removed completely; this is the eucentric position.

position is also the position for which the objective lens current generates the smallest possible spherical aberration coefficient.

(vii) *Voltage centering*. To ensure that the electron beam travels along the optical axis of the microscope, we can superimpose an oscillating voltage on the acceleration voltage. This will result in a sinusoidally varying electron energy and hence wavelength. This, in turn, means that the total rotation angle of the electron trajectory will change and the resulting image will rotate around a point that does not necessarily coincide with the optical axis. Use the beam tilt controls (in bright field mode) to bring the center of rotation onto the optical axis (center of the screen). Alternatively, one could superimpose an alternating current on the objective lens current and bring the apparent rotation center to the center of the screen; this is known as *current centering*. Current and voltage centering should in principle give rise to the same position of the rotation center, but in practice there may be a small difference between the two.

(viii) *Diffraction centering*. When you switch to diffraction mode the central beam may not be located precisely in the middle of the viewing screen. You can use the projector lens alignment to bring the central beam to the center. In older microscope models the projector alignment is a mechanical alignment, using two pairs of screws near the bottom of the column. In more modern instruments the projector alignment uses deflection coils. Since the projector lens is the final lens in the microscope, changing its alignment will only cause a lateral movement of the entire image or diffraction pattern.

Now the microscope is ready to begin observations. Insert the sample into the beam path if it is not already there, and use the specimen translation controls to locate the edge of the thin foil. In the next section we will discuss a few basic techniques to obtain images and diffraction patterns. Then we will record a series of micrographs illustrating contrast features which one can expect to observe in any defect-free crystalline material.

4.4.2 Basic observation modes

4.4.2.1 Going from image to diffraction mode

Assuming that you have prepared a foil from one of the four study materials, now is the time to insert it in the column. When the alignment procedure has been completed, you should perform the following steps.

- Make sure that both diffraction and selected area apertures are removed from the column; set the magnification to somewhere between 5000× and 10 000×. Do *not* move or remove the condensor aperture, as you have just completed its alignment!
- Use the specimen translation controls to bring the edge of the sample (or any other area of interest) to the center of the screen. You may need to reduce the magnification to find the edge first.
- Change the height of the sample (using the z-control) to bring it to the eucentric position; this can be done by tilting the sample back and forth around the primary tilt axis, and minimizing the lateral shift of the sample on the screen.
- Insert the selected area aperture and select an aperture size that covers the area of interest in the sample. Center the aperture on the screen.
- If necessary, adjust the position of the sample so that the area of interest is visible inside the aperture.
- Focus the image inside the aperture.[†] This ensures that the specimen is conjugate to the SA aperture plane.
- Change to diffraction mode.

The procedure described above is one of the basic procedures for transmission electron microscopy. You will perform this routine over and over again, every time you switch from image to diffraction mode. The first three steps may not always be necessary, but the remaining ones should become second nature to you. It is a bit like changing gears in a car with manual transmission; this involves a number of coordinated moves, but after a while you do not really have to think about them. Going from image to diffraction mode (and then back, which is described in the next section) should become part of your *microscope driving skills*, so you may want to spend some time practicing the individual moves.

Before moving on with the next step we should point out a few image characteristics that you may have observed while carrying out the above routine. Before you inserted the SA aperture you may have noticed that the image contrast was rather poor. If you go back to image mode and remove the aperture, you will only observe contrast if you change the OL focus (either over-focus or under-focus). At the in-focus condition, you actually have the lowest contrast, and you may not even be able to see the sample at all. When you change the OL focus, you may also notice bright features moving back and forth with changing focus. There may

[†] Some older microscopes have a special focus knob to focus the edge of the SA aperture independently.

be several of those features with a nearly identical shape, or you may not see any at all. When you change the orientation of the sample, you may see their number change. When you approach the in-focus condition several of them will converge to one location (at which you will generally see a dark version of the same feature), and when they all merge, very little or no contrast will remain.

An example is shown in Fig. 4.10(a): the center image is an in-focus image of a few grains near the edge of a Ti foil. When the image is defocused by ± 2 and ± 4 μm,

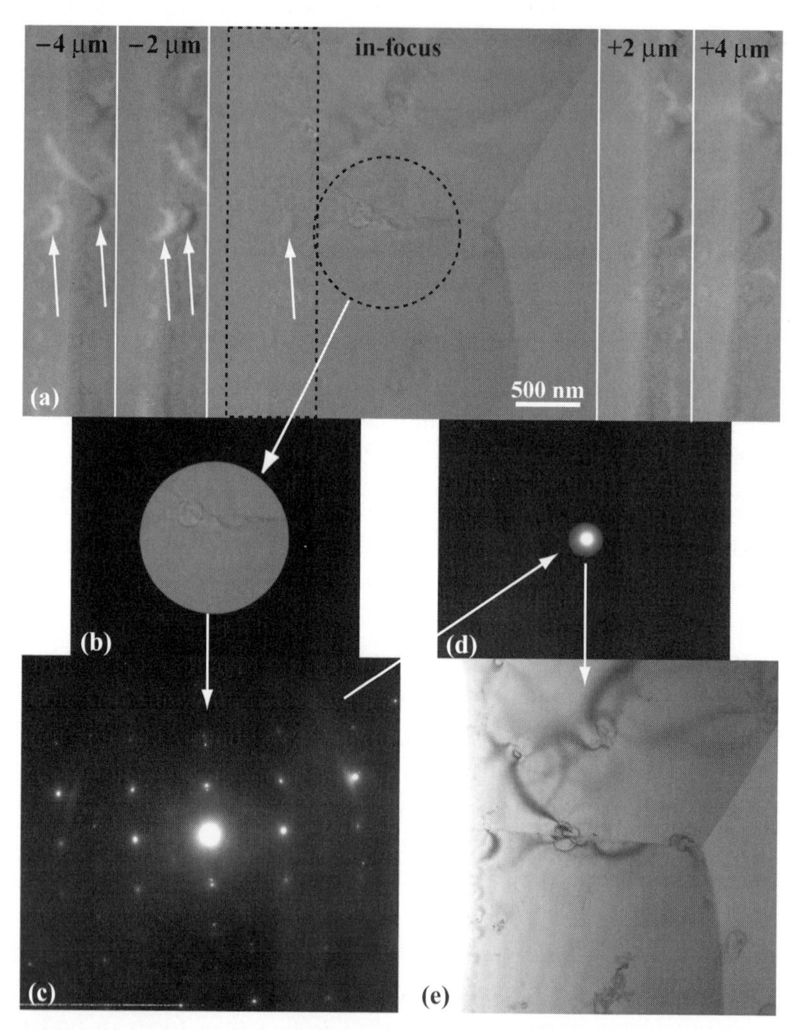

Fig. 4.10. (a) Experimental illustration of in-focus and out-of-focus images of a Ti foil, recorded with no apertures inserted in the column; (b) the selected area aperture is placed at the dashed circle in the center image of (a), and the image inside the aperture is focused; (c) switching to diffraction mode, a diffraction pattern is obtained (in this case near [01.0] zone axis orientation); (d) the diffraction aperture is inserted around the central beam, and (e) in image mode a high-contrast bright field image is observed.

we obtain the images to the left and right of the center image (only the rectangular region outlined in the center image is shown). There are clearly several bright features visible inside the hole of the sample (negative defocus), and the distance between those features and the sample edge increases with increasing defocus (white arrows). When you approach the in-focus condition, all bright features merge together with darker features that do not change position with defocus, and the resulting image has very little contrast. Then you can insert the selected area aperture into the column and position it at a region of interest, say the dashed circle in the center image of Fig. 4.10(a); after you focus the image inside the aperture (Fig. 4.10b) you switch to diffraction mode to obtain a selected area diffraction pattern, similar to that shown in Fig. 4.10(c).

The origin of those "ghost images" is easy to understand. Consider the following geometry: a sample, shown in Fig. 4.11, contains a rectangular particle which strongly diffracts electrons into a number of diffracted beams. This means that, generally speaking, the transmitted beam will have fewer electrons coming from that particular location in the sample, whereas the diffracted beams will travel in various directions predicted by the Bragg equation. Let us assume that every part of this region scatters electrons equally strongly and that the surrounding matrix does not scatter at all. If the imaging lenses are focused on the OL image plane, then on the viewing screen all electrons that leave the particle will be recombined into a magnified image of that region (assuming that we have a reasonably good objective lens). Since most electrons leaving the sample are recombined at the proper location

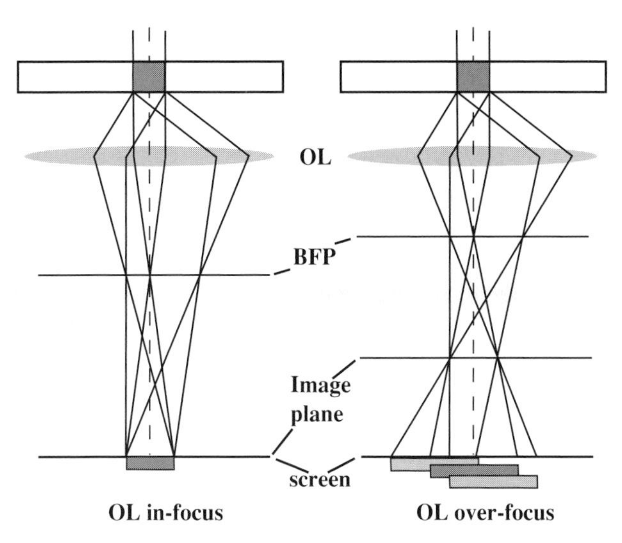

Fig. 4.11. Illustration of the origin of "ghost-images" when the objective lens is not properly focused.

in the image, we will not observe a lot of contrast. If, on the other hand, we focus the imaging lenses onto a plane slightly *above* the OL image plane, then the resulting image is quite different: instead of having one rectangular region where all electrons merge into an image, we now have several of those regions in the image plane, each corresponding to a different diffracted beam. On the viewing screen, each of those regions will be separately imaged, and the spacing between the individual images will vary with the amount of defocus. In principle, there should not be any contrast in the in-focus image. In practice, spherical aberration will affect the trajectories of the diffracted electrons, and the various images may not merge perfectly in the in-focus image.

What we observe on the viewing screen when both selected area and diffraction aperture are removed from the column is the result of all combined scattering processes, both elastic and inelastic. By the time the electrons leave the sample in various directions, nearly all of the relevant physics is over, and all we can do as microscope operators is attempt to extract information from this complex pattern of scattered electrons. Using the terminology introduced on page 106, what we observe when we remove both post-specimen apertures from the microscope column is essentially the function $\mathcal{R}''$; when we change the focus of the objective lens we are actually changing the point spread function $\mathcal{T}$, and we observe "different versions" of the modified signal $\mathcal{R}'$. It is now your task, as a microscope operator, to cleverly manipulate this function $\mathcal{T}$ (by twiddling knobs and moving apertures around), so as to end up with a fairly accurate representation (image or diffraction pattern) of the information that is present in the modified signal $\mathcal{R}'$. The in-focus image just happens to be the one that is easiest to interpret. The out-of-focus images contain the same information, but the human brain is simply not capable of separating the information content and the point spread function.

4.4.2.2 Working in diffraction mode

At the end of the routine discussed in the previous section, you should have a diffraction pattern on the viewing screen. You can now change the magnification of the diffraction pattern (i.e. the camera length L), and you can focus the pattern (i.e. ensure that the back focal plane of the objective lens is conjugate to the viewing screen). While in diffraction mode, you should also change the second condensor lens current back and forth and observe what happens. Note that the diffraction spots will change from spots to disks and back to spots as you change the C2 current. When C2 is under-focused (the knob turned counter-clockwise on most microscopes), the electron beam will be more or less parallel when it enters the sample. The OL will then focus this parallel beam into a point in the back focal plane. When C2 is in-focus we observe the smallest possible beam diameter on the sample, but now the beam is converged into a point and the OL will produce a diffraction disk in the back focal plane. While this particular mode of sample illumination is of importance for convergent beam electron diffraction (CBED) observations, for now we will not use it; instead, you should always obtain regular diffraction patterns with the C2 lens under-focused, i.e. with a reasonably parallel incident beam.

As part of the alignment procedure, you used the projector lens controls to bring the central beam (the transmitted beam) onto the center of the viewing screen. You should periodically ensure that this is still the case. In particular, when you change the camera length, the central beam may wander off the center of the screen, and you can simply bring it back using the projector alignment. You should also ensure that the diffraction pattern is properly focused; this can be done by inserting the diffraction aperture and making sure that the edge of the aperture is as sharp as possible. You can then remove the aperture again to observe the entire diffraction pattern.

At this point you may also want to try changing the orientation of the sample with respect to the incident beam. Since you have already made sure that the sample is in the eucentric position, the sample region illuminated by the beam should not move laterally while tilting the sample. Conversely, if you did not properly adjust the eucentric position of the sample, then the area illuminated by the electron beam will move back and forth as you change the sample orientation, and you cannot be certain that the diffraction pattern you observe actually comes from the sample region you selected while in image mode. This illustrates the importance of a proper eucentric adjustment. On modern microscopes, the specimen attitude (position and orientation) is controlled by a computer program that will attempt to maintain the eucentric position at all times, so the operator should not have to worry about this particular adjustment. The algorithms controlling the eucentric specimen positioning are quite accurate, but you should be aware that small shifts may still occur, in particular at higher magnifications.

Continue to tilt the specimen around one or both of the tilt axes until you observe many diffracted beams on the screen. Adjust the specimen tilts so that you obtain a symmetric pattern with respect to the center of the screen (this is explained in more detail in Section 4.6.2). We already know from the previous chapter that when the crystal is oriented such that a zone axis is parallel to the incident beam, then the planes belonging to that zone will all be close to the Bragg orientation. Their reciprocal lattice vectors all lie in a plane which is tangent to the Ewald sphere. Therefore, we expect to see a two-dimensional array of spots in the diffraction pattern. Such a symmetric pattern is known as a *zone axis diffraction pattern*. When the sample is tilted over a small angle, the intersection of the reciprocal lattice plane and the Ewald sphere will become a circle (the *Laue circle*) along which the reciprocal lattice vectors are close to Bragg orientation, and this is clearly visible in the diffraction pattern (see Fig. 4.20 on page 283). This is your first experimental proof that the Ewald sphere is indeed a valid tool for the description of electron diffraction!

4.4.2.3 Obtaining a bright field image

Once you have placed the sample in a suitable orientation (while working in diffraction mode), you may want to go back to image mode to see what things look like. Here are a few steps to get back to the image (you may already have carried out a few of them in the previous section).

- Select a convenient camera length and, if necessary, bring the central beam to the center of the viewing screen using the projector alignment controls.
- Partially converge the incident beam using the C2 lens (i.e. make sure you observe disks rather than spots). If your sample is very sensitive to the incident electron beam you may want to skip this step. It is not an essential step, it only makes it easier to center the aperture in the next step.
- Insert the diffraction aperture, select a suitable diameter (slightly larger than the diffraction disk diameter), and center the aperture on the optical axis around the central diffraction disk.
- Return to image mode.

At this point you will most likely not see very much contrast on the screen because the SA aperture is still inserted in the column. Remove the SA aperture from the column and adjust the C2 lens to obtain an even illumination across the entire viewing screen. Congratulations! You have just obtained your first *bright field image*! The last two steps above are illustrated in Figs 4.10(d) and (e) on page 258.

A bright field image (or BF image) is defined as an image obtained with electrons from the transmitted beam only, i.e. electrons which left the sample in the same direction they entered the sample. This does not mean that those electrons were not scattered by the crystal. We will see in Chapter 5 that electrons can be scattered many times on their way through the crystal foil, and some may end up traveling in the original direction of the incident beam. Furthermore, many electrons will lose energy as they are scattered; there are various inelastic scattering processes that can give rise to such energy losses, and some of those electrons may also leave the sample in the transmitted beam. The bright field image is thus created by electrons that leave the sample in the direction of the incident beam; these electrons may have undergone multiple elastic and inelastic scattering processes.

As mentioned before, switching back and forth between image and diffraction mode will become second nature to you, so that you no longer have to think about the details of the procedure. At this point in time, however, you may be a bit confused about apertures and various image, object, and focal planes. Do not worry, things will get easier with time. The following brief summary might help.

- In image mode, you will nearly always have an aperture inserted in the back focal plane of the objective lens. This aperture selects which beam is used to create the image.
- In diffraction mode, you will nearly always have the SA aperture inserted into the column, so that the diffraction pattern arises from a "selected area" on the sample.
- You will rarely use the microscope with both apertures taken out of the column. There is very little contrast in such images (see Fig. 4.10a) at the lower end of the magnification range, and the corresponding diffraction patterns are usually rather messy. At the highest magnifications you may use the microscope without apertures for high-resolution observations, although typically a large-diameter diffraction aperture would be inserted in the column. If you use a small spot size with a converged incident beam instead of a parallel beam, then you can obtain so-called *convergent beam electron diffraction* patterns with both diffraction and selected area apertures removed from the column.
- You will never use the microscope with both apertures inserted in the column. There is not much information to be gained from such a setup.

In other words, the one aperture (other than the condensor aperture) that should be in the column is the one located in the plane that you are not looking at.

4.4.2.4 Obtaining a dark field image

To obtain the BF image we inserted the diffraction aperture in the path of the transmitted beam, centered along the optical axis. Nothing will keep you from moving this aperture to another location in the back focal plane to select one of the diffracted beams. The routine is exactly the same as for obtaining a BF image, except that *you should not use the projector lens alignment controls to bring the diffracted beam to the center of the screen*. This would be a rather inefficient way of doing things, and the projector controls may not even have a sufficiently wide range to do this. There is another way to bring the relevant diffracted beam to the center and we shall introduce it in a little while. First, you should take a look at the image obtained when selecting an off-center diffracted beam with the diffraction aperture.

Most likely you will observe a region on the sample that is very bright, while the surroundings are dark. In particular, if you are looking close to the edge of the thin foil you will find that the hole in the sample is completely dark. The reason for this is obvious: vacuum does not diffract electrons, and since you are only using electrons scattered over an angle β the hole in the sample will not become bright. This is known as a *dark field* (DF) image. Any image created with a diffracted beam is known as a dark field image. Since there may be many diffracted beams, you should always specify which diffracted beam you used to obtain the DF image, and we would typically speak of, for instance, a "two-oh-oh dark field image" when the operative reflection corresponds to the (200) lattice planes. We will see in a later section how this information should be represented on an electron micrograph.

All of the diffracted beams in the pattern correspond to electrons traveling at an angle β with respect to the optical axis, which means that those electrons will suffer from spherical aberration (remember that spherical aberration goes with the cube of the angle β). It would therefore be unwise to record images corresponding to an off-center diffraction aperture. You can easily convince yourself that the image quality deteriorates when you use diffracted beams at increasingly larger distances from the optical axis. In addition, the electrons in a high-order diffracted beam will come from a region on the sample that does not completely coincide with the area selected by the SA aperture. This is again a consequence of spherical aberration. Indeed, there will nearly always be a shift between a diffraction pattern and the corresponding selected area. In Fig. 4.12(a) an object *ab* is centered on the optical axis. Two sets of electrons leave the object: the transmitted beam (parallel to the optical axis) and one diffracted beam (at an angle $\beta = 2\theta_B$). If an aperture is introduced in the image plane, such that the image from the transmitted beam exactly fills the aperture (between points A and B), then the image formed by the diffracted beam will be shifted by an amount $MC_S\beta^3$, according to the definition

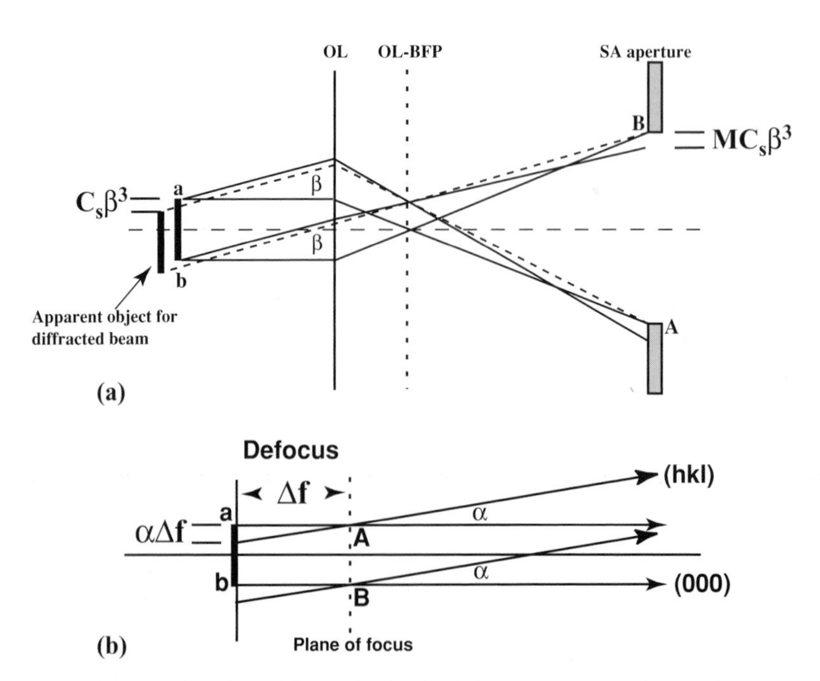

Fig. 4.12. (a) Illustration of the effect of spherical aberration on the image for two electron beams. (b) Effect of overfocusing of the objective lens on the image for selected area imaging.

of spherical aberration in Chapter 3. Assuming that the two images coincide, the diffracted beam may be traced back (dotted lines) to the object plane, which results in an object shift of $C_S\beta^3$; in other words, the diffracted electrons arise from a region that is not completely identical to the region imaged by the transmitted electrons.

Since diffraction angles in electron diffraction are rather small (of the order of milliradians), this shift effect is small, except for high magnifications. There is a much more important image shift effect: that caused by under- or over-focusing of the objective lens. Figure 4.12(b) shows what happens when the image plane of the objective lens does not coincide with the object plane of the imaging system. The diffracted beam traveling at an angle β will arise from an area displaced by a distance $\Delta f\beta$. The distance Δf is known as the *defocus*. The total displacement *in the object plane* caused by defocusing and spherical aberration is then given by

$$y = C_S\beta^3 + \Delta f\beta. \tag{4.1}$$

The out-of-focus images in Fig. 4.10(a) illustrate this image shift. For higher-order reflections the image shift becomes increasingly more important. It is hence important to correctly focus the image (so that $\Delta f \approx 0$) and to use only low-order beams when obtaining DF images with an off-center diffraction aperture. We recommend that you always use the *centered dark field* method introduced in the next section to minimize the effects of spherical aberration.

4.4.2.5 Centered dark field images

Since the off-center dark field image suffers from lens aberrations, we must find a way to bring the diffracted beam onto the optical axis, so that $\beta = 0$. Fortunately, the microscope is equipped with pre-specimen beam deflection coils that allow us to do just that. If your microscope has been aligned properly, you should be able to focus the beam onto a small region (using C2), and then use the beam tilt controls to change the orientation of the incident beam, without lateral motion of the beam. You should try the following:

- in image mode, focus the beam using C2 so that you obtain the smallest possible spot size;
- use the beam translation controls to bring the spot to the center of the viewing screen;
- now change the beam tilt controls; you should not observe any beam shift on the screen (although the contrast in your image will change);
- if you do observe a beam shift, then that means that the beam deflection system is misaligned, and you should refer to the standard procedures manual for the alignment procedure (this is essentially the iterative alignment procedure for the condensor lenses discussed in Section 4.4.1).

Once the beam tilt has been aligned properly, switch to diffraction mode, underfocus the C2 lens, and change the beam tilt. You will find that the diffraction pattern shifts laterally. In other words, *a tilt in direct space corresponds to a shift in reciprocal space*. This is not too difficult to understand since the position of a diffraction spot in the back focal plane is determined by the angle of the electron trajectory with respect to the optical axis; if we tilt the entire beam, then all electron trajectories will be changed by the same angle and thus the entire diffraction pattern is translated in the direction opposite to the beam tilt direction.

We have seen before that we can also translate the diffraction pattern by means of the projector lens controls. There is one important difference between those two translations: shifting the pattern by means of the projector lens controls is a post-specimen operation, and the intensity distribution in the diffraction pattern does not change at all. Tilting the incident beam, on the other hand, is a pre-specimen operation, and the intensity distribution in the back focal plane will change in addition to the pattern shift.

Next, tilt your specimen while in diffraction mode to an orientation for which one of the beams closest to the transmitted beam is very intense. It does not matter whether other beams are intense as well. Switch the microscope to dark field mode.[†] Then use the beam tilt to bring this intense reflection to the center of the screen. Do not change the sample orientation. Insert an aperture around this beam and switch to image mode. You will find that the image has a rather low intensity. You may not see anything at all! In fact, you may have noticed that the intensity of the diffracted beam which you moved to the screen center *decreased* as you brought it closer to the center.

To understand why this happens we need to take a close look at the Ewald sphere construction for this situation. Consider the initial situation as shown in Fig. 4.13(a): the reciprocal lattice point **g** is on the Ewald sphere and a high-intensity diffraction spot is observed on the screen. The higher-order reflections 2**g** and 3**g** are at increasingly larger excitation errors and will have correspondingly lower intensities. The electrons in the beam **g** travel at an angle 2θ with respect to the optical axis. The corresponding lattice planes are parallel to the perpendicular bisector of the vector **g**, as indicated in the drawing. To bring the reflection **g** onto the optical axis, we must tilt the incident beam towards the right by an angle 2θ, as shown in Fig. 4.13(b). The crystal remains in the same orientation as before. Since Bragg angles are quite small (of the order of a few milliradians) the electronic beam tilt controls will be quite sufficient to change the orientation of the incident beam with respect to the optical axis. The angle between the incident beam and the

[†] Most microscopes have multiple settings or memories for the beam tilt. It is a good idea to use the primary setting or memory for the bright field beam tilt condition, and other memory slots for various dark field beam tilts.

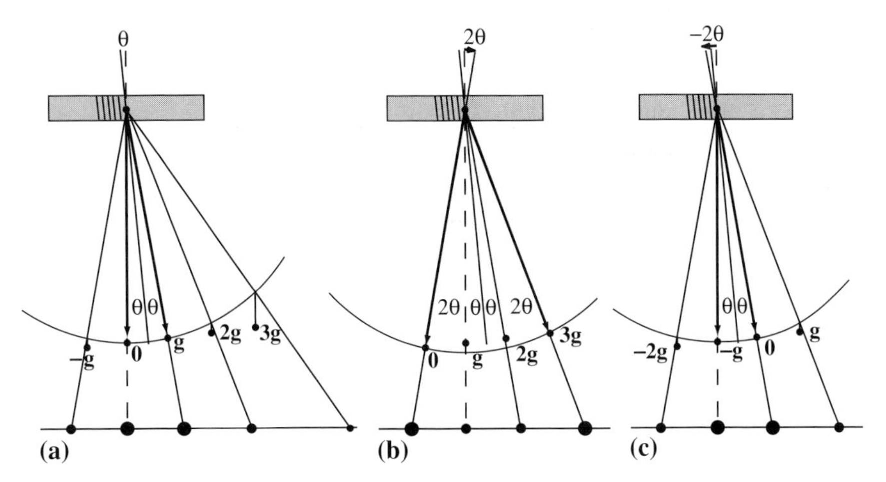

Fig. 4.13. (a) Bright field configuration with **g** in Bragg orientation; (b) a beam tilt of 2θ will bring **g** onto the optical axis but only 3**g** will be in Bragg orientation; (c) a beam tilt of -2θ will bring $-\mathbf{g}$ to the optical axis and simultaneously in Bragg orientation. This is known as the centered dark field mode.

planes **g** is now 3θ, which is rather far away from the exact Bragg orientation. This means that the reciprocal lattice point **g** has a large positive excitation error and hence a weak diffraction spot. Since the incident angle is 3θ, the Bragg equation will be satisfied for the reflection 3**g**, and the corresponding diffraction spot will become bright. In other words, our attempt to bring the reflection **g** onto the optical axis has changed the diffraction conditions to the extent that now the planes 3**g** are strongly diffracting. We could now tilt the crystal over an angle 2θ to restore the Bragg condition for the planes **g**. This would work fine, but tilting a sample over an angle of a few milliradians is not an easy task, especially when only mechanical tilting controls are available. In addition, as is verified readily from a drawing of the transmitted and diffracted beams, the bright field image would then change from what it was before the sample tilt.

There is a much easier way to obtain a dark field image from the planes **g**, with the operative reflection located on the optical axis of the microscope. You should now try to bring the reflection $-\mathbf{g}$ onto the optical axis, using only the beam tilt controls. Note that the intensity of the $-\mathbf{g}$ reflection increases and becomes maximal when the beam is at the screen center. When you insert the diffraction aperture and switch to image mode, a portion of your image will be rather bright, and you can easily focus the image. This is known as a *centered dark field image* (CDF image). To understand why this works we need to go back to the Ewald sphere drawing in Fig. 4.13. We have tilted the incident beam by an angle -2θ, to bring the reflection $-\mathbf{g}$ onto the optical axis, as shown in Fig. 4.13(c). In other words, we have satisfied the Bragg condition for the other side of the planes **g**, i.e. for $-\mathbf{g}$. In Chapter 5 we will see that the dark field images for **g** and −**g** are identical. Summarizing, to obtain a centered dark field image for the reflection **g** we bring −**g** onto the optical axis using the beam tilt controls. After insertion of the diffraction aperture, we can simply switch from BF to the CDF image by pressing a single button on the microscope console. When you switch back and forth in diffraction mode you will see the diffraction pattern jump between the two positions; when you switch back and forth in image mode the image will switch from BF to DF and back. Both images should be simultaneously in focus, provided all alignments have been carried out properly.

We will see later on in this book that the first method of bringing **g** to the optical axis is very useful, and is the basis of the so-called *weak beam imaging method*, also known as **g**–3**g** imaging. This is a very powerful method for observing crystal defects with a high spatial resolution. The dark field images thus obtained do not suffer from spherical aberration since the electrons travel along the optical axis, as they do in the CDF technique. As you have discovered, such weak beam images have a very low overall intensity and require long exposure times.

4.4.2.6 Astigmatism and BF/DF images

Astigmatism is a parasitic lens aberration, which means that it is caused by imperfections in the lens surfaces, inhomogeneities in the lens coils, or possibly by small magnetic fields caused by the sample itself. Astigmatism correction is perhaps the most difficult aspect of electron microscopy; it takes practice and a skilled microscopist can correct astigmatism about as well as a computerized system, which uses discrete Fourier transforms of digital images.

The easiest procedure for astigmatism correction relies on the use of *Fresnel fringes*; these fringes arise wherever the specimen has an edge or any other discontinuity. Electrons scattered from the edge (in a spherically scattered wave) interfere with the transmitted electron beam and form an interference pattern. The number of fringes in this pattern is a direct measure of the *beam coherence*. For conventional TEM one can usually distinguish only one fringe; for a field emission machine as many as 100 fringes have been observed. As shown in the through-focus series of gold particles on a holey carbon support film in Figs 4.14(a)–(c), this fringe is bright in under-focus and dark in over-focus. By changing the current through the objective lens we can switch from one to the other. The fringe must have the same distance to the sample edge all around the hole; if this distance is not constant, or if part of the fringe is bright and part dark, then astigmatism is present in the image, as ilustrated in Figs 4.14(d) and (e). Correction of the astigmatism either involves obtaining a circular Fresnel fringe around the entire hole, or, alternatively, no preferential direction in the image. The images in Figs 4.14(d) and (e) clearly

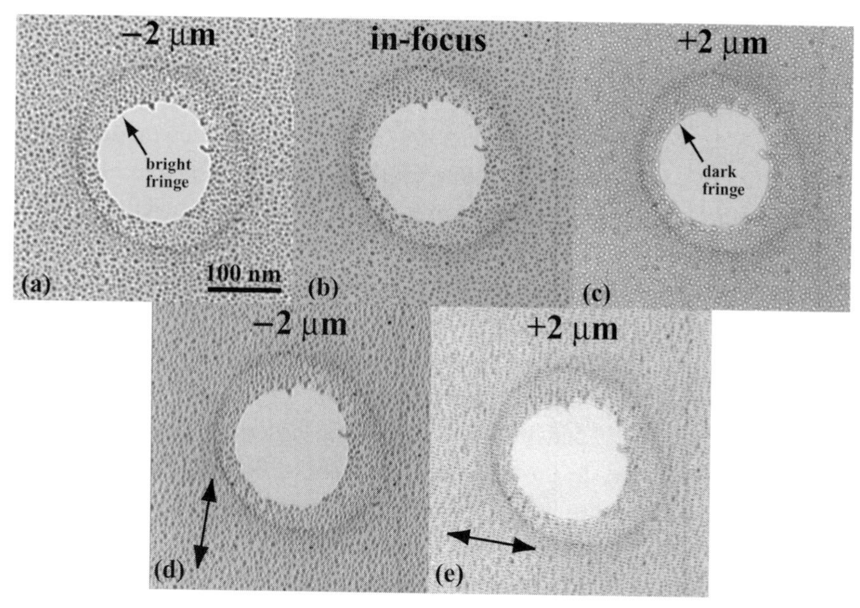

Fig. 4.14. Illustration of the effect of astigmatism on Fresnel fringes around a circular hole in a holey carbon film with gold particles. Parts (a)–(c) are a through-focus series with prefectly corrected astigmatism. Parts (d) and (e) show the clear presence of astigmatism, indicated by the absence of a Fresnel fringe around the entire hole, and the presence of a preferential direction (arrowed) in the image. (JEOL 2000EX operated at 200 kV.)

show a preferential direction (arrowed). Astigmatism can be corrected by removing this preferential direction from the images for all focus values. Astigmatism correction is easiest when there are small holes in the sample. If there are no holes, then you may have to move to the edge of the sample, specifically where the edge is not perfectly straight.† At the highest magnifications one can often observe a narrow amorphous area at the edge of a sample (even a crystalline sample) and this amorphous area can be used to correct the astigmatism (see Chapter 10 for examples).

4.4.2.7 Taking good pictures

Now you are ready to take a picture of what you see on the viewing screen. This is a crucial step and you should learn how to properly expose an electron micrograph, either the standard photographic negative or the electronic (digital) version. The exposure time τ should be such that the relevant information will be recorded in the nearly linear part of the characteristic curve (see page 229). If τ is too short,

† You cannot use a straight Fresnel fringe to correct astigmatism. Why not?

portions of your image may be invisible (intensity below the fog level of the film or the background level of the CCD camera); if τ is too long, the film will be overexposed and saturated.

We know from Chapter 3 that the density of a developed negative is related to the number of electrons per unit area that hit the emulsion. We have estimated this number using the concept of beam brightness on page 186. Typical current densities are in the range of tenths to hundreds of pA cm^{-2}, which translates into exposure times in the range of 0.1 to several tens of seconds, considerable longer than the times needed for most conventional light photography.

Every TEM is equipped with a sensitive exposure meter, which measures the current density at the screen. It is important to realize that this is an average number, and that local current densities can be several orders of magnitude larger or smaller than the average, in particular for electron diffraction patterns. The exposure meter signal is automatically converted into an exposure time τ, and the microscope operator can choose to use this value as the exposure time (automatic exposure mode) or to set a different value (manual exposure mode). The automatic exposure mode has one adjustable parameter known as *film sensitivity* (related to the electron speed of the film). This parameter is set once and for all after a series of calibration exposures, and should be changed only if the exposure meter is replaced or if the user switches to a different brand or type of negatives.

The automatic exposure mode can be used for bright field images that do not have significant contrast variations (i.e. rapid intensity changes from bright to dark regions). Depending on the microscope magnification you may have exposure times in the range 0.1–2 s. It is usually not a good idea to focus the incident beam using the C2 lens so that maximum intensity is observed on the screen. A converged incident beam has a lower beam coherence than a parallel beam and you may want to record image details (such as Fresnel fringes) that are only visible for a reasonably coherent beam. So, even if you have more electrons available to reduce the exposure time, sometimes you will need to use a longer exposure time. For many dark field images the image contrast will vary strongly from one location to the next, and the automatic exposure meter will typically underestimate the exposure time (large portions of your dark field image may be dark). You will have to manually increase the exposure time by a few clicks but only experience will tell you the correct number of clicks.

For all other images we recommend that you use the manual exposure mode. Estimating the exposure time requires some experience, so until you have acquired this expertise you should always record more than one micrograph at different exposure times, and make a note (both mental and written) of the exposure times that work best for a given type of image. For instance, for a zone axis diffraction pattern on which you plan to perform measurements (indexing of the pattern) you could use the following procedure:

- place the sample in the proper orientation for the pattern that you wish to record;
- under-focus the C2 lens (in diffraction mode) to obtain a more parallel incident beam and therefore smaller diffraction spots;
- reduce the C2 lens current until you can barely see the weakest diffraction spots on the screen (make sure the room is completely dark, so that you have good night vision);
- focus the diffraction spots (you may have to use the binocular to see the spots);
- use the projector lens alignment to bring the transmitted beam to the center of the screen;
- set the manual exposure time to somewhere in the range 22–45 s;
- place the viewing screen cover over the screen and take a picture.

Most of the zone axis patterns shown in this book were recorded using this procedure. If diffuse features are present in the diffraction pattern (thermal diffuse scattering, short-range order scattering, etc.), then you will probably have to use either a longer exposure time, or increase the C2 current a bit. Again, it might be a good idea to record multiple images with different C2 settings.

If you are about to record a convergent beam electron diffraction pattern, then you can obviously not change the C2 current, since this would destroy the CBED pattern (see Section 4.6.4). You will then have to record multiple images with different exposure times to record all the image details present in the pattern. Weak beam dark field images require very long exposure times, in the range of 22–90 s, and again you will need to take multiple recordings.

Digitally acquired images have one significant advantage over conventional photographic images in that they are immediately available for inspection. If an image is over-exposed, you can simply discard it, adjust the exposure time, and record another one before saving it to disk. The best digital images are those that use the entire available dynamic range of intensities (e.g., 0 to $2^N - 1$, with $N = 12, 14$, or 16 for modern CCD cameras).

Photographic negatives must be developed and printed or digitized if they are to be used for analysis or publication. The standard procedure for developing electron micrographs is as follows (all steps to be carried out in a darkroom with only red-filtered safelight present; latex or lint-free gloves should be worn whenever the negatives or negative holders are manipulated):

- remove the exposed negatives from the negative holders and place them in a metal or plexiglass carrier;
- place the carrier in the developer tank; develop for 3–5 min, depending on the type and strength of the developer;
- agitate the developer either manually or by means of nitrogen bursts once every 10 s or so;
- remove the carrier from the developer and rinse in room-temperature running water for 90 s;

- place the carrier in the fixing bath and fix for 8–10 min, again depending on the type and age of the fixing solution; use frequent agitation;
- rinse in room-temperature running water for 20–30 min, apply a soapy solution to minimize drying marks (optional), and dry in a warm and dust-free place.

The details of this procedure may vary depending on the type of chemicals and negatives used. For nearly all micrographs reproduced in this book we used Kodak Electron Image Film SO-163 with 2:1 diluted D19 developer.

If you plan to digitize your negative, you should use a good flatbed scanner with a transparency adaptor, and scan the negative at a resolution of 1200 or 2400 dpi. The grain size in the photographic emulsion is then about equal to the pixel size of the digitized image. There is little point in using a higher resolution, unless you use special film with a smaller grain size. Once you have a digital version of the image, you can use the full power of image analysis programs to enhance the contrast, scale the brightness linearly or non-linearly, select portions of the image, place a micron bar on the image, and so on. If you alter the image in any non-standard way (i.e. adding or removing information, enhancing features, etc.), then you should explicitly state this in the figure caption (assuming that you are going to use the figure in a report or paper).

4.5 Microscope calibration

As with any other scientific instrument, a transmission electron microscope must be calibrated. The image magnification and diffraction camera length are perhaps the most important calibrations for conventional TEM observations. In addition, the relative orientation of the image and the corresponding diffraction pattern must be calibrated. The following two subsections will describe briefly how these calibrations can be performed. It is likely that calibrations will have already been carried out for the microscope(s) you are using, so it may not be necessary to repeat them yourself. However, you should make sure that you understand the imaging conditions used for the calibrations, so that you can reproduce them in case you wish to make accurate measurements.

Calibrations must be carried out not only for conventional micrographs, but also for digital and analog cameras. In all cases it is important to know the absolute orientation of the camera/negative with respect to the column. For a negative it is rather easy to find out how it is oriented inside the column (the emulsion side faces the electron beam, and there is usually a fiducial marker along the side of the negative which indicates its orientation in the camera). Calibration measurements should always be carried out with the negative placed in the same orientation as it was inside the microscope. Note that it is not sufficient to calibrate magnifications for electron micrographs alone; if your microscope is equipped with a digital camera, a magnification calibration must be carried out since the image plane of the camera does not in general coincide with the image plane of the negative. There are several other microscope parameters that may need to be calibrated. We will expand on the topic of calibration in Chapter 10.

4.5.1 Magnification and camera length calibration

The diffraction camera length can be calibrated using a polycrystalline sample of a known material; commonly used calibration samples consist of evaporated aluminum or gold on a holey carbon grid. Figure 4.15 shows a diffraction pattern for polycrystalline aluminum, obtained with the nominal camera lengths of 60, 80, and 100 cm on a JEOL 2000EX microscope (the nominal camera length is what is indicated on the microscope console). The first six ring diameters were measured in two different ways: by hand using a ruler, and numerically, by first digitizing the negatives at a resolution of 1200 dots per inch and then fitting a circle to each ring. Example results from the second method are shown in Table 4.5. The measured average values for L are close to the nominal values. The standard deviations are simply based on the number of measured values. A more realistic standard deviation would take into account the accuracy of the digital measurement, which is estimated to be slightly better than 1%.

The orientation of the diffraction pattern with respect to the corresponding image must also be calibrated. The most straightforward method involves the use of α-MoO_3, an orthorhombic oxide with lattice parameters $a = 0.3963$, $b = 1.3856$, and

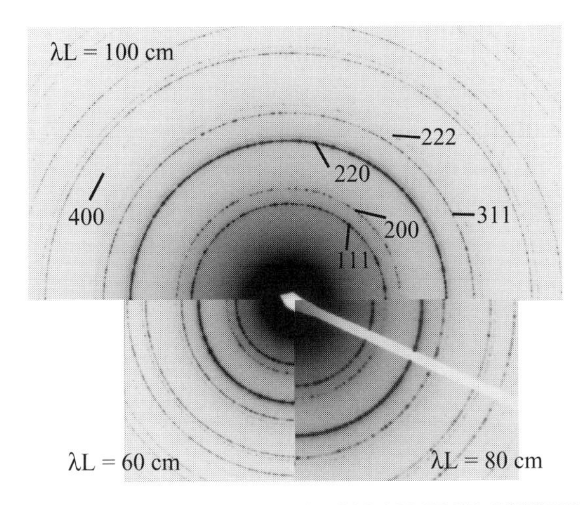

Fig. 4.15. Calibration of the camera length of a 200 kV JEOL 2000EX microscope using an evaporated aluminum film. The diffraction rings used for the calibration are indicated; the (400) ring is just barely visible. The bottom quarters represent $\lambda L = 60$ and 80 cm, and the top corresponds to 100 cm.

Table 4.5. *Camera length calibration results for a JEOL 2000EX microscope operated at* 200 *kV, using an evaporated aluminum film.* λ *is measured in nm; L is computed by dividing by the electron wavelength* $\lambda = 0.002\,508$ *nm.*

(hkl)	d_{hkl} [nm]	λL_{60}	λL_{80}	λL_{100}
111	0.234	1.537	2.004	2.493
200	0.202	1.531	1.989	2.484
220	0.143	1.529	1.994	2.487
311	0.122	1.524	1.994	2.487
222	0.117	1.528	1.995	2.494
400	0.101	1.528	1.983	2.491
$\langle \lambda L \rangle$		1.529	1.993	2.489
Standard deviation		0.004	0.007	0.004
$\langle L \rangle$		60.96	79.46	99.24

$c = 0.3697$ nm, space group *Pbnm*.[†] This oxide grows in elongated platelet form, with the normal to the platelet along the [010] direction, and the platelet length axis along [001]. The [010] zone axis pattern is therefore rectangular, with the longest axis along [001]. Figure 4.16(a) shows a double exposure of a single platelet along with its diffraction pattern. The 100 and 001 reflections are labeled. The longest axis in reciprocal space corresponds to the length axis of the platelet, and the angle between the two directions is measured to be nearly 6°.

Since the diffraction pattern in Fig. 4.16(a) is symmetric, the angle between the image and the diffraction pattern is either 6° or 186°. We can determine which of the two angles is the correct one by defocusing the diffraction pattern (decrease the current through the diffraction lens, so that its focal length increases). The resulting image is shown in Fig. 4.16(b); a dark field image of the platelet is visible in each diffraction spot. The orientation of the platelet is clearly rotated by 180°, as can be seen by comparing the circled regions in Figs 4.16(a) and (b). The diffraction pattern can be brought into the proper orientation with respect to the image by a counterclockwise rotation of 186°.

When the camera length is changed the diffraction pattern may also rotate around the origin. To determine whether or not this rotation is present we can simply obtain a multiple exposure of the diffraction pattern at different camera lengths, as shown in Fig. 4.16(c). Between exposures the diffraction pattern was re-centered using the projector lens shift controls. It is clear that there is no rotation between

[†] This is not a standard space group setting. *The International Tables* give the setting *Pnma*, which can be obtained from *Pbnm* by cyclic permutation of the unit cell axes.

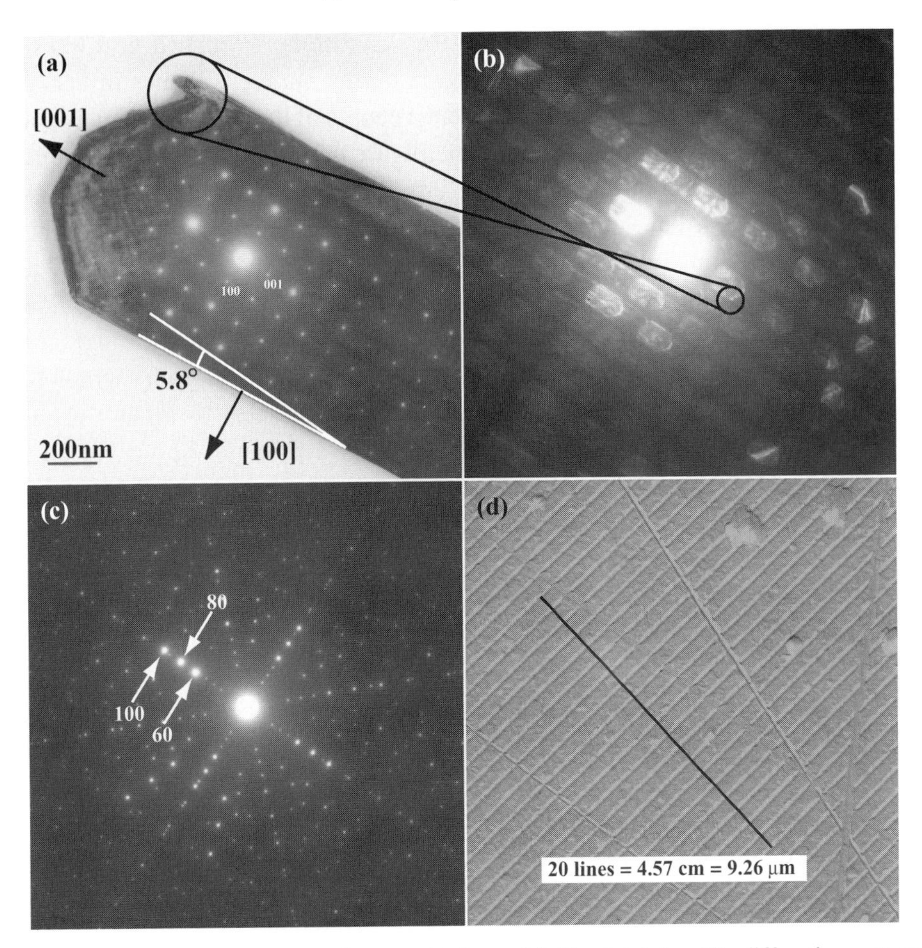

Fig. 4.16. (a) Double-exposure micrograph of an α-MoO_3 platelet and its diffraction pattern; (b) defocused diffraction pattern showing a dark field image of the platelet in each reflection; (c) multiple exposure diffraction pattern of [010] α-MoO_3 reveals no rotation between different camera lengths; (d) image at 5000× (indicated on the microscope console) of a 2160 lines per millimeter calibration grating. All images obtained on a JEOL 2000EX microscope.

different camera lengths for this particular microscope (JEOL 2000EX). For other microscopes there may be a significant rotation, which can simply be measured from the multiple exposure diffraction pattern.

Calibration gratings for magnification calibration (image mode) are available from many of the microscopy supply companies (listed on the website). Typically a calibration grid with around 2000 lines per millimeter should be sufficient to calibrate all but the highest magnifications. It is important that the sample should be at the eucentric position throughout all calibration steps. For all microscope magnifications record a bright field image of the calibration grid. After developing the micrographs, use a ruler to measure the distance between two grid lines on opposite sides of the micrograph and divide by the number of lines in between them. You should perform this measurement a couple of times for each micrograph and take the average of all measurements. Figure 4.16(d) shows a typical image for the magnification 5000× for a JEOL 2000EX microscope, operated at 200 kV. The calibration grid has 2160 lines per mm, so that the spacing per grid line is equal to $d = 463$ nm. Twenty lines correspond to a measured distance of 45.7 mm (Fig. 4.16d), which should also be equal to $20d = 9260$ nm. The magnification is therefore equal to $45.7 \times 10^6/9260 = 4935$, close to the 5000× indicated on the microscope console. If you make multiple measurements for each micrograph then you can compute error bars for the measured magnifications.

For the highest magnifications, only a single grid line will be visible on the micrograph. Measure the distance between two easily recognizable features, and compare it with the same distance at the largest magnification for which two grid lines are present. The ratio between those two distances can then be used as a multiplicative factor to compute the actual magnification. Typically, magnifications above about 60k× are obtained in that way. For the highest magnifications (about 200k× and above) it is customary to use lattice fringes with a known spacing.

If you are going to publish micrographs, then you will need to convert the calibrated magnifications to "micron bars". If your images are conventional micrographs on photographic negatives, you will need to know how many millimeters in the image correspond to a typical distance on the object (1 μm, 10 μm, etc.). If you digitized your micrograph, or the image was digitally acquired, then a more convenient parameter may be the number of pixels per unit image length. We leave it to the reader to derive the appropriate conversion relations.

4.5.2 Image rotation

In many older microscopes, the image rotates around the optical axis as the magnification is changed. This is due to the Lorentz force on the electrons, which causes a helical path around the optical axis, as described in Chapter 3. Modern microscopes may have an additional lens in the magnification stage, so that this image rotation is minimized or, ideally, eliminated. For such machines, the image does not rotate when the magnification is changed, and the diffraction pattern is in the correct relative orientation with respect to the image.

If the image does rotate, then we must calibrate the rotation angles as a function of magnification. First of all, we must determine the relation between the physical orientation of the primary tilt axis and the projected orientation of this axis. This can be done in the following way: lower the sample using the z-control so that the sample holder is near its lowest point. Adjust the image focus and bring a feature to the center of the screen. Then tilt the sample along the primary tilt axis *by a positive rotation angle* until this feature is a few centimeters away from the center. Record a double-exposure micrograph of these two images (i.e. with the feature at the center and tilted away). The line connecting the two images of this feature on the micrograph is normal to the primary tilt axis. Since the tilt angle was positive, the direction of the image shift also indicates the absolute orientation of the primary tilt axis with respect to the negative.

This procedure is illustrated in Fig. 4.17: the first image on the top left shows a double exposure of the calibration line grid at a nominal magnification of 8200×. The image was acquired using a Gatan DualView camera on a Philips EM420 microscope operated at 120 kV. The arrowed feature moved from right to left for a positive (counterclockwise) rotation around the primary tilt axis. The tilt axis is then oriented at right angles to the line connecting two identical features in the double-exposure image. The image rotation at other magnifications (two representative ones are shown in Fig. 4.17) can then be measured with respect to the tilt axis. At the highest magnifications only a single grid line may be visible in the image, making it difficult to obtain an accurate measure of the image rotation.

It is convenient to represent rotation data in graphical form. The drawing in the lower half of Fig. 4.17 represents the orientations of the bottom edge of the image with respect to the primary tilt axis. The orientation of three diffraction patterns (different camera lengths) are also added to the figure. They were measured using the double-exposure technique shown in Fig. 4.16(a). We will use this graphical calibration chart later on, when we describe how to determine the orientation of a crystal with respect to the microscope column. Regardless of the type of microscope you use, it is important to understand the relative orientation of tilt axes and images/diffraction patterns. We encourage the reader to perform a series of calibration measurements similar to that described in this chapter.

4.6 Basic CTEM observations

Now we are finally ready to start some basic TEM observations. In the following sections we will describe typical image and diffraction features that are easily observed in the four study materials. We will emphasize those features that are common to all four, and which therefore appear to be relatively independent of the particular details of the crystal structure. We will discuss basic imaging techniques, and illustrate how information from a zone axis electron diffraction pattern can be transposed onto a stereographic projection.

The images shown in the following sections are easy to obtain, and we encourage the reader to spend time on the microscope with one or more of the study materials. You should feel free to experiment with the microscope (within reasonable limits

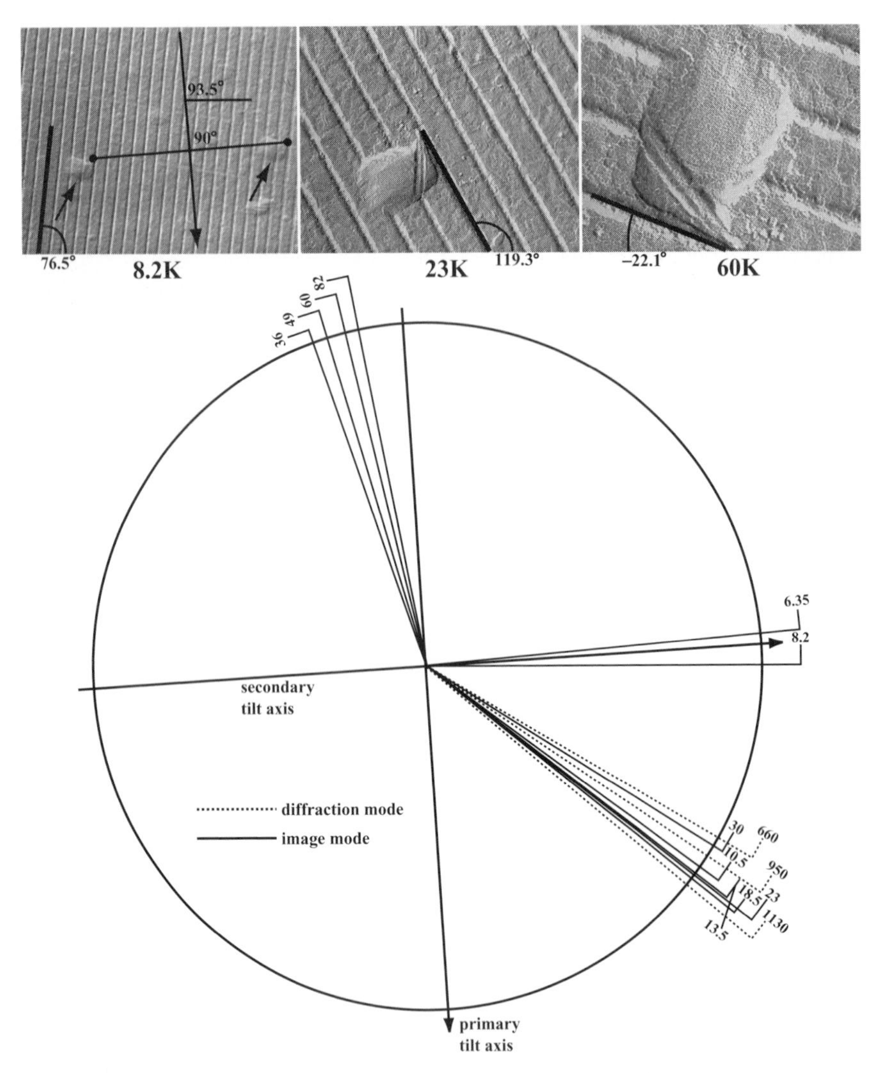

Fig. 4.17. Illustration of the rotation calibration for a Philips EM420; images were recorded on a Gatan DualView camera. The top row shows three images of a calibration line grid with 2160 lines per mm (463 nm line spacing). The bottom drawing shows the rotation angles of image and diffraction mode with respect to the primary and secondary tilt axes.

of course!). When you observe a certain image feature, say, a broad dark band running across the bright field image, try to determine how this band behaves when you change the orientation of the sample. Or, in diffraction mode, observe how the diffraction pattern changes when you move the selected area aperture across one of those bands. It is only by experimenting with various imaging conditions that you will begin to see how the image contrast depends on the microscope parameters and settings *which you as an operator can select*. Once you have familiarized yourself

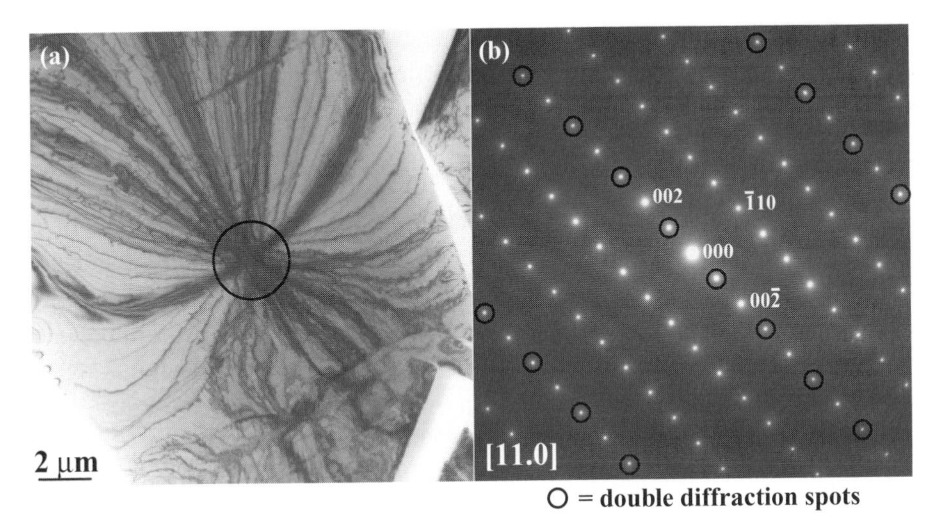

Fig. 4.18. Example of a bend center in a Ti thin foil. The image in (a) was obtained on a JEOL 2000EX, operated at 200 kV. The diffraction pattern in (b) was obtained using a selected area aperture centered on the intersection of bend contours (circle in a). The resulting zone axis pattern can be indexed as the [01.0] zone axis pattern of Ti. The reflections circled with a black circle have a zero structure factor, but appear because of double diffraction.

with the basic image features common to all four study materials, it will be time to proceed to the next chapter and explain these observations, starting from the relativistically corrected Schrödinger equation.

4.6.1 Bend contours

In this section, we will assume that you have a polycrystalline metallic thin foil, either Cu-15 at% Al or Ti. Since most metallic thin foils are locally bent, it should be easy to find a region (in the bright field image mode) that looks somewhat similar to the region shown in Fig. 4.18(a). We observe a large number of dark contours that intersect each other in a small region. When you tilt the sample, you will find that all of these contours move across the grain. They are known as *bend contours*, and their intersection point is a *bend center*. Place the selected area aperture over the bend center and switch to diffraction mode. You will observe a symmetric diffraction pattern, not unlike the one shown in Fig. 4.18(b).[†] This pattern corresponds to a plane in reciprocal space, tangent to the Ewald sphere, which means that the incident beam direction is along a crystallographic direction or a zone axis direction. The pattern is known as a *zone axis diffraction pattern*. Using the methods discussed in

[†] The pattern you obtain will not necessarily be the same as the one shown in the figure, but it should be a symmetric pattern with a significant number of reflections.

Section 9.2.1 we can assign Miller indices to each of the reflections, and using the cross product of any two independent reflections we find the indices of the zone axis. For the case shown in Fig. 4.18(b) we have used a Ti thin foil, and the zone axis is [11.0] in 4-index notation.

Several reflections have been circled in Fig. 4.18(b); the structure factor vanishes for all of these reflections, but we clearly have significant intensity in each of them. This is our first evidence that electron diffraction is not simply the same as x-ray diffraction with a shorter wavelength. For conventional x-ray diffraction experiments, a vanishing structure factor would mean zero intensity in the corresponding diffracted beam. The circled reflections in Fig. 4.18(b) will not give rise to diffraction peaks in an x-ray powder pattern, because the presence of the 6_3 screw axis in the non-symmorphic space group of Ti forbids these reflections from occurring. Yet, the same reflections are clearly present in the electron diffraction pattern. This means that the structure factor is not sufficient to determine the intensity of diffracted beams in an electron diffraction experiment. Intensity calculations based on the structure factor are commonly known as *kinematical calculations*. Calculations which account for the presence of the "forbidden" reflections are known as *dynamical calculations*, and we will spend a significant portion of this book discussing the dynamical theory (Chapters 5–7). The forbidden reflections are present through a mechanism known as *double diffraction*, and we will return to this concept in Section 9.2.3.

After you have convinced yourself that the intersection of bend contours corresponds to a point where the incident beam travels along a zone axis orientation of the crystal, you should return to image mode (bright field) and move the SA aperture onto one of the bend contours, away from the bend center. Then switch back to diffraction mode; you will find that one single row of reflections is prominent, while reflections that do not belong to this row are significantly weaker than they were for the zone axis orientation. Since a row of reflections corresponds to a single set of planes $n\mathbf{g}_{hkl}$ we also find that *a single bend contour is associated with a particular set of planes*. We could therefore label the bend contour with the Miller indices hkl of the corresponding reciprocal lattice point. All along the bend contour the planes (hkl) are strongly diffracting, which means that the Bragg condition is satisfied along the bend contour. We will determine in Section 7.2.1 exactly where along the contour the Bragg condition is satisfied. If a single row of strongly excited reflections is present in the diffraction pattern, then we say that the crystal is in a *systematic row orientation*. Let us consider this case in somewhat more detail.

Consider the zone axis pattern shown in Fig. 4.19(a). This pattern can be indexed as the [01.0] zone axis pattern of Ti. If we tilt the crystal by about 8° around the row $n\mathbf{g}_{002}$ then we obtain the systematic row pattern shown in Fig. 4.19(b).

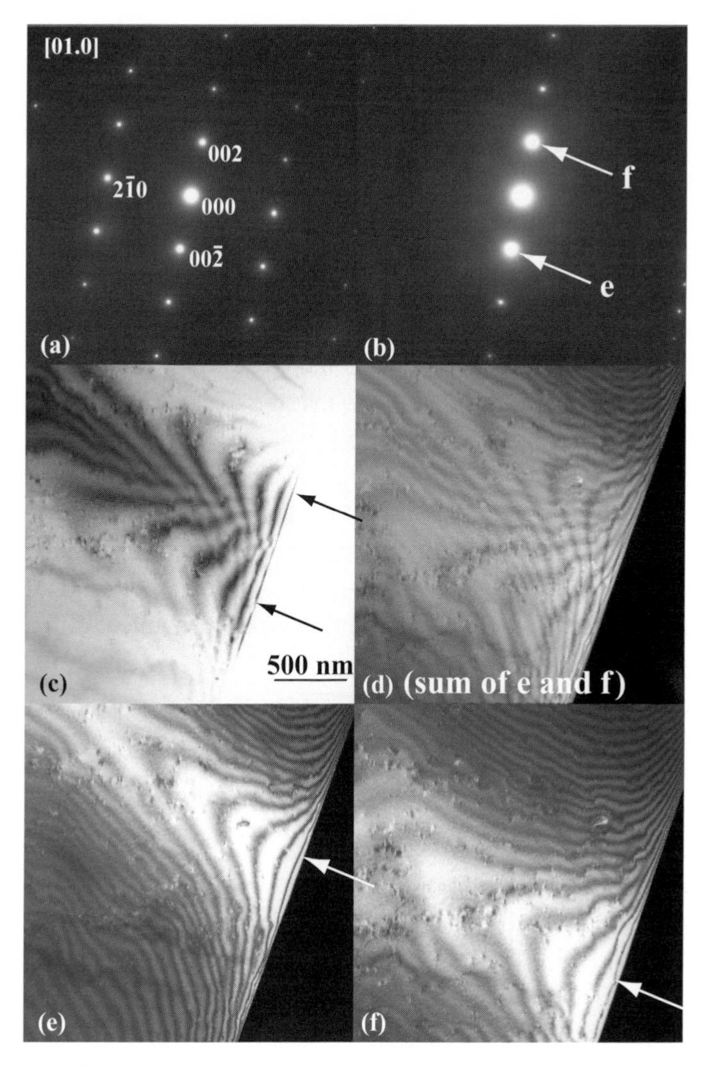

Fig. 4.19. (a) [01.0] zone axis pattern of Ti; (b) same but tilted about 8° around the $\mathbf{g}_{002}$ systematic row axis; (c) bright field image corresponding to (b); (e) and (f) are centered dark field images for the reflections labeled in (b); (d) is the sum of images (e) and (f), and is complementary to (c).

The corresponding bright field image is shown in Fig. 4.19(c); we observe a bend contour that is nearly normal to the edge of the foil.[†] Where the contour intersects

[†] The local shape of the foil determines the geometry of the bend contour, and it is not possible to tell from the systematic row diffraction pattern what the orientation of the bend contour will be with respect to other features in the foil. To find a configuration similar to that shown in Fig. 4.19(c), you should search in bright field mode until you find a bend contour nearly normal to the sample edge; you may need to tilt the sample back and forth until you find such a bend contour. Then switch to diffraction mode and you will observe a systematic row pattern.

the foil edge, intensity oscillations are present, and the first fringe is nearly parallel to the edge. We will have to wait until Chapters 6 and 7 to explain the details of this image contrast.

Images 4.19(e) and (f) were obtained by means of the centered dark field method explained in Section 4.4.2.5. The corresponding reflections are labeled in the systematic row pattern 4.19(b). It is important to note that the sample orientation was not changed at all with respect to the bright field orientation. Only beam tilts were used to obtain the two dark field images. We observe a large number of fringes that are nearly parallel to the sample edge close to the edge, but move away from the edge with increasing sample thickness. At the arrowed locations the fringes are at a maximum distance from the edge. Note that the locations of the two arrows do not coincide at the same location in the foil.

If we add the two dark field images together we obtain the computed image shown in Fig. 4.19(d). Note that there is some similarity between this sum and the bright field image; the bright field image appears to be the inverse of the sum image. We say that these two images are *complementary* images. The details of this fringe contrast will have to wait until we cover multi-beam dynamical diffraction theory in Chapter 7.

4.6.2 Tilting towards a zone axis pattern

If you insert a sample in the microscope, turn on the beam, select a random point on the sample and switch to diffraction mode, you will almost never end up with a perfectly symmetric zone axis diffraction pattern. Instead, you will most likely obtain a pattern similar to that shown in Fig. 4.20(a). The reflections visible in this pattern lie along a segment of the Laue circle, corresponding to the intersection of the Ewald sphere with a plane in reciprocal space. The corresponding zone axis orientation can be obtained by "closing the circle". Figs 4.20(b)–(d) show how you can accomplish this.

It is easy to say "close the circle", but you will actually have to learn how to do this. In rare situations, the foil will be oriented such that a simple rotation around either the primary or secondary tilt axis will do the trick. In most cases you will have to use both tilt axes simultaneously to close the circle. This requires first of all an understanding of the orientation of the tilt axes with respect to the diffraction pattern; this you can derive from the calibration measurements described earlier. Then you will need to keep the sample in the same location during tilting, so that you always look at the diffraction pattern corresponding to the location you selected in the image. This can also be tricky, especially when the sample is far from the eucentric position. If your sample has a small grain size, then it will be even more difficult to keep the selected grain under the beam during tilting. There are also

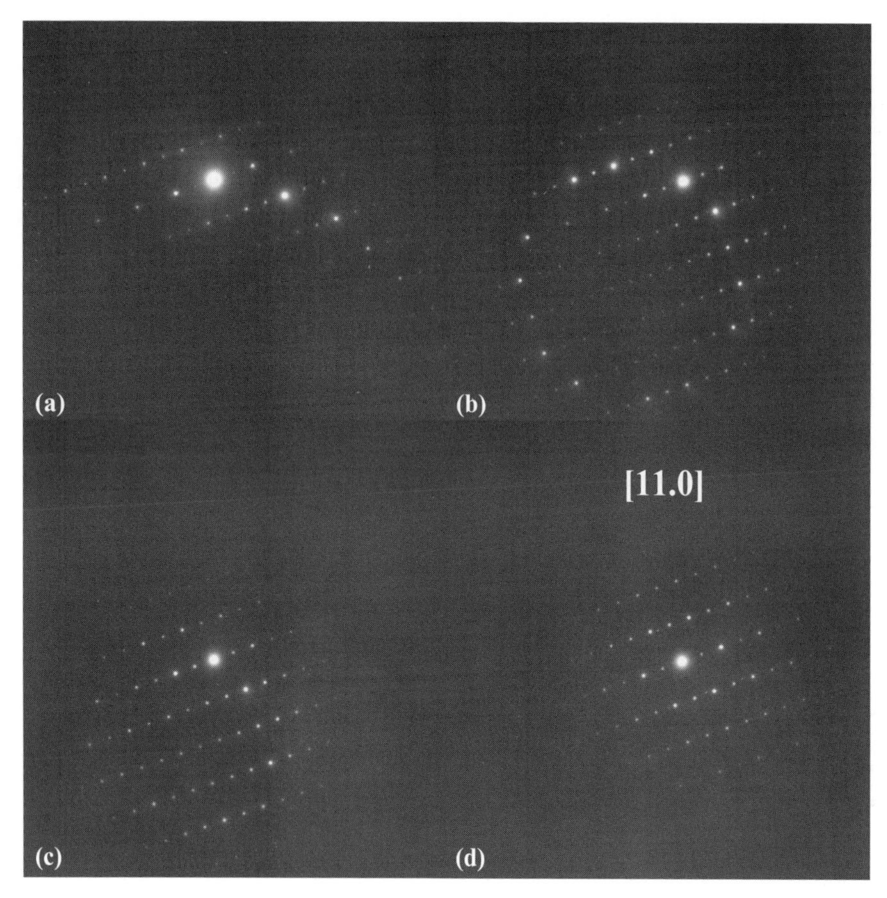

Fig. 4.20. Schematic illustration of tilting a foil by closing the "Laue circle", the intersection of the Ewald sphere with the reciprocal plane normal to the zone axis direction. The [11.0] pattern was obtained from Ti, using a JEOL 2000EX microscope.

significant differences between the tilting of a double-tilt holder and a rotation-tilt holder.

This is where bend contours can come in handy. Assuming that you can see a bend contour near the area you selected in the image, you can tilt the sample while looking at the image rather than the diffraction pattern. All you need to do is keep the bend contour at the same location in the sample during tilting. When you approach the zone axis orientation, a bend center should appear and when you position this bend center at the selected location, the corresponding zone axis diffraction pattern will be symmetric, as shown in Fig. 4.18. If the sample is not significantly bent, or the grain size is too small for bend contours to be visible, then you should tilt until the selected grain (in image mode) becomes very dark with respect to its neighbors, indicating that a large number of electrons are diffracted out of the transmitted beam.

Tilting from one zone axis orientation to another one is similar to the method described above. In image mode you would keep a single bend contour (corresponding to the plane common to both zone axis orientations) in view while you tilt the sample. In diffraction mode you need to keep the reflection corresponding to this common plane as bright as possible while tilting. This requires a coordinated movement of one or both feet (to operate the tilt controls) and both hands (to keep the selected area under the beam). Since tilting often results in a change of the specimen height, the image will go out of focus during tilting and you will need to adjust the image focus periodically. Do not forget to set your sample back to the eucentric specimen height after you have tilted to the correct orientation! All of this requires a lot of practice and you should take your time to learn how to tilt a sample in a controlled way. On modern computer-controlled microscopes some of the tediousness of maintaining eucentric position can be taken over by the microscope computer. However, the chances are that, as a microscopy student, you will not have access to such modern machines, so that it will be necessary to practice, and practice, and practice.

4.6.3 Sample orientation determination

The zone axis diffraction pattern contains significant geometrical information about your crystal. If you obtain two or more zone axis diffraction patterns of a particular region in your foil, then you can transfer the geometrical information onto a stereographic projection which will allow you to index all diffraction patterns in a consistent way. Consider the four bright field images shown in Fig. 4.21, obtained

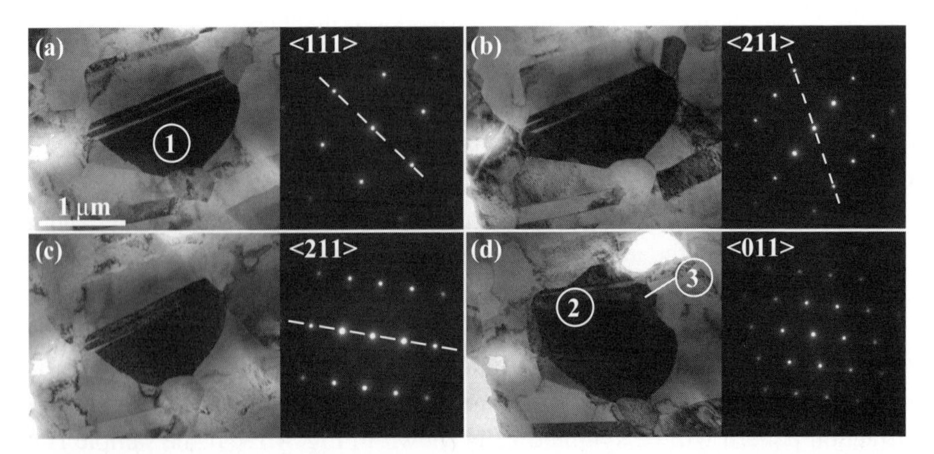

Fig. 4.21. Four different zone axis diffraction patterns and corresponding bright field images of a small grain in a Cu thin foil. The region labeled 1 in (a) was used to obtain all four diffraction patterns. The images were obtained in a Philips EM420, operated at 120 kV, using a Gatan DualView digital camera. The image orientations are as-acquired and image rotations have not been compensated for.

from a fine-grained Cu foil on a Philips EM420 microscope operated at 120 kV. The central grain in (a) is dark and in zone axis orientation, as shown in the corresponding diffraction pattern. By comparison with precomputed zone axis patterns from the xtalinfo.f90 or zap.f90 programs or, alternatively, using the indexing program indexZAP.f90 described in Chapter 9, it is easy to show that this pattern belongs to a zone axis from the $\langle 111 \rangle$ family. The sample orientation angles were read from the goniometer and are equal to $\alpha = 8.2^\circ$ and $\beta = 6.1^\circ$. Bright field images were acquired at a nominal magnification of 30 000×, and the diffraction patterns were recorded at a nominal camera length of 660 mm. The relative orientation of image and diffraction pattern can be derived from Fig. 4.17.

When the sample was tilted around the dashed line in Fig. 4.21(a) we obtained a pattern (b) of the $\langle 211 \rangle$ family at an orientation $\alpha = 28^\circ$ and $\beta = 12.5^\circ$. A second rotation around the axis labeled in Fig. 4.21(b) resulted in another $\langle 211 \rangle$ pattern at $\alpha = -4.7^\circ$ and $\beta = 20.4^\circ$. And a final rotation around the labeled axis resulted in the $\langle 011 \rangle$ pattern at $\alpha = -27.0^\circ$ and $\beta = -6.0^\circ$, as shown in Fig. 4.21(d). The question we will answer next is: how can we index all patterns and reflections in a consistent way? In other words, how do we select the actual family members from the four direction families?

Consider the stereographic projection in Fig. 4.22; the original drawing for this pattern was made on a transparent foil, mounted on top of a Wulff net as shown in Fig. 1.11 on page 32. The primary tilt axis is placed horizontally from **A** to **B**, the arrow indicating the direction in which the sample holder is inserted into the column. From the calibration measurements in Section 4.5.2 we find that the angle between the diffraction pattern obtained at a nominal camera length of 660 mm and the primary tilt axis is 57°; this is indicated to the left of the center in Fig. 4.22. If we take the zone axis pattern of Fig. 4.21(d), rotate it so that its bottom line is parallel to the line on the stereographic projection and the central beam is positioned at the center of the projection, then we can draw lines through each row of reflections (i.e. from $-\mathbf{g}$ to $+\mathbf{g}$ for all reflections close to the origin). If we extend those lines across the stereographic projection we obtain (in this case) six intersections with the projection circle; those points are indicated by small filled circles. The zone axis corresponds to the point at the center of the stereographic projection. The stereographic projections of the reciprocal lattice points in the $\langle 011 \rangle$ pattern in Fig. 4.21(d) are therefore given by the six points on the projection circle in Fig. 4.22, in the proper relative orientation with respect to the primary tilt axis.

Since the sample was tilted from the $\alpha = 0^\circ$, $\beta = 0^\circ$ reference orientation to obtain the $\langle 011 \rangle$ zone axis pattern we must now correct for this tilt. First, we correct for the α tilt angle by aligning the M'–M'' axis of the Wulff net (see Fig. 1.10) with the primary tilt axis. Then we rotate all six projection points along the arcs centered on the points M' and M'' *by an angle opposite to the tilt angle*. In this case, the α tilt

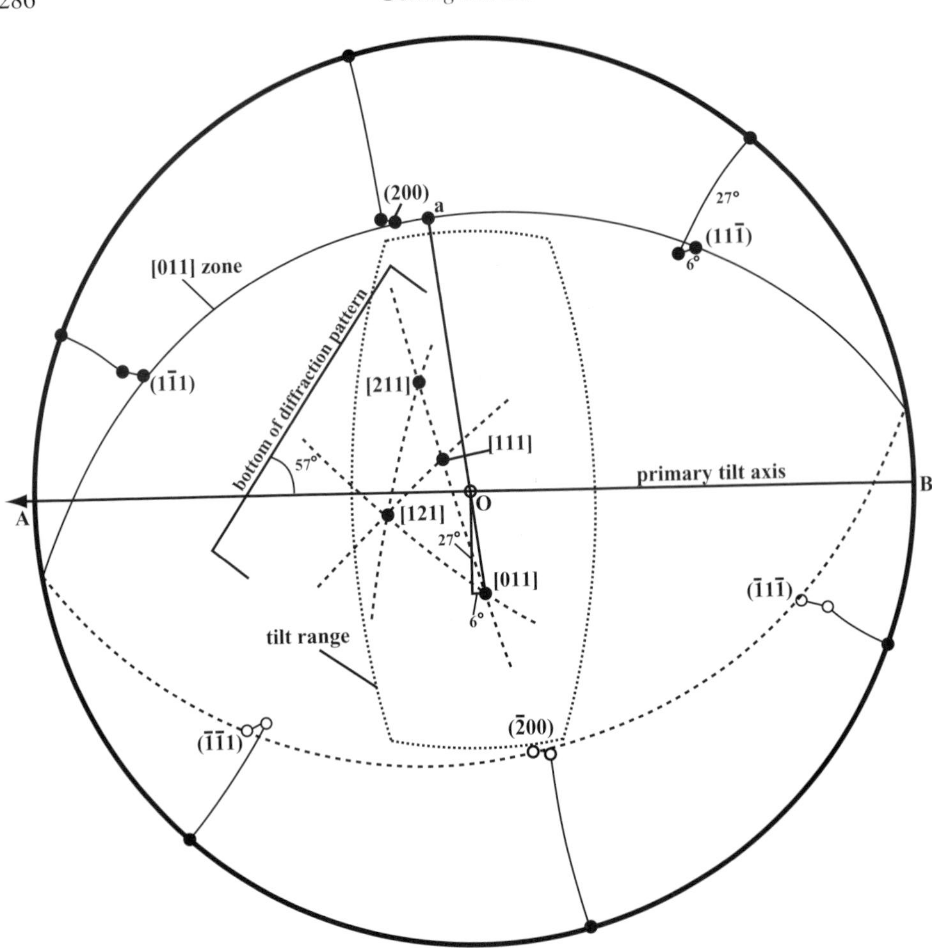

Fig. 4.22. Construction of the stereographic projection of the $\langle 011 \rangle$ zone axis pattern of Fig. 4.21(d).

angle is -27°, so we tilt by $+27^\circ$ and obtain six new points. We also tilt the central point of the projection by $+27^\circ$. Points in the lower hemisphere are indicated by open circles. The second tilt can be corrected by first rotating the Wulff net by 90°, and then performing the same operation on all points. This time the rotation is $-\beta = 6.0^\circ$.

The six points obtained in this way lie on a great circle, and we can draw this circle on the projection, using the Wulff net. The pole of this great circle is the projection of the zone axis, labeled [011]. If we repeat this procedure for the other three zone axis patterns of Fig. 4.21 we obtain the four zone axis projections shown near the center of Fig. 4.22. Next we select the zone axis direction of the $\langle 111 \rangle$ family to be the [111] direction. Comparison of the locations of the other poles with a standard stereographic projection (for instance, you could use the stereo.f90 program to draw

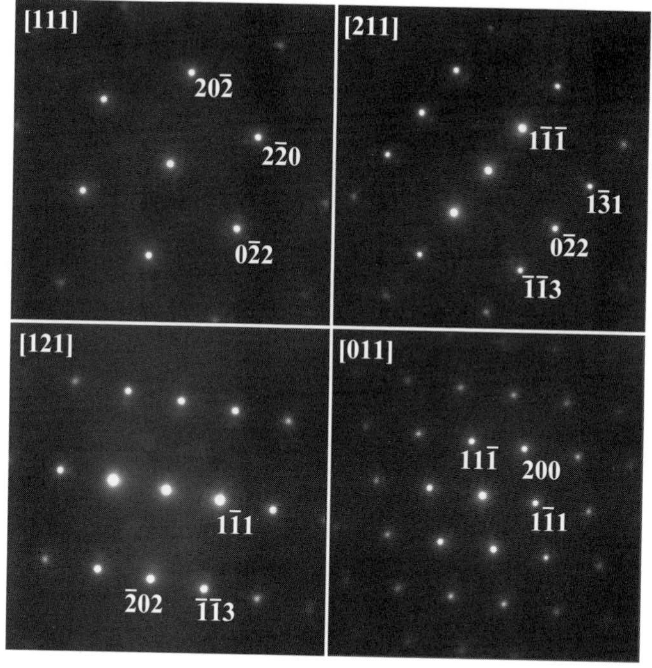

Fig. 4.23. Consistently indexed diffraction patterns of Fig. 4.21(a)–(d).

a [111] projection) shows that the other poles are consistently indexed by selecting [111], [211], [121], and [011] for the patterns of Figs. 4.21(a)–(d), respectively.

The last step involves determination of the Miller indices of all reflections of the four patterns. Since we know which reflections were used to tilt from one zone axis orientation to the next one, we can simply use the cross product of directions to determine the Miller indices of the reflection on the rotation axis. As an example, the plane common to both the [111] and [211] zones is the $(0\bar{2}2)$ plane. Figure 4.23 shows the four diffraction patterns again, this time with the proper consistent Miller and direction indices indicated. Once the $0\bar{2}2$ reflection is labeled, the other reflections in this pattern can be identified readily by comparison with a precomputed pattern. The same cross product procedure identifies that the $\bar{1}\bar{1}3$ reflection lies on the rotation axis between the [211] and [121] zones, and the $1\bar{1}1$ reflection is common to [121] and [011]. This allows us to unambiguously and consistently index every reflection of the four patterns.

This completes the indexing procedure. The grain orientation is completely and unambiguously determined with respect to the primary tilt axis, and hence with respect to the microscope column. With care you should be able to determine grain orientations to within a degree or so. The procedure outlined in this section is important for defect identification. As we will see in Chapter 8, for many defect types it is necessary to obtain a series of two-beam bright field–dark field image pairs for three or more different reciprocal lattice vectors **g**. It is not sufficient to simply know the family {**g**} of the reflection; we must determine *which member* of the family corresponds to the active reflection, so that all **g**-vectors are known unambiguously. The reader is encouraged to practice the procedure outlined in this section; first, start on a large-grained material, or even on the single-crystal GaAs study material, so that you do not have to worry about losing the region of interest when tilting the sample. Then, when you have mastered the procedure, you can work on more difficult situations, for instance a small-grained material, such as the one used for Fig. 4.21.

4.6.4 Convergent beam electron diffraction patterns

All images and diffraction patterns obtained so far used a parallel incident beam, which means that the second condensor lens was used in an under-focus condition. It is also possible to use a small spot size† and a large condensor aperture to obtain a conical incident beam. Observations made with such a conical beam are known as *convergent beam* observations.

We will describe the full geometry and symmetry of the convergent beam electron diffraction (CBED) pattern in Chapters 6 and 7. For now, we will simply illustrate how CBED patterns can be obtained for systematic row and zone axis orientations. We will assume that you have inserted one of the study materials in your microscope, and that you have obtained either a systematic row orientation or a zone axis orientation. Most likely you obtained this pattern using a large spot size (i.e. a small spot size number), and an under-focused second condensor lens, to generate a parallel incident beam. Next, carry out the following steps:

- in image mode, set the first condensor lens to one of the smallest spot sizes, and correct the condensor astigmatism;
- focus the image using the objective focus control (you may have to use the binoculars to do this, since the image might be quite dim);
- focus the second condensor lens to cross-over (i.e. the smallest possible illumination spot), and center the spot on the region of interest;
- switch to diffraction mode and remove the diffraction aperture.

The resulting pattern is a convergent beam diffraction pattern. Next, try each of the following modifications.

- Change the diameter of the condensor aperture; the radius of the diffraction disks will change, because you are changing the range of incident beam angles (see Fig. 3.27b). If

† On the microscope console this usually means a high spot size number. On a JEOL 2000EX microscope for instance, spot size 8 results in a much narrower beam than spot size 2, as illustrated in Fig. 4.8(c).

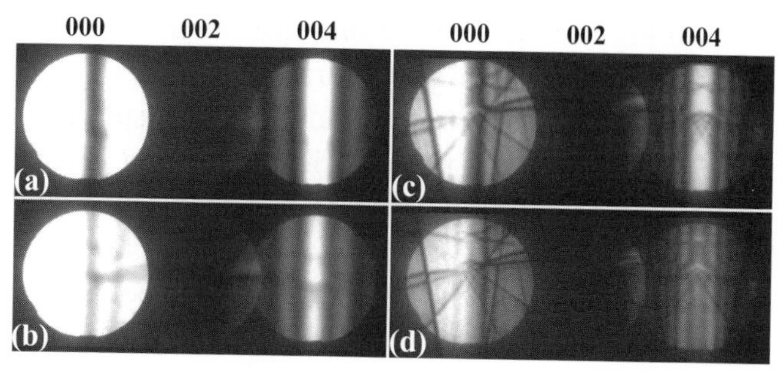

Fig. 4.24. $\mathbf{g}_{002}$ systematic row CBED pattern for GaAs, obtained on a Philips EM420 microscope operated at 120 kV. The selected condensor aperture corresponds to a beam convergence angle of $\theta_c = 6.6$ mrad, slightly larger than the Bragg angle $\theta_{002} = 5.92$ mrad. The foil thickness increases from (a) to (d).

you completely remove the condensor aperture from the beam path,† you will observe a complex pattern of lines and intensity variations inside a large-diameter central disk. You may need to reduce the image intensity to observe this pattern, in particular if you use a digital camera to record the image, but you should not use the second condensor lens control to do this! Why not? In CBED mode, image intensity should be changed by means of the filament emission control.

- Move the condensor aperture using the aperture translation controls; note that the location of the diffraction disks changes, but the intensity distribution remains in the same location. If you have a zone axis diffraction pattern you can use this method to precisely position the central disk at the zone center.
- Change the beam tilt (you may want to do this using the dark field memory setting); both the location of the diffraction disks and the intensity distribution inside them changes. You can use this to precisely align the beam along a given direction, and then return the central beam to the center of the screen using the projector alignment.
- Change the sample orientation using the primary tilt axis; this has the same result as changing the beam tilt, except that the diffraction disks remain in the same location during tilting.
- Change the second condensor lens current or the objective lens current and observe how that changes the CBED pattern. There is only one current setting for which the resulting pattern is a true CBED pattern, and you should experiment with both lens currents to discover for which setting the pattern is obtained.

If you started from a systematic row orientation, then your CBED pattern may be similar to the one shown in Fig. 4.24, which shows a section of the $\mathbf{g}_{002}$ systematic row of GaAs for four different thicknesses. The patterns were obtained on a Philips EM420, operated at 120 kV, equipped with a Gatan DualView camera system. The Bragg condition is satisfied for the $\mathbf{g}_{004}$ reflection, and the $\mathbf{g}_{002}$ reflection is

† It is not always possible to completely remove the condensor aperture from the column. This depends on the microscope model. Without a condensor aperture, a large number of x-ray photons are generated in the column.

barely visible. From Fig. 4.5 (part 2) we find that this is in agreement with the relative magnitudes of the corresponding Fourier coefficients $V_{002}/V_{004} = 0.127$. The number of vertical fringes in the 004 diffraction disks increases with increasing sample thickness. We also note that the intensity distribution in the central disk is asymmetric: the left half is significantly brighter than the right half, in particular for the thickest foil. Chapters 6 and 7 will explain the details of the contrast variations in these CBED patterns. For now, you should familiarize yourself with obtaining and recording these patterns.

If you started from a zone axis orientation, then your pattern might be similar to those shown in Fig. 4.25. The patterns correspond to a [110] zone axis of

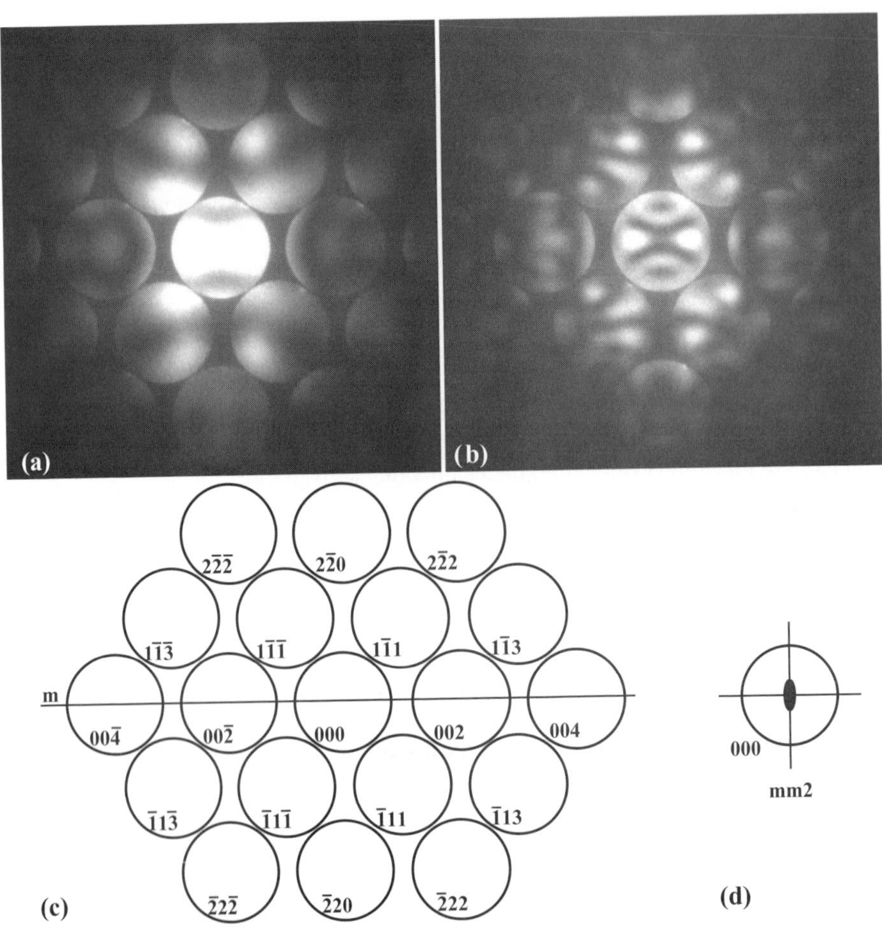

Fig. 4.25. (a) and (b) Zone axis CBED patterns of the [110] orientation of GaAs, recorded at 120 kV on a Philips EM420, for two different foil thicknesses. Part (c) shows a schematic of the patterns with Miller indices labeling each disk; (d) shows the internal symmetry of the central disk.

GaAs, again obtained at 120 kV in a Philips EM420, for two different sample thicknesses. The beam convergence angle is 5.1 mrad, just below the Bragg angle $\theta_B =$ 5.13 mrad for the $\mathbf{g}_{111}$ reflections. With increasing foil thickness, more details become visible within the disks. The disks are indexed in Fig. 4.25(c). From the pattern in Fig. 4.25(b) we see that the overall pattern has a horizontal mirror plane, and therefore has planar point group symmetry **m**. The central disk itself has an additional vertical mirror plane, as shown schematically in Fig. 4.25(d), and hence has planar point group symmetry **mm2**. The intensity distributions inside the $1\bar{1}\bar{1}$ and $1\bar{1}1$ disks are not identical and this difference is due to the non-centrosymmetric nature of the GaAs crystal structure.

4.7 Lorentz microscopy: observations on magnetic thin foils

4.7.1 Basic Lorentz microscopy (classical approach)

As described in equation (3.7) on page 146, an electron moving through a region of space with an electrostatic field **E** and a magnetic field **B** experiences a velocity-dependent force, commonly known as the *Lorentz force* $\mathbf{F}_L$:

$$\mathbf{F}_L = -e\,(\mathbf{E} + \mathbf{v} \times \mathbf{B}). \tag{4.2}$$

We will assume that all fields are static. Whereas Chapter 3 dealt with the behavior of electrons in the magnetic field of round lenses, deflectors, and stigmators, in this section we will describe how the electron interacts with the magnetic field caused by the sample itself.

Since the magnetic component of the Lorentz force acts normal to the travel direction of the electron, a *deflection* will occur. Only the component of **B** normal to **v** will contribute to the deflection, and we will refer to this component as the *in-plane magnetic induction* $\mathbf{B}_\perp$, i.e.

$$\mathbf{B} = \mathbf{B}_\perp + B_z\mathbf{n}, \tag{4.3}$$

where **n** is a unit vector parallel to the beam direction. Since only two out of the three components of **B** act on the electron trajectory, it is clear *a priori* that a complete determination of all three components will require the use of at least two independent incident beam directions **n**.

While it is possible to mount the sample in a field-free region (see Section 4.7.2) it is in general not possible to completely remove the electrostatic field **E**, since the crystal structure itself generates an electrostatic lattice potential. The magnetic deflection caused by the Lorentz force does not change the electron energy, whereas the (positive) electrostatic component causes the electron to accelerate which changes its kinetic energy (refraction).

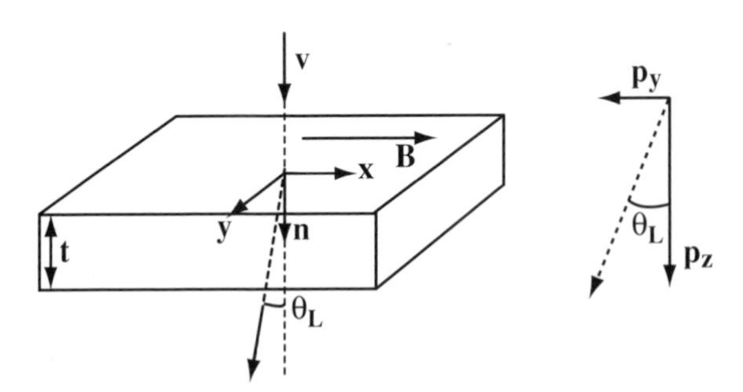

Fig. 4.26. Schematic of a magnetic thin foil and the resulting deflection of an incident electron beam.

Consider a planar thin foil of thickness t, normal to the incident beam. If the in-plane magnetic induction is directed along the x-axis (see Fig. 4.26) with magnitude $B_\perp$, then the momentum transferred to the electron is given by

$$p_y = \int_0^\tau F_L \,\mathrm{d}\tau, \tag{4.4}$$

with τ the time it takes for the electron to traverse the sample. In the absence of electrostatic fields, the magnitude of the Lorentz force along the y direction is given by

$$F_L = ev_z B_\perp. \tag{4.5}$$

The integral can be rewritten in terms of the sample thickness $t = v\tau$ as

$$p_y = e\int_0^t B_\perp \,\mathrm{d}z = eB_\perp t. \tag{4.6}$$

The deflection angle θ_L is then to a good approximation (Fig. 4.26) given by the ratio of the y and z momentum components:

$$\theta_L = \frac{p_y}{p_z} = \frac{eB_\perp t}{mv}, \tag{4.7}$$

where m is the relativistic electron mass. This equation can be rewritten in a more useful form by means of the de Broglie relation between particle momentum and wavelength ($p = h/\lambda$), and we find for the Lorentz deflection angle θ_L:

$$\theta_L = \frac{e\lambda}{h} B_\perp t = C_L(E) B_\perp t. \tag{4.8}$$

The constant $C_L(E)$ is determined by the acceleration voltage E of the microscope, and is given by

$$C_L(E) = \frac{9.377\,83}{\sqrt{E + 0.974\,85 \times 10^{-3} E^2}} \quad \mu\mathrm{rad\,T^{-1}\,nm^{-1}};$$

(to obtain the Lorentz angle in μrad, $B_\perp$ must be stated in tesla, t in nm and E in kV). For the commonly used acceleration voltages we have: $C_L(100) = 0.895\,018$, $C_L(200) = 0.606\,426$, $C_L(300) = 0.476\,050$, $C_L(400) = 0.397\,511$, and $C_L(1000) = 0.210\,834$ μrad T^{-1} nm^{-1}. A 100 nm thin foil with an in-plane magnetic induction of $B_\perp = 1$ T, will give rise to a beam deflection of $\theta_L = 39.75$ μrad at 400 kV. For comparison, typical Bragg angles for electron diffraction are in the range of a few *milli*radians, i.e. two to three orders of magnitude larger than Lorentz deflection angles. Note that the sample thickness and the in-plane induction component both appear in the expression for the deflection angle. Since the sample thickness is notoriously difficult to quantify experimentally, it should be clear that any measurement based on the Lorentz deflection angle will at best be qualitative and provide only the product $B_\perp t$. Localized thickness variations and local out-of-plane excursions of the magnetic induction will produce identical changes in the Lorentz deflection angle θ_L. Only when an independent thickness measurement is available can Lorentz methods provide a direct map of the in-plane induction component.

4.7.2 Experimental methods

Direct observations of the magnetic substructure of a material in a TEM require that a thin foil be made, using appropriate thinning procedures. One must always be aware that the magnetic structure of a thin foil may be different from the bulk magnetic structure. In particular, in soft magnetic materials, the introduction of two free surfaces may completely change the energetics of magnetic domains, and, indeed, the very nature of the magnetic domain walls.

In order to preserve the magnetic microstructure of the thin foil, the sample must be mounted in a field-free region in the microscope column. Since the objective lens in a standard TEM is an immersion-type lens, the requirement of a field-free region has profound consequences on the electron optical properties of the microscope. The low-field sample environment and consequent increase in the focal length of the objective lens result in a reduced final image magnification compared with that of conventional transmission electron microscopy. This represents a serious limitation in the quantitative study of nanoscale magnetic structures. Furthermore, inelastic scattering in the sample contributes noise to the images, which further limits the attainable resolution of the standard Lorentz modes.

A number of different sample + lens configurations can be used.

- Mount the sample *above* the main objective lens. This generally requires the introduction of a second goniometer stage in between the condensor and objective lenses, and is therefore only attempted in dedicated instruments.
- Use a dedicated low-field Lorentz pole piece. This is perhaps the most efficient and reliable method, but may require a pole piece change, which is not always an easy thing

to do. Furthermore, if the microscope is used for many different types of materials, the loss of magnification due to the reduced field may be unacceptable to other users. A permanent Lorentz pole piece downstream from the sample holder is therefore preferable, assuming it does not deteriorate the electron optical properties of the column when used in conventional or high-resolution observation modes.

- Turn the objective lens off and use an objective minilens (if present) to obtain a back focal plane at the location of the selected area aperture. If a post-column energy filter is available, the loss of image magnification may be partially compensated by the internal magnification of the filter [DDG97].

All images shown in this section were obtained on a JEOL 4000EX top-entry high-resolution TEM, operated at 400 kV in *Low Mag* mode (with the main objective lens turned off). The objective minilens provides a cross-over near the selected area aperture plane, at a maximum magnification of 3000×. A Gatan imaging filter (GIF) is then used to remove most inelastically scattered electrons (using an energy selecting slit width of 20 eV), and to magnify the image by an additional factor of about 20×. All images were acquired on a 1K×1K CCD camera, and are gain normalized and background subtracted. Unless mentioned otherwise, no additional image processing was carried out on the experimental images.

4.7.2.1 Fresnel mode

The Fresnel or *out-of-focus* Lorentz mode is perhaps the easiest observation method to carry out experimentally, since it only requires a change in the lens current for the main imaging lens (either a Lorentz lens or an objective minilens). The method is schematically shown in Fig. 4.27. Consider a sample with three magnetic domains, separated from each other by 180° magnetic domain walls. The character of the walls is unimportant at this point. The outer domains have their magnetization point

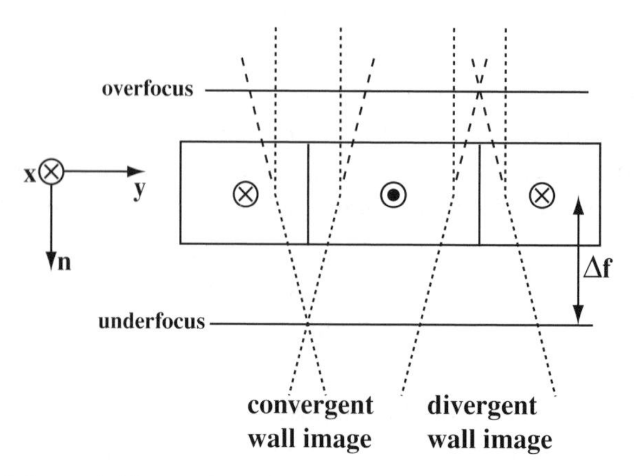

Fig. 4.27. Schematic illustration of the Fresnel or out-of-focus imaging mode.

into the plane of the drawing, and the central domain has the opposite magnetization direction. The resulting trajectory deflections are towards the positive y-direction for the outer domains.

A standard bright field image, obtained by selecting only the transmitted beam with a sufficiently large aperture in the back focal plane of the imaging lens, does not show any contrast (other than diffraction or absorption contrast) for an *in-focus* condition. In other words, when the object plane of the imaging lens is located at the sample position, no magnetic contrast is observed. If the current through the imaging lens is reduced, the focal length increases (causing a reduction in the lateral image magnification), and the object plane is located at a distance Δf *below* the sample. Because of the small Lorentz deflection angles, a rather large *defocus* Δf is needed to obtain an overlap between the electrons deflected on either side of the left domain wall in Fig. 4.27. In the overlap region, an excess of electrons will be detected and this will show up in the image as a bright line feature at the projected position of the domain wall. The domain wall on the right will show up as a dark feature, since electrons are deflected away from the domain wall position. The bright domain wall image is known as the *convergent wall*, the dark wall is the *divergent wall*.

If the imaging lens current is increased, so that the focal length is reduced and the image magnification increased, then an over-focus image is obtained. The image contrast of convergent and divergent domain walls reverses, as can be seen from the extended electron trajectories in Fig. 4.27.

Figure 4.28 shows a set of experimental Fresnel images for a cross-tie wall arrangement in a 50 nm thick Ni-20 wt% Fe Permalloy film. The in-focus image is shown in (a), while the under-focus and over-focus images are shown in (c) and (d), respectively. A schematic of the magnetization configuration around the cross-tie wall is shown in (b). Note that this configuration can not be deduced from the Fresnel images; the Foucault images shown in the next section (Fig. 4.32) were used to derive the magnetization arrangement shown in the schematic. The bright points along the main domain wall (Fig. 4.28c) correspond to the centers of regions around which the magnetization has a circular pattern (vortices); this arrangement acts as a lens which focuses the electrons into a single spot. The magnetization at the vortex is predominantly out-of-plane (see e.g. [Chi78, pages 402–408]), and opposite to the magnetization at the intersection of the main wall and the divergent walls in Fig. 4.28(c). This divergent wall is known as a *cross-tie wall*. Such walls gradually disappear with increasing distance from the main 180° domain wall.

Since the Fresnel image mode highlights the domain walls in the thin foil, it should be possible to derive from the images an estimate of the domain wall width δ. This parameter plays a central role in the energetics of magnetic materials, and a direct experimental measurement method would be an extremely valuable tool.

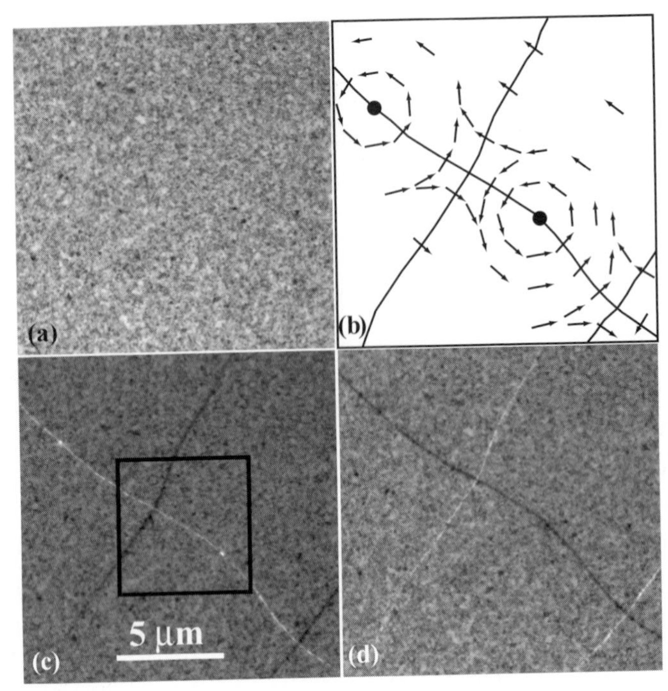

Fig. 4.28. (a) In-focus, (c) under-focus, and (d) over-focus images (zero-loss filtered with 20 eV slit width) of cross-tie domain walls in a 50 nm thick Ni-20 wt% Fe Permalloy film. Note that the image magnification changes slightly with defocus. The magnetization pattern in (b) is derived from the Foucault images in Fig. 4.31. (Sample courtesy of Chang-Min Park.)

One can use the divergent wall image to obtain an estimate of the domain wall width δ; a reasonable estimate of δ is obtained when the measured width of a divergent wall is plotted against the defocus value and extrapolated to zero defocus (e.g. [SWP68, SW69, Woh71]). This extrapolation method works well for large domain wall widths, but overestimates δ for the narrow walls found in hard magnetic materials. A discussion of the accuracy of the extrapolation method can be found in [GC87].

The classical Lorentz deflection theory provides a good first approximation for the understanding of image contrast in the Fresnel imaging mode. However, when a more coherent electron beam is used (for instance, by under-focusing the second condensor lens), then the convergent wall image reveals the presence of interference fringes which cannot be explained by the classical model. Figure 4.29 shows a sequence of under-focus images of a 180° domain wall (upper right-hand corner) in a permalloy thin film at increasing defocus Δf; all images were taken with an under-focused second condensor lens, or, equivalently, a reasonably coherent electron beam. The traces below the micrographs are averages over 40 pixels along

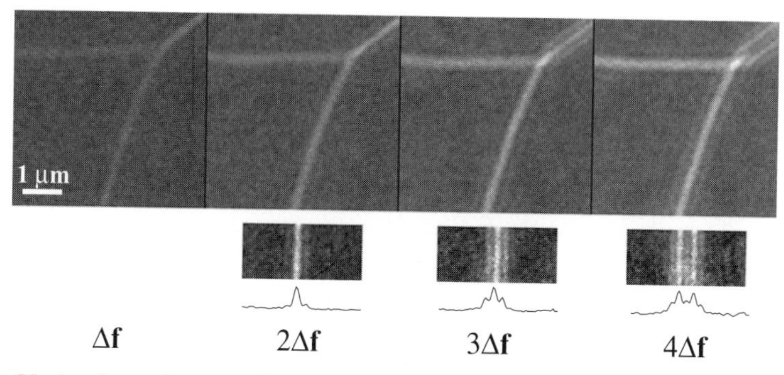

Fig. 4.29. Under-focus images of a 180° domain wall (upper right) which splits into two 90° domain walls in a 50 nm Ni-20 wt% Fe Permalloy thin film. The images were obtained with a strongly under-focused second condensor lens, for increasing under-focus value Δf. The fringe contrast is clearly visible; intensity profiles across the 180° domain wall are shown at the bottom.

the wall profile. While the number of fringes increases with increasing defocus, the spacing between the fringes appears to be constant. The classical model for the Fresnel observation mode fails to explain the presence of such fringes. We will return to the origin of the fringes in Section 7.5.2; image simulations for coherent Fresnel images will be discussed in Section 10.4.1.

While the Fresnel imaging mode is easy to use and provides a direct, qualitative image of the domain wall configurations in the sample, its quantitative usefulness is rather limited because of the sometimes large magnification changes caused by the defocus. However, in Chapter 10 we will introduce a method, based on the Fresnel mode, that can be used to determine quantitatively the magnetic microstructure.

4.7.2.2 Foucault mode

A second commonly used observation mode employs an in-focus image and is somewhat similar to the *dark field* imaging mode. As illustrated in Fig. 4.30 an aperture is introduced in the back focal plane of the imaging lens (an objective minilens or a dedicated Lorentz lens), and by translating the aperture normal to the beam a section of the split central beam is cut off. The regions in the sample that give rise to deflected electrons, which are passed by the aperture, will appear bright in the image. This observation mode produces high-contrast in-focus images of magnetic domain configurations, which are known as *Foucault images*. Since a typical aperture stage has two translation controls which are normal to each other, one would acquire typically either two or four images, one or two for each main aperture translation direction. An example of Foucault images for the same sample region as shown in Fig. 4.28 is shown in Fig. 4.31; the approximate aperture shift directions are indicated by white arrows in the upper left-hand image. Since we

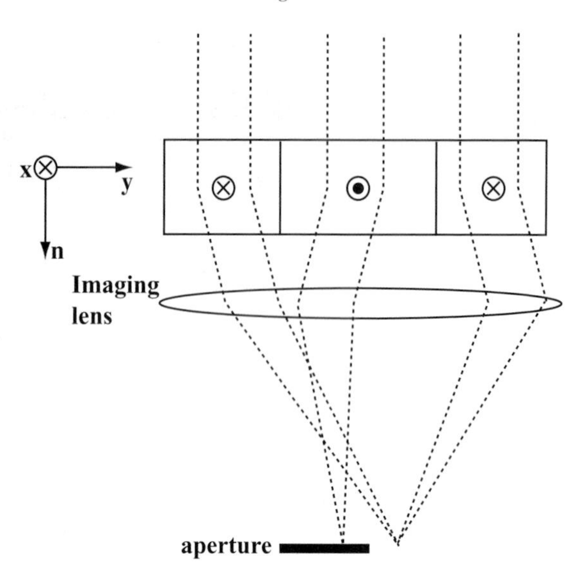

Fig. 4.30. Schematic illustration of the Foucault imaging mode.

know that the direction of the Lorentz deflection is normal to the corresponding in-plane induction direction, we can use the four images in Fig. 4.31 to obtain a schematic (qualitative) drawing of the magnetization configuration, which is shown in Fig. 4.28(b).

When more coherent illumination is used to obtain the Foucault images, fringes are often observed parallel to the domain walls, inside the bright portions of the image (e.g. [DDG97, Fig. 7]). The origin of these fringes is the same as for the Fresnel mode and will be addressed in more detail in Section 7.5.2.

The Foucault mode is a highly qualitative observation mode, because it is notoriously difficult to reproducibly position the aperture. Apertures are often not perfectly circular, or may have debris at the aperture edge. When a set of four Foucault images is acquired, it is difficult if not impossible to manually (for mechanically controlled apertures) position the aperture at equal shifts from the central beam position. This would be a prerequisite for quantitative work. It is also a nontrivial task to determine the location of the optical axis, i.e. the position of the beam in the absence of a magnetic sample. This position would be the "origin" for all Foucault images, and one would have to specify the magnitude of the aperture shift (typically in nm^{-1}) with respect to this origin.

4.7.2.3 Diffraction mode

Figure 4.32(a) shows an electron diffraction pattern obtained from the divergent part of a cross-tie wall in a 50 nm Permalloy thin film. Since the saturation induction

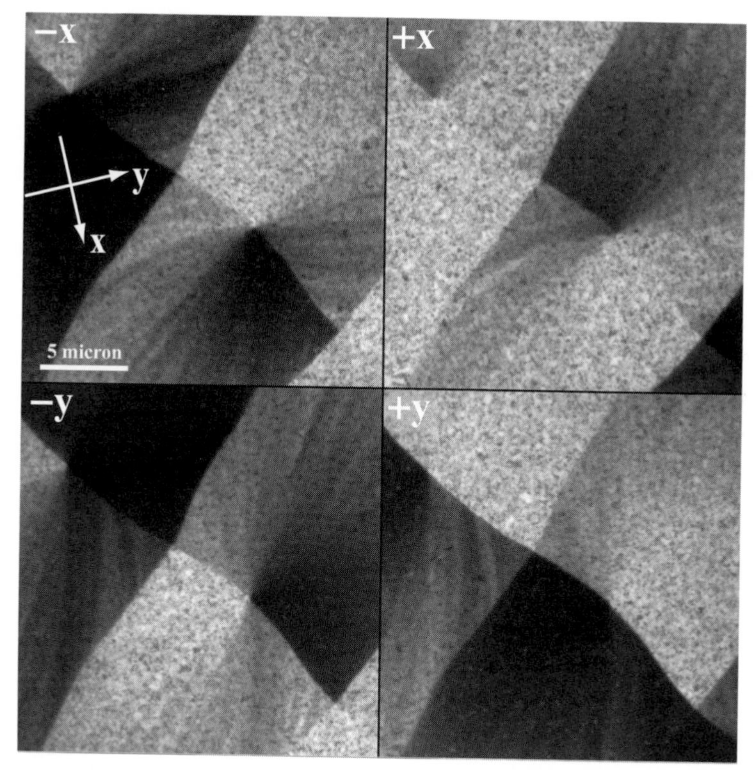

Fig. 4.31. Foucault images (zero-loss filtered with 20 eV slit width, 400 kV) for four aperture shifts for the same sample region as Fig. 4.28 ($\pm x$ and $\pm y$ directions, with respect to the aperture directions, indicated by the white arrows). The magnitude of the aperture shift is approximately equal for all four images.

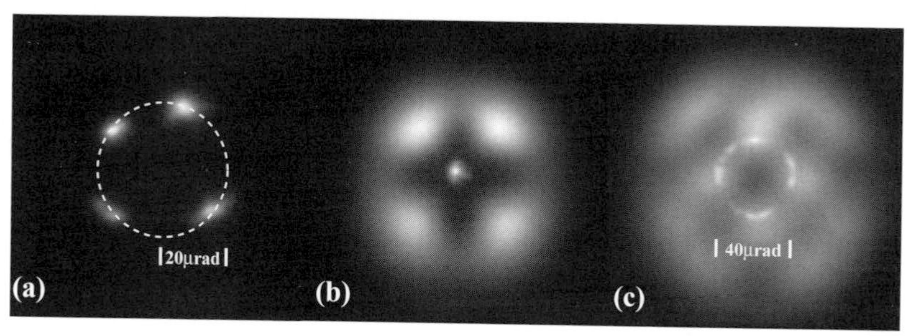

Fig. 4.32. (a) Electron diffraction pattern for the beam centered on the divergent part of a cross-tie wall; each spot corresponds to the deflection from one of the quadrants of the magnetization pattern. (b) Image of the undersaturated filament tip (LaB_6), for the beam passing through a single magnetic domain; (c) same as (b), but now the beam passes through the convergent part (vortex) of a cross-tie wall.

of this film is about 1 T, the Lorentz deflection angle at 400 kV is about 20 μrad. The pattern was acquired directly on the CCD camera of the energy filter, using an under-saturated filament to prevent over-exposure of the camera. An exposure time of 0.1 s was used, along with an energy selecting slit width of 20 eV. The pattern is obtained by first focusing the second condensor lens so that the gun cross-over is conjugate to the specimen, and then adjusting the objective minilens current to make the viewing screen (CCD plane) conjugate to the specimen. The lateral magnification of the diffraction pattern can be adjusted by using the free-lens control on the microscope, or the magnification reduction options on the energy filter.

Figure 4.32(b) shows an image of the filament tip (LaB_6) in an under-saturated condition, when the beam passes through a single magnetic domain. The beam is deflected from where it would be in the absence of a magnetic sample, but that precise location is difficult to determine experimentally. In Fig. 4.32(c) the incident beam is focused on the vortex part of a cross-tie wall, and a nearly continuous circular intensity distribution is obtained, along with a complex distribution from the outer parts of the filament emission pattern. This pattern contains all the information about the momentum transfer due to the sample magnetization, but it is complicated due to the fact that all magnetization directions around the vortex contribute *simultaneously* to the image. If a fine electron probe were scanned across the region containing the cross-tie vortex, then a synchronized measurement of both magnitude and direction of the deflection angle would provide a direct magnetization map of the sample (actually, a map of the product $B_{\perp}t$). This is essentially the so-called *differential phase contrast* (DPC) acquisition mode [CBWF78].

4.8 Recommended additional reading

Before proceeding with the theory of dynamical elastic scattering, we conclude this chapter with a brief list of additional reading material that the reader may find interesting. The following books (some of them are quite old and out-of-print but you may find them in your library) contain information on basic microscopy techniques and sample preparation:

- *Transmission Electron Microscopy and Diffractometry of Materials*, by B. Fultz and J.M. Howe, 2001 [FH01].
- *Transmission Electron Microscopy, a Textbook for Materials Science*, by D.B. Williams and C.B. Carter, 1996 [WC96].
- *Transmission Electron Microscopy: Physics of Image Formation and Microanalysis*, by L. Reimer, 1993 [Rei93].
- *Specimen Preparation for Transmission Electron Microscopy of Materials*, by P.J. Goodhew, 1984 [Goo84].

- *Electron Microscopy of Thin Crystals*, by P.B. Hirsch, A. Howie, R.B. Nicholson, D.W. Pashley, and M.J. Whelan, 1977 [HHN+77].
- *Practical Electron Microscopy in Materials Science*, by J.W. Edington, 1976 [Edi76].
- *Computed Electron Micrographs and Defect Identification*, by A.K. Head, P. Humble, L.M. Clarebrough, A.J. Morton, and C.T. Forwood, 1973 [HHC+73].
- *Introduction to Electron Microscopy*, by C.E. Hall, 1953 [Hal53].

At this point we leave the experimental observations, and we set out to compute and explain the details of the bright field and dark field images observed thus far. This will require a fair number of mathematical calculations, but as a result we will find that many contrast features can be explained satisfactorily in terms of only a small number of parameters.

Exercises

4.1 Ask one of your colleagues to intentionally misalign the microscope that you plan to use. Then, carefully restore the alignment to what it is supposed to be. Note that you shouldn't do this on a machine that is heavily used by more experienced users.

4.2 Derive the scaling factor needed to convert a length in the sample to a number of pixels on a digital image of this sample. In other words, how many pixels of the image correspond to a given distance on the sample? You can use this relation to create a table which expresses the number of pixels corresponding to different sample lengths at all commonly used microscope magnifications.

4.3 Compute the structure factor for Ti and verify that the reflections circled in Fig. 4.18(b) are indeed forbidden reflections.

4.4 Perform a magnification and rotation calibration for the microscope that you plan to work on for the foreseeable future. Estimate the accuracy of the magnification calibration. Create a rotation chart similar to the one constructed in Fig. 4.17.

4.5 Perform a calibration of the camera lengths on the same microscope. Determine the relative orientation of the diffraction pattern and corresponding image, including the 180° ambiguity. Add this calibration data to the rotation chart that you constructed in the previous exercise.

4.6 The relation between the distance measured on a diffraction pattern and the corresponding distance in reciprocal space does not take into account the fact that the Ewald sphere is curved. Derive a corrected expression and determine when the corrections become important (e.g. when the corrections are larger than 5% of the measured values).

4.7 Record a series of micrographs similar to that shown in Fig. 4.19. Use a small selected area aperture to obtain systematic row diffraction patterns for different aperture locations with respect to the bend contour. Determine which portions of the bend contour correspond to the so-called *two-beam* conditions (i.e. only two strongly excited beams present in the systematic row).

4.8 Use the xtalinfo.f90 program (or any equivalent program) to compute low-index zone axis patterns for the study material Cu-15 at% Al. Then obtain a low-index zone axis orientation in the microscope, write down the tilt angles, record both a bright field image and a diffraction pattern. Use either bend contours or Kikuchi lines (see Section 9.4.3) to tilt to at least two other zone axis orientations for the same area of the sample. Record diffraction patterns and write down the tilt angles. Then, after you develop the negatives, index the zone axis patterns using the precomputed patterns, and transfer the information to a stereographic projection. Determine the poles of the $\langle 100 \rangle$ family. Determine the direction indices of the foil normal (hint: if your foil is not too strongly bent, you may assume that at zero tilt the incident beam is parallel to the foil normal).

4.9 Record a zone axis diffraction pattern for any one of the four study materials. Then tilt the sample by 1° or 2° around the primary tilt axis and record the pattern. After you develop the patterns, find a way to calculate the tilt angle between the two patterns. (Hint: use the position of the Laue center to find an expression for the tilt angle.)

4.10 Obtain a $\mathbf{g}_{111}$ systematic row CBED pattern for materials Cu-15 at% Al and GaAs. Convince yourself that the symmetry of the $\mathbf{g}$ and $-\mathbf{g}$ disks is different for the non-centrosymmetric structure, but identical for Cu-15 at% Al.

4.11 Use study material Cu-15 at% Al, Ti, or $BaTiO_3$ to determine its average grain size (use the calibrated magnification values). (You may wish to read about the ASTM (American Society for Testing and Materials) standards for grain size measurements [fTM98] before you start this exercise.)

5

Dynamical electron scattering in perfect crystals

5.1 Introduction

In the previous chapter, we have seen that electron micrographs can display a rich variety of contrast features: thickness fringes, bend contours, intersections of bend contours, defects, and so on. Now that we know the basic experimental techniques to obtain the electron micrographs, we must take the next step and attempt to *explain* all of these contrast features, for both images and electron diffraction patterns. We have prepared the foundation for a theoretical description of electron scattering through the derivation of the relevant Schrödinger equation in Chapter 2. Now we must solve the equation for an arbitrary electrostatic lattice potential and an arbitrary electron wavelength.

We know from the experiments in the preceding chapter that there are two basic observation modes: two-beam observations, whereby only one diffracted beam has appreciable intensity, and multi-beam observations, where the electron beam is oriented close to a crystal zone axis, such that a plane in reciprocal space is nearly tangent to the Ewald sphere. We will begin this series of theoretical chapters by first solving the equations for the most general case (i.e. the multi-beam case). Then we will describe the two-beam case as a special case. While it is possible to approach the elastic scattering problem in the opposite way, i.e. first deal with the two-beam case and then generalize the results to the multi-beam case, the method followed in this and the next two chapters systematically deals with all important aspects of multi-beam scattering, and the two-beam equations follow naturally from the results.

In this chapter, we will rephrase the governing equation (2.37) in several different ways, and work out formal solution methods for the N-beam elastic scattering problem. In Chapter 6, we simplify the results of this chapter and discuss the two-beam case in great detail. Finally, in Chapter 7, we deal with the numerical solution of the N-beam case for the *systematic row* and *zone axis* orientations. We will find that the methods described in these three theoretical chapters can be applied to a large number of different experimental situations, among others: convergent beam electron diffraction (CBED), high-resolution TEM, dynamical diffraction patterns, Eades patterns, defect contrast analysis, and so on. While all three chapters rely rather heavily on the mathematical formalism, the reader is encouraged to persist and work through the equations. A good theoretical understanding of contrast mechanisms will facilitate experimental work, since there is a direct relation between the mathematical formalism and what the microscope operator actually does during a microscope session. Throughout the next three chapters, we will attempt to highlight this relation whenever possible.

On the website, the reader will find a significant amount of Fortran source code dealing with the numerical solution of the N-beam elastic electron scattering problem. Some of this code is described in the text, but space restrictions do not permit an extensive discussion of each routine or algorithm. The reader is encouraged once more to study the source code and carry out some of the example runs suggested on the website. Most multi-beam simulations are too demanding computationally to be performed in real-time, in particular through a website. Nevertheless, several interactive **ION** routines are made available for limited real-time simulations. The software will be discussed in Chapters 6 and 7. The theoretical derivations in the present chapter will be limited to *formal solutions* of the elastic scattering problem; actual numerical solutions will be treated in Chapter 7.

In Section 5.8, we will analyze the intrinsic symmetry of the dynamical scattering process, through the introduction of the *reciprocity theorem* and the *diffraction groups*. We conclude this chapter with a brief historical review of the development of the theory of elastic electron scattering.

5.2 The Schrödinger equation for dynamical electron scattering

The general equation for relativistic electron scattering is given by equation (2.37) in Chapter 2:

$$\Delta\Psi + 4\pi^2 K_0^2 \Psi = -4\pi^2 U(\mathbf{r})\Psi = -\frac{8\pi^2 me}{h^2} V_c(\mathbf{r})\Psi. \tag{5.1}$$

Note that we have replaced the electrostatic lattice potential V by the complex potential $V_c = V + \mathrm{i}W$, introduced in Section 2.6.5. The Fourier transform of this complex potential is given by

$$V_c(\mathbf{r}) = V_0 + V'(\mathbf{r}) + \mathrm{i}W(\mathbf{r}) = V_0 + \sum_{\mathbf{g}\neq\mathbf{0}} V_{\mathbf{g}} e^{2\pi \mathrm{i}\mathbf{g}\cdot\mathbf{r}} + \mathrm{i}\sum_{\mathbf{g}} W_{\mathbf{g}} e^{2\pi \mathrm{i}\mathbf{g}\cdot\mathbf{r}}, \tag{5.2}$$

where we have separated the zero-frequency component V_0 from the rest of the Fourier series. V_0 is the positive *mean inner potential*, usually on the order of 5–50 V.

Since it is positive, this potential *accelerates* the electrons when they enter the crystal, and hence the relativistic acceleration potential of the electron increases to

$$\hat{\Psi}_c = \hat{\Psi} + \gamma V_0. \tag{5.3}$$

This acceleration is known as *refraction*, and is equivalent to the change of the velocity of light when a photon enters a transparent medium with a refractive index different from that of vacuum.† We define the *wave number* k_0 as follows:

$$k_0 \equiv \frac{\sqrt{2me\hat{\Psi}_c}}{h}. \tag{5.4}$$

We introduce the following shorthand notation:

$$U(\mathbf{r}) + \mathrm{i}U'(\mathbf{r}) \equiv \frac{2me}{h^2}[V'(\mathbf{r}) + \mathrm{i}W(\mathbf{r})].$$

In other words,

$$U(\mathbf{r}) = \frac{2me}{h^2}\sum_{\mathbf{g}\neq\mathbf{0}} V_{\mathbf{g}} \mathrm{e}^{2\pi\mathrm{i}\mathbf{g}\cdot\mathbf{r}};$$

$$U'(\mathbf{r}) = \frac{2me}{h^2}\sum_{\mathbf{g}} W_{\mathbf{g}} \mathrm{e}^{2\pi\mathrm{i}\mathbf{g}\cdot\mathbf{r}}.$$

Note that U' contains the term $W_{\mathbf{0}}$, whereas U does not have a zero-frequency component. This will become important later on. Using equation (5.4) we obtain:‡

$$\boxed{\Delta\Psi + 4\pi^2 k_0^2 \Psi = -4\pi^2[U + \mathrm{i}U']\Psi,} \tag{5.5}$$

which is the starting equation for this chapter.

In the absence of a crystal potential other than the mean inner potential V_0, and ignoring absorption, we have $U + \mathrm{i}U' = 0$. This is known as the *empty crystal approximation*. The solution to the resulting homogeneous equation is given by

$$\Psi_0(\mathbf{r}) = \mathrm{e}^{2\pi\mathrm{i}\mathbf{k}_0\cdot\mathbf{r}}, \tag{5.6}$$

i.e. a plane wave, corrected for refraction, as is verified easily by substitution into equation (5.5). In what follows we will choose the positive z-direction along the electron propagation direction, i.e. $\mathbf{k}_0 = |k_0|\mathbf{e}_z$. We will assume that the $\mathbf{e}_x$-axis lies along the projection of one of the reciprocal lattice vectors, say $\mathbf{g}$, onto the plane normal to the optical axis. Mathematically, this is represented by

$$\mathbf{e}_x \parallel (\mathbf{k}_0 \times \mathbf{g}) \times \mathbf{k}_0.$$

† In the case of light, of course, the velocity *decreases* as light enters a medium with a refractive index larger than unity. For electrons, the velocity *increases* upon entering the crystal.

‡ From here on we will no longer write the $\mathbf{r}$ dependence, except where confusion could arise.

Furthermore, we will assume that the crystal is planar, with parallel top and bottom surfaces, and that it is defect-free. The bottom surface of the crystal, i.e. on the opposite side of the electron source, will be known as the *exit plane* of the crystal.

5.3 General derivation of the Darwin–Howie–Whelan equations

Bragg's law (equation 2.41) tells us that electrons are diffracted in directions $\mathbf{k}'$ given by $\mathbf{k}_0 + \mathbf{g}$, for all reciprocal lattice points $\mathbf{g}$ on the Ewald sphere. Hence, we anticipate that the total wave function at the exit plane of the crystal will be a superposition of plane waves, one in each of the directions predicted by the Bragg equation. All that remains to be computed is the complex amplitude of each of these diffracted waves. The general solution to equation (5.5) must then be of the form:

$$\Psi(\mathbf{r}) = \sum_{\mathbf{g}} \psi_{\mathbf{g}} e^{2\pi i(\mathbf{k}_0+\mathbf{g})\cdot\mathbf{r}}. \tag{5.7}$$

Note that this is a rather important assumption: instead of writing the total wave function as a general superposition of momentum eigenfunctions, as we did in Chapter 2, we limit the allowed momentum states to those satisfying the Bragg equation. In Section 5.7, we will remove this assumption and *compute* the allowed wave vectors inside the crystal, rather than *imposing* them from the beginning. At this point in time we are not interested in what happens inside the crystal; we are only interested in the behavior of the scattered waves far from the crystal. The geometry for which all diffracted waves emanate from the back surface of the crystal is commonly known as the *Laue case*; the other geometry, with scattered waves on the same side of the crystal as the incident beam, is known as the *Bragg case*.

After substitution of $\Psi(\mathbf{r})$ above into equation (5.5) and some straightforward computations, we find:

$$\sum_{\mathbf{g}} \left\{ \Delta\psi_{\mathbf{g}} + i4\pi(\mathbf{k}_0+\mathbf{g})\cdot\nabla\psi_{\mathbf{g}} + 4\pi^2\left[k_0^2 - (\mathbf{k}_0+\mathbf{g})^2\right]\psi_{\mathbf{g}} \right\} e^{2\pi i(\mathbf{k}_0+\mathbf{g})\cdot\mathbf{r}}$$
$$= -4\pi^2 \sum_{\mathbf{g}} [U + iU']\psi_{\mathbf{g}} e^{2\pi i(\mathbf{k}_0+\mathbf{g})\cdot\mathbf{r}}.$$

It can be shown [VD76] that the first term (Laplacian derivative) in this equation is negligible for high-energy electrons; this is the so-called *high-energy approximation*, or, equivalently, the *forward scattering approximation*. The argument for the validity of this approximation involves the formal solution of the complete second-order differential equation. Van Dyck [VD76] has shown that the high-energy approximation is a good approximation for crystal thicknesses up to a few hundred nanometers, depending on the average atomic number of the crystal. First-order correction terms† to the high-energy approximation can be found in [VD76].

For a perfect planar slab of material, the wave function component $\psi_{\mathbf{g}}$ cannot depend on the lateral coordinates x and y, so the gradient, $\nabla\psi_{\mathbf{g}}$, contains only a derivative with respect to z. Consequently, using the general definition of the dot product (1.3), we can rewrite the second term as

$$(\mathbf{k}_0 + \mathbf{g}) \cdot \nabla\psi_{\mathbf{g}} = |\mathbf{k}_0 + \mathbf{g}| \cos\alpha \frac{\mathrm{d}\psi_{\mathbf{g}}}{\mathrm{d}z},$$

with α the angle between the vectors $\mathbf{e}_z$ and $\mathbf{k}_0 + \mathbf{g}$. Often, $\cos\alpha$ can be approximated by 1, since the diffraction angle for high energy electrons is of the order of a few milliradians. Rearranging terms, we find:

$$\sum_{\mathbf{g}} \left[\frac{\mathrm{d}\psi_{\mathbf{g}}}{\mathrm{d}z} - \mathrm{i}2\pi \frac{k_0^2 - (\mathbf{k}_0 + \mathbf{g})^2}{2|\mathbf{k}_0 + \mathbf{g}| \cos\alpha} \psi_{\mathbf{g}} \right] \mathrm{e}^{2\pi \mathrm{i}(\mathbf{k}_0 + \mathbf{g})\cdot\mathbf{r}}$$
$$= \sum_{\mathbf{g}} \mathrm{i}\pi \frac{(U + \mathrm{i}U')}{|\mathbf{k}_0 + \mathbf{g}| \cos\alpha} \psi_{\mathbf{g}} \mathrm{e}^{2\pi \mathrm{i}(\mathbf{k}_0 + \mathbf{g})\cdot\mathbf{r}}. \qquad (5.8)$$

At this point, we must remove the $\mathbf{r}$ dependence of the potential $U + \mathrm{i}U'$, so that the only $\mathbf{r}$ dependence remaining will be that of the plane waves. We introduce the Fourier expansion:

$$U + \mathrm{i}U' = \sum_{\mathbf{q}} \left(U_{\mathbf{q}} + \mathrm{i}U'_{\mathbf{q}}\right) \mathrm{e}^{2\pi \mathrm{i}\mathbf{q}\cdot\mathbf{r}}. \qquad (5.9)$$

Note that $U_{\mathbf{0}} = 0$, since $V_{\mathbf{0}}$ was separated out earlier on. The right-hand side of equation (5.8) can be rewritten as follows (we represent the left-hand side by LHS):

$$\mathrm{LHS} = \mathrm{i}\pi \sum_{\mathbf{g}} \sum_{\mathbf{q}} \frac{\left(U_{\mathbf{q}} + \mathrm{i}U'_{\mathbf{q}}\right)}{|\mathbf{k}_0 + \mathbf{g}| \cos\alpha} \psi_{\mathbf{g}} \mathrm{e}^{2\pi \mathrm{i}(\mathbf{k}_0 + \mathbf{q} + \mathbf{g})\cdot\mathbf{r}}.$$

Since $\mathbf{q}$ is an arbitrary reciprocal lattice vector, we can replace it by any other reciprocal lattice vector, say $\mathbf{q} = \mathbf{g}' - \mathbf{g}$. This results in

$$\mathrm{LHS} = \mathrm{i}\pi \sum_{\mathbf{g}} \sum_{\mathbf{g}'} \frac{\left(U_{\mathbf{g}'-\mathbf{g}} + \mathrm{i}U'_{\mathbf{g}'-\mathbf{g}}\right)}{|\mathbf{k}_0 + \mathbf{g}| \cos\alpha} \psi_{\mathbf{g}} \mathrm{e}^{2\pi \mathrm{i}(\mathbf{k}_0 + \mathbf{g}')\cdot\mathbf{r}}.$$

Finally, since both $\mathbf{g}$ and $\mathbf{g}'$ are summation variables, we can interchange them without changing the result of the summation:

$$\mathrm{LHS} = \mathrm{i}\pi \sum_{\mathbf{g}'} \sum_{\mathbf{g}} \frac{\left(U_{\mathbf{g}-\mathbf{g}'} + \mathrm{i}U'_{\mathbf{g}-\mathbf{g}'}\right)}{|\mathbf{k}_0 + \mathbf{g}'| \cos\alpha} \psi_{\mathbf{g}'} \mathrm{e}^{2\pi \mathrm{i}(\mathbf{k}_0 + \mathbf{g})\cdot\mathbf{r}}.$$

† The correction factor is proportional to the cube of the electron wavelength.

When we substitute this back into equation (5.8) we find:

$$\sum_{\mathbf{g}}\left[\frac{\mathrm{d}\psi_{\mathbf{g}}}{\mathrm{d}z}-\mathrm{i}2\pi\frac{k_0^2-(\mathbf{k}_0+\mathbf{g})^2}{2|\mathbf{k}_0+\mathbf{g}|\cos\alpha}\psi_{\mathbf{g}}\right]\mathrm{e}^{2\pi\mathrm{i}(\mathbf{k}_0+\mathbf{g})\cdot\mathbf{r}}$$
$$=\sum_{\mathbf{g}}\left[\mathrm{i}\pi\sum_{\mathbf{g}'}\frac{\left(U_{\mathbf{g}-\mathbf{g}'}+\mathrm{i}U'_{\mathbf{g}-\mathbf{g}'}\right)}{|\mathbf{k}_0+\mathbf{g}'|\cos\alpha}\psi_{\mathbf{g}'}\right]\mathrm{e}^{2\pi\mathrm{i}(\mathbf{k}_0+\mathbf{g})\cdot\mathbf{r}}. \tag{5.10}$$

Let us briefly summarize what we have done so far: starting from equation (5.5) we have introduced a plane wave expansion for the wave function, based on Bragg's law; we have introduced a Fourier expansion for the complex lattice potential $U+\mathrm{i}U'$; we have applied the high-energy approximation; and we have assumed that we have a perfect crystal. Equation (5.10) must be valid for every location $\mathbf{r}$, therefore the equality must be valid for each of the individual terms in the summation:

$$\frac{\mathrm{d}\psi_{\mathbf{g}}}{\mathrm{d}z}-\mathrm{i}2\pi\frac{k_0^2-(\mathbf{k}_0+\mathbf{g})^2}{2|\mathbf{k}_0+\mathbf{g}|\cos\alpha}\psi_{\mathbf{g}}=\mathrm{i}\pi\sum_{\mathbf{g}'}\frac{\left(U_{\mathbf{g}-\mathbf{g}'}+\mathrm{i}U'_{\mathbf{g}-\mathbf{g}'}\right)}{|\mathbf{k}_0+\mathbf{g}'|\cos\alpha}\psi_{\mathbf{g}'}. \tag{5.11}$$

This set of equations, one for each $\mathbf{g}$, states that the change of the amplitude of the diffracted wave $\psi_{\mathbf{g}}$ with depth z in the crystal (first term on the left-hand side) is determined by: (1) the current value of the amplitude (second term on the left-hand side) and (2) the amplitudes of all other waves $\psi_{\mathbf{g}'}$ (summation on the right-hand side). The *strength* of the interaction between the waves is determined by the Fourier coefficients of the complex electrostatic lattice potential. In other words, the amplitude of any given diffracted wave is determined by the amplitudes of all other waves, and the electrostatic lattice potential determines how strong these interactions or *couplings* are.

We have seen in Chapter 2 (on page 119) that for a finite crystal, diffraction may occur even when the reciprocal lattice point is not precisely located on the Ewald sphere. We have introduced a geometrical parameter $s_{\mathbf{g}}$, the *excitation error* or *deviation parameter*, which measures the distance between the reciprocal lattice point and the Ewald sphere, along the direction of the foil normal (or, equivalently, along the relrod). Replacing K_0 by k_0 to allow for refraction, we recognize the expression for $s_{\mathbf{g}}$ in the second term of the left-hand side of equation (5.11), and we write:

$$\frac{\mathrm{d}\psi_{\mathbf{g}}}{\mathrm{d}z}-2\pi\mathrm{i}s_{\mathbf{g}}\psi_{\mathbf{g}}=\mathrm{i}\pi\sum_{\mathbf{g}'}\frac{\left(U_{\mathbf{g}-\mathbf{g}'}+\mathrm{i}U'_{\mathbf{g}-\mathbf{g}'}\right)}{|\mathbf{k}_0+\mathbf{g}'|\cos\alpha}\psi_{\mathbf{g}'}. \tag{5.12}$$

From our discussion of the relation between reciprocal space and the objective lens back focal plane in Chapter 2, we remember that we can move the reciprocal lattice points through the Ewald sphere (i.e. change their distance to the Ewald sphere) by changing the orientation of the crystal, or by tilting the incident beam. Therefore, the parameters $s_{\mathbf{g}}$ *are experimental variables, and their values can be set by the microscope operator.* One should note that the various excitation errors $s_{\mathbf{g}}$ are not independent of each other; in fact, when the excitation errors of two non-collinear vectors $\mathbf{g}$ and $\mathbf{g}'$ are selected (by tilting the crystal in a particular orientation), all other excitation errors are fixed for a given electron wavelength.

Since the first term in equation (5.12) has the dimensions of reciprocal length, the right-hand side must have the same dimensions. Therefore, we introduce two new parameters with the dimensions of length: the *extinction distance* $\xi_{\mathbf{g}}$ and the *absorption length* $\xi'_{\mathbf{g}}$. The Fourier coefficients of the real and imaginary part of the electrostatic lattice potential can be written in terms of their modulus and phase angle, as follows:

$$U_{\mathbf{g}} = |U_{\mathbf{g}}|\mathrm{e}^{\mathrm{i}\theta_{\mathbf{g}}};$$
$$U'_{\mathbf{g}} = |U'_{\mathbf{g}}|\mathrm{e}^{\mathrm{i}\theta'_{\mathbf{g}}}.$$

For a centrosymmetric crystal (which must have real Fourier coefficients), the phase factor $\theta_{\mathbf{g}}$ can only take on the values 0 and π, corresponding to positive and negative values for the Fourier coefficients $U_{\mathbf{g}}$. There is no *a priori* reason why, for a non-centrosymmetric crystal, the phase factors of $U_{\mathbf{g}}$ and $U'_{\mathbf{g}}$ should be the same. We will address this issue in some detail in Chapter 6.

We can use the moduli of these Fourier coefficients to define the extinction distance and the absorption length:

$$(\text{extinction distance})^{-1} \rightarrow \quad \frac{1}{\xi_{\mathbf{g}}} \equiv \frac{|U_{\mathbf{g}}|}{|\mathbf{k}_0 + \mathbf{g}|\cos\alpha}; \tag{5.13}$$

$$(\text{absorption length})^{-1} \rightarrow \quad \frac{1}{\xi'_{\mathbf{g}}} \equiv \frac{|U'_{\mathbf{g}}|}{|\mathbf{k}_0 + \mathbf{g}|\cos\alpha}. \tag{5.14}$$

Noting that, to a very good approximation, we have $|\mathbf{k}_0 + \mathbf{g}| \approx |\mathbf{k}_0 + \mathbf{g}'| \approx |\mathbf{k}_0|$, so we can write

$$\frac{(U_{\mathbf{g}-\mathbf{g}'} + \mathrm{i}U'_{\mathbf{g}-\mathbf{g}'})}{|\mathbf{k}_0|\cos\alpha} = \frac{\mathrm{e}^{\mathrm{i}\theta_{\mathbf{g}-\mathbf{g}'}}}{\xi_{\mathbf{g}}} + \mathrm{i}\frac{\mathrm{e}^{\mathrm{i}\theta'_{\mathbf{g}-\mathbf{g}'}}}{\xi'_{\mathbf{g}}}.$$

For notational purposes it is convenient, following [GBA66], to introduce the following complex quantity:

$$\frac{1}{q_{\mathbf{g}}} \equiv \frac{1}{\xi_{\mathbf{g}}} + \mathrm{i}\frac{\mathrm{e}^{\mathrm{i}\beta_{\mathbf{g}}}}{\xi'_{\mathbf{g}}}, \tag{5.15}$$

with

$$\beta_{\mathbf{g}} = \theta'_{\mathbf{g}} - \theta_{\mathbf{g}}. \tag{5.16}$$

Note that, since $U_\mathbf{0} = 0$ (and hence $\xi_\mathbf{0} = \infty$), and $U'_\mathbf{0}$ is real, we also have

$$\frac{1}{q_\mathbf{0}} = \frac{1}{\xi_\mathbf{0}} + \mathrm{i}\frac{\mathrm{e}^{\mathrm{i}\beta_\mathbf{0}}}{\xi'_\mathbf{0}} = \frac{\mathrm{i}}{\xi'_\mathbf{0}}. \tag{5.17}$$

Finally, we can rewrite equation (5.12) as

$$\boxed{\frac{\mathrm{d}\psi_\mathbf{g}}{\mathrm{d}z} - 2\pi\mathrm{i}s_\mathbf{g}\psi_\mathbf{g} = \mathrm{i}\pi\sum_{\mathbf{g}'}\frac{\mathrm{e}^{\mathrm{i}\theta_{\mathbf{g}-\mathbf{g}'}}}{q_{\mathbf{g}-\mathbf{g}'}}\psi_{\mathbf{g}'}.} \tag{5.18}$$

Equations (5.18) are known as the *multibeam Darwin–Howie–Whelan equations* of dynamical electron diffraction theory [HW61]. They are valid for an arbitrary number of diffracted beams, and they describe how the amplitude of a diffracted beam changes with depth in the crystal. These changes depend on how well the Bragg condition is satisfied, and on the strengths of the interactions with the other diffracted beams.

Using the Dirac delta-function we may rewrite the Darwin–Howie–Whelan (or DHW) equations as follows:

$$\frac{\mathrm{d}\psi_\mathbf{g}}{\mathrm{d}z} = \mathrm{i}\pi\sum_{\mathbf{g}'}\left[2s_\mathbf{g}\delta(\mathbf{g}-\mathbf{g}') + \frac{\mathrm{e}^{\mathrm{i}\theta_{\mathbf{g}-\mathbf{g}'}}}{q_{\mathbf{g}-\mathbf{g}'}}\right]\psi_{\mathbf{g}'}. \tag{5.19}$$

This form of the system of equations will prove to be useful in the following section, where we convert the dynamical scattering equations to a single matrix equation. The numbers $(q_{\mathbf{g}-\mathbf{g}'})^{-1}$ are related to the probability that an electron will be scattered from the beam $\mathbf{g}'$ into the beam $\mathbf{g}$. In particular, the number $(q_{\mathbf{g}-\mathbf{0}})^{-1} = (q_\mathbf{g})^{-1}$ refers to scattering from the incident beam into the diffracted beam $\mathbf{g}$. The number $(q_{\mathbf{0}-\mathbf{g}})^{-1} = (q_{-\mathbf{g}})^{-1}$ describes the reverse process, i.e. scattering from the diffracted beam $\mathbf{g}$ back into the incident beam direction. In the absence of absorption ($\xi'_\mathbf{g} = \infty$), and for a centrosymmetric crystal, we have $(q_{-\mathbf{g}})^{-1} = (q_\mathbf{g})^{-1}$.

In Section 6.5.1, we will take a closer look at the numerical computation of the extinction distances and absorption lengths for a given crystal structure. For now, we assume that we know all relevant parameters, and then the only "free" parameters are the crystal orientation, through the excitation errors $s_\mathbf{g}$, the crystal thickness z_0, and the electron wavelength λ.

In principle, we could now use any differential equation solver algorithm to compute the solutions to these equations, for an arbitrary crystal. At this point, it is far more useful, however, to rewrite the equations in a number of different forms, so that we may discover other alternative (and possibly faster) ways to solve the dynamical scattering equations.

5.4 Formal solution of the DHW multibeam equations

We begin with equations (5.19). Instead of labeling every diffracted beam with the corresponding reciprocal lattice vector **g**, we will simply number all the beams from $n = 1$ to N. It is common practice, but not necessary, to label the transmitted beam with $n = 1$. The DHW system of equations then becomes

$$\frac{d\psi_n}{dz} = i\pi \sum_{n'=1}^{N} \left[2s_{n'}\delta(n - n') + \frac{e^{i\theta_{n-n'}}}{q_{n-n'}} \right] \psi_{n'} \qquad n = 1, \ldots, N. \tag{5.20}$$

Using equation (5.17), these equations can be simplified substantially by the following substitution:

$$\psi_n = S_n \, e^{i(\theta_n + \pi z/q_0)} = S_n \, e^{i\theta_n} e^{-\pi z/\xi_0'}. \tag{5.21}$$

The intensity of the nth scattered beam is given by

$$I_n = |\psi_n|^2 = e^{-2\pi z/\xi_0'} |S_n|^2,$$

i.e. an exponentially damped function. The parameter ξ_0' is commonly known as the *normal absorption length*, and it is common to *all* beams, including the transmitted beam. For a sample thickness equal to the normal absorption length, the exponential factor equals $e^{-2\pi} = 0.001\,8674$, which means that only just under 0.2% of the electrons make it through a sample of this thickness. Since this factor is common to all scattered beams, it is appropriate at this point to remove it from the equations. Substitution of equation (5.21) into the DHW equations above, results in the following system of equations:

$$\frac{dS_n}{dz} = i\pi \sum_{n'=1}^{N} \left[2s_{n'}\delta(n - n') + \frac{1 - \delta(n - n')}{q_{n-n'}} \right] S_{n'}. \tag{5.22}$$

These equations cannot be solved analytically, and we could use, say, the fourth-order Runge–Kutta method to obtain a numerical solution; we will discuss numerical solution methods in Chapter 7. For now, we will proceed analytically and rewrite the system of equations in matrix form. Consider the square matrix $\mathcal{A}$, defined by

$$\left. \begin{aligned} \mathcal{A}_{nn} &= 2\pi s_n; \\ \mathcal{A}_{nn'} &= \tfrac{\pi}{q_{n-n'}} \qquad n \neq n'. \end{aligned} \right\} \tag{5.23}$$

In other words, the matrix $\mathcal{A}$ contains along its diagonal the excitation errors of all the beams contributing to the scattering process. The off-diagonal elements describe the interactions between the various beams through the complex numbers $q_{n-n'}$. There is, hence, a clean separation between the diffraction geometry and the coupling between the diffracted beams. We will see later on that we can make good use of this separation (Section 5.5).

If we write the diffracted amplitudes S_n as components of a column vector $\mathbf{S}$, the DHW equations can be rewritten in matrix form as:[†]

$$\frac{\mathrm{d}\mathbf{S}}{\mathrm{d}z} = \mathrm{i}\mathcal{A}\mathbf{S}. \tag{5.24}$$

This equation can be solved formally for a crystal of thickness z_0:

$$\mathbf{S}(z_0) = \mathrm{e}^{\mathrm{i}\mathcal{A}z_0}\mathbf{S}(0) \equiv \mathcal{S}\mathbf{S}(0), \tag{5.25}$$

as can be verified easily by substitution in the above equation. The matrix $\mathcal{A}$ is the so-called *crystal transfer matrix* or *structure matrix*, and $\mathcal{S}$ is known as the *scattering matrix* [Stu62].

The exponential of a matrix can be defined from the standard Taylor expansion for the exponential function (e.g. [AS77]):

$$\mathrm{e}^{ax} = \sum_{n=0}^{\infty} \frac{a^n}{n!}x^n = 1 + \frac{a}{1!}x + \frac{a^2}{2!}x^2 + \cdots + \frac{a^n}{n!}x^n + \cdots. \tag{5.26}$$

For the scattering matrix $\mathcal{S}$ we can write:

$$\mathcal{S} = \mathrm{e}^{\mathrm{i}\mathcal{A}z_0} = \mathbf{1} + \mathrm{i}\frac{\mathcal{A}}{1!}z_0 - \frac{\mathcal{A}^2}{2!}z_0^2 + \cdots + (\mathrm{i})^n\frac{\mathcal{A}^n}{n!}z_0^n + \cdots, \tag{5.27}$$

where $\mathbf{1}$ is the identity matrix with the same number of rows and columns as $\mathcal{A}$. We will discuss an algorithm for the numerical computation of $\mathcal{S}$ in Chapter 7.

At this point, it is useful to take a closer look at the two new matrices $\mathcal{S}$ and $\mathcal{A}$. From the theory of differential equations we know that the general solution to a system of N first-order coupled differential equations can be written as a linear combination of N special solutions. The special solutions can be obtained by stating N independent boundary conditions at the entrance surface of the crystal. The most convenient choice for the independent conditions is

$$\mathbf{S}(0) = \begin{pmatrix} 1 \\ 0 \\ 0 \\ \vdots \\ 0 \end{pmatrix}; \quad \begin{pmatrix} 0 \\ 1 \\ 0 \\ \vdots \\ 0 \end{pmatrix}; \quad \begin{pmatrix} 0 \\ 0 \\ 1 \\ \vdots \\ 0 \end{pmatrix}; \quad \ldots; \quad \begin{pmatrix} 0 \\ 0 \\ 0 \\ \vdots \\ 1 \end{pmatrix}.$$

Solving the DHW equations for each of those initial conditions results in N column vectors $\mathbf{S}_n(z_0)$, $n = 1, \ldots, N$, which can be placed in an $N \times N$ square matrix $\mathcal{S}$.

[†] The order of the entries in $\mathbf{S}$ is irrelevant, as long as the same order is used for the matrix $\mathcal{A}$.

If a general incident amplitude is described by the column vector $\mathbf{S}(0)$, then the general solution at the exit plane of the crystal is given by

$$\mathbf{S}(z_0) = \mathcal{S}\mathbf{S}(0).$$

Hence, the scattering matrix $\mathcal{S}$ can be computed by explicitly solving the DHW equations for N independent initial conditions. However, the exponential expression for the matrix $\mathcal{S}$ in terms of the structure matrix $\mathcal{A}$ shows that $\mathcal{S}$ depends in a non-linear way on the crystal thickness z_0. This means that the previous equation should actually read as

$$\mathbf{S}(z_0) = \mathcal{S}(z_0)\mathbf{S}(0).$$

The scattering matrix must, therefore, be recalculated for every crystal thickness. In Section 5.7 on Bloch waves we will describe how this matrix can be written as the product of three matrices, with only one of them being dependent on the crystal thickness.

In the absence of absorption, the sum of the intensities of the beams leaving the crystal must equal the incident intensity, and the matrix relation above expresses a *unitary transformation*, i.e. a transformation that leaves the norm of the vector $\mathbf{S}$ unchanged. Consequently, the dynamical scattering process can be regarded as a rotation in an N-dimensional space, and $\mathcal{S}$ is a *hermitian matrix*. When absorption occurs, the transformation is no longer unitary, and the total intensity decreases with increasing crystal thickness.

5.5 Slice methods

The exponential of a matrix, as defined in equation (5.27), converges after a small number of terms if the crystal thickness z_0 is sufficiently small. We can make use of the properties of exponential functions and rewrite relation (5.25) as follows:

$$\begin{aligned} \mathbf{S}(z_0) &= \mathcal{S}(z_0)\mathbf{S}(0); \\ &= \mathcal{S}(n\epsilon)\mathbf{S}(0); \\ &= [\mathcal{S}(\epsilon)]^n\,\mathbf{S}(0). \end{aligned}$$

We have made use of the fact that $(x)^{ab} = (x^a)^b$. The total crystal thickness equals $z_0 = n\epsilon$, and, if we take ϵ sufficiently small, then the series expansion for $\mathcal{S}(\epsilon)$ converges rapidly; the scattering matrices for thicknesses that are a multiple of ϵ are obtained by repeated multiplication of $\mathcal{S}(\epsilon)$ with itself, and matrix multiplication can be carried out efficiently on fast processors.

We have subdivided the crystal into a stack of *slices* of equal thickness ϵ, as illustrated in Fig. 5.1(a). Now it becomes apparent how we can use the matrix

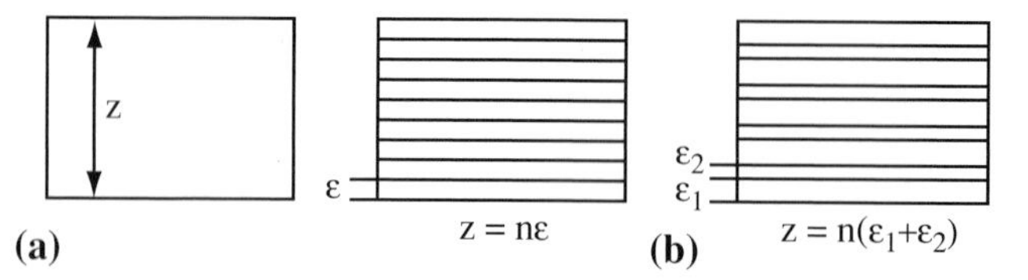

Fig. 5.1. (a) Illustration of a crystal divided into parallel slices of equal thickness; (b) more complex stack of slices, consisting of two different materials. The boundary conditions at each of the interfaces must be properly taken into account in the computation of the multibeam scattering process.

formulation to our advantage. Consider a crystal consisting of an alternating sequence of layers of different thickness, normal to the electron beam. The layers could consist of different materials, or identical material with a different orientation. To compute the amplitudes at the exit plane we would then only need to compute two scattering matrices $\mathcal{S}_1(\epsilon_1)$ and $\mathcal{S}_2(\epsilon_2)$, and multiply them together as many times as needed to obtain the final amplitudes:

$$\mathbf{S}(z_0) = [\mathcal{S}_2(\epsilon_2)\mathcal{S}_1(\epsilon_1)]^n\, \mathbf{S}(0), \tag{5.28}$$

where n is the number of bilayers in the stack. The matrix formalism turns this type of computation into a very simple sequence of matrix multiplications;† it would be very difficult to employ analytical methods to obtain the same results. For a practical implementation of the matrix method we refer to Chapter 7.

The structure matrix $\mathcal{A}$ can be written as the sum of two matrices, one diagonal, and one with zeros along the diagonal (recall that the term $1/q_0$ was separated out in (5.21) and therefore does not contribute to $\mathcal{V}$):

$$\begin{aligned} \mathcal{T}_{nn} &= 2\pi s_n; \\ \mathcal{V}_{nn'} &= \frac{\pi}{q_{n-n'}} \qquad n \neq n'. \end{aligned} \tag{5.29}$$

The formal solution to the DHW equations is given by

$$\mathbf{S}(z_0) = \left[\mathrm{e}^{\mathrm{i}(\mathcal{T}+\mathcal{V})\epsilon}\right]^n \mathbf{S}(0). \tag{5.30}$$

Matrix multiplication is in general non-commutative and therefore we *cannot write*:

$$\mathrm{e}^{\mathrm{i}(\mathcal{T}+\mathcal{V})\epsilon} = \mathrm{e}^{\mathrm{i}\mathcal{T}\epsilon}\mathrm{e}^{\mathrm{i}\mathcal{V}\epsilon} \qquad \leftarrow \textbf{this is incorrect!} \tag{5.31}$$

It is easy to verify that this statement is incorrect by writing down and comparing the first few terms in the Taylor expansions of left- and right-hand sides. Instead,

† One must take care to specify the proper boundary conditions at each interface plane.

one has to use the so-called *Zassenhaus theorem* (from the theory of Lie algebras) which states the following [Suz77]:

For every pair of square matrices $\mathcal{A}$ and $\mathcal{B}$ of dimension N there exists a series of unique matrices $\mathcal{C}_m$ of degree m in $\mathcal{A}$ and $\mathcal{B}$ such that

$$\mathrm{e}^{\mathcal{A}+\mathcal{B}} = \prod_m e^{\mathcal{C}_m}$$

and the product series converges. The first three elements are given by $\mathcal{C}_0 = \mathcal{A}$, $\mathcal{C}_1 = \mathcal{B}$ and $\mathcal{C}_2 = -\frac{1}{2}[\mathcal{A}, \mathcal{B}]$, where $[a, b]$ indicates the commutator $ab - ba$.

Hence we can write

$$\mathrm{e}^{\mathrm{i}(\mathcal{T}+\mathcal{V})\epsilon} = \mathrm{e}^{\mathrm{i}\mathcal{T}\epsilon} \times \mathrm{e}^{\mathrm{i}\mathcal{V}\epsilon} \times \mathrm{e}^{-\frac{1}{2}(\mathcal{T}\mathcal{V}-\mathcal{V}\mathcal{T})\epsilon^2} \times \cdots .$$

For small values of the slice thickness ϵ, higher-order terms can be ignored and the product may be truncated after the second term. We find for the approximate wave function components at the exit plane of a crystal of thickness $z_0 = n\epsilon$:

$$\mathbf{S}(z_0) = \mathrm{e}^{\mathrm{i}\mathcal{T}\epsilon}\mathrm{e}^{\mathrm{i}\mathcal{V}\epsilon} \cdots \mathrm{e}^{\mathrm{i}\mathcal{T}\epsilon}\mathrm{e}^{\mathrm{i}\mathcal{V}\epsilon}\mathrm{e}^{\mathrm{i}\mathcal{T}\epsilon}\mathrm{e}^{\mathrm{i}\mathcal{V}\epsilon}\mathbf{S}(0), \tag{5.32}$$

where the product of the pair of exponentials is executed n times. It appears that we have not really gained anything by splitting the structure matrix into two parts. However, we know that the matrix $\mathcal{T}$ is a diagonal matrix; it is straightforward to show that the exponential of a diagonal matrix is equal to the diagonal matrix containing the exponentials of the original diagonal elements. Therefore we have

$$\left(\mathrm{e}^{\mathrm{i}\mathcal{T}\epsilon}\right)_{nn'} = \mathrm{e}^{2\pi \mathrm{i} s_{n'}\epsilon}\delta(n - n'), \tag{5.33}$$

and this is rather straightforward to compute numerically. The exponential of the second matrix, $\mathcal{V}$, is much more demanding computationally, and we will next derive a method to convert the exponential of $\mathcal{V}$ into an expression that is much easier to compute.

5.6 The direct space multi-beam equations

We start again from the basic Schrödinger equation (5.5) on page 305. This time, we will not make use of Fourier expansions at all, and describe the dynamical scattering process entirely in direct space, following [VDC84]. The incident electron is represented by a plane wave with free-space wave vector $\mathbf{K}_0$, and after passing through the crystal we assume that the wave function can be written as a *modulated plane wave*:

$$\Psi(\mathbf{r}) = \psi(\mathbf{r})e^{2\pi i\mathbf{K}_0\cdot\mathbf{r}}. \tag{5.34}$$

Substitution into equation (5.5) results in (dropping the argument $\mathbf{r}$):

$$\Delta\psi + i4\pi\mathbf{k}_0\cdot\nabla\psi = -4\pi^2(U + iU')\psi,$$

where we have once again included refraction in the wave vector $\mathbf{k}_0$. Since the z-component of the wave vector $\mathbf{k}_0$ is much longer than the x and y components, we can separate the terms in z from the others, and we find

$$\frac{\partial^2\psi}{\partial z^2} + i4\pi k_{0z}\frac{\partial\psi}{\partial z} + \Delta_{xy}\psi + i4\pi\mathbf{k}_{xy}\cdot\nabla_{xy}\psi = -4\pi^2(U + iU')\psi.$$

The differential operators Δ_{xy} and ∇_{xy} are two-dimensional operators, and contain only x and y derivatives. Once again we can make use of the *high-energy approximation* (see page 306) and ignore the second-order derivative along the beam direction. That this is again a valid approximation has been shown in [VDC84]. This results in

$$\frac{\partial\psi}{\partial z} = \frac{i}{4\pi k_{0z}}\left[\left(\Delta_{xy} + i4\pi\mathbf{k}_{xy}\cdot\nabla_{xy}\right)\psi + 4\pi^2(U + iU')\psi\right].$$

Next, we introduce a shorthand notation for the terms in this equation:

$$\bar{\Delta} \equiv \frac{i}{4\pi k_{0z}}\left(\Delta_{xy} + i4\pi\mathbf{k}_{xy}\cdot\nabla_{xy}\right);$$

$$\bar{V} \equiv \frac{i\pi}{k_{0z}}(U + iU').$$

The resulting equation is

$$\frac{\partial\psi}{\partial z} = \left(\bar{\Delta} + \bar{V}\right)\psi. \tag{5.35}$$

This equation has the same form as the reciprocal space equation (5.24) after splitting the crystal transfer matrix $\mathcal{A}$ into its two component matrices:

$$\frac{d\mathbf{S}}{dz} = i(\mathcal{T} + \mathcal{V})\mathbf{S}.$$

Again we have separated the geometry of the problem ($\bar{\Delta}$) from the electron interactions ($\bar{V}$). Let us take a closer look at the two components of this equation.

5.6.1 The phase grating equation

Let us assume first that $\bar{\Delta} = 0$, which means that there is no lateral scattering of the electron; the main equation then reduces to

$$\frac{\partial\psi}{\partial z} = \bar{V}\psi. \tag{5.36}$$

For a crystal of thickness z_0, this equation is integrated readily to give

$$\psi(x, y, z_0) = \mathrm{e}^{\int_0^{z_0} \bar{V}(x,y,z)\,\mathrm{d}z},$$

where we have written the explicit coordinate dependence. The argument of the exponential can be rewritten in more familiar terms as follows:

$$\bar{V} = \frac{\mathrm{i}\pi}{k_{0z}} \frac{2me}{h^2} V_c = \mathrm{i}\sigma V_c,$$

where we have used the definition of the interaction constant σ (2.35) and $k_{0z} = 1/\lambda$. The solution to the first component equation is then rewritten as

$$\psi(x, y, z_0) = \mathrm{e}^{\mathrm{i}\sigma \int_0^{z_0} V_c(x,y,z)\,\mathrm{d}z} \equiv \mathrm{e}^{\mathrm{i}\sigma V_p};$$

V_p is commonly known as the *projected potential* and is defined by

$$V_p \equiv \int_0^{z_0} V_c(x, y, z)\,\mathrm{d}z. \tag{5.37}$$

Equation (5.36) can be interpreted easily if we consider a thin crystal with $z_0 = \epsilon$; in the absence of absorption, the projected potential V_p changes only the phase of the electron wave function, not the amplitude, and hence the thin slice of crystal acts as a phase shifting layer or a *phase grating*. If the phase change is small (much smaller than a radian), then the exponential can be approximated by its Taylor expansion

$$\mathrm{e}^{\mathrm{i}\sigma V_p} \approx 1 + \mathrm{i}\sigma V_p + \cdots$$

and the crystal is referred to as a *weak phase object* . We will see in Chapter 10 that ferromagnetic thin foils give rise to large phase shifts, and are hence *strong phase objects*.

5.6.2 The propagator equation

Next, we assume that $\bar{V} = 0$ (this is similar to the empty crystal approximation) and we solve the following equation:

$$\frac{\partial \psi}{\partial z} = \bar{\Delta}\psi. \tag{5.38}$$

This equation is first order in z, and second order in x and y, hence it belongs to the class of *diffusion equations*, with z taking on the role of time t. The main difference between this equation and a standard diffusion equation is the fact that $\bar{\Delta}$ is a *complex differential operator*. If the incident (refraction-corrected) wave vector does not have x and y components, i.e. if $\mathbf{k}_{xy} = \mathbf{0}$, then the differential operator is purely imaginary. The formal solution for a slice of thickness ϵ is, as before, given by

$$\psi(x, y, \epsilon) = \mathrm{e}^{\epsilon\bar{\Delta}}\psi(x, y, 0).$$

This equation describes how the amplitude in the plane $z = 0$ is modified in the plane $z = \epsilon$; in other words, the equation describes how the electrons *propagate* from one slice to the next, and hence this equation is known as the *propagator equation*. The exponential of the differential operator $\bar{\Delta}$ is rather difficult to compute analytically. However, following [CVDOdBVL97] or [Kir98], for example, we can make use of the properties of Fourier transforms to derive a much simpler expression. For a beam incident along the z-direction, the second term in the definition of $\bar{\Delta}$ vanishes, and the operator is written as

$$\bar{\Delta} = \frac{\mathrm{i}\lambda}{4\pi}\Delta_{xy}.$$

Consider the following relation (using the definition of the inverse Fourier transform):

$$\bar{\Delta}\psi(\mathbf{R}) = \bar{\Delta}\iint \psi(\mathbf{q})\mathrm{e}^{2\pi\mathrm{i}\mathbf{q}\cdot\mathbf{R}}\,\mathrm{d}\mathbf{q},$$

where $\mathbf{R}$ is a two-dimensional direct space vector, and $\mathbf{q}$ is a 2D reciprocal space vector. Since the Laplacian operator acts only on the direct space coordinates, we can bring it inside the integral and we find:

$$\bar{\Delta}\psi(\mathbf{R}) = \iint (-\mathrm{i}\pi\lambda q^2)\psi(\mathbf{q})\mathrm{e}^{2\pi\mathrm{i}\mathbf{q}\cdot\mathbf{R}}\,\mathrm{d}\mathbf{q}. \tag{5.39}$$

Next, we use the Taylor expansion (5.26):

$$\mathrm{e}^{\epsilon\bar{\Delta}}\psi = \sum_n \frac{\epsilon^n}{n!}(\bar{\Delta})^n\psi.$$

The higher-order derivatives of ψ can be calculated by repeated use of equation (5.39), and we find

$$\begin{aligned}\mathrm{e}^{\epsilon\bar{\Delta}}\psi(\mathbf{R}) &= \iint \left\{\sum_n \frac{\epsilon^n}{n!}(-\mathrm{i}\pi\lambda q^2)^n\right\}\psi(\mathbf{q})\mathrm{e}^{2\pi\mathrm{i}\mathbf{q}\cdot\mathbf{R}}\,\mathrm{d}\mathbf{q};\\ &= \iint \mathrm{e}^{-\pi\mathrm{i}\lambda\epsilon q^2}\psi(\mathbf{q})\mathrm{e}^{2\pi\mathrm{i}\mathbf{q}\cdot\mathbf{R}}\,\mathrm{d}\mathbf{q}.\end{aligned}$$

Finally, we find the important result:

$$\mathcal{F}\left[\mathrm{e}^{\epsilon\bar{\Delta}}\psi(\mathbf{R})\right] = \mathrm{e}^{-\pi\mathrm{i}\lambda\epsilon q^2}\psi(\mathbf{q}). \tag{5.40}$$

The complicated exponential of a differential operator in direct space is equivalent to multiplication by a simple phase factor in reciprocal space. We have already seen that in reciprocal space the geometry-dependent factor is given by $\mathrm{e}^{2\pi\mathrm{i}s_g\epsilon}$, and this in turn means that

$$-\pi i\lambda\epsilon q^2 = 2\pi i s_{\mathbf{q}}\epsilon \quad \rightarrow \quad s_{\mathbf{q}} = -\frac{\lambda q^2}{2}.$$

The second equality is valid for exact zone axis orientation and we leave it to the reader to show that this relation is indeed correct.

We can gain some additional insight into the physical meaning of the propagator term by using the convolution theorem of equation (5.40):

$$e^{\bar{\Delta}\epsilon}\psi = \mathcal{F}^{-1}\left[e^{-\pi i\lambda\epsilon q^2}\right] \otimes \psi(\mathbf{R}). \tag{5.41}$$

Propagation from one slice to the next is thus equivalent mathematically to a convolution operation. The first factor on the right-hand side can be rewritten as

$$\mathcal{F}^{-1}\left[e^{-\pi i\lambda\epsilon q^2}\right] = e^{i\pi R^2/\epsilon\lambda} = e^{\pi i k R^2/\epsilon} = \mathcal{P}_\epsilon(R),$$

and we recognize the Fresnel propagator, which was introduced in Section 3.4.7.

5.6.3 Solving the full direct-space equation

If we subdivide the crystal into n layers or slices of thickness ϵ, then in the limit of vanishing slice thickness we have

$$\psi(x, y, z_0) = e^{\bar{\Delta}\epsilon}e^{i\sigma V_p^n} \cdots e^{\bar{\Delta}\epsilon}e^{i\sigma V_p^2}e^{\bar{\Delta}\epsilon}e^{i\sigma V_p^1}\psi(x, y, 0), \tag{5.42}$$

where the number of pairs of exponentials increases with decreasing ϵ, such that the product $n\epsilon = z_0$ is constant. This relation is remarkably similar to equation (5.32). Each slice can, in principle, have a different projected potential, reflected in the use of the superscript n. The interpretation of this relation is rather simple: the scattering process of a fast electron is broken up into two steps per crystal slice. The first involves *multiplication* with the phase grating term. The entire potential of the slice is projected onto a plane normal to the beam direction, and, for a sufficiently thin slice, the phase grating term changes only the phase of the electron wave function. Then, the wave *propagates* to the next slice, as illustrated in Fig. 5.2. From the starting point at (x, y, z) the electron can reach any point within a certain distance $w/2$ from the point $(x, y, z + \epsilon)$; the magnitude of this distance is determined by the electron wavelength and the slice thickness in the form of the complex differential operator $\epsilon\bar{\Delta}$. We will see in Chapter 7 that equations (5.32) and (5.42) form the basis for fast numerical algorithms to compute the solution to the dynamical multi-beam electron diffraction equations.

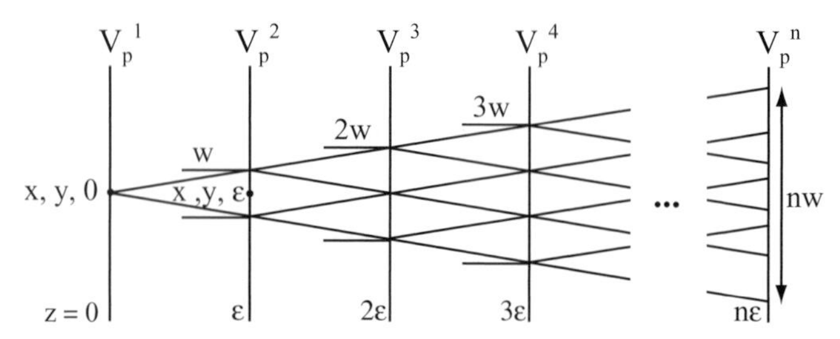

Fig. 5.2. Illustration of the propagation of an electron through consecutive slices of a crystal.

5.7 Bloch wave description

In Section 5.3, we have approached the electron scattering problem assuming that electrons will travel only in the directions $\mathbf{k}_0$ and $\mathbf{k}_0 + \mathbf{g}$. While this is a valid assumption for an external observer (who is interested mainly in the travel direction of the electrons *after they have left the crystal*), it is not necessarily valid inside the crystal. Let us consider a simple analogy: a sound wave – a compressive wave in the medium "air" – enters a solid on one side and emerges from that solid on the opposite side, ignoring reflection of the wave. We know that the only sound waves possible inside a crystal with 3D periodicity are *phonons* (or quantized lattice vibrations); thus, the sound wave will propagate through the crystal as a linear superposition of phonons, each with its own frequency and energy. The relation between frequency (or wave vector) and energy is called a *dispersion relation*. It is possible that there are *gaps* in the dispersion function; this indicates that certain frequencies are not transmitted by the crystal.

A similar situation occurs for high-energy electrons traveling through matter. Before entering the crystal, the electron can travel with any wave vector.[†] Inside the crystal, the allowed directions depend on the crystal structure and on the electron energy. The problem of finding the allowed wave vectors inside a crystal is closely related to another important solid-state problem: the band-structure of a material. Band theory addresses the question: which electron energies are allowed for a given wave vector (or electron momentum)? When the allowed energies are plotted versus the wave vector, the so-called *band-structure* is obtained; certain energy levels may be inaccessible to the electrons and form *band-gaps*. The width of a band-gap is determined by the Fourier coefficients of the lattice potential. We will show in this section that the problem of electron diffraction is closely related to the band-structure of a material. We will ask the question: for a given *incident energy*, what are the allowed wave vectors inside the crystal? The incident energy is several

[†] Wavelength and energy are related to each other by equation (2.33) of Chapter 2, which is therefore the dispersion relation in vacuum, and the direction of the wave is arbitrary.

orders of magnitude larger than the typical energies of electrons in a solid, which will lead to some qualitative differences between the two theories.

Most of the theory presented in this section is based on two important review articles of the Bloch wave method: *Diffraction of Electrons by Perfect Crystals*, by A.J.F. Metherell [Met75], and *The Scattering of Fast Electrons By Crystals*, by C.J. Humphreys [Hum79]. We have adapted slightly the notation used in those papers to conform with the remainder of this book, and also with the notation used in the book *Electron Microdiffraction*, by J.C.H. Spence and J.M. Zuo [SZ92].

We start from equation (2.37): we expand the lattice potential in a Fourier series,

$$V(\mathbf{r}) = \sum_{\mathbf{h}} V_{\mathbf{h}} \mathrm{e}^{2\pi \mathrm{i}\mathbf{h}\cdot\mathbf{r}}.$$

Once again, we separate the V_0 term from the rest of the potential and we will use the refraction-corrected wave vector $\mathbf{k}_0$ (see equation 5.4) throughout this section. We do not include absorption at this point, so that the potential above is real, not complex. The effect of absorption will be described in Section 5.7.3.

The electron wave inside the crystal is written as the product of a wave with arbitrary wave vector $\mathbf{k}$ (to be determined later) and a function $C(\mathbf{r})$ with the periodicity of the lattice. This type of wave function is known as a *Bloch wave*:

$$\Psi(\mathbf{r}) = C(\mathbf{r})\mathrm{e}^{2\pi \mathrm{i}\mathbf{k}\cdot\mathbf{r}} = \sum_{\mathbf{g}} C_{\mathbf{g}} \mathrm{e}^{2\pi \mathrm{i}(\mathbf{k}+\mathbf{g})\cdot\mathbf{r}}, \tag{5.43}$$

where we have introduced a Fourier expansion for $C(\mathbf{r})$. Bloch waves were first introduced into quantum mechanics by Felix Bloch in 1929 [Blo29], for the electron theory of ferromagnetism. Note that, for a perfect crystal, the coefficients $C_{\mathbf{g}}$ are *independent of position*; this is different from expression (5.7), where the coefficients $\psi_{\mathbf{g}}$ do depend upon position. The coefficients $C_{\mathbf{g}}$ do, however, depend on the wave vector $\mathbf{k}$. A second important difference, as mentioned above, is that an *arbitrary* wave vector $\mathbf{k}$ is used, instead of $\mathbf{k}_0$. The solution method described below will enable us to determine the allowed wave vector(s) $\mathbf{k}$ for a given incident beam energy. The coefficients $C_{\mathbf{g}}$ are called *Bloch wave coefficients*.

Substitution of $\Psi(\mathbf{r})$ into equation (5.5) leads to

$$\sum_{\mathbf{g}} \left\{ \left[k_0^2 - (\mathbf{k}+\mathbf{g})^2\right] C_{\mathbf{g}} + \sum_{\mathbf{h}\neq\mathbf{g}} U_{\mathbf{g}-\mathbf{h}} C_{\mathbf{h}} \right\} \mathrm{e}^{2\pi \mathrm{i}(\mathbf{k}+\mathbf{g})\cdot\mathbf{r}} = 0, \tag{5.44}$$

which must be valid for any position $\mathbf{r}$ in the crystal. Therefore, we can write

$$\boxed{\left[k_0^2 - (\mathbf{k}+\mathbf{g})^2\right] C_{\mathbf{g}} + \sum_{\mathbf{h}\neq\mathbf{g}} U_{\mathbf{g}-\mathbf{h}} C_{\mathbf{h}} = 0.} \tag{5.45}$$

This set of equations, one for each $\mathbf{g}$, is *exact*, i.e. no approximations have been used so far. They relate the wave vector $\mathbf{k}$ of the Bloch wave to the energy of the incident electron (through k_0), hence they are *dispersion relations*.

5.7.1 General solution method

We now proceed to solve equations (5.45) for a multi-beam situation. We note that, as before, the scattering problem naturally separates into two contributions: the geometry of the problem is described in the first term of (5.45), and the interaction with the crystal potential in the second summation. The first term can again be represented by a diagonal matrix, while the second term only gives rise to off-diagonal matrix elements. For convenience, we select the transmitted beam to be the first entry in the column vector of Bloch wave coefficients, and the governing equations (5.45) can then be rewritten in matrix form as

$$\begin{pmatrix} k_0^2 - k^2 & U_{\mathbf{0}-\mathbf{g}} & \cdots & U_{\mathbf{0}-\mathbf{h}} & \cdots & U_{\mathbf{0}-\mathbf{m}} \\ U_{\mathbf{g}-\mathbf{0}} & k_0^2 - (\mathbf{k}+\mathbf{g})^2 & \cdots & U_{\mathbf{g}-\mathbf{h}} & \cdots & U_{\mathbf{g}-\mathbf{m}} \\ \vdots & \vdots & \ddots & \vdots & & \vdots \\ U_{\mathbf{h}-\mathbf{0}} & U_{\mathbf{h}-\mathbf{g}} & \cdots & k_0^2 - (\mathbf{k}+\mathbf{h})^2 & \cdots & U_{\mathbf{h}-\mathbf{m}} \\ \vdots & \vdots & & \vdots & \ddots & \vdots \\ U_{\mathbf{m}-\mathbf{0}} & U_{\mathbf{m}-\mathbf{g}} & \cdots & U_{\mathbf{m}-\mathbf{h}} & \cdots & k_0^2 - (\mathbf{k}+\mathbf{m})^2 \end{pmatrix} \begin{pmatrix} C_{\mathbf{0}} \\ C_{\mathbf{g}} \\ \vdots \\ C_{\mathbf{h}} \\ \vdots \\ C_{\mathbf{m}} \end{pmatrix} = 0. \tag{5.46}$$

Non-trivial solutions can occur only when the determinant of this matrix is equal to zero. It is easy to see that the leading term in this determinant is the product of the main diagonal elements, and that this term is proportional to k^{2N}, where N is the number of beams. All other contributions to the determinant contain only lower powers of k. The resulting *characteristic equation* is then a polynomial equation of order $2N$, which must have $2N$ roots. We thus find the important result that for an N-beam case, the *total number of allowed wave vectors inside the crystal is equal to* $2N$. We will denote the different wave vectors by a superscript (j), as in $\mathbf{k}^{(j)}$. The most general wave solution inside the crystal therefore consists of a superposition of $2N$ Bloch waves, each with its own amplitude $\alpha^{(j)}$:

$$\Psi(\mathbf{r}) = \sum_j \alpha^{(j)} \sum_{\mathbf{g}} C_{\mathbf{g}}^{(j)} e^{2\pi i(\mathbf{k}^{(j)}+\mathbf{g})\cdot\mathbf{r}} = \sum_j \alpha^{(j)} C^{(j)}(\mathbf{r}) e^{2\pi i\mathbf{k}^{(j)}\cdot\mathbf{r}} \tag{5.47}$$

and the coefficient $\alpha^{(j)}$ is the *excitation amplitude* of the jth Bloch wave.

It may appear that we have introduced many more unknowns than equations; in equation (5.47) we have $2N$ coefficients $\alpha^{(j)}$, $2N$ wave vectors $\mathbf{k}^{(j)}$, and $2N \times N$ Bloch wave coefficients $C_{\mathbf{g}}^{(j)}$, for a total of $2N^2 + 4N$ (complex) unknowns. In the following paragraphs, we will show that we do have enough information to determine all the unknown parameters unambiguously.

Since the total energy of the incident electron is constant, all $2N$ Bloch waves must have the same total energy; we say that the Bloch waves are *degenerate*. Since the $2N$ wave vectors $\mathbf{k}^{(j)}$ are, in general, different from each other, each Bloch wave must correspond to a different kinetic energy, and therefore also a different potential energy. This, in turn, means that *different Bloch waves must travel at different locations through the crystal.* This is a very important observation, and we will return to it in the next chapter.

The $2N$ wave vectors $\mathbf{k}^{(j)}$ are ranked according to decreasing kinetic energy. This means that the Bloch wave with the highest kinetic energy is number (1), and that corresponding to the lowest kinetic energy is number $(2N)$. The reader should be warned that in some of the earlier literature on the Bloch wave theory the wave vector with the lowest kinetic energy was taken to be $\mathbf{k}^{(1)}$.

We define the unit vector $\mathbf{n}$ as the surface normal to the sample, in the direction opposite† to the incident beam, i.e. along the $-z$-direction. Since both the incident wave and the wave inside the crystal are solutions to the Schrödinger equation, which is a second-order equation in the spatial coordinates, both the functions and their first-order derivatives must be continuous across the entrance plane of the crystal. It can be shown quite generally (e.g. [Met75]) that this continuity condition implies that the *tangential component* of the wave vector must be conserved across the entrance plane. The only component of the wave vector that can change upon entering the crystal is, therefore, the normal component. It is then convenient to write the solution vectors $\mathbf{k}^{(j)}$ as

$$\mathbf{k}^{(j)} = \mathbf{k}_0 + \gamma^{(j)}\mathbf{n}. \tag{5.48}$$

This expression guarantees that all wave vectors $\mathbf{k}^{(j)}$ have the same component normal to the vector $\mathbf{n}$. Let us use this expression to rewrite equation (5.45):

$$\begin{aligned} k_0^2 - \left(\mathbf{k}^{(j)} + \mathbf{g}\right)^2 &= k_0^2 - \left(\mathbf{k}_0 + \gamma^{(j)}\mathbf{n} + \mathbf{g}\right)^2; \\ &= -\mathbf{g}\cdot(2\mathbf{k}_0 + \mathbf{g}) - 2\mathbf{n}\cdot(\mathbf{k}_0 + \mathbf{g})\gamma^{(j)} - \left(\gamma^{(j)}\right)^2; \\ &= 2k_0 s_{\mathbf{g}} - 2\mathbf{n}\cdot(\mathbf{k}_0 + \mathbf{g})\gamma^{(j)} - \left(\gamma^{(j)}\right)^2. \end{aligned} \tag{5.49}$$

We have used the definition (2.89) of the excitation error $s_{\mathbf{g}}$ in the last step. At this point, we will make the assumption that the coefficients $\gamma^{(j)}$ are small compared to k_0, so that the quadratic term in $\gamma^{(j)}$ can be ignored. This is equivalent to the *high-energy approximation* used in the previous sections, and as a result only N of the $2N$ wave vectors $\mathbf{k}^{(j)}$ remain.‡ The N wave vectors which we chose to ignore

† Note that some authors select the foil normal to point *along* the incident beam. The reader should be aware that this leads to sign differences for the coefficients $\gamma^{(j)}$.

‡ Equation (5.49) is now linear instead of quadratic in $\gamma^{(j)}$.

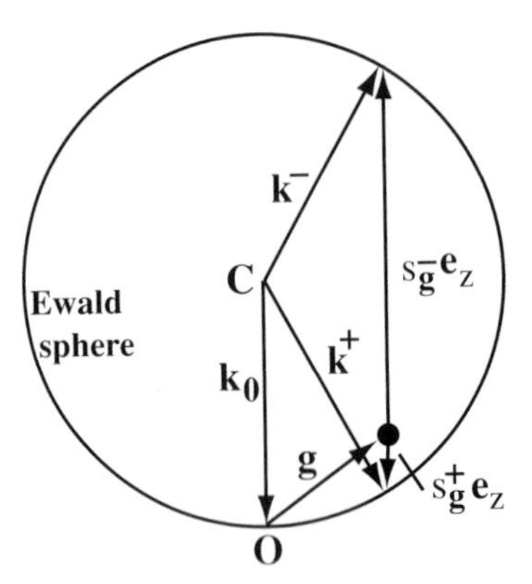

Fig. 5.3. Schematic illustration of positive and negative excitation errors for a single reciprocal lattice point. For large acceleration voltages, the negative excitation error is always significantly larger than the positive one, justifying the use of the high-energy approximation.

correspond to electrons which travel in the direction opposite to the incident beam, i.e. they correspond to *reflected* waves. For the acceleration voltages used in typical TEM experiments this is a good approximation, as is shown explicitly in [KS82].

A simple geometrical argument can be used to qualitatively justify the high-energy approximation. Consider the Ewald sphere construction in Fig. 5.3: the reciprocal lattice point **g** is just inside the Ewald sphere, and if we assume that the sample surfaces are normal to the incident beam $\mathbf{k}_0$ (i.e. the corresponding relrod is parallel to $\mathbf{k}_0$), then relation (2.89) on page 122 has two solutions for the excitation error $s_\mathbf{g}$. The positive solution is small in magnitude and corresponds to the excitation error used in the previous sections. The negative solution has a large magnitude and corresponds to a beam traveling in nearly the opposite direction, i.e. a reflected beam. The two scattered wave vectors are given by

$$\mathbf{k}^+ = \mathbf{k}_0 + \mathbf{g} + s_\mathbf{g}^+ \mathbf{e}_z;$$
$$\mathbf{k}^- = \mathbf{k}_0 + \mathbf{g} + s_\mathbf{g}^- \mathbf{e}_z.$$

The excitation error $s_\mathbf{g}^-$ will increase rapidly with increasing acceleration voltage (i.e. with increasing Ewald sphere radius), so that the probability of Bragg scattering in the direction $\mathbf{k}^-$ will decrease with decreasing wavelength. For the acceleration voltages used in modern microscopes, it is a good approximation to consider only electrons which were scattered in the forward direction (remember that the angle between $\mathbf{k}_0$ and $\mathbf{k}^+$ is of the order of milliradians), and this is essentially the high-energy approximation.

Elimination of half of the allowed wave vectors decreases the total number of unknowns from $2N^2 + 4N$ to $N^2 + 2N$. If we denote the normal components of

$\mathbf{k}_0$ and $\mathbf{g}$ by $\mathbf{n} \cdot \mathbf{g} = g_n$ and $\mathbf{n} \cdot \mathbf{k}_0 = k_n$, then we find for equations (5.45):

$$2k_0 s_\mathbf{g} C_\mathbf{g}^{(j)} + \sum_{\mathbf{h} \neq \mathbf{g}} U_{\mathbf{g}-\mathbf{h}} C_\mathbf{h}^{(j)} = 2k_n \left(1 + \frac{g_n}{k_n}\right) \gamma^{(j)} C_\mathbf{g}^{(j)}. \tag{5.50}$$

If $\mathbf{k}_0$ is very nearly antiparallel to $\mathbf{n}$, then we also have $g_n \ll k_n$ and the equations reduce to

$$2k_0 s_\mathbf{g} C_\mathbf{g}^{(j)} + \sum_{\mathbf{h} \neq \mathbf{g}} U_{\mathbf{g}-\mathbf{h}} C_\mathbf{h}^{(j)} = 2k_n \gamma^{(j)} C_\mathbf{g}^{(j)}. \tag{5.51}$$

These equations will serve as the starting point for the remainder of this section. It is easy to see that (5.51) is an eigenvalue equation of the form

$$\boxed{\bar{\mathcal{A}} \mathbf{C}^{(j)} = 2k_n \gamma^{(j)} \mathbf{C}^{(j)},} \tag{5.52}$$

where the column vectors $\mathbf{C}^{(j)}$ contain the Bloch wave coefficients $C_\mathbf{g}^{(j)}$. For the N-beam case, the matrix $\bar{\mathcal{A}}$ has $N \times N$ entries, and therefore N eigenvalues and N eigenvectors. Solution of this equation will determine all wave vectors $\mathbf{k}^{(j)}$ and all Bloch wave coefficients $C_\mathbf{g}^{(j)}$. The Bloch wave excitation amplitudes $\alpha^{(j)}$ can then be determined by applying the appropriate boundary conditions at the entrance plane, as will be shown below. Since there are N eigenvalues, and N^2 components of the N eigenvectors, and we also determine the N coefficients $\alpha^{(j)}$ from the boundary conditions, we have effectively determined all $N^2 + 2N$ unknowns, and therefore the problem is, at least formally, solved.

It is important to consider some general properties of the eigenvalues and eigenvectors of $\bar{\mathcal{A}}$. For a non-centrosymmetric crystal we have $U_\mathbf{g} = U_{-\mathbf{g}}^*$, and therefore $\bar{\mathcal{A}}$ is a *hermitian matrix* in the absence of absorption. The diagonal of $\bar{\mathcal{A}}$ contains information about the orientation of the crystal, through the excitation errors $s_\mathbf{g}$, and the off-diagonal elements contain information on the interactions between the beams. The eigenvalues of a hermitian matrix are real, the eigenvectors are complex and form a *unitary matrix*. In other words, when we consider the eigenvectors $\mathbf{C}^{(j)}$ as columns of a matrix $\mathcal{C}$, then this matrix satisfies the relation:

$$\mathcal{C}^{-1} = \tilde{\mathcal{C}}^*, \tag{5.53}$$

where the tilde indicates the transposition operator. If we consider the equation $\mathcal{C}\mathcal{C}^{-1} = \mathbf{1}$, with $\mathbf{1}$ the $N \times N$ unit matrix, then we can easily show that

$$\sum_\mathbf{g} C_\mathbf{g}^{(i)} C_\mathbf{g}^{(j)*} = \delta_{ij}, \tag{5.54}$$

and

$$\sum_j C_\mathbf{g}^{(j)} C_\mathbf{h}^{(j)*} = \delta_{\mathbf{gh}}. \tag{5.55}$$

This means that the eigenvectors of $\bar{A}$ form an *orthonormal set*. Standard matrix theory then states that $\bar{A}$ can be written as

$$\bar{A} = C\Lambda\tilde{C}, \tag{5.56}$$

where Λ is a diagonal matrix containing the eigenvalues. This decomposition of $\bar{A}$ is known as the *spectral factorization* of $\bar{A}$.

5.7.2 Determination of the Bloch wave excitation coefficients

The Bloch wave excitation coefficients $\alpha^{(j)}$ can be determined from the boundary conditions at the crystal entrance plane. We will rewrite the Bloch wave expansion in the so-called *Darwin representation*; this is essentially the expansion (5.7), which formed the starting point for the derivation of the Darwin–Howie–Whelan equations on page 306.

Using (5.48) we have

$$\begin{aligned}\Psi(\mathbf{r}) &= \sum_{\mathbf{g}}\left[\sum_j \alpha^{(j)} C_{\mathbf{g}}^{(j)} e^{2\pi i\gamma^{(j)}z}\right] e^{2\pi i(\mathbf{k}_0+\mathbf{g})\cdot\mathbf{r}};\\ &= \sum_{\mathbf{g}} \psi_{\mathbf{g}}(z) e^{2\pi i(\mathbf{k}_0+\mathbf{g})\cdot\mathbf{r}},\end{aligned}$$

with

$$\psi_{\mathbf{g}}(z) = \sum_j \alpha^{(j)} C_{\mathbf{g}}^{(j)} e^{2\pi i\gamma^{(j)}z}. \tag{5.57}$$

This can be rewritten in matrix form as

$$\begin{pmatrix}\psi_{\mathbf{0}}(z)\\ \psi_{\mathbf{g}}(z)\\ \vdots\\ \psi_{\mathbf{h}}(z)\end{pmatrix} = \begin{pmatrix} C_{\mathbf{0}}^{(1)} & C_{\mathbf{0}}^{(2)} & \dots & C_{\mathbf{0}}^{(N)}\\ C_{\mathbf{g}}^{(1)} & C_{\mathbf{g}}^{(2)} & \dots & C_{\mathbf{g}}^{(N)}\\ \vdots & \vdots & \ddots & \vdots\\ C_{\mathbf{h}}^{(1)} & C_{\mathbf{h}}^{(2)} & \dots & C_{\mathbf{h}}^{(N)}\end{pmatrix}$$

$$\times \begin{pmatrix} e^{2\pi i\gamma^{(1)}z} & 0 & \dots & 0\\ 0 & e^{2\pi i\gamma^{(2)}z} & \dots & 0\\ \vdots & \vdots & \ddots & \vdots\\ 0 & 0 & \dots & e^{2\pi i\gamma^{(N)}z}\end{pmatrix}\begin{pmatrix}\alpha^{(1)}\\ \alpha^{(2)}\\ \vdots\\ \alpha^{(N)}\end{pmatrix}. \tag{5.58}$$

At the entrance plane of the crystal we have $z = 0$, and the diagonal matrix reduces to the identity matrix. The resulting equation is

$$\begin{pmatrix} \psi_{\mathbf{0}}(0) \\ \psi_{\mathbf{g}}(0) \\ \vdots \\ \psi_{\mathbf{h}}(0) \end{pmatrix} = \begin{pmatrix} C_{\mathbf{0}}^{(1)} & C_{\mathbf{0}}^{(2)} & \dots & C_{\mathbf{0}}^{(N)} \\ C_{\mathbf{g}}^{(1)} & C_{\mathbf{g}}^{(2)} & \dots & C_{\mathbf{g}}^{(N)} \\ \vdots & \vdots & \ddots & \vdots \\ C_{\mathbf{h}}^{(1)} & C_{\mathbf{h}}^{(2)} & \dots & C_{\mathbf{h}}^{(N)} \end{pmatrix} \begin{pmatrix} \alpha^{(1)} \\ \alpha^{(2)} \\ \vdots \\ \alpha^{(N)} \end{pmatrix} = \begin{pmatrix} 1 \\ 0 \\ \vdots \\ 0 \end{pmatrix}. \tag{5.59}$$

This, in turn, means that the Bloch wave excitation coefficients $\alpha^{(j)}$ are given by the first column of the inverse of the eigenvector matrix. In the case of a centrosymmetric crystal without absorption, this matrix is Hermitian, which means (using 5.53) that $\alpha^{(j)} = C_{\mathbf{0}}^{(j)*}$.

Note that this derivation also produces an explicit expression for the scattering matrix $\mathcal{S}$, introduced on page 312. If we denote the diagonal matrix $e^{2\pi i\gamma^{(j)}z}\delta_{ij}$ by $\mathcal{E}(z)$, then we have

$$\Psi(z) = \mathcal{C}\mathcal{E}(z)\mathcal{C}^{-1}\Psi(0) = \mathcal{S}\Psi(0). \tag{5.60}$$

The Bloch wave formalism thus enables us to compute the scattering matrix directly for an arbitrary crystal thickness. Only the diagonal matrix $\mathcal{E}(z)$ depends on the crystal thickness, in a simple exponential way. Later we will see that this results in a fast algorithm to compute bright field and dark field images, in particular for the N-beam case. It is clear from the relations in this and the previous section that the Bloch wave problem is completely solved if the spectral factorization (5.56) of $\bar{A}$ is known. From the eigenvalues $\gamma^{(j)}$ and the eigenvector matrix $\mathcal{C}$ we can then also compute the scattering matrix $\mathcal{S}$. We will discuss an algorithm for spectral factorization in Chapter 7.

5.7.3 Absorption in the Bloch wave formalism

Absorption is taken into account by adding an imaginary part to the electrostatic lattice potential. The eigenvalues $\gamma^{(j)}$ then become complex, and we will denote the new eigenvalues by $\Gamma^{(j)} = \gamma^{(j)} + iq^{(j)}$. Consequently, the wave vectors inside the crystal are given by

$$\mathbf{k}^{(j)} = \mathbf{k}_0 + \left(\gamma^{(j)} + iq^{(j)}\right)\mathbf{n}.$$

Starting from equation (5.5), and proceeding along the same path as before, we find for the eigenvalue equation (including absorption):

$$\begin{pmatrix} \mathrm{i}U'_{\mathbf{0}} & U_{-\mathbf{g}}+\mathrm{i}U'_{-\mathbf{g}} & \dots & U_{-\mathbf{h}}+\mathrm{i}U'_{-\mathbf{h}} \\ U_{\mathbf{g}}+\mathrm{i}U'_{\mathbf{g}} & 2k_0s_{\mathbf{g}}+\mathrm{i}U'_{\mathbf{0}} & \dots & U_{\mathbf{g}-\mathbf{h}}+\mathrm{i}U'_{\mathbf{g}-\mathbf{h}} \\ \vdots & \vdots & \ddots & \vdots \\ U_{\mathbf{h}}+\mathrm{i}U'_{\mathbf{h}} & U_{\mathbf{h}-\mathbf{g}}+\mathrm{i}U'_{\mathbf{h}-\mathbf{g}} & \dots & 2k_0s_{\mathbf{h}}+\mathrm{i}U'_{\mathbf{0}} \end{pmatrix} \begin{pmatrix} C^{(j)}_{\mathbf{0}} \\ C^{(j)}_{\mathbf{g}} \\ \vdots \\ C^{(j)}_{\mathbf{h}} \end{pmatrix} = 2k_n\Gamma^{(j)} \begin{pmatrix} C^{(j)}_{\mathbf{0}} \\ C^{(j)}_{\mathbf{g}} \\ \vdots \\ C^{(j)}_{\mathbf{h}} \end{pmatrix}. \tag{5.61}$$

We have made use of the high-energy approximation (ignoring factors of order $(\Gamma^{(j)})^2$), and also $U'_{\mathbf{0}} \neq 0$. The resulting eigenvalue problem is a general complex eigenvalue problem, which is a non-trivial numerical problem. We will return to this equation in Chapter 6 when we discuss the two-beam case, and in Chapter 7 where we will introduce a standard numerical procedure for determining the eigenvalues and eigenvectors of this equation.

5.8 Important diffraction geometries and diffraction symmetry

5.8.1 Diffraction geometries

In Chapter 4, we have seen that interesting image features are observed for a number of different sample orientations. The most important orientations are as follows.

- **The two-beam orientation**. In this orientation a single diffracted beam is strongly excited, while all other beams are weak. While this orientation is somewhat artificial in that it is nearly impossible to obtain for crystal structures with a large unit cell, it is of importance from a theoretical point of view; the dynamical diffraction equations for the two-beam case have an analytical solution. The two-beam orientation is important for defect characterization and we will discuss the experimental and theoretical details in Chapters 6 and 8.
- **The systematic row orientation**. The *systematic row orientation* is similar to the two-beam orientation, except that now additional reflections along the row $\mathbf{g}$ are included in the image formation process. In an actual experiment, we can, of course, not avoid contributions from higher-order reflections, such as $2\mathbf{g}$ and $-\mathbf{g}$. For crystals with a relatively small unit cell, it is rather easy to orient the sample in a two-beam orientation; this is no longer the case for materials with larger unit cells, or for observations at higher acceleration voltages, for which the Ewald sphere is flatter. For a two-beam sample orientation far away from any zone axis, an entire row of reflections may be visible in the diffraction pattern. If the most intense reflections in a diffraction pattern can be expressed as multiples of a single reciprocal lattice vector $\mathbf{g}$, then we say that the sample is in a *systematic row orientation*. The reciprocal lattice points can be written as $n\mathbf{g}$, with $n_{min} \leq n \leq n_{max}$.

The number of active reflections depends on the unit cell size, the sample orientation and the microscope acceleration voltage.

- **The zone axis orientation**. If the incident beam is parallel to a crystal zone axis, then the diffraction pattern in the objective lens back focal plane will be a two-dimensional pattern. This is the so-called *zone axis orientation*, and the diffraction pattern is known as a *zone axis pattern* or ZAP. If the zone axis is given by $[uvw]$, then we can employ the duality between real space and reciprocal space and the diffraction pattern will correspond to the plane $(uvw)^*$ in reciprocal space.

  Each reflection in a ZAP can be indexed as an integer linear combination of two short reciprocal lattice vectors, $\mathbf{g}_1$ and $\mathbf{g}_2$. The number of contributing reflections need not be the same along these two directions, and hence the general diffracted beam is described by

$$\mathbf{g} = m\mathbf{g}_1 + n\mathbf{g}_2$$

  with $m_{\min} \leq m \leq m_{\max}$ and $n_{\min} \leq n \leq n_{\max}$. In exact zone axis orientation we usually have $m_{\max} = -m_{\min}$ and similarly for n, so that the total number of diffraction spots is equal to $N = (2m_{\max} + 1)(2n_{\max} + 1)$. For materials with a large unit cell, there may well be several hundred diffracted beams.

  We can use the magnitude of the excitation error as a selection criterion for which beams contribute to the multiple scattering process (i.e. all reflections for which $|s_{\mathbf{g}}| \leq s_{max}$ contribute). A general multi-beam simulation program must then allow for the number of beams to depend on the actual beam orientation. This, in turn, obviates the need for an adaptive algorithm that automatically determines which beams need to be included in the simulation.

  For both systematic row orientation and zone axis orientation, the relative orientation of the beam and the crystal is commonly described by the *Laue center*, i.e. the projection of the center of the Ewald sphere onto the reciprocal plane $(uvw)^*$.

In the following chapters, we will discuss the elastic scattering theory for all three sample orientations, and, when possible, provide analytical expressions for the beam intensities. Before we attempt to solve the general dynamical equations, however, we must first discuss the symmetry of the diffraction process. As was the case with crystallography, discussed in Chapter 1, the use of symmetry may provide useful insights into the diffraction process, and may also allow us to create fast algorithms to solve the dynamical equations.

The following sections are based on two landmark papers:

(i) Pogany and Turner [PT68]: *Reciprocity in Electron Diffraction and Microscopy*;
(ii) Buxton, Eades, Steeds, and Rackham [BESR76]: *The Symmetry of Electron Diffraction Zone Axis Patterns.*

The reader is encouraged to read these original papers, as they provide the foundation for all diffraction symmetry work. We will closely follow both papers, with minor changes in notation where appropriate.

5.8.2 Thin-foil symmetry

We have seen in Chapter 2 that the finite nature of a thin foil has consequences for the shape of the reciprocal lattice points. The reciprocal lattice point is no longer a mathematical point with zero volume, but, instead, is related to the Fourier transform of the thin-foil shape function. A second important consequence is that the symmetry of a thin-foil sample *is no longer equal to the space group symmetry of the crystal structure!* This is surprising at first, but it is rather easy to understand. Space groups are *infinite* groups, and the full space group symmetry, in particular the translational symmetry of the underlying Bravais lattice, cannot apply to a planar foil with a finite thickness. Let us consider the various symmetry elements (we will refer to the plane through the center of the foil and parallel to both surfaces as the *center plane*).

- *Bravais lattice translations.* Only translation vectors which lie in the plane of the sample can survive the introduction of the surfaces of the thin foil. All translation vectors with a component along the foil normal are lost.
- *Rotation axes.* All rotation axes which are inclined with respect to the foil normal disappear. Only rotations around the foil normal, and two-fold rotations around an axis in the center plane survive.
- *Mirror planes.* All mirror planes which are inclined with respect to the foil normal disappear. Only mirror planes which contain the foil normal, or the mirror plane coinciding with the center plane survive.
- *Inversion operation.* Only inversion centers in the center plane can survive the introduction of the sample surfaces.
- *Glide planes and screw axes.* Since the only surviving translations are those with a vanishing component along the foil normal, nearly all glide planes and screw axes will vanish. The only surviving glide planes are those for which the mirror plane coincides with the center plane, or glide planes which contain the beam direction and have glide components in the center plane. The only surviving screw axis is the two-fold screw in the center plane (and higher-order screw axes which reduce to the same projected symmetry).

Out of the surviving symmetry elements, there are only two which interchange the top and bottom surfaces of the foil: the mirror plane coinciding with the center plane, and the inversion center in the center plane. These two symmetry elements will become important in the following section.

If the foil is inclined with respect to the beam direction, then there are even fewer surviving symmetry elements; we leave it to the reader to verify that the highest possible point group symmetry of such an inclined parallel foil is $\mathbf{2/m}$, with the two-fold axis normal to the beam direction. For a wedge-shaped foil, the only surviving symmetry element is a mirror plane containing the beam and normal to the wedge edge (point group symmetry **m**).

While the symmetry reductions imposed by the presence of the sample surfaces are significant, things are not as bad as they appear to be. There is a strong probability for electron scattering, so that even a small number of unit cells along the beam direction will impose some of the crystal symmetry on the scattering process. In many cases, the presence of the surfaces can be considered as a perturbation of the perfect space group symmetry. Similarly, internal defects in the crystal lattice will change the space group symmetry, but the perfect material surrounding the defect will impose the space group symmetry on the scattering process. The space group symmetry is therefore a good starting point, even in the case of a thin foil with non-parallel surfaces.

A simple argument, based on equation (5.35), can illustrate† how the crystal symmetry influences the symmetry of the electron wave function at the crystal exit plane. Consider a symmetry operator $\mathcal{O}$ that belongs to the space group of the electrostatic lattice potential $V(\mathbf{r})$, i.e. $\mathcal{O}[V(\mathbf{r})] = V(\mathbf{r})$. In the derivation of the phase grating equation (Section 5.6.1) we have seen that the phase of the electron wave is modified by the *projected potential* V_p (equation 5.37). The projected potential is a two-dimensional function which exhibits the *projected space group symmetry*. This means that only a subset of the space group symmetry operators $\mathcal{O}$ survives the projection operation. We will denote such operators by $\mathcal{O}_p$. If the wave function ψ satisfies equation (5.35), and we apply the projected operator $\mathcal{O}_p$ we find

$$\begin{aligned}\mathcal{O}_p\psi' &= [\mathcal{O}_p(\bar{\Delta}\psi) + \mathcal{O}_p(\bar{V}\psi)],\\ &= (\bar{\Delta}\mathcal{O}_p\psi + \mathcal{O}_p\bar{V}\mathcal{O}_p\psi),\\ &= (\bar{\Delta} + \mathcal{O}_p\bar{V})\mathcal{O}_p\psi,\\ &= (\bar{\Delta} + \bar{V})\mathcal{O}_p\psi.\end{aligned}$$

Both ψ and $\mathcal{O}_p\psi$ satisfy the same equation, which means that the wave function has the symmetry of the projected crystal space group.

5.8.3 The reciprocity theorem

It has been known for a long time that the *principle of reciprocity* holds for Maxwell's theory of the electromagnetic field (e.g. [Wan79, page 532]). Figure 5.4

† This argument is intuitive and not very rigorous. For a rigorous approach to crystal symmetry and its effect on the diffraction process we refer the interested reader to Section 2b in [BESR76].

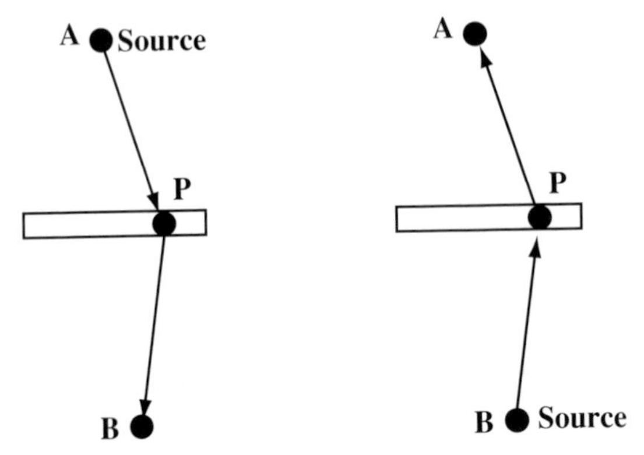

Fig. 5.4. Schematic illustration of the reciprocity theorem.

illustrates the principle:

> The amplitude at B of a wave originating from a source at A, and scattered by P, is equal to the scattered amplitude at A due to the same source placed at B.

Following von Laue's pioneering work on x-ray fluorescence in single crystals in 1935 [vL35], Kainuma [Kai55] and Fukuhara [Fuk66] applied the reciprocity principle to the theory of Kikuchi lines and to multi-beam electron diffraction, respectively. Pogany and Turner [PT68] then formally stated the *reciprocity theorem* as it applies to electron diffraction.

The meaning of the reciprocity theorem can be made clear by the following example [PT68]. Consider a sample with parallel top and bottom surfaces. A beam is incident at an angle α with respect to the foil normal, such that the reciprocal lattice point $\mathbf{g}_{hkl}$ is in exact Bragg orientation. Multiples of $\mathbf{g}_{hkl}$, labeled by the integer n, will also give rise to diffracted beams, as shown schematically in Fig. 5.5(a). The reciprocity theorem then states that the amplitude of the hkl scattered beam for a given incident amplitude along $\mathbf{k}_0$ is the same as the amplitude that would be measured in the $-\mathbf{k}_0$-direction for an incident beam along $-(\mathbf{k}_0 + \mathbf{g}_{hkl})$, shown schematically in Fig. 5.5(b). The amplitudes of the other beams will in general be very different for these two incident beam directions.

The reciprocity theorem thus states that for a given diffracted beam $\psi_\mathbf{g}$, there are two different incident beam directions which will give rise to the same amplitude. The two incident beam directions are on opposite sides of the crystal, which makes it a bit inconvenient to check the validity of the theorem in the microscope. It can be shown (e.g. [PT68,Wan95]) that the reciprocity theorem is also valid for intensities, even in the presence of inelastic scattering processes.

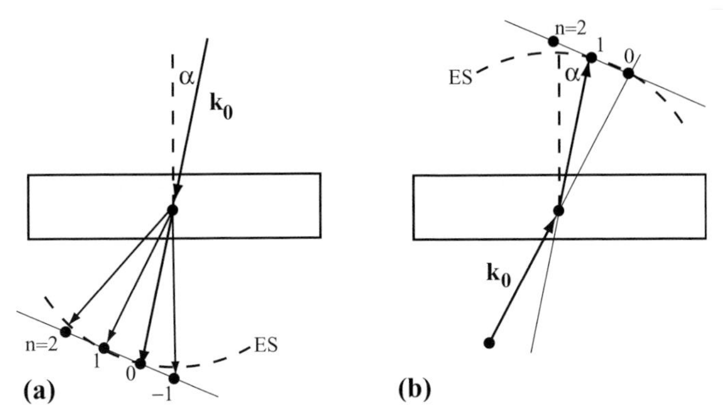

Fig. 5.5. (a) Schematic beam configuration for an incident beam and $\mathbf{g}_{hkl}$ reciprocal lattice point in exact Bragg orientation; (b) configuration equivalent to (a) by reciprocity.

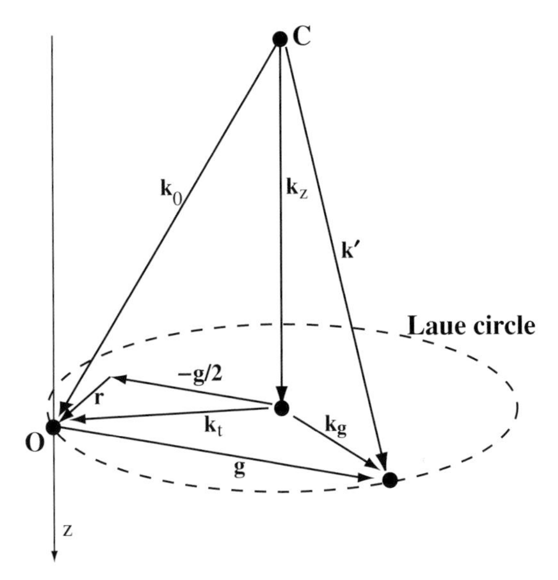

Fig. 5.6. Graphical definition of the various vectors used to derive the effect of the reciprocity theorem on the symmetry of the amplitudes of diffracted beams.

It is useful to formalize the reciprocity theorem, and to study its relationship to other symmetry properties of the crystalline sample. Following Pogany and Turner [PT68], but with slightly different notation, we will write the incident wave vector $\mathbf{k}_0$ as the sum of two vectors; $\mathbf{k}_z$ along the z-direction (which we will assume to be normal to the foil surfaces), and $\mathbf{k}_t$ which is tangential to the foil surface (and normal to the z-direction). Figure 5.6 shows a graphical representation of the decomposition.

We also introduce the vector $\mathbf{k_g}$, which is the tangential component of the scattered wave vector $\mathbf{k}'$. The following relations are now valid:

$$\mathbf{k}_0 = \mathbf{k}_t + \mathbf{k}_z;$$
$$\mathbf{k}' = \mathbf{k}_0 + \mathbf{g} = \mathbf{k}_z + \mathbf{k_g};$$
$$\mathbf{k_g} = \mathbf{k}_t + \mathbf{g}.$$

Since the length of the wave vectors is constant for elastic scattering processes, we only need to keep track of the magnitude of the tangential component; the normal component is then calculated easily from the wave number and the tangential component. The only information we need to keep about the normal component is whether the beam is incident from above or from below the sample. We will denote a wave vector incident from above the sample by the components $(\otimes, \mathbf{k}_t)$, and a wave vector incident from below by the components $(\odot, \mathbf{k}_t)$. The amplitude of the diffracted beam $\mathbf{g}$ is then written as

$$\psi_{\mathbf{g}} = \psi_{\mathbf{g}}(\otimes, \mathbf{k}_t, C),$$

where the last argument C indicates the crystal itself; the meaning of this last argument will be made clear shortly.

Application of the reciprocity theorem is equivalent to reversing the beam direction to be along the vector $-\mathbf{k}'$, which can be decomposed into $-\mathbf{k}_z - (\mathbf{k}_t + \mathbf{g}) = (\odot, -(\mathbf{k}_t + \mathbf{g}))$. The corresponding amplitude of the diffracted beam from *the same set of planes* $\mathbf{g}$ is then given by

$$\psi_{\mathbf{g}}(\odot, -(\mathbf{k}_t + \mathbf{g}), C).$$

The second argument of this expression is the tangential component, which does not necessarily have the same magnitude as $\mathbf{k}_t$. We recall from Chapter 2 that in exact Bragg orientation the tangential component of the incident wave vector must be equal to $-\mathbf{g}/2$, and therefore we introduce a new vector $\mathbf{r}$, which describes the deviation from the Bragg condition. Decomposing the tangential component $\mathbf{k}_t$ into $-\mathbf{g}/2 + \mathbf{r}$ (as defined in Fig. 5.6), and applying the reciprocity theorem, we have for the diffracted amplitudes:

$$\boxed{\psi_{\mathbf{g}}\left(\otimes, -\frac{\mathbf{g}}{2} + \mathbf{r}, C\right) = \psi_{\mathbf{g}}\left(\odot, -\frac{\mathbf{g}}{2} - \mathbf{r}, C\right).} \tag{5.62}$$

This equation states explicitly the reciprocity relation for a given set of diffracting planes, and is valid regardless of the underlying crystal symmetry. It is also valid regardless of the number of diffracted beams involved.

The schematic drawing in Fig. 5.7 illustrates the meaning of the reciprocity theorem: the configurations in *a* and *b* are equivalent by reciprocity. The reciprocity

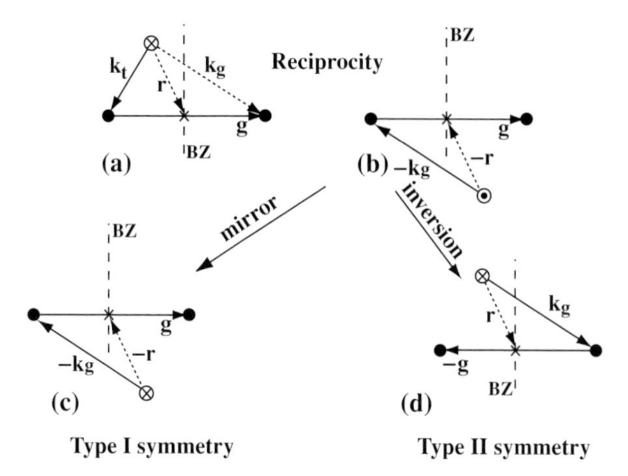

Fig. 5.7. Parts (a) and (b) are equivalent by reciprocity; (c) is derived from (b) by application of a mirror operation in the plane of the drawing. Part (d) follows from (b) by application of an inversion operation.

theorem is particularly useful if the crystal foil has symmetry elements which relate the top and bottom surfaces to each other. On page 330 we have found that the only crystal symmetry elements that survive the introduction of the foil surfaces and which interchange the top and bottom of the crystal foil are:

- a mirror plane through the center of the foil, parallel to the foil surfaces, and
- an inversion center located in the middle of the foil.

We will denote the crystal foil obtained from C by application of the mirror plane by $C^{(m)}$. The mirror operation changes the sign of the normal component of every vector. Similarly, the crystal foil obtained from C by application of the inversion operation is denoted by $C^{(-)}$. The inversion operation replaces every vector by its opposite.

A crystal with a mirror plane through the center, normal to the z-axis, is said to have *type I symmetry*; the reciprocity relation (5.62) then reads as

$$\psi_{\mathbf{g}}\left(\otimes, -\frac{\mathbf{g}}{2} + \mathbf{r}, C\right) = \psi_{\mathbf{g}}\left(\otimes, -\frac{\mathbf{g}}{2} - \mathbf{r}, C^{(m)}\right). \tag{5.63}$$

This relation is depicted graphically in Fig. 5.7(c). A crystal with an inversion center in the center plane is said to exhibit *type II symmetry*, and application of the reciprocity relation results in

$$\psi_{\mathbf{g}}\left(\otimes, -\frac{\mathbf{g}}{2} + \mathbf{r}, C\right) = \psi_{-\mathbf{g}}\left(\otimes, \frac{\mathbf{g}}{2} + \mathbf{r}, C^{(-)}\right). \tag{5.64}$$

This relation is depicted graphically in Fig. 5.7(d).

Let us consider these two symmetry types in more detail. For a crystal with type I symmetry (i.e. $C = C^{(m)}$), the relation (5.63) reads as

$$\psi_{\mathbf{g}}\left(\otimes, -\frac{\mathbf{g}}{2} + \mathbf{r}, C\right) = \psi_{\mathbf{g}}\left(\otimes, -\frac{\mathbf{g}}{2} - \mathbf{r}, C\right).$$

This relation states that, in the presence of a center mirror plane, the diffracted amplitudes (and hence intensities) corresponding to the deviations from Bragg orientation $\mathbf{r}$ and $-\mathbf{r}$ must be equal. This implies that the diffraction disk for the beam corresponding to $\mathbf{g}$ has at least this internal symmetry. Since an inversion operation in two dimensions is actually a rotation of 180° around the normal to the plane of the drawing, we find that in the presence of a horizontal mirror, each diffraction disk must have a two-fold rotation axis through the location of the exact Bragg orientation. The operation which rotates a diffraction disk around its own center by 180° is commonly denoted by the symbol R [BESR76].

As an example, consider a crystal with point group symmetry **1**, i.e. no symmetry elements. If the incident beam lies along a zone axis, then a plane of reciprocal lattice points will be tangent to the Ewald sphere. If only the reciprocal lattice points in this plane contribute to the scattering process, then the plane itself becomes a mirror plane (all points not in the plane are ignored, and therefore the mirror symmetry is valid). This is known as the *projection approximation*. For a converged incident beam, each diffraction spot will be a disk with radius determined by the beam convergence angle. The presence of the mirror plane (both in reciprocal and direct space) implies that each diffraction disk has the 180° symmetry around the location of exact Bragg orientation. The resulting symmetry is written as $\mathbf{1}_R$, and this is one of the 31 so-called *diffraction groups*. A diffraction group combines the point group symmetry of the crystal with the type I and type II symmetry operations of the crystal foil.

The graphical depiction of the diffraction group $\mathbf{1}_R$ can be constructed by combining Figs 5.7(a) and (c) into a single drawing, as shown in Fig. 5.8(a). We can then simplify this figure by only drawing those components which are essential: the origin of reciprocal space is indicated by a plus-sign, the diffraction disk by a circle. The direction of the vector $\mathbf{g}$ is along the line between the origin and the center of the disk, and is represented by a solid line across the disk. The disk center, corresponding to the exact Bragg orientation, is indicated by a short line segment. Points in the disk which are related to each other by a symmetry operation are indicated by small open circles. The resulting schematic drawing for diffraction group $\mathbf{1}_R$ is then shown in Fig. 5.8(b).

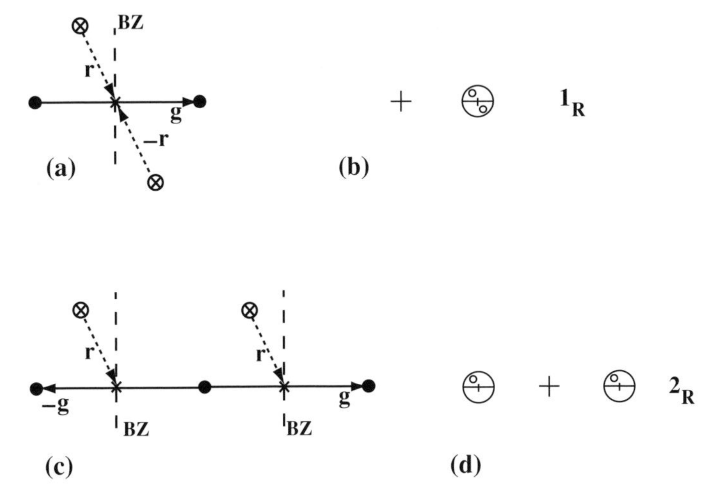

Fig. 5.8. Parts (a) and (b) derivation of the $\mathbf{1}_R$ diffraction group drawing; (c) and (d) the same for the diffraction group $\mathbf{2}_R$.

Next, we consider a crystal foil with type II symmetry ($C = C^{(-)}$). The relation (5.64) reads as

$$\psi_{\mathbf{g}}\left(\otimes, -\frac{\mathbf{g}}{2} + \mathbf{r}, C\right) = \psi_{-\mathbf{g}}\left(\otimes, \frac{\mathbf{g}}{2} + \mathbf{r}, C\right).$$

This implies a relation between amplitudes (and hence intensities) in opposite diffraction disks $\mathbf{g}$ and $-\mathbf{g}$. We can again determine the resulting symmetry by combining Figs. 5.7(a) and (d) into a single drawing, shown in Fig. 5.8(c). The resulting diffraction group is $\mathbf{2}_R$, i.e. a rotation of 180° around the origin followed by a 180° rotation around the disk center. An important implication of this particular diffraction symmetry is that, if the $\mathbf{g}$ and $-\mathbf{g}$ disks *do not* show this relation, then the crystal structure can not have inversion symmetry.

We can repeat this same procedure for all possible orientations of all 32 crystallographic point groups, and it can be shown (e.g. [BESR76]) that this leads to 31 *diffraction groups*. Alternatively, one can use the 10 two-dimensional point groups† and combine each of them with the crystal foil symmetries of type I and type II. The graphical representations of the 31 diffraction groups are given in Figs. 5.9 (adapted from Table 1 in [BESR76]). The diffraction groups derived from a given two-dimensional point group are displayed in 10 sets of two, three or four groups. The short straight line segments in some of the drawings indicate mirror planes; if the reciprocal lattice point lies on one of those mirror planes, then the internal

† The two-dimensional crystallographic point groups (planar groups) are **1**, **2**, **m**, **2mm**, **4**, **4mm**, **3**, **3m**, **6**, and **6mm**.

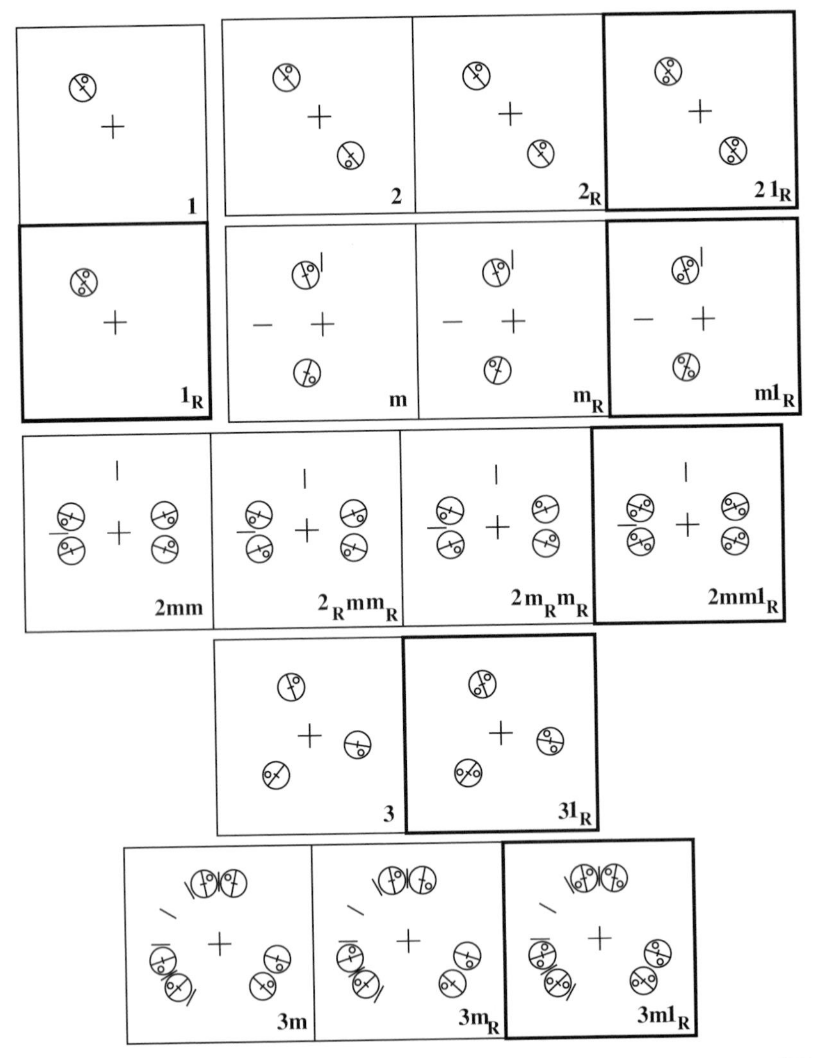

Fig. 5.9. Graphical representation of the 31 diffraction groups (based on [BESR76]).

symmetry of the corresponding disk is higher than that of the general disk. It is clear from the figures that there are only four possible symmetries for an individual diffraction disk: **1**, **2**, **m**, and **2mm**, as shown schematically in Fig. 5.10. The symmetry of an individual diffraction disk is generally known as the *dark field symmetry*, and we distinguish between the *general* dark field symmetry (**1** or **2**), and the *special* dark field symmetry (**m** or **2mm**), which can only occur in those diffraction groups that contain mirror planes.

If the projection approximation is valid, then the highest-order diffraction group in each set must be used. This is the so-called *projection diffraction group*; the

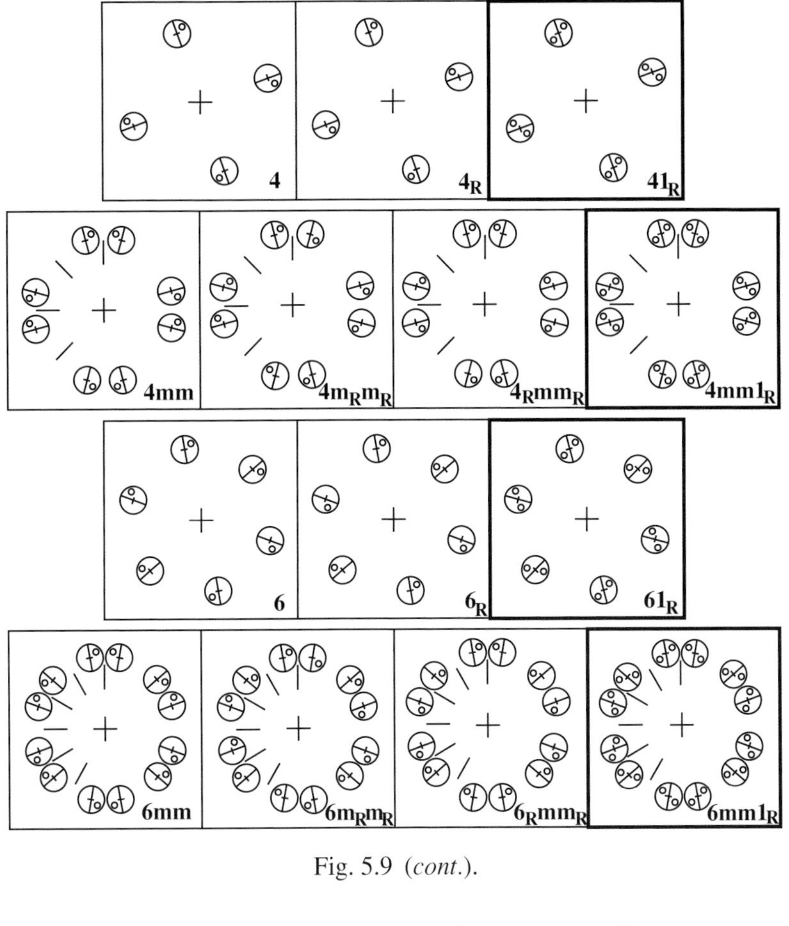

Fig. 5.9 (*cont.*).

Fig. 5.10. The four possible symmetries of an individual diffraction disk. Symmetries **m** and **2mm** are only possible if the diffraction disk lies on a mirror plane of the parent crystal structure.

schematic for this group is surrounded by a thicker frame in Figs. 5.9 and is marked by the letter [P] in Table 5.1. All projection diffraction group symbols end in $\mathbf{1}_R$, reflecting the mirror symmetry of the projection approximation.

It is now a simple, but somewhat lengthy task to determine the diffraction group for every possible zone axis orientation for each crystallographic point group. This was first done by Buxton et al. [BESR76], and their Table 3 is reproduced in Fig. 5.11. The table lists for each crystallographic point group which diffraction groups are possible. We will make use of this table when we discuss convergent

Table 5.1. *Diffraction pattern symmetries (from Table 2 in [BESR76]).*

Diffraction group		Bright field	Whole pattern	Dark field General	Dark field Special
1		**1**	**1**	**1**	None
$\mathbf{1}_R$	[P]	**2**	**1**	**2**	None
2		**2**	**2**	**1**	None
$\mathbf{2}_R$		**1**	**1**	**1**	None
$\mathbf{21}_R$	[P]	**2**	**2**	**2**	None
$\mathbf{m}_R$		**m**	**1**	**1**	**m**
m		**m**	**m**	**1**	**m**
$\mathbf{m1}_R$	[P]	**2mm**	**m**	**2**	**2mm**
$\mathbf{2m}_R\mathbf{m}_R$		**2mm**	**2**	**1**	**m**
2mm		**2mm**	**2mm**	**1**	**m**
$\mathbf{2}_R\mathbf{mm}_R$		**m**	**m**	**1**	**m**
$\mathbf{2mm1}_R$	[P]	**2mm**	**2mm**	**2**	**2mm**
4		**4**	**4**	**1**	None
$\mathbf{4}_R$		**4**	**2**	**1**	None
$\mathbf{41}_R$	[P]	**4**	**4**	**2**	None
$\mathbf{4m}_R\mathbf{m}_R$		**4mm**	**4**	**1**	**m**
4mm		**4mm**	**4mm**	**1**	**m**
$\mathbf{4}_R\mathbf{mm}_R$		**4mm**	**2mm**	**1**	**m**
$\mathbf{4mm1}_R$	[P]	**4mm**	**4mm**	**2**	**2mm**
3		**3**	**3**	**1**	None
$\mathbf{31}_R$	[P]	**6**	**3**	**2**	None
$\mathbf{3m}_R$		**3m**	**3**	**1**	**m**
3m		**3m**	**3m**	**1**	**m**
$\mathbf{3m1}_R$	[P]	**6mm**	**3m**	**2**	**2mm**
6		**6**	**6**	**1**	None
$\mathbf{6}_R$		**3**	**3**	**1**	None
$\mathbf{61}_R$	[P]	**6**	**6**	**2**	None
$\mathbf{6m}_R\mathbf{m}_R$		**6mm**	**6**	**1**	**m**
6mm		**6mm**	**6mm**	**1**	**m**
$\mathbf{6}_R\mathbf{mm}_R$		**3m**	**3m**	**1**	**m**
$\mathbf{6mm1}_R$	[P]	**6mm**	**6mm**	**2**	**2mm**

beam electron diffraction in Chapter 9. For completeness we also reproduce in Table 5.1 a portion of a related table (Table 2, [BESR76]), which lists the various symmetries for each diffraction group. This table will be discussed in more detail in Chapter 9.

5.9 Concluding remarks and recommended reading

The literature on the dynamical theory of electron diffraction is extensive. The theory can be traced back to the initial Bloch wave treatment of electron diffraction by Bethe [Bet28] in 1928, followed in 1940 by a more detailed theory by

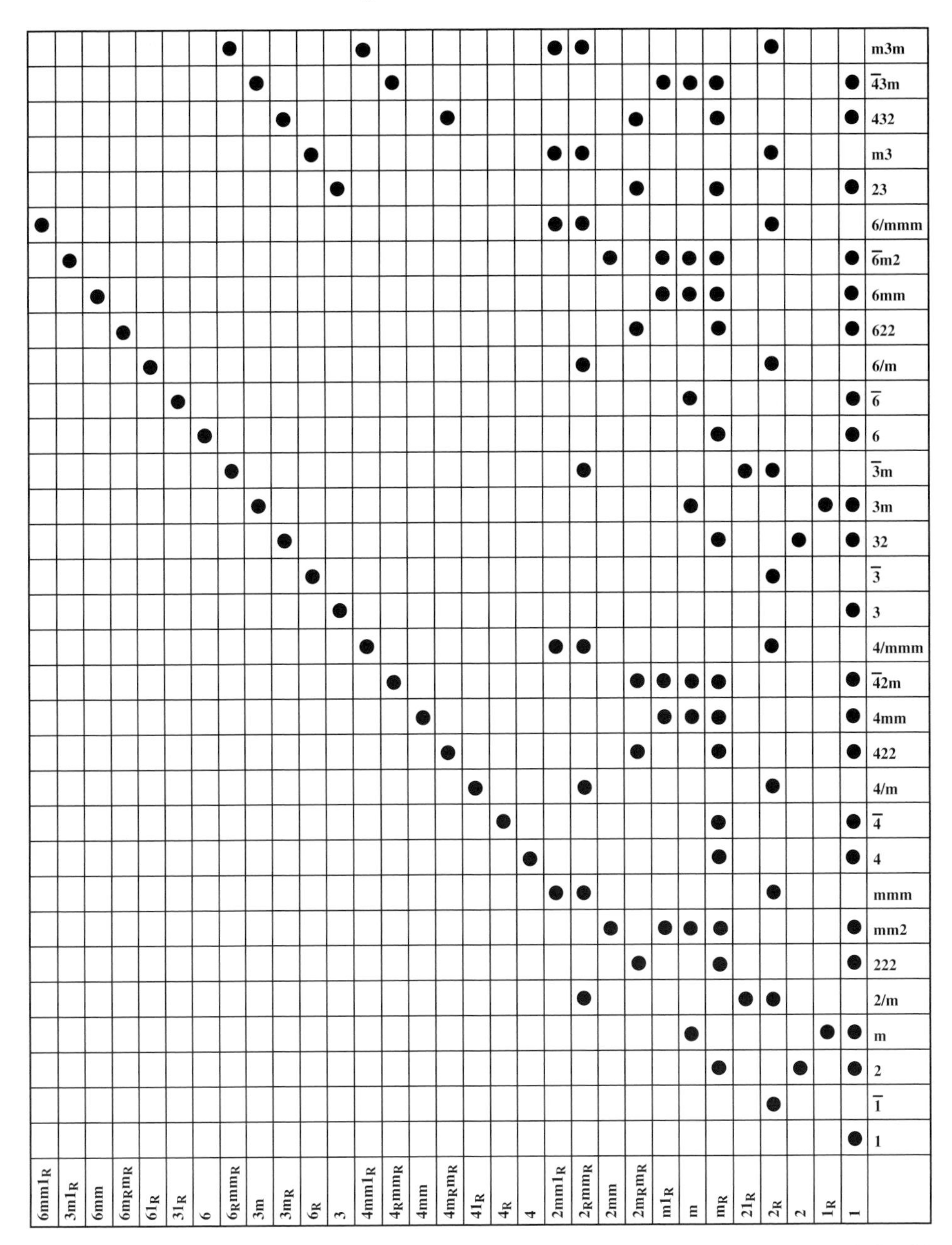

$6mm1_R$	$3m1_R$	6mm	$6m_Rm_R$	61_R	31_R	6	6_Rmm_R	3m	$3m_R$	6_R	3	$4mm1_R$	4_Rmm_R	4mm	$4m_Rm_R$	41_R	4_R	4	$2mm1_R$	2_Rmm_R	2mm	$2m_Rm_R$	$m1_R$	m	m_R	21_R	2_R	2	1_R	1	
							●					●							●	●							●				m3m
								●					●										●	●	●					●	$\bar{4}3m$
									●						●							●			●					●	432
										●									●	●							●				m3
											●											●			●					●	23
●																			●	●							●				6/mmm
	●																				●		●	●	●					●	$\bar{6}m2$
		●																					●	●	●					●	6mm
			●																			●			●					●	622
				●																●							●				6/m
					●																			●						●	$\bar{6}$
						●																			●					●	6
							●													●						●	●				$\bar{3}m$
								●																●					●	●	3m
									●																●			●		●	32
										●																	●				$\bar{3}$
											●																			●	3
												●							●	●							●				4/mmm
													●									●	●	●	●					●	$\bar{4}2m$
														●									●	●	●					●	4mm
															●							●			●					●	422
																●				●							●				4/m
																	●								●					●	$\bar{4}$
																		●							●					●	4
																			●	●							●				mmm
																					●		●	●	●					●	mm2
																						●			●					●	222
																				●						●	●				2/m
																								●					●	●	m
																									●			●		●	2
																											●				$\bar{1}$
																														●	1

Fig. 5.11. Table representing the relation between the 31 diffraction groups and the 32 crystallographic point groups. (Table 3 in [BESR76].)

MacGillavry [Mac40]. In 1949, Heidenreich [Hei49] published an important paper on the theory and application of electron diffraction. This paper can be considered to be the first didactical exposition, and although the notation is somewhat different from that used in this book, the reader is strongly encouraged to study this publication. In the 1950s, electron microscopes became more widely available; several research groups worked on the development of the dynamical theory and the formalism became firmly rooted in quantum mechanics. Among the most important advances or reviews are the papers by: Cowley and Moodie (introduction of slice methods, 1957, [CM57]); Hirsch, Howie, and Whelan (kinematical theory of defect images, 1960, [HHW60]); Howie and Whelan (first derivation of DHW equations (5.18), 1961, [HW61]); Tournarie (matrix solution similar to equation (5.24), 1961, [Tou61]); Sturkey (scattering matrix formalism, 1962, [Stu62]); Gevers, Blank, and Amelinckx (extension of DHW equations to non-centrosymmetric crystals, 1966, [GBA66]); Metherell (review of the Bloch wave method, 1975, [Met75]); Humphreys (review of Bloch wave method, 1979, [Hum79]); Bird (review of theory and application to convergent beam electron diffraction, 1989, [Bir89]). This list is by no means exhaustive, and the reader should consult the references in these papers for additional publications.

The relativistic nature of high-energy electron diffraction was analyzed in detail by Fujiwara in 1961 [Fuj61]. The equivalence of the various theories developed in the 1950s and 1960s was, at first, not obvious; in 1974 Goodman and Moodie [GM74] reviewed the different formalisms and described their equivalence in detail. In 1975 Van Dyck [VD75] showed that all of the dynamical theory formulations could be derived from a path integral formulation of the scattering process.

The theory presented in this chapter is based entirely on *plane waves*. The electron wave functions are written as linear combinations of plane waves, with wave vectors given by the Bragg equation (for the DHW formulation), or determined by the solution of an eigenvalue problem (for the Bloch wave method). Since the high-energy elastic scattering problem is equivalent to a standard band-structure problem in solid-state physics, we might consider changing from a plane wave description to a description in terms of other basis functions. We have already seen in Chapter 2 that a large number of plane waves (Fourier coefficients) is needed to describe accurately the variations of the electrostatic lattice potential near atoms. Basis functions which are better suited to describing such variations are well known in the solid-state physics community, and several alternative basis sets have been proposed in the literature, among which are the following.

- Combined basis algorithm (CB). Tochilin and Whelan [TW93] introduced a so-called *combined basis* approach to reduce the size of the dynamical matrix. The basis functions

consist of a combination of plane waves and localized functions describing electrons channeling along atomic strings. Such a description results in a fast algorithm and sometimes allows for an analytical solution of a multi-beam configuration.
- Orthogonalized plane waves (OPW). Barreteau and Ducastelle [BD95] have suggested the use of an OPW expansion to reduce the size of the scattering matrix. For zone axis orientations this method, combined with the use of symmetry considerations (see Chapter 7), results in a significantly faster algorithm.
- Atomic column approximation (ACA). Van Dyck and Chen [VDC99] have proposed an analytical solution for the exit wave function, based on an atomic column approximation. They show that the contribution of each column of atoms parallel to a zone axis can be parameterized with a single parameter, which depends on the "projected weight" of the column.

While all of these alternative methods do succeed in reducing the size of the computational problem, they are not commonly available as public-domain or commercial algorithms, nor have they been implemented for the most general crystal structure case. It will be up to the interested and adventurous reader to experiment with these and other alternative approaches.

To conclude this chapter on the multi-beam theory of elastic electron scattering, we list a few other sources which the reader may wish to consult for additional information:

- *Modern Diffraction and Imaging Techniques in Material Science*, edited by S. Amelinckx, R. Gevers, G. Renault, and J. Van Landuyt [AGRVL70].
- *Electron Microscopy of Thin Crystals*, P. Hirsch, A. Howie, R.B. Nicholson, D.W. Pashley, and M.J. Whelan [HHN+77].
- *Electron Microdiffraction*, J.C.H. Spence and J.M. Zuo [SZ92].
- *Transmission Electron Microscopy, Physics of Image Formation and Microanalysis*, L. Reimer [Rei93].
- *Principles of Electron Optics. Volume 3: Wave Optics*, P.W. Hawkes and E. Kasper [HK94].
- *Transmission Electron Microscopy: a Textbook for Materials Science*, D.B. Williams and C.B. Carter [WC96].
- *Advanced Computing in Electron Microscopy*, E.J. Kirkland [Kir98].

Exercises

5.1 Derive equations (5.54) and (5.55).

5.2 Write down the explicit equations (5.24) for a three-beam case with the following beams:

(a) $-\mathbf{g}, \mathbf{0}, \mathbf{g}$;

(b) $\mathbf{0}, \mathbf{g}, 2\mathbf{g}$.

Next, write down the corresponding equations (5.61) for the Bloch wave formulation.

5.3 Show that equation (5.31) is indeed incorrect.

5.4 Show that a wedge-shaped foil can have at the most the point group symmetry **m**.

5.5 For a crystal with point group symmetry **4/mmm** determine which zone axis orientations $[uvw]$ correspond to each of the allowed diffraction groups.

6

Two-beam theory in defect-free crystals

6.1 Introduction

Most x-ray diffraction experiments are designed to *increase* the probability that a reciprocal lattice point will fall on the Ewald sphere. This is usually not a problem for electron diffraction, since the radius of the Ewald sphere is much larger, as we have seen in Chapter 2. It is, therefore, ironic that the most basic TEM technique involves *minimizing* the number of lattice points on the Ewald sphere! Indeed, the most popular conventional TEM technique relies on the use of only two electron beams, the transmitted beam and a single scattered beam. In many cases, we have to work hard at getting only one diffracted beam. The shorter the wavelength, the more reflections will fall on or close to the Ewald sphere for any given crystal orientation. Furthermore, if the direct space lattice has a large unit cell, then reciprocal space will be filled densely with lattice points and hence two-beam microscopy becomes progressively more difficult with increasing unit cell size.

There is a good reason for the use of two-beam techniques: the theory becomes mathematically more tractable and closed-form solutions to the dynamical scattering equations are known. Intuitively, it makes a lot of sense to study a crystal by looking at one set of lattice planes at a time. For instance, when studying dislocations, the displacement field around the dislocation core causes lattice planes to distort. For a given dislocation type, certain lattice planes may remain undistorted and the two-beam technique allows identification of these undistorted lattice planes, thus providing valuable information concerning the displacement field of the dislocation. Before we can begin to study defects, however, we need to understand elastic electron scattering in a defect-free crystal.

We will see in this chapter that perfect, defect-free crystals give rise to image contrast features (in two-beam mode), which are *essentially independent of the crystal structure*. In other words, the contrast features observed for a perfect crystal are the same whether we study metals, semiconductors, ceramics, or any other crystalline material. In addition, the general image features are qualitatively the same at all acceleration voltages, for all microscopes. We will assume that we have a defect-free crystal for which we can compute the electrostatic lattice potential, as described in Section 2.6. We will start from the general multi-beam elastic scattering equations derived in the previous chapter, and apply them to the non-centrosymmetric two-beam case. This approach will permit us to derive explicit mathematical expressions for the intensities of transmitted and diffracted beams. Once the general expressions are derived, it will be straightforward to apply them to specific cases.

All theoretical computations will be compared with experimental images obtained from the four study materials. The reader should by now have prepared thin foils of one or more of the study materials, following the instructions in Chapter 4. Several of the experimental images of Chapter 4 will be analyzed in the present chapter and compared with theoretical predictions. First, we will simplify the DHW equations (5.18) to the two-beam situation. After deriving the solutions for a perfect defect-free crystal, we will discuss different ways to compute the image contrast numerically and compare the simulations with the observations on the study materials. We will introduce the *column approximation* and apply it to the two-beam case.

In the second half of this chapter we will repeat the two-beam analysis, this time using the Bloch wave formalism (equation 5.51). We will introduce the dispersion surface construction, which is the dynamical equivalent to the kinematical Ewald sphere construction. The Bloch wave formalism will provide physical insight into the phenomenon of *anomalous absorption*. We will conclude the chapter with examples of two-beam image simulations, along with Fortran-90 source code and **ION**-routines available from the website.

6.2 The column approximation

Since scattering angles for electron diffraction are small compared with the angles in x-ray diffraction, electron diffraction is essentially a *forward scattering* process. This means that, even after being scattered several times, most electrons will travel nearly parallel to the incident electron beam. Let us assume that the maximum cumulative scattering angle is, say, 2°; then an electron which travels through a 100 nm foil will end up at most 3.5 nm away from the projection of the entrance point on the exit plane. The higher the acceleration voltage, the smaller the diffraction angle, which means that the lateral spread of the electron wave decreases. It is then a good approximation to assume that an electron which enters the foil at one point will never leave a cylindrical column centered on this point; this column is parallel to the incident beam direction, and has a diameter of only a few nanometers (depending on the foil thickness and the acceleration voltage). For numerical purposes it is more

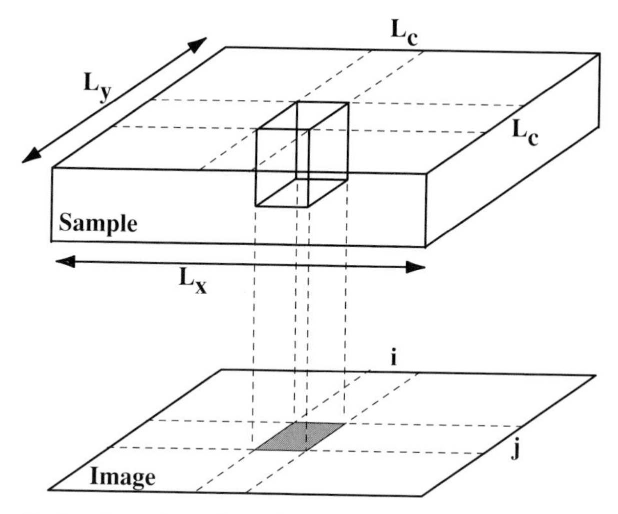

Fig. 6.1. Schematic drawing of a column in a crystal, and how that column relates to the image pixels.

convenient to consider *rectangular* columns of width $L_c = 5$–10 nm, as shown in Fig. 6.1. This is called the *column approximation*, first introduced around 1957 by M. Whelan and coworkers [WHHB57]. It is assumed that neighboring columns do not interact, i.e. *no electrons are exchanged between columns*, so that the dynamical equations can be solved for each column in turn. The column approximation is essentially equivalent to the high-energy approximation (see page 306).

Consider a rectangular sample area (perpendicular to the electron beam) of dimensions L_x by L_y (distances are measured in nanometers). A schematic drawing is shown in Fig. 6.1. We can subdivide this area into square elements with edge length L_c (typically 5–10 nm). The total number of squares N_t is then given by

$$N_t = \frac{L_x}{L_c} \times \frac{L_y}{L_c} = N_x \times N_y,$$

where N_x (N_y) is the number of squares along the x (y) direction. Each square can be considered to be the top of a column, and is labeled by the pair of integers (i, j). The length of the column (the local sample thickness $z_0(i, j)$) and the crystal orientation along the column (the local excitation error $s_{\mathbf{g}}(i, j)$) are known, and then it is a simple matter of solving the dynamical diffraction equations for those parameters to obtain the transmitted and scattered intensities at the exit plane of the column. These intensities can then be arranged into arrays of $N_x \times N_y$ pixels, where each pixel is labeled by two integers i and j, as shown in Fig. 6.1. In this way, we can calculate numerically the bright field and multiple dark field images for a given sample geometry. We will make extensive use of this basic image simulation method throughout this and the following chapter.

6.3 The two-beam case: DHW formalism

6.3.1 The basic two-beam equations

We will consider only the transmitted beam $\psi_{\mathbf{0}}$ and one diffracted beam $\psi_{\mathbf{g}}$. We have two equations of the type (5.18), one for each beam. The summation on the right-hand side of equation (5.18) also contains only two terms, $\mathbf{g}' = \mathbf{0}$ and $\mathbf{g}' = \mathbf{g}$. This results in the following equations (using equation 5.17):

$$\left.\begin{array}{llllllll}
\text{beam} & \text{left-hand side} & & & \mathbf{g}' = \mathbf{0} & & \mathbf{g}' = \mathbf{g} \\
\Downarrow & \Downarrow & & & \Downarrow & & \Downarrow \\
\mathbf{0}: & \frac{d\psi_{\mathbf{0}}}{dz} & - & 2\pi i s_{\mathbf{0}} \psi_{\mathbf{0}} & = & -\frac{\pi}{\xi_0'}\psi_{\mathbf{0}} & + & i\pi \frac{e^{i\theta_{-\mathbf{g}}}}{q_{-\mathbf{g}}}\psi_{\mathbf{g}}; \\
\mathbf{g}: & \frac{d\psi_{\mathbf{g}}}{dz} & - & 2\pi i s_{\mathbf{g}} \psi_{\mathbf{g}} & = & i\pi \frac{e^{i\theta_{\mathbf{g}}}}{q_{\mathbf{g}}}\psi_{\mathbf{0}} & - & \frac{\pi}{\xi_0'}\psi_{\mathbf{g}}.
\end{array}\right\} \quad (6.1)$$

These equations will be the starting point for two different derivations: the kinematical two-beam theory and the dynamical two-beam theory.

6.3.2 The two-beam kinematical theory

As a first step towards the general solution of the dynamical equations (6.1), let us make a few simplifying assumptions.

(i) Ignore absorption. This means that all factors $1/\xi_{\mathbf{g}}'$ (including the normal absorption term $1/\xi_0'$) must vanish. Equivalently, we have

$$\frac{1}{q_{\mathbf{g}-\mathbf{g}'}} = \frac{1}{\xi_{\mathbf{g}-\mathbf{g}'}}.$$

(ii) The *single scattering approximation*. The electron either remains in the transmitted beam and is never scattered, or it is scattered elastically into a diffracted beam and remains there. Only scattering *from* the transmitted beam *into* a diffracted beam is allowed, which can be expressed mathematically as

$$\frac{1}{\xi_{\mathbf{g}-\mathbf{g}'}} = 0 \quad \text{except for} \quad \frac{1}{\xi_{\mathbf{g}-\mathbf{0}}}.$$

This is known as the *kinematical approximation*.

Since the excitation error $s_{\mathbf{0}}$ is equal to zero (the origin of reciprocal space always lies on the Ewald sphere), we can simplify the two-beam equations (6.1) to their kinematical form:

$$\frac{\mathrm{d}\psi_\mathbf{0}}{\mathrm{d}z} = 0; \tag{6.2}$$

$$\frac{\mathrm{d}\psi_\mathbf{g}}{\mathrm{d}z} = 2\pi \mathrm{i} s_\mathbf{g}\psi_\mathbf{g} + \mathrm{i}\pi\frac{\mathrm{e}^{\mathrm{i}\theta_\mathrm{g}}}{\xi_\mathbf{g}}\psi_\mathbf{0}. \tag{6.3}$$

The first equation follows because both the extinction distance and the excitation error of the transmitted beam vanish. It is clear that under the kinematical approximation the incident beam has a *constant amplitude*, and therefore a constant intensity which is usually taken to be $I_T = |\psi_\mathbf{0}|^2 = 1$. It is convenient to also take $\psi_\mathbf{0} = 1$.

The above assumptions can also be inserted directly into the DHW equations (6.1), and we find

$$\boxed{\frac{\mathrm{d}\psi_\mathbf{g}}{\mathrm{d}z} - 2\pi \mathrm{i} s_\mathbf{g}\psi_\mathbf{g} = \mathrm{i}\pi\frac{\mathrm{e}^{\mathrm{i}\theta_\mathrm{g}}}{\xi_\mathbf{g}}.} \tag{6.4}$$

These equations (one for each set of lattice planes **g**) form the starting point for the *kinematical multi-beam theory*.

Equation (6.3), which is the same for *every* diffracted beam in the multi-beam case, can be solved with the following substitution:

$$\psi_\mathbf{g} = S\mathrm{e}^{2\pi \mathrm{i} s_\mathbf{g} z},$$

which results in

$$\frac{\mathrm{d}S}{\mathrm{d}z} = \frac{\mathrm{i}\pi}{\xi_\mathbf{g}}\mathrm{e}^{\mathrm{i}\theta_\mathrm{g}}\mathrm{e}^{-2\pi \mathrm{i} s_\mathbf{g} z}.$$

Integration from $z = 0$, with $S(z = 0) = 0$ (i.e. there is no diffracted amplitude at the entrance surface) results in

$$S = \frac{\mathrm{i}\pi}{\xi_\mathbf{g}}\mathrm{e}^{\mathrm{i}\theta_\mathrm{g}}\int_0^{z_0}\mathrm{e}^{-2\pi \mathrm{i} s_\mathbf{g} z}\,\mathrm{d}z, \tag{6.5}$$

from which the solution follows:

$$\psi_\mathbf{g} = \mathrm{e}^{\mathrm{i}\theta_\mathrm{g}}\mathrm{e}^{\mathrm{i}\pi s_\mathbf{g} z_0}\frac{\mathrm{i}\sin\left(\pi s_\mathbf{g} z_0\right)}{s_\mathbf{g}\xi_\mathbf{g}}. \tag{6.6}$$

The kinematical diffracted intensity $I_S = |\psi_\mathbf{g}|^2$ for a crystal with total thickness z_0 is given by

$$\boxed{I_S = \frac{\sin^2\left(\pi s_\mathbf{g} z_0\right)}{\left(s_\mathbf{g}\xi_\mathbf{g}\right)^2}.} \tag{6.7}$$

Note that this expression can also be derived by means of the so-called *first Born approximation*. If we replace the wave function Ψ on the right-hand side of equation (5.5) by the incident plane wave Ψ_0, the resulting equation can be solved for Ψ. The mathematics is similar to that for the dynamical theory, except for the fact that the resulting differential equations for the diffracted beams are independent of each other. For a detailed derivation and discussion of the kinematical diffraction theory we refer to [Gev70]. Although it is intuitively clear that the kinematical approximation must break down – the contrast variations in the experimental bright field observations in Section 4.6 demonstrate that the transmitted beam cannot have constant intensity – it is instructive to take a closer look at the solution for the diffracted intensity.

It is common practice to introduce a dimensionless parameter w, defined by

$$w \equiv s_{\mathbf{g}}\xi_{\mathbf{g}}. \tag{6.8}$$

The crystal thickness is often also expressed in units of the extinction distance, and we introduce the following notation for the dimensionless thickness:

$$z_\xi \equiv \frac{z_0}{\xi_{\mathbf{g}}}. \tag{6.9}$$

We will refer to z_ξ as the *reduced thickness*. Equation (6.7) takes the following form:

$$I_S = \pi^2 z_\xi^2 \frac{\sin^2(\pi w z_\xi)}{(\pi w z_\xi)^2} = \pi^2 z_\xi^2 \text{sinc}^2(\pi w z_\xi).$$

Apart from the constant prefactor π^2, this expression has the same mathematical form as the square of the Fourier transformed shape function for a thin foil, derived on page 119. The deviation parameter $s_{\mathbf{g}}$ takes on the role of the variable q_z. The meaning of the shape function now becomes clear: as we tilt the crystal, different parts of the relrod intersect the Ewald sphere, and the diffracted intensity varies according to (6.7).

The dimensionless parameters w and z_ξ simplify the task of representing the scattered intensity as a function of crystal thickness and orientation. For all illustrations in this and the following sections, we will consider a wedge-shaped, bent thin foil, similar to the top sketch in Fig. 6.2(a). For simplicity, we take the curvature of the foil to be such that the excitation error $s_{\mathbf{g}}$ varies linearly from negative on the left to positive on the right. This is consistent with the orientation of the diffracting plane indicated in the bottom part of Fig. 6.2(a). The lattice planes are assumed to have a constant orientation with respect to the foil top and bottom surfaces. The dimensionless parameter w is then taken in the range -4 to $+4$. The reduced thickness varies linearly from the wedge edge ($z_\xi = 0$) to

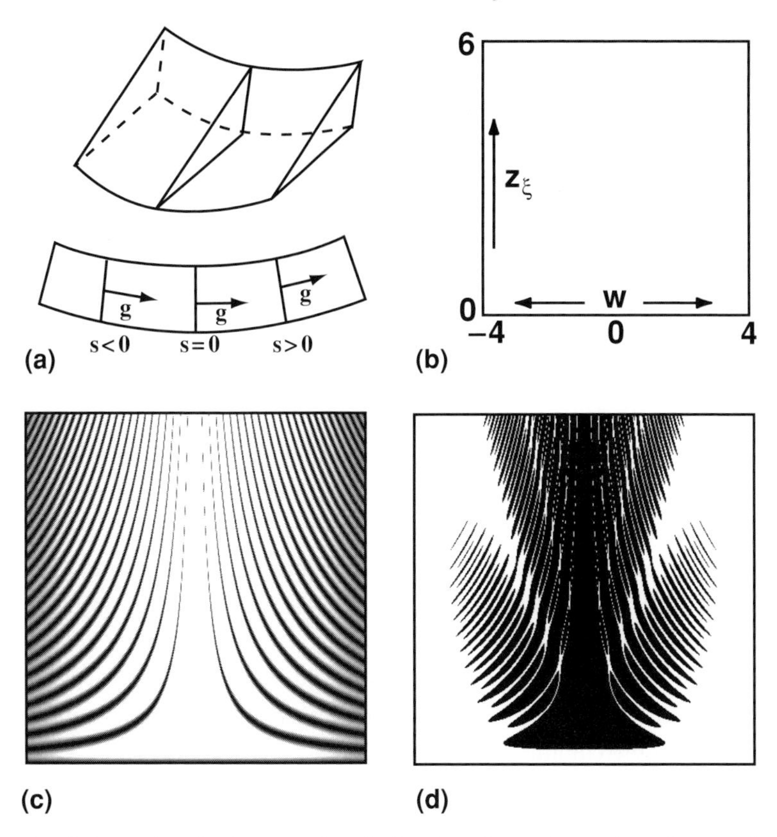

Fig. 6.2. (a) Schematic of a bent wedge, indicating the regions of positive and negative excitation error; (b) schematic of the (w, z_ξ) parameter space used for several of the figures in this chapter; (c) grayscale representation of the kinematical diffracted intensity for the intensity range [0.0, 0.1] and (d) black regions indicate those parts of the parameter space where the kinematical and dynamical theory differ by more than 5%.

the thickest part ($z_\xi = 6$). The parameter ranges are shown in Fig. 6.2(b). We are essentially making use of the *column approximation*, which we introduced in Section 6.2.

Figure 6.2(c) shows the kinematical diffracted intensity I_S for an array of 400×400 points (or columns) sampling the (w, z_ξ) parameter space. The grayscale ranges from intensity 0.0 (black) to 0.1 (white); all values larger than 0.1 are also represented by white pixels. The value of 0.1 has no particular importance and only serves to delineate the regions of parameter space where the diffracted intensity is less than 10% of the (constant) transmitted intensity. Those regions of the (w, z_ξ) parameter space for which the difference between the kinematical expression for I_S and the correct, dynamical expression derived in the next section (equation 6.22) is larger than 5% are represented by black pixels in the image in Fig. 6.2(d). It is clear that the kinematical approximation is only valid (with less than 5% error) for small reduced thicknesses z_ξ and for large values of $|w|$, or, equivalently, far from the Bragg orientation.

The kinematical theory is a poor approximation near the exact Bragg orientation for nearly all thicknesses (as is obvious from the dark central band in Fig. 6.2d) because the diffracted intensity is then proportional to the square of the reduced thickness,

$$I_S(s_{\mathbf{g}} = 0) = \pi^2 z_\xi^2, \tag{6.10}$$

and this intensity can increase without bound, clearly an unphysical situation. It can be shown that the thickness must be of the order of one-tenth of the extinction distance near Bragg orientation, otherwise dynamical corrections are needed [Gev70].

Note that the excitation error (or equivalently, the dimensionless parameter w) can be interpreted as a parameter probing the relrod corresponding to the vector **g**; when we tilt the crystal back and forth, i.e. change the value of $s_{\mathbf{g}}$, we are actually "rocking" the reciprocal lattice point through the Ewald sphere. Any curve representing the intensity as a function of w is therefore known as a *rocking curve*.

Summarizing, the kinematical theory is only valid when: (a) the crystal thickness is very small and/or (b) the crystal is far from Bragg orientation. Note that within the kinematical approximation there is no distinction between centrosymmetric and non-centrosymmetric crystals, since the phase $\theta_{\mathbf{g}}$ does not appear in the final intensity expression. The kinematical theory can also be extended to include absorption effects and it can be shown that this does not substantially alter the validity range in the (w, z_ξ) parameter space. A correct theory for thicker samples and orientations close to the Bragg orientation can only be derived by allowing multiple scattering events for each electron, or, equivalently, by avoiding the first Born approximation. This will be done in the following sections.

6.3.3 The two-beam dynamical theory

6.3.3.1 General derivation

The terms on the right-hand side of equations (6.1) are of two different types: one term is real and depends on the normal absorption length ξ_0' and the other term represents the interaction with the other beam. We can remove the dependence on the normal absorption length from the equations (as we did on page 311) and simultaneously render them more symmetric by means of the following substitution:

$$\psi_{\mathbf{0}} = T(z)\,\mathrm{e}^{\alpha z}; \tag{6.11}$$

$$\psi_{\mathbf{g}} = S(z)\,\mathrm{e}^{\mathrm{i}\theta_{\mathbf{g}}}\mathrm{e}^{\alpha z}, \tag{6.12}$$

with

$$\alpha = \pi \mathrm{i}\left(\frac{\mathrm{i}}{\xi_0'} + s_{\mathbf{g}}\right) = -\frac{\pi}{\xi_0'} + \pi \mathrm{i} s_{\mathbf{g}}. \tag{6.13}$$

After some simple mathematics we arrive at the following final equations:

$$\frac{\mathrm{d}T}{\mathrm{d}z} + \pi \mathrm{i} s_{\mathbf{g}} T = \frac{\mathrm{i}\pi}{q_{-\mathbf{g}}} S; \tag{6.14}$$

$$\frac{\mathrm{d}S}{\mathrm{d}z} - \pi \mathrm{i} s_{\mathbf{g}} S = \frac{\mathrm{i}\pi}{q_{\mathbf{g}}} T. \tag{6.15}$$

These are the dynamical two-beam equations, including absorption, for a perfect crystal of arbitrary symmetry. For a centrosymmetric crystal, the factor $q_{-\mathbf{g}}$ in equation (6.14) is replaced by $q_{\mathbf{g}}$.

Since one can only observe intensities in an experiment, we have from equations (6.11) and (6.12):

$$\text{bright field} \quad \rightarrow \quad I_T = |\psi_{\mathbf{0}}|^2 = |T|^2\,\mathrm{e}^{-(2\pi/\xi_0')z_0}; \tag{6.16}$$

$$\text{dark field} \quad \rightarrow \quad I_S = |\psi_{\mathbf{g}}|^2 = |S|^2\,\mathrm{e}^{-(2\pi/\xi_0')z_0}. \tag{6.17}$$

In the following sections we will consider the solutions of equations (6.14) and (6.15), and write them in a form that is suitable for the numerical computation of image contrast.

6.3.3.2 Solutions of the dynamical two-beam equations

We will derive the solutions to the dynamical two-beam equations for the most general situation, including absorption. Then we will make various approximations to obtain analytical expressions for the transmitted and diffracted intensities.

From equation (6.14) we find

$$S = \frac{q_{-\mathbf{g}}}{\mathrm{i}\pi}\frac{\mathrm{d}T}{\mathrm{d}z} + s_{\mathbf{g}} q_{-\mathbf{g}} T;$$

and

$$\frac{\mathrm{d}S}{\mathrm{d}z} = \frac{q_{-\mathbf{g}}}{\mathrm{i}\pi}\frac{\mathrm{d}^2 T}{\mathrm{d}z^2} + s_{\mathbf{g}} q_{-\mathbf{g}} \frac{\mathrm{d}T}{\mathrm{d}z}.$$

Substitution in equation (6.15) and rearranging terms leads to

$$\frac{\mathrm{d}^2 T}{\mathrm{d}z^2} + \pi^2\sigma^2 T = 0, \tag{6.18}$$

with

$$\sigma = \sqrt{s_{\mathbf{g}}^2 + \frac{1}{q_{\mathbf{g}} q_{-\mathbf{g}}}}. \tag{6.19}$$

This is the differential equation for a *harmonic oscillator* and the general solution is given by

$$T(z) = A\cos(\pi\sigma z) + B\sin(\pi\sigma z).$$

From the boundary condition at the entrance surface we have $T(z=0) = 1$ and hence $A = 1$. The coefficient B can be found by computing $\mathrm{d}T/\mathrm{d}z$, substituting into the equation for S above, and applying the boundary condition $S(z=0) = 0$, which leads to $B = -\mathrm{i}s_{\mathbf{g}}/\sigma$, and thus

$$T(s,z) = \cos(\pi\sigma z) - \frac{\mathrm{i}s_{\mathbf{g}}}{\sigma}\sin(\pi\sigma z); \tag{6.20}$$

$$S(s,z) = \frac{\mathrm{i}}{q_{\mathbf{g}}\sigma}\sin(\pi\sigma z). \tag{6.21}$$

These are the general solutions to the dynamical two-beam equations, including absorption. Note that the parameter σ is complex. From these expressions we can now compute the transmitted and diffracted intensities; first, we will take the simplest case without absorption, then we will derive the general expressions.

6.3.3.3 Dynamical intensities, no absorption

In the absence of absorption, the factors $q_{\mathbf{g}}$ become real and equal to $\xi_{\mathbf{g}}$. In addition, $\xi_{\mathbf{g}} = \xi_{-\mathbf{g}}$ for all crystal structures, since the extinction distance depends only on the modulus of the Fourier coefficient of the electrostatic lattice potential, not on its phase, and hence

$$\sigma^2 = s_{\mathbf{g}}^2 + \frac{1}{\xi_{\mathbf{g}}^2} = \frac{1}{\xi_{\mathbf{g}}^2}\left(1 + w^2\right),$$

where we have again introduced the dimensionless parameter $w = s_{\mathbf{g}}\xi_{\mathbf{g}}$. σ is now a real number and it is then straightforward to compute the intensities for the transmitted and diffracted beams for a crystal of thickness z_0:

$$I_S = |S|^2 = \frac{\sin^2\left(\pi\sigma z_0\right)}{1+w^2} = \frac{\sin^2\left(\pi\sqrt{1+w^2}z_\xi\right)}{1+w^2}; \tag{6.22}$$

$$I_T = 1 - I_S = \frac{w^2 + \cos^2\left(\pi\sigma z_0\right)}{1+w^2} = \frac{w^2 + \cos^2\left(\pi\sqrt{1+w^2}z_\xi\right)}{1+w^2}. \tag{6.23}$$

The last equality in both equations is included to facilitate comparison of these expressions with the corresponding scattered intensity for the kinematical theory, equation (6.7). The scattered intensity differs from the kinematical solution by the fact that the denominator no longer vanishes upon approaching the Bragg condition, and by the replacement of $s_{\mathbf{g}}$ by σ in the argument of the sin-function. The transmitted intensity is no longer constant; in the absence of absorption, I_T and I_S are

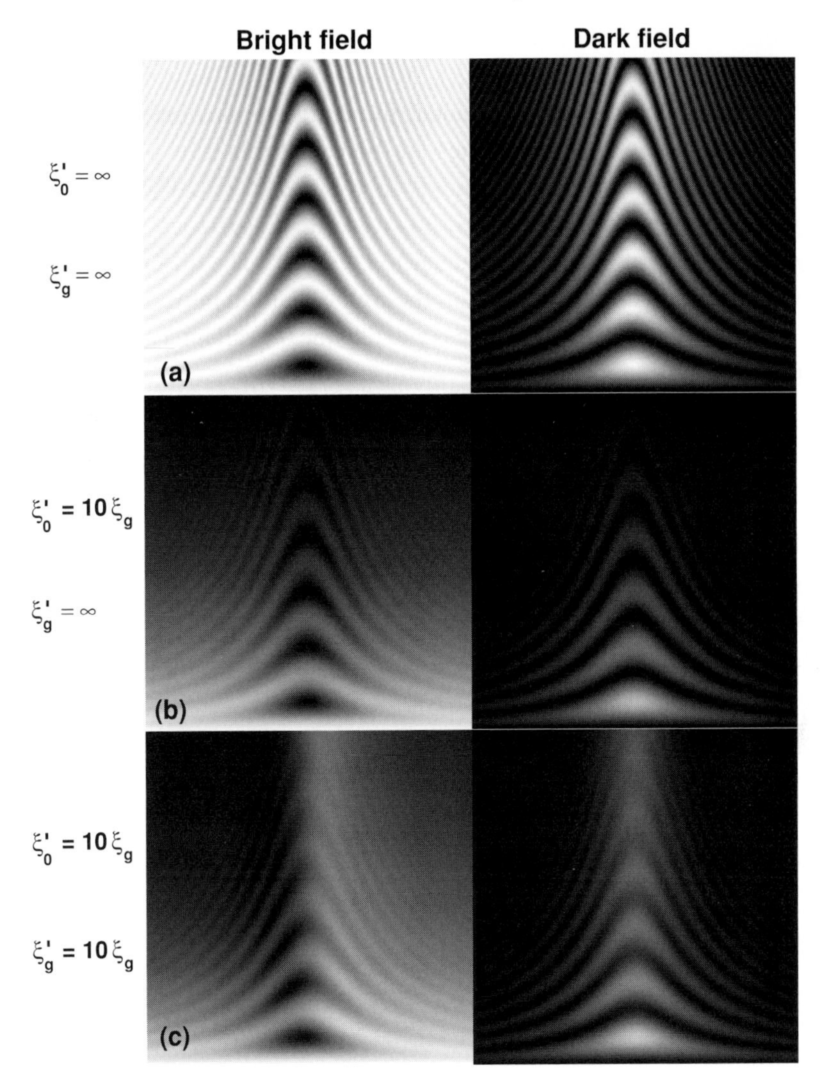

Fig. 6.3. Dynamical rocking curves (bright field and dark field): (a) in the absence of absorption (infinite absorption lengths); (b) in the presence of normal absorption; and (c) in the presence of both normal and anomalous absorption. All computations are carried out for the standard parameter space.

complementary functions (i.e. they add up to 1) and hence bright field and dark field images are complementary. For large deviations from the Bragg orientation we can ignore factors of order 1 compared to w^2 (this includes the cos-function in the transmitted intensity), and we recover equation (6.7) and the fact that $I_T = 1$, i.e. the kinematical expressions.

Figure 6.3(a) shows the dynamical bright field and dark field intensities for the standard parameter space (w, z_ξ) introduced in Figs 6.2(a) and (b). Note that both bright field and dark field rocking curves are symmetric in w (because they only depend on w^2).

When normal absorption is included, the computations are identical to those above, except that now the intensities are both multiplied by the normal absorption exponential, as shown in equations (6.16) and (6.17). The effect of normal absorption on the image intensities is shown in Fig. 6.3(b), for $\xi'_0 = 10\xi_g$, or equivalently, an absorption exponential equal to $e^{-2\pi z_\xi/10}$. The absorption length ξ'_g is taken to be infinite. It is obvious from the figure that the intensity oscillations decrease with increasing crystal thickness and that the overall intensity level decreases exponentially.

6.3.3.4 *Dynamical intensities including absorption*

In this section, we will derive analytical expressions for the intensities of transmitted and diffracted beams for the most general situation: a non-centrosymmetric crystal with absorption. Approximate expressions, which are obtained by ignoring factors of order $O[1/\xi_g'^2]$, can be found in [GBA66]. We start from equations (6.20) and (6.21). We can make use of the following relations (e.g. [AS77]):

$$\begin{aligned}
2\cos z_1 \cos z_2 &= \cos(z_1 - z_2) + \cos(z_1 + z_2);\\
2\sin z_1 \sin z_2 &= \cos(z_1 - z_2) - \cos(z_1 + z_2);\\
2\sin z_1 \cos z_2 &= \sin(z_1 - z_2) + \sin(z_1 + z_2);\\
\cos z &= \cosh \mathrm{i}z;\\
\sin z &= -\mathrm{i}\sinh \mathrm{i}z;\\
\sigma - \sigma^* &= 2\mathrm{i}\sigma_i;\\
\sigma + \sigma^* &= 2\sigma_r,
\end{aligned}$$

where z_1 and z_2 are general complex numbers, and $\sigma = \sigma_r + \mathrm{i}\sigma_i$, to derive explicit expressions for the transmitted and scattered intensities. The computations are somewhat lengthy but straightforward, and we find:

$$I_T = e^{-\frac{2\pi}{\xi'_0} z_0}\left[\frac{1}{2}\cosh(2\pi\sigma_i z_0)\left(1 + \frac{s_g^2}{|\sigma|^2}\right) + \frac{s_g\sigma_r}{|\sigma|^2}\sinh(2\pi\sigma_i z_0)\right.$$
$$\left. + \frac{1}{2}\cos(2\pi\sigma_r z_0)\left(1 - \frac{s_g^2}{|\sigma|^2}\right) - \frac{s_g\sigma_i}{|\sigma|^2}\sin(2\pi\sigma_r z_0)\right]; \qquad (6.24)$$

$$I_S = \frac{e^{-(2\pi/\xi'_0)z_0}}{2|q_g|^2\,|\sigma|^2}\left[\cosh(2\pi\sigma_i z_0) - \cos(2\pi\sigma_r z_0)\right]. \qquad (6.25)$$

The parameter $|\sigma|^2 = \sigma_r^2 + \sigma_i^2$ is given by

$$|\sigma|^2 = \sqrt{\left(s_{\mathbf{g}}^2 + \frac{1}{\xi_{\mathbf{g}}^2} - \frac{1}{\xi_{\mathbf{g}}'^2}\right)^2 + \frac{4\cos^2\beta_{\mathbf{g}}}{\left(\xi_{\mathbf{g}}\xi_{\mathbf{g}}'\right)^2}}.$$

The factor $|q_{\mathbf{g}}|^2$ can be computed from

$$\frac{1}{|q_{\mathbf{g}}|^2} = \left(\frac{1}{\xi_{\mathbf{g}}} + \mathrm{i}\frac{\mathrm{e}^{\mathrm{i}\beta_{\mathbf{g}}}}{\xi_{\mathbf{g}}'}\right)\left(\frac{1}{\xi_{\mathbf{g}}} - \mathrm{i}\frac{e^{-\mathrm{i}\beta_{\mathbf{g}}}}{\xi_{\mathbf{g}}'}\right) = \frac{1}{\xi_{\mathbf{g}}^2} + \frac{1}{\xi_{\mathbf{g}}'^2} - \frac{2\sin\beta_{\mathbf{g}}}{\xi_{\mathbf{g}}\xi_{\mathbf{g}}'}.$$

It is verified easily that, in the limit of no absorption ($\sigma_i \rightarrow 0$), both expressions (6.24) and (6.25) converge to the simpler expressions (6.23) and (6.22). The hyperbolic functions give rise to a thickness-dependent background intensity, onto which a thickness-dependent oscillatory term is superimposed. Both intensities are subject to normal absorption. The resulting bright field and dark field rocking curves are shown in Fig. 6.3(c). Note the asymmetry of the bright field image with respect to the excitation error; the asymmetry increases with increasing foil thickness.

The explicit expressions for the transmitted and diffracted intensities are quite useful for computer implementations: not every programming language has built-in support for complex number arithmetic (in particular, the relations between trigonometric and hyperbolic functions of complex arguments), whereas equations (6.24) and (6.25) no longer contain complex quantities and are straightforward to implement in any programming language.

Next, we will take a closer look at the symmetry of expressions (6.24) and (6.25).

6.3.3.5 *Symmetry of the dynamical two-beam amplitudes*

In the kinematical approximation, the probability of scattering from the transmitted beam into a diffracted beam $\mathbf{g}$ is described by the inverse of the extinction distance (i.e. by $1/\xi_{\mathbf{g}}$), or, equivalently, by the Fourier coefficient $V_{\mathbf{g}}$ of the lattice potential, which, in turn, is related to the structure factor $F_{\mathbf{g}}$. The intensity of the scattered beam (equation 6.7) is proportional to the modulus squared of this structure factor, or

$$I_{\mathbf{g}} = |F_{\mathbf{g}}|^2 = F_{\mathbf{g}}F_{\mathbf{g}}^*.$$

It is easy to show that $F_{\mathbf{g}}^* = F_{-\mathbf{g}}$, and therefore also $F_{\mathbf{g}} = F_{-\mathbf{g}}^*$. Combining all relations we find

$$I_{\mathbf{g}} = F_{\mathbf{g}}F_{\mathbf{g}}^* = F_{-\mathbf{g}}^*F_{-\mathbf{g}} = I_{-\mathbf{g}}. \tag{6.26}$$

This equation states that *a kinematical diffraction pattern always shows inversion symmetry*, even when the crystal is non-centrosymmetric. This is commonly known as *Friedel's law* (e.g. [MM86, page 199]). As a direct consequence of Friedel's law, a kinematical diffraction pattern must belong to one of the 11 centrosymmetric point groups. All point groups which become a given centrosymmetric point group when an inversion element is added belong to a so-called *Laue class*; there are 11 Laue classes, and the point group dependences are described in Appendix A4. We will now see that under dynamical electron diffraction conditions, Friedel's law may be violated.

The symmetry of the dynamical two-beam scattering process can be analyzed in a number of different ways. First of all, let us consider equations (6.20) and (6.21). The transmitted amplitude depends on both the excitation error $s_{\mathbf{g}}$ and the lattice potential through the quantity σ. We have already seen that the parameters $(q_{\mathbf{g}})^{-1}$ and $(q_{-\mathbf{g}})^{-1}$ describe the scattering from $\mathbf{0}$ to $\mathbf{g}$, and from $\mathbf{g}$ back to $\mathbf{0}$, respectively. The parameter σ depends on the *product* of both qs. This product expresses the fact that electrons in the transmitted beam are either not scattered at all, or scattered twice, or four times, and so on. Thus, electrons in the transmitted beam have been scattered *an even number of times*. Therefore, we expect that the transmitted amplitude $\psi_{\mathbf{0}}$ for a given excitation error $s_{\mathbf{g}}$ will be equal to the amplitude $\psi_{\mathbf{0}}$ for the opposite orientation $s_{-\mathbf{g}} = s_{\mathbf{g}}$. Friedel's law is therefore valid for the transmitted beam, for both centrosymmetric and non-centrosymmetric crystals.

Electrons in the scattered beam have been scattered an *odd* number of times, which is reflected in the presence of an extra factor $(q_{\mathbf{g}})^{-1}$ in equation (6.21). Since for a non-centrosymmetric crystal the numbers $(q_{\mathbf{g}})^{-1}$ and $(q_{-\mathbf{g}})^{-1}$ are not necessarily equal, we find that the scattered amplitudes for the beams $\mathbf{g}$ and $-\mathbf{g}$ for the same excitation error *may be different for a non-centrosymmetric crystal.* This means that Friedel's law is not valid for the scattered beam under dynamical diffraction conditions. For a centrosymmetric crystal, for which $q_{\mathbf{g}} = q_{-\mathbf{g}}$, Friedel's law is not violated.

This behavior can also be derived from the reciprocity theorem (equation 5.62), as applied to the transmitted beam $\mathbf{0}$. In the two-beam case, the non-centrosymmetric crystal must have type I symmetry, since there are no other reflections and the projection approximation is valid. This means that we must have (from equation 5.63)

$$\psi_{\mathbf{0}}(\otimes, \mathbf{r}, C) = \psi_{\mathbf{0}}(\otimes, -\mathbf{r}, C). \tag{6.27}$$

We find that the transmitted amplitude (and therefore intensity) is symmetric in the excitation error $s_{\pm\mathbf{g}}$, and hence Friedel's law is valid. This can also be seen from the explicit equation (6.24), when $s_{\mathbf{g}}$ is replaced by $s_{-\mathbf{g}}$.

The situation is quite different when we consider the symmetry with respect to the sign of the excitation error $\pm s_{\mathbf{g}}$. The scattered amplitude (6.25) only depends on

Table 6.1. *Intensity symmetry relations for centrosymmetric and non-centrosymmetric crystal structures under dynamical two-beam conditions with absorption.*

Centrosymmetric	Non-centrosymmetric	Comment
$I_0(+s_+) = I_0(+s_-)$	$I_0(+s_+) = I_0(+s_-)$	Friedel's law
$I_\mathbf{g}(+s_+) = I_{-\mathbf{g}}(+s_-)$	$I_\mathbf{g}(+s_+) \neq I_{-\mathbf{g}}(+s_-)$	Friedel's law
$I_0(+s_+) \neq I_0(-s_+)$	$I_0(+s_+) \neq I_0(-s_+)$	Asymmetric bright field
$I_\mathbf{g}(+s_+) = I_\mathbf{g}(-s_+)$	$I_\mathbf{g}(+s_+) = I_\mathbf{g}(-s_+)$	Symmetric dark field

$s_\mathbf{g}^2$, and hence the scattered rocking curve is symmetric in the excitation error. The transmitted intensity (6.24) depends linearly on the excitation error, and therefore is not symmetric in $\pm s_\mathbf{g}$. Both of these statements are valid for centrosymmetric and non-centrosymmetric crystal structures.

Table 6.1 summarizes the various symmetry relations; the shorthand notation $\pm s_\pm$ is used to denote the excitation error (positive or negative according to the first $\pm$ sign) for either the **g** or $-$**g** reflection (indicated by the second $\pm$ sign). The magnitude of the excitation error is the same for all entries in the table. The transmitted intensity is hence asymmetric in the excitation error, but satisfies Friedel's law for all crystal structures; the scattered intensity is symmetric in the excitation error, but violates Friedel's law in non-centrosymmetric crystals. In the absence of absorption ($\sigma_i \rightarrow 0$), the bright field rocking curve becomes symmetric in the excitation error.

The BF–DF rocking curves for a centrosymmetric crystal ($\beta_\mathbf{g} = 0$) are shown in Fig. 6.4 for a range of crystal thicknesses and orientations. The BF asymmetry is clearly present. It appears that there is a second type of absorption process which is only visible for negative excitation errors; this process depends on the value of the absorption length $\xi_\mathbf{g}'$ and is known as *anomalous absorption*.† When we talk about Bloch waves in Section 6.4, we will discuss the physics underlying anomalous absorption.

For a non-centrosymmetric crystal, there is no a priori reason why the phases of the real and imaginary potentials should be identical, and $\beta_\mathbf{g}$ can take on any value. This has profound consequences. Figure 6.5 shows how the bright field and dark field contrast for a foil thickness of $z_0 = 2\xi_\mathbf{g}$ (see Fig. 6.4) varies with the parameter $\beta_\mathbf{g}$. The bright field contrast reverses completely when $\beta_\mathbf{g} = \pi$, and is symmetric in the excitation error for $\beta_\mathbf{g} = \pi/2$ and $3\pi/2$. The bright field contrast

† One can also think of this phenomenon in terms of *enhanced transmission* for the opposite sign of the excitation error. Both terminologies are in use.

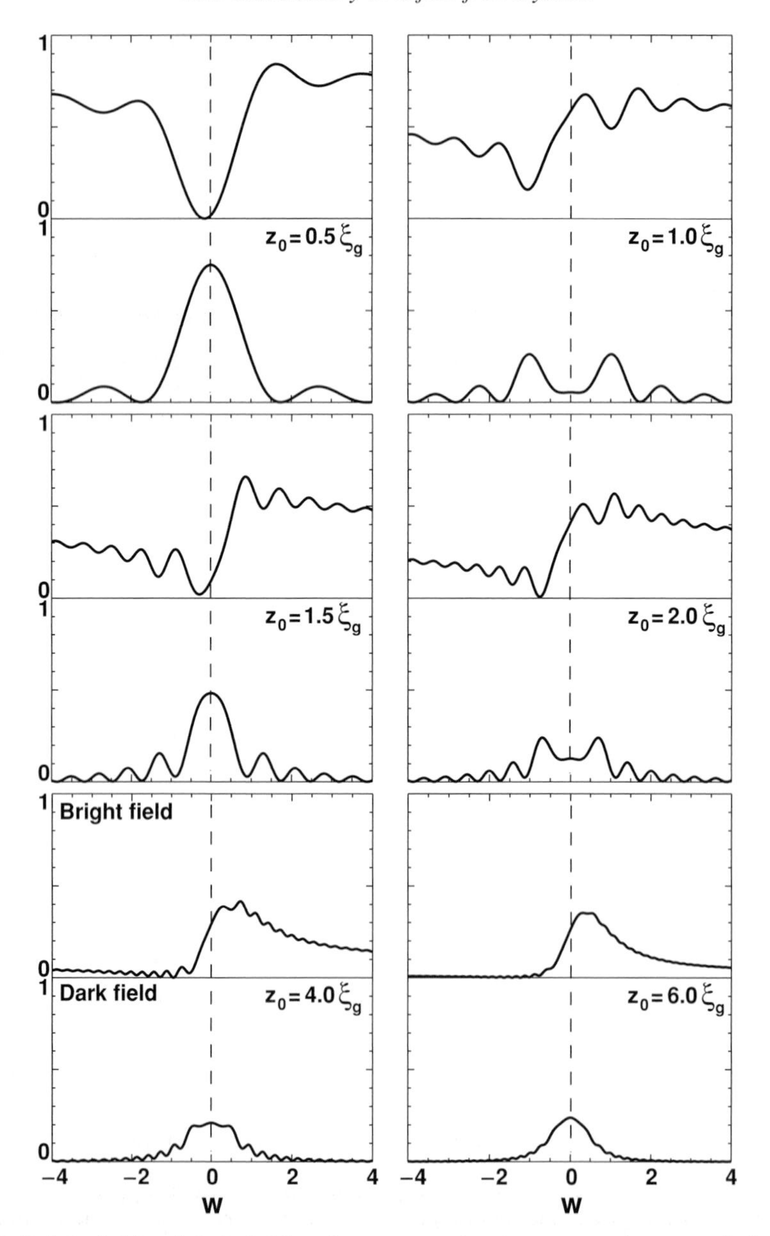

Fig. 6.4. Bright field and dark field rocking curves for the range $-4 \leq w \leq 4$, for crystal thicknesses $z_\xi = 0.5, 1.0, 1.5, 2.0, 4.0$, and 6.0. Note the asymmetric bright field curve. Other parameters are: $\beta_g = 0$, $\xi_g' = 10\xi_g = \xi_0$.

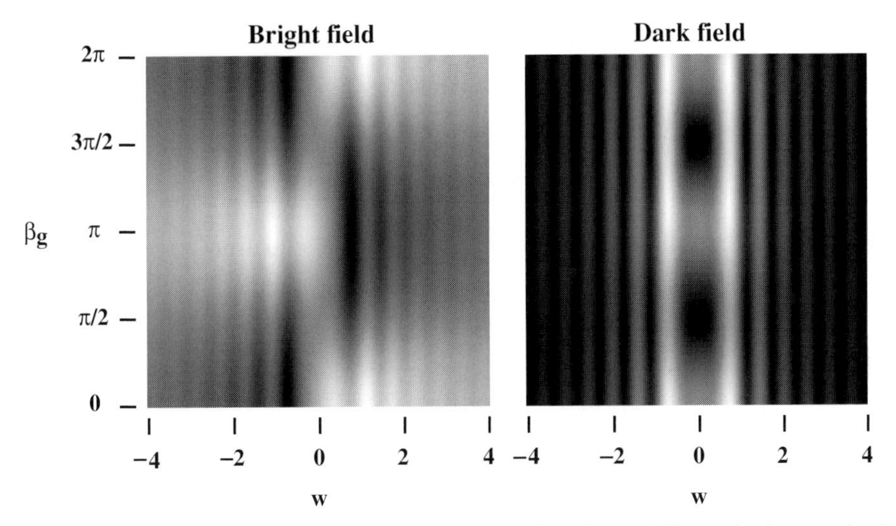

Fig. 6.5. Bright field and dark field intensities as a function of dimensionless excitation error w and $\beta_{\mathbf{g}}$, for the case $z_0 = 2\xi_{\mathbf{g}}$ of Fig. 6.4.

is also symmetric in $\pm\beta_{\mathbf{g}}$. The dark field contrast remains symmetric in w for all values of $\beta_{\mathbf{g}}$, but is asymmetric in $\pm\beta_{\mathbf{g}}$.

The reader may find it useful to enter various parameters into the **ION**-routine betag.pro on the website. This routine displays the rocking curves for a non-centrosymmetric crystal, based on equations (6.20) and (6.21). The absorption ratio is defined as ξ'/ξ, so that low absorption implies a large absorption ratio. A second **ION**-routine called rock.pro is also available on the website; this routine displays the bright field and dark field rocking curves for the standard parameter space for a centrosymmetric crystal; the user can then click on any point in the bright field image to generate a curve similar to those shown in Fig. 6.4.

6.4 The two-beam case: Bloch wave formalism

The purpose of this section is to compare the Bloch wave approach to the differential equation approach of the previous section, for a case where the entire computation can be performed analytically. We will rederive the solution of the two-beam dynamical scattering problem for the case of a non-centrosymmetric crystal with absorption. We will also introduce a graphical method to solve the problem; this is known as the *dispersion surface construction*, and is equivalent to the Ewald sphere construction of the kinematical scattering problem. Where appropriate, we will compare the results with those from the differential equation approach.

6.4.1 Mathematical solution

For the two-beam case, the eigenvalue equation (5.51) reads as

$$\begin{pmatrix} 0 & U_{-\mathbf{g}} \\ U_{\mathbf{g}} & 2k_0 s_{\mathbf{g}} \end{pmatrix} \begin{pmatrix} C_{\mathbf{0}}^{(j)} \\ C_{\mathbf{g}}^{(j)} \end{pmatrix} = 2k_n \gamma^{(j)} \begin{pmatrix} C_{\mathbf{0}}^{(j)} \\ C_{\mathbf{g}}^{(j)} \end{pmatrix}. \tag{6.28}$$

This results in the following quadratic determinantal equation:

$$\left[2k_n\gamma^{(j)}\right]^2 - (2k_0 s_{\mathbf{g}})\left[2k_n\gamma^{(j)}\right] - U_{-\mathbf{g}}U_{\mathbf{g}} = 0, \tag{6.29}$$

with solutions

$$2k_n\gamma^{(j)} = k_0 s_{\mathbf{g}} \pm \sqrt{k_0^2 s_{\mathbf{g}}^2 + U_{-\mathbf{g}}U_{\mathbf{g}}}.$$

For a centrosymmetric crystal, we have $U_{-\mathbf{g}} = U_{\mathbf{g}}$, and also $k_0/|U_{\mathbf{g}}| = \xi_{\mathbf{g}}$, and we can rewrite the solutions as

$$2k_n\gamma^{(j)} = |U_{\mathbf{g}}|\left(w \pm \sqrt{1+w^2}\right), \tag{6.30}$$

where $w = s_{\mathbf{g}}\xi_{\mathbf{g}}$ is the dimensionless excitation error. The difference between the two eigenvalues $\gamma^{(1)}$ and $\gamma^{(2)}$ is given by

$$\Delta\gamma = \gamma^{(1)} - \gamma^{(2)} = \frac{|U_{\mathbf{g}}|}{k_n}\sqrt{1+w^2} = \sigma,$$

where we have used the definition of the variable σ on page 354. We find that in the exact Bragg orientation, when $w = 0$, the difference between the two eigenvalues is precisely equal to the inverse of the extinction distance $\xi_{\mathbf{g}}$. Figure 6.6 shows the graphical representation of the two eigenvalues $\xi_{\mathbf{g}}\gamma^{(j)}$ in equation (6.30), for w from -3 to $+3$.

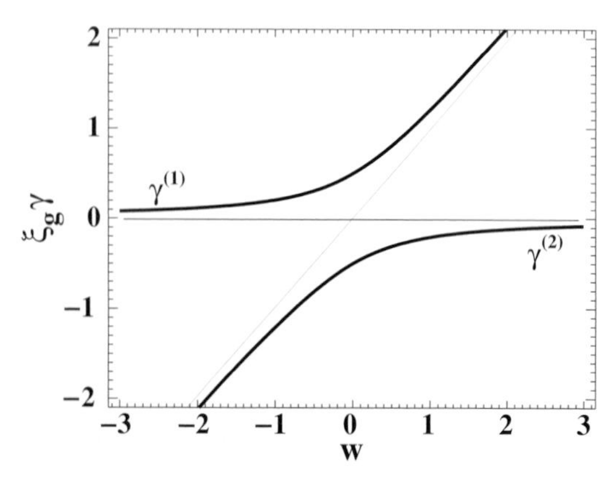

Fig. 6.6. Graphical representation of the two-beam eigenvalues $\gamma^{(1)}$ and $\gamma^{(2)}$ (multiplied by the extinction distance $\xi_{\mathbf{g}}$) as a function of the dimensionless parameter $w = s_{\mathbf{g}}\xi_{\mathbf{g}}$. The straight lines represent the asymptotes of the empty crystal approximation.

The *gap* between the two dispersion surface branches is the inverse of the extinction distance and is, therefore, proportional to the potential Fourier coefficient. This is very similar to solid-state band theory, where the band-gap is also proportional to the Fourier coefficient. Furthermore, whereas band theory assumes *parabolic* branches close to the Brillouin zone boundary, electron diffraction is described by *hyperbolic* branches.

At this point, we have sufficient information to determine the eigenvectors. From equation (6.28), we find that

$$\frac{C_{\mathbf{g}}^{(1)}}{C_{\mathbf{0}}^{(1)}} = \left(w + \sqrt{1+w^2}\right) e^{-i\theta_{\mathbf{g}}};$$
$$\frac{C_{\mathbf{g}}^{(2)}}{C_{\mathbf{0}}^{(2)}} = \left(w - \sqrt{1+w^2}\right) e^{-i\theta_{\mathbf{g}}}.$$

If we introduce the notation $\cot\beta = w$, then we can make use of the following relations [AS77]:

$$\frac{\cos\beta + 1}{\sin\beta} = \frac{\cos(\beta/2)}{\sin(\beta/2)} \quad \text{and} \quad \frac{\cos\beta - 1}{\sin\beta} = -\frac{\sin(\beta/2)}{\cos(\beta/2)}. \tag{6.31}$$

Combining these expressions with the normalization relations (5.54) and (5.55) we find for the components of the eigenvectors:

$$\left.\begin{array}{ll} C_{\mathbf{0}}^{(1)} = \sin\frac{\beta}{2}; & C_{\mathbf{0}}^{(2)} = \cos\frac{\beta}{2} e^{i\theta_{\mathbf{g}}}; \\ C_{\mathbf{g}}^{(1)} = \cos\frac{\beta}{2} e^{-i\theta_{\mathbf{g}}}; & C_{\mathbf{g}}^{(2)} = -\sin\frac{\beta}{2}. \end{array}\right\} \tag{6.32}$$

For the two-beam case we then have

$$\psi_{\mathbf{0}}(z_0) = C_{\mathbf{0}}^{(1)} e^{2\pi i\gamma^{(1)} z_0} \alpha^{(1)} + C_{\mathbf{0}}^{(2)} e^{2\pi i\gamma^{(2)} z_0} \alpha^{(2)};$$
$$\psi_{\mathbf{g}}(z_0) = C_{\mathbf{g}}^{(1)} e^{2\pi i\gamma^{(1)} z_0} \alpha^{(1)} + C_{\mathbf{g}}^{(2)} e^{2\pi i\gamma^{(2)} z_0} \alpha^{(2)}.$$

For the general case with absorption, we must explicitly invert the dynamical matrix to obtain the excitation amplitudes $\alpha^{(j)}$. In the absence of absorption, the two-beam Bloch wave excitation amplitudes are given by (using $\alpha^{(j)} = C_{\mathbf{0}}^{(j)*}$):

$$\alpha^{(1)} = \sin\frac{\beta}{2};$$
$$\alpha^{(2)} = \cos\frac{\beta}{2} e^{-i\theta_{\mathbf{g}}}.$$

The excitation amplitudes are shown in Fig. 6.7(a) as a function of the dimensionless parameter w. When the reciprocal lattice point $\mathbf{g}$ is inside the Ewald sphere (positive excitation error), then the second Bloch wave contributes most to the diffracted

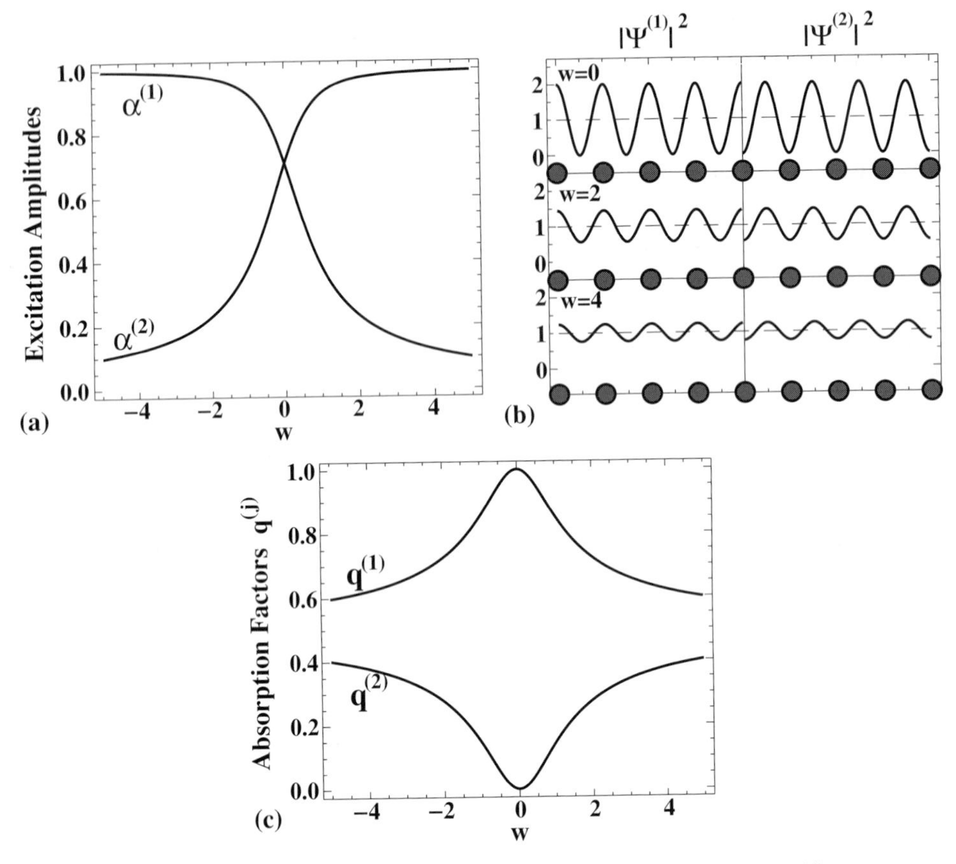

Fig. 6.7. Two-beam Bloch wave parameters: (a) excitation amplitudes $\alpha^{(j)}$; (b) type I and II Bloch wave intensities for $w = 0, 2, 4$, relative to the atom locations (spheres); (c) Bloch wave absorption parameters versus dimensionless parameter w.

intensity. When the reciprocal lattice point is outside the Ewald sphere, then the first Bloch wave is more strongly excited.

It is now straightforward to compute the intensities of the transmitted and diffracted beams for the two-beam case without absorption. We will only compute the dark field intensity as an example. Inserting all eigenvectors and Bloch wave excitation amplitudes into the expression for $\psi_{\mathbf{g}}(z)$ we find:

$$\begin{aligned}
\psi_{\mathbf{g}}(z_0) &= \frac{\sin\beta}{2}\mathrm{e}^{-\mathrm{i}\theta_{\mathbf{g}}}\left[\mathrm{e}^{2\pi\mathrm{i}\gamma^{(1)}z_0} - \mathrm{e}^{2\pi\mathrm{i}\gamma^{(2)}z_0}\right];\\
&= \mathrm{i}\mathrm{e}^{-\mathrm{i}\theta_{\mathbf{g}}}\mathrm{e}^{2\pi\mathrm{i}\gamma^{(1)}z_0}\mathrm{e}^{-\pi\mathrm{i}\sigma z}\sin\beta\sin(\pi\sigma z_0);\\
|\psi_{\mathbf{g}}|^2 &= \sin^2\beta\sin^2(\pi\sigma z_0);\\
&= \frac{\sin^2(\pi\sigma z_0)}{1+w^2},
\end{aligned}$$

which is identical to the solution (6.22) on page 354. We leave it to the reader to verify the expression for the transmitted intensity.

One particular advantage of the Bloch wave formulation is that it provides the answer to the question: where in the crystal do the electrons travel? Since there are two Bloch wave vectors inside the crystal, each of those waves must be associated with a different potential energy, and must therefore travel at a different location in the crystal. The two Bloch waves for the two-beam case are given by

$$\Psi^{(j)}(\mathbf{r}) = C_{\mathbf{0}}^{(j)} e^{2\pi i \mathbf{k}^{(j)}\cdot\mathbf{r}} + C_{\mathbf{g}}^{(j)} e^{2\pi i(\mathbf{k}^{(j)}+\mathbf{g})\cdot\mathbf{r}}.$$

It is a simple exercise to show that the intensities associated with the Bloch waves are given by

$$|\Psi^{(1)}(\mathbf{r})|^2 = 1 + \sin\beta\cos\left(2\pi\mathbf{g}\cdot\mathbf{r} - \theta_{\mathrm{g}}\right); \tag{6.33}$$

$$|\Psi^{(2)}(\mathbf{r})|^2 = 1 - \sin\beta\cos\left(2\pi\mathbf{g}\cdot\mathbf{r} - \theta_{\mathrm{g}}\right). \tag{6.34}$$

If we take $\theta_{\mathrm{g}} = 0$, and the x-direction is normal to the planes $\mathbf{g}$, then $2\pi\mathbf{g}\cdot\mathbf{r} = 2\pi g x$, and we find:

$$|\Psi^{(1)}(\mathbf{r})|^2 = 1 + \frac{\cos 2\pi g x}{\sqrt{1+w^2}};$$

$$|\Psi^{(1)}(\mathbf{r})|^2 = 1 - \frac{\cos 2\pi g x}{\sqrt{1+w^2}}.$$

The resulting intensity profiles are shown in Fig. 6.7(b): the wave $\Psi^{(1)}$ on the left has its maximum intensity on the planes $x = nd$, where $d = 1/g$ is the interplanar spacing. The second Bloch wave has its maxima in between the atomic planes, at $x = (n + \frac{1}{2})d$.

We shall call a Bloch wave with maxima at the atom positions a type I wave; a Bloch wave with maxima in between the atom positions is then a type II wave. Electrons in the type I wave spend much of their time near the atom cores, and are hence likely to undergo inelastic scattering events, which essentially means that they are absorbed according to the phenomenological description of absorption on page 122. Electrons in a type II wave spend most of their time *in between* the atom columns and are less likely to undergo inelastic scattering events. We have already seen that for negative excitation errors, the first Bloch wave has the strongest excitation amplitude $\alpha^{(1)}$ (Fig. 6.7a). Conversely, for a positive excitation error the second Bloch wave is more strongly excited. Since the first Bloch wave is more strongly absorbed, this generates an asymmetry between the transmitted intensities for positive and negative excitation errors. We have called this asymmetry *anomalous absorption* (see page 359).

Two sound waves with a slightly different wavelength will give rise to an interference sound (a "beating" sound) at the difference frequency. A similar thing happens for electrons in a solid: for a two-beam situation, the two allowed waves interfere and give rise to interference fringes, which we can observe as *thickness fringes* or *Pendellösung fringes*. Note that the variable *time* (for sound waves) is replaced by *depth* in the crystal (for electrons).

It is rather straightforward to include absorption explicitly into the two-beam equations. The resulting eigenvalue equation reads as

$$\begin{pmatrix} \mathrm{i}U'_{\mathbf{0}} & U_{-\mathbf{g}} + \mathrm{i}U'_{-\mathbf{g}} \\ U_{\mathbf{g}} + \mathrm{i}U'_{\mathbf{g}} & 2k_0 s_{\mathbf{g}} + \mathrm{i}U'_{\mathbf{0}} \end{pmatrix} \begin{pmatrix} C_{\mathbf{0}}^{(j)} \\ C_{\mathbf{g}}^{(j)} \end{pmatrix} = 2k_n \Gamma^{(j)} \begin{pmatrix} C_{\mathbf{0}}^{(j)} \\ C_{\mathbf{g}}^{(j)} \end{pmatrix}. \tag{6.35}$$

For a centrosymmetric crystal we have $U_{\mathbf{g}} = U_{-\mathbf{g}}$ and $U'_{\mathbf{g}} = U'_{-\mathbf{g}}$, and the determinantal equation becomes

$$\left[2k_n\Gamma^{(j)}\right]^2 - 2k_n\Gamma^{(j)}\left(2k_0 s_{\mathbf{g}} + 2\mathrm{i}U'_{\mathbf{0}}\right) - (U_{\mathbf{g}} + \mathrm{i}U'_{\mathbf{g}})^2 = 0.$$

Ignoring quadratic terms in U'/U we find for the solutions:

$$2k_n\Gamma^{(j)} \approx |U_{\mathbf{g}}| \left[w + \mathrm{i}\frac{U'_{\mathbf{0}}}{|U_{\mathbf{g}}|} \pm \sqrt{1 + w^2 + \mathrm{i}\frac{2}{|U_{\mathbf{g}}|}(wU'_{\mathbf{0}} + U'_{\mathbf{g}})} \right].$$

The imaginary components of these equations are small, and we can use the following Taylor expansion to separate the real and imaginary components of the eigenvalues:

$$\sqrt{x + \mathrm{i}y} \approx \sqrt{x} + \frac{1}{1!} \left.\frac{\partial\sqrt{x+\mathrm{i}y}}{\partial y}\right|_{y=0} y = \sqrt{x} + \frac{\mathrm{i}y}{2\sqrt{x}}.$$

Introducing the expressions for the absorption distances we find

$$\gamma^{(j)} = \frac{1}{2\xi_{\mathbf{g}}}\left(w \pm \sqrt{1 + w^2}\right); \tag{6.36}$$

$$q^{(j)} = \frac{1}{2}\left(\frac{1}{\xi'_{\mathbf{0}}} \pm \frac{1}{\xi'_{\mathbf{g}}\sqrt{1 + w^2}}\right). \tag{6.37}$$

The real part of the eigenvalues is identical to the case without absorption. The imaginary components for the type I and type II Bloch waves in exact Bragg orientation $w = 0$ are given by

$$q^{(1)} = \frac{1}{2}\left(\frac{1}{\xi'_{\mathbf{0}}} + \frac{1}{\xi'_{\mathbf{g}}}\right); \tag{6.38}$$

$$q^{(2)} = \frac{1}{2}\left(\frac{1}{\xi'_0} - \frac{1}{\xi'_{\mathbf{g}}}\right). \tag{6.39}$$

The scaled absorption parameters $\xi_{\mathbf{g}} q^{(j)}$ are shown as a function of w in Fig. 6.7(c); the ratio ξ/ξ' was taken to be equal to 1. Since the type I wave is more strongly excited (Fig. 6.7a) for negative excitation errors, the preferential absorption will be more noticeable than in the case of a positive excitation error and results in anomalous absorption.

Equation (6.39) provides us with an important restriction on the allowed values of the normal and anomalous absorption lengths. Since $q^{(2)}$ must be positive,† we must have

$$\frac{1}{\xi'_0} \geq \frac{1}{\xi'_{\mathbf{g}}},$$

which means that $\xi'_{\mathbf{g}}$ must be greater than ξ'_0 [HHW62]. Experiments show that typically $\xi'_{\mathbf{g}} \approx 2\xi'_0$ and also that $\xi_{\mathbf{g}} \approx 0.1\xi'_0$ (e.g. [Has64]). We have seen in Fig. 6.5 that, for a non-centrosymmetric crystal, anomalous absorption may occur for positive excitation errors, as a consequence of the parameter $\beta_{\mathbf{g}}$, which expresses the phase difference between the Fourier coefficients of the real and imaginary parts of the electrostatic lattice potential. We leave it to the reader to derive the explicit equations (6.24) and (6.25) from the Bloch wave formalism.

6.4.2 Graphical solution

In this section, we will derive a graphical method to determine the Bloch wave vectors inside the crystal. The method is the dynamical equivalent of the kinematical Ewald sphere construction discussed in Chapter 2, and is known as the *dispersion surface construction*. This section is based on [Met75] and [SZ92]. Unless stated otherwise, a centrosymmetric crystal is assumed.

Consider the drawing in Fig. 6.8(a); it shows the Ewald sphere construction for a reciprocal lattice point **g**, which falls somewhat inside the sphere (i.e. with a positive excitation error $s_{\mathbf{g}}$). The diffracted wave vector is given by $\mathbf{k}' = \mathbf{k}_0 + \mathbf{g} + s_{\mathbf{g}}\mathbf{e}_z$, where $\mathbf{e}_z$ is a unit vector along the incident beam direction. This is a purely kinematical description, which only deals with the geometry of Bragg's law. We can now use equation (6.29) to derive the eigenvalues $\gamma^{(j)}$ for the so-called *empty crystal approximation*. In this approximation, all Fourier coefficients $U_{\mathbf{g}}$ vanish, except for the mean inner potential which is taken into account in the refraction-corrected wave vector $\mathbf{k}_0$. The solutions of the determinantal equation are

† A negative value for $q^{(2)}$ would give rise to an *increase* with crystal thickness z_0 in the number of electrons in Bloch wave (2).

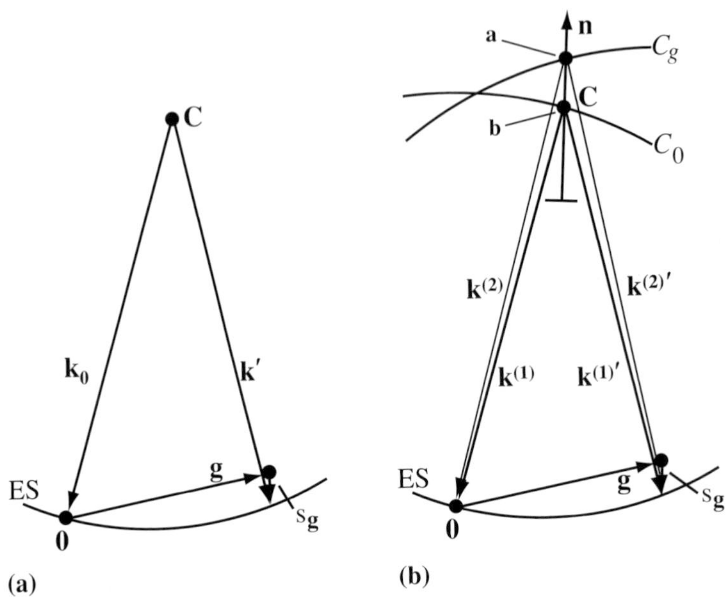

Fig. 6.8. (a) Standard Ewald sphere construction and (b) an equivalent construction for the empty crystal approximation.

then given by

$$\gamma^{(1)} = 0;$$

$$\gamma^{(2)} = \frac{k_0}{k_n} s_{\mathbf{g}}.$$

These eigenvalues are shown as thin straight lines in Fig. 6.6. The corresponding wave vectors inside the crystal are:

$$\mathbf{k}^{(1)} = \mathbf{k}_0;$$

$$\mathbf{k}^{(2)} = \mathbf{k}_0 + \frac{k_0}{k_n} s_{\mathbf{g}}\mathbf{n} \approx \mathbf{k}_0 - s_{\mathbf{g}}\mathbf{n}.$$

The minus sign arises because the vector $\mathbf{n}$ points in a direction opposite to the incident beam direction $\mathbf{k}_0$. In Bragg orientation, both vectors are equal since $s_{\mathbf{g}} = 0$. The second equality is only valid when the incident beam is nearly parallel to the foil normal. The vectors $\mathbf{k}^{(j)}$ can be represented graphically as follows: on the Ewald sphere construction of Fig. 6.8(a), superimpose two circles (spheres) with centers at $\mathbf{0}$ and $\mathbf{g}$, and radius equal to $|\mathbf{k}_0|$. The circles are labeled C_0 and C_g, respectively in Fig. 6.8(b). Then, draw a line through the center $\mathbf{C}$ of the Ewald sphere, parallel to the foil normal $\mathbf{n}$. This line intersects the two circles in the points $\mathbf{a}$ and $\mathbf{b}$. The point $\mathbf{b}$ coincides with the center of the Ewald sphere. Next, connect both points with the origin of reciprocal space; these are the vectors $\mathbf{k}^{(j)}$ inside the "empty crystal". The scattered wave vectors $\mathbf{k}^{(j)\prime}$ can be drawn by connecting the points **a** and **b** with the endpoint of $\mathbf{k}_0 + \mathbf{g} + s_\mathbf{g}\mathbf{e}_z$.

Next, we can "turn on" the lattice potential and increase the Fourier coefficient $U_\mathbf{g}$ from zero to its proper value. As illustrated in Fig. 6.6, the eigenvalues are no longer linear, and, more importantly, they no longer intersect each other at Bragg orientation. We can now transfer the information in this figure to a drawing similar to that in Fig. 6.8(b). The two spheres around **0** and **g** must now be replaced by two surfaces that do not intersect each other. The surfaces approach the spheres for large excitation errors, and in Bragg orientation they are separated by a distance $1/\xi_\mathbf{g}$. The resulting graphical construction is shown in Fig. 6.9(a) for exact Bragg orientation and in Fig. 6.9(b) for a positive excitation error. The wave vectors $\mathbf{k}^{(j)}$ are constructed in the same way as before, and now neither of them is equal to the incident wave vector $\mathbf{k}_0$. The drawings in Fig. 6.9 are not to scale. It is straightforward, and left as an exercise for the interested reader, to draw the dispersion surface construction for the case where the foil normal **n** is not perpendicular to the reciprocal lattice vector **g**.

The **ION**-routine tbbloch.pro on the website can be used to calculate the eigenvalues $\gamma^{(j)}$ for a given lattice potential Fourier coefficient $V_\mathbf{g}$ (in volts). The routine produces two drawings, one using the same format as Fig. 6.6 and the second using a different representation in terms of the z-component of the wave vectors. We know that the tangential components of both wave vectors $\mathbf{k}^{(j)}$ are equal to k_t. It is

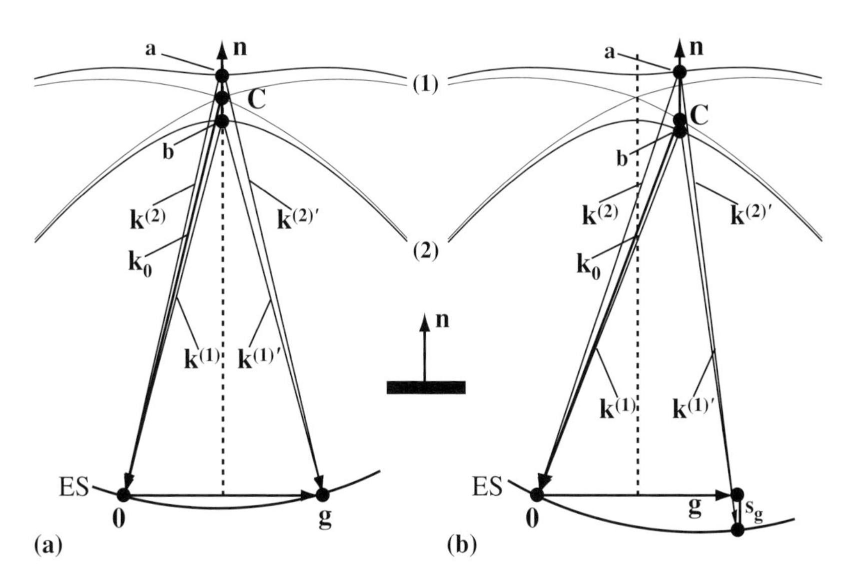

Fig. 6.9. Dispersion surface construction for the two-beam case: (a) exact Bragg orientation and (b) positive excitation error.

easy to show that the tangential component is also equal to

$$k_t = -\frac{g}{2} - k_0\frac{s_{\mathbf{g}}}{g}. \tag{6.40}$$

Now we have two expressions for the wave vector $\mathbf{k}^{(j)}$:

$$\mathbf{k}^{(j)} = \mathbf{k}_0 + \gamma^{(j)}\mathbf{n};$$
$$\mathbf{k}^{(j)} = \mathbf{k}_t + \mathbf{k}_z^{(j)},$$

with $\mathbf{k}_z^{(j)}$ normal to the tangential component. Equating the lengths of both vectors we find:

$$k_z^{(j)} = \left|\mathbf{k}_n + \gamma^{(j)}\mathbf{n}\right|, \tag{6.41}$$

with $\mathbf{k}_n = \mathbf{k}_0 - \mathbf{k}_t$. It is then common to plot the value of $k_z^{(j)} - k_0$ versus the dimensionless orientation parameter $-k_t/g$. In Bragg orientation we have $k_t = -g/2$; the minus sign stems from the fact that the wave vector points towards the origin, not towards **g**. Example output of the tbbloch **ION** routine is shown in Fig. 6.10; the computation was performed for $V_{\mathbf{g}} = 5$ V, a mean inner potential of

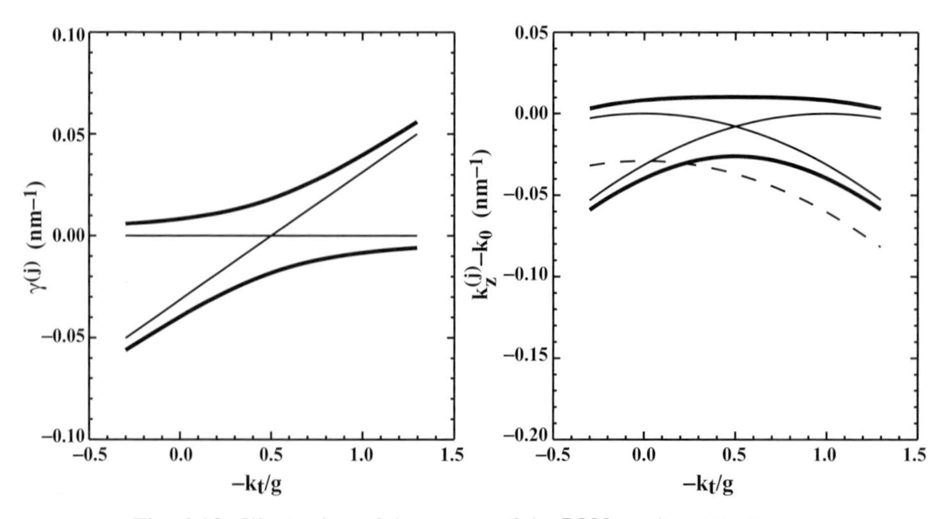

Fig. 6.10. Illustration of the output of the **ION** routine tbbloch.pro.

25 V, and a microscope acceleration voltage of 200 kV. The dashed curve on the right-hand side represents the sphere centered on the origin, with the vacuum wave number as the radius (i.e. $|K|$). The thin curves represent the empty crystal results, and the thicker curves form the dispersion surfaces.

The second representation method ($k_z^{(j)} - k_0$ versus $-k_t/g$) has the advantage that it can be extended readily to multi-beam representations, in particular to the systematic row case which we will address in the next chapter. For the two-beam case, the dispersion surfaces are surfaces of revolution around the reciprocal lattice vector **g**;† because of the limited validity of the two-beam approximation, only the region of the dispersion surface close to the Bragg orientation (i.e. close to the Brillouin zone boundary at $\mathbf{g}/2$) is of practical importance.

6.5 Numerical two-beam image simulations

In this section, we will discuss the basic principles of two-beam image simulations. We start with an extensive discussion of the numerical computation of the various parameters affecting the image contrast (excitation error, extinction distance, and absorption length). Then we discuss the numerical implementation of the scattering matrix formalism, and the two-beam Bloch wave formalism. We conclude this section with example image simulations, for a variety of sample geometries, followed by a brief discussion of a two-beam CBED method to determine the local sample thickness.

6.5.1 Numerical computation of extinction distances and absorption lengths

The parameter q_g is given by

$$\frac{1}{q_\mathrm{g}} = \frac{1}{\xi_\mathrm{g}} + \mathrm{i}\frac{\mathrm{e}^{\mathrm{i}\beta_\mathrm{g}}}{\xi'_\mathrm{g}} = \left(\frac{1}{\xi_\mathrm{g}} - \frac{\sin\beta_\mathrm{g}}{\xi'_\mathrm{g}}\right) + \mathrm{i}\frac{\cos\beta_\mathrm{g}}{\xi'_\mathrm{g}}, \tag{6.42}$$

where we have written explicitly the real and imaginary parts. From the definition of the extinction distance (equation 5.13) and using equation (2.38) on page 95 we have

$$\begin{aligned}\frac{1}{\xi_\mathrm{g}} &= \frac{|U_\mathrm{g}|}{|\mathbf{k}_0 + \mathbf{g}|\cos\alpha} \\ &= \frac{\sigma}{\pi\lambda}\frac{|V_\mathrm{g}|}{|\mathbf{k}_0 + \mathbf{g}|\cos\alpha}.\end{aligned}$$

† This is analogous to the conical surface on which the Bragg condition is satisfied, as illustrated in Fig. 2.2(b) on page 96.

Since, to a very good approximation, we have $|\mathbf{k}_0 + \mathbf{g}| \approx 1/\lambda$ and $\cos\alpha \approx 1$, we find

$$\frac{1}{\xi_{\mathbf{g}}} \approx \frac{\sigma}{\pi}|V_{\mathbf{g}}|. \tag{6.43}$$

The reader should be careful not to confuse the interaction constant σ in this equation with the parameter σ introduced in equation (6.19).

Numerical computation of the extinction distance is now straightforward, since we already have routines for the computation of both σ (CalcWaveLength on page 94) and $V_{\mathbf{g}}$ (CalcUcg, Section 2.6.7 on page 128).

The computation of the absorption length presents a whole new set of difficulties. We have seen in Section 2.6.5 that the effects of absorption can be taken into account in a phenomenological way, by adding an imaginary part to the electrostatic lattice potential:

$$V_c(\mathbf{r}) = V(\mathbf{r}) + \mathrm{i}W(\mathbf{r}). \tag{6.44}$$

In the early days of TEM research, it was common practice to assume that the two potential functions are *in phase*, with a constant ratio $a = W/V$. The resulting complex potential can then be written as

$$V_c(\mathbf{r}) = (1 + \mathrm{i}a)V(\mathbf{r}).$$

It was found by comparison with experiment that reasonable values of a are in the range 0.05–0.15 for monatomic solids (i.e. for the pure elements) [HHN+77].

For more complex unit cells, in particular non-primitive cells, this simple approximation does not hold as well, and the absorption potential is no longer simply proportional to the electrostatic potential. There is, in fact, no a priori reason why the two potential functions should even be "in phase" for a non-centrosymmetric crystal structure [GBA66]. We will follow Bird and King [BK90], and Weickenmeier and Kohl [WK91], and assume that the absorption part of the potential is mostly due to electron–phonon scattering, also known as *thermal diffuse scattering* (TDS). This means that we ignore plasmon scattering and energy losses due to core excitations. The derivation in this section serves as an example of how inelastic scattering phenomena can be incorporated into the dynamical scattering theory. For a more detailed discussion of inelastic scattering we refer to the book by Z.L. Wang [Wan95].

We must further simplify the problem, since it is quite difficult to compute the electron–phonon interactions in general for an arbitrary crystal structure. The simplest approximation to the phonon spectrum of a material is the *Einstein dispersion model*, which assumes that the thermal vibrations of each atom are statistically independent. We have already made use of this approximation in the derivation of the Debye–Waller factor (equation 2.101 in Chapter 2).

We follow Hall and Hirsch [HH65] and calculate the probability amplitude for a Bloch wave in Bragg orientation to be scattered in the direction $\mathbf{k}$ by an assembly of N identical atoms at positions $\mathbf{r}_i$. The electrostatic potential in the notation of Chapter 2 is given by

$$V(\mathbf{r}) = V^a(\mathbf{r}) \otimes \sum_{i=1}^{N} \delta(\mathbf{r} - \mathbf{r}_i).$$

Using the two-beam approximation for the incident Bloch wave in Bragg orientation (see page 365), we arrive at the following expressions for the two Bloch waves:†

$$\Psi^{(j)}(\mathbf{r}) = \frac{1}{\sqrt{2}}\left[e^{2\pi i\mathbf{k}_0\cdot\mathbf{r}} + (-1)^{j-1}e^{2\pi i(\mathbf{k}_0+\mathbf{g})\cdot\mathbf{r}}\right].$$

The probability amplitude P_j for scattering of this wave in the direction $\mathbf{k}$ is given by the *bra-ket*:

$$P_j = \left\langle e^{2\pi i\mathbf{k}\cdot\mathbf{r}}\left|V(\mathbf{r})\right|\Psi^{(j)}(\mathbf{r})\right\rangle.$$

If we denote the scattering vector $\mathbf{k} - \mathbf{k}_0$ by $\mathbf{S}$, then it is easy to show, using the definition of the direct Fourier transform on page 104, that

$$P_j = \frac{1}{\sqrt{2}}\left\{\mathcal{F}_{\mathbf{S}}[V(\mathbf{r})] + (-1)^{j-1}\mathcal{F}_{\mathbf{S}-\mathbf{g}}[V(\mathbf{r})]\right\},$$

where the subscript on the Fourier transform operator indicates the reciprocal space vector of the transform. Using the convolution theorem (2.54), we can rewrite the Fourier transforms as

$$\begin{aligned}\mathcal{F}_{\mathbf{q}}[V(\mathbf{r})] &= \mathcal{F}_{\mathbf{q}}\left[V^a(\mathbf{r})\right]\mathcal{F}_{\mathbf{q}}\left[\sum_{i=1}^{N}\delta(\mathbf{r} - \mathbf{r}_i)\right];\\ &= f_{\mathbf{q}}\sum_{i=1}^{N} e^{2\pi i\mathbf{q}.\mathbf{r}_i}.\end{aligned}$$

We have used the definition of the atomic scattering factor f (equation 2.70, dropping the superscript e), and equation (2.50). The subscript to f indicates the scattering vector. In the kinematical approximation, the intensity scattered in the direction

† The theory in this section is also valid for orientations away from the Bragg orientation, and the reader is encouraged to work through the full derivation.

k is proportional to the modulus squared of P_j, and we find

$$|P_j|^2 = \frac{1}{2}\left[f_{\mathbf{S}}^2 \sum_{i=1}^{N}\sum_{j=1}^{N} e^{2\pi i \mathbf{S}\cdot(\mathbf{r}_i-\mathbf{r}_j)} + f_{\mathbf{S}-\mathbf{g}}^2 \sum_{i=1}^{N}\sum_{j=1}^{N} e^{2\pi i(\mathbf{S}-\mathbf{g})\cdot(\mathbf{r}_i-\mathbf{r}_j)} \right.$$
$$\left. + 2(-1)^{j-1} f_{\mathbf{S}} f_{\mathbf{S}-\mathbf{g}} \sum_{i=1}^{N}\sum_{j=1}^{N} \cos\{2\pi[\mathbf{S}\cdot(\mathbf{r}_i-\mathbf{r}_j)+\mathbf{g}\cdot\mathbf{r}_j]\} \right].$$

Next, let us assume that the atoms vibrate independently of each other, and that the instantaneous displacements are given by the vectors $\mathbf{u}_i(t)$; we will not write the time argument explicitly, as it is clear that only the $\mathbf{u}_i$ depend on time. The first term on the right-hand side of the above equation then reads as

$$\sum_{i=1}^{N}\sum_{j=1}^{N} e^{2\pi i \mathbf{S}\cdot(\mathbf{r}_i-\mathbf{r}_j)} e^{2\pi i \mathbf{S}\cdot(\mathbf{u}_i-\mathbf{u}_j)} = N + \sum_{i,j}^{N}\sum_{i\neq j}^{N} e^{2\pi i \mathbf{S}\cdot(\mathbf{r}_i-\mathbf{r}_j)} e^{2\pi i \mathbf{S}\cdot(\mathbf{u}_i-\mathbf{u}_j)},$$

with similar expressions for the other terms. The time average can be computed in the same way as the Debye–Waller factor in Chapter 2:

$$\left\langle e^{2\pi i \mathbf{S}\cdot(\mathbf{u}_i-\mathbf{u}_j)} \right\rangle = e^{-2M_S},$$

with $M_S = 2\pi^2 S^2 \langle u^2 \rangle$. Noting that

$$\langle \cos(2\pi \mathbf{g}\cdot\mathbf{u}) \rangle = 1 - \frac{4\pi^2 g^2 \langle u^2 \rangle}{2!} + \cdots = 1 - \frac{2\pi^2 g^2 \langle u^2 \rangle}{1!} + \cdots = e^{-M_g},$$

and working out all time averages we find for the intensity:

$$|P_j|^2 = \frac{1}{2}\left[f_{\mathbf{S}}^2 \left[N + e^{-2M_S} \sum_{i,j}^{N}\sum_{i\neq j}^{N} e^{2\pi i \mathbf{S}\cdot(\mathbf{r}_i-\mathbf{r}_j)} \right] \right.$$
$$+ f_{\mathbf{S}-\mathbf{g}}^2 \left[N + e^{-2M_{S-g}} \sum_{i,j}^{N}\sum_{i\neq j}^{N} e^{2\pi i(\mathbf{S}-\mathbf{g})\cdot(\mathbf{r}_i-\mathbf{r}_j)} \right]$$
$$+ 2(-1)^{j-1} f_{\mathbf{S}} f_{\mathbf{S}-\mathbf{g}} \left[N e^{-M_g} + e^{-(M_S+M_{S-g})} \right]$$
$$\left. \times \sum_{i,j}^{N}\sum_{i\neq j}^{N} \cos\{2\pi[\mathbf{S}\cdot(\mathbf{r}_i-\mathbf{r}_j)+\mathbf{g}\cdot\mathbf{r}_j]\} \right].$$

To determine the thermal diffuse scattering component of this expression, we must subtract the Bragg scattering part, which can be obtained by repeating the computation for a crystal with atomic scattering factors equal to $f_{\mathrm{S}} e^{-M_s}$. The resulting expression for the thermal diffuse scattering intensity is

$$I_{TDS}(\mathbf{k}) = \frac{N}{2}\left[f_{\mathbf{S}}^{2}\left(1-\mathrm{e}^{-2M_S}\right)+f_{\mathbf{S}-\mathbf{g}}^{2}\left(1-\mathrm{e}^{-2M_{S-g}}\right)\right.$$
$$\left.+\,2(-1)^{j-1}f_{\mathbf{S}}f_{\mathbf{S}-\mathbf{g}}\left(\mathrm{e}^{-M_g}+\mathrm{e}^{-(M_S+M_{S-g})}\right)\right]. \qquad (6.45)$$

At this point, we should briefly review what we have accomplished: our final goal is to come up with a method to compute the absorption length $\xi'_{\mathbf{g}}$, or, equivalently, the imaginary potential coefficients $V'_{\mathbf{g}}$. There are several contributions to the absorption potential, and experience shows that thermal diffuse scattering is, for most materials, the dominant contribution at room and elevated temperature. We have assumed a crystal in the Bragg orientation, and computed the probability for the corresponding Bloch waves to be scattered in an *arbitrary* direction $\mathbf{k}$. After introducing atomic vibrations and time-averaging the resulting expression, we ended up with the TDS intensity in equation (6.45). Now we must connect this expression to the absorption potential.

Electrons are absorbed when they no longer contribute to the Bragg-diffracted beams, whatever the origin of the absorption process may be. If we integrate $I_{TDS}(\mathbf{k})$ over all possible orientations of $\mathbf{k}$, then we will have calculated the total number of electrons that have undergone thermal diffuse scattering events. Those electrons no longer contribute to the Bragg-diffracted intensity and therefore the integrated TDS expression provides an estimate of the overall absorption. Consider the Bloch wave absorption factor $q^{(j)}$ in exact Bragg orientation, defined on page 366:

$$q^{(j)} = \frac{1}{2}\left[\frac{1}{\xi'_{\mathbf{0}}}+(-1)^{j-1}\frac{1}{\xi'_{\mathbf{g}}}\right].$$

If we compare this expression with equation (6.45), we find that they have the same structure: the first two terms in (6.45) contribute to $\xi'_{\mathbf{0}}$, and the last term in (6.45) contributes to $\xi'_{\mathbf{g}}$.

We can use an expression for the absorption Fourier coefficient $V'_{\mathbf{g}}$, *which is formally similar to that of* $V_{\mathbf{g}}$ (equation 2.102 on page 129):

$$V'_{\mathbf{g}} = \frac{0.047\,878\,01}{\Omega}\sum_{j=1}^{N_a}\left[\mathrm{e}^{-B_j s^2}f'_j(s)\right]\sum_{(\mathbf{D}|\mathbf{t})}\mathrm{e}^{-2\pi i\mathbf{g}\cdot(\mathbf{D}|\mathbf{t})[\mathbf{r}_j]}. \qquad (6.46)$$

In this expression, $f'_j(s)$ are the so-called *absorptive form factors*, which are atomic quantities similar to the electron scattering factors $f^e_j(s)$. The expansion (2.76) of $f^e_j(s)$ in terms of exponential functions is now advantageous because it turns out that an analytical expression for the absorptive form factors $f'_j(s)$ can be obtained; this allows them to be calculated for every crystal structure, using an algorithm that is not significantly more complicated than that for the computation of $f^e_j(s)$ [WK91].

The explicit expression is too long to include here and we refer to the appendix of Weickenmeier and Kohl's paper for the details. The procedure involves integration of the TDS intensity over all scattering vectors **S**:

$$f'(g) = \frac{\lambda}{2\pi} \iint f_{\mathbf{S}} f_{\mathbf{S}-\mathbf{g}} \left(\mathrm{e}^{-M_g} + \mathrm{e}^{-(M_S + M_{S-g})} \right) \mathrm{d}\mathbf{S}. \qquad (6.47)$$

The numerical computation of the absorptive form factor has been implemented in the Fortran routine CalcUcg. The absorptive form factors for individual atoms can be plotted using the **ION** routine scatfac.pro, available from the website, and discussed in more detail in Chapter 2.

The reader should realize that many assumptions are involved in the computation of both the extinction distances and the absorption lengths. The Fourier coefficients of the electrostatic lattice potential are calculated on the basis of an isolated spherical atom approximation for the electron scattering factors; since there are quite a few parameterizations available for the scattering factors, the precise value of a Fourier coefficient $V_{\mathbf{g}}$ will depend on the particular model used. In addition, the parameterizations ignore bonding contributions.

The Debye–Waller factors are unknown for most crystal structures. They can be calculated if a model for the phonon dispersion relation is available. In most cases this is not feasible, and the parameters will need to be estimated. This results in some uncertainty in the values of the Fourier coefficients of the electrostatic lattice potential. Moreover, it is probably not a good idea to use an isotropic Debye–Waller factor in all situations.

If experimental values for the Debye–Waller factors are unavailable, then one could use the elemental values as rough estimates. Peng and coworkers [PRDW96] list the Debye–Waller factors as a function of temperature for 42 different elemental crystals. From their Table 1 it can be seen that, in general, the Debye–Waller factor is larger for elements in the leftmost columns of the periodic table. At room temperature, values around 0.1 nm^2 would be quite reasonable for first column elements *(Li, Na, K, Rb, Cs)*, whereas for second column elements *(Be, Mg, Ca, Sr, Ba)* values around 0.01–0.03 nm^2 are reasonable. For most other elements, values in the range 0.003–0.007 are acceptable. The Debye–Waller factors for the elemental solids decrease down a column of the periodic table. If experimental values for $B(T)$ are available, then those should be used instead of the estimated values. At liquid nitrogen temperature, the Debye–Waller factors are typically about one-third of their room-temperature values.

Because of these uncertainties, one should treat the calculated values for $V'_{\mathbf{g}}$ (using CalcUcg or any other similar routine) as first-order estimates. Refinements based on experimental observations should then lead to more accurate values; this is currently an active area of research and we refer the reader to [SZ92] for a more detailed description. In recent years, there has been a dramatic increase of interest in *quantitative* TEM observations, and the measurement of V_g and related parameters certainly falls under this umbrella. Since this is not the topic of this book, we will assume hence forth that the extinction distance and absorption length can be calculated in one way or the other.

The Fortran program qg.f90 can be used to compute the real and imaginary parts of $1/q_g$ for an arbitrary crystal structure. The program calls the CalcUcg routine to compute both the real and imaginary parts of the potential Fourier coefficients for a given plane (hkl). As an example, Table 6.2 shows the results for three face-centered cubic materials (Al, Cu, and Au), along with the non-centrosymmetric study material GaAs.

Table 6.3 shows computed parameters for the non-centrosymmetric tetragonal form of $BaTiO_3$, and the hexagonal materials Ti, ZnS, and BeO, along with experimental (refined) values for the (002) reflection of BeO [SZ92]. The information in this table can be generated for individual planes by means of the qg.f90 program, or for all independent families of planes using the xtalinfo.f90 program introduced in Chapter 4.

6.5.2 The two-beam scattering matrix

Matrix methods were introduced in 1961 by Howie and Whelan [HW61] to facilitate contrast computations, and they have since been used by many researchers, e.g. [GVLA65, Man62, Thö70]. We will closely follow [Thö70], but adapt the notation to that used throughout this chapter. The theory described below will prove to be very useful for the description of defect image contrast.

In obtaining the solutions (6.20) and (6.21) to the two-beam equations we have made use of the following initial conditions:

$$T(s, z = 0) = 1;$$
$$S(s, z = 0) = 0.$$

A second special solution can be obtained by using the following independent initial conditions:

$$T'(s, z = 0) = 0;$$
$$S'(s, z = 0) = 1;$$

in other words, all the incident amplitude is in the direction of the *scattered* beam rather than the transmitted beam. Proceeding in the same way as in Section 6.3.3.2

Table 6.2. *Modulus and phase of the real and imaginary part of the electrostatic lattice potential and several derived parameters for a few low index planes in face-centered cubic materials. All parameters were computed using the Weickenmeier–Kohl scattering factor parameterization for* 100 *kV electrons. The Debye–Waller factors are (at* 270 *K, in* nm^2*):* $B_{Al} = 0.007\,194$, $B_{Cu} = 0.005\,073$, $B_{Au} = 0.005\,714$, $B_{Ga} = 0.005\,844$, $B_{As} = 0.006\,288$ *(sources [VC85, PRDW96, GP99]).*

	hkl	V_g (V)	V'_g (V)	ξ_g (rad)	ξ'_g (nm)	ξ'_g/ξ_g (nm)	$1/q_g$ (nm^{-1})
Al	000	20.270	0.895	–	454.0	–	i0.002 20
	111	7.051	0.238	57.6	1703.7	29.6	0.017 35 + i0.000 59
	200	5.828	0.216	69.7	1882.6	27.0	0.014 34 + i0.000 53
	220	3.534	0.167	115.0	2428.4	21.1	0.008 69 + i0.000 41
	311	2.767	0.147	146.9	2758.5	18.8	0.006 81 + i0.000 36
	222	2.584	0.142	157.3	2862.3	18.2	0.006 36 + i0.000 35
	400	2.049	0.124	198.3	3265.4	16.5	0.005 04 + i0.000 31
	331	1.773	0.114	229.3	3566.7	15.6	0.004 36 + i0.000 28
Cu	000	26.965	1.994	–	203.8	–	i0.004 91
	111	13.49	0.966	30.1	420.6	14.0	0.033 19 + i0.002 38
	200	11.94	0.908	34.0	447.7	13.1	0.029 37 + i0.002 23
	220	8.556	0.761	47.5	533.8	11.2	0.021 05 + i0.001 87
	311	7.098	0.689	57.3	589.9	10.3	0.017 46 + i0.001 70
	222	6.707	0.668	60.6	608.1	10.0	0.016 50 + i0.001 64
	400	5.443	0.597	74.7	680.8	9.1	0.013 39 + i0.001 47
	331	4.722	0.552	86.1	736.3	8.5	0.011 62 + i0.001 36
Au	000	35.734	4.300	–	94.5	–	i0.010 58
	111	22.26	3.516	18.3	115.6	6.3	0.054 77 + i0.008 65
	200	19.93	3.403	20.4	119.4	5.9	0.049 05 + i0.008 37
	220	14.38	3.060	28.3	132.8	4.7	0.035 38 + i0.007 53
	311	11.97	2.859	33.9	142.2	4.2	0.029 46 + i0.007 03
	222	11.34	2.798	35.8	145.3	4.0	0.027 89 + i0.006 88
	400	9.317	2.579	43.6	157.6	3.6	0.022 92 + i0.006 34
	331	8.182	2.433	49.7	167.0	3.4	0.020 13 + i0.005 99
GaAs	000	18.116	1.178	–	345.1	–	i0.002 90
	111	$8.348e^{-i0.726}$	$0.517e^{-i0.729}$	48.7	785.4	16.1	0.020 55 + i0.001 27
	200	0.659	0.040	616.4	10136.9	16.4	0.001 62 + i0.000 10
	220	7.645	0.615	53.2	661.3	12.4	0.018 81 + i0.001 51
	311	$4.559e^{i0.750}$	$0.407e^{i0.726}$	89.1	999.6	11.2	0.011 24 + i0.001 00
	222	0.191	0.034	2122.4	12103.2	5.7	0.000 47 + i0.000 08
	400	5.201	0.523	78.1	776.3	9.9	0.012 80 + i0.001 29
	331	$3.315e^{-i0.771}$	$0.354e^{-i0.726}$	122.6	1149.46	9.4	0.008 12 + i0.000 87

Table 6.3. *Modulus and phase of the real and imaginary part of the electrostatic lattice potential and derived parameters for several non-cubic materials. All parameters were computed using the Weickenmeier–Kohl scattering factor parameterization for* 100 *kV electrons, except for the BeO data which was computed for* 80 *kV. The Debye–Waller factors are (at* 270 *K, in nm*^{2}*):* $B_{Ba} = 0.004$ *(estimated),* $B_{Ti} = 0.004$ *(estimated) in* $BaTiO_3$ *and* 0.004 735 *in pure Ti,* $B_O = 0.002$ *(estimated) in* $BaTiO_3$ *and* 0.0028 *in BeO,* $B_{Zn} = 0.007\,844$*,* $B_S = 0.006\,787$*,* $B_{Be} = 0.003\,55$*. (sources [VC85,PRDW96,GP99,SZ92]).*

	hkl	V_g (V)	V'_g (V)	ξ_g (rad)	ξ'_g (nm)	ξ'_g/ξ_g (nm)	$1/q_g$ (nm^{-1})
$BaTiO_3$	000	29.336	1.494	–	272.0	–	i0.003 68
	111	$6.237e^{i0.108}$	$0.430e^{i0.044}$	65.2	944.9	14.5	0.015 41 + i0.001 06
	200	10.525	0.616	38.6	659.7	17.1	0.025 90 + i0.001 52
	220	7.240	0.527	56.1	770.8	13.7	0.017 82 + i0.001 30
	311	$3.211e^{i0.102}$	$0.333e^{i0.035}$	126.5	1220.1	9.6	0.007 96 + i0.000 82
	222	$5.439e^{i0.010}$	$0.472e^{-i0.023}$	74.7	861.2	11.5	0.013 42 + i0.001 16
	400	4.380	0.434	92.8	936.6	10.1	0.010 78 + i0.001 07
	331	$2.109e^{i0.100}$	$0.285e^{i0.033}$	192.7	1426.7	7.4	0.005 24 + i0.000 70
Ti	000	28.334	1.086	–	374.3	–	i0.002 67
	$\bar{1}1.0$	−5.825	−0.230	69.8	1770.9	25.4	0.014 33 − i0.000 56
	11.0	6.320	0.340	64.3	1196.1	18.6	0.015 55 + i0.000 84
	$2\bar{2}.0$	−2.600	−0.155	156.3	2625.0	16.8	0.006 40 − i0.000 38
	$\bar{1}1.1$	8.820	−0.372	46.1	1091.3	23.7	0.021 70 + i0.000 92
	$\bar{2}2.1$	4.274	0.262	95.1	1553.3	16.3	0.010 52 + i0.000 64
	00.2	−10.649	−0.440	38.2	924.7	24.2	0.026 20 + i0.001 08
	$\bar{1}1.2$	3.823	0.187	106.3	2178.7	20.5	0.009 41 + i0.000 46
ZnS	000	16.248	1.106	–	367.6	–	i0.002 72
	$\bar{1}1.0$	−5.594	−0.295	72.7	1377.7	18.9	0.013 76 − i0.000 73
	11.0	6.853	0.467	59.3	870.9	14.7	0.016 86 + i0.001 15
	$2\bar{2}.0$	−2.890	−0.216	140.6	1877.7	13.3	0.007 11 − i0.000 53
	$\bar{1}1.1$	$3.452e^{-i0.562}$	$0.274e^{-i1.213}$	117.7	1485.2	12.6	0.008 90 + i0.000 54
	$\bar{2}2.1$	$1.940e^{-i0.721}$	$0.219e^{-i1.267}$	209.5	1851.7	8.8	0.005 05 + i0.000 46
	00.2	$7.634e^{-i0.716}$	$0.446e^{-i0.372}$	53.2	911.4	17.1	0.018 42 + i0.001 03
	$\bar{1}1.2$	$2.923e^{-i0.701}$	$0.198e^{-i0.348}$	139.0	2056.8	14.8	0.007 02 + i0.000 46
BeO	000	20.169	1.614	–	223.2	–	i0.004 48
[80 kV]	$\bar{1}1.0$	−4.788	−0.083	75.2	4317.2	57.4	0.013 29 − i0.000 23
	11.0	4.809	0.090	74.9	3985.5	53.2	0.013 35 + i0.000 25
	$2\bar{2}.0$	−1.924	−0.039	187.2	9205.1	49.1	0.005 34 − i0.000 11
	$\bar{1}1.1$	$2.975e^{i0.037}$	$0.068e^{i0.433}$	121.1	5274.1	43.6	0.008 19 + i0.000 17
	$\bar{2}2.1$	$1.582e^{i0.344}$	$0.039e^{i0.509}$	227.7	9204.7	40.4	0.004 37 + i0.000 11
	$\bar{1}1.2$	$2.273e^{-i1.031}$	$0.043e^{-i1.191}$	158.5	8462.6	53.4	0.006 33 + i0.000 12
	00.2	$6.505e^{-i0.918}$	$0.120e^{-i1.155}$	55.4	3002.0	54.2	0.018 13 + i0.000 32
Exp.	00.2	$5.955e^{-i0.885}$	$0.110e^{-i1.100}$	52.3	2857.1	54.6	0.019 19 + i0.000 34

we find that this special solution $(T'(s,z), S'(s,z))$ can be written as

$$T'(s,z) = S(-s,z);$$

$$S'(s,z) = T(-s,z).$$

In other words, the expressions for transmitted and scattered amplitudes are swapped, and the sign of the excitation error is changed. From equations (6.20) and (6.21) we know that $S(-s,z) = S(s,z)$, i.e. the dark field rocking curve is symmetric in the excitation error. We will denote $T(-s,z)$ by $T^{(-)}$, $T(s,z)$ by T, and $S(s,z)$ by S. The amplitude at the exit plane of a crystal with thickness z_0 (including absorption) is then given by

$$\begin{pmatrix} \psi_\mathbf{0} \\ \psi_\mathbf{g} \end{pmatrix}_{z=z_0} = \mathrm{e}^{-(\pi/\xi_0')z_0} \begin{pmatrix} T & S\mathrm{e}^{-\mathrm{i}\theta_\mathbf{g}} \\ S\mathrm{e}^{\mathrm{i}\theta_\mathbf{g}} & T^{(-)} \end{pmatrix} \begin{pmatrix} \psi_\mathbf{0} \\ \psi_\mathbf{g} \end{pmatrix}_{z=0}. \tag{6.48}$$

In Section 5.4, we introduced the scattering matrix $\mathcal{S}$. In the centrosymmetric two-beam case we have

$$\mathcal{S}(s, z_0) \equiv \mathrm{e}^{-(\pi/\xi_0')z_0} \begin{pmatrix} T & S\mathrm{e}^{-\mathrm{i}\theta_\mathbf{g}} \\ S\mathrm{e}^{-\mathrm{i}\theta_\mathbf{g}} & T^{(-)} \end{pmatrix}, \tag{6.49}$$

and equation (6.48) reads as

$$\Psi(z_0) = \mathcal{S}(s, z_0)\Psi(0). \tag{6.50}$$

The scattering matrix $\mathcal{S}(s,z)$ can be written as

$$\mathcal{S}(s,z) = \mathcal{S}_r(s,z) + \mathrm{i}\mathcal{S}_i(s,z). \tag{6.51}$$

It is easy to show that the explicit elements of the two matrices are given by

$$\begin{aligned} \mathcal{S}_{r,11} &= \mathrm{e}^{-(\pi/\xi_0')z} \Big\{ \cos(\pi\sigma_r z)\cosh(\pi\sigma_i z) \\ &\quad - \frac{s_\mathbf{g}}{|\sigma|^2}\left[\sigma_i \sin(\pi\sigma_r z)\cosh(\pi\sigma_i z) - \sigma_r \cos(\pi\sigma_r z)\sinh(\pi\sigma_i z)\right]\Big\}; \\ \mathcal{S}_{r,22} &= \mathcal{S}_{r,11}(-s,z); \\ \mathcal{S}_{r,21} &= \frac{\mathrm{e}^{-(\pi/\xi_0')z}}{|\sigma|^2} \left\{ \left[\frac{\sigma_i}{\xi_\mathbf{g}} - \frac{1}{\xi_\mathbf{g}'}(\sigma_r \cos\beta_\mathbf{g} + \sigma_i \sin\beta_\mathbf{g}) \right] \sin(\pi\sigma_r z)\cosh(\pi\sigma_i z) \right. \end{aligned}$$

$$
-\left[\frac{\sigma_r}{\xi_{\mathbf{g}}}-\frac{1}{\xi'_{\mathbf{g}}}(\sigma_r\sin\beta_{\mathbf{g}}-\sigma_i\cos\beta_{\mathbf{g}})\right]\cos(\pi\sigma_r z)\sinh(\pi\sigma_i z)\Bigg\} = \mathcal{S}_{r,12};
$$

$$
\begin{aligned}
\mathcal{S}_{i,11} &= -\mathrm{e}^{-(\pi/\xi'_0)z}\Bigg\{\sin(\pi\sigma_r z)\sinh(\pi\sigma_i z)\\
&\quad+\frac{s_{\mathbf{g}}}{|\sigma|^2}\left[\sigma_r\sin(\pi\sigma_r z)\cosh(\pi\sigma_i z)+\sigma_i\cos(\pi\sigma_r z)\sinh(\pi\sigma_i z)\right]\Bigg\};\\
\mathcal{S}_{i,22} &= \mathcal{S}_{i,11}(-s,z);\\
\mathcal{S}_{i,21} &= \frac{\mathrm{e}^{-(\pi/\xi'_0)z}}{|\sigma|^2}\Bigg\{\left[\frac{\sigma_i}{\xi_{\mathbf{g}}}-\frac{1}{\xi'_{\mathbf{g}}}(\sigma_r\cos\beta_{\mathbf{g}}+\sigma_i\sin\beta_{\mathbf{g}})\right]\cos(\pi\sigma_r z)\sinh(\pi\sigma_i z)\\
&\quad+\left[\frac{\sigma_r}{\xi_{\mathbf{g}}}-\frac{1}{\xi'_{\mathbf{g}}}(\sigma_r\sin\beta_{\mathbf{g}}-\sigma_i\cos\beta_{\mathbf{g}})\right]\sin(\pi\sigma_r z)\cosh(\pi\sigma_i z)\Bigg\} = \mathcal{S}_{i,12}.
\end{aligned}
$$

While these expressions look rather complicated, they actually contain many repeating factors which need only be computed once. The expressions above are implemented in the subroutine TBCalcSM; the source code has been written to have the smallest possible number of computations and function evaluations. The routine returns the real and imaginary parts of the two-beam scattering matrix $\mathcal{S}$. A second routine, TBCalcdz, is provided which applies the transfer matrix to the column vector Ψ_1 and returns the updated column vector $\Psi_2(z+\mathrm{d}z)=\mathcal{S}\Psi_1(z)$. Finally, the routine TBCalcInten returns the intensities of the transmitted and diffracted beams computed directly from equations (6.25) and (6.24). All images shown in the next section were computed using either the intensity expressions or the matrix formalism.

The main advantage of the matrix formalism is that matrix multiplication is much faster computationally than numerically solving the dynamical diffraction differential equations by Runge–Kutta-type algorithms. The matrix method is even faster than the direct application of equations (6.25) and (6.24). As an example, consider the computation of the transmitted and diffracted intensities as a function of crystal thickness along the centerline of the standard parameter space, i.e. for $w=0$. The thickness varies from $z=0$ to $z=6\xi_{\mathbf{g}}$. We can subdivide this interval into, say, N equidistant steps $\mathrm{d}z=6\xi_{\mathbf{g}}/N$. Equations (6.25) and (6.24) would then require evaluation of $2N$ trigonometric, $2N$ hyperbolic, and N exponential functions, a potentially expensive computation in terms of processor clock cycles. The matrix method on the other hand would evaluate the matrix $\mathcal{S}(0,\mathrm{d}z)$ once (with only seven calls to trigonometric, hyperbolic, and exponential functions), and then multiply this matrix $N-1$ times by itself to obtain the amplitudes for all thicknesses. A possible drawback of the matrix method is that only multiples of the thickness dz can be computed; for intermediate values we must either use a different value of dz and recompute the matrix product, or interpolate intensities for thicknesses bracketing the desired thickness.

The bright field and dark field images for the standard parameter space in Figs 6.3 and 6.4 were computed using the TBSM.f90 program. This program implements both the scattering matrix algorithm and the analytical solutions, and compares the execution time of the algorithms for a BF–DF image pair of 512×512 pixels. The scattering matrix approach is about five times faster than the direct evaluation of the analytical solutions.

6.5.3 Numerical (two-beam) Bloch wave calculations

Numerical implementation of a general complex eigenvalue problem is a non-trivial task, and is best left to mathematicians and experts in linear algebra. For all Bloch wave computations in this book, we will make use of version 3.0 of the public domain package LAPACK, which is available from `www.netlib.org` for a large number of computer platforms. LAPACK is a library of Fortran-77 routines *for solving the most commonly occurring problems in numerical linear algebra* [ABB$^+$99]. LAPACK is also available in Fortran-95, Java, and C++. Pre-compiled libraries for DEC-ALPHA, Linux (RedHat), IBM RS/6000, Solaris, and Windows NT (Visual Fortran) are available; alternatively, the package can be installed from scratch using the extensive installation instructions available from the `www.netlib.org` website.†

LAPACK replaces the older LINPACK [DBMS79] and EISPACK [GBDM77] libraries, and makes extensive use of the BLAS routines or *basic linear algebra subprograms* [DDCHH88]. Documentation for LAPACK is available on-line, or in book form [ABB$^+$99]. We will assume that the reader has access to the LAPACK libraries, so that all Fortran source code for Bloch wave computations can be compiled and executed.

From a numerical point of view, the Bloch wave formalism requires the solution of an eigenvalue problem for a complex non-symmetric matrix in the most general case. Eigenvalues $\Gamma^{(j)}$ and eigenvectors $\mathbf{C}^{(j)}$ must be computed, and the eigenvector matrix $\mathcal{C}$ must be inverted to obtain the Bloch wave excitation amplitudes $\alpha_{\mathbf{g}}^{(j)}$ for a given initial condition at the sample entrance surface.

The basic structure of a Bloch wave program is independent of the number of beams in the calculation. We will, in fact, use only one subroutine for the two-beam case, the systematic row case and the zone axis case (the latter two will be discussed in the next chapter). The relevant LAPACK routine is CGEEV, which computes the

† The complete LAPACK package consists of about 805 000 lines of Fortran source code.

Pseudo Code PC-14 Core of the Bloch wave computation.

Input: dynamical matrix
Output: eigenvalues and eigenvectors
initialize auxiliary arrays
compute eigenvalues and eigenvectors for complex non-symmetric matrix {CGEEV (LAPACK)}
rank eigenvalues from most positive to most negative; also rank eigenvectors in same order {SPSORT (SLATEC)}
compute LU decomposition of eigenvector matrix {CGETRF (LAPACK)}
compute inverse of eigenvector matrix {CGETRI (LAPACK)}
return eigenvalues, eigenvector matrix and its inverse

Pseudo Code PC-15 Outline of complete Bloch wave computation.

Input: crystal data file
Output: Bloch wave eigenvalues, eigenvectors, excitation amplitudes
ask user for range of contributing reciprocal lattice points
ask user for incident beam range
compute off-diagonal components of dynamical matrix {CalcVcg}
for each incident beam orientation **do**
complete diagonal of dynamical matrix (excitation errors) {Calcsg}
solve the eigenvalue problem {Pseudo Code PC-14}
compute amplitudes of all beams {Equation 5.58}
store relevant information in file
end for

eigenvalues and eigenvectors of an $N \times N$ complex non-symmetric matrix. The eigenvector matrix is then inverted using the routines CGETRF, which computes the LU decomposition of the matrix, and CGETRI, which computes the inverse of the matrix using the LU decomposition. The eigenvalues and eigenvectors are then ranked so that the largest real part corresponds to $\gamma^{(1)}$. The sorting can be done with a variety of different algorithms, and we use the SPSORT routine available in the SLATEC library. The main flow of the N-beam Bloch wave algorithm BWsolve is shown in pseudo code PC-14.[†] The outline of a complete Bloch wave program is shown in pseudo code PC-15.

[†] If the LAPACK package is not available, any other eigenvalue solver may be substituted; the BWsolve subroutine is then the only routine that must be modified.

Note that all the off-diagonal elements of the dynamical matrix $\bar{\mathcal{A}}$ can be precomputed, since they do not depend on the sample orientation. The imaginary part of the diagonal contains only the normal absorption contribution and can also be precomputed. The real part of the diagonal is computed for each sample (or incident beam) orientation. After matrix inversion of the eigenvector matrix, the amplitudes of all beams can be computed using equation (5.58).

For the two-beam case, we have an analytical solution for both the Bloch wave and the DHW formalisms, so that a numerical Bloch wave solution is not really needed. Nevertheless, the reader will find the program TBBW.f90 on the website. The program produces an output file with eigenvalues, eigenvector matrices, and excitation amplitudes for all Bloch waves and for each beam orientation. The program BWshow.f90 can then be used to produce bright field and dark field images, along with plots of eigenvalues, excitation amplitudes, and absorption parameters versus beam direction. The resulting drawings are similar to the ones shown in Figs 6.6 and 6.7(a) and (c).

6.5.4 Example two-beam image simulations

In this section, we will discuss a number of different sample geometries and illustrate the computation of two-beam bright and dark field images. The reader is encouraged to study and modify the Fortran source code used to compute the images in Figs 6.11–6.14. The source code can be found in the file TBBFDF.f90 and is described in pseudo code PC-16. A second file, BFDF.routines, is available from which the relevant lines can be selected and pasted into the TBBFDF.f90 program to simulate a number of different sample geometries. All images in the following four subsections were obtained using the TBBFDF.f90 program. The same image

---

Pseudo Code PC-16 Computation of two-beam bright field and dark field images

Input: crystal data file [*.xtal]
Output: PostScript file(s)

```
get acceleration voltage and operative reflection
compute excitation and absorption length                      {Calcqgi}
compute thickness and excitation error arrays
        for a given foil geometry         {source code from BFDF.routines}
compute intensities using analytical expressions          {TBCalcInten}
generate PostScript output
```

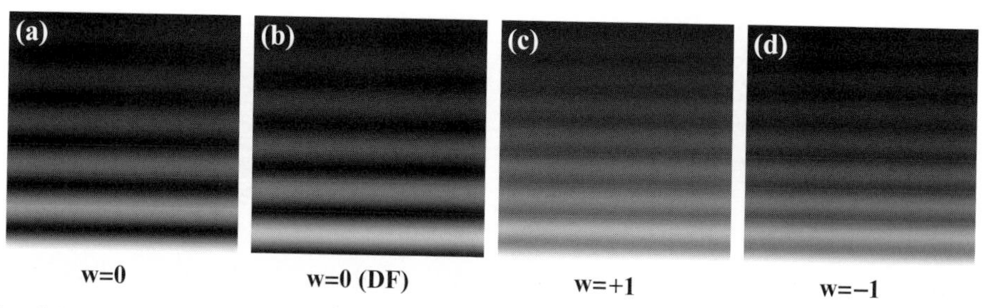

Fig. 6.11. Simulated bright field images for the parameters stated in the text. (a) Bright field and (b) dark field images for exact Bragg orientation; (c) bright field image for $w = +1$, and (d) $w = -1$. The effect of anomalous absorption is clearly seen in (d).

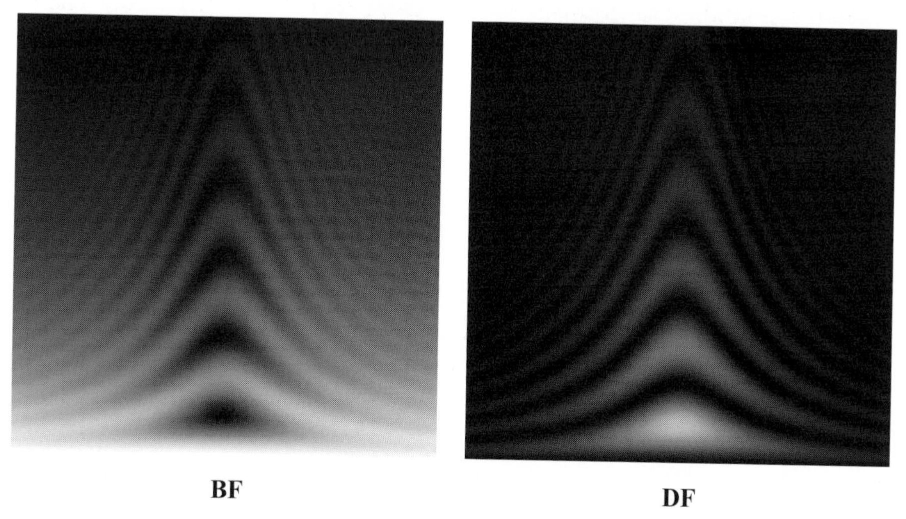

Fig. 6.12. Simulated BF/DF pair for a bent wedge in aluminum for the same parameters stated in Section 6.5.4.1.

simulation examples are also available in the interactive **ION** routine wedge.pro on the website.

6.5.4.1 Wedge-shaped foil

Figures 6.11(a) and (b) show a bright field/dark field image pair for a straight wedge in aluminum. The microscope acceleration voltage is 200 kV, and the 111 reflection is in the Bragg orientation. The thickness varies from 0 to 400 nm, the extinction distance is $\xi_{111} = 73.1$ nm, the anomalous absorption parameter is $\xi'_{111} = 2742.6$ nm, and the normal absorption length is $\xi'_0 = 697.2$ nm (all based on the Weickenmeier–Kohl scattering and atomic form factors).

The bright field image shows intensity fringes running parallel to the wedge edge. The fringe spacing can be derived from equation (6.22) (ignoring absorption). In exact Bragg orientation $w = 0$, the scattered and transmitted intensities are identical

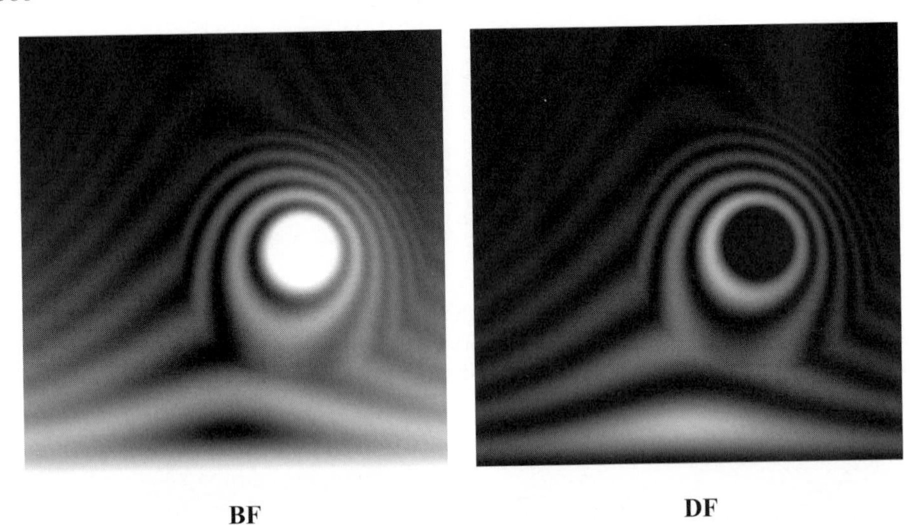

Fig. 6.13. Simulated BF/DF pair for a bent aluminum foil with a hole near the edge.

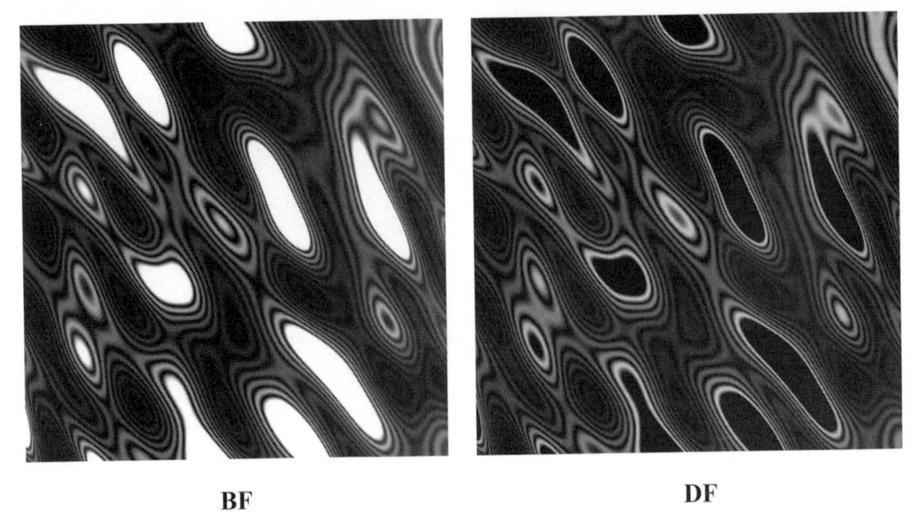

Fig. 6.14. Simulated BF/DF pair for a titanium foil with average thickness 100 nm. Random thickness variations are present, with negative thickness values corresponding to holes in the foil. The dimensionless orientation parameter w varies from -1 to $+1$ (left to right).

when z_ξ increases from n to $n+1$, which indicates that the periodicity of the fringes is equal to $\xi_{\mathbf{g}}$. In other words, the crystal thickness increases by one extinction distance $\xi_{\mathbf{g}}$ for every thickness fringe. This fact may be used to *estimate* the local thickness of the foil. Care must be taken that the foil is indeed in exact Bragg orientation, and that there are no other strongly excited reflections in the diffraction pattern. Furthermore, the first fringe near the edge may not correspond to $z_\xi = 1$. Thicknesses obtained using this method are at best rough estimates of the true thickness. We refer to Section 6.5.5 for a more accurate local thickness determination procedure.

Since the fringes follow the contours of constant thickness, they are known as *thickness fringes*. When the excitation error increases in magnitude, the fringe spacing becomes smaller, since the periodicity is now determined by $\sqrt{1+w^2}z_\xi$. When $w = 1$, the effective extinction distance decreases by a factor of $\sqrt{2}$, and the fringes are closer together, as seen in the simulated bright field image of Fig. 6.11(c). The effect of anomalous absorption can be seen clearly in the asymmetry between the $w = +1$ and -1 (Fig. 6.11d) bright field images.

6.5.4.2 Wedge-shaped bent foil

Figure 6.12 illustrates the BF–DF images for a bent foil with imaging parameters stated in Section 6.5.4.1. This is the standard two-beam rocking curve discussed earlier in this chapter; the reader may access the rock.pro routine on the website to display the rocking curve for a specific reduced thickness z_ξ.

6.5.4.3 Wedge-shaped bent foil with hole near edge

During specimen preparation, one occasionally produces many small holes in the foil, in addition to the main hole. Often those smaller holes are located near the edge of the foil. Figure 6.13 illustrates how a circular hole with sample thickness increasing radially away from the edge of the hole perturbs the thickness fringes around the Bragg orientation. As stated for the first example, the fringes represent contours of constant $\sqrt{1+w^2}z_\xi$. For a straight wedge, such contours would coincide with constant thickness contours, but in the more general case of a bent foil, the interpretation of the fringes is somewhat more difficult. The imaging parameters are identical to those for Fig. 6.12.

6.5.4.4 Planar foil with random thickness variations

The previous examples all dealt with wedge-shaped foils. With the modern specimen preparation techniques now available, we can often obtain large thin areas with a nearly uniform thickness, and it becomes important to understand how random thickness variations affect the image contrast. Using the TBBFDF.f program, we can explore the image contrast caused by such thickness variations by randomly superimposing a series of cosine waves with random orientations and phases on the average thickness.

Figure 6.14 shows a simulated BF/DF pair for a titanium foil with an average thickness of 100 nm. The minimum thickness has been set to a negative value, which effectively means that regions with a negative thickness are holes in the foil.

The thickness variations are generated through the superposition of 50 cosine waves with random amplitude and phase. The microscope acceleration voltage is 200 kV, $\xi_{002} = 48.4$ nm, $\xi'_{002} = 1489.2$ nm, and $\xi'_0 = 576.7$ nm. The foil orientation varies from $w = -1$ to $+1$ from left to right. This type of image contrast is sometimes observed in thin foils, and is usually caused by specimen preparation artifacts.

The preceding examples, which took less than 0.3 s each to compute on a 512×512 grid of pixels, show that it is quite straightforward to simulate two-beam images. For a given reflection **g**, we only need to compute the relevant imaging parameters (ξ_g, ξ'_g, ξ'_0, and β_g), provide an array of local thickness and excitation error values, and feed this information to the TBCalcInten routine. The reader is encouraged to modify the sample geometries provided in the BFDF.routines file, and/or to create additional geometries.

6.5.5 Two-beam convergent beam electron diffraction

In Chapter 4 (Section 4.6.4), we have seen how to obtain a two-beam CBED pattern. Such a pattern is essentially a *parallel recording* of many diffraction patterns for a continuous range of incident beam directions. Consider the schematic drawing in Fig. 6.15; the transmitted disk is centered on the optical axis, and each point in the disk corresponds to a different *tangential component* $\mathbf{k}_t$ of the incident wave vector $\mathbf{k}_0$. We have also seen that a beam tilt is equivalent to a specimen tilt in the opposite direction, so that a two-beam CBED pattern contains roughly the same information as the bright field image of a bend contour. Electrons which are scattered into the diffracted beam **g** experience an elastic momentum transfer equal

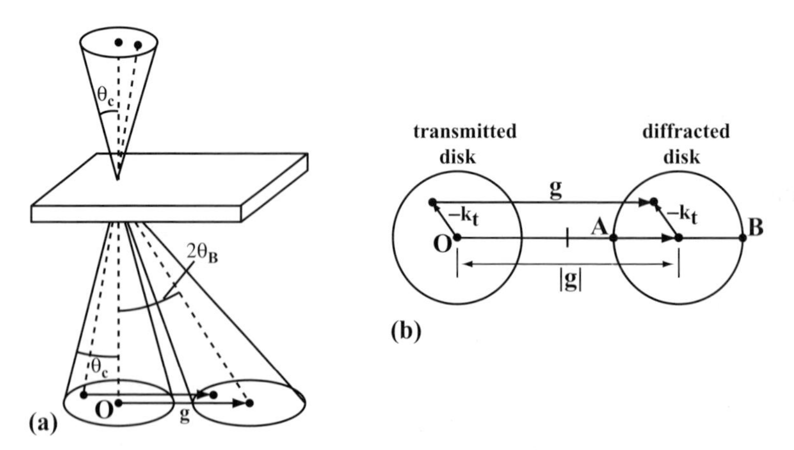

Fig. 6.15. (a) Schematic illustration of the geometry of a two-beam convergent beam electron diffraction pattern; (b) relation between equivalent locations in the bright field and dark field diffraction disks.

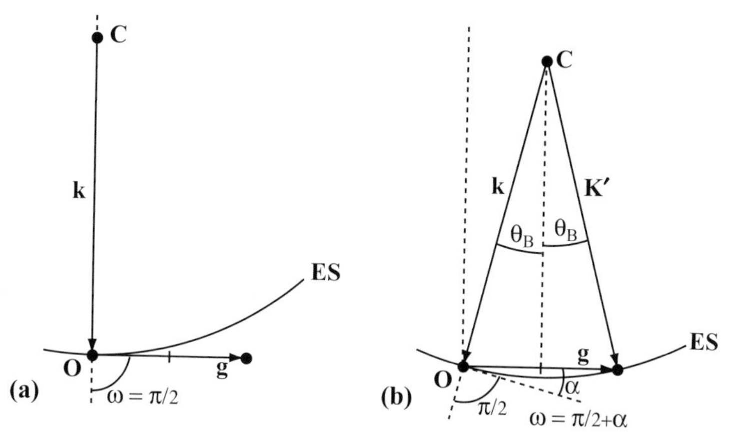

Fig. 6.16. (a) Incident beam normal to **g**; (b) for the Bragg orientation a beam tilt (or sample tilt) of α is needed to bring the reciprocal lattice point onto the Ewald sphere.

to $h\mathbf{g}$, which means that the point in the diffracted disk equivalent to the point $-\mathbf{k}_t$ in the transmitted disk is at the location $\mathbf{g} - \mathbf{k}_t$.

The opening angle of the incident beam is the beam divergence angle θ_c; it is clear from the drawing in Fig. 6.15 that if $\theta_c = \theta_B$, with θ_B the Bragg angle, then the transmitted and diffracted disks will touch each other at the point $\mathbf{g}/2$. From the Bragg equation in reciprocal space, we also know that the Bragg condition is satisfied when the center of the Ewald sphere lies on the perpendicular bisector plane of **g**, which is equivalent to saying that the tangential component of the incident wave vector must be equal to $\mathbf{k}_t = -\mathbf{g}/2$. If the initial incident beam direction **k** is perpendicular to **g**, as shown in Fig. 6.16(a), then **g** is not in Bragg orientation. A beam tilt of $\alpha = \theta_B$ milliradians is needed to bring the reciprocal lattice point **g** onto the Ewald sphere. The angles θ_c and α are therefore independent variables, both of which can be set by the microscope operator. Depending on the microscope model, the angle θ_c may have either a discrete range of settings or a continuous range, whereas α can be continuously adjusted using the beam tilt controls.

The intensity distribution inside the transmitted and diffracted disks can be understood easily from the theoretical derivations in this chapter. Since we are considering scattering from only one set of planes, described by a single **g**, we have rotational symmetry around **g**; in other words, the angle α is sufficient to define the orientation of the sample. The incident beam directions relative to the sample therefore fall in the range $[\alpha - \theta_c, \alpha + \theta_c]$, which corresponds to the line segment AB in Fig. 6.15(b). The distance between A and B is determined by θ_c, and the microscope operator can change the orientation of the sample (i.e. α) independently. It is an easy experiment to convince yourself that the intensity distribution inside the disks can be changed by tilting the sample or the incident beam orientation. The size and relative location of the disks do not change when α is changed.

In Exercise 2.9 on page 134 we have derived an expression for the excitation error $s_{\mathbf{g}}$ as a function of the angle ω between the transmitted beam direction and $\mathbf{g}$. Since $\omega = \pi/2 + \alpha$, we can use this expression to determine the angle α for which $s_{\mathbf{g}} = 0$:

$$\cos\omega = \frac{\lambda|\mathbf{g}|}{2}, \tag{6.52}$$

from which follows

$$\alpha = \sin^{-1}\left(\frac{\lambda|\mathbf{g}|}{2}\right) \approx \frac{\lambda|\mathbf{g}|}{2} = \theta_B. \tag{6.53}$$

For this particular value of α, $s_{\mathbf{g}}$ will be equal to zero at the center of the diffraction disk. Moving away from the center towards A or B (Fig. 6.15b) the excitation error becomes negative (towards B) or positive (towards A). The intensity profile along AB is therefore the dark field rocking curve profile derived earlier in this chapter. The excitation error varies from $s_{\mathbf{g}}(\theta_B - \theta_c)$ at A to $s_{\mathbf{g}}(\theta_B + \theta_c)$ at B. When θ_c, and hence the disk radius, is increased, a larger portion of the rocking curve profile becomes visible inside the diffraction disk.

Numerical simulation of a two-beam CBED pattern is now a straightforward exercise. The **ION** routine tbcbed.pro displays two-beam CBED patterns for six different sample thicknesses (multiples of $\xi_{\mathbf{g}}/2$), for a given microscope acceleration voltage V, beam divergence angle θ_c, and beam tilt angle α. The user can also change the length of the reciprocal lattice vector $\mathbf{g}$. Example output of the tbcbed routine is shown in Fig. 6.17 for the following parameters: $d = 0.2$ nm, $V = 120$ kV, $\xi_{\mathbf{g}} = 100$ nm, and the absorption ratio is equal to 15. Fig. 6.17 shows the output of three separate runs of the tbcbed routine, for the following pairs of (α, θ_c): $(5, 5)$, $(7, 6)$, and $(11, 7)$ mrad. The Bragg angle for this configuration is $\theta_B = 8.373$ mrad.

It is apparent from the examples in Fig. 6.17 that the details of the CBED pattern are quite sensitive to the sample thickness. It is possible to use two-beam CBED to estimate the local sample thickness z_0 and the extinction distance $\xi_{\mathbf{g}}$. This was first suggested by MacGillavry in 1940 [Mac40] and subsequently applied by, among others, Kelly et al. [KJBN75], Allen [All81] and Allen and Hall [AH82].

From equation (6.22), we see that the scattered intensity vanishes (in the absence of absorption) whenever

$$\sigma z_0 = \sqrt{1 + w^2} z_\xi = n, \tag{6.54}$$

with n a positive integer. If there are N intensity minima on one side of the Bragg orientation in the scattered disk, then there will also be N different values for n,

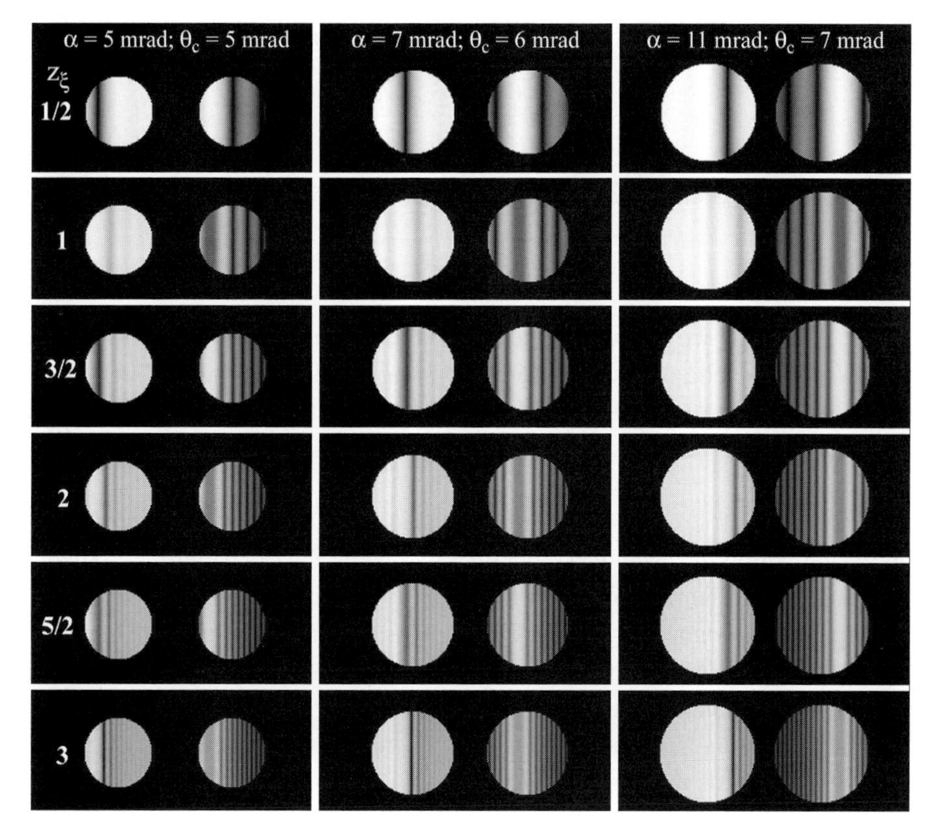

Fig. 6.17. (a) Example output of the tbcbed.pro **ION**-routine for the imaging parameters stated in the text.

starting with $n_1 = n$, $n_2 = n + 1$, and so on. For each minimum, we also have a corresponding excitation error s_i. This results in

$$z_\xi^2 + z_0^2 s_i^2 = (n + i)^2. \tag{6.55}$$

The parameters z_ξ, z_0, and n are unknown and must be determined. This can be done in a number of different ways. Perhaps the most straightforward method is to write down a χ^2-type expression:

$$\sum_{i=0}^{N-1} \left[z_\xi^2 + z_0^2 s_i^2 - (n + i)^2 \right]^2 = \chi^2, \tag{6.56}$$

and minimize it with respect to the three parameters. In order for this method to work there must be at least three intensity minima in the diffraction disk. Alternatively, we could plot s_i^2 as a function of $(n + i)^2$ and through linear regression analysis determine the parameters which provide the best fit for the following relation:

$$s_i^2 = \frac{(n+i)^2}{z_0^2} - \frac{1}{\xi_{\mathbf{g}}^2}. \tag{6.57}$$

It is easy to show geometrically that the parameters s_i are given by the expression

$$s_i = \frac{X_i}{X} g^2 \lambda, \tag{6.58}$$

where X is the distance between the centers of the two disks (conveniently measured from the edge of one disk to the corresponding edge of the other disk), and X_i is the distance between the ith minimum and the Bragg orientation. Since the dark field rocking curve is always symmetric, the Bragg orientation is readily identifiable.

In practice, the following procedure may be used to estimate the local thickness and the extinction distance. First, obtain a two-beam CBED pattern, as explained in Section 4.6.4. Then, measure on the electron micrograph the distances X and X_i along the line connecting the disk centers, as shown in Fig. 6.18. Compute the length of **g** from the crystallographic parameters. Measure the distance between the center of the dark field disk and the location of the Bragg orientation, and the radius of the diffraction disks. Distances are negative on the outside of the dark field disk, and positive towards the central disk, as shown in Fig. 6.18. The reader will find the tbthick.pro **ION** routine on the website. This routine takes all the experimental

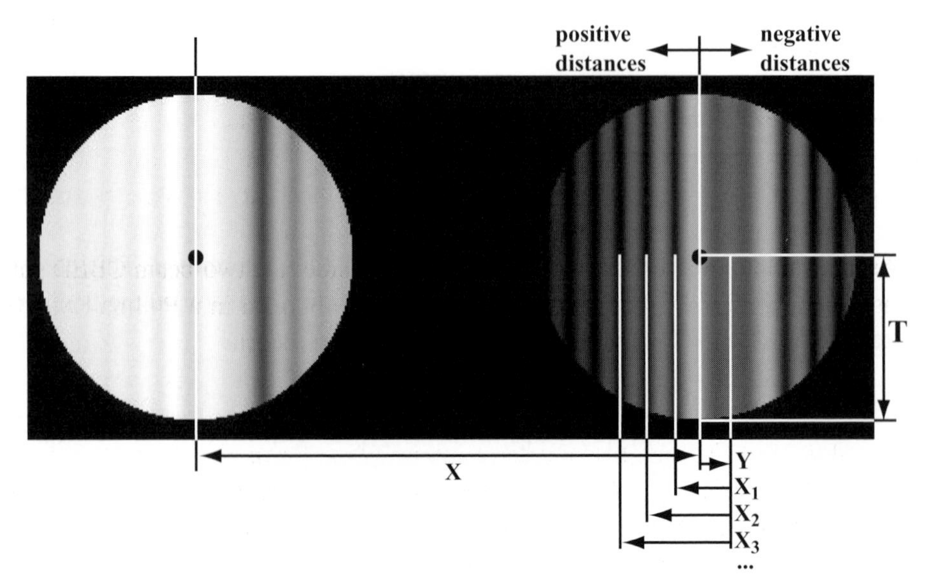

Fig. 6.18. Illustration of the input parameters for the tbthick.pro **ION** routine.

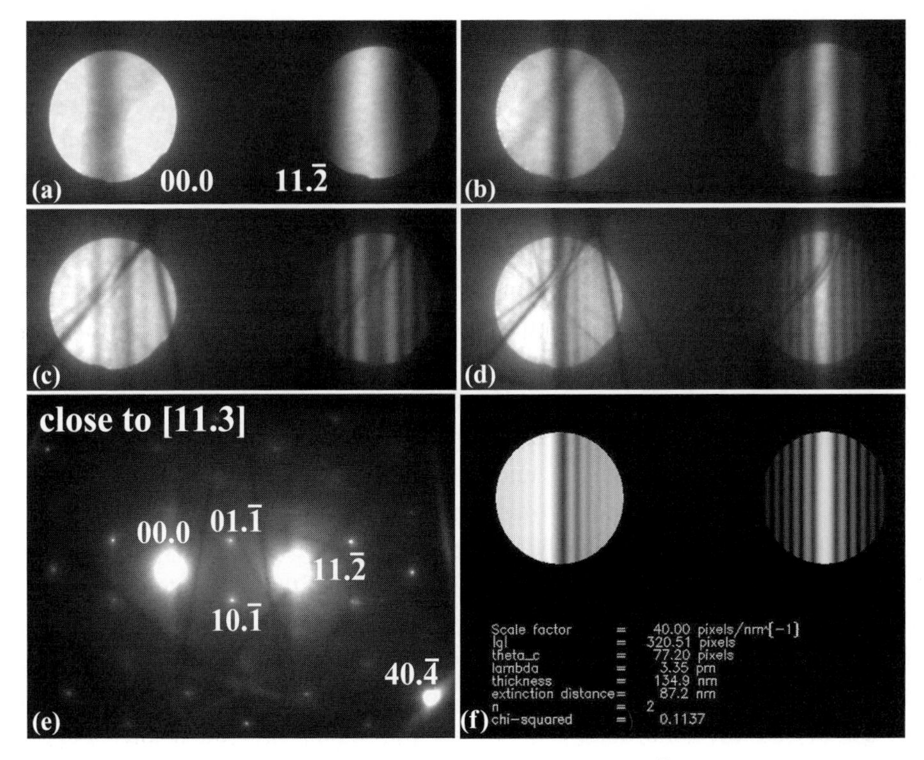

Fig. 6.19. Experimental two-beam CBED patterns for the $(11.\bar{2})$ reflection of Ti (Philips EM420 operated at 120 kV). The thickness increases from (a) to (d). The diffraction pattern in (e) shows that in addition to the two main reflections, the $(40.\bar{4})$ reflection is also strongly excited; this gives rise to the diagonal dark lines running through both bright field and dark field disks in (a)–(d). The pattern in (f) is the output of the tbthick.pro routine on the website for the parameters of (d).

parameters and minimizes the χ^2 expression to obtain $\xi_\mathbf{g}$, z_0, and n. The routine also displays the computed two-beam CBED pattern for the fitted parameters, so that a comparison with the experimental image may be carried out. The source-code for the algorithm may be found on the website.

As an example of a thickness determination, consider the two-beam CBED patterns shown in Figs 6.19(a)–(d); the patterns were obtained in a Philips EM420, operated at 120 kV, using a Ti foil oriented close to the [11.3] zone axis. Figure 6.19(e) shows the corresponding diffraction pattern. The thickness of the foil increases from (a) to (d). The parameters defined in Fig. 6.18 were measured on an enlarged print of Fig. 6.19(d), and are given by (all in mm): $X = 68.5$, $T = 16.5$, $Y = 0.0$, $X_1 = 3.0$, $X_2 = 6.0$, $X_3 = 8.8$, and $X_4 = 11.2$. The d-spacing of the $(11.\bar{2})$ planes is 0.1248 nm. The fitting algorithm of the tbthick.pro routine then returns the following parameters: $z_0 = 134.9$ nm, $\xi_\mathbf{g} = 87.2$ nm, and $n = 2$. This compares reasonably well with the theoretical extinction distance for the $\{2\bar{1}.2\}$

family of $\xi_g = 86.47$ nm (at 120 kV; computed using the xtalinfo.f90 program). The simulated CBED pattern is shown in Fig. 6.19(f).

The two-beam CBED method provides a simple way to estimate the local thickness of the thin foil and the extinction distance. Comparison of the measured ξ_g with a calculated value may reveal significant differences between the two values. The measured values are only reliable to the same extent that the two-beam approximation is reliable. In other words, if a crystal structure has a large lattice parameter along the direction $\mathbf{g}$, then the two-beam approximation will fail since the intensity in other reflections of the type $n\mathbf{g}$ may not be negligible. The accuracy of the thickness and extinction distance determination improves as the length of $\mathbf{g}$ increases, or, equivalently, when higher-order planes are selected for the acquisition of the CBED pattern. The accuracy also improves with decreasing microscope acceleration voltage, since the increased curvature of the Ewald sphere improves the validity of the two-beam approximation. Two-beam CBED thickness determinations should therefore always be regarded as rough estimates. A more detailed and accurate thickness determination is only possible if the complete intensity profile and possibly the effects of other diffracted beams are taken into account. For a detailed comparison of various two-beam methods to determine the local foil thickness and the extinction error we refer to the paper by Delille et al. [DPVC01]. We will discuss the contributions of multi-beam scattering in the next chapter.

Exercises

6.1 Repeat the dispersion surface construction of Fig. 6.9 for the case where the foil normal $\mathbf{n}$ is not perpendicular to the reciprocal lattice vector $\mathbf{g}$.

6.2 The TBBFDF.f90 program is provided on the website along with a list of possible sample geometries (BFDF.routines). Create a new sample geometry for a sample that has facets on its top and bottom surfaces. The facets correspond to surface steps of a few nanometers, and can be triangular, rectangular, or lozenge-shaped. Compute bright field and dark field images, and determine for which step height the steps still give rise to detectable image contrast.

6.3 Derive the explicit equations (6.24) and (6.25) from the Bloch wave formalism.

6.4 Show geometrically that the excitation error corresponding to a minimum intensity in the dark field CBED disk is given by equation (6.58).

7

Systematic row and zone axis orientations

7.1 Introduction

In the previous chapter, we analyzed in great detail how the bright field and dark field image contrast for the two-beam case depend on the sample thickness and the crystal orientation. We defined two dimensionless parameters (w and z_ξ) in terms of which the intensities can be expressed in the absence of absorption. Two additional parameters appear in the presence of absorption: the normal absorption length ξ_0' and the anomalous absorption length $\xi_{\mathbf{g}}'$. The former causes an exponential damping with increasing sample thickness, common to both bright field and dark field intensities; the latter produces an asymmetry between the bright field intensities for positive and negative excitation errors. We used the Bloch wave approach to explain the origin of this anomalous absorption, and we introduced thermal-diffuse scattering as one of the sources of the absorption process.

In the present chapter, we will consider the complete N-beam dynamical diffraction equations in several of the formalisms described in Chapter 5, still under the assumption of a perfect crystal. We will discuss a number of numerical techniques for solving the dynamical equations, and provide illustrations of various contrast features which are commonly observed in experiments but cannot be explained by the two-beam theory. We will discuss matrix methods, the differential equation approach, the Bloch wave approach, the multi-slice method, and the real-space method. For each of those methods we will introduce source code in the form of pseudo code descriptions. While many of the multi-beam simulations are too computationally demanding to perform in real-time, the website has several **ION** routines which illustrate various aspects of the multi-beam simulations. We will, along the way, introduce a number of techniques to speed up the computations, mostly through the use of symmetry arguments. For instance, for a beam direction slightly inclined to a zone axis there are usually additional beam directions, related to the first one by the symmetry operations of the crystal structure, which will give rise to an identical intensity distribution at the exit plane of the crystal (identical apart from a symmetry operation). It is then computationally more efficient to first identify equivalent beam orientations, and then carry out the calculation for only one of the equivalent orientations. A detailed description of the application of diffraction group symmetry to multi-beam dynamical calculations is available as a web-based appendix to this book.

7.2 The systematic row case

7.2.1 The geometry of a bend contour

The solutions to the two-beam dynamical diffraction equations can be expressed in terms of the two dimensionless parameters w and z_ξ, as illustrated in Section 6.3.3.2. This effectively means that two-beam image contrast *is independent of the particular material being studied.* A visual comparison of the bright field images for the four study materials shown in Chapter 4 immediately shows that there are significant differences in the details of bend contour and bend center contrast, and that therefore different materials *do* give rise to different image contrast. The two-beam theory represents the idealized case, and, with very few exceptions, it is also the only case that can be solved analytically.†

One of the most striking differences between, for instance, the bend contour in Fig. 4.19 (page 281) and the bright field image for the standard parameter space shown in Fig. 6.3 (page 355) is the fact that in the experimental contour there are two regions near the foil edge where the bright field contrast shows intensity oscillations (arrows in Fig. 4.19). In addition, the darker image contrast is confined to a narrow band (the actual bend contour), whereas in the standard parameter space image *all* orientations with negative w display reduced contrast due to anomalous absorption. The two-beam theory, therefore, cannot explain all of the details of bend contour contrast.

It is, however, not very difficult to understand the image contrast associated with a bend contour. Consider the schematic in Fig. 7.1. A bent foil of constant thickness z_0 is oriented such that on the left-hand side of the diagram the planes with plane normal $-\mathbf{g}$ are close to the Bragg orientation; on the right-hand side diffraction occurs from the opposite side of the planes $-\mathbf{g}$, or, equivalently, from the planes $\mathbf{g}$. Consider the excitation error for the planes $-\mathbf{g}$: towards the left of the drawing, there is a location in the foil for which $s_{-\mathbf{g}} = 0$ (location 2). At this point in the foil, $\mathbf{g}$ lies rather far outside of the Ewald sphere ($s_\mathbf{g} < 0$), so it is a good approximation to use the two-beam theory for the image contrast. The two-beam bright field and dark

† There are certain multi-beam cases that also have analytical solutions (e.g. [Fuk66]), but the general multi-beam dynamical diffraction equations have no known analytical solutions.

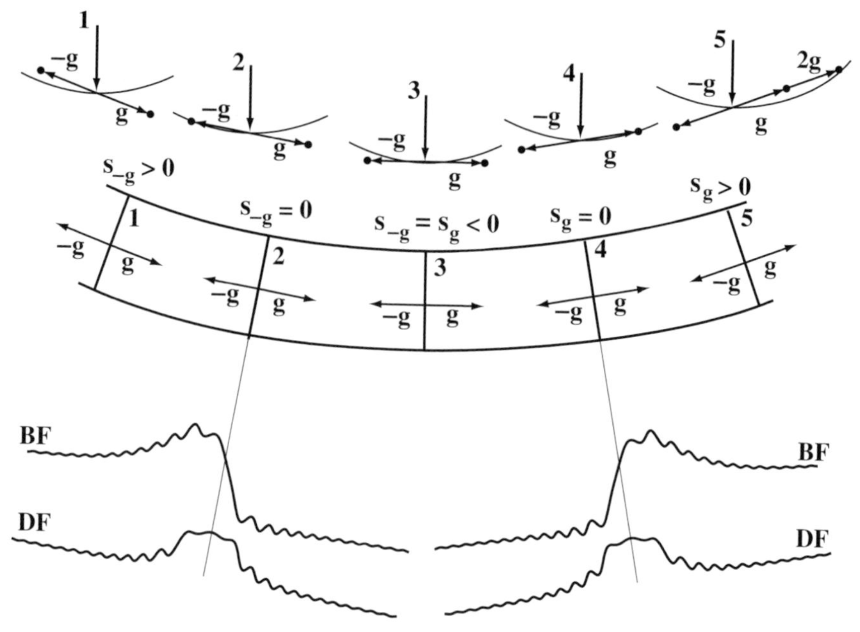

Fig. 7.1. Illustration of the geometry giving rise to a bend contour. The orientation of the lattice planes $-\mathbf{g}$ (1) varies such that the reflection $-\mathbf{g}$ approaches the Ewald sphere (2), falls outside of the Ewald sphere (3), and then the situation repeats for the lattice planes $\mathbf{g}$ (4) and (5).

field contrast around the location $s_{-\mathrm{g}} = 0$ is shown on the lower left-hand corner of the figure, for an arbitrary reduced thickness $z_\xi = 4$. When $s_{-\mathrm{g}} < 0$, anomalous absorption occurs and the bright field image will have a rather low intensity in this region.

The same situation occurs on the other side of the foil: the planes $\mathbf{g}$ are in exact Bragg orientation at location 4. Since $s_{-\mathrm{g}}$ is large and negative, we can again use the two-beam theory to describe the image contrast, as shown on the lower right-hand side. Note that this time the bright field curve is reversed, and the anomalous absorption occurs on the side for which $s_\mathrm{g} < 0$. In the central region *both* $s_{-\mathrm{g}}$ *and* s_g *are negative*, which means that anomalous absorption will occur for scattering from both sides of the planes. The situation depicted at location 3 will thus give rise to a low transmitted intensity and this is what makes a bend contour look dark in bright field images.

At location 3, both $-\mathbf{g}$ and $\mathbf{g}$ have the same excitation error, and the two-beam approximation must break down. A computation of the image contrast at this location will therefore involve at least three beams. We say "at least" because there may be more than three beams involved; consider location 5 in Fig. 7.1. The reciprocal lattice point $\mathbf{g}$ is well inside the Ewald sphere, but the point $2\mathbf{g}$ is close to the Ewald

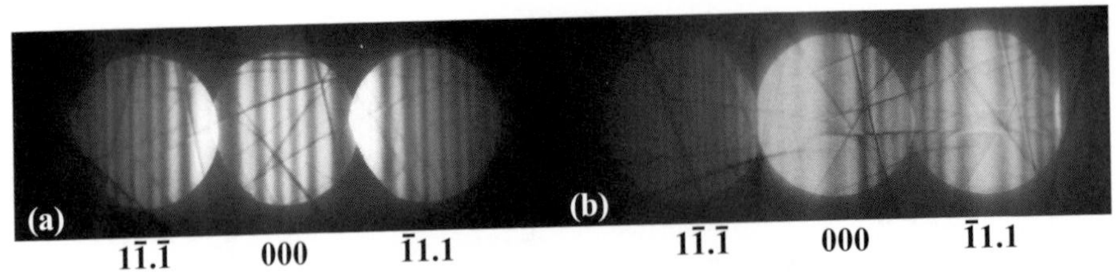

Fig. 7.2. ($\bar{1}$1.1) systematic row CBED pattern for Ti acquired at 200 kV with a beam convergence angle of 5.5 mrad. Pattern (a) is taken in the symmetric orientation and in (b) the ($\bar{1}$1.1) reflection is in the Bragg orientation.

sphere, indicating that there is a substantial probability for electrons in the incident beam to be scattered into the beam 2**g**. How many beams are involved depends on the acceleration voltage used for the observations and on the local orientation of the foil. A higher voltage means a flatter Ewald sphere, which increases the probability that many reciprocal lattice points are simultaneously close to the Ewald sphere. Consequently, even a contrast feature as simple as a bend contour requires a multi-beam computation rather than a two-beam computation.

This is also shown in the systematic row convergent beam pattern of Fig. 7.2. This pattern was obtained using the ($\bar{1}$1.1) systematic row of Ti at 200 kV. The beam convergence angle is 5.5 mrad. It is clear that the symmetric orientation shown in (a) would require at least three beams, whereas the Bragg orientation pattern in (b) could be reasonably well approximated by a two-beam simulation. In Section 7.2.2.7 we will see how systematic row CBED patterns can be simulated.

7.2.2 Theory and simulations for the systematic row orientation

7.2.2.1 The N-beam dynamical equations

When it comes to the computation of diffracted amplitudes, there is no intrinsic difference between the systematic row case and the zone axis case (which we will treat in Section 7.3). The crystal transfer matrix $\mathcal{A}$, introduced in Chapter 5, has one row entry for each contributing reciprocal lattice point **g**, so that we can simply number all the contributing beams, starting from 1 for the transmitted beam.

In Section 5.4, we have seen that the Darwin–Howie–Whelan equations can be written as

$$\frac{dS_n}{dz} = i\pi \sum_{n'} \left[2s_{n'}\delta(n-n') + \frac{1-\delta(n-n')}{q_{n-n'}} \right] S_{n'}, \tag{7.1}$$

where n labels the individual beams. In matrix notation we have

$$\frac{d\mathbf{S}}{dz} = i\mathcal{A}\mathbf{S}. \tag{7.2}$$

A general multi-beam simulation program will then proceed along the following lines.

(i) *Setup*. Read the crystal data file, compute all symmetry operations, compute all atom positions, and determine the electron wavelength, corrected for refraction.
(ii) *Diffraction geometry*. Ask the user for the incident beam direction **k** and the foil normal **F**; for a systematic row, also ask for **g**.
(iii) *Determine contributing beams*. For the systematic row case, these are simply multiples of **g**. For the zone axis orientation, we will need an algorithm based on the symmetry of the zone axis. Either way, we will end up with a list of N contributing beams.
(iv) *Compute the off-diagonal part of* $\mathcal{A}$. We will use the subroutine CalcUcg (introduced in Chapter 2) to compute the real and imaginary parts of the off-diagonal elements of $\mathcal{A}$. The routine CalcDynMat is provided to compute the dynamical matrix for either the Darwin–Howie–Whelan approach or the Bloch wave approach.
(v) *Beam orientations*. Determine the range of incident beam directions, the number of pixels in the final image (this will determine the number of individual multi-beam computations to be performed), and the thickness of the foil (either constant thickness or thickness gradient). For each incident beam direction, we will need to compute the diagonal of $\mathcal{A}$. We will also make use of symmetry relations to reduce the total number of beam directions for which the computation will be carried out.
(vi) *Solve the equations*. Use your favorite solution method to obtain the amplitudes S_n of all N beams, for each incident beam direction and/or foil thickness. Conversion to bright field and dark field images is then straightforward.

In the following subsections, we will provide a more detailed description of the most important steps and algorithms for the systematic row case. The reader may download the program SR.f90 from the website; this program implements various solution methods for the general systematic row case, as described below.

7.2.2.2 The differential equation approach

We will use the numerical package rksuite90, available from `www.netlib.org`, to solve the system of first-order differential equations. We have already used this package when we discussed the lens.f90 program in Chapter 3. The reader may consult the source code of the SRIntegrate routine in the SR.f90 program on the website. Pseudo code for the systematic row case is shown in PC-17. The calling sequence of the rksuite90 routines is as follows:

```
call setup(comm,t_start,YS,t_end,tol,thres,method='M', &
          task='R',message=.FALSE.)
call range_integrate(comm,f,t_want,t_got,y_got,yderiv_got,
                      & flag=flag)
call statistics(comm,num_succ_steps=steps)
call collect_garbage(comm)
```

Pseudo Code **PC-17** Solution of the dynamical scattering equations for the N-beam systematic row case SR.f90.

Input: PC-2
Output: multi-beam wave function
  ask user for beam direction, foil normal, and **g**-vector
  ask for extent of integration (z_0 and $k_{t,max}$)
  precompute off-diagonal part of dynamical matrix {CalcUcg}
  for each incident beam orientation **do**
    complete diagonal of dynamical matrix {Calcsg}
    *integrate the system of equations
    compute transmitted and scattered intensities for required thickness(es)
  end for
  prepare PostScript/TIFF output

The first routine (`setup`) defines the integration parameters: starting thickness (`t_start`), ending thickness (`t_end`), initial values (`YS`), integration tolerance (`tol`), and error threshold (`thres`); the method (`M`) indicates that either fourth- or fifth-order Runge–Kutta integration will be used. The task parameter identifies that the next call will be to the `range_integrate` subroutine, and messages are turned off, to prevent all but the most serious of errors and warnings being shown. The `range_integrate` subroutine does the actual work: the variable `f` is a function that implements the actual differential equation (7.2); `t_want` is the requested thickness, `t_got` the actual ending thickness, and the resulting vector is returned in `y_got`, along with its derivative in `yderiv_got`. The `flag` parameter provides information about the success or failure of the integration. The remaining two procedure calls are optional: `statistics` can be used to retrieve statistical information about the integration, and `collect_garbage` cleans up all dynamically allocated memory. For more details on each of these routines we refer the interested reader to the rksuite90 manual [BG94].

7.2.2.3 The scattering matrix approach

The scattering matrix approach, introduced on page 312, uses a Taylor expansion for the matrix $\mathcal{S}$ of the following form:

$$\mathcal{S}(z_0) = \mathbf{1} + \sum_{n=1}^{\infty} \frac{(\mathrm{i}\mathcal{A})^n}{n!} z_0^n.$$

Table 7.1. *Comparison of the computation times for the* SR.f90 *program for a $(2N+1)$-beam systematic row using the* $\mathbf{g}_{200}$ *reflection of aluminum. The microscope acceleration voltage is 200 kV, the beam orientation varies from* $\mathbf{k}_t = -3\mathbf{g}$ *to* $+3\mathbf{g}$*, and the foil thickness increases from zero to 200 nm. The computational array consists of* 512×512 *pixels. All computations were performed on a 666 MHz Compaq Digital UNIX workstation, computational times are scaled relative to the five-beam* $(N=2)$ *Bloch wave computation.*

Computational method	$N=2$	4	6	8
Runge–Kutta integration	61.1	211.7	483.6	1031.5
Runge–Kutta computation for $\mathcal{S}(\epsilon)$	3.0	13.6	29.9	60.0
Taylor expansion $\mathcal{S}$	1.9	9.9	22.1	40.1
Bloch wave approach	1.0	1.4	2.6	4.1

For a small thickness $z_0 = \epsilon$, the series expansion will converge after only a few terms. The general term in the expansion can be written as

$$\mathcal{A}_n = \mathcal{A}_{n-1} \times \frac{\mathrm{i}\mathcal{A}}{n}\epsilon, \tag{7.3}$$

and this relation can be used in an efficient algorithm (CalcSMslice and SMloop) implemented in the SR.f90 program. The algorithm computes the term on the right-hand side of equation (7.3) and multiplies it with itself until the largest change in any element of $\mathcal{A}_n$ is smaller than a preset threshold. The slice thickness ϵ is typically of the order of the unit cell size, i.e. a few tenths of a nanometer. The larger the slice thickness, the more terms are needed in the Taylor expansion. The wave function $\mathbf{S}$ at any thickness $n\epsilon$ can be computed by multiplication of $\mathcal{S}(\epsilon)^n$ with the incident wave $\mathbf{S}(0)$.

A combination of the differential equation and scattering matrix methods is also implemented in the SR.f90 program; the scattering matrix for the first slice of thickness ϵ is computed using the rksuite90 routines instead of the Taylor expansion above, and then the resulting scattering matrix is multiplied with itself as before. Table 7.1 compares the computation times of the pure differential equation approach, the scattering matrix approach, and the mixed approach, along with the Bloch wave approach to be discussed next. It is clear that the differential equation approach is significantly slower than any of the other methods.

7.2.2.4 The Bloch wave approach

The Bloch wave approach, using the BWsolve routine introduced in the previous chapter, is implemented in the program SR.f90. In addition to the bright field and dark field images, which obviously are identical to those computed with the other three methods, the Bloch wave approach provides much more detailed information about the scattering process. In Section 7.2.2.6 we will discuss the representation of Bloch wave eigenvalues, excitation amplitudes, and Bloch wave intensities with respect to the atom positions in the unit cell. First, we will analyze the symmetry of the systematic row case.

7.2.2.5 Symmetry of the systematic row case

In Section 5.8.3, where we introduced the reciprocity theorem, we have seen that a thin foil can have two different symmetries, type I or type II symmetry, corresponding to a mirror plane through the center of the foil, parallel to the foil surfaces, and an inversion center located in the middle of the foil, respectively. For the systematic row case the entire crystal is considered to be one-dimensional (only one row of reciprocal lattice vectors is important), so that all crystals must have at least type I symmetry in the systematic row orientation. This means that the diffracted amplitude $\psi_{\mathbf{g}}$ satisfies the following relation:

$$\psi_{\mathbf{g}}\left(\otimes, -\frac{\mathbf{g}}{2}+\mathbf{r}, C\right) = \psi_{\mathbf{g}}\left(\otimes, -\frac{\mathbf{g}}{2}-\mathbf{r}, C\right),$$

using the notation of Chapter 5. This in turn means that the diffracted beam $\mathbf{g}$ belongs to the diffraction group $\mathbf{1}_R$. The dark field bend contour image must therefore have this internal symmetry. If the crystal is non-centrosymmetric, then the $-\mathbf{g}$ dark field bend contour need not have the same intensity distribution as the $+\mathbf{g}$ contour, provided both do not belong to the same family.†

For a centrosymmetric space group the systematic row has, in addition to the type I symmetry, also type II symmetry, which means that

$$\psi_{\mathbf{g}}\left(\otimes, -\frac{\mathbf{g}}{2}+\mathbf{r}, C\right) = \psi_{-\mathbf{g}}\left(\otimes, \frac{\mathbf{g}}{2}+\mathbf{r}, C\right).$$

This, in turn, means that the diffraction group is $\mathbf{21}_R$, so that the $\mathbf{g}$ and $-\mathbf{g}$ dark field bend contour images are identical.

A careful analysis of bend contour bright field and dark field images may in principle be used to determine whether a crystal structure has inversion symmetry or not. However, the reader should bear in mind that the corresponding asymmetries in the dark field intensities may be very small, so that they are not detected easily

† In a non-centrosymmetric space group, the reciprocal lattice vectors $\mathbf{g}$ and $-\mathbf{g}$, in general, do not belong to the same family. It is possible, however, that certain families of planes contain both $\mathbf{g}$ and $-\mathbf{g}$. As an example, consider the non-centrosymmetric space group Pm (# 6): all reciprocal lattice vectors normal to the mirror plane belong to families of the type $\{\mathbf{g}, -\mathbf{g}\}$, whereas all other vectors belong to families with only one single member $\{\mathbf{g}\}$.

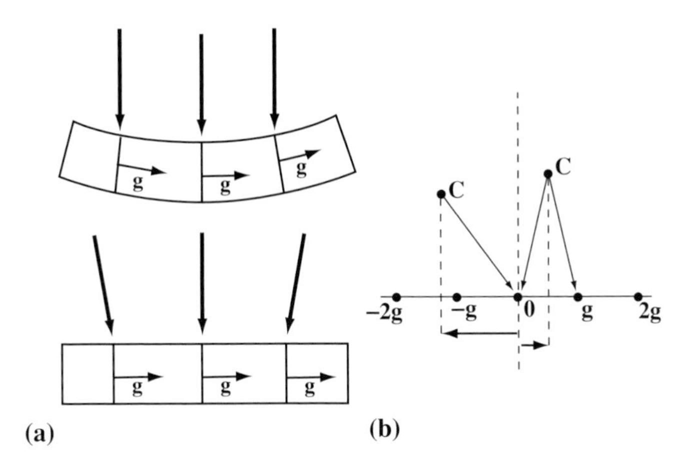

Fig. 7.3. (a) Within the column approximation a bent foil with parallel incident illumination is equivalent to a planar foil with conical incident illumination; (b) schematic illustrating the use of the Laue center to determine the orientation of the incident beam.

in experimental images. Furthermore, experimental bend contours are rarely as straight and perfect as simulated ones, so that it may become even more difficult to analyze small intensity differences between **g** and −**g** dark field images.

7.2.2.6 Simulation examples

It is common practice to compute the excitation errors of the reflections in the systematic row by tilting the incident wave vector, rather than changing the foil orientation. Within the column approximation, these two descriptions are equivalent, as shown in Fig. 7.3(a). It is much easier computationally to change the orientation of the incident wave vector; this corresponds to specifying the location of the *Laue center*, the projection of the center of the Ewald sphere onto the systematic row. The Laue center position is measured in units of $|\mathbf{g}|$, as shown in Fig. 7.3(b). If we denote the component of $\mathbf{k}_0$ along the systematic row by k_t, then the reflection $N\mathbf{g}$ is in exact Bragg orientation when $\mathbf{k}_t = -\frac{1}{2}N\mathbf{g}$. The Laue center positions for the two incident beam directions in Fig. 7.3(b) would then be −1.2 for the beam on the left and 0.5 for the beam on the right. All the Fortran routines in this chapter employ this method to determine the relative orientation of crystal foil and incident beam.

Figure 7.4 shows the bright field and five dark field images for the (00.2) systematic row in titanium, for a microscope acceleration voltage of 200 kV, a thickness varying from 0 to 200 nm, an incident beam tilt between $\mathbf{k}_t = -3\mathbf{g}$ and $+3\mathbf{g}$ and a total of 11 beams ($-5\mathbf{g} \cdots +5\mathbf{g}$). The images were computed on a grid of 512 × 512 pixels using the transfer matrix series expansion (SR.f90 program). Since the Ti structure is centrosymmetric, the $\pm\mathbf{g}$ dark field images are identical. The images are

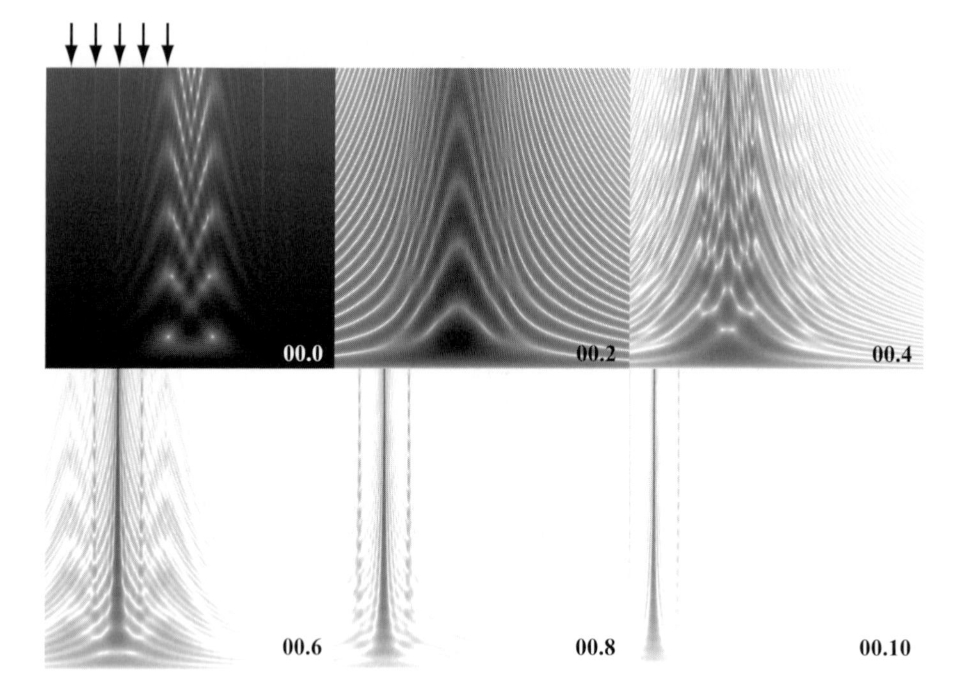

Fig. 7.4. Titanium (00.2) systematic row for the parameters stated in the text. The arrows at the top indicate the locations for which one of the reflections $N\mathbf{g}_{00.2}$ is in Bragg orientation. The images are shown as optical densities, i.e. as $-\log(I)$, with $I = 0.0001$ corresponding to white and $I = 1.0$ to black.

displayed as optical densities to bring out small intensity variations; this display mode also makes it easier to compare them directly to electron negatives. It is obvious that with increasing order of reflection, the dark field bend contour contrast is confined to a narrower region. The straight lines in the bright field image (arrowed) correspond to the locations where the $n\mathbf{g}_{00.2}$ reflections are in exact Bragg orientation. Those lines are known as *Bragg lines* [Ead92].

Figure 7.5 shows a comparison between the calculated bright field and (00.2) dark field images, displayed as optical densities, and experimental images obtained on a JEOL 2000EX microscope and displayed with inverted contrast (i.e. as negatives). The simulated images used a perfect linear wedge-shaped foil, whereas the foil shape for the experimental images is not a perfectly linear wedge. Instead, the thickness increases faster than the theoretical wedge close to the edge of the foil, resulting in more closely spaced fringes. There is good qualitative agreement between the two sets of images; in the dark field image, there are several locations at which two fringes merge, and these merge points can also be observed in the experimental dark field image (black arrows). The vertical white lines in the simulated bright field image, labeled 1 and 2, can also be found in the experimental image.

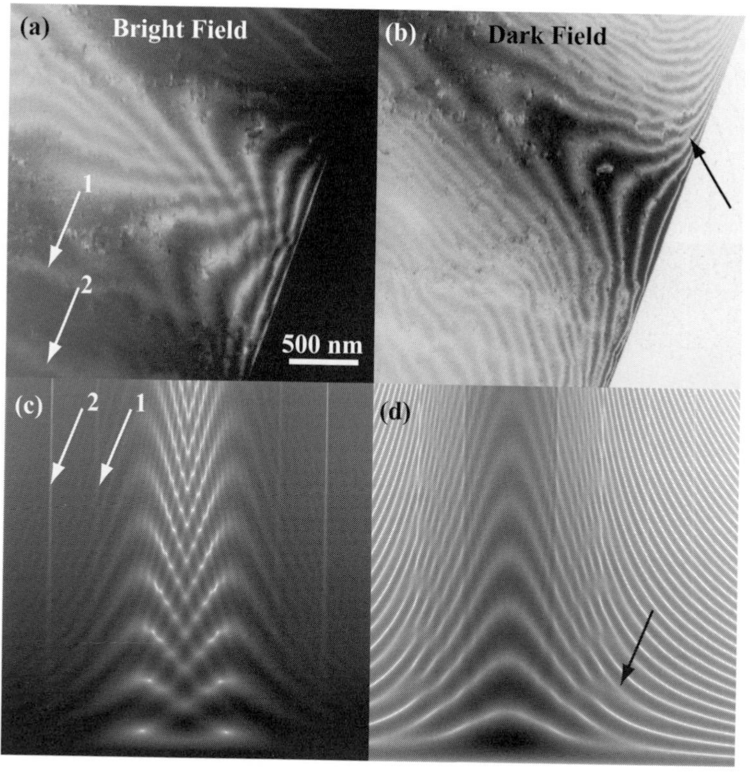

Fig. 7.5. Comparison of experimental and computed bright field and (00.2) dark field images for a Ti foil (inverted contrast, same parameters as in Fig. 7.4, except for the maximum thickness, which was taken to be 300 nm).

Because of the local shape of the foil, the lines in the experimental image are not perfectly straight.

It is rather straightforward to compute systematic row bright and dark field images for a foil with an irregular shape. The only input parameters that are needed are the local thickness and the local value of k_t/g for each column (i.e. pixel) of the image. Figure 7.6 shows an example of the results of such a simulation: the local value of k_t/g is shown as a contour plot in Fig. 7.6(a), along with a contour plot of the local thickness in (b). The irregular shape on the lower right is a hole in the foil. The other simulation parameters are: Ti foil, 200 kV, nine beams, $\mathbf{g}_{00.2}$ systematic row, Bloch wave formalism. The resulting bright and dark field images are shown in (c) and (e)–(l), respectively. The corresponding diffraction pattern is shown in Fig. 7.6(d). The simulation was carried out for a grid of 512×512 columns, and took about 2 min on a 666 MHz workstation. We leave it to the interested reader to adapt the SR.f90 program for the simulation of systematic row images similar to those shown in Fig. 7.6.

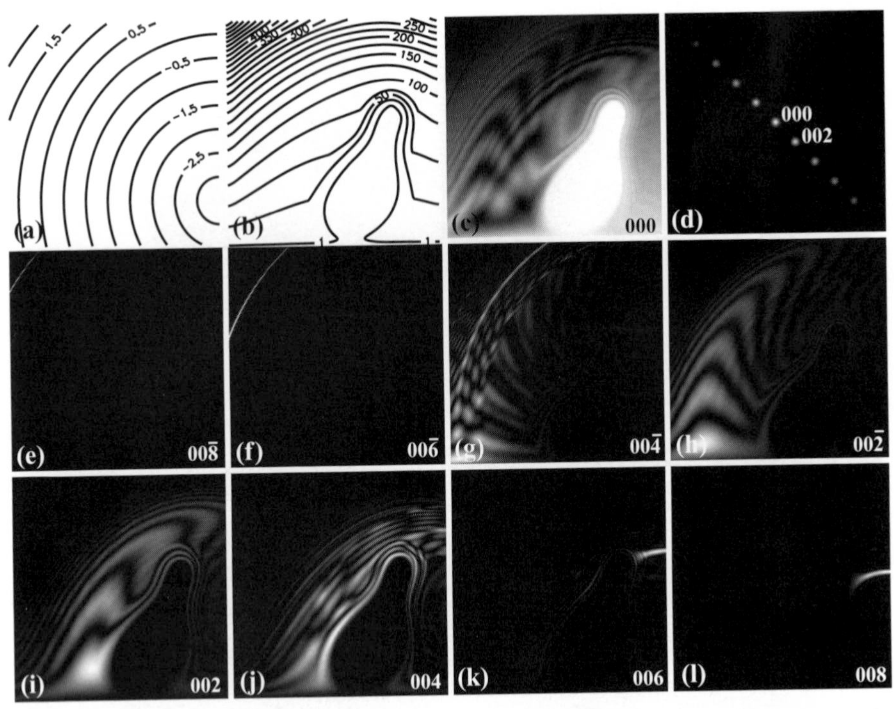

Fig. 7.6. Example systematic row simulation for an irregularly shaped foil with a k_t/g contour map (a) and thickness contour map (b). The simulation used a Ti foil, 200 kV, nine-beam $\mathbf{g}_{00.2}$ systematic row Bloch wave calculation. The bright field image is shown in (c), along with the diffraction pattern (d) and all eight dark field images.

The Bloch wave approach has a significant advantage over the other simulation methods in that more detailed information about the scattering process is available. Bloch wave eigenvalues $\Gamma^{(j)}$ and excitation amplitudes $\alpha^{(j)}$ can be displayed as a function of incident beam orientation, and the position of the individual Bloch waves with respect to the crystal reference frame can also be determined. Let us consider two examples: the $\mathbf{g}_{202}$ systematic row of copper and the $\mathbf{g}_{101}$ systematic row of $BaTiO_3$. In both cases we will use 200 kV electrons, nine beams, and the incident beam orientation varies in the range $-3 < k_t/g < 3$.

Figures 7.7(a) and 7.8(a) show the eigenvalues $\gamma^{(j)}$ of the dynamical matrix for both Cu and $BaTiO_3$ systematic rows. An alternative representation of the eigenvalues in terms of $k_z^{(j)} - k_0$ is shown in Figs 7.7(b) and 7.8(b). The eigenvalues inside the first Brillouin zone, which runs from $k_t/g = -0.5$ to $+0.5$, are repeated periodically in the so-called *extended zone representation*. For the largest beam tilts the periodicity is no longer obvious from the figure, and this indicates that the computation did not take into account a sufficient number of beams; the beam tilt is so large that higher-order reflections are close enough to the Ewald sphere to give

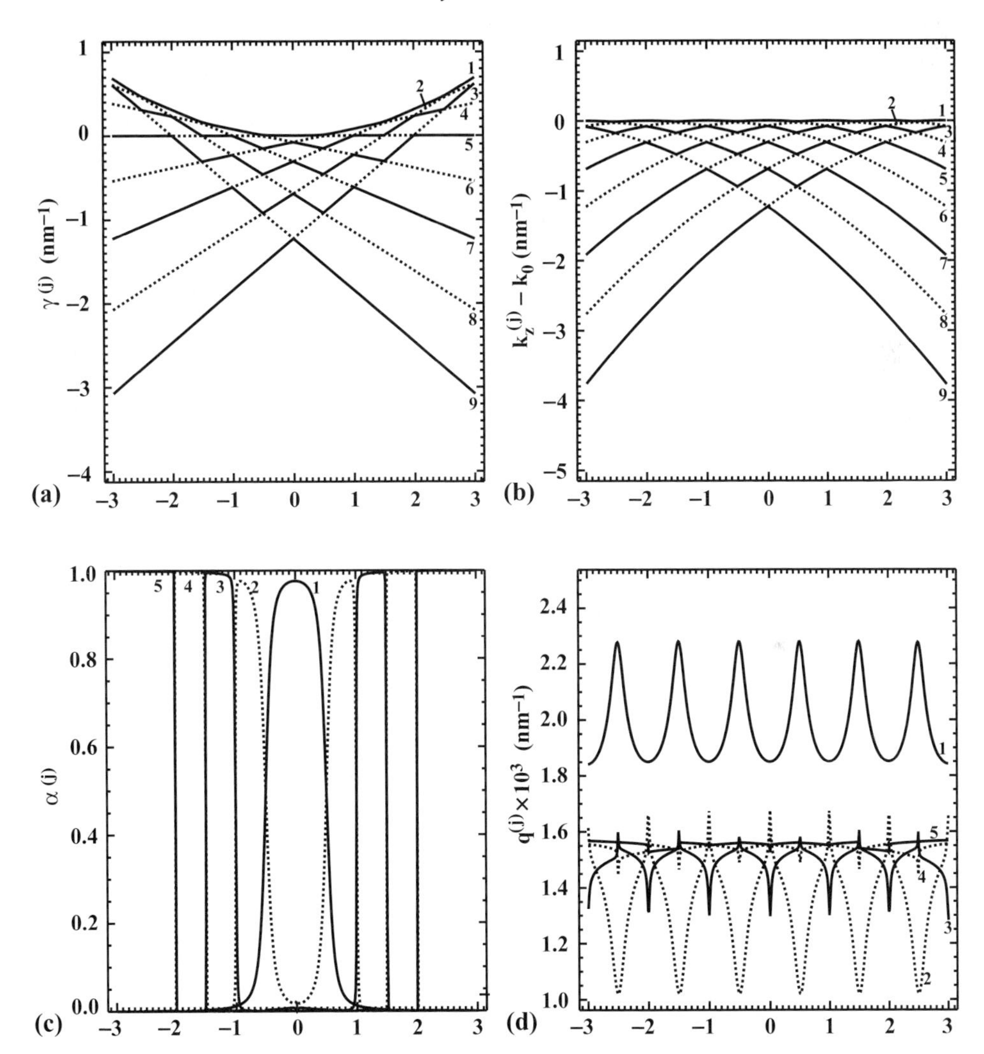

Fig. 7.7. (a) and (b) eigenvalues $\gamma^{(j)}$ of the dynamical matrix for pure Cu, 200 kV, nine-beam $\mathbf{g}_{202}$ systematic row simulation. (c) Bloch wave excitation coefficients and (d) absorption coefficients $((j) = 1\text{–}5)$.

rise to significant scattering. When we repeat the computation with a larger number of beams the periodicity is fully restored.

The modulus of the Bloch wave excitation coefficients $\alpha^{(j)}$ is shown in Figs 7.7(c) and 7.8(c). The excitation coefficients are not periodic along the systematic row, since they are determined by the boundary conditions at the entrance plane of the crystal, i.e. by the incident beam direction. These curves are quite similar to the ones for the two-beam case in Fig. 6.7(a): at the Bragg orientation of a particular reflection the amplitude of two of the excitation coefficients changes rapidly. In

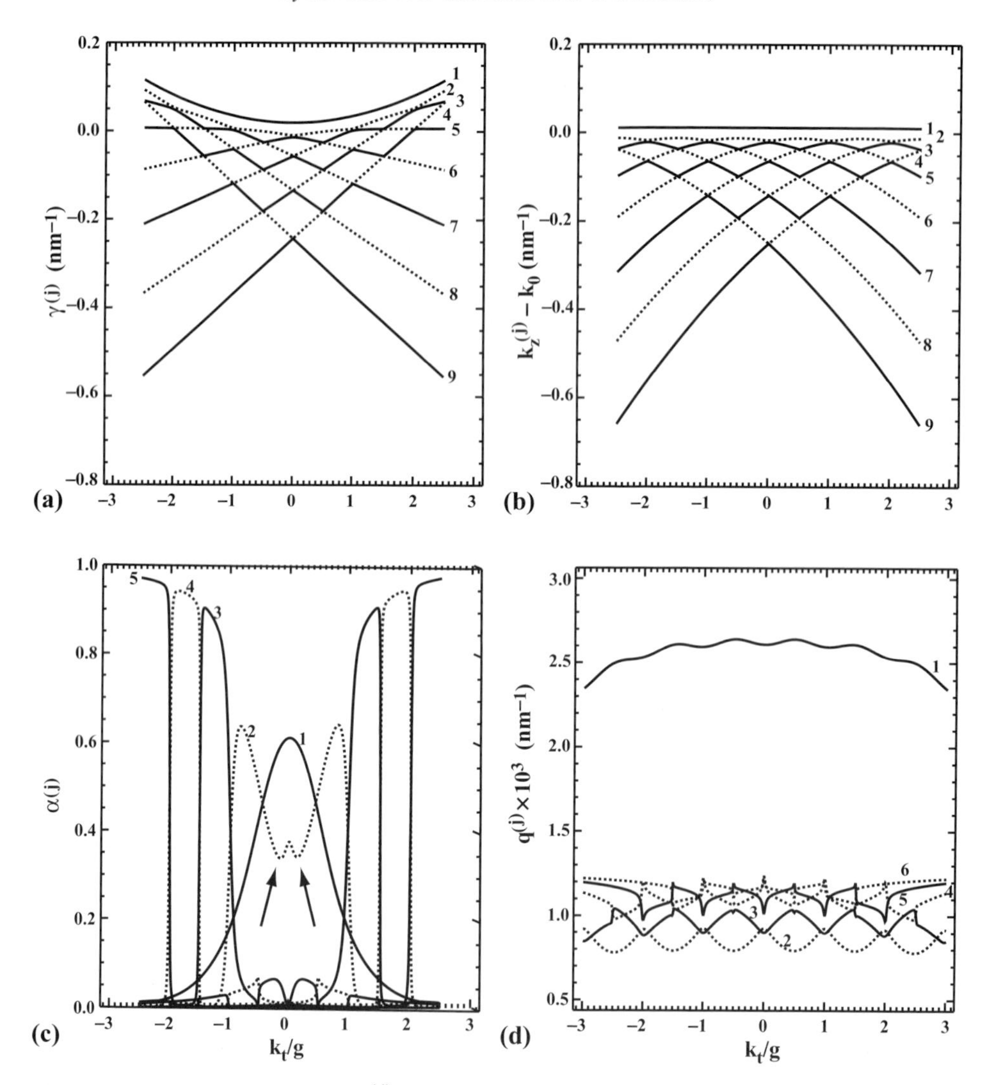

Fig. 7.8. (a) and (b) eigenvalues $\gamma^{(j)}$ of the dynamical matrix for $BaTiO_3$, 200 kV, nine-beam $\mathbf{g}_{101}$ systematic row simulation. (c) Bloch wave excitation coefficients and (d) absorption coefficients ($(j) = 1$–6).

$BaTiO_3$, the second Bloch wave excitation amplitude, labeled 2 in Fig. 7.8(c), shows a small asymmetry near the center (arrows); this asymmetry is due to the non-centrosymmetric nature of the structure. If we move the Ti atom to the center of the cell, then the asymmetry disappears. Conversely, if we move the Ti atom further away from the center, the asymmetry increases. The excitation amplitudes for the Cu systematic row do not show any asymmetry since the structure is centrosymmetric.

The absorption components $q^{(j)}$ are shown in Figs 7.7(d) and 7.8(d); they are again periodic along the systematic row since they are the imaginary parts of the

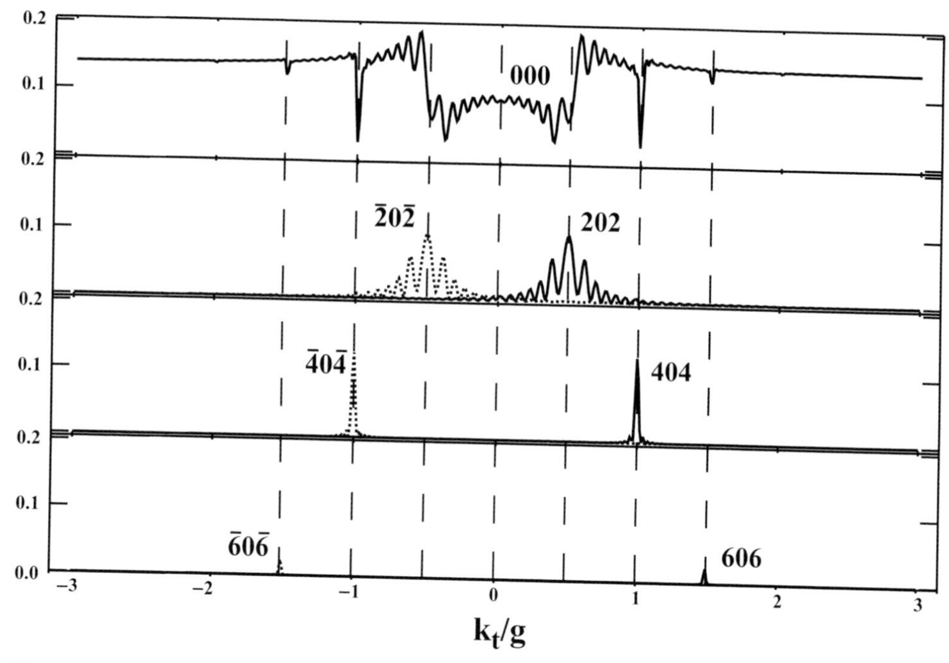

Fig. 7.9. Systematic row bright field and dark field intensities for a 100 nm thick Cu foil, oriented along the $\mathbf{g}_{202}$ systematic row (see Fig. 7.7).

complex eigenvalues $\Gamma^{(j)}$. The lowest index absorption coefficients vary strongly with beam orientation while the higher index coefficients are nearly constant. The absorption coefficients contain both normal absorption and anomalous absorption contributions.

Figures 7.9 and 7.10 show the bright field and dark field intensities for both systematic rows for a 100 nm thick foil. The bright field curve shows a lower intensity between $k_t/g = -\frac{1}{2}$ and $+\frac{1}{2}$, and this would correspond to a dark bend contour in a bright field image. The $+\mathbf{g}$ and $-\mathbf{g}$ reflections give rise to symmetric dark field contours, and the width of the contour decreases rapidly with increasing Miller indices. This can also be seen in the simulated bend contours of Fig. 7.4. For $BaTiO_3$, the 101 and $\bar{1}0\bar{1}$ dark field curves are asymmetric (arrows), again a signature of the non-centrosymmetric nature of the crystal structure. A careful comparison of experimental 101 and $\bar{1}0\bar{1}$ bend contours may reveal this asymmetry.

Using equation (5.43) on page 321, we can determine what each of the individual Bloch waves looks like, similarly to the analysis that led to the type I and type II Bloch waves for the two-beam case in Chapter 6. Figure 7.11(a) shows the modulus squared of Bloch waves $C^{(1)}$ to $C^{(8)}$ for the $\mathbf{g}_{202}$ systematic row in Cu, for identical simulation parameters as in Figs 7.7 and 7.9. Two sets of curves are shown in relation

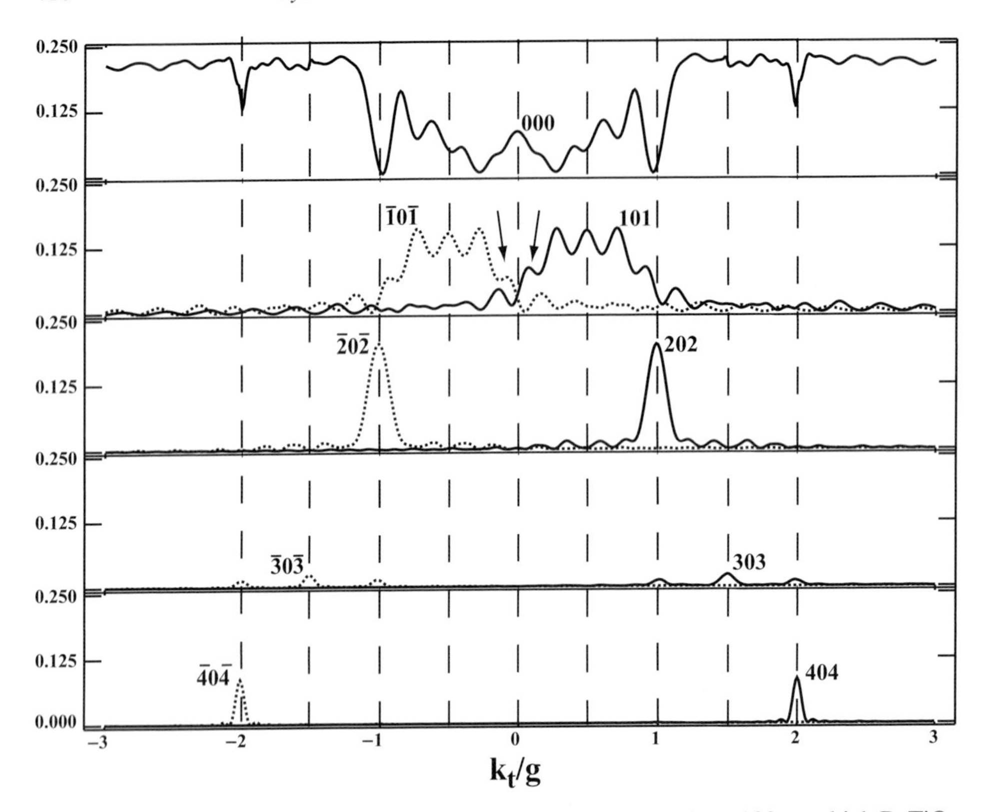

Fig. 7.10. Systematic row bright field and dark field intensities for a 100 nm thick $BaTiO_3$ foil, oriented along the $\mathbf{g}_{101}$ systematic row (see Fig. 7.8).

to the lattice planes in the lower half of the figure:[†] the solid lines correspond to the Bloch wave intensities for incident beam directions of the type $k_t/g = n$, with $n = 0, \pm 1, \pm 2, \ldots$, i.e. for the Bragg orientation of the even multiples of $\mathbf{g}_{202}$. The dotted curves show the same curves for the orientation $k_t/g = n/2$, i.e. for the Bragg orientation of the odd multiples of $\mathbf{g}_{202}$. These curves repeat themselves because of the periodicity of the Bloch wave coefficients $C_{\mathbf{g}}^{(j)}$ and the eigenvalues $\Gamma^{(j)}$.

The Bloch waves for $BaTiO_3$ are shown in Fig. 7.11(b). The asymmetry caused by the displacement of the Ti atom from the center of the unit cell is clearly visible for the Bloch waves (2), (3), and (4). It is no longer straightforward to define a type I and type II wave, but it is clear from the simulations in Fig. 7.11 that several waves have their maxima on the projected atom positions, whereas others have maxima in between the atoms.

[†] Since the systematic row is a 1D section of the reciprocal lattice, the structures in Figs 7.11(a) and (b) must be projected onto the [101] direction.

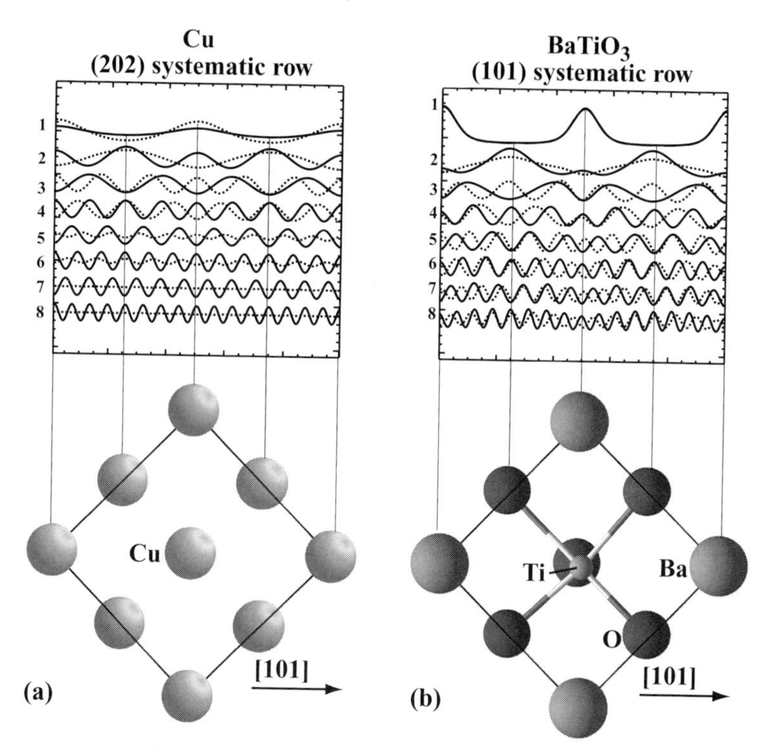

Fig. 7.11. Bloch wave intensities of the first eight Bloch waves for (a) the $\mathbf{g}_{202}$ systematic row in Cu and (b) the $\mathbf{g}_{101}$ systematic row in $BaTiO_3$, both at 200 kV, nine-beam simulation. The waves are drawn in the correct position relative to the structure drawings in the lower half.

7.2.2.7 Systematic row convergent beam electron diffraction

Figure 7.2 shows a typical systematic row convergent beam electron diffraction (SRCBED) pattern. Several diffraction disks have significant intensity in them, mostly in the form of parallel fringes, similar to those discussed for the two-beam case in Section 6.5.5. The simulation of a systematic row CBED pattern proceeds along similar lines as for the two-beam CBED pattern: for a given incident beam direction (represented by k_t), beam divergence angle θ_c, and **g**-vector, the N-beam rocking curve is computed and displayed across diffraction disks located at the proper relative positions. Pseudo code for the algorithm is shown in PC-17.

The algorithm is implemented in the SRCBED.f90 program; example output is shown in Fig. 7.12 for the (200) systematic row in copper, 200 kV, nine beams. The figure shows the intensity distributions for the reflections ±(400), ±(200) and (000), for five different incident beam directions. The beam direction changes from $k_t/g_{200} = 1.0$, for which $\bar{4}00$ is in Bragg orientation, to $k_t/g_{200} = -1.0$, corresponding to $s_{400} = 0$. The SRCBED pattern contains the same information as

Pseudo Code PC-17 Systematic row CBED simulation.

```
Input: PC-2
Output: systematic row CBED pattern
  ask user for number of beams
  repeat
    ask for k_t, α, g and θ_c
    compute geometrical parameters
    set up crystal structure matrix A
    use Taylor expansion to compute S
    compute scattered amplitudes
    create individual diffraction disks
    check for disk overlap
    if coherent illumination and disk overlap then
      add amplitudes to row image
    else
      add intensities to row image
    end if
    dump image to PostScript file
  until no further patterns requested
```

Fig. 7.9, but represented in a different form. Note that each diffraction disk acts as a "window" through which a portion of the corresponding bright field or dark field rocking curve can be observed. Increasing the diameter of the condensor aperture† results in a larger window; moving the aperture reveals different portions of the rocking curves. Moving the aperture is equivalent to tilting the conical incident beam, so we can also use the beam tilt to translate the disks in the plane of the pattern.

Any of the systematic row algorithms discussed in the previous sections may be used for this computation; the SRCBED.f90 implementation available from the website uses the scattering matrix (Taylor expansion) approach. It is left as an exercise for the reader to implement the SRCBED algorithm using the Bloch wave approach.

7.2.3 *Thickness integrated intensities*

To conclude this section on systematic row multi-beam dynamical scattering we will apply the Bloch wave formalism to the computation of the orientation

† Some microscopes, like the JEOL 3000 series, have an optional α-selector, which permits a finer control of the beam convergence angle, so that the discrete condensor aperture diameters no longer determine the only disk diameters that can be obtained.

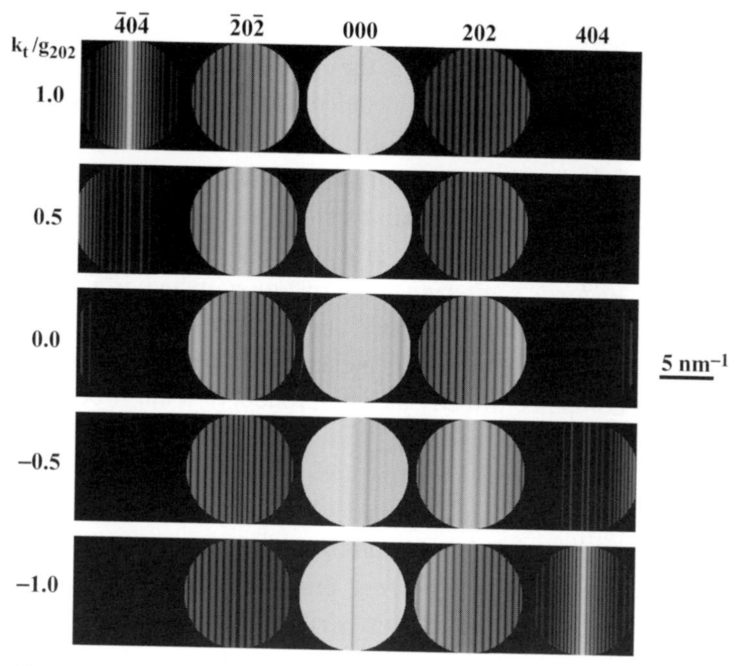

Fig. 7.12. Simulated systematic row CBED patterns for the (202) row in copper; 200 kV, foil thickness 100 nm, nine-beam simulation (these conditions are identical to those of Fig. 7.9). The radius of the diffraction disks corresponds to a convergence angle of 9.0 mrad.

dependence of electron-induced x-ray emission *from selected sites in the unit cell.* Although this book does not deal with the multitude of analytical methods based on inelastic scattering processes, it is important for the microscope user who wishes to employ these analytical techniques to understand how dynamical elastic scattering processes can affect the results of analytical observations.

7.2.3.1 Theoretical derivation

Consider a cubic crystal structure with two atom types, say A and B. For simplicity we take the structure to be the CsCl structure, i.e. atom A is located in the origin and atom B in the cell center. We know from structure factor considerations that for the disordered structure all reflections for which $h + k + l = 2n + 1$ are forbidden. For the ordered structure, these reflections have $|F_{hkl}| = |f_A - f_B|$. We can now ask the question: *what is the probability that the beam electrons will induce x-ray emission by A (or B) atoms, and how does this probability depend on the sample orientation?*

The answer to this question is not straightforward. We need to compute the probability that an incident electron will transfer energy to an atom at a certain site in the unit cell, for a given orientation of the sample with respect to the incident beam. We do not concern ourselves with the computation of the inelastic scattering probability itself; that would require detailed considerations of various inelastic scattering mechanisms and is beyond the scope of this book. We are, however, in a position to compute a second factor of importance to the inelastic scattering probability: the probability that a beam electron will actually travel near a given atom type. The total inelastic scattering probability from an atom A is then given by the product of the *inelastic scattering cross-section* σ_i, which is a quantity derived from time-dependent quantum mechanics, and the probability that a beam electron is actually in a position to undergo this inelastic scattering event. It is this second factor that can be computed directly as a by-product of a multi-beam Bloch wave simulation.

The probability that a beam electron will be found at a certain site $\mathbf{r}_i$ in the unit cell is given by the modulus squared $|\Psi(\mathbf{r}_i)|^2$ of the electron wave function $\Psi(\mathbf{r})$ evaluated at that site. Since this probability depends upon the thickness of the sample, we will compute the *average* probability over the sample thickness. Mathematically, this is done as follows [Kri88, Wan95]: if $\mathcal{S}$ denotes the set of equivalent points in the unit cell for which we wish to evaluate the probability of inelastic scattering, then the number of electrons per unit thickness which may undergo inelastic scattering events by atoms in the set $\mathcal{S}$ is given by

$$\frac{N}{z_0} = \sum_{i\in\mathcal{S}} \left[\frac{1}{z_0} \int_0^{z_0} \Psi^*(\mathbf{r}_i)\Psi(\mathbf{r}_i)\,\mathrm{d}z \right], \tag{7.4}$$

where z_0 is the foil thickness. Using the Bloch wave formalism, the electron wave function can be expressed as

$$\begin{aligned}\Psi(\mathbf{r}) &= \sum_j \alpha^{(j)} \sum_{\mathbf{g}} C_{\mathbf{g}}^{(j)} \mathrm{e}^{2\pi\mathrm{i}[\mathbf{k}^{(j)}+\mathbf{g}]\cdot\mathbf{r}};\\ &= \sum_j \alpha^{(j)} \sum_{\mathbf{g}} C_{\mathbf{g}}^{(j)} \mathrm{e}^{2\pi\mathrm{i}[\mathbf{k}_0+\gamma^{(j)}\mathbf{n}+\mathrm{i}q^{(j)}\mathbf{n}+\mathbf{g}]\cdot\mathbf{r}}.\end{aligned}$$

If we only consider normal incidence, i.e. $\mathbf{n}\cdot\mathbf{r} = z$, then the wave function can be rewritten as

$$\Psi(\mathbf{r}) = \sum_j \alpha^{(j)} \mathrm{e}^{-2\pi q^{(j)} z} \sum_{\mathbf{g}} C_{\mathbf{g}}^{(j)} \mathrm{e}^{2\pi\mathrm{i}\gamma^{(j)} z} \mathrm{e}^{2\pi\mathrm{i}(\mathbf{k}_0+\mathbf{g})\cdot\mathbf{r}}.$$

Substitution in equation (7.4) (exercise) results in

$$\frac{N}{z_0} = \sum_{\mathbf{g}}\sum_{\mathbf{h}} \left[\sum_{i\in\mathcal{S}} \mathrm{e}^{2\pi\mathrm{i}(\mathbf{h}-\mathbf{g})\cdot\mathbf{r}_i} \right] \sum_j\sum_k \alpha^{(i)*}\alpha^{(k)} C_{\mathbf{g}}^{(j)*} C_{\mathbf{h}}^{(k)} \mathcal{I}_{jk},$$

where

$$\mathcal{I}_{jk} = \frac{1}{z_0}\int_0^{z_0} e^{-2\pi(\alpha_{jk}+i\beta_{jk})z}\, dz,$$

and

$$\alpha_{jk} = q^{(j)} + q^{(k)};$$
$$\beta_{jk} = \gamma^{(j)} - \gamma^{(k)}.$$

The integration is elementary and results in

$$\mathcal{I}_{jj} = \frac{1 - e^{-4\pi q^{(j)} z_0}}{4\pi q^{(j)} z_0}; \tag{7.5}$$

$$\mathcal{I}_{jk} = \frac{1 - e^{-2\pi(\alpha_{jk}+i\beta_{jk})z_0}}{2\pi(\alpha_{jk}+i\beta_{jk})z_0} \qquad (j \neq k). \tag{7.6}$$

The thickness-averaged intensity can then be expressed as the product of two factors, one determined by the atom positions and the active reflections, and one by the crystal orientation through the Bloch wave coefficients:

$$\left(\frac{N}{z_0}\right)_{\mathcal{S}} = \sum_{\mathbf{g}}\sum_{\mathbf{h}} S_{\mathbf{gh}} L_{\mathbf{gh}}, \tag{7.7}$$

with

$$S_{\mathbf{gh}} \equiv \sum_{i\in\mathcal{S}} e^{2\pi i(\mathbf{h}-\mathbf{g})\cdot\mathbf{r}_i};$$

$$L_{\mathbf{gh}} \equiv \sum_j\sum_k C_{\mathbf{g}}^{(j)*}\alpha^{(j)*}\mathcal{I}_{jk}\alpha^{(k)}C_{\mathbf{h}}^{(k)}.$$

The matrix elements $S_{\mathbf{gh}}$ can be precomputed once for a given subset $\mathcal{S}$ of atom positions in the unit cell, whereas the matrices $L_{\mathbf{gh}}$ must be calculated for each crystal orientation, or, equivalently, for each incident beam direction $\mathbf{k}_0$.

It is interesting to note that the off-diagonal terms of the matrix $\mathcal{I}$ (equation 7.6) have an oscillatory character; with increasing crystal thickness, the off-diagonal elements will become less important, and only the diagonal elements $\mathcal{I}_{jj}$ will remain. It is then easy to show that

$$L_{\mathbf{gh}} \approx \sum_j |\alpha^{(j)}|^2 C_{\mathbf{g}}^{(j)*} C_{\mathbf{h}}^{(j)} \frac{1 - e^{-4\pi q^{(j)} z_0}}{4\pi q^{(j)} z_0} \qquad \text{(for large } z_0\text{)}. \tag{7.8}$$

We can apply the above equation to the two-beam situation for our model A–B crystal structure. Using the two-beam Bloch wave results from the previous

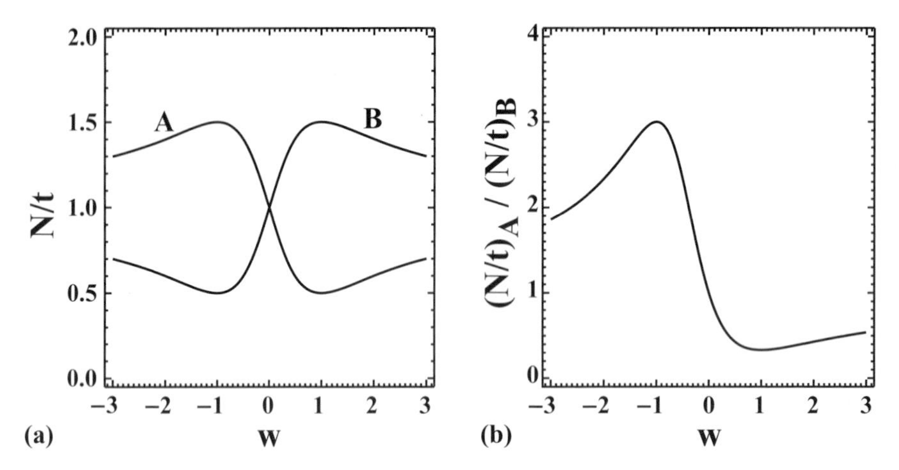

Fig. 7.13. (a) Thickness-integrated intensity (two-beam case) for the A and B sites of the CsCl structure; (b) ratio of the two curves in (a).

chapter we find for the thickness averaged expression above (in the thick crystal approximation without absorption):

$$\left(\frac{N}{z_0}\right)_A = 1 - \frac{1}{2}\sin(2\beta) = 1 - \frac{w}{1+w^2}; \tag{7.9}$$

$$\left(\frac{N}{z_0}\right)_B = 1 - \frac{1}{2}\sin(2\beta)\cos\pi(h+k+l) = 1 - \frac{w\cos\pi(h+k+l)}{1+w^2}. \tag{7.10}$$

The resulting curves are shown in Fig. 7.13; we find that, for negative s_{100}, the induced x-ray photons are more likely to come from the A-sites than from the B-sites. The opposite is true for a positive excitation error. Figure 7.13 also shows the ratio of the A-curve to the B-curve. Note that this ratio is only dependent on the excitation error for the superlattice reflections, for which $h+k+l = 2n+1$; for fundamental reflections the thickness integrated scattering probability for A and B sites is identical for all excitation errors (since $\cos(2\pi n) = +1$), and the ratio is always equal to 1. At $w = -1$, the probability of electron-induced x-ray emission from the A-sites is three times as large as for the B-sites, and a simple change of the orientation to $w = +1$ changes the ratio to $\frac{1}{3}$.

The importance of this orientation dependence of the electron-induced x-ray emission probability can not be overstated. The geometrical component of the inelastic scattering probability for AB compounds with the CsCl structure depends in a sensitive way on the foil orientation when superlattice reflections are close to (two-beam) Bragg orientation. Nearly all analytical observation methods (energy dispersive x-ray spectroscopy (EDAX or EDS), electron energy loss spectroscopy (EELS), etc.) performed on crystalline samples aim to determine the local chemistry and electronic structure. It is crucial that the effect of sample orientation on analytical TEM observations be taken into account properly, otherwise the analytical results may be incorrectly interpreted. In fact, one can make good use of the channeling phenomena that give rise to this orientation dependence; in 1982, Taftö and Liliental [TL82] used the orientation dependence of the electron-induced x-ray emission to determine the cation distribution in $ZnCr_xFe_{2-x}O_4$ spinels. This technique was then nicknamed *ALCHEMI* (Atom Location by CHanneling Enhanced MIcro analysis) by Spence and Taftö [ST83]. The ALCHEMI method is now used by many researchers to determine site occupations and substitutional atom types in a large variety of materials. The basic computation underlying the ALCHEMI technique uses equation (7.7).

For more details on the ALCHEMI method and related theory we refer the interested reader to the following references: [MR84, RM84, Kri88, Pen88, RTW88, AR93, JHJF99]. An early discussion of the channeling effect can be found in [KLF74]. Since the two-beam approach of Fig. 7.13 is only a crude approximation to the complete multi-beam scattering process, we will next perform more realistic multi-beam computations for the systematic row orientation.

7.2.3.2 Thickness integrated intensities: systematic row case

As an example of the thickness integrated intensity for the systematic row case, let us consider three **g** vectors belonging to the $[1\bar{1}0]$ zone of Si and GaAs. Figure 7.14(a) shows the Si structure projected along $[1\bar{1}0]$. The (111), (004), and (440) planes are indicated by solid lines. Using the SRpot.f90 program, we can compute the potential of the systematic row for each of these planes; the resulting potential curves are shown in Fig. 7.14(a). The curves were computed for a 29-beam case, and are displayed for one repeat period (i.e. the distance x is drawn in units of the interplanar spacing d_{hkl}). The potentials were computed using the Fourier coefficients $V_{n\mathbf{g}}$ of the systematic row. Figure 7.14(b) shows the same curves for the $[1\bar{1}0]$ projection of GaAs. In this case, the structure is non-centrosymmetric, and the (220) and (002) reflections are allowed. The potential curves clearly show the difference between the Ga and As positions.

Next, we apply equations (7.5)–(7.7) to all six systematic rows of Fig. 7.14. The equations are implemented in the SRTII.f90 program, along with the systematic row Bloch wave code of the SR.f90 program. The resulting thickness integrated intensities for a 100 nm thick foil, 200 kV acceleration voltage, and $-6 < k_t/g < 6$ are shown in Fig. 7.15. The first column shows the results for Si, the center column for GaAs, and the last column shows the ratio of the

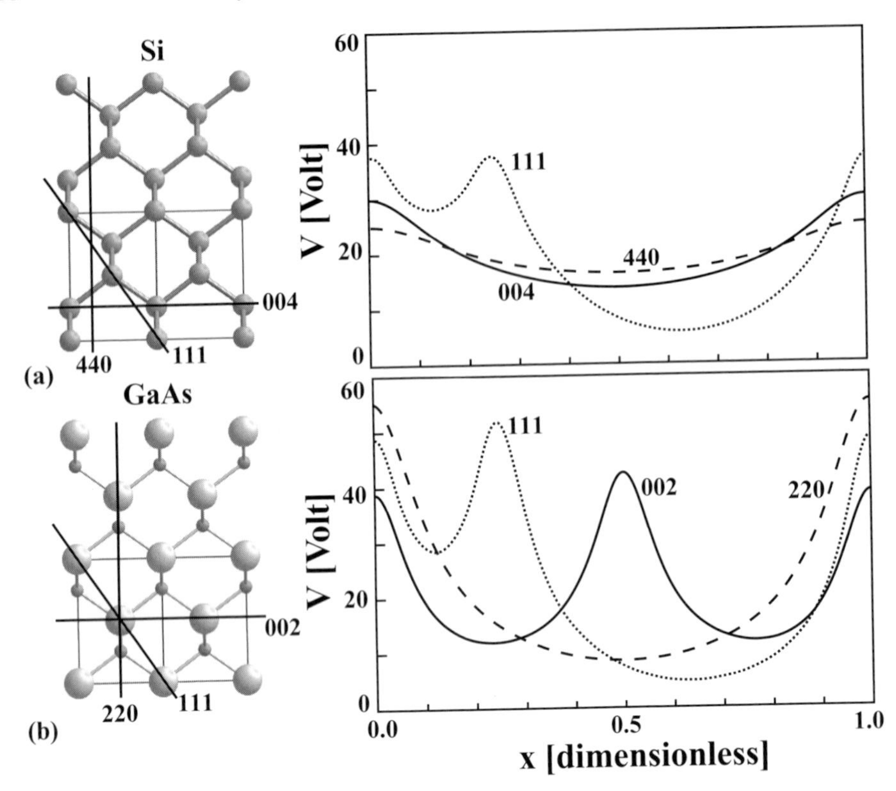

Fig. 7.14. $[1\bar{1}0]$ projection of the structures of Si (a) and GaAs (b), along with the systematic row potentials for three different rows.

intensities for As and Ga, i.e. $(N/z_0)_{As}/(N/z_0)_{Ga}$. The following observations can be made:

- for large beam tilts k_t/g, the integrated intensity approaches a constant level for all systematic row orientations;
- since the (220) GaAs potential curve in Fig. 7.14(b) is symmetric, there is no difference between the integrated intensities for the Ga and As sites; conversely, both (002) and (111) GaAs potential curves show a difference between the Ga and As sites, and therefore the integrated intensities for both sites are also different;
- for the (111) GaAs systematic row, there are virtually no beam tilts for which the ratio equals unity; if the thickness integrated intensity varies strongly with incident beam orientation, then the corresponding systematic row orientation can be used for ALCHEMI-like observations.

We conclude that electron channeling near systematic row orientations may significantly alter the probability of electron-induced x-ray emission from particular

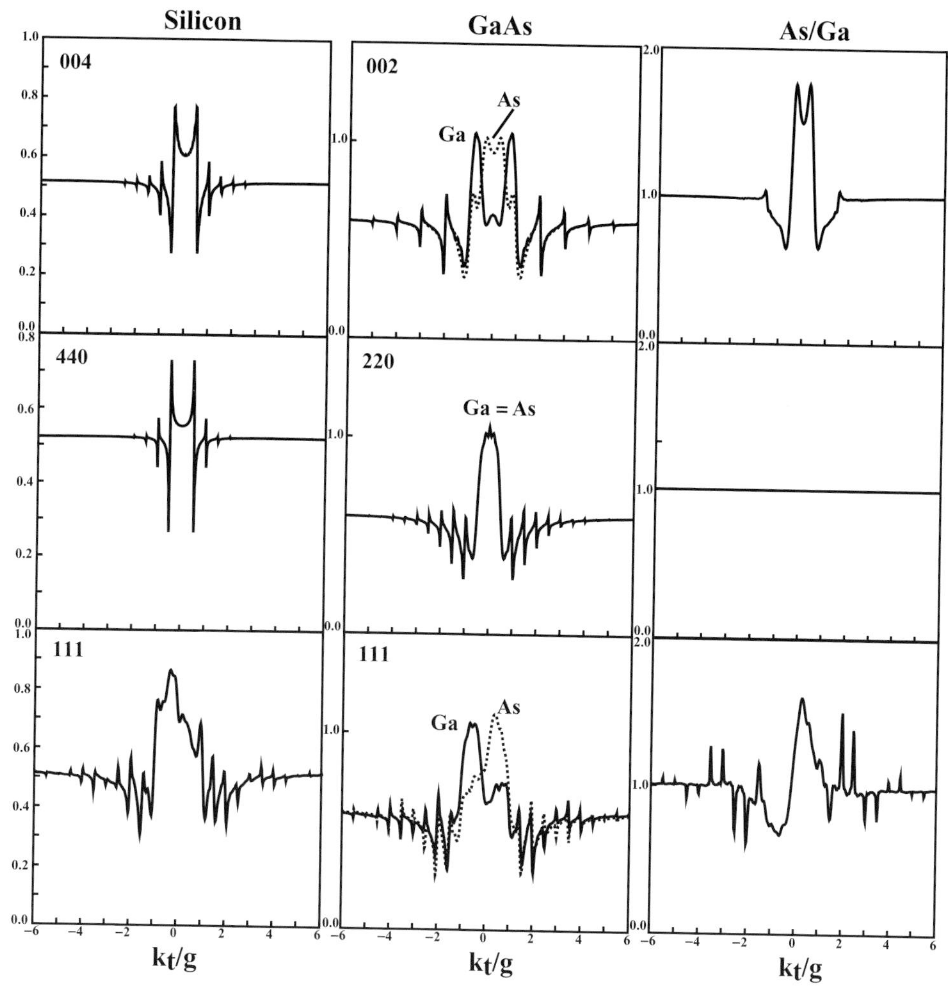

Fig. 7.15. Thickness integrated intensities for the systematic row orientations of Fig. 7.14, for a 100 nm thick foil, 200 kV, and $-6 < k_t/g < 6$ (29 beams). The last column shows the ratio of the As curve to the Ga curve.

sites within the unit cell. For quantitative analytical observations, this orientation dependence must be taken into account.

7.3 The zone axis case

In this section, we will review how the multi-beam dynamical diffraction problem can be solved for the zone axis orientation. First, we will describe the geometry of the problem, and to what extent it differs from the systematic row case. Then we will introduce several simulation methods, equivalent to the SR.f90 program described for the systematic row case. We will describe how the lattice symmetry can be used to reduce the number of multi-beam simulations needed for bend-center and convergent beam computations. We will describe briefly how channeling effects determine the thickness-averaged electron-induced x-ray emission probabilities for the zone axis case.

7.3.1 The geometry of the zone axis orientation

In Chapter 4, we have seen several examples of images or diffraction patterns obtained in the zone axis orientation. Figure 4.18 shows a bend center in Ti; multiple bend contours intersect at the location of the exact zone axis orientation. The resulting diffraction pattern is symmetric with respect to the central beam. Figure 4.25 shows a convergent beam electron diffraction pattern for GaAs, again for the exact zone axis orientation.

Near a zone axis orientation, many reciprocal lattice points lie close to the Ewald sphere. We can always select two short reciprocal lattice vectors $\mathbf{g}_1$ and $\mathbf{g}_2$, and express every other reciprocal lattice point in the same zone as a linear combination of these two vectors:

$$\mathbf{g} = m\mathbf{g}_1 + n\mathbf{g}_2,$$

where m and n are integers.† For computational purposes, we will use a single index to label all the diffracted beams. We will either number the reciprocal lattice points one row at a time, or number them by family $\{hkl\}$, with increasing numbers indicating increasing $|\mathbf{g}_{hkl}|$.

The basic multi-beam simulation proceeds as described on page 399, with essentially no differences between the systematic row case and the zone axis case, except for:

(i) the determination of the contributing beams (explained in the next paragraph), and
(ii) the graphical representation of the resulting intensities or Bloch wave coefficients and eigenvalues, which are now 2D functions of the incident beam orientation.

The reader will find a zone axis multi-beam program MB.f90 on the website; this program implements the zone axis case using the scattering matrix approach, the differential equation approach, and the Bloch wave formalism.

For a given incident beam direction $\mathbf{t}_{uvw}$, it is rather straightforward to determine two independent reciprocal lattice vectors which belong to the zone $[uvw]$. We can

† The selection of the two shortest vectors must be made carefully. As we will see in Section 9.2.3 it is possible for diffracted beams corresponding to "forbidden reflections" to have a non-zero intensity. This means that those reciprocal lattice points must be taken into account, even when they correspond to a vanishing structure factor.

simply select the following two vectors:

$$\mathbf{g}_a = v\mathbf{a}_1^* - u\mathbf{a}_2^*;$$
$$\mathbf{g}_b = w\mathbf{a}_1^* - u\mathbf{a}_3^*.$$

It is easy to verify that both vectors satisfy the zone equation. If one or more indices of the direction $[uvw]$ are zero, then the corresponding reciprocal basis vector can be selected as one of the two vectors. As an example, consider the direction [201]: the second index is zero and therefore we select $\mathbf{g}_a = \mathbf{a}_2^*$. The second vector can then be obtained from the second relation above as $\mathbf{g}_b = \mathbf{a}_1^* - 2\mathbf{a}_3^*$. The algorithm is straightforward to implement and the subroutine ShortestG in the library file symmetry.f90 implements the general and all special cases.

The vectors obtained from this procedure are not necessarily the shortest possible reciprocal lattice vectors. If we denote the shortest vectors by $\mathbf{g}_1$ and $\mathbf{g}_2$, then we have

$$\mathbf{g}_a = n_a\mathbf{g}_1 + m_a\mathbf{g}_2;$$
$$\mathbf{g}_b = n_b\mathbf{g}_1 + m_b\mathbf{g}_2,$$

where n_a, n_b, m_a, and m_b are integers. It is then straightforward to determine these integers such that the lengths of $\mathbf{g}_1$ and $\mathbf{g}_2$ are minimized. This is also implemented in the subroutine ShortestG. The vector $\mathbf{g}_1$ is selected to fall on a symmetry element, such as a mirror plane, if one is present. This is accomplished by computing the families $\{\mathbf{g}_1\}$ and $\{\mathbf{g}_2\}$ and selecting as the final $\mathbf{g}_1$ the vector with the *smallest* multiplicity.

If we do not use symmetry arguments to reduce the number of incident beam directions for which the dynamical equations are to be solved, then we can simply define a square grid of points, each point corresponding to an image pixel and also to a particular value of $\mathbf{k}_t$:

$$\mathbf{k}_t(i, j) = i\Delta\frac{\mathbf{g}_1}{|\mathbf{g}_1|} + j\Delta\frac{\mathbf{g}_\perp}{|\mathbf{g}_\perp|}, \tag{7.11}$$

with Δ the grid spacing (in nm^{-1}). The vector $\mathbf{g}_\perp$ is normal to both $\mathbf{g}_1$ and the incident beam direction $\mathbf{t}$, i.e.

$$\mathbf{g}_\perp = \mathbf{g}_1 \times \mathbf{t}. \tag{7.12}$$

Since directions are expressed with respect to the *direct* basis vectors, we must make sure that the components of $\mathbf{t}$ are converted to reciprocal space before computation of the cross product.

Once the tangential wave vector component is known, it is straightforward to determine the complete wave vector, since we know that its length should be equal to λ^{-1}. For a convergent beam simulation, the range of the indices i and j in equation (7.11) is determined by the illumination cone angle θ_c and the grid spacing Δ:

$$i^2 + j^2 \leq \left(\frac{\theta_c}{\lambda\Delta}\right)^2. \tag{7.13}$$

For a bend center computation, we can define the range of i and j by means of the maximum length of $\mathbf{k}_t(i, j)$ along the vector $\mathbf{g}_1$. If we represent the maximum projection of $\mathbf{k}_t$ onto $\mathbf{g}_1$ by k_m, and the number of grid points along $\mathbf{g}_1$ by N_a, then the grid spacing (in nm^{-1}) is given by

$$\Delta = \frac{2k_m|\mathbf{g}_1|}{N_a}.$$

The total number of beam orientations along the positive and negative $\mathbf{g}_1$ vector is then equal to $N_1 = 2N_a + 1$. The number of grid points N_2 along $\mathbf{g}_\perp$ is taken to be equal to N_1, so as to obtain a square grid.

The routine Calckvectors in the MB.f90 program implements the computation of the incident wave vectors. The algorithm employs dynamical memory allocations using *pointers* to store all incident beam directions in a so-called *linked list*. This list is then passed on to the routine which computes the actual multi-beam solution for each member of the list. While such a memory allocation scheme is not really needed for this type of computation, it will become useful when we incorporate symmetry arguments in the algorithm; the number of independent incident beam directions is then not known a priori, and dynamical memory allocations are then the only efficient means to implement the multi-beam computation. Once the incident beam directions are known, it is also a trivial matter to compute the excitation error of each scattered beam using the Calcsg routine. The computation of all factors $s_\mathbf{g}$ completes the diagonal of the crystal structure matrix $\mathcal{A}$. The off-diagonal elements can be precomputed before the iteration over all beam directions is started.

7.3.2 Example simulations for the zone axis case

In this section, we will discuss several representative multi-beam computations for the zone axis case. Once the exit plane wave function is known for all incident beam directions, we can use several representation methods for the images: conventional bright and dark field images, convergent beam electron diffraction patterns, and Eades patterns. Examples of each will be provided below. All multi-beam computations are carried out with the MB.f90 program, and image output is generated with MBshow.f90.

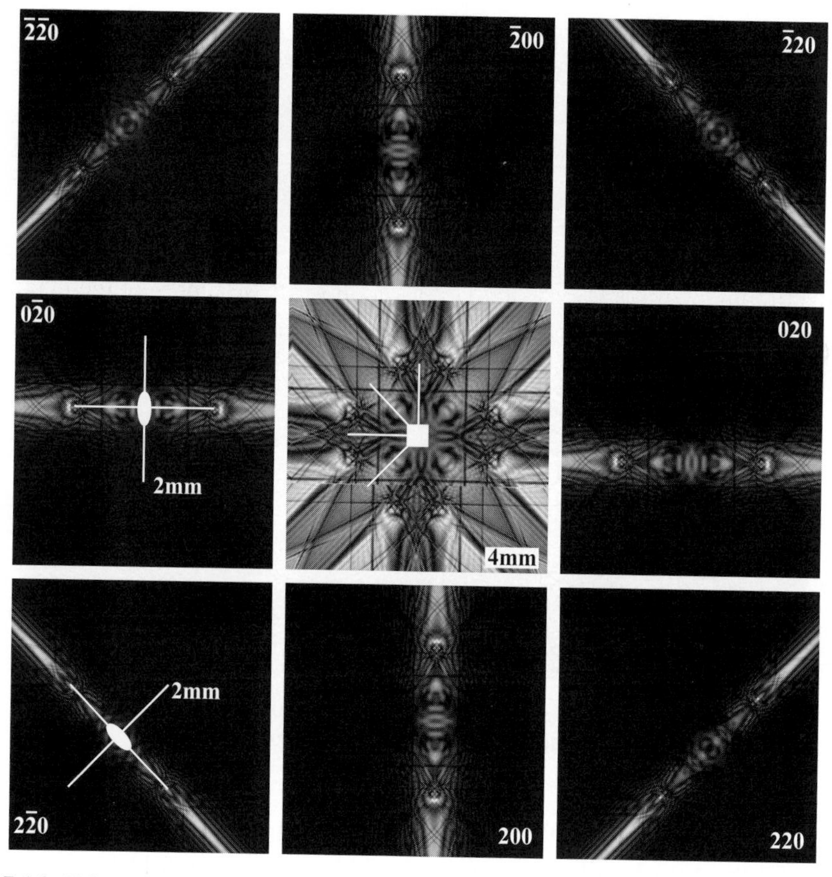

Fig. 7.16. [001] zone axis bright field and dark field images for a spherically bent copper foil at 200 kV; the foil is 100 nm thick, and the Bloch wave computation was performed for 49 beams.

7.3.2.1 Bend center simulations

Figure 7.16 shows a bright field (center) and eight dark field images for a spherically bent copper foil oriented along the [001] zone axis. The computation was performed using the MB.f90 program for 49 beams, 512 × 512 pixels, a crystal thickness of 100 nm, and a microscope acceleration voltage of 200 kV. The Laue center position ranges from −3 to +3 along both $\mathbf{g}_{200}$ and $\mathbf{g}_{020}$ systematic rows. The bright field image shows a rich variety of fringes and contrast variations. Note that the dark field images for opposite reflections, e.g. 200 and $\bar{2}00$, are displaced with respect to the center of the corresponding bend contour. This corresponds to the scenario described in Fig. 7.1, where the Bragg orientation is satisfied for each individual reflection at locations with opposite excitation errors. The symmetry of the patterns is also indicated in Fig. 7.16: the bright field image has an obvious **4mm** symmetry,

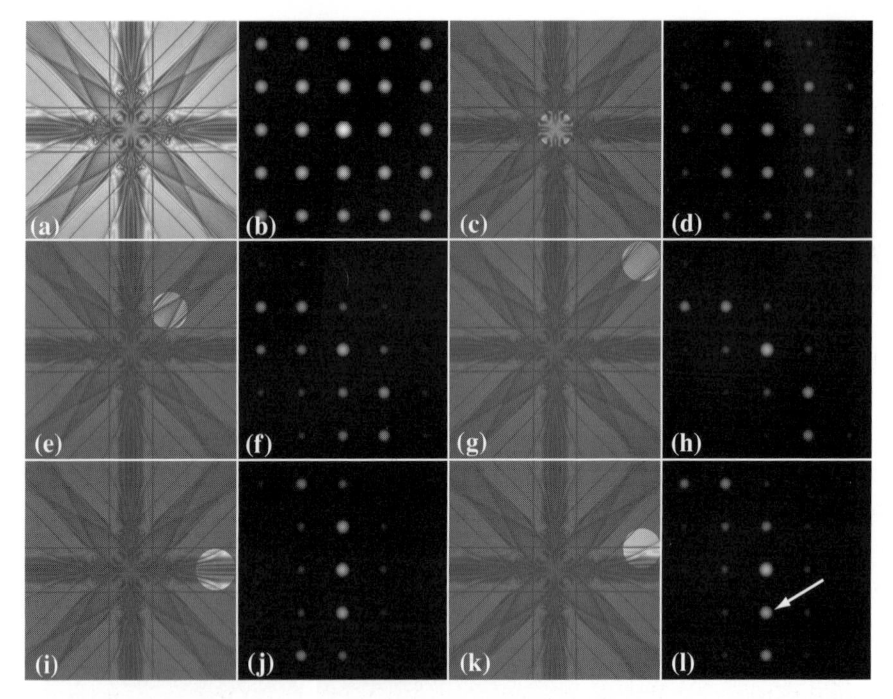

Fig. 7.17. Bright field zone axis patterns corresponding to Fig.7.16, for various locations of the selected area aperture, and the resulting diffraction patterns.

whereas both {200} and {220} patterns have **2mm** symmetry, in agreement with the entries for the $\mathbf{4mm1}_R$ diffraction group in Table 5.1 and Fig. 5.11.†

One might wonder about the usefulness of this type of image simulation. After all, bend contours are not usually considered to be of great interest in conventional microscopy observations. There is, however, a distinct advantage in understanding the nature and origin of bend contours, and how they relate to local diffraction conditions. Suppose that you have a foil of copper which is bent and oriented in such a way that you observe an intersection of bend contours, similar to the one shown in Fig. 7.17(a). If you were to insert a selected area aperture which would enclose the entire region, the corresponding diffraction pattern would be that shown in Fig. 7.17(b), where the intensity in each individual beam is an average over the entire area included in the SA aperture. There is not much difference between the intensities of different beams, other than perhaps a general decrease for higher-order reflections. For a smaller aperture centered on the zone axis, (c), the diffraction pattern, (d), shows an average over a smaller area, and intensity differences between

† Cu has point group $\mathbf{m\bar{3}m}$, which corresponds to the diffraction group $\mathbf{4mm1}_R$ according to Fig. 5.11. For this diffraction group the bright field image has symmetry **4mm**, whereas the special **G** reflections have **2mm** symmetry. Both {200} and {220} are special reflections because they lie on mirror planes.

diffracted beams are more pronounced. If we move the aperture along one of the individual bend contours, towards the upper right-hand corner, (e) and (g), then the corresponding diffraction patterns, (f) and (h), show an off-axis pattern, since the Laue center of the selected area no longer coincides with the optical axis. Note that when the aperture is centered on an individual bend contour, the diffraction pattern will reveal symmetric intensities for **g** and −**g** reflections, as is apparent from the systematic row patterns in (f), (h), and (j). If the aperture is moved away from the center of the bend contour, as in (k), then one reciprocal lattice point will move closer to the Ewald sphere, and the diffraction pattern approaches that of a true two-beam situation; the arrowed reflection in (l) is in nearly exact Bragg orientation.

The bend center image depends strongly on the foil thickness, as shown in Fig. 7.18 for a 49-beam Bloch wave computation for pure copper (200 kV, [001]). The thickness increases in steps of 25 nm. For small foil thicknesses, the contrast features are a bit diffuse, and intensity oscillations are broad. For larger thicknesses, the number of intensity oscillations increases, similar to the increase of the number of thickness fringes in the two-beam case (see Fig. 6.4). Experimental observations of bend centers are often not as detailed as the simulated images. There are several possible reasons for this difference: a high defect density may blur the smallest features of the image, and the foil is not necessarily symmetrically bent, giving rise to various distortions of the bend center image. In addition, inelastic scattering often causes a significant amount of blurring of the finest details. Zero-loss energy filtering may remove most of the inelastic scattering, and it may be necessary to

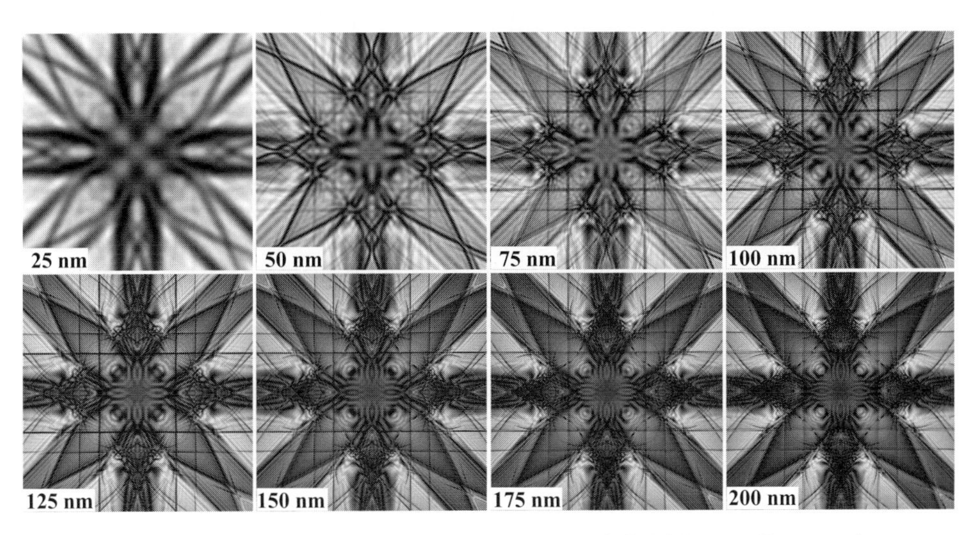

Fig. 7.18. Bright field zone axis patterns for increasing foil thickness. Computations were carried out for the same conditions as Fig. 7.16.

cool the sample down to a low temperature to enable observation of the smallest details in Fig. 7.18.

It is relatively easy to incorporate more realistic foil shapes into the computations. The figure on the cover of this book was computed for the same foil shape as Fig. 7.16, and in addition the foil thickness increases linearly from 0 to 300 nm (bottom to top). The computation used a 77-beam Bloch wave simulation ($|\mathbf{g}| \leq 28.5\ \mathrm{nm}^{-1}$), 200 kV, [001] zone axis orientation, 2100×3000 image pixels. A second example of a bend center simulation is shown in Fig. 7.19: an experimental

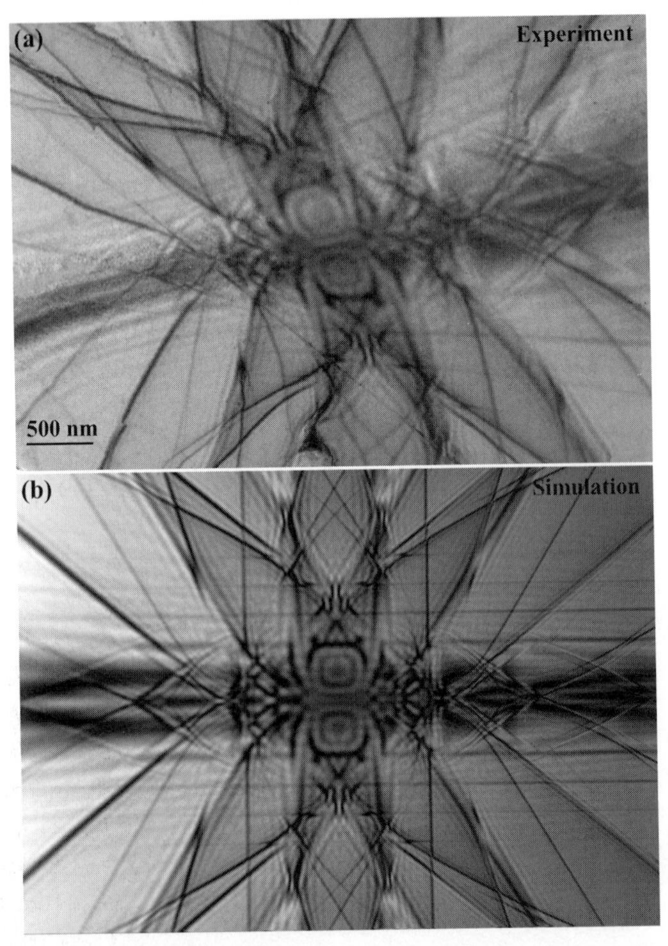

Fig. 7.19. Experimental (a) and simulated (b) bend center for [211] copper at 200 kV. The foil thickness increases from 40 (left) to 110 nm (right). The incident beam tangential component varies from $-7 \leq k_t/g_{0\bar{1}1} \leq +7$ horizontally and vertically. A 47-beam Bloch wave simulation was carried out for 750×525 image pixels. The normal absorption parameter ξ_0' was reduced by a factor of 10 to enhance the contrast in the thicker region of the foil and a γ brightness correction of 1.5 was applied. (Experimental image courtesy of S. Tandon.)

[211] bend center in a thin copper foil (a) can be compared with a simulated image (b). The simulation uses 47 beams ($|\mathbf{g}| \leq 25.0\ \text{nm}^{-1}$), an acceleration voltage of 200 kV, and the thickness increases from 40 nm on the left to 110 nm on the right. The tangential component of the wave vector ranges from $-7|\mathbf{g}_{0\bar{1}1}|$ to $+7|\mathbf{g}_{0\bar{1}1}|$ along both vertical and horizontal directions. The horizontal bend contour is the $(1\bar{1}\bar{1})$ contour, the vertical contour corresponds to $(0\bar{2}2)$. The agreement with the experimental image is remarkably good, down to the smallest details, even allowing for the distortions due to the local foil shape. Along the horizontal bend contour, the experimental image is somewhat blurred due to the presence of dislocations which destroy the local lattice symmetry.

The reader should be warned that this type of image simulation can be very time and memory consuming; the cover page illustration took about 76 hours of CPU time on a 660 MHz Compaq TRU64 UNIX workstation, using the cover.f90 program, available from the website. This type of computation lends itself to implementation on parallel computers, or on distributed workstation clusters, since each image pixel is independent of all the others. In addition, the use of symmetry arguments can significantly speed up the computation. In the case of the cover figure, we used the vertical mirror plane to reduce the number of computations by a factor of 2. Furthermore, points at equal distance from the horizontal center line have Bloch wave solutions related to each other by symmetry; only the foil thickness is different for these pairs of points. This provided a further reduction by a factor of 2.

7.3.2.2 Bloch wave simulations

The results of a Bloch wave simulation can also be displayed as 2D images, 3D surfaces or contour plots. Figure 7.20 shows a 3D rendering of the dispersion surface for the Cu simulation of Fig. 7.16. The tangential component of the wave vector varies from $0 \leq k_t/g_{200} \leq \frac{3}{2}$ along both 200 and 020 systematic rows. The vertical bar on the left indicates the position of the bend center (i.e. the incident beam is parallel to the zone axis). In the absence of the electrostatic lattice potential, the *empty crystal approximation*, the dispersion surface would consist of a set of intersecting spheres, centered on each of the contributing reciprocal lattice points. In the presence of a lattice potential the various branches of the dispersion surface are asymptotic to these spheres. While the spheres for the systematic row are all aligned along the row and give rise to a relatively simple dispersion surface drawing (e.g. Fig. 7.7b), for the zone axis case the geometry of the dispersion surface is significantly more complicated. The dispersion surface construction of Fig. 6.9 on page 369 remains valid for the zone axis case.

The Bloch wave excitation amplitudes $\alpha^{(j)}$ can also be represented as intensity distributions. The first eight coefficients are shown in Fig. 7.21(a) for the wave

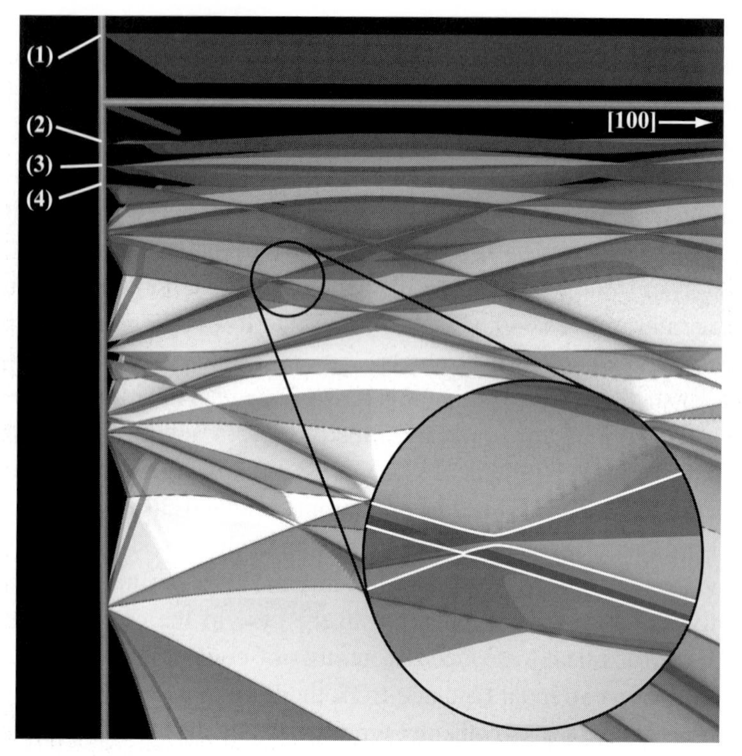

Fig. 7.20. Three-dimensional representation of the 25 dispersion surfaces $\gamma^{(j)}$ of Cu for the conditions of Fig. 7.16. Only one quadrant of the surfaces is shown.

vector range $-1 \leq k_t/g_{200} \leq +1$ and $-1 \leq k_t/g_{020} \leq +1$. Close to the zone axis orientation, the first and second Bloch waves are strongly excited. When the beam tilt is increased, higher order Bloch waves begin to contribute, in a manner similar to that shown in Figs 7.7(c) and 7.8(c). Figure 7.21(b) shows the corresponding absorption coefficients q^j for the first eight Bloch waves. The absorption coefficients vary smoothly for the first Bloch wave, and show rapid variations for the higher-order waves.

Finally, the position of each Bloch wave with respect to the atom positions can also be computed, similar to the systematic row results in Fig. 7.11. Figure 7.22 shows the first 14 Bloch waves for the [001] zone axis orientations of copper, along with the projected atom positions (upper left). It is clear that the first and second Bloch waves are centered on the atom positions. Close to the zone axis orientation, these two waves have the strongest excitation coefficients, as can be seen in Fig. 7.21(a). Most of the beam electrons spend most of their time in the crystal close to the projected atom positions (for orientations close to the zone axis orientation) and one can draw an analogy with bound electron states in atoms [BLS78].

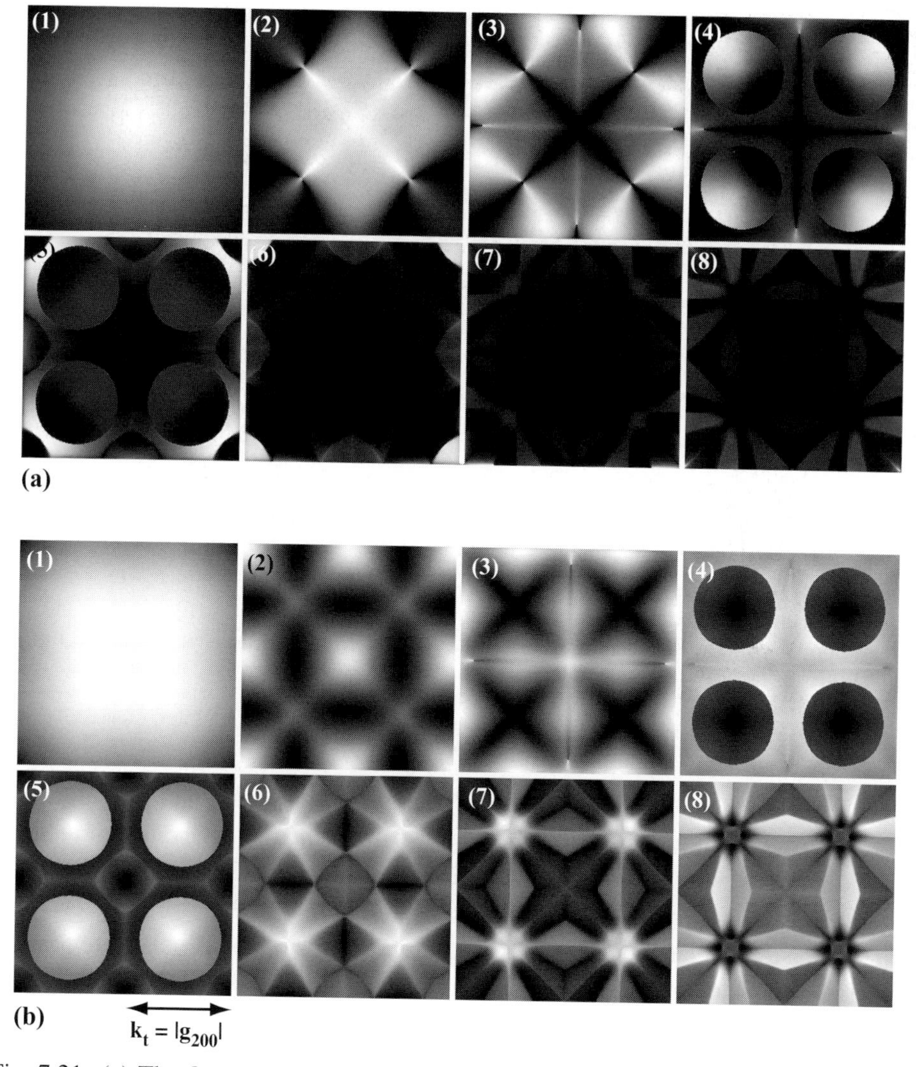

Fig. 7.21. (a) The first eight Bloch wave excitation coefficients $|\alpha^{(j)}|^2$; (b) the first eight Bloch wave absorption coefficients $q^{(j)}$ for the conditions of Fig. 7.16.

This analogy lies at the basis of an alternative description of elastic electron scattering in terms of the *atomic column approximation* introduced by Van Dyck and Chen [VDC99]. Each projected atom position represents a column of atoms parallel to the beam direction. Electrons travel through the foil and "channel" along these columns. The probability of finding an electron inside the column oscillates with depth in the crystal, and different Bloch waves have a different oscillation period (i.e. a different extinction distance). We will return to this oscillation behavior in Section 7.4.1 when we discuss the *multi-slice method*.

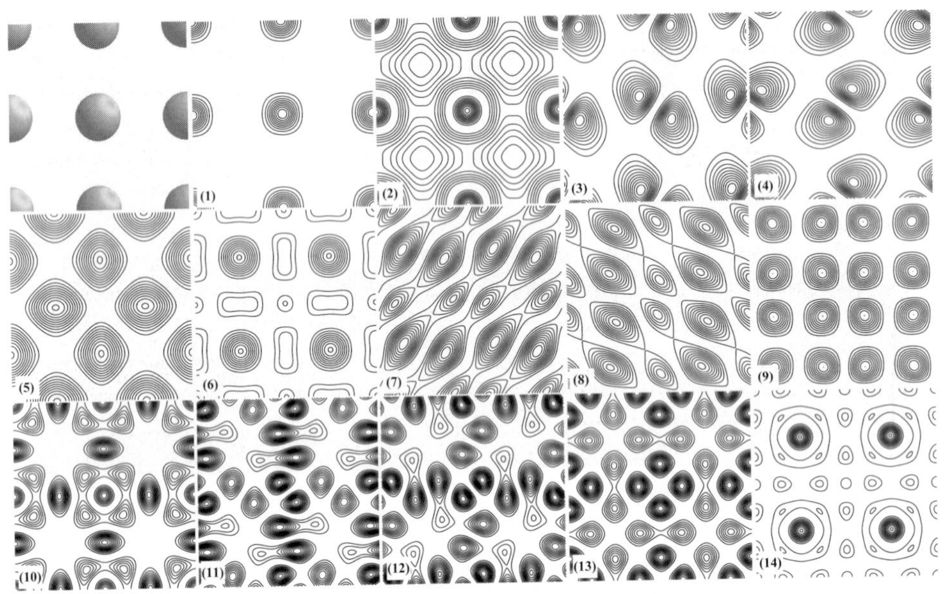

Fig. 7.22. Bloch wave contour plots for the first 14 Bloch waves for the [001] zone axis orientation of copper. The computations used 25 Bloch waves and were carried out for 200 kV electrons.

7.3.2.3 CBED patterns

Zone axis convergent beam patterns can be computed using the same algorithms discussed in the previous sections; the only two differences are the range of incident beam directions, which is defined according to equation (7.13), and the representation of the result as a set of (possibly overlapping) disks. The MBCBED.f90 program can be used to simulate zone axis CBED patterns based on the Bloch wave formalism. The input parameters are the beam convergence angle θ_c, the camera length L, the zone axis direction $[uvw]$, the location of the Laue center, and the microscope acceleration voltage E. The output is a series of CBED patterns for increasing foil thickness. The total number of incident beam directions to be considered depends on the camera length and on the desired printing resolution. For a printing resolution of 300 dpi and a camera length of $L = 1000$ mm, a beam convergence angle of $\theta_c = 8$ mrad would correspond to a disk radius of 8 mm, which requires 189 pixels or incident beam directions along a disk diameter. There is no point in computing more beam directions than will be displayed in the final pattern.

The results of two example calculations are shown in Figs 7.23 and 7.24. Figure 7.23 shows two GaAs CBED patterns computed using the Bloch wave approach of the MBCBED.f90 program at 120 kV; 35 beams were used for foil thicknesses of 50 and 120 nm. These images are to be compared with the experimental images of Fig. 4.25 on page 290. The agreement between simulations and

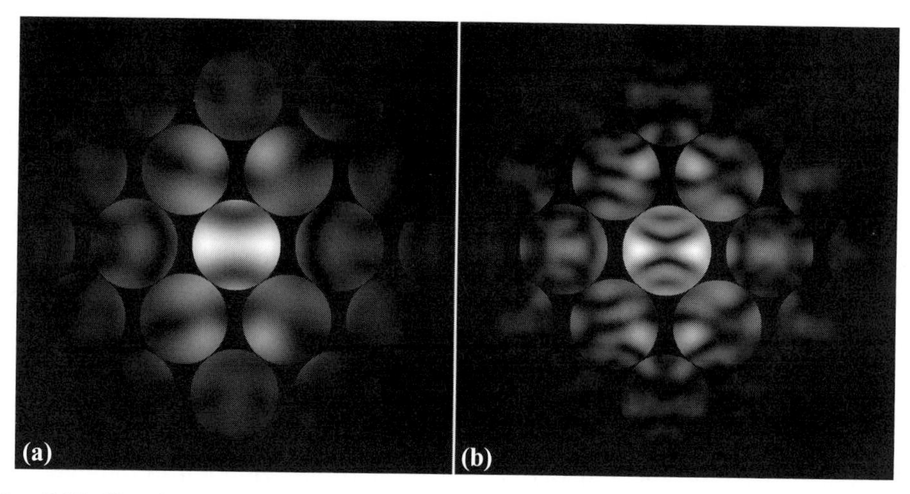

Fig. 7.23. Simulated CBED patterns for the [110] zone axis orientation of GaAs. The simulation parameters are 120 kV, $\theta_c = 5.1$ mrad, and 35 beams were used (Bloch wave formalism). The foil thicknesses are 50 nm (left) and 120 nm (right). The patterns are to be compared with the experimental patterns in Fig. 4.25 on page 290.

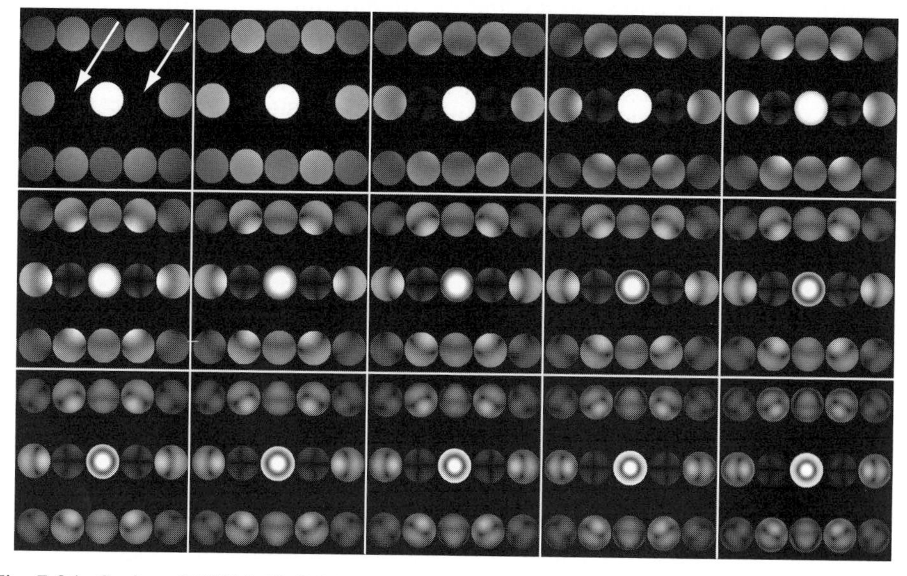

Fig. 7.24. Series of 15 [11.0] CBED patterns for Ti (200 kV, $\theta_c = 2.6$ mrad, 57 beam Bloch wave simulation). The kinematically forbidden $\pm(00.1)$ reflections are arrowed in the first pattern. The thickness increases from 10 nm in steps of 10 nm from left to right and top to bottom. (Intensities have been scaled exponentially to emphasize the weaker features.)

experimental observations is rather good. The absence of a vertical mirror plane for the whole pattern symmetry is clear for the thicker foil and indicates that the structure is non-centrosymmetric.

While it is easy to compare experimental and simulated CBED patterns visually, it is possible to do a lot better than that. CBED patterns, in particular energy-filtered CBED patterns in which most of the inelastically scattered electrons have been removed, can be used to determine with very high accuracy the amplitude and phase of electron structure factors. For a detailed description of the procedures involved we refer the interested reader to the book *Electron Microdiffraction* by Spence and Zuo [SZ92]; a review of quantitative CBED can be found in [Spe93].

Figure 7.24 shows a series of 15 computed CBED patterns for the $[11\bar{2}0]$ zone axis orientation of Ti. 57 beams were used at 200 kV, for a beam convergence angle of 2.6 mrad. The thickness increases in steps of 10 nm from left to right and top to bottom; the first image on the top left has $z_0 = 10$ nm. A total of 6625 incident beam directions were used for the central disk. The kinematically forbidden $(00.l)$ reflections ($l = 2n + 1$) are arrowed in the first pattern. As the thickness increases a faint intensity can be observed in the forbidden disks. The intensity distribution shows a thin dark horizontal line for all thicknesses. Such a line in a kinematically forbidden reflection is known as a *Gjönnes–Moodie line* or *G-M line* [GM65]; the origin of such lines will be discussed in Section 9.2.3. G-M lines can be used to distinguish between different space groups, as illustrated in Section 9.5.

HOLZ reflections can also be taken into account in this type of CBED simulations. This would turn the diffraction problem into a true three-dimensional problem, instead of simply using only the ZOLZ reflections. Computationally, HOLZ reflections would increase significantly the size of the dynamical matrix and hence the computation time. If HOLZ reflections are to be included in the CBED computation, the method of *Bethe potentials*, introduced in Section 7.3.3, should be employed to reduce the size of the dynamical matrix.

7.3.2.4 Eades patterns

We have seen in Fig. 7.3, that a parallel beam incident on a bent foil is equivalent to a convergent beam incident on a planar foil. If we had a method to acquire a diffraction pattern for every incident beam orientation separately (somewhat like serial acquisition of a convergent beam pattern), then we would be able to image directly patterns similar to those shown in Fig. 7.16. In practice, this can be done by using a narrow but parallel incident beam, similar to that used in microdiffraction observations, and rocking the beam in two orthogonal directions. For each incident orientation we measure the bright field intensity and the dark field intensity for a particular Bragg angle (or range of angles). If your microscope has a scanning attachment (STEM attachment) then you can set up the illumination conditions to obtain such *double-rocking patterns* or *Eades patterns* [Ead80]. If your microscope has an *annular dark field detector*, a ring-shaped detector that measures electron current scattered into a given angular range, then you can determine for each incident beam direction the total intensity scattered in between two conical surfaces. Plotting the bright field and dark field signals as a function of beam position then results in the Eades patterns.

Figure 7.25 shows a simulated Eades pattern for the [001] zone axis orientation of Cu. The acceleration voltage is 200 kV, crystal thickness is 200 nm, and the annular range includes the following families: {200}, {220}, {400}, and {420} (not shown as

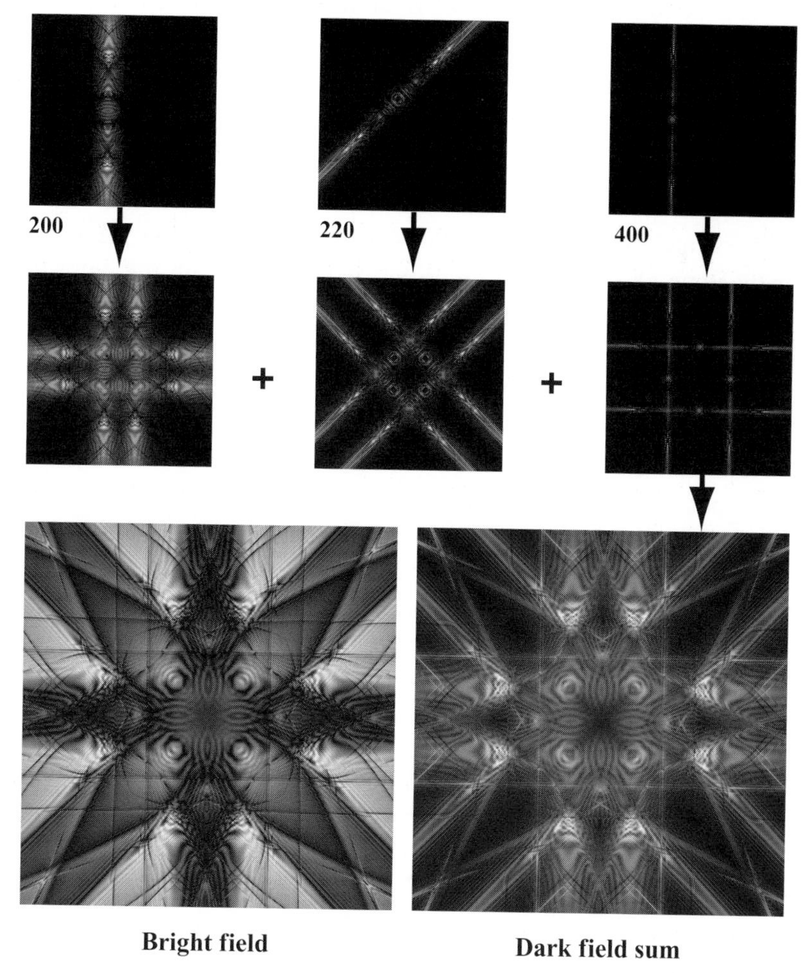

Fig. 7.25. Eades pattern simulation for the [001] zone axis orientation in Cu (200 kV, 200 nm thickness). The dark field images of the 200, 220, and 400 reflections are copied into the symmetrized patterns of the second row, which are then added together to form the dark field image.

separate images, but included in the dark field sum). The dark field image is the sum of the symmetrized dark field images of the individual reflections. For instance, after application of the **4mm** symmetry to the 200 dark field image, the symmetrized pattern on the second row is obtained. Summation of all such patterns for the reflections inside the annular range (in this case from 10 to 32 mrad) then results in the complete dark field image on the bottom row. Simulations of Eades patterns are identical to the simulations of CBED and zone axis bend center patterns. It is only the final representation for the dark field image that is different. The dark field images for all reflections detected by the annular dark field detector are added together in the proper relative orientation and position. Other experimental techniques can be used to obtain the same information; we refer the reader to the beautiful volumes on convergent beam electron diffraction by Tanaka and coworkers [TT85], [TTK88], and [TTT94]. The computation of Eades patterns has been implemented as part of the MBshow.f90 program.

7.3.2.5 Thickness-integrated intensities: zone axis case

The thickness-integrated intensity expression derived in Section 7.2.3 can also be used for the zone axis case. The algorithm is identical, with the only difference being that the resulting intensities are represented as 2D plots instead of 1D curves. The computation has been implemented in the MBTII.f90 program. An example of the output of this program is shown in Fig. 7.26 for the [110] zone axis orientation of GaAs. The acceleration voltage is 200 kV, foil thickness 50 nm, and 117 beams (up to 18 nm^{-1}) were taken into account for a total of 66 049 ($= 257^2$) incident beam directions (resulting in a computation time of slightly more than 5 days on a 666 MHz workstation). The maximum incident beam tilt is $\mathbf{k}_t = -\mathbf{g}_{006}$. The image on the left shows the thickness-integrated intensity for the Ga positions; the intensity varies between 0.223 and 2.296. The intensity profile for As is very nearly the mirror image (vertical mirror) of that for Ga, with an intensity range between 0.237 and 2.433. The third image shows the ratio of As to Ga; the intensity range is from

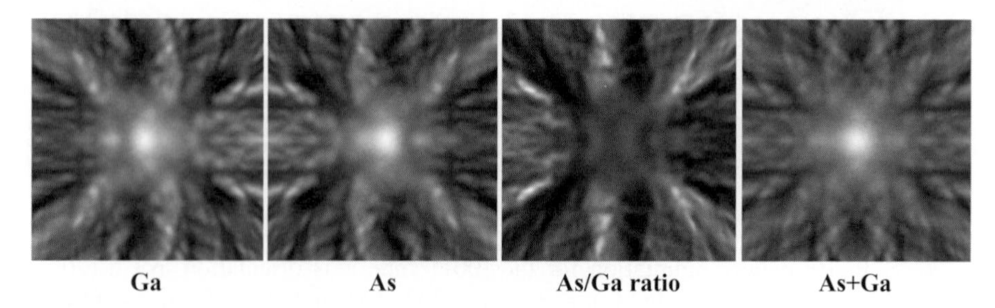

Fig. 7.26. Thickness-averaged integrated intensities for Ga and As in the [110] zone axis orientation of GaAs. Simulation parameters are stated in the text.

0.227 to 4.931. The ratio varies rapidly from one location to another, indicating that channeling along the [110] zone axis orientation is strong and that this zone axis should be avoided for quantitative EDS or EELS experiments. The sum of the Ga and As integrated intensities is shown as the final image and resembles the [110] bright field bend center image.

It should be clear by now that this type of multi-beam simulation can be rather time-consuming, in particular when large numbers of beams are taken into account. In the following two sections, we will introduce briefly two different ways to increase the speed of the computations (other than running them on parallel computers): Bethe potentials, which can reduce the size of the dynamical matrix through the use of perturbation theory, and diffraction group symmetry, which can be used to reduce the number of independent incident beam directions for which the computations need to carried out.

7.3.3 Bethe potentials

The bend center computation in Fig. 7.19 shows that certain reflections only contribute to a small region of the bright field image (Bragg lines). Everywhere else in the image their contribution is small. It is therefore not strictly necessary to take exact account of the contributions of all these weak reflections to every pixel in the image. We can use the following criterion to determine whether a reflection is "strong" or "weak": if the reciprocal lattice point is far enough from the Ewald sphere, so that $|s_{\mathbf{g}}|$ is significantly larger than $\lambda|U_{\mathbf{g}}|$, then we can consider the contribution of that reflection to be a *perturbation* to the overall scattered amplitude. The criterion for a beam to be weak is derived from equation (5.51) in Chapter 5:

$$2k_0 s_{\mathbf{g}} C_{\mathbf{g}}^{(j)} + \sum_{\mathbf{h}\neq\mathbf{g}} U_{\mathbf{g}-\mathbf{h}} C_{\mathbf{h}}^{(j)} = 2k_n \gamma^{(j)} C_{\mathbf{g}}^{(j)}.$$

If the prefactor of the first term ($k_0|s_{\mathbf{g}}|$) is significantly larger than the prefactor of the second term ($|U_{\mathbf{g}-\mathbf{h}}|$), then the coupling of $\mathbf{g}$ to $\mathbf{h}$ will be weak. Let us rewrite this eigenvalue equation as follows:

$$\eta_{\mathbf{g}}^{(j)} C_{\mathbf{g}}^{(j)} + \sum_{\mathbf{h}\neq\mathbf{g}} U_{\mathbf{g}-\mathbf{h}} C_{\mathbf{h}}^{(j)} = 0,$$

with

$$\eta_{\mathbf{g}}^{(j)} \equiv 2\left[k_0 s_{\mathbf{g}} - k_n \gamma^{(j)}\right].$$

We will denote weak beams by primed reciprocal lattice vectors, e.g. $\mathbf{g}'$. We will also assume that weak beams do not interact with each other, only with strong beams. In that case, the eigenvalue equation for a weak beam becomes

$$\eta_{\mathbf{g}'}^{(j)} C_{\mathbf{g}'}^{(j)} + \sum_{\mathbf{h}} U_{\mathbf{g}'-\mathbf{h}} C_{\mathbf{h}}^{(j)} = 0,$$

where the summation runs *only over the strong beams*. This equation can now be solved for the Bloch wave coefficients of the weak beam:

$$C_{\mathbf{g}'}^{(j)} = -\frac{1}{\eta_{\mathbf{g}'}^{(j)}} \sum_{\mathbf{h}} U_{\mathbf{g}'-\mathbf{h}} C_{\mathbf{h}}^{(j)}. \tag{7.14}$$

If the Bloch wave coefficients of all strong beams are known, then this equation can be used to determine the coefficients of the weaker beams.

Next, we rewrite the eigenvalue equation for the strong beams by splitting the summation over strong and weak beams:

$$\eta_{\mathbf{g}}^{(j)} C_{\mathbf{g}}^{(j)} + \sum_{\mathbf{h} \neq \mathbf{g}} U_{\mathbf{g}-\mathbf{h}} C_{\mathbf{h}}^{(j)} + \sum_{\mathbf{h}'} U_{\mathbf{g}-\mathbf{h}'} C_{\mathbf{h}'}^{(j)} = 0.$$

The Bloch wave coefficients $C_{\mathbf{h}'}^{(j)}$ of the weak beams can be replaced by equation (7.14). This is left as an exercise for the reader. After rearranging the terms in the resulting equation we arrive at

$$\left[\eta_{\mathbf{g}}^{(j)} - \sum_{\mathbf{h}'} \frac{|U_{\mathbf{g}-\mathbf{h}'}|^2}{\eta_{\mathbf{h}'}^{(j)}}\right] C_{\mathbf{g}}^{(j)} + \sum_{\mathbf{h} \neq \mathbf{g}} \left[U_{\mathbf{g}-\mathbf{h}} - \sum_{\mathbf{h}'} \frac{U_{\mathbf{g}-\mathbf{h}'} U_{\mathbf{h}'-\mathbf{h}}}{\eta_{\mathbf{h}'}^{(j)}}\right] C_{\mathbf{h}}^{(j)} = 0, \tag{7.15}$$

where the factors $\eta_{\mathbf{h}'}^{(j)}$ in the denominators of the sums over the weak beams must be approximated by $2k_0 s_{\mathbf{h}'}$. If we introduce the following notation:

$$\bar{\eta}_{\mathbf{g}}^{(j)} \equiv \eta_{\mathbf{g}}^{(j)} - \sum_{\mathbf{h}'} \frac{|U_{\mathbf{g}-\mathbf{h}'}|^2}{\eta_{\mathbf{h}'}^{(j)}};$$

$$\bar{U}_{\mathbf{g}-\mathbf{h}} \equiv U_{\mathbf{g}-\mathbf{h}} - \sum_{\mathbf{h}'} \frac{U_{\mathbf{g}-\mathbf{h}'} U_{\mathbf{h}'-\mathbf{h}}}{\eta_{\mathbf{h}'}^{(j)}},$$

then the eigenvalue equation becomes

$$\bar{\eta}_{\mathbf{g}}^{(j)} C_{\mathbf{g}}^{(j)} + \sum_{\mathbf{h} \neq \mathbf{g}} \bar{U}_{\mathbf{g}-\mathbf{h}} C_{\mathbf{h}}^{(j)} = 0, \tag{7.16}$$

which has the same form as the original eigenvalue equation. We have replaced the excitation errors of the strong beams by effective excitation errors; the Fourier coefficients of the electrostatic lattice potential have also been modified to take the weaker beams into account. The coefficients $\bar{U}_{\mathbf{g}-\mathbf{h}}$ are known as *Bethe potentials* or *dynamical potentials*. This approximation was first introduced by Bethe using first-order perturbation theory [Bet28].

In all multi-beam computations described thus far, we have always explicitly taken all beams into account. The advantage of doing this is that the off-diagonal elements of the dynamical matrix need only be computed once. The disadvantage is that the dynamical matrix has a large number of entries and perhaps only a small number of beams with appreciable amplitude. We can reduce the size of the dynamical matrix by incorporating all the weakest reflections by means of Bethe potential coefficients. The disadvantage of this approach would be that the off-diagonal elements of the dynamical matrix must be recomputed for every incident beam orientation, but this is more than offset by the reduced size of the Bloch wave eigenvalue problem. Bethe potentials have been implemented in the CalcDynMat routine, which is available from the website. A detailed description of the implementation is also available in PDF format as a web appendix (appbethe.pdf) to this book.

The Bethe potential can be rewritten as follows:

$$\bar{U}_{\mathbf{g}} = U_{\mathbf{g}} - \lambda \sum_{\mathbf{h}'} \frac{U_{\mathbf{g}-\mathbf{h}'} U_{\mathbf{h}'}}{2 s_{\mathbf{h}'}}.$$

From this relation, we see that the Bethe potential can become equal to zero for a particular wavelength λ, or, equivalently, for a particular acceleration voltage E. The voltage for which a Bethe potential coefficient vanishes is known as a *critical voltage* E_c, and the vanishing of $\bar{U}_{\mathbf{g}-\mathbf{h}}$ is known as the *critical voltage effect*. If the Bethe potential coefficient for a particular reflection vanishes, then that reflection will have zero intensity for all crystal thicknesses, even when the structure factor for that reflection does not vanish. The voltage for which an allowed reflection vanishes for all crystal thicknesses can be determined by varying the acceleration voltage over a large range (e.g. from 100 to 1000 kV) and this, in turn, allows for the accurate determination of the potential coefficients $U_{\mathbf{g}-\mathbf{h}}$. There is extensive literature on the critical voltage effect and we refer the reader to [LHMF72] and [SZ92] for more details.

7.3.4 Application of symmetry in multi-beam simulations

We have seen in Chapter 5 that the symmetry of the dynamical diffraction process is described by the 31 diffraction groups. The diffraction groups relate the intensity at a particular point of a diffraction disk to the intensity at one or more equivalent points on other diffraction disks of the same family. It is possible to use this group theoretical information in numerical computations to reduce the number of beam directions for which the dynamical calculations must be performed. The implementation of the diffraction groups is rather technical and would interrupt the main line of this text, so we refer the interested reader to a web appendix (MBsym.pdf), which deals with all of the details. For completeness we simply list in Table 7.2,

Table 7.2. *Zone axis symmetries for several independent zone axis directions in all 32 crystal point groups [based on Table 4 in [BESR76]].*

Point groups	$\langle 111\rangle$	$\langle 100\rangle$	$\langle 110\rangle$	$\langle uv0\rangle$	$\langle uuw\rangle$		$[uvw]$
$\mathbf{m\bar{3}m}$	$\mathbf{6_R mm_R}$	$\mathbf{4mm1_R}$	$\mathbf{2mm1_R}$	$\mathbf{2_R mm_R}$	$\mathbf{2_R mm_R}$		$\mathbf{2_R}$
$\mathbf{\bar{4}3m}$	$\mathbf{3m}$	$\mathbf{4_R mm_R}$	$\mathbf{m1_R}$	$\mathbf{m_R}$	$\mathbf{m}$		$\mathbf{1}$
$\mathbf{432}$	$\mathbf{3m_R}$	$\mathbf{4m_R m_R}$	$\mathbf{2m_R m_R}$	$\mathbf{m_R}$	$\mathbf{m_R}$		$\mathbf{1}$
Point groups	$\langle 111\rangle$	$\langle 100\rangle$		$\langle uv0\rangle$			$[uvw]$
$\mathbf{m3}$	$\mathbf{6_R}$	$\mathbf{2mm1_R}$		$\mathbf{2_R mm_R}$			$\mathbf{2_R}$
$\mathbf{23}$	$\mathbf{3}$	$\mathbf{2m_R m_R}$		$\mathbf{m_R}$			$\mathbf{1}$
Point groups	$[00.1]$	$\langle 11.0\rangle$	$\langle 1\bar{1}.0\rangle$	$[uv.0]$	$[uu.w]$	$[u\bar{u}.w]$	$[uv.w]$
$\mathbf{6/mmm}$	$\mathbf{6mm1_R}$	$\mathbf{2mm1_R}$	$\mathbf{2mm1_R}$	$\mathbf{2_R mm_R}$	$\mathbf{2_R mm_R}$	$\mathbf{2_R mm_R}$	$\mathbf{2_R}$
$\mathbf{\bar{6}m2}$	$\mathbf{3m1_R}$	$\mathbf{m1_R}$	$\mathbf{2mm}$	$\mathbf{m}$	$\mathbf{m_R}$	$\mathbf{m}$	$\mathbf{1}$
$\mathbf{6mm}$	$\mathbf{6mm}$	$\mathbf{m1_R}$	$\mathbf{m1_R}$	$\mathbf{m_R}$	$\mathbf{m}$	$\mathbf{m}$	$\mathbf{1}$
$\mathbf{622}$	$\mathbf{6m_R m_R}$	$\mathbf{2m_R m_R}$	$\mathbf{2m_R m_R}$	$\mathbf{m_R}$	$\mathbf{m_R}$	$\mathbf{m_R}$	$\mathbf{1}$
Point groups	$[00.1]$			$[uv.0]$			$[uv.w]$
$\mathbf{6/m}$	$\mathbf{61_R}$			$\mathbf{2_R mm_R}$			$\mathbf{2_R}$
$\mathbf{\bar{6}}$	$\mathbf{31_R}$			$\mathbf{m}$			$\mathbf{1}$
$\mathbf{6}$	$\mathbf{6}$			$\mathbf{m_R}$			$\mathbf{1}$
Point groups	$[00.1]$	$[11.0]$				$[u\bar{u}.w]$	$[uvw]$
$\mathbf{\bar{3}m}$	$\mathbf{6_R mm_R}$	$\mathbf{21_R}$				$\mathbf{2_R mm_R}$	$\mathbf{2_R}$
$\mathbf{3m}$	$\mathbf{3m}$	$\mathbf{1_R}$				$\mathbf{m}$	$\mathbf{1}$
$\mathbf{32}$	$\mathbf{3m_R}$	$\mathbf{2}$				$\mathbf{m_R}$	$\mathbf{1}$
Point groups	$[00.1]$						$[uv.w]$
$\mathbf{\bar{3}}$	$\mathbf{6_R}$						$\mathbf{2_R}$
$\mathbf{3}$	$\mathbf{3}$						$\mathbf{1}$
Point groups	$[001]$	$\langle 100\rangle$	$\langle 110\rangle$	$[u0w]$	$[uv0]$	$[uuw]$	$[uvw]$
$\mathbf{4/mmm}$	$\mathbf{4mm1_R}$	$\mathbf{2mm1_R}$	$\mathbf{2mm1_R}$	$\mathbf{2_R mm_R}$	$\mathbf{2_R mm_R}$	$\mathbf{2_R mm_R}$	$\mathbf{2_R}$
$\mathbf{\bar{4}2m}$	$\mathbf{4_R mm_R}$	$\mathbf{2m_R m_R}$	$\mathbf{m1_R}$	$\mathbf{m_R}$	$\mathbf{m_R}$	$\mathbf{m}$	$\mathbf{1}$
$\mathbf{4mm}$	$\mathbf{4mm}$	$\mathbf{m1_R}$	$\mathbf{m1_R}$	$\mathbf{m}$	$\mathbf{m_R}$	$\mathbf{m}$	$\mathbf{1}$
$\mathbf{422}$	$\mathbf{4m_R m_R}$	$\mathbf{2m_R m_R}$	$\mathbf{2m_R m_R}$	$\mathbf{m_R}$	$\mathbf{m_R}$	$\mathbf{m_R}$	$\mathbf{1}$
Point groups	$[001]$				$[uv0]$		$[uvw]$
$\mathbf{4/m}$	$\mathbf{41_R}$			$\mathbf{2_R mm_R}$			$\mathbf{2_R}$
$\mathbf{\bar{4}}$	$\mathbf{4_R}$			$\mathbf{m_R}$			$\mathbf{1}$
$\mathbf{4}$	$\mathbf{4}$			$\mathbf{m_R}$			$\mathbf{1}$
Point groups	$[001]$	$\langle 100\rangle$		$[u0w]$	$[uv0]$		$[uvw]$
$\mathbf{mmm}$	$\mathbf{2mm1_R}$	$\mathbf{2mm1_R}$		$\mathbf{2_R mm_R}$	$\mathbf{2_R mm_R}$		$\mathbf{2_R}$
$\mathbf{mm2}$	$\mathbf{2mm}$	$\mathbf{m1_R}$		$\mathbf{m}$	$\mathbf{m_R}$		$\mathbf{1}$
$\mathbf{222}$	$\mathbf{2m_R m_R}$	$\mathbf{2m_R m_R}$		$\mathbf{m_R}$	$\mathbf{m_R}$		$\mathbf{1}$
Point groups	$[010]$			$[u0w]$			$[uvw]$
$\mathbf{2/m}$	$\mathbf{21_R}$			$\mathbf{2_R mm_R}$			$\mathbf{2_R}$
$\mathbf{m}$	$\mathbf{1_R}$			$\mathbf{m}$			$\mathbf{1}$
$\mathbf{2}$	$\mathbf{2}$			$\mathbf{m_R}$			$\mathbf{1}$
Point groups							$[uvw]$
$\mathbf{\bar{1}}$							$\mathbf{2_R}$
$\mathbf{1}$							$\mathbf{1}$

derived from Table 4 in [BESR76], all possible zone axis symmetries for different zone axis directions for all 32 point groups.

7.4 Computation of the exit plane wave function

In Chapter 10, we will discuss various types of phase contrast microscopy, in particular high-resolution TEM (HRTEM) and Lorentz microscopy. The main goal of the HRTEM method is to obtain *structural* information about the material at the length scale of the interatomic distances. As for every other observation method that we have described so far, the resulting experimental high-resolution images must be compared with theoretical, simulated images. Such a simulation requires two types of input: the wave function of the electrons at the exit plane of the crystal, and the transfer function of the microscope. We postpone a discussion of the transfer function until Chapter 10; in this section, we will describe briefly how the exit plane wave function can be computed in an efficient way.

As we have seen in Chapter 5, there are three major computational methods which can be used to obtain this wave function: Bloch waves, the multi-slice method, and the real-space method. We will assume that the thin foil is oriented so that a low-index zone axis $[uvw]$ is parallel to the electron beam, and that the foil is not bent. This means that we are effectively only considering the exact zone axis orientation, i.e. the point at the center of the bend center. For many crystal structures, parallel projection drawings along low-index zone axes can be interpreted readily in terms of the structural elements, such as atom columns, coordination polyhedra, and so on.

On the website we provide a single program, BWEW.f90, to compute the exit wave using the Bloch wave formalism. For a numerical implementation of the multi-slice formalism we refer the interested reader to the book *Advanced Computing in Electron Microscopy* by E.J. Kirkland [Kir98]; this text describes programming details for the technique and all the underlying theory. The book also provides extensive C-source code that implements the basic algorithms of the multi-slice method. The source code for the third method, the real-space method, has been made available by its authors and can also be downloaded from the website. For a critical review of several simulation methods we refer the reader to [SOBS83].

7.4.1 The multi-slice and real-space approaches

We begin with the basic equation (5.35):

$$\frac{\partial \psi}{\partial z} = \left(\bar{\Delta} + \bar{V}\right)\psi. \tag{7.17}$$

If we subdivide the crystal into slices of thickness ϵ normal to the beam direction, then the formal solution to this equation is given by

$$\psi(x, y, z = n\epsilon) = \hat{B}_n \hat{B}_{n-1} \dots \hat{B}_1 \psi(x, y, z = 0), \tag{7.18}$$

with

$$\hat{B}_j(x, y) \equiv \mathrm{e}^{\epsilon(\bar{\Delta}+\bar{V}_j)}, \tag{7.19}$$

and

$$\bar{V}_j = \frac{\mathrm{i}\sigma}{\epsilon} \int_{(j-1)\epsilon}^{j\epsilon} V_c(x, y, z)\,\mathrm{d}z. \tag{7.20}$$

This solution is exact in the limit of vanishing slice thickness ϵ. In the simplest case, the slice thickness is equal to the lattice periodicity c along the z-axis and all slices are identical:

$$\bar{V}_j = \bar{V} = \frac{\mathrm{i}\sigma}{c} \int_0^c V_c(x, y, z)\,\mathrm{d}z. \tag{7.21}$$

In Chapter 5, we have seen that the exponential of a sum of operators is not simply equal to the product of the exponentials. If we apply the Zassenhaus theorem to second order in the slice thickness we find for the operator $\hat{B}_j$:

$$\begin{aligned} \hat{B}_j = \mathrm{e}^{\epsilon(\bar{\Delta}+\bar{V}_j)} &\approx \mathrm{e}^{\epsilon\bar{\Delta}} \mathrm{e}^{\epsilon\bar{V}_j} \mathrm{e}^{-\frac{1}{2}(\bar{\Delta}\bar{V}_j - \bar{V}_j\bar{\Delta})\epsilon^2}; \\ &= \mathrm{e}^{\frac{1}{2}\epsilon\bar{\Delta}} \mathrm{e}^{\epsilon\bar{V}_j} \mathrm{e}^{\frac{1}{2}\epsilon\bar{\Delta}}. \end{aligned} \tag{7.22}$$

The second expression can be shown to be equal to the first by means of a simple Taylor expansion to second order in ϵ. The exit wave can then be written as

$$\psi(x, y, z = n\epsilon) = \mathrm{e}^{-\frac{1}{2}\epsilon\bar{\Delta}} \left[\mathrm{e}^{\epsilon\bar{\Delta}} \mathrm{e}^{\epsilon\bar{V}_j}\right]^n \mathrm{e}^{\frac{1}{2}\epsilon\bar{\Delta}} \psi(x, y, z = 0). \tag{7.23}$$

The first and last terms simply propagate the wave function by half a slice thickness. The terms between the square brackets are identical to the expression (5.42) on page 319. This expression is correct to second order in the slice thickness.

Now that we have established a theoretical equation for the exit wave we must answer the following questions:

- how many image points (pixels) do we need to have an accurate representation of the exit wave in both direct and reciprocal space?
- how do we compute the phase grating term?
- how do we compute the propagator term?

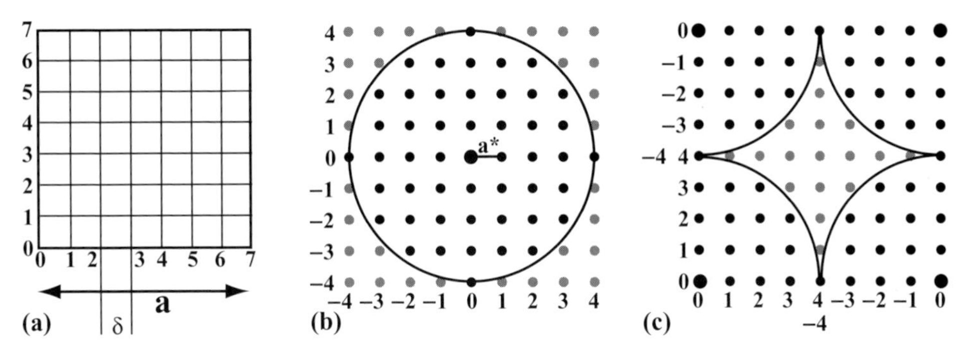

Fig. 7.27. Schematic representation of the relation between real-space (a) and reciprocal space (b) sampling for a square unit cell. In (c) the origin of the reciprocal lattice is shifted to the lower left-hand corner.

7.4.1.1 The computational grid

The sampling theorem, introduced in Section 3.11.1, states that a continuous function can be represented by a discrete sample, provided the sampling frequency is sufficiently high and the function is bandwidth limited. A continuous function $f(\mathbf{r})$ with an infinite domain has a Fourier transform $f(\mathbf{q})$ that also exists on an infinite domain. If we select a finite region of the domain of $f(\mathbf{r})$, and sample it discretely, then its discrete Fourier transform no longer has an infinite domain but becomes periodic. Typically, the finite domain will be the projected unit cell of the structure of interest. If the sampling grid has a sampling interval δ, and we use N points (or pixels) to cover the side of the unit cell, i.e. $a = N\delta$ (see Fig. 7.27a), then the Nyquist frequency q_c is given by

$$q_c = \frac{1}{2\delta} = \frac{N}{2a} \equiv g_{max}.$$

This means that the maximum reciprocal lattice point that must be taken into account in the computation is $\frac{1}{2}N\mathbf{a}^*$. The same argument can be made for any direction in the real-space unit cell, so that the relevant portion of reciprocal space is a circle, as shown in Fig. 7.27(b). There is no point in taking more reciprocal lattice points because the sampling interval δ is too large to correctly represent the contributions of these higher spatial frequencies. The reciprocal lattice points indicated in gray in Fig. 7.27(b) should therefore have zero contribution. In numerical work it is customary to shift the origin of reciprocal space to the lower left-hand corner of the array, as shown in Fig. 7.27(c). Each corner of the resulting array is then identical as a consequence of the periodicity of the Fourier transform. The points with frequency $N/2$ and $-N/2$ are also identical.

It is customary to select the number of grid points N_i along each axis to be a power of 2, i.e. $N_i = 2^{n_i}$, with n_i integers. The main reason for this choice is the fact that the

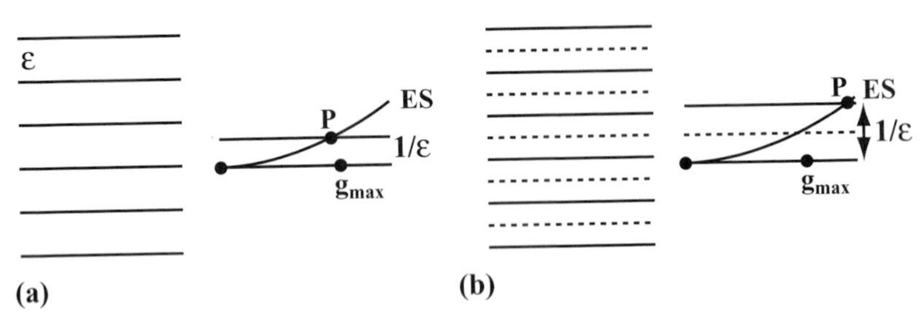

Fig. 7.28. Graphical representation of the relation between slice thickness and g_{max}. The slice thickness should be sufficiently small, so that the corresponding FOLZ layer intersects the Ewald sphere far from g_{max}.

HRTEM image simulation makes extensive use of the fast Fourier transform (FFT), introduced in Section 2.5.4, and the traditional FFT algorithms require a power of 2 for all dimensions. This is no longer a necessary restriction, since modern FFT algorithms can deal with any number N_i, not just powers of 2 or other small integers. The fftw-package used for the computations in this textbook can handle transforms of arbitrary N_i; when N_i is a large prime number the computational speed will be somewhat slow, but it is rather easy to avoid such numbers. This means that we can select the grid spacing δ to be the same along both axes, even when the axes have different lengths.

The number of sampling points is important because it also determines the maximum slice thickness that can be considered for multi-slice-type simulations. Figure 7.28 shows schematically how the slice thickness in real-space relates to reciprocal space. A slice thickness of ϵ creates a periodic structure along the beam direction which has a FOLZ layer at a distance $1/\epsilon$ from the ZOLZ layer (e.g. [CVD84a]). To avoid such situations, the point P in Fig. 7.28 must be kept far enough from the origin so that the excitation error of the $\mathbf{g}_{max}$ reflection will be significantly smaller than $1/\epsilon$:

$$\frac{\lambda}{2} g_{max}^2 \ll \frac{1}{\epsilon}, \tag{7.24}$$

or

$$\epsilon \ll \frac{2}{\lambda g_{max}} = \frac{8\delta^2}{\lambda}. \tag{7.25}$$

If the sampling conditions are selected so that this condition is satisfied, then HOLZ effects will be negligible. If the unit cell size along the beam direction is too large for a given sampling interval δ, then the unit cell must be cut into thinner slices and the computation time increases, proportional to the number of slices needed.

The value of g_{max} to be used for a multi-slice-type computation depends on the particular atom type(s) present in the unit cell. Heavier atoms scatter more strongly and will need more reciprocal lattice points to represent their scattering potentials. Coene and Van Dyck [CVD84a] have proposed that g_{max} should be somewhere between Z (the atomic number) and $Z^{1/3}$ (expressed in nm^{-1}), where Z is taken for the heaviest atom in the unit cell. For pure gold a value of 50 nm^{-1} is suggested, whereas for Cu $g_{max} \approx 39$ nm^{-1}. Repeating a computation for several values of g_{max} may give an indication as to the minimum value of this parameter (or equivalently, the maximum sampling interval δ).

7.4.1.2 Computation of the phase grating term

The phase grating exponential can be computed in a number of different ways. We will only discuss the convolution method introduced by D.F. Lynch [Lyn74]. Other more advanced methods can be found in [SOBS83] and [dJCVD87]. Consider the phase grating exponential $e^{\epsilon \bar{V}_j}$. It is easy to show by means of Taylor expansions that the following equality holds:

$$e^{\epsilon \bar{V}_j} = \lim_{s \to \infty} \left(1 + \frac{\epsilon \bar{V}_j}{s} \right)^s . \tag{7.26}$$

Replacing $\bar{V}_j$ by $i\sigma V_p$ we have

$$e^{i\sigma V_p} = \lim_{s \to \infty} \left(1 + i\frac{\sigma\epsilon}{s} V_p \right)^s . \tag{7.27}$$

This is a slowly converging product and typically we have $s \approx 10^3$ to have a reasonable accuracy of 10^{-5} [Lyn74]. If we take the Fourier transform of the s-fold product, we end up with an s-fold convolution product of the Fourier transforms:

$$\mathcal{F}\left[e^{i\sigma V_p}\right]_{hk} = \left[\delta(h,k) + i\sigma V_{hk}\frac{\epsilon}{s}\right] \otimes \left[\delta(h,k) + i\sigma V_{hk}\frac{\epsilon}{s}\right] \otimes \cdots . \tag{7.28}$$

While the actual product series converges slowly, the convolution product can be computed in a reasonably short time by taking s to be a power of 2, i.e. $s = 2^r$ with typically $r = 10$–12. If we represent a single term between square brackets on the right-hand side of the above equation by Q_1, then the first convolution produces $Q_2 = Q_1 \otimes Q_1$, the second produces $Q_3 = Q_2 \otimes Q_2$ and so on until $r - 1$ convolutions have been carried out.

The coefficients V_{hk} are the Fourier coefficients of the lattice potential in the plane normal to the beam direction. The range of coefficients is again limited by $\mathbf{g}_{max}$, so that Q_1 is a bandwidth-limited function. Convolutions of bandwidth-limited functions can be computed by means of the convolution theorem, but require special care to avoid aliasing artifacts. It is not difficult to show that a doubling of the number of sampling points from N to $2N$ in combination with a scaled Nyquist limit of $\frac{2}{3}q_c = \frac{2}{3}\frac{2N}{2a} = \frac{2N}{3a}$ will completely eliminate aliasing. This means that Q_1 must be bandwidth-limited to two-thirds of the Nyquist frequency for the double-sampled function. The resulting convolution product Q_2 will have amplitude in all frequencies out to $\frac{4}{3}q_c$; application of a mask can then set all frequencies between $\frac{2}{3}$ and $\frac{4}{3}$ of q_c to zero before the next convolution is carried out. For more details we refer to Section 6.7 in [Kir98]. At the end of the r-fold convolution product, we simply Fourier transform Q_r (sampled over N points instead of $2N$) to real-space to obtain the phase grating term. If we represent the mask by $\mathcal{M}$, then a general step of the algorithm can be summarized as follows:

$$Q_{i+1} \equiv \mathcal{F}\left[\left(\mathcal{F}^{-1}[\mathcal{M}Q_i]\right)^2\right]. \tag{7.29}$$

This recipe should be carried out $r - 1$ times to obtain Q_r. The implementation of this algorithm is left as an exercise for the reader.

The computation of the phase grating term makes the implicit assumption that the atoms inside a slice are located at the center of the slice. This may not be a realistic assumption, and more accurate algorithms for the computation of the projected potential have been proposed in the literature. Van Dyck and Coene defined the so-called *potential eccentricity* [VDC84], which explicitly allows the potential inside a slice to be asymmetric with respect to the center of the slice. This leads to a slightly modified solution (7.23), and therefore a slightly slower but more accurate multi-slice/real-space computation. We refer the interested reader to [KOK87] for a detailed comparison of various multi-slice implementations.

7.4.1.3 Computation of the propagator term

The difference between the multi-slice and real-space methods lies principally in the way both methods deal with the computation of the propagator $e^{\epsilon\bar{\Delta}}$. Both methods attempt to circumvent the direct computation of this term.

- **The multi-slice method.** In the multi-slice method, the propagator is evaluated in reciprocal space, using the following relation (derived in Chapter 5):

$$e^{\epsilon\bar{\Delta}}\psi = \mathcal{F}^{-1}\left[e^{-i\pi\lambda\epsilon q^2}\mathcal{F}[\psi]\right]. \tag{7.30}$$

This expression assumes that the incident beam travels along the zone axis. For a tilted beam the following expression can be derived:

$$e^{\epsilon\bar{\Delta}}\psi = \mathcal{F}^{-1}\left[e^{-i\pi\lambda\epsilon(q^2+2\mathbf{q}\cdot\mathbf{k}_{xy})}\mathcal{F}[\psi]\right], \tag{7.31}$$

with $\mathbf{k}_{xy}$ the normal component of the incident wave vector.

The multi-slice algorithm then proceeds as follows: first, select a slice thickness ϵ and compute the phase grating (in real-space) and the propagator (in reciprocal space). We will denote these two factors by p and q, respectively. Then initialize the incident wave $\phi(0)$ in real-space by assigning a unit complex amplitude to each sampling point. Propagate this initial wave by half a slice thickness to obtain $\phi'(0)$. The wave amplitude at $z = \epsilon$ is then given by

$$\phi(\epsilon) = \mathcal{F}^{-1}[q\mathcal{F}[p\phi'(0)]]. \tag{7.32}$$

Repeated use of this single step then results in the wave amplitude at multiples of the slice thickness ϵ. The general step is given by

$$\phi_n = \mathcal{F}^{-1}\left[q\mathcal{F}\left[p_n\phi_{n-1}\right]\right], \tag{7.33}$$

where p_n is the phase grating term for the nth slice. At the end, the wave function must be propagated for half a slice again, as described in equation (7.23). As long as the slice thickness is constant, the propagator term q must be computed only twice, once for the half-slice thickness and once for the whole slice thickness. In principle, each slice could have a different projected potential V_p, which is reflected in the use of p_n. For a perfect crystal structure with a small unit cell one would typically use a single phase grating term, but more complex defect arrangements with a structure which changes along the beam direction require the use of possibly a large number of different phase grating terms.

There are many implementations of the multi-slice algorithm. The EMS suite of programs is particularly well suited for simulations that require many different phase grating terms [Sta87].† The C source code accompanying the book *Advanced Computing in Electron Microscopy* [Kir98] provides a good starting point for the reader who wishes to write specialized multi-slice code. A basic multi-slice implementation (multis.f, in Fortran-77) can be found in [SZ92];‡ on the website the reader can find an **ION**-routine ms.pro which implements the essential components of the multis.f program (see examples in Section 7.4.3).

- **The real-space method.** The real-space method [VDC84, CVD84a, CVD84b] assumes that the value of the wave function in a given point in a slice is influenced only by the values of a small neighborhood of points in the previous slice. In other words, the propagator is considered in real-space (hence the name of the method) and is written as a small convolution kernel. The explicit derivation of this kernel is somewhat complicated and we refer the reader to the original paper for all details [CVD84b]. If we represent the real-space propagator by q_{RS}, then the general step of the real-space method is given by

$$\phi_n = q_{RS} \otimes \left[p_n\phi_{n-1}\right]; \tag{7.34}$$

there are no Fourier transforms, and the propagator kernel typically is a small circular array (15–20 pixels across). The convolution product is carried out entirely in real-space, using periodic boundary conditions. The Fortran-77 source code of the real-space method can be downloaded from the website; it is available as a compressed UNIX archive realspace.tar.gz.§

† An on-line version of EMS is available at `http://cimesg1.epfl.ch/CIOL/ems.html`.

‡ The multis.f program was written by J.M. Zuo; the source code is listed on pages 311–322 in *Electron Microdiffraction* by Spence and Zuo [SZ92].

§ The author would like to thank Professor D. Van Dyck for permission to distribute the 1990 version of the real-space source code, written by W. Coene in standard Fortran-77 and modified by M. Op de Beeck.

An important advantage of the real-space method is the fact that the finite propagator allows for the patching together of many neighboring images into a larger region. This would be difficult to achieve with the other methods without the need for very large arrays. An example of the patching method can be found in [CVDVTVL85].

7.4.2 The Bloch wave approach

We know from Section 5.7 that the exit wave function can be expressed as a sum of Bloch waves, each with its own wave vector $\mathbf{k}^{(j)}$ and excitation amplitude $\alpha^{(j)}$. The individual Bloch waves for both the systematic row case and the zone axis case have already been analyzed earlier in this chapter. All that remains to be done is to combine the individual Bloch waves $C^{(j)}(\mathbf{r})$ with the proper relative weights to obtain the total wave function for each grid point. The grid of points at which the exit wave function is to be evaluated is chosen in the same way as for the multi-slice and real-space methods. Bloch waves are typically used for smaller unit cells for which the number of beams is relatively small (of the order of a few hundred beams). The zone axis Bloch wave program BWEW.f90 is available from the website; the program computes the exit wave for a series of foil thicknesses.

7.4.3 Example exit wave simulations

Figure 7.29 shows the exit wave amplitude and phase for the [100] zone axis orientation of silicon at 200 kV, computed with the BWEW.f90 program. The computation includes 373 beams (all beams for which $|\mathbf{g}| < 40$ nm^{-1}; note that this is significantly smaller than $g_{max} \approx 235$ nm^{-1}) and was carried out for 12 thicknesses (multiples of 10 nm). The advantage of the Bloch wave method is that the thickness can be substituted at the end of the computation; this means that for large numbers of beams the results of the eigenvalue computation can simply be stored on disk and used again should other thicknesses be required.

It is apparent from Fig. 7.29 that the wave function remains sharply peaked at the atom positions for the entire thickness range. We have seen earlier in this chapter that, for the exact zone axis orientation, typically only one Bloch wave is excited. For the copper example in Figs 7.21 and 7.22, the first Bloch wave is centered on the atom positions and is strongly excited. The same happens for most crystal structures in which the projected structure consists of columns of identical atoms. The lowest-order Bloch wave(s) have the highest excitation coefficients in exact zone axis orientation and the beam electrons "channel" along the atomic column. From $\gamma^{(j)}$ dispersion surfaces similar to those shown in Fig. 7.20, we find that one (or a few) Bloch waves correspond to bound states, i.e. the electrons in the beam "fall" into the potential well created by the atom column and remain there throughout the thickness of the foil.

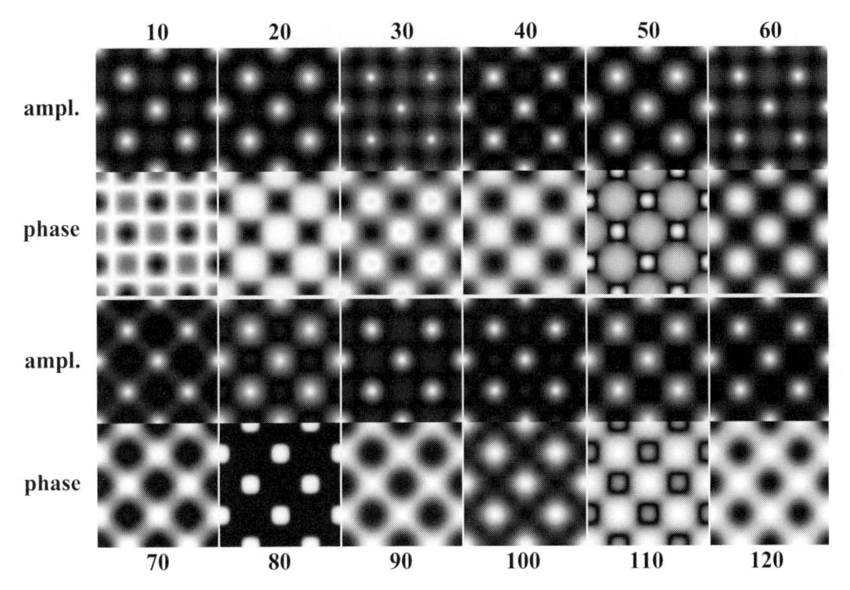

Fig. 7.29. Bloch wave exit wave simulation for the [100] zone axis orientation of silicon (373 beams, 200 kV, 256 × 256 pixels). Phase and amplitude are shown for thicknesses from 10 to 120 nm. The white spots in the amplitude at 10 nm coincide with the projected atom positions for a single unit cell.

For an isolated column of atoms the bound states are similar to the radially symmetrical states of the hydrogen atom, i.e. similar to s states. We can use the ms.pro **ION**-program on the website to study this channeling behavior. The ms.pro program takes a limited number of input parameters. The crystal structure is tetragonal, with $a = 0.4$ nm, and c is variable. Two identical atoms can be placed anywhere in the unit cell. The cell is sampled at 64 × 64 pixels. The user can select the slice thickness ϵ, the number of slices, the acceleration voltage E, the value of g_{max}, and the c lattice parameter. By varying these parameters, the reader can obtain a good understanding of how channeling works. An example simulation is shown in Fig. 7.30. The unit cell is shown in Fig. 7.30(a) and consists of two W atoms at 0.142 nm spacing. The unit cell thickness is $c = 0.2$ nm and a slice thickness of $\epsilon = 0.05$ nm is used (200 kV). The projected potential and the propagator (origin shifted to lower left-hand corner) are shown in Figs 7.30(b) and (c), respectively.

Figure 7.30(d) shows the amplitude (top row) and $\sin(\phi)$ (bottom row) for 10 thicknesses. There are three different intensity features, indicated by small arrows in the 40 nm amplitude image, which occur for all images, though not always all at the same time. The smallest feature is a bright spot at the atom location. The second feature is ring-shaped and surrounds the atom position. The third feature is more diffuse and surrounds each atom spot at a larger distance; this feature is

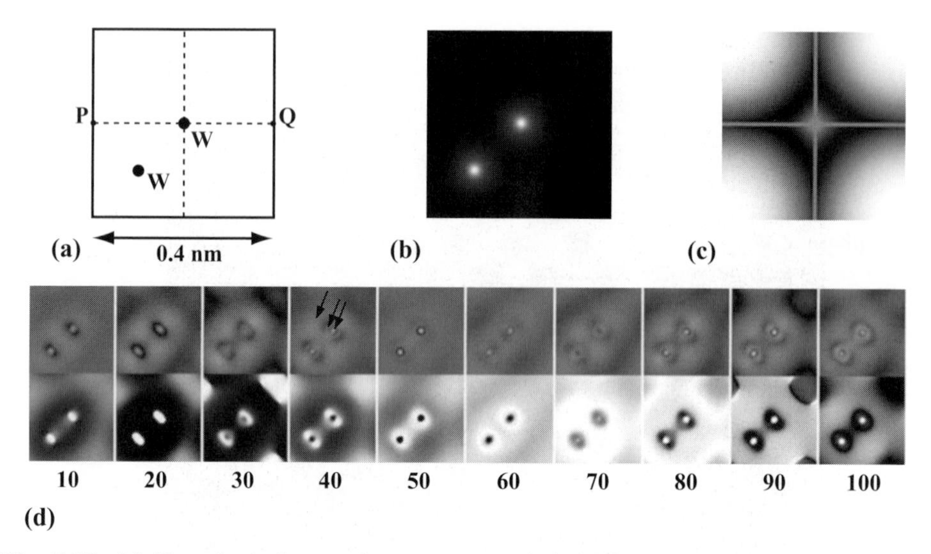

Fig. 7.30. (a) Hypothetical crystal structure consisting of two W atoms; (b) projected potential for one slice of thickness 0.05 nm; (c) real part of the propagator (200 kV); (d) thickness series of amplitude (top row) and sin (phase) (bottom row).

also affected by the neighboring atom. The central feature corresponds to a tightly bound 1*s* state, the ring is a 2*s* state and the more diffuse feature is a combination of the 3*s* state and other Bloch waves. At a particular depth in the crystal each of the Bloch waves has a different excitation amplitude so that the intensity of the various features oscillates with depth in the crystal, similar to the intensity oscillations in the two-beam bright field–dark field case.

The probability for a beam electron to be in a bound 1*s* state oscillates with depth in the crystal. The oscillation period depends on the electron energy, the type of atoms in the column, and the distance between the atoms along the column. Figure 7.31 shows this oscillatory behavior for 12 selected elements. The computations were carried out with a modified ms.pro program for the following simulation parameters: one atom in the center of unit cell, 200 kV, $g_{max} = 80\,\text{nm}^{-1}$, slice thickness 0.05 nm, 2000 slices, tetragonal axis $c = 0.2$ nm, and no absorption. The thickness increases from left to right, and each vertical intensity profile corresponds to the line PQ in Fig. 7.30(a). The bound 1*s* state is centered on the atom itself and the intensity oscillates with a periodicity which generally increases with increasing atomic number. With increasing atomic number the 2*s* state becomes occupied and oscillates with a different period. For the heaviest elements the 3*s* bound state is occupied as well. The oscillation periods are not simply related to the atomic number, but instead vary in a way similar to the variations in electronic structure across the periodic table of the elements.

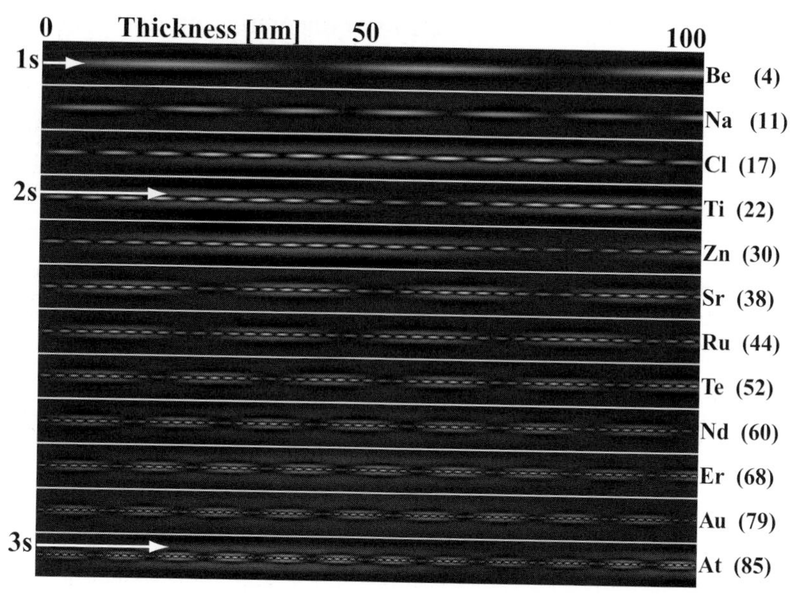

Fig. 7.31. Radial intensity profile as a function of foil thickness (indicated at the top) for 12 selected elements (multi-slice computation, slice thickness 0.1 nm, 200 kV, 3205 beams or $g_{max} = 80\,\text{nm}^{-1}$, Doyle–Turner scattering factors, no absorption). The oscillatory character of the bound Bloch states of $1s$, $2s$, and $3s$ type is clearly visible.

What we learn from this type of simulation is that the propagation of a plane incident wave along a zone axis orientation gives rise to bound states at each atom column. If the columns are sufficiently far away from each other, then the oscillatory behavior of the symmetrical lowest-order Bloch waves dominates the scattering behavior and the wave function will be peaked at the column positions. If columns are close together, as they are in many projected crystal structures, then there will be a finite probability for electrons from one column to scatter into a neighboring column, which means that the non-symmetrical Bloch waves will be excited (see Fig. 7.30d). Which particular waves are excited will depend on the configuration of columns. For many crystal structures we can therefore describe the zone axis scattering process as essentially a collection of bound s-type states; the interaction between the bound states could then be viewed as a perturbation. It is possible to work out analytical solutions for scattering from an isolated column; this lies at the basis of the *atomic column approximation* described in [VDOdB96]. An extensive study of the channeling behavior in terms of Bloch waves can be found in [BLS78].

The reader is encouraged to experiment with the ms.pro routine, in particular to study what happens when the illumination is tilted slightly away from the zone axis orientation. The routine returns an animated GIF file for up to 400 slices.

7.5 Electron exit wave for a magnetic thin foil

7.5.1 The Aharonov–Bohm phase shift

In 1959 Aharonov and Bohm published a revolutionary paper on the importance of the phase of the electron wave function in the presence of electrostatic and magnetic potentials [AB59]. While the existence of such phase shifts was known to others at that time, their paper provided the first detailed discussion of the effect and drew a lot of attention from the physics community. They found that even in regions of space where all the fields vanish, the wave function of a charged particle could still experience changes due to the corresponding electromagnetic potentials. It has taken several decades for the scientific community to accept the notion that the wave function of a particle can be affected by something other than a force, and an extensive review of the quantum effects of electromagnetic fluxes can be found in [OP85]. There is now ample evidence that the Aharonov–Bohm (or A–B) effect is indeed real, and a direct experimental proof based on electron holography was provided by A. Tonomura and coworkers [TOM$^+$86] in 1986. The existence of the A–B phase shift is related to the fact that the electric and magnetic fields do not appear directly in the Schrödinger equation, but only in the form of the potentials; while the electromagnetic potentials are not observables in the classical theory, they do become the fundamental quantities in the quantum mechanical framework.

The A–B phase shift imparted on an electron wave with relativistic wavelength λ by the presence of electromagnetic potentials V and $\mathbf{A}$ is given by [AB59]:

$$\phi(\mathbf{r}_\perp) = \frac{\pi}{\lambda E_t}\int_L V(\mathbf{r}_\perp, z)\,\mathrm{d}z - \frac{e}{\hbar}\int_L \mathbf{A}(\mathbf{r}_\perp, z)\cdot \mathrm{d}\mathbf{r} \qquad (7.35)$$

where E_t is the total energy of the beam electron and the integrals are carried out along a straight line L parallel to the incident beam direction (i.e. crossing the plane of the sample at the point $(x, y, 0)$). It is left as an exercise for the reader to show that the prefactor of the electrostatic component of the phase shift is equal to the interaction constant σ introduced in Chapter 2. A 100 nm foil with a mean inner potential V_0 of 30 V will hence give rise to an electrostatic phase shift of 18.36 radians at 400 kV (using Table 2.2 on page 93). The prefactor of the magnetic component of the A–B phase shift equals $e/\hbar = 0.001\,519\,27\ \mathrm{T}^{-1}\,\mathrm{nm}^{-2}$ and is independent of the electron energy.

Since it is rather difficult to measure absolute phases, we usually work in terms of the phase difference $\Delta\phi(\mathbf{r}_\perp)$ between a given electron trajectory crossing the sample at the location $\mathbf{r}_\perp$ and a reference trajectory which is conveniently chosen to coincide with the optical axis of the microscope, as schematically indicated in Fig. 7.32. In the absence of an electrostatic potential, the magnetic component of

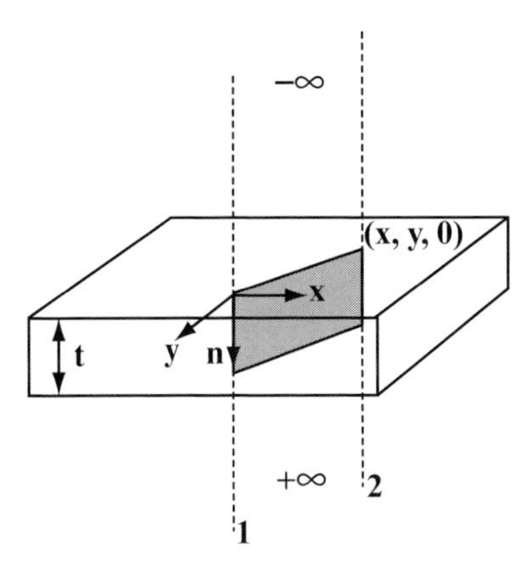

Fig. 7.32. Schematic of the two trajectories involved in the formulation of the Aharonov–Bohm phase shift.

the phase difference $\phi_2 - \phi_1$ between the trajectories 1 and 2 is equal to

$$\Delta\phi_m(\mathbf{r}_\perp) = \frac{e}{\hbar}\left(\int_{-\infty}^{+\infty} \mathbf{A}\cdot \mathrm{d}\mathbf{r}_2 - \int_{-\infty}^{+\infty} \mathbf{A}\cdot \mathrm{d}\mathbf{r}_1\right) = \frac{e}{\hbar}\oint \mathbf{A}\cdot \mathrm{d}\mathbf{r}.$$

This last line integral can be converted into a double integral using *Stokes' theorem* (e.g. [Jac75]):

$$\Delta\phi_m(\mathbf{r}_\perp) = \frac{e}{\hbar}\iint \nabla\times\mathbf{A}\cdot \mathrm{d}\mathbf{S}, \tag{7.36}$$

with $\mathbf{S}$ a unit vector normal to the integration surface. This integral is equal to the *magnetic flux* Φ_m enclosed between the two trajectories:

$$\Delta\phi_m(\mathbf{r}_\perp) = \frac{e}{\hbar}\Phi_m(\mathbf{r}_\perp) = \pi\frac{\Phi_m(\mathbf{r}_\perp)}{\Phi_0}, \tag{7.37}$$

where $\Phi_0 = h/2e$ is the flux quantum. The phase shift is therefore proportional to the enclosed flux, measured in units of the flux quantum. If we take trajectory 1 to be the reference phase (i.e. "zero phase"), then the phase at any other point $\mathbf{r}_\perp$ can be calculated if either the vector potential $\mathbf{A}$ or the magnetic induction $\mathbf{B}$ are known. In Section 7.5.3 we will discuss an example of the use of this relation.

Consider a thin foil of thickness t with parallel top and bottom surfaces. The foil contains three magnetic domains, as illustrated in Fig. 7.33, separated from each other by 180° domain walls. The origin of the reference frame is taken to be in the center of the foil, with a right-handed Cartesian reference frame oriented as shown in the figure. The z-axis coincides with the optical axis of the electron microscope, with positive z in the direction of the beam. The magnetic induction $B_\perp$ in the domains is uniform and parallel to the x-direction, so that the Lorentz deflection will be in the $\pm y$-direction. In the absence of fringe fields we can calculate the phase

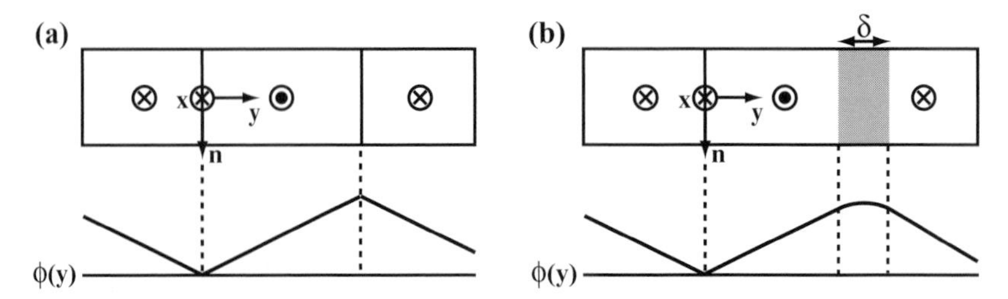

Fig. 7.33. Schematic illustration of the phase computation for a set of three alternating magnetic domains (magnetization normal to the plane of the drawing). The domain wall profile is taken to be discontinuous (black line) and with a finite domain wall width δ (gray line).

shift of the electron wave function along the line 0–y from the A–B equation (7.36):

$$\phi(y) = \frac{e}{\hbar}\int_{-t/2}^{+t/2}\int_0^y B_\perp \,\mathrm{d}y\,\mathrm{d}z = \frac{e}{\hbar}B_\perp t y. \tag{7.38}$$

If the domain wall is infinitely narrow, then the slope of the phase changes discontinuously across the domain wall, as indicated in Fig. 7.33. For a more realistic domain wall profile, the slope change is continuous. The range over which the slope changes corresponds to the domain wall width. The phase function is a linear function inside magnetic domains with a uniform magnetization and a constant foil thickness. We leave it to the reader to show that the phase becomes a quadratic function of y for a linear wedge-shaped foil. Small local thickness variations will cause non-linearities in the phase as well, but generally do not change the sign of the slope. The same is true for magnetization ripple, which causes fluctuations in the orientation of the gradient vector of the phase function.

The Lorentz deflection angle, equation (4.8), can be converted into a phase gradient:

$$\nabla\phi = \frac{2\pi}{\lambda}\theta_L \mathbf{e}, \tag{7.39}$$

where $\mathbf{e}$ is a suitably oriented unit vector. For a uniformly magnetized foil of constant thickness t this results in a phase gradient of

$$\nabla\phi = \frac{e}{\hbar}B_\perp t\mathbf{e} = -\frac{e}{\hbar}(\mathbf{B}\times\mathbf{n})t. \tag{7.40}$$

This expression is independent of the electron energy. Since the electrostatic component of the phase is proportional to the interaction constant σ, and σ decreases with increasing microscope acceleration voltage E, it follows that magnetic contributions to the phase shift are larger relative to the electrostatic contribution for higher-voltage microscopes. Since the Lorentz deflection angle is proportional to the electron wavelength, it is clear that intermediate voltages in the range 200–400 kV provide the best compromise between a strong relative magnetic contribution and reasonably large Lorentz deflection angles.

7.5.2 Direct observation of quantum mechanical effects

The most readily observable quantum effect in Lorentz microscopy is the appearance of interference fringes at convergent wall images in Fresnel mode (see Fig. 4.29). We will show that there is one fringe per fluxon $h/2e$ in the image, independent of defocus, magnification, and electron energy [Woh71].

Consider again a film of thickness t, with an infinitely narrow domain wall centered at $y = 0$ (Fig. 7.34, based on Fig. 5 in [Woh71]). A point source S at a distance s from the sample illuminates the area containing the domain wall. In Fresnel mode the image plane is located at a distance Δf below the sample. The Lorentz deflection of electrons passing through the domain on the left is $-\theta_L$ and at the image plane the electrons appear to originate from an effective source S', located at a distance $s\theta_L$ from the actual source (measured at the source plane). Similarly, electrons deflected by the angle $+\theta_L$ on the right-hand side appear to originate from the point source S”. The two sources will give rise to a region containing interference fringes with width δ_i. The angle subtended by the two sources at the

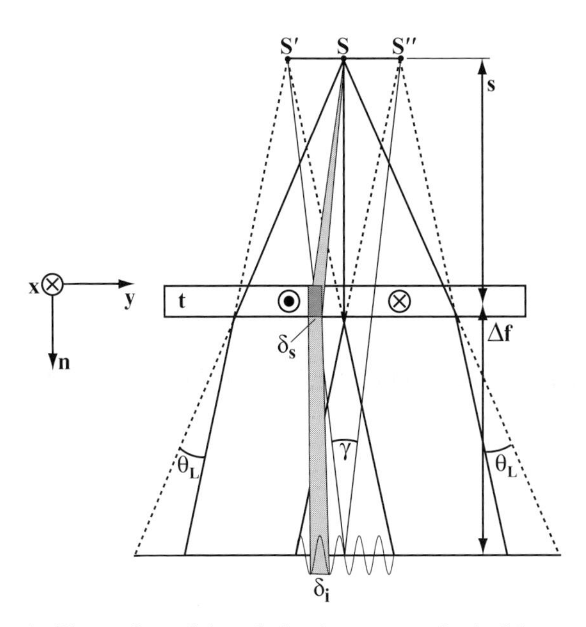

Fig. 7.34. Schematic illustration of the relation between a single fringe period in a Fresnel image (defocus Δf), and the corresponding region in the sample (gray rectangle). The flux through this region is equal to one flux quantum (based on Fig. 5 in [Woh71]).

image plane is equal to $\gamma = \frac{2s\theta_L}{s+\Delta f}$. The width of a single fringe in the image plane is then given by

$$\delta_i = \frac{\lambda}{\gamma};$$
$$= \frac{h(s+\Delta f)}{2etsB_\perp},$$

where we have made use of equation (4.8). Next, we refer this fringe width back to the width δ_s in the sample plane:

$$\delta_s = \delta_i \frac{s}{s+\Delta f};$$
$$= \frac{h}{2etB_\perp};$$
$$= \frac{\Phi_0}{tB_\perp}.$$

We find that $\delta_s t B_\perp = \Phi_0$, or, that the flux contained in a region in the sample corresponding to a single projected fringe width in the image is equal to one flux quantum. The fringe width depends inversely on both film thickness and magnetic induction.

7.5.3 Numerical computation of the magnetic phase shift

Magnetization configurations in thin foils rarely have an analytical representation of either the vector potential or the magnetic induction. For such cases the magnetic phase shift imparted on the beam electron may be calculated by assuming that the magnetization has in-plane periodicity. It is usually not too difficult to create a periodic continuation of any given magnetization pattern; one can either pad the region of interest with a zero magnetization edge (making sure that the magnetization goes to zero smoothly as the edge is approached), or, if the spatial derivatives of the magnetization vanish at the edge, then one can readily use the pattern as a single unit cell. We will hence assume in this section that a two-dimensional (2D) magnetization unit cell can be found, and to each pixel (i, j) in the cell we assign a three-dimensional (3D) magnetization vector $\mathbf{M}(i, j)$. The magnetization is assumed to be constant along the beam direction and we ignore the demagnetization or fringe field outside the foil. An example of such a magnetization configuration is shown in Fig. 7.35(a). This configuration commonly occurs in materials which have the $\langle 111 \rangle$ directions as the so-called "easy axes", i.e. low-energy directions for the magnetization vector. The incident beam is normal to the plane of the drawing along the [110] direction.

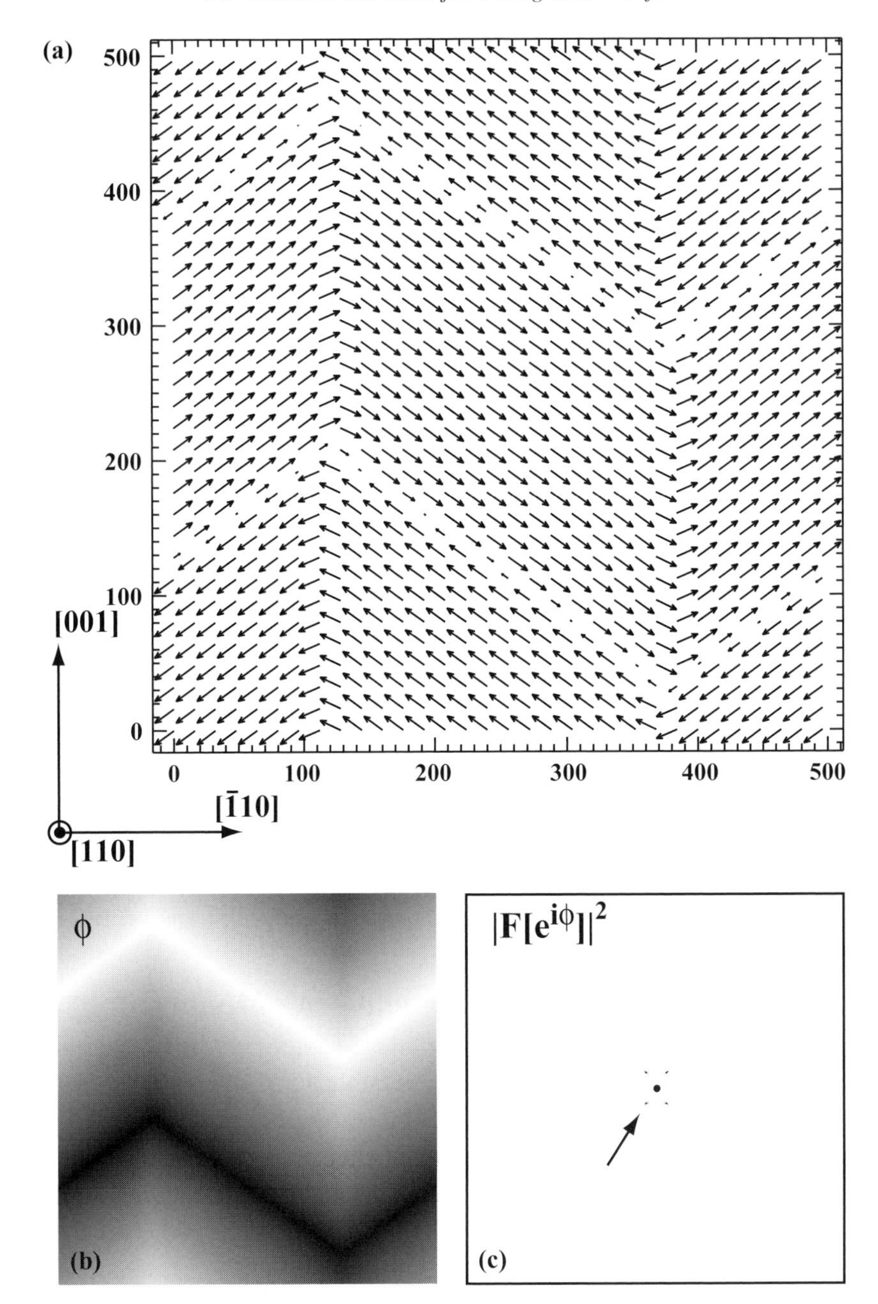

Fig. 7.35. (a) Periodic 2D magnetization configuration with four different domain orientations; magnetization vectors are along ⟨111⟩-type directions. 71° (vertical) and 180° (zig-zag) domain walls with widths $\delta_{71} = 24$ nm and $\delta_{180} = 48$ nm are present. (b) The calculated Aharonov–Bohm phase shift ϕ for the configuration in (a); (c) is the power spectrum of a wave $e^{i\phi}$; the central spot is split into four components, one for each magnetic domain.

The angle between two $\langle 111 \rangle$ directions is either 70.53°, 109.47°, or 180°. Therefore the possible magnetic domain walls are known as 71°, 109° and 180° domain walls. The first set of domain walls is normal to the $[\bar{1}10]$ direction and is of the 71° type, i.e. the magnetization vector rotates by 71° upon crossing the domain wall. The magnetization vectors are at an angle of 35.3° with respect to the plane of the drawing. The second set of domain walls is parallel to the magnetization vectors and lie along $\{11\bar{2}\}$-type planes. These walls are 180° walls. Each wall has an associated domain wall width δ, typically a few tens to a few hundreds of nanometers. The magnetization changes direction smoothly across a domain wall. The domain wall widths in Fig. 7.35(a) are $\delta_{71} = 24$ nm and $\delta_{180} = 48$ nm. This magnetization configuration is periodic in both x and y. The configuration in Fig. 7.35(a) is defined for an array of $P \times Q = 512 \times 512$ pixels, and the pixel size is equal to $D = 1$ nm. The routine PhaseExample in the lorentz.f90 module can be used to create the magnetization vectors corresponding to Fig. 7.35(a).

It has been shown by Mansuripur [Man91] that for such a periodic configuration the magnetic A–B phase shift may be written as a 2D Fourier series. We refer the reader to the original paper for the lengthy proof of this statement, and we will continue by stating the explicit result. We represent the discrete Fourier components of the magnetization over a unit cell of $P \times Q$ pixels with pixel spacing D by $\mathbf{M}_{mn}$, with $m = 1, \ldots, P$ and $n = 1, \ldots, Q$; in other words, we have

$$\mathbf{M}_{mn} = \sum_{i}^{P} \sum_{j}^{Q} \mathbf{M}(i, j) \exp\left[-2\pi \mathrm{i}\left(\frac{mi}{P} + \frac{nj}{Q}\right)\right]. \tag{7.41}$$

The A–B phase shift is then given by the discrete 2D Fourier transform:

$$\phi(\mathbf{r}) = \frac{2e}{\hbar} \sum_{m=0}^{P} \sideset{}{'}\sum_{n=0}^{Q} \mathrm{i}\frac{t}{|\mathbf{q}|} G_{\mathbf{p}}(t|\mathbf{q}|)(\hat{\mathbf{q}} \times \mathbf{e}_z) \cdot [\mathbf{p} \times (\mathbf{p} \times \mathbf{M}_{mn})]\, e^{2\pi \mathrm{i}\mathbf{r}\cdot\mathbf{q}}, \tag{7.42}$$

where the prime on the summation indicates that the term $(m, n) = (0, 0)$ does not contribute to the summation, $\mathbf{q} = \frac{m}{P}\mathbf{e}_x^* + \frac{n}{Q}\mathbf{e}_y^*$ is the frequency vector, $\mathbf{e}_x^*$ and $\mathbf{e}_y^*$ are the reciprocal unit vectors, $\mathbf{p}$ is the beam direction expressed in the orthonormal reference frame, t is the sample thickness, a hat (ˆ) indicates a unit vector, and the function $G_{\mathbf{p}}(t|\mathbf{q}|)$ is given by

$$G_{\mathbf{p}}(t|\mathbf{q}|) = \frac{1}{(\mathbf{p} \cdot \hat{\mathbf{q}})^2 + p_z^2} \operatorname{sinc}\left(\pi t|\mathbf{q}|\frac{\mathbf{p} \cdot \hat{\mathbf{q}}}{p_z}\right), \tag{7.43}$$

where $\operatorname{sinc}(x) \equiv \sin(x)/x$. For normal beam incidence ($\mathbf{p} \parallel \mathbf{e}_z$) the function $G_{\mathbf{p}}(t|\mathbf{q}|)$ takes on the value 1. Numerical implementation of this equation is straightforward and the corresponding source code can be found on the website as the PhaseMap routine in the lorentz.f90 module.

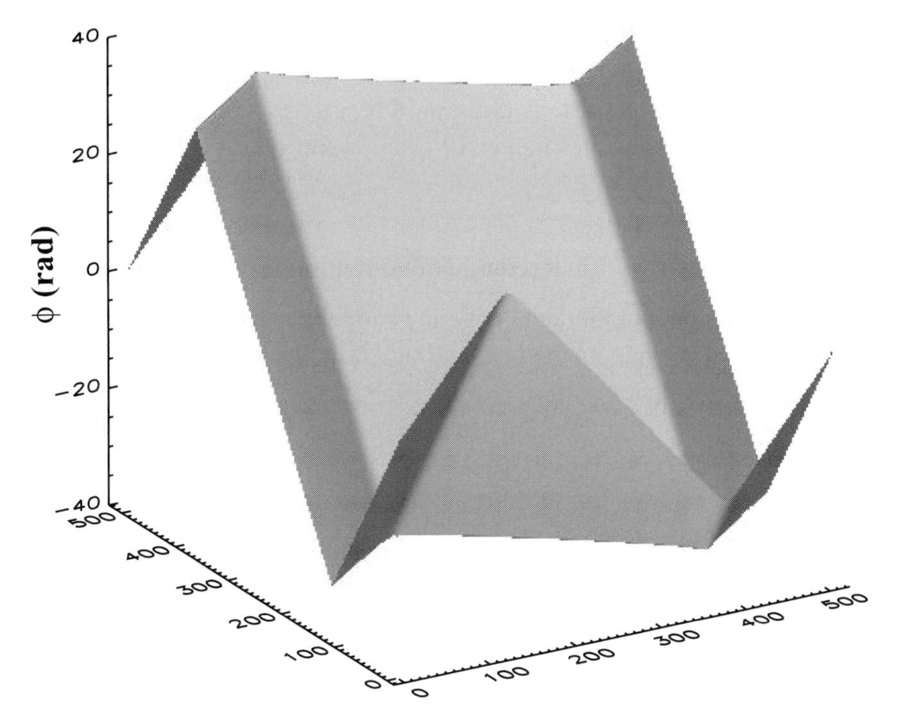

Fig. 7.36. Shaded surface representation of the calculated phase shift of Fig. 7.35(b).

The magnetic phase shift ϕ for the magnetization configuration of Fig. 7.35(a) (with $t = 1$ and $|\mathbf{M}(i, j)| = 1$ for all i, j) is shown in Fig. 7.35(b) as a grayscale image. The minimum and maximum phase shifts are ± 0.325 rad. This phase shift must be multiplied by the sample thickness (t, in nm) and the saturation magnetization (M_0, in tesla). An alternative representation as a shaded surface (Fig. 7.36, $M_0 t = 100$) clearly shows that the magnetic phase shift inside a domain is a linear (planar) function of position, whereas the slope of the phase changes upon crossing a magnetic domain wall.

A plane wave which travels through a thin foil with the magnetization configuration of Fig. 7.35(a) will experience a position-dependent phase shift such that the exit wave has the phase ϕ. In the back focal plane of the objective lens we can observe the modulus-squared of the Fourier transform of the wave function $|\mathcal{F}[e^{iM_0 t\phi}]|^2$. For $M_0 t = 100$ nm T the resulting diffraction pattern is shown in Fig. 7.35(c). The central beam, represented by a small circle, is split into four separate beams, one for each magnetic domain.

Note that the magnetic phase shift does not affect the amplitude of the wave. Since the phase shift can be large, several radians, a magnetic thin foil is known as a so-called *strong phase object*. If there are thickness variations across the foil, then the phase computation becomes much more involved, since we can no longer simply multiply by the thickness. In addition, electrostatic contributions to the phase shift will become important for the non-uniform thickness case. In the case of a foil with uniform thickness the electrostatic phase shift is simply a constant. We will return to the magnetic phase shift in Chapter 10 when we introduce image formation for phase objects.

7.6 Recommended reading

There is a large amount of literature covering various aspects of multi-beam theory and simulations. The slice methods are all based on the physical-optics approach first introduced by Cowley and Moodie in 1957 [CM57]. The first use of the FFT for multi-beam electron diffraction simulations was described by Ishizuka and Uyeda in 1977 [IU77]. Various elastic scattering formalisms have been reviewed by Goodman and Moodie [GM74]. Much of the initial literature (1970–1975) on these simulation methods has appeared in journals such as *Acta Crystallographica A* and *Ultramicrocopy*. Several simulation packages have then emerged, among them SHRLI [OBI74], EMS [Sta87], NCEMS [OK89], MacTempas [Kil87], and the real-space package [VDC84, CVD84a, CVD84b]. While these packages are sufficient for routine applications, the reader may find it necessary to develop new algorithms and code for more specialized situations. The interested reader should also consult [KO89] for more information on simulation algorithms and methods.

Source code for many applications can be found on the website of the *Microscopy Society of America* at the URL `http://www.msa.microscopy.com`. In many cases one can make good use of routines and algorithms written by others (no need to re-invent the wheel); a good starting point for general purpose numerical procedures is the collection at `http://www.netlib.org`, maintained by the Oak Ridge National Laboratory. Combined with the *Numerical Recipes* book [PFTV89] these resources should be sufficient to tackle most computational problems.

Exercises

7.1 Consider a bright field image containing a dark bend contour of width W_1 (this width could be measured by locating two points on either side of the center of the contour that have equal intensity). If the acceleration voltage of the microscope is increased from V_1 to V_2, what happens to the width of this contour (keeping all other parameters constant)?

7.2 Show that $\pi/\lambda E_t$, with E_t the total energy of the electron, is equal to the interaction constant σ.

7.3 Use the SRBW.f90 program as a starting point to write a new Bloch wave program that will compute images similar to those shown in Fig. 7.6.

7.4 The SRCBED.f90 program uses the scattering matrix approach for its dynamical intensity computations. Rewrite this program using the Bloch wave approach. (You will most likely be able to use significant portions of the SRBW.f90 program.)

7.5 Derive equations (7.9) and (7.10) for the two-beam case.

7.6 Implement equation (7.29) for the phase grating computation.

7.7 Show that for a wedge-shaped sample with constant slope the magnetic phase shift depends on the square of the coordinate variable y (see equation 7.38).

8

Defects in crystals

8.1 Introduction

In this chapter, we will continue the discussion of the dynamical theory of electron diffraction. We will cover several of the most important defects: dislocations, stacking faults, and other planar boundaries, and we will discuss the diffraction contrast caused by various types of interfaces, e.g. domain walls and twins. Finally, some aspects of the weak beam technique will be covered. In all cases, we will describe how image simulations can be carried out.

The survey of defects in this chapter is by necessity limited to only a few defects. The reader will find that the methods are sufficiently general, so that they may be applied to other defects with only minimal modifications to the experimental procedures or numerical algorithms.

8.2 Crystal defects and displacement fields

One of the most important applications of conventional TEM is the study of defects in crystalline materials. As we have seen in Chapter 1, a crystal is characterized by a periodic 3D lattice, decorated with a motif or asymmetric unit consisting of one or more atoms. From space group theory, we know that the symmetry elements of a crystal structure are either rotational,[†] translational, or a combination of both. Since a space group is a group of infinite order, and since *real* crystals are always of finite size, we find that a crystal cannot completely satisfy all elements of a particular space group. In other words, space group symmetry must be *broken* at one or more points in space. The most obvious case of broken symmetry is found at the bounding surfaces of the crystal and we have already described the importance of this type of symmetry breaking in Sections 2.6.3 (page 119) and 5.8 (page 328). The surface truncates the infinite crystal into a finite size and shape which can be

[†] This includes both proper and improper rotations.

represented by a *shape function* $D(\mathbf{r})$, as described in Section 2.6.3. The existence of *relrods* is a direct consequence of this type of symmetry breaking.

In addition to the free surfaces of a crystal, there are numerous other ways in which space group symmetry may be broken. The broken symmetry may involve only translational symmetry elements, as will be the case for the defects discussed in this chapter, or it may involve rotational symmetry elements (a much rarer situation in crystalline solids). Symmetry may be broken on a very local scale, such as a missing atom (vacancy), or on a much more extended scale, as is the case for *planar defects*.

When the translational symmetry of a crystal is broken, this generally means that atoms are not found at the positions predicted by the Bravais lattice translation vectors. The atoms are displaced from those positions by a vector $\mathbf{R}$, which is in general a function of position $\mathbf{r}$. The vector field $\mathbf{R}(\mathbf{r})$ is known as the *displacement field*. It relates the displaced position $\mathbf{r}'$ of the atom to the ideal (or reference) position $\mathbf{r}$ by

$$\mathbf{r}' = \mathbf{r} + \mathbf{R}(\mathbf{r}). \tag{8.1}$$

From a purely mathematical point of view, the vector field $\mathbf{R}$ may be continuous (and differentiable) or discontinuous. We will discuss examples of both in this chapter.

Using the results of Section 2.5 (page 103) on Fourier transforms, it is easy to see that a translation of a function in crystal space is equivalent to a phase shift in reciprocal space. Mathematically, this is expressed as

$$f(\mathbf{r} - \mathbf{a}) = \mathcal{F}^{-1}\left[\mathrm{e}^{-2\pi i \mathbf{a}\cdot\mathbf{k}} f(\mathbf{k})\right].$$

In the case of a defect in a crystal, the translation vector $\mathbf{a}$ is replaced by the position-dependent displacement field $\mathbf{R}(\mathbf{r})$. The electrostatic lattice potential, $V(\mathbf{r})$, takes the place of the function f, and, hence, we expect to find that the Fourier coefficients $V_{\mathbf{g}}$ of the electrostatic lattice potential experience a phase shift near a translational lattice defect. Let us cast this intuitive observation into more precise mathematical terms.

Assume that the perfect crystal potential is given by $V(\mathbf{r})$. A defect with displacement field $\mathbf{R}(\mathbf{r})$ deforms the crystal and the new potential *at the deformed position*, $V'(\mathbf{r} + \mathbf{R}(\mathbf{r}))$, is to a good approximation equivalent to the original potential at the undistorted location. In other words, if displacements are small, then the local environment of each atom remains approximately the same, or

$$V'(\mathbf{r} + \mathbf{R}(\mathbf{r})) = V(\mathbf{r}). \tag{8.2}$$

This is equivalent to the statement

$$\begin{aligned} V'(\mathbf{r}) &= V(\mathbf{r}-\mathbf{R}),\\ &= V_0 + \sum_{\mathbf{g}} V_{\mathbf{g}}\,\mathrm{e}^{2\pi i\mathbf{g}\cdot(\mathbf{r}-\mathbf{R})},\\ &= V_0 + \sum_{\mathbf{g}} V_{\mathbf{g}}\mathrm{e}^{-2\pi i\mathbf{g}\cdot\mathbf{R}}\,\mathrm{e}^{2\pi i\mathbf{g}\cdot\mathbf{r}}. \end{aligned} \tag{8.3}$$

As anticipated, we find that the defect modifies the Fourier coefficients of the electrostatic lattice potential by a phase factor

$$V_{\mathbf{g}} \rightarrow V_{\mathbf{g}}\,\mathrm{e}^{-i\alpha_{\mathbf{g}}(\mathbf{r})} \quad \text{with} \quad \alpha_{\mathbf{g}}(\mathbf{r}) \equiv 2\pi\mathbf{g}\cdot\mathbf{R}(\mathbf{r}). \tag{8.4}$$

It is straightforward to incorporate this additional phase factor in the fundamental equations of electron diffraction: it is left as an exercise for the reader to show that the modified Darwin–Howie–Whelan equations (5.18) on page 310 are given by

$$\boxed{\frac{\mathrm{d}\psi_{\mathbf{g}}}{\mathrm{d}z} - 2\pi i s_{\mathbf{g}}\psi_{\mathbf{g}} = i\pi\sum_{\mathbf{g}'}\frac{\mathrm{e}^{i\theta_{\mathbf{g}-\mathbf{g}'}}}{q_{\mathbf{g}-\mathbf{g}'}}\mathrm{e}^{-i\alpha_{\mathbf{g}-\mathbf{g}'}(\mathbf{r})}\psi_{\mathbf{g}'}.} \tag{8.5}$$

If the displacement field is continuous and differentiable, then it is convenient to remove the $\mathbf{r}$ dependence from the phase factor on the right-hand side of equations (8.5). This can be achieved by the substitution:

$$\psi_{\mathbf{g}} = \phi_{\mathbf{g}}\,\mathrm{e}^{-i\alpha_{\mathbf{g}}(\mathbf{r})}.$$

This substitution does not change the intensity, since the modulus squared of a pure phase factor equals 1. Using

$$\alpha_{\mathbf{g}-\mathbf{g}'}(\mathbf{r}) = \alpha_{\mathbf{g}}(\mathbf{r}) - \alpha_{\mathbf{g}'}(\mathbf{r}),$$

this leads, after some simple mathematics, to the modified fundamental equations of electron diffraction for a crystal containing a defect with a *differentiable* displacement field $\mathbf{R}(\mathbf{r})$:

$$\boxed{\frac{\mathrm{d}\phi_{\mathbf{g}}}{\mathrm{d}z} - 2\pi i\left(s_{\mathbf{g}} + \mathbf{g}\cdot\frac{\mathrm{d}\mathbf{R}}{\mathrm{d}z}\right)\phi_{\mathbf{g}} = i\pi\sum_{\mathbf{g}'}\frac{\mathrm{e}^{i\theta_{\mathbf{g}-\mathbf{g}'}}}{q_{\mathbf{g}-\mathbf{g}'}}\phi_{\mathbf{g}'}.} \tag{8.6}$$

Comparing this equation with that for a perfect crystal (5.18), we notice only one difference: the excitation error $s_{\mathbf{g}}$ has been replaced by an *effective, position-dependent excitation error*:

$$s_{\mathbf{g}}^{eff}(\mathbf{r}) = s_{\mathbf{g}} + \mathbf{g}\cdot\frac{\mathrm{d}\mathbf{R}(\mathbf{r})}{\mathrm{d}z} = s_{\mathbf{g}} + \frac{\mathrm{d}[\mathbf{g}\cdot\mathbf{R}(\mathbf{r})]}{\mathrm{d}z}, \tag{8.7}$$

where the derivative is taken parallel to the incident electron beam. Equation (8.7) deserves some careful analysis.

- If the displacement field is constant in the z-direction, then the derivative vanishes and *the defect is invisible*. In other words, what we can observe in the TEM is not the actual displacement field but *the variation of the displacement field with depth in the crystal.*
- If $\mathbf{g} \cdot \mathbf{R}(\mathbf{r})$ vanishes, then the effective excitation error $s_{\mathbf{g}}^{eff}(\mathbf{r})$ is equal to the excitation error for the perfect crystal, $s_{\mathbf{g}}$, and therefore *the defect will be invisible,* since the equations reduce to those for the perfect crystal. When $\mathbf{g} \cdot \mathbf{R}$ vanishes, this means that the displacements are confined to the planes represented by $\mathbf{g}$ (i.e. *in-plane* displacements).
- The excitation error $s_{\mathbf{g}}$ describes the orientation of the planes $\mathbf{g}$ with respect to the Ewald sphere. A local bending of the lattice planes due to the defect is then described by the second term in equation (8.7). Conversely, we can say that the presence of a defect changes the *shape* of the reciprocal lattice point (or relrod).

In summary, a (translational) lattice defect described by a displacement field $\mathbf{R}(\mathbf{r})$ will only be visible for a particular two-beam condition if $\mathbf{g} \cdot \mathbf{R}$ varies along the beam direction. This is known as the *visibility criterion.*

The visibility criterion lies at the basis of defect identification and suggests a simple experimental procedure: look for two-beam conditions *for which the defect contrast vanishes*. For these $\mathbf{g}$ vectors, the dot product $\mathbf{g} \cdot \mathbf{R}$ vanishes or is constant, which provides a number of linear equations from which the displacement vector can be determined. This is especially useful when the displacement field is only weakly dependent upon position, or when the field can be defined in terms of a single vector, e.g. the Burgers vector of a dislocation, or the displacement vector of a planar fault.

We have seen in Chapter 6 that the strongest bright field image contrast occurs for excitation errors close to zero. This remains true for the effective excitation error, $s_{\mathbf{g}}^{eff}$; the strongest image contrast for defects will also occur in those regions of the foil where the effective excitation error is close to zero. If we orient a foil such that a bend contour is visible in the bright field image, then we know from Chapters 6 and 7 that the edges of the bend contour, where the contrast varies rapidly, correspond to $s_{\pm\mathbf{g}} \approx 0$ (Fig. 8.1a). If we move away from the bend contour to a region with positive $s_{\mathbf{g}}$, and a defect is present in this region, then the local, effective excitation error will become small wherever the second term in equation (8.7) becomes comparable in magnitude to $s_{\mathbf{g}}$, but opposite in sign, as illustrated in Fig. 8.1(a). An experimental dark field image for the $\bar{2}0.1$ bend contour in Ti is shown in Figs 8.1(b) and (c). This foil has a rather high defect density, and many dislocations can be observed near the bend contour. The dislocations close to the bend contour have broad contrast (black arrows), whereas dislocations slightly removed from the contour (white arrows) are visible as much narrower white lines.

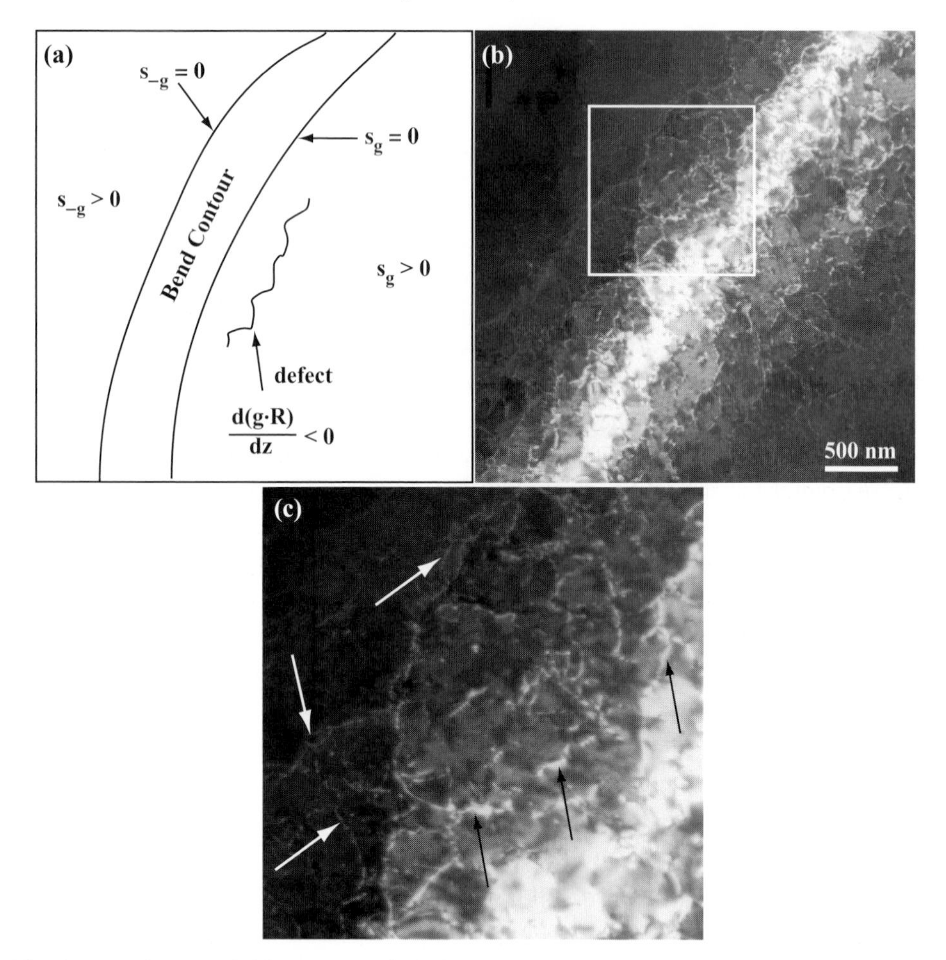

Fig. 8.1. (a) Schematic illustration of a defect near a bend contour; the defect displacement field may locally contribute a negative $d(\mathbf{g} \cdot \mathbf{R})/dz$ to the positive s_g, resulting in a near-zero effective excitation error and hence strong defect image contrast. The titanium $(\bar{2}1.0)$ dark field image, obtained on a JEOL 2000EX microscope operated at 200 kV, clearly shows dislocation contrast well outside of the bend contour. Part (c) is a magnified view of the region outlined in (b).

This variation of the apparent width of the dislocation contrast can be understood as follows: as we move away from the bend contour, the excitation error s_g increases to larger positive values. Only those regions near the defect where the second term of equation (8.7) is equally large but negative will give rise to strong image contrast. Large derivatives of $\mathbf{g} \cdot \mathbf{R}$ only occur near the core of the defect, where the atom displacements are largest, and this core region is rather narrow.[†] The resulting

[†] We will see in Section 8.4.2 that the displacement field of an edge dislocation in an elastically isotropic medium depends on $\ln r$, with r the distance to the dislocation line. The derivative of the logarithmic function is largest for small r-values.

image (bright field or dark field) will show a narrow contrast feature located close to the actual defect core. You can easily verify this behavior yourself, by looking at the image contrast of a defect while tilting the sample such that the bend contour moves away from the defect; the region of strong image contrast will become narrower.

In Fig. 4.13(b) we have seen that, if we bring the reflection **g** to the optical axis to obtain a dark field image, then the reflection 3**g** will be in Bragg orientation (assuming that **g** is in Bragg orientation before we tilt the beam). If $s_{3\mathbf{g}} \approx 0$, then $s_{\mathbf{g}}$ must be large and positive, which is exactly the diffraction condition necessary to obtain a narrow dislocation image. The **g**–3**g** diffraction condition, also known as the *weak beam imaging condition*, can therefore be used to obtain defect images which reveal the location(s) around the defect where the derivative of $\mathbf{g} \cdot \mathbf{R}$ is large; i.e. the core of the defect. Continuing this intuitive argument along the same lines, we also find that a gradual increase of $s_{\mathbf{g}}$ (i.e. by tilting the crystal such that $s_{3\mathbf{g}}$ becomes positive) will lead to contrast in an increasingly narrower region around the dislocation core. However, simultaneously the overall intensity of the image will decrease so that a longer exposure time is needed.

8.3 Numerical simulation of defect contrast images

In this section, we will discuss general principles of image contrast simulations for a crystal containing a defect with displacement field $\mathbf{R}(\mathbf{r})$. It is assumed throughout that the displacement field is available in analytical or numerical form; in other words, that its components can be computed at an arbitrary point in the crystal.

Nearly all defect contrast simulations make use of the column approximation, introduced in Chapter 6. While explicit formal expressions are available for the two-beam transmitted and scattered amplitudes in a perfect crystal, analytical solutions are no longer possible in the general case of a crystal with an arbitrary defect. Equations (8.5) must be integrated numerically along each column, taking into account the variation of the displacement field along the center line of the column. Each column again corresponds to one pixel in the computed bright or dark field image.

In the late 1950s and throughout the 1960s, when the first defect image simulations were carried out, computers were primitive and slow compared with contemporary machines. Random access memory (RAM) was expensive and limited, to the extent that the memory of most computers in those days was measured in kilobytes rather than megabytes. This meant that image simulations (or any other type of numerical simulation) had to be optimized to run in the limited available memory. Furthermore, there were no graphics terminals and laser printers, so that the reproduction quality of computed images was rather poor: grayscale images were often printed on line-printers, where the number and type of characters printed on one single location would determine the gray level of that location.

Nowadays, RAM is no longer expensive and even laptop computers can be equipped with hundreds of megabytes of RAM. This means that image simulations can now be carried out easily on a laptop. It is interesting to note, however, that the "older" programs, those written when the available memory was rather small, still work fine on modern machines, and that they tend to run much faster than programs written more recently. Perhaps memory restrictions forced programmers to optimize their code to run in the available space, whereas nowadays we rely more on the optimization capabilities of compilers and we need not worry too much about how much memory is allocated for a particular simulation.

For the numerical simulation of defect contrast, in both two-beam and multi-beam cases, we must perform two basic operations: (1) set up the geometry of the problem, and (2) solve the dynamical equations. We will discuss both topics in the following subsections.

8.3.1 Geometry of a thin foil containing a defect

We will assume that the image simulation is carried out for an array of pixels, each pixel corresponding to a single "column" through the thin foil, parallel to the beam direction (see Fig. 6.1). The number of columns is then not important for our discussion, and is usually limited by the available computing power. All columns have the same width and a square cross-section normal to the electron beam. The position of a column in the image is determined by the integer pair (i, j). The width of a column is described by L (expressed in nanometers).

With every column (i, j) we can associate a number of parameters which fully describe the crystal geometry and properties in the absence of a defect.

(i) **Foil thickness**. The local sample thickness can change from one column to the next, and is described by an array of values (or a mathematical function, if available) $z_0(i, j)$. The local foil thickness is not necessarily equal to the length of the column, since the entire foil may be inclined with respect to the beam direction. We will refer to the crystal thickness along the beam direction by the symbol $z_b(i, j)$.

(ii) **Diffraction condition**. The local diffraction condition may also vary from one column to the next. In the two-beam case, this would simply require an array of values for the excitation error $s_{\mathbf{g}}(i, j)$. In the multi-beam case, we must compute the excitation error of each diffracted beam based on the actual orientation of the crystal with respect to the Ewald sphere (i.e. the tangential component of the incident wave vector).

(iii) **Lattice potential**. The local electrostatic lattice potential can also differ from one column to the neighboring one. This could be due to the presence of an interface between the two columns, an interface separating two regions with the same crystal structure but different orientation. Alternatively, the local chemistry might change gradually (i.e. a composition gradient), which would also give rise to a position-dependent potential and therefore position-dependent parameters $q_{\mathbf{g}}(i, j)$.

It is clear that the description of these parameters requires a number of independent reference frames. We have introduced the crystal reference frames in Chapter 1: the direct and reciprocal reference frames (or Bravais reference frames, described by the direct and reciprocal metric tensors g_{ij} and g^*_{ij}), and the corresponding Cartesian reference frame (using the direct and reciprocal structure matrices a_{ij} and b_{ij}). To distinguish between the various reference frames introduced in the following paragraphs, we will identify the reference frame by a single lowercase superscript. The Cartesian reference frame defined in Chapter 1 is then labeled by the superscript c. We will continue to use an asterisk $*$ to indicate quantities in the reciprocal reference frame, and quantities in the direct Bravais reference frame will have no superscript.

Next we have the *foil reference frame*, which is fixed with respect to the thin foil. For convenience we select the foil normal ($\mathbf{F}$ with respect to the direct Bravais reference frame) to lie along the z-direction of the foil reference frame. The x- and y-directions are both normal to $\mathbf{F}$ and to each other. This means that they must belong to the zone $\mathbf{F}$. We will assume that the foil normal for the *untilted* foil is parallel to the incident beam direction $\mathbf{k}_0$, but points towards the electron gun, i.e.

$$\hat{\mathbf{F}} = -\frac{\mathbf{k}_0}{|\mathbf{k}_0|},$$

where the ^ hat indicates a unit vector. Then we select for the x-direction a reciprocal space vector $\mathbf{q}$ that lies in the foil plane, such that $\mathbf{F} \cdot \mathbf{q} = 0$, and which points towards the goniometer airlock along the primary tilt axis.

The vector $\mathbf{q}$ has non-integer components in the general case, and can be determined based on the calibration procedure described in Section 4.5.2. Since the foil normal may vary from one point of the foil to another, we will take the foil normal $\mathbf{F}$ to be for the central pixel of the image simulation. The Cartesian foil reference frame, labeled by the superscript f, is then constructed as follows:

$$\left.\begin{aligned} \mathbf{e}^f_x &= \frac{\mathbf{q}}{|\mathbf{q}|} = \frac{b_{ij}q_j}{|\mathbf{q}|}\mathbf{e}^c_i; \\ \mathbf{e}^f_z &= \frac{\mathbf{F}}{|\mathbf{F}|} = \frac{a_{ij}F_j}{|\mathbf{F}|}\mathbf{e}^c_i; \\ \mathbf{e}^f_y &= \mathbf{e}^f_z \times \mathbf{e}^f_x. \end{aligned}\right\} \tag{8.8}$$

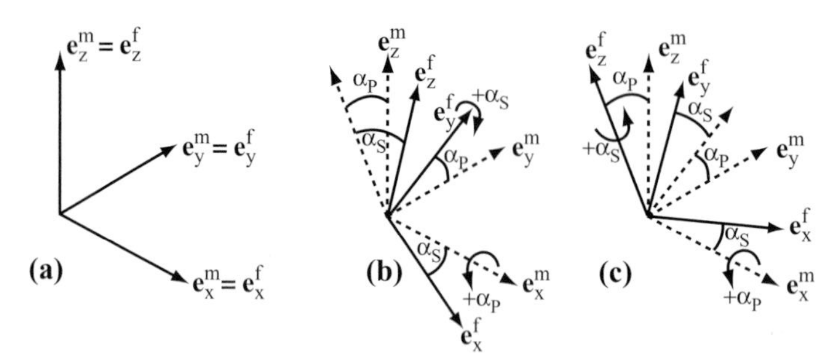

Fig. 8.2. (a) Definition of the microscope and foil reference frames for the sample reference orientation (no tilts); (b) sign conventions for positive rotations around the primary and secondary tilt axes (for a double-tilt holder); and (c) similarly for a rotation holder.

The equations above define the transformation matrix α^{fc} from the Cartesian crystal reference frame (old) to the foil reference frame (new):

$$\mathbf{e}_i^f = \alpha_{ij}^{fc}\mathbf{e}_j^c. \tag{8.9}$$

This matrix satisfies all the transformation rules described in Table 1.6 on page 51.

The next reference frame describes how the foil is oriented with respect to the microscope column. There are a number of ways in which this reference frame can be defined. We will take the z-direction of the microscope reference frame, labeled with a superscript m, as opposite to the beam direction. The x-axis is taken to be along the primary tilt axis, pointing towards the specimen airlock.† If the goniometer tilt angles are equal to zero, i.e. the sample holder is in its reference orientation, then we can define two rotation matrices, which express the microscope basis vectors in terms of the foil basis vectors for a double-tilt and rotation-tilt holder, respectively. Consider the reference frames shown in Fig. 8.2. In the reference sample holder orientation, the microscope and foil reference frames coincide, as shown in Fig. 8.2(a). For a double-tilt holder, the two rotation axes have positive rotations (counterclockwise when viewing towards the origin), α_P (primary axis) and α_S (secondary tilt angle). The sign conventions are shown in Fig. 8.2(b). Since the primary tilt axis is attached to the microscope and the secondary tilt axis is attached to the sample holder (i.e. changes orientation with the primary tilt), the order of the two rotations is unimportant. The rotation matrix α^{fm} connecting the microscope and foil reference frames is given by

$$\begin{pmatrix} \mathbf{e}_x^f \\ \mathbf{e}_y^f \\ \mathbf{e}_z^f \end{pmatrix} = \begin{pmatrix} \cos\alpha_S & \sin\alpha_S\sin\alpha_P & -\sin\alpha_S\cos\alpha_P \\ 0 & \cos\alpha_P & \sin\alpha_P \\ \sin\alpha_S & -\cos\alpha_S\sin\alpha_P & \cos\alpha_S\cos\alpha_P \end{pmatrix} \begin{pmatrix} \mathbf{e}_x^m \\ \mathbf{e}_y^m \\ \mathbf{e}_z^m \end{pmatrix}. \tag{8.10}$$

† This assumes that the microscope is equipped with a side-entry goniometer. For a top-entry goniometer a different x-direction must be selected.

A similar rotation matrix can be defined for a rotation-tilt holder. The secondary rotation angle is defined as α_R, measured from a suitable reference direction.

$$\begin{pmatrix} \mathbf{e}_x^f \\ \mathbf{e}_y^f \\ \mathbf{e}_z^f \end{pmatrix} = \begin{pmatrix} \cos\alpha_R & \sin\alpha_R\cos\alpha_P & -\sin\alpha_S\cos\alpha_P \\ -\sin\alpha_R & \cos\alpha_R\cos\alpha_P & \cos\alpha_R\sin\alpha_P \\ 0 & -\sin\alpha_P & \cos\alpha_P \end{pmatrix} \begin{pmatrix} \mathbf{e}_x^m \\ \mathbf{e}_y^m \\ \mathbf{e}_z^m \end{pmatrix}. \quad (8.11)$$

The resulting transformation matrix α_{fm} is defined by

$$\mathbf{e}_i^f = \alpha_{ij}^{fm}\mathbf{e}_j^m. \quad (8.12)$$

We will need an additional reference frame to describe the defect itself. The nature of this reference frame, labeled d, depends on the defect, and may simply be the crystal reference frame in the case of an isotropic defect. For instance, for a straight dislocation it appears natural to select the line direction $\mathbf{u}$ to be one of the reference directions, say, $\mathbf{e}_z^d$. For a planar fault we could select the normal to the fault plane to be the $\mathbf{e}_z^d$-axis. Whatever the choice might be, we will need to define another rotation matrix α^{dc}, which transforms the Cartesian crystal reference basis vectors to the defect basis vectors:

$$\mathbf{e}_i^d = \alpha_{ij}^{dc}\mathbf{e}_j^c. \quad (8.13)$$

The final reference frame is that of the simulated image itself, which locates each pixel in terms of the integer pair (i, j). If we define the image basis vectors $\mathbf{e}_{x,y}^i$ to lie in the lower left-hand corner along the horizontal and vertical directions of the image (i.e. of the electron micrograph), then the relation between the microscope reference frame $\mathbf{e}_{x,y,z}^m$ and the image can be expressed in terms of the angle β_P between the primary tilt axis and, say, the bottom edge of the negative:

$$\begin{pmatrix} \mathbf{e}_x^m \\ \mathbf{e}_y^m \\ \mathbf{e}_z^m \end{pmatrix} = \begin{pmatrix} \cos\beta_P & -\sin\beta_P & 0 \\ \sin\beta_P & \cos\beta_P & 0 \\ 0 & 0 & 1 \end{pmatrix} \begin{pmatrix} \mathbf{e}_x^i \\ \mathbf{e}_y^i \\ \mathbf{e}_z^i \end{pmatrix}. \quad (8.14)$$

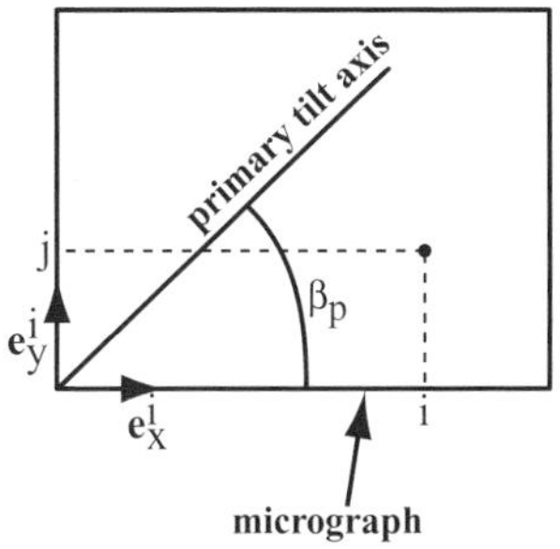

It may seem that the introduction of several reference frames makes things more complicated than necessary. However, it is useful to incorporate the microscope reference frame in the simulations, because it then becomes a trivial matter to simulate what happens to the image contrast when the crystal foil is tilted by a small angle around the primary or secondary tilt axis, which is what one would do experimentally. Furthermore, the explicit introduction of the microscope reference frame allows one to compute the image contrast for situations where multiple defects in different orientations with respect to the crystal lattice are present in the field of view. Each defect j would then be described in its own reference frame, the origin of which is located at a position vector $\mathbf{r}^i_j$ with respect to the *main origin*, which is conveniently chosen to be the lower left-hand corner of the image. The displacement fields due to all defects can then be added together (assuming a linear elastic description holds), so that images of arrays of defects or overlapping defects can be computed.

8.3.2 Example of the use of the various reference frames

A concrete example will illustrate the various coordinate transformations introduced in the previous section. Consider a face-centered cubic material, which contains a perfect edge dislocation with Burgers vector $\mathbf{b} = \frac{1}{2}[110]$ and line direction $\mathbf{u} = [1\bar{1}2]$. The dislocation lies in the $(\bar{1}11)$ glide plane. The foil normal is taken to be along $\mathbf{F} = [\bar{1}\bar{1}3]$, and the foil thickness is uniform and equal to z_0. The center of the dislocation line (at the center of the foil) corresponds to the origin of the defect reference frame $\mathbf{e}^d_i$, and the $\mathbf{e}^d_z$ direction is taken to be parallel to the Burgers vector. The geometry is shown schematically in the stereographic projection of Fig. 8.3.

For a cubic crystal, the Cartesian crystal reference frame coincides with the Bravais reference frame, but is normalized, so that we have

$$\mathbf{e}^c_i = \frac{\mathbf{a}_i}{a},$$

with a the lattice parameter. The direct structure matrix is equal to $a_{ij} = a\delta_{ij}$, and the reciprocal structure matrix is given by $b_{ij} = \frac{1}{a}\delta_{ij}$. The transformation matrix, α^{dc}, relating the crystal and defect reference frames, is derived from a definition of the defect reference frame: the dislocation line direction is taken along the $\mathbf{e}^d_z$-direction. The $\mathbf{e}^d_y$-direction lies in the plane formed by the beam direction $\mathbf{B}$ and the line direction $\mathbf{u}$, and $\mathbf{e}^d_x$ completes the right handed reference frame. We leave it as an exercise for the reader to show that, for the configuration of Fig. 8.3, we have

$$\begin{pmatrix} \mathbf{e}^d_x \\ \mathbf{e}^d_y \\ \mathbf{e}^d_z \end{pmatrix} = \begin{pmatrix} 0.082\,51 & 0.907\,27 & 0.412\,38 \\ -0.909\,13 & -0.100\,98 & 0.404\,08 \\ 0.408\,25 & -0.408\,25 & 0.816\,50 \end{pmatrix} \begin{pmatrix} \mathbf{e}^c_x \\ \mathbf{e}^c_y \\ \mathbf{e}^c_z \end{pmatrix}. \quad (8.15)$$

Each row is normalized, so that this matrix is a *unitary* matrix.

Next, we consider the orientation of the thin foil in the microscope column. We will assume that the $\mathbf{g}_{031}$ plane normal lies along the primary tilt axis, which itself is at an angle of, say, $\beta_P = +96^\circ$ with respect to the $\mathbf{e}^i_x$-axis; i.e. the bottom edge

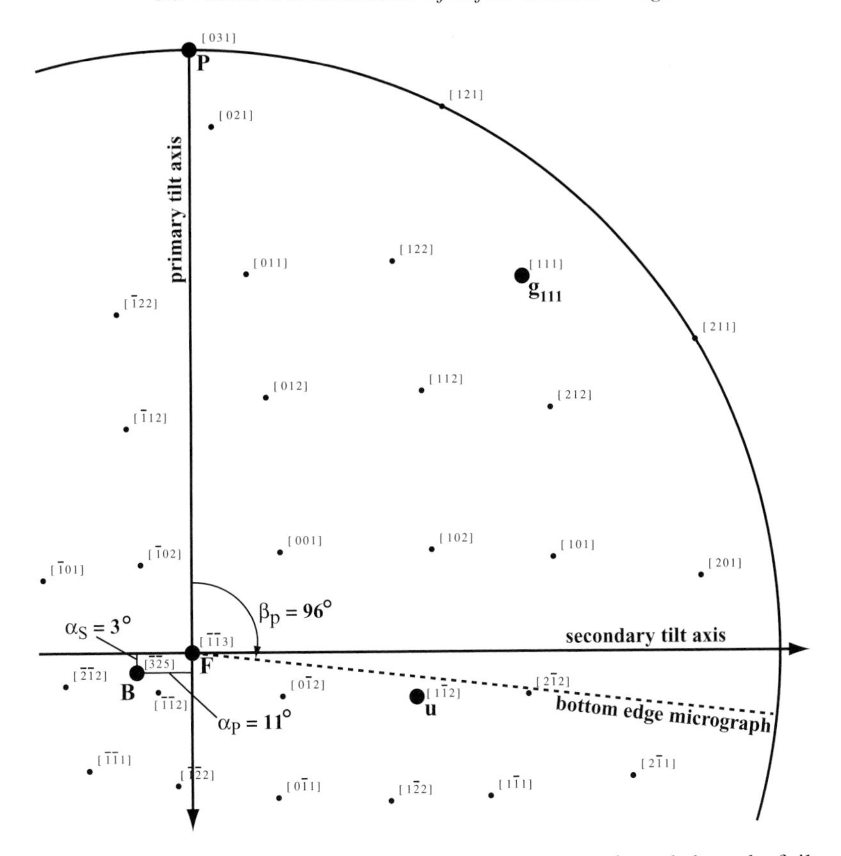

Fig. 8.3. Partial stereographic projection of a cubic structure, projected along the foil normal **F**. Other vectors indicated are: dislocation line direction **u**, diffraction vector **g**, and the beam direction **B**. The tilt angles α_P and α_S, along with β_P, are also indicated.

of the computed micrograph. This means that

$$\alpha^{mi} = \begin{pmatrix} -0.104\,528 & -0.994\,522 & 0 \\ 0.994\,522 & -0.104\,528 & 0 \\ 0 & 0 & 1 \end{pmatrix}. \tag{8.16}$$

The foil reference frame transformation matrix α^{fc} is given by

$$\alpha^{fc} = \begin{pmatrix} 0 & \frac{3}{\sqrt{10}} & \frac{1}{\sqrt{10}} \\ \frac{-10}{\sqrt{110}} & \frac{1}{\sqrt{110}} & \frac{-3}{\sqrt{110}} \\ \frac{-1}{\sqrt{11}} & \frac{-1}{\sqrt{11}} & \frac{3}{\sqrt{11}} \end{pmatrix}, \tag{8.17}$$

with the $\mathbf{e}_y^f$-direction along $[\bar{1}0\,1\bar{3}]$. Again the middle row follows from the cross product between the third and first rows, and all rows are normalized. Both defect and foil reference frames are now completely defined.

Assume, next, that the foil is mounted in a double-tilt holder, and that an image is obtained for the orientation $\alpha_P = 11.6°, \alpha_S = 3°$.[†] The following questions must then be answered as part of the image simulation.

- What is the length of each image column (referring to the column approximation)? This length determines the integration interval for the dynamical equations.
- What is the defect displacement field along the centerline of the column (i, j)?
- What is the derivative $\mathrm{d}\mathbf{R}/\mathrm{d}z$ along an arbitrary direction in the defect reference frame (for the case of a defect with a continuous displacement field)?

The first question is easy to answer: the foil normal $\mathbf{e}_z^f$ is rotated by the angles α_P and α_S, and, therefore, the crystal thickness measured along the beam direction, which is described by the vector $\mathbf{e}_z^m$, is given by

$$z_b = \frac{z_0}{\mathbf{e}_z^f \cdot \mathbf{e}_z^m} = \frac{z_0}{\cos\alpha_P \cos\alpha_S} = 1.022 z_0.$$

A point along the column corresponding to pixel (i, j) has coordinates (Li, Lj, z) with respect to the main origin (lower left-hand corner of the simulated image). The defect origin expressed in the image reference frame is given by (Li_d, Lj_d, z_d); the components i_d and j_d are not necessarily integers. The transformation matrix from image space to defect space, α^{di}, is given by the product

$$\alpha^{di} = \alpha^{dc}\alpha^{cf}\alpha^{fm}\alpha^{mi}, \tag{8.18}$$

and, since all matrices are known, the product can be worked out. The matrix α^{cf} is equal to the transpose of α^{fc}, since rotation matrices are unitary matrices. The result is

$$\alpha^{di} = \begin{pmatrix} 0.007\,94 & 0.999\,97 & 0.000\,00 \\ -0.596\,01 & 0.004\,73 & 0.802\,96 \\ 0.802\,93 & -0.006\,38 & 0.596\,03 \end{pmatrix}.$$

Now that we have the transformation matrix relating the image and defect basis vectors

$$\mathbf{e}_k^d = \alpha_{kl}^{di}\mathbf{e}_l^i,$$

we can apply the transformation rules listed in Table 1.6 to convert image space coordinates to defect space coordinates. A vector with components r_k^i in image space is converted to defect space[‡] by post-multiplication with the inverse

[†] These tilt angles will bring the beam direction $\mathbf{B} = [\bar{3}\bar{2}5]$ to the center of the stereographic projection in Fig. 8.3.
[‡] Remember that the superscripts indicate the reference frame with respect to which the components are expressed.

matrix $(\alpha^{di})^{-1} = \alpha^{id}$, or

$$r_l^d = r_k^i \alpha_{kl}^{id}. \tag{8.19}$$

The column (Li, Lj, z) in image space, corresponding to pixel (i, j) of the simulated image, corresponds to the straight line of length z_b between the top (t) and bottom (b) points with defect space coordinates

$$\left(x_t^d, y_t^d, z_t^d\right) = \left(L\,(i - i_d)\,, L\,(j - j_d)\,, \frac{z_b}{2}\right)\alpha^{id};$$
$$\left(x_b^d, y_b^d, z_b^d\right) = \left(L\,(i - i_d)\,, L\,(j - j_d)\,, -\frac{z_b}{2}\right)\alpha^{id}.$$

Finally, it is rather easy to derive an explicit expression for the derivative $\mathrm{d}\mathbf{R}/\mathrm{d}z$ at an arbitrary location $\mathbf{r}$ along a unit direction vector $\hat{\mathbf{k}}^d$ expressed in the defect reference frame. This direction vector is parallel to the incident beam direction, which is given by $[00\bar{1}]$ in the image reference frame, or, more conventionally, by the components k_m^* in the crystal reciprocal reference frame. In the defect reference frame the incident beam direction can be expressed by first transforming the components k_m^* to the crystal Cartesian reference frame, using the reciprocal structure matrix b_{im}, and subsequently transforming to the defect reference frame using the matrix α_{dc}:

$$k_j^d = \alpha_{ji}^{dc} b_{im} k_m^*. \tag{8.20}$$

The incident beam direction is then described by the unit vector $\hat{\mathbf{k}}^d$.

The directional derivative of a vector field $\mathbf{R}(\mathbf{r})$ along the direction $\hat{\mathbf{k}}^d$ is given by [BT79, page 165]:

$$\frac{\mathrm{d}\mathbf{R}(\mathbf{r})}{\mathrm{d}z} = \left(\hat{\mathbf{k}}^d \cdot \nabla^d\right)\mathbf{R}(\mathbf{r}), \tag{8.21}$$

where the differential operator ∇^d operates in the defect reference frame. This derivative can be rewritten in a form useful for numerical implementation:

$$\frac{\mathrm{d}\mathbf{R}^d}{\mathrm{d}z} = k_i^d \left.\frac{\partial R_j^d}{\partial x_i}\right|_{\mathbf{r}} \mathbf{e}_j^d. \tag{8.22}$$

The dot products $\mathbf{g} \cdot \mathbf{R}$ and $\mathbf{g} \cdot \mathrm{d}\mathbf{R}/\mathrm{d}z$ require that the directional derivative be expressed with respect to the direct space basis vectors $\mathbf{a}_m$, which can be computed using the inverse of the direct structure matrix (see Chapter 1):

$$\frac{\mathrm{d}\mathbf{R}(\mathbf{r})}{\mathrm{d}z} = k_i^d \left.\frac{\partial R_j^d}{\partial x_i}\right|_{\mathbf{r}} \alpha_{jl}^{dc} b_{lm} \mathbf{a}_m. \tag{8.23}$$

where we have used the fact that $b^T = a^{-1}$. The dot products of **g** and **R** or its derivative along the beam direction are then given by

$$\mathbf{g}\cdot\mathbf{R}(\mathbf{r}) = R_j^d(\mathbf{r})\alpha_{jl}^{dc}b_{lm}g_m;$$

$$\mathbf{g}\cdot\frac{\mathrm{d}\mathbf{R}(\mathbf{r})}{\mathrm{d}z} = k_i^d \left.\frac{\partial R_j^d}{\partial x_i}\right|_{\mathbf{r}} \alpha_{jl}^{dc}b_{lm}g_m.$$

This completes the computation of the defect geometry. We can now proceed with the solution of the dynamical equations for every image column (i, j); it is clear from the above discussion that we will need to have an efficient algorithm to compute the displacement vector **R** or its derivative at any point along the column center line. Depending on the defect being considered, this may be an analytical expression, or perhaps a discrete set of displacement vectors obtained from a finite element computation.

If more than one defect is present in the field of view, then the image contrast due to the combined defects may be calculated. In this case, we must change the definition of the defect origin to coincide with one of the defects, and introduce a translation vector relating each defect to the one at the origin. This is not difficult and is left as an exercise for the reader. If the foil thickness is position dependent, then the top and bottom points of each column may be different, and the computation becomes slightly more involved; this is again left as an exercise.

8.3.3 Dynamical multi-beam computations for a column containing a displacement field

In this section, we use the modified DHW equations as a starting point for the numerical simulation of defect contrast. Equations (8.5) are valid for the general multi-beam case, and for any defect. For numerical purposes, it is useful to distinguish two cases.

- **Displacement field R(r) is not differentiable**

  The following substitution can be used to rewrite the modified DHW equations in a form useful for numerical computations:

$$\psi_{\mathbf{g}} = S_{\mathbf{g}}\mathrm{e}^{\mathrm{i}\theta_{\mathbf{g}}}\mathrm{e}^{\alpha z} \qquad \text{with } \alpha = -\frac{\pi}{\xi_0'}. \tag{8.24}$$

  A simple computation then results in

$$\frac{\mathrm{d}S_{\mathbf{g}}}{\mathrm{d}z} = 2\pi\mathrm{i}s_{\mathbf{g}}S_{\mathbf{g}} + \mathrm{i}\pi{\sum_{\mathbf{g}'}}' \frac{\mathrm{e}^{-\mathrm{i}\alpha_{\mathbf{g}-\mathbf{g}'}(\mathbf{r})}}{q_{\mathbf{g}-\mathbf{g}'}}S_{\mathbf{g}'}, \tag{8.25}$$

  where the prime on the summation indicates that the term $\mathbf{g}' = \mathbf{g}$ is excluded.

- **Displacement field R(r) is differentiable**
  We have already derived the relevant equation in Section 8.2; if we also include substitution (8.24), then the resulting equations are given by

$$\frac{\mathrm{d}S_{\mathbf{g}}}{\mathrm{d}z} = 2\pi\mathrm{i}\left[s_{\mathbf{g}} + \mathbf{g}\cdot\frac{\mathrm{d}\mathbf{R}(\mathbf{r})}{\mathrm{d}z}\right]S_{\mathbf{g}} + \mathrm{i}\pi\sum_{\mathbf{g}'}{}^{\prime}\frac{1}{q_{\mathbf{g}-\mathbf{g}'}}S_{\mathbf{g}'}. \tag{8.26}$$

Now we are set to compute the image contrast for any arrangement of defects, with mixed differentiable and non-differentiable displacement fields.† A simple example of such a combination would be a pair of partial dislocations (continuous and differentiable **R**) separated by a stacking fault (discontinuous **R**). All continuous displacement fields contribute to the $\mathrm{d}\mathbf{R}/\mathrm{d}z$ part of the effective excitation error, while the discontinuous displacement fields contribute to the right-hand side of the equation in the phase shift term. The modified DHW equations for a configuration of N_c defects with continuous displacement fields and N_d defects with discontinuous displacement fields are then given by

$$\boxed{\frac{\mathrm{d}S_{\mathbf{g}}}{\mathrm{d}z} = \left[2\pi\mathrm{i}s_{\mathbf{g}} + \mathrm{i}\frac{\mathrm{d}\alpha^c_{\mathbf{g}}(\mathbf{r})}{\mathrm{d}z}\right]S_{\mathbf{g}} + \mathrm{i}\pi\sum_{\mathbf{g}'}{}^{\prime}\frac{\mathrm{e}^{-\mathrm{i}\alpha^d_{\mathbf{g}-\mathbf{g}'}(\mathbf{r})}}{q_{\mathbf{g}-\mathbf{g}'}}S_{\mathbf{g}'},} \tag{8.27}$$

with

$$\alpha^c_{\mathbf{g}}(\mathbf{r}) \equiv 2\pi\mathbf{g}\cdot\sum_{j=1}^{N_c}\mathbf{R}_j(\mathbf{r});$$

$$\alpha^d_{\mathbf{g}}(\mathbf{r}) \equiv 2\pi\mathbf{g}\cdot\sum_{k=1}^{N_d}\mathbf{R}_k(\mathbf{r}).$$

In most defect image contrast simulations, only one single defect must be considered, and the equations above are simplified somewhat. It is useful, however, to have the general equations, because they open the way to the simulation of more complex defect images, such as faults which change orientation, or intersecting faults. In the remainder of this chapter, we will make use of the equations derived above to compute the image contrast of a number of defects. We will introduce analytical expressions for defect displacement fields whenever they are available. In some cases, analytical solutions to the modified DHW equations are possible for the two-beam case, in particular for planar defects such as stacking faults and anti-phase boundaries. We will see that the crystal transfer matrix formalism introduced in Chapter 5 is useful to compute rapidly the contrast of defects and defect configurations.

† It is assumed throughout that linear elasticity theory is valid, which means that the total displacement due to an arrangement of defects is equal to the sum of the individual displacements.

The solution to equation (8.27) can always be obtained numerically by means of an appropriate differential equation solver. We have seen in Chapter 7 that such a solution method is typically slower than other algorithms, such as the scattering matrix approach or the Bloch wave formalism. It is possible, however, to use the scattering matrix formalism for the systematic row case in the presence of defects, and the algorithm is described in pseudo code PC-19. Since most defect contrast observations are carried out for systematic row conditions, there is no real need for a defect simulation program for the zone axis case.

The simulation method relies on the fact that all defects, with continuous or discontinuous displacement fields, give rise to phase shifts of the Fourier coefficients of the electrostatic lattice potential and that such phase shifts are periodic functions. This effectively means that it is not the magnitude of $\mathbf{g}\cdot\mathbf{R}$ that matters, but only $\mathbf{g}\cdot\mathbf{R}$ mod 1. Consider the phase factor

$$\theta_{\mathbf{gg}'}(\mathbf{r}) \equiv \mathrm{e}^{-2\pi\mathrm{i}(\mathbf{g}-\mathbf{g}')\cdot\mathbf{R}(\mathbf{r})}.$$

For a systematic row, all vectors $\mathbf{g}$ are multiples of a single reciprocal lattice vector $\mathbf{G}$, and we can write

$$\theta_{nn'}(\mathbf{r}) = \mathrm{e}^{-2\pi\mathrm{i}(n-n')\mathbf{G}\cdot\mathbf{R}(\mathbf{r})}.$$

Since n and n' are integers, it is easy to see that only values $\mathbf{G}\cdot\mathbf{R}$ mod 1 are of importance and this suggests that we could precompute the possible phase factors $\theta_{nn'}$ and employ the scattering matrix formalism to implement systematic row defect simulations. This would include two-beam and weak-beam ($\mathbf{g}$–$3\mathbf{g}$) simulations.

The scattering matrix formalism relies on the crystal transfer matrix $\mathcal{A}$, as defined in Chapter 5. In the presence of a defect the crystal transfer matrix depends on location in the crystal. In particular, only the off-diagonal members of $\mathcal{A}$ depend on position:

$$\mathcal{A}_{nn'} = \frac{\pi}{q_{nn'}}\theta_{nn'}(\mathbf{r}).$$

From $\mathcal{A}$ we can compute $\mathcal{S}(\epsilon)$ for a thin slice of thickness ϵ, using the Taylor expansion algorithm described in Section 7.2.2.3. We can repeat this computation for an array of values for the dot product $\mathbf{g}\cdot\mathbf{R}$. The memory requirements for storing, say, $N = 10\,000$ scattering matrices are not unreasonable: for a nine-beam systematic row the total number of complex numbers that need to be stored is $9^2 \times 10\,000$, which amounts to about 13 Mbytes of storage space for double-precision arithmetic (16 bytes per complex number).

The selection of the appropriate scattering matrix is straightforward: for a given location in the foil the sum of all local defect displacement vectors $\mathbf{R}(\mathbf{r}) = \sum_{i=1}^{M}\mathbf{R}_i(\mathbf{r})$ is computed. The number of the corresponding scattering matrix is then given by the integer nearest to $N(\mathbf{G} \cdot \mathbf{R} \bmod 1)$. The program SRdefect.f90, available from the website, implements the entire algorithm. The number of slices needed for each column will depend on how fast the displacement field varies along the column; the simplest approach to determining the number of necessary slices would be to repeat the computation for half the slice thickness and compare the resulting images. If there is no significant difference between the two images, then the larger slice thickness is sufficiently small. All systematic row defect image simulations in this chapter were performed with the SRdefect.f90 program.

Pseudo Code **PC-19** Systematic row defect contrast computation.

Input: PC-2, PC-3, elastic data
Output: bright field and dark field defect contrast images

```
ask for microscope voltage and systematic row indices
define slice thickness, beam orientation
define foil and defect configuration
initialize the perfect crystal dynamical matrix
for each value of G · R do
   compute and store scattering matrix S(ε)
end for
for each image pixel do
   determine contributing defects
   initialize incident wave
   determine the number of slices
   for each slice in column (top to bottom) do
      compute total G · R mod 1
      select appropriate slice matrix and multiply wave function
   end for
   convert to intensities
end for
display bright field and dark field images
```

It is straightforward to include a bent foil shape in this systematic row simulation by including an additional displacement field. A bent foil can be described by a position-dependent excitation error $s_{\mathbf{G}}(x, y)$. If we introduce a displacement field of the form

$$\mathbf{R}(\mathbf{r}) = z s_{\mathbf{G}}(x, y)\mathbf{G}^*, \tag{8.28}$$

with $\mathbf{G}^*$ a vector such that $\mathbf{G}\cdot\mathbf{G}^* = 1$,[†] then it is easy to see that the contribution of this "defect" to the effective excitation error is given by

$$\mathbf{G}\cdot\frac{\mathrm{d}\mathbf{R}}{\mathrm{d}z} = s_{\mathbf{G}}(x, y). \tag{8.29}$$

This is similar to the technique used to obtain Fig. 7.6 in Chapter 7. This method approximates the foil excitation error, and its accuracy increases with increasing acceleration voltage. At lower acceleration voltage, this approach overestimates the excitation error of the higher-order reflections.

Weak beam images can also be simulated using the systematic row program. All that is required is to provide the correct orientation of the systematic row with respect to the incident beam direction. This is most easily accomplished by entering the value of the excitation error $s_{\mathbf{G}}$. The following cases are of importance and can all be derived from the solution to problem 2.9, Chapter 2:

$$s_{n\mathbf{G}} \approx -n\frac{|\mathbf{G}|}{2}\left(\frac{2\cos\omega + n\lambda|\mathbf{G}|}{1 + n\lambda|\mathbf{G}|\cos\omega}\right), \tag{8.30}$$

with ω the angle between the incident beam and $\mathbf{G}$ (see Fig. 2.13).

- *Symmetric systematic row orientation*: $\omega = \pi/2$, so that

$$s_{\mathbf{G}} = -\frac{\lambda}{2}|\mathbf{G}|^2. \tag{8.31}$$

- *Weak beam orientation*: if $s_{3\mathbf{G}} = 0$, we have $\cos\omega = -\frac{3}{2}\lambda|\mathbf{G}|$ and

$$s_{\mathbf{G}} = \frac{2\lambda|\mathbf{G}|^2}{2 - 3\lambda^2|\mathbf{G}|^2}. \tag{8.32}$$

- $-\mathbf{G}$ *in Bragg orientation*: if $s_{-\mathbf{G}} = 0$, we have $\cos\omega = \frac{\lambda}{2}|\mathbf{G}|$ and

$$s_{\mathbf{G}} = -\frac{2\lambda|\mathbf{G}|^2}{2 + \lambda^2|\mathbf{G}|^2}. \tag{8.33}$$

8.4 Image contrast for selected defects

In the following sections, we will describe two-beam and systematic row image contrast for selected defects: coherent precipitates, dislocations, stacking faults, anti-phase boundaries, inversion boundaries, permutation boundaries, and grain boundaries. Most of the simulations will involve the two-beam theory. For some defects we will also describe weak beam contrast, which involves the systematic row

[†] In other words, $\mathbf{G}^* = \mathbf{G}/G^2$.

solution. For dislocations and planar defects the website has several **ION** routines that the reader can use to obtain bright field and dark field images for a variety of input parameters.

8.4.1 Coherent precipitates and voids

8.4.1.1 Crystallography

Consider a cubic second phase particle which has a lattice parameter a_p that is nearly equal to that of the matrix, a_m, in which the particle is embedded. The *lattice misfit* δ is then defined by

$$\delta = \frac{a_p - a_m}{a_m}. \tag{8.34}$$

When $a_p < a_m$, the particle puts the surrounding matrix in tension, and $\delta < 0$. The displacement field for a spherical particle with radius r_0 and the associated diffraction contrast were first derived by Ashby and Brown [AB63a, AB63b]:

$$\mathbf{R}(\mathbf{r}) = C(\delta)\frac{(|\mathbf{r}| < r_0)^3}{|\mathbf{r}|^3}\mathbf{r}, \tag{8.35}$$

where the symbol $(a < b)$ means that the smaller one of a and b must be used. $C(\delta)$ is a constant that depends on the elastic properties of the isotropic matrix and the lattice misfit, and is given by

$$C = \frac{3K\delta(1+\nu)}{3K(1+\nu)+2E}, \tag{8.36}$$

where K is the bulk modulus of the particle, and E and ν are Young's modulus and Poisson's ratio of the matrix, respectively. If $\nu = \frac{1}{3}$ and $K \approx E$, then we have $C = \frac{2}{3}\delta$. Hence, inside the particle the displacement is proportional to $\mathbf{r}$, whereas the displacements outside the particle decay as $1/r^2$.

The displacement field (8.35) is valid for small isotropic coherent inclusions in an isotropic matrix. Small means that the particle diameter should be significantly smaller than one extinction distance. For larger, misfitting particles, Mader [Mad87] has described the diffraction contrast based on a Bloch wave approach.

For a void or a gas bubble, it is a good approximation to assume that the cavity is empty, which means that the scattering potential $U + iU'$ vanishes. The only components of the scattering matrix that are non-zero are, hence, the diagonal components, and it is easy to see that the scattering matrix of a thin empty slice is equal to the vacuum propagator, described in Section 5.6.2.

8.4.1.2 Image contrast

It is straightforward to implement equation (8.35). The SRdefect.f90 program includes this displacement field in its standard set of displacement fields. An example simulation for an array of nine inclusions of radius 10 nm in a 200 nm thick Cu foil is shown in Fig. 8.4. The top illustration shows the distance of the center of each inclusion to the top surface, as a percentage of the total foil thickness. The simulations were carried out for the $\mathbf{g}_{200}$ systematic row with seven beams (200 kV) and for three different systematic row conditions: (a) symmetric orientation, (b) $s_{\mathbf{g}} = 0$ Bragg orientation, and (c) **g**–3**g** weak beam orientation. The images on each individual row have a common intensity scale.

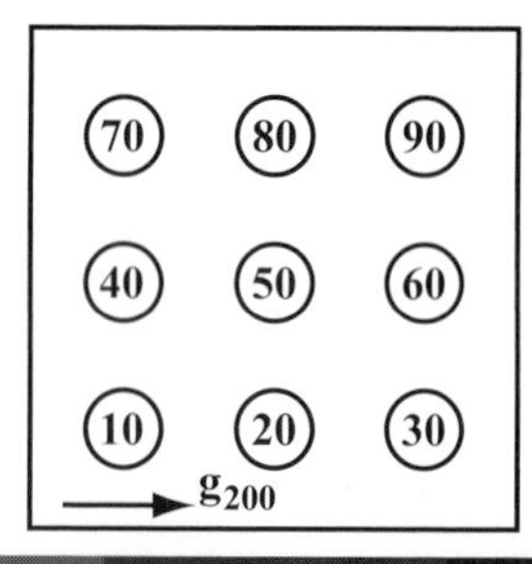

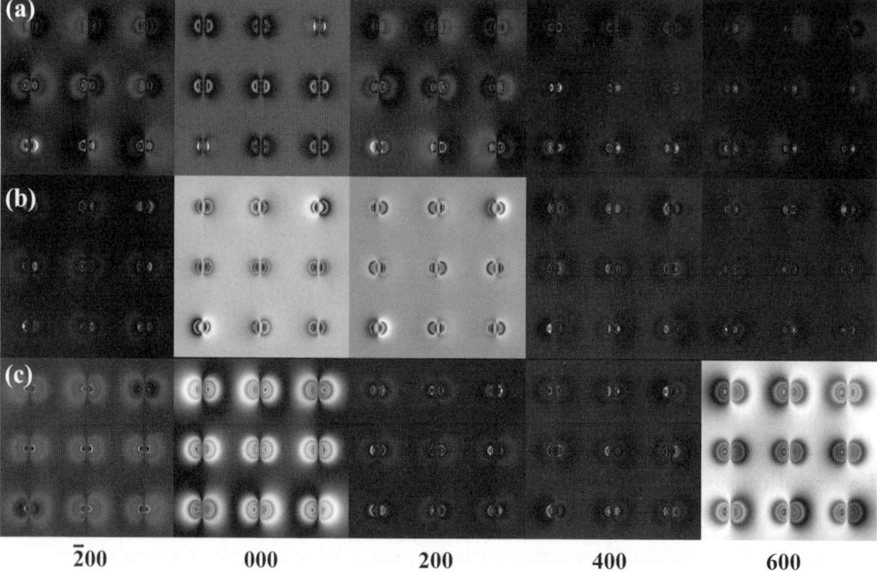

Fig. 8.4. Seven-beam systematic row simulation (200 kV) for nine spherical inclusions in a 200 nm Cu foil. The operating reflection is $\mathbf{g}_{200}$. (a) Systematic orientation ($s_{\mathbf{g}} = -0.0383$ nm^{-1}), (b) Bragg orientation ($s_{\mathbf{g}} = 0.0$ nm^{-1}), (c) weak beam orientation ($s_{\mathbf{g}} = 0.076\,75$ nm^{-1}).

The displacement field of equation (8.35) is strictly only valid for the case where the inclusion sits in the middle of the foil. Inclusions closer to the foil surfaces must have contributions from so-called *mirror-inclusions*, to guarantee that the surfaces are stress-free. Such mirror-inclusions were not taken into account for the simulations in Fig. 8.4.

The bright field contrast on the central row displays characteristic diffraction contrast: the inclusion at the center of the foil has a symmetrical dark-lobed image, with a straight no-contrast line normal to the $\mathbf{g}_{200}$ vector. The inclusion near the top (lower left) displays an asymmetric lobe contrast, which is identical to the contrast in the dark field image. For the inclusion near the bottom foil surface, this bright field/dark field symmetry is reversed. This contrast reversal remains even when the mirror-inclusions are taken into account. From a measurement of the extent of the contrast lobes, and a knowledge of the extinction distance and particle diameter, the sign and magnitude of the misfit parameter δ can be derived. We refer the reader to [AB63a] for further details.

There are several other lattice defects that give rise to nearly identical diffraction contrast. Small coherent precipitates, small dislocation loops, voids, and vacancy clusters can be difficult to distinguish from each other, and the reader should consult [Edi76] and [JK01] for a detailed discussion of experimental procedures.

8.4.2 Line defects

The study of dislocations forms an important part of a more general study of the mechanical properties of a material. Identification of line directions, Burgers vectors, slip planes, dislocation interaction mechanisms, pinning sites, etc., is needed in order to explain plasticity, work hardening, creep, and many other mechanical phenomena. In this section, we will first describe the geometry and crystallography of dislocations and then introduce image simulation methods to compute dislocation image contrast.

8.4.2.1 Crystallography

A dislocation is a linear lattice defect and is characterized by a Burgers vector $\mathbf{b}$ and a line direction $\mathbf{u}$. The line direction may be constant along the line, in which case we refer to the dislocation as a *straight dislocation*, or it may vary randomly for a *curved dislocation* or change along a closed loop for a *dislocation loop*. A dislocation line segment has *edge* character when $\mathbf{b} \cdot \mathbf{u} = 0$, *screw* character when $\mathbf{b} \parallel \mathbf{u}$, and *mixed* character in all other cases. For an extensive introduction to dislocations we refer the reader to the books by Hull and Bacon [HB84], and Hirth and Lothe [HL68].

In addition to the Burgers vector and line direction, the nature of the displacement field of a dislocation is also determined by the *elastic properties* of the material. The second-order elastic moduli tensor c_{ijkl} relates strain ϵ_{kl} to stress σ_{ij} (assuming that Hooke's law is valid):

$$\sigma_{ij} = c_{ijkl}\epsilon_{kl},$$

and the strain is related to the displacement through

$$\epsilon_{ij} = \frac{1}{2}\left(\frac{\partial R_i}{\partial x_j} + \frac{\partial R_j}{\partial x_i}\right).$$

It can be shown, and we refer the reader to [HL68], that for an elastically isotropic medium, the displacement field of a straight dislocation is given by

$$\mathbf{R} = \frac{1}{2\pi}\left\{\mathbf{b}\theta + \mathbf{b}_e\frac{\sin 2\theta}{4(1-\nu)} + \mathbf{b}\times\mathbf{u}\left[\frac{1-2\nu}{2(1-\nu)}\ln r + \frac{\cos\theta}{4(1-\nu)}\right]\right\}, \tag{8.37}$$

where ν is Poisson's ratio and (r,θ) are polar coordinates in a plane perpendicular to the line direction $\mathbf{u}$. The angle θ is conventionally measured from the slip plane. $\mathbf{b}_e$ is the edge component of the Burgers vector.

In the case of an elastically anisotropic medium, the situation is more complicated and a general analytical solution does not exist. We simply state the result and refer the reader to [HL68] and [Str58] for more details. The dislocation line direction is taken to be the $z = x_3$-direction; the displacement field then only depends on the coordinates (x_1, x_2) normal to the dislocation line.

$$R_k = \Re\left[\frac{1}{2\pi}\sum_{\alpha=1}^{3} A_{k\alpha}M_{\alpha j}H_{ji}b_i\ln(x_1 + p_\alpha x_2)\right], \tag{8.38}$$

where $\Re[]$ denotes the real part of the enclosed expression and

$$\begin{aligned}
L_{i\alpha} &= (c_{i2k1} + p_\alpha c_{i2k2})A_{k\alpha};\\
M_{ij} &= L_{ij}^{-1};\\
B_{ij} &= \frac{\mathrm{i}}{2}\sum_\alpha (A_{i\alpha}M_{\alpha j} - A_{i\alpha}^{*}M_{\alpha i}^{*});\\
H_{ij} &= B_{ij}^{-1}.
\end{aligned}$$

The vectors $A_{k\alpha}$ are the solution vectors to the following system of equations (summation over k):

$$[c_{i1k1} + (c_{i1k2} + c_{i2k1})p + c_{i2k2}p^2]A_k = 0 \qquad (i = 1, \ldots, 3).$$

This gives rise to a sextic determinantal equation for the coefficients p:

$$|c_{i1k1} + (c_{i1k2} + c_{i2k1})p + c_{i2k2}p^2| = 0,$$

and one can show that the six solutions occur in three complex conjugate pairs $p_\alpha, \alpha = 1, \ldots, 3$. For each solution p_α there is also a corresponding solution vector $A_{k\alpha}$. The computation of the displacement components does not depend on the absolute values of the elastic moduli; only relative values are of importance. The expressions for the displacement field have been implemented in the routine makedislocation, which is part of the SRdefect.f90 program.

8.4.2.2 Image contrast

For a screw dislocation in an isotropic medium, the edge component $\mathbf{b}_e$ vanishes and the Burgers vector is parallel to the line direction $\mathbf{u}$. Hence, the displacement field (8.37) reduces to

$$\mathbf{R} = \frac{\mathbf{b}\theta}{2\pi}. \tag{8.39}$$

The visibility criterion for pure screw dislocations is $\mathbf{g} \cdot \mathbf{R} = \mathbf{g} \cdot \mathbf{b} = 0$, or, alternatively, planes containing the line direction remain undistorted. This leads to a simple recipe for the determination of the *direction* of the Burgers vector of a screw dislocation: find two vectors $\mathbf{g}_i$ for which the dislocation contrast vanishes. Then take the cross product between the two $\mathbf{g}$-vectors to find the direction of $\mathbf{b}$.†

For an edge dislocation in an isotropic medium the situation is slightly more complex: the edge component $\mathbf{b}_e = \mathbf{b}$ and the vector $\mathbf{b} \times \mathbf{u}$ never vanishes. A pure edge dislocation is hence only invisible if simultaneously $\mathbf{g} \cdot \mathbf{b} = 0$ *and* $\mathbf{g} \cdot (\mathbf{b} \times \mathbf{u}) = 0$. This only happens for one single reflection, i.e. only for $\mathbf{g}$ parallel to the line direction $\mathbf{u}$. This is again consistent with the intuitive observation that, for an edge dislocation, planes perpendicular to the line direction remain undistorted.

For a general dislocation in an anisotropic matrix, neither of the above cases is exactly satisfied and no generally valid visibility criterion can be used. One has to solve the DHW differential equations, using the expression (8.38) for the displacement field. In addition, the observed contrast will depend on other factors, such as the beam direction $\mathbf{B}$, the foil normal $\mathbf{F}$, the extinction distance, the overall excitation error, the crystal thickness, and the anomalous absorption factor. We conclude that dislocation contrast is a complex function of many variables and detailed image simulations may be needed to unambiguously identify a dislocation.

† The length of $\mathbf{b}$ is more difficult to determine from images; often, crystallographic arguments can be used to list possible values for the length.

For a dislocation in an elastically *anisotropic* matrix, *all* lattice planes are distorted and hence the visibility criterion should be used with caution, even when pure screw or edge dislocations are considered.

8.4.2.3 Experimental example

Figure 8.5(a)–(f) show seven sets of bright field/dark field images, along with diffraction patterns, for the two-beam orientations indicated. The images were

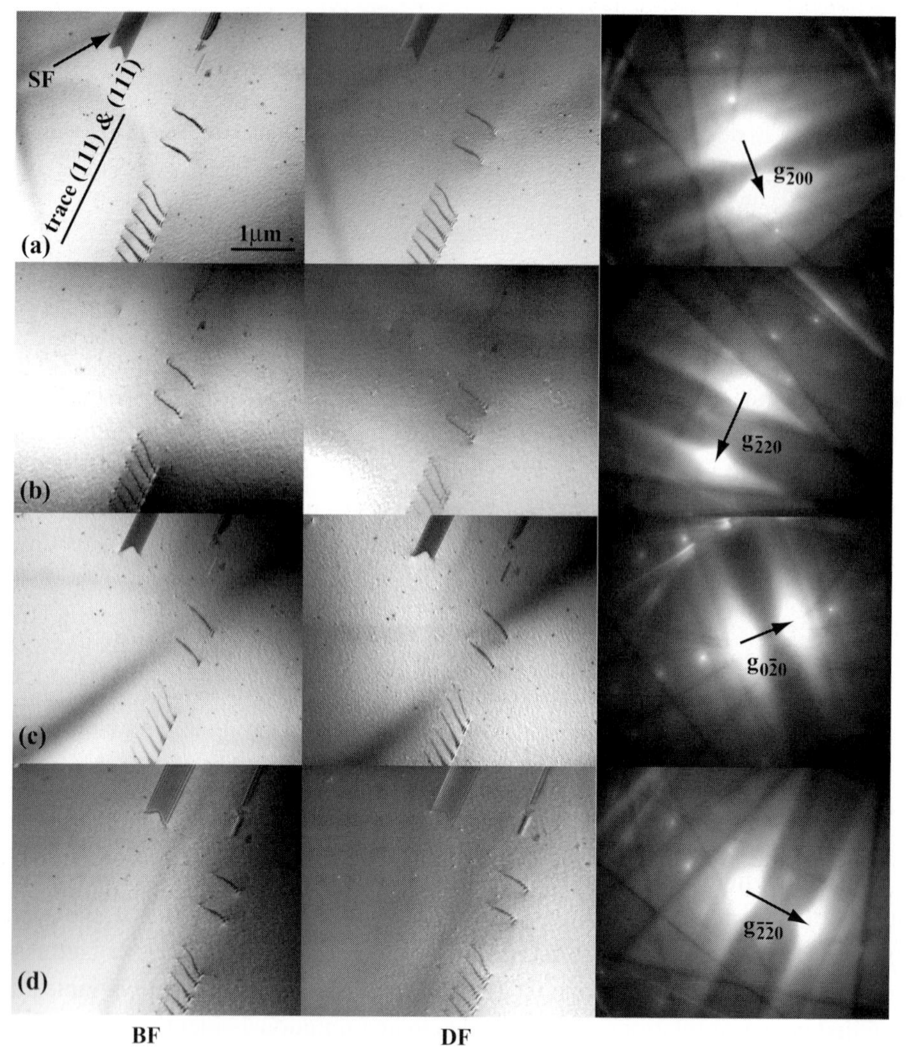

Fig. 8.5. BF-DF image pairs along with two-beam diffraction patterns of a series of dislocations and a stacking fault in Cu-15 at% Al. The images were recorded at 200 kV in a Philips CM20.

Fig. 8.5 (*cont.*).

obtained at 200 kV, using a thick foil of Cu-15 at% Al. Two-beam conditions were obtained by tilting from the [001], [$\bar{1}$14], and [013] zone axis orientations. All dark field images are centered dark field (CDF) images. The images show a sequence of dissociated perfect dislocations (lower center), and a stacking fault (top, arrowed in a). The trace of the stacking fault is indicated by a solid line, and corresponds to either the (111) or the (11$\bar{1}$) plane.

From the contrast rules for stacking faults, described in Section 8.4.3.3, we find that the left-hand side of the stacking fault corresponds to the bottom of the foil, and hence the fault plane is the (111) plane. The row of dislocations then lies in the (11$\bar{1}$) glide plane. There are three two-beam conditions for which the stacking fault is invisible: ($\bar{2}$20), (3$\bar{1}$1), and ($\bar{1}\bar{3}$1). This gives rise to three equations for the components of the displacement vector:

$$-2R_x + 2R_y = 0, \pm1, \ldots;$$
$$3R_x - R_y + R_z = 0, \pm1, \ldots;$$
$$-R_x - 3R_y + R_z = 0, \pm1, \ldots,$$

where the right-hand side should be an integer (but not necessarily the same integer for every equation). It is straightforward to solve these equations, and we find that $R_x = R_y = 1$ and $R_z = -2$. The resulting displacement vector is then $\mathbf{R} = \frac{1}{6}[11\bar{2}]$, which corresponds to the Burgers vector of the partial dislocation that created this fault (the *leading partial*). The trailing partial is then either a $\mathbf{b}_1 = \frac{1}{6}[2\bar{1}\bar{1}]$ or a $\mathbf{b}_2 = \frac{1}{6}[\bar{1}2\bar{1}]$ Schockley partial dislocation. Analysis of the image contrast in Fig. 8.5(g) shows that the trailing partial is visible for the $\mathbf{g}_{\bar{1}\bar{3}1}$ reflection; this leaves only the Burgers vector $\mathbf{b}_2$, since $\mathbf{g}_{\bar{1}\bar{3}1} \cdot \mathbf{b}_1 = 0$. The original dislocation which dissociated into the two partials and created the stacking fault is therefore a $\frac{1}{2}[01\bar{1}]$ dislocation.

The analysis in the preceding paragraphs is straightforward for the particular images shown in Fig. 8.5, since sufficient information was available concerning the orientation of the foil, and the various active reflections. In general, the identification of dislocations can be a complex problem, in particular if there are many possibilities to choose from. The more crystallographic information can be determined, the easier identification will become. Several worked out examples of dislocation characterization can be found in Chapters 5 and 8 in [HHC+73]. The reader is encouraged to study these methods, and to apply them to dislocation characterization in the four study materials. In many cases, image simulations will be needed to distinguish between several possible dislocation models. In the next section, we will describe the basics of dislocation contrast simulations for the two-beam and systematic row cases.

8.4.2.4 Image simulations

Dislocation image simulations have a long history, starting with the well-known ONEDIS and TWODIS programs, developed at the University of Melbourne, and described in detail in the book *Computed Electron Micrographs and Defect Identification* [HHC+73]. On the website, the reader will find the Fortran-77 source code for the program hh.f, which is based on the original dislocation programs. The program was modified extensively by P. Skalicky and coworkers of the Technical University of Vienna, Austria, and is made available as *public domain* source code with their permission. The modified program has been augmented to deal with up to four parallel dislocations and three stacking faults, and also takes into account the effects of piezoelectric charges along the dislocation line.

The program is written in standard Fortran-77, and makes use of the so-called *generalized cross-section*. Up to four parallel dislocations connected by stacking faults can be simulated. The generalized cross-section reduces the total number of integrations that need to be carried out to compute the complete dislocation image. Consider the straight dislocation in Fig. 8.6: the dislocation line *PQ* runs from the bottom left to the top right of the thin foil, which itself is tilted with respect to

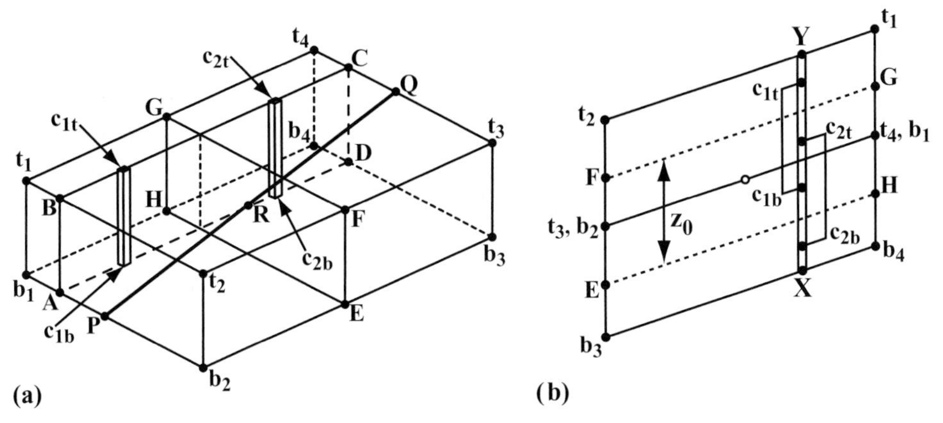

Fig. 8.6. (a) Schematic illustration of a straight dislocation in a tilted foil; neighboring columns in the plane *ABCD* have identical displacements on lines parallel to the dislocation line; (b) generalized cross-section. (Figure based on Figs 4.2, 4.4, and 4.5 in [HHC+73].)

the incident electron beam. The integration columns are parallel to the beam. If we consider two columns in the plane *ABCD* then the displacements are identical along a line parallel to the dislocation line. This means that we only need to consider one single plane, normal to the plane *ABCD*: the plane *EFGH*. If we draw the dislocation line normal to the plane of the drawing (central circle in Fig. 8.6b), then each vertical line corresponds to an entire plane of the type *ABCD*. The column with bottom c_{1b} and top c_{1t} corresponds to a segment of the line XY in Fig. 8.6(b). The column c_{2b}–c_{2t} corresponds to a different segment of the line XY, so that we only need to evaluate the displacements along the entire line XY to cover the entire plane *ABCD*. This leads to a significant reduction of the number of integration steps to be carried out. For a detailed description of the generalized cross-section we refer the interested reader to Chapter 4 in [HHC+73].

Because of the geometry of the generalized cross-section, the hh.f program cannot deal with dislocations for which $\mathbf{u} \perp \mathbf{B}$ or $\mathbf{u} \parallel \mathbf{B}$. The expression (8.38) for the displacement components was efficiently encoded in Fortran-77 in the routine ANCALC of the original "Head–Humble" code. On the website, the hh.f program is available in two different forms. The standard form takes input in namelist format† and produces an ASCII output file which contains the bright field and dark field intensities for the entire image; this file must then be converted to a more standard image file format (GIF, TIFF, JPEG, PostScript, etc.) for display. The second format of the program links the hh executable directly into the **ION** framework, and all variables are directly passed from IDL to the executable; the resulting images are

† Namelist format is a standard Fortran input format in which all variables in a common block can be read in from a textfile. Examples of the correct namelist format for the hh.f program are available on the website.

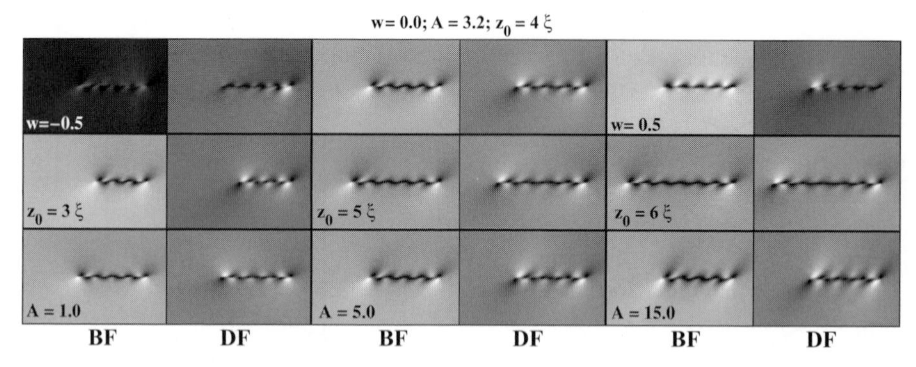

Fig. 8.7. Image simulations using the parameters in Table 8.1 and the hh.f program. The following parameters are varied: excitation error (top row), foil thickness (center row), and elastic anisotropy factor (bottom row).

then passed back to the **ION** routine for display as a JPEG image. The source code for the second version is available as the hhion.f program.†

In 1973, the ONEDIS program required about 40–50 s of computation time per image (CDC 3600 computer with 32 Kbyte of random access memory) [HHC+73, page 310]. This computation time refers to a single image (bright field *or* dark field) on a grid of 128 × 80 columns. The computational power has increased significantly during the past three decades, and it is now possible to run dislocation image simulations almost in real-time; a *pair* of bright field-dark field images on a grid of 256 × 160 columns takes about 0.1 s on a Compaq TRU64 666 MHz workstation, so that a tilt series of images (i.e. dislocation images as a function of excitation error) can be computed in a matter of a few seconds. Examples of such tilt series "movies" can be found on the website.

An example image simulation for the dislocation configuration described in Section 8.3.2 and Fig. 8.3 is shown in Fig. 8.7, for a range of computational parameters. The main parameters are summarized in Table 8.1. The elastic moduli of pure copper are $C_{11} = 168.4$ GPa, $C_{12} = 121.4$ GPa, and $C_{44} = 75.5$ GPa [Edi76], resulting in an elastic anisotropy factor A of 3.21. In a cubic crystal system, A is defined by

$$A = \frac{C_{11} - C_{12}}{2C_{44}}.$$

For an elastically isotropic material, we have $A = 1$. The simulations in Fig. 8.7 were carried out with the hh.f program. The top center image pair shows the bright field and dark field images computed on a grid of 256 × 160 pixels, for the indicated

† Implementation of this version requires the IDL programming environment.

Table 8.1. *Computational parameters for the dislocation simulation examples. The geometry is based on Fig. 8.3. The extinction distance and absorption lengths were computed for copper at 200 kV, using the Weickenmeier–Kohl scattering and absorption form factors.*

Description	Symbol	Value
Dislocation line direction	$\mathbf{u}$	$[1\bar{1}2]$
Dislocation Burgers vector	$\mathbf{b}$	$\frac{1}{2}[110]$
Foil normal	$\mathbf{F}$	$[\bar{1}\bar{1}3]$
Beam direction	$\mathbf{B}$	$[\bar{3}\bar{2}5]$
Diffracting planes	$\mathbf{g}$	(111)
Extinction distance [nm]	ξ_{111}	38.3
Normal absorption length [nm]	ξ'_0	316.6
Anomalous absorption length [nm]	ξ'_{111}	679.5
Foil thickness [nm]	z_0	$4\xi_{111} = 153.3$
Thickness along **B** [nm]	z_b	156.7

parameters. To the left and right of these images, the sample excitation error is changed, and this is expressed by the dimensionless parameter $w = s_{\mathbf{g}}\xi_{\mathbf{g}} = \pm 0.5$. For a negative excitation error, the bright field image becomes rather dark, in agreement with the conclusions of Chapter 6. The fine details of the dislocation contrast are highly sensitive to the precise sample orientation.

Dislocation contrast often has the shape of a "wiggly line", and there is a simple relation between the effective crystal thickness z_b, the two-beam extinction distance $\xi_{\mathbf{g}}$, and the number of wiggles or oscillations n: $z_b = n\xi_{\mathbf{g}}$. If the thickness is decreased or increased, the number of wiggles changes accordingly. This is illustrated in the second line of Fig. 8.7, for $n = 3$, 5, and 6. The contrast near the ends of the line does not depend strongly on the foil thickness. The bottom row shows the effect of changing the elastic anisotropy factor A, for $A = 1.0$, 5.0, and 15.0. There are not many materials with such large anisotropy factors; materials that undergo martensitic phase transformations will typically have a large elastic anisotropy.

Dislocation image simulations can also be carried out with the SRdefect.f90 program. Since this program does not use the generalized cross-section geometry, the computations take significantly longer than those of the hh.f program. It is, however, possible to include multiple dislocations in arbitrary orientations, including parallel or normal to the incident beam. It is also possible to compute weak-beam dislocation images, since the complete systematic row dynamical interactions are

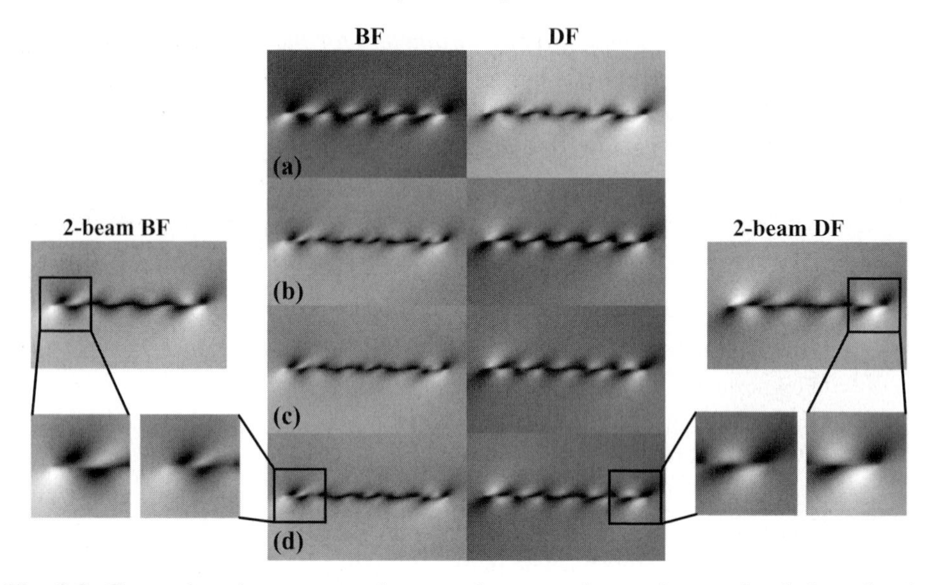

Fig. 8.8. Comparison between two-beam and systematic row image simulations for the computational parameters of Table 8.1. The central row shows n-beam simulations, with $n = 3$ (a), $n = 5$ (b), $n = 7$ (c) and $n = 9$ (d). The ends of the dislocation line are compared for the two-beam and multi-beam cases.

taken into account. Alternatively, one could write a simulation program based on the kinematical theory, which is valid for the reflection **g** in the **g**–3**g** orientation.

Figure 8.8 shows a comparison between the two-beam computation using the hh.f program, and the systematic row computations of SRdefect.f90 (the program used 1 nm thick slices and the scattering matrix formalism). The parameters are the same as in Table 8.1. The two-beam bright field and dark field images are shown on the left and right. The central column contains multi-beam simulations using the beams $-n\mathbf{g}\ldots+n\mathbf{g}$, for $n = 1$ (a), $n = 2$ (b), $n = 3$ (c), and $n = 4$ (d). The main difference between the two simulations is the number of wiggles along the dislocation line. It is clear from the $n = 4$ case that there are five wiggles along the line, as opposed to four for the two-beam case. The foil thickness was identical for the two-beam and systematic row simulations. The reason for this difference lies in the fact that the extinction distance ξ_{111} describes the complete scattering process for the two-beam case, but not for the systematic row case. Other interactions occur for the systematic row case, which decrease the value of ξ_{111} to a smaller, effective value ξ^e_{111}. The number of wiggles is then related to the foil thickness by $z_b = n\xi^e_{111}$, instead of $z_b = n\xi_{111}$. For a more detailed analysis of this effect, we refer the reader to [HHC+73, Section 9.3].

If we compare the ends of the dislocation images, then we see that there is very little difference between the two-beam and systematic row simulations. This gives

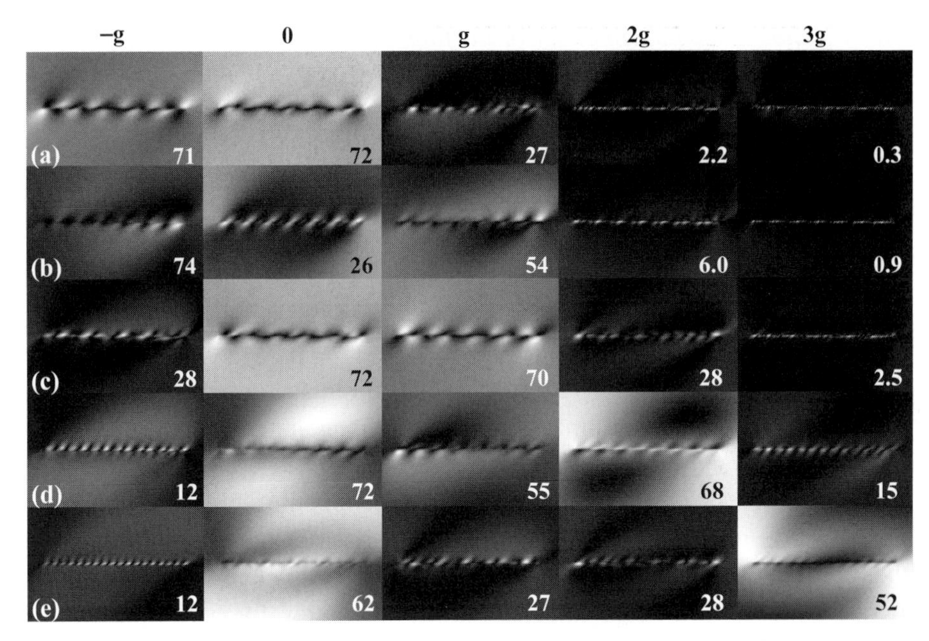

Fig. 8.9. Nine-beam systematic row simulations for the following orientations: (a) $s_{-\mathbf{g}} = 0$, (b) $s_{-\mathbf{g}} = s_{\mathbf{g}}$, (c) $s_{\mathbf{g}} = 0$, (d) $s_{2\mathbf{g}} = 0$, and (e) $s_{3\mathbf{g}} = 0$. The numbers in the lower right-hand corners indicate 1000 times the maximum intensity level in each image.

us some confidence that two-beam image simulations of dislocation contrast can be useful and reliable, as long as effective extinction distances are used. A distinct advantage of the systematic row simulation is that larger deviations from the Bragg orientation can be reliably simulated. This includes the symmetric systematic row orientation and the weak beam **g**–3**g** orientation. Figure 8.9 shows the results of a nine-beam computation for five different foil orientations. The excitation errors of the reflection $\mathbf{g}_{111}$ are given by (in nm $^{-1}$): (a) $-0.057\,57$ ($s_{-\mathbf{g}} = 0$); (b) $-0.028\,79$ ($s_{-\mathbf{g}} = s_{\mathbf{g}}$); (c) 0.0; (d) $+0.028\,79$ ($s_{2\mathbf{g}} = 0$); and (e) $+0.057\,57$ ($s_{3\mathbf{g}} = 0$). The number in the lower right-hand corner indicates the maximum intensity level in each individual image, multiplied by 1000. All images were scaled between 0 (black) and the maximum intensity (white).

A comparison of the first and third row bright field images shows that the contrast near the ends of the dislocation line is complementary. The right-hand side of the dislocation line corresponds to the top of the thin foil. This can be derived from the image, by comparing the bright field and dark field images in rows (a) or (c). The right-hand side end contrast is the same for both −**g** and **0** images in row (a), and also for **0** and **g** in row (c). The contrast lobes have opposite contrast for the other side of the dislocation line. It is also clear that the weak beam image in row (e) is significantly narrower that the conventional dark field image, so that the weak

Table 8.2. *Elastic moduli (in GPa) for the four study materials [Edi76]. (Material Cu-15 at% Al is approximated by pure copper.)*

	C_{11}	C_{12}	C_{44}	C_{13}	C_{33}	C_{66}
Cu-15 at% Al	168.4	121.4	75.5			
Ti	162.0	92.0	46.7	69.0	181.0	35.0
GaAs	119.2	59.9	59.4			
$BaTiO_3$	282.6	186.5	80.6	141.6	178.1	113.1

beam observation mode results in finer dislocation image contrast. For a detailed discussion of the weak beam imaging mode and image interpretation we refer the reader to [Sto70] and Chapter 26 in [WC96].

For completeness, we list in Table 8.2 the elastic moduli for the four study materials, so that the interested reader can carry out dislocation image simulations using either the hh.f or the SRdefect.f90 program. The **ION**-implementation of the hhion.f program is limited to the simulation of dislocation contrast for cubic materials. The elastic moduli cannot be independently varied, and only the anisotropy factor A can be specified.

8.4.3 Planar defects

In this section, we turn our attention to planar defects with essentially discontinuous displacement fields $\mathbf{R}(\mathbf{r})$; these include stacking faults, anti-phase boundaries, twins, and inversion boundaries. We will first discuss the crystallographic features of these defects; then we introduce a simple method to compute their two-beam diffraction contrast. After several experimental and computational examples, we conclude this section with a brief discussion of the systematic row contrast of planar faults.

8.4.3.1 Crystallography

Consider a compound with composition A_3B, which forms a structure with Bravais lattice cF; the two different atomic species A and B are placed randomly on the lattice sites. This generates a *disordered* face-centered cubic structure (Fig. 8.10a). The Bravais lattice cF is described by the infinite group

$$\mathcal{T} = \{\mathbf{a}, \mathbf{b}, \mathbf{c}, \mathbf{A}, \mathbf{B}, \mathbf{C}, \ldots\},$$

using the notation from Chapter 1. The infinite lattice is formed from integer linear combinations of the basis vectors and the centering vectors. The structure factor of each allowed reflection is proportional to $|f_B + 3f_A|$.

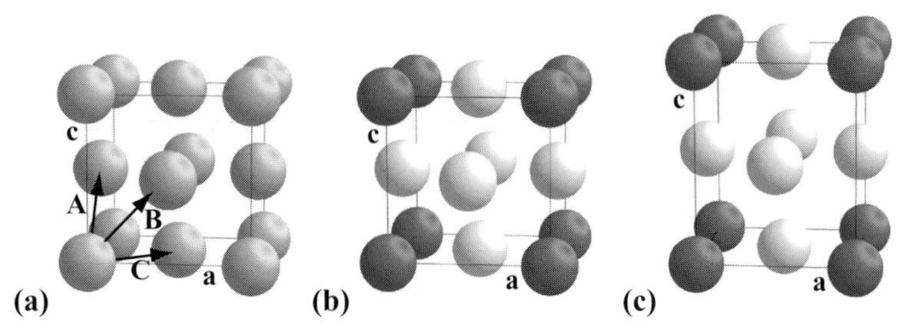

Fig. 8.10. (a) Disordered face-centered cubic structure; (b) $L1_2$ ordered Cu_3Au structure; (c) tetragonally distorted version of (b).

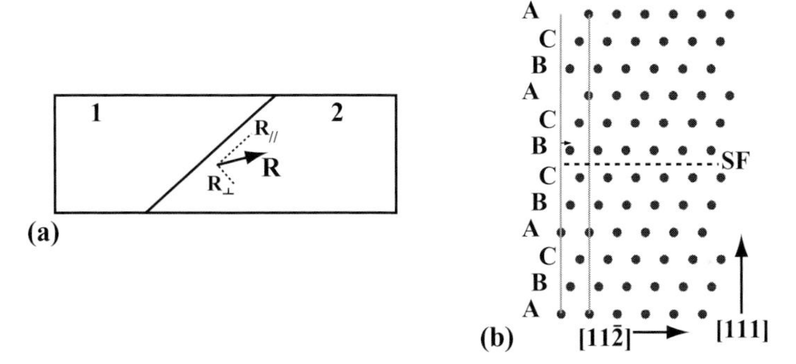

Fig. 8.11. (a) Schematic illustration of a planar defect inclined with respect to the foil normal. The displacement vector **R** has components parallel ($\mathbf{R}_{\|}$) and normal ($\mathbf{R}_{\perp}$) to the defect interface plane. (b) An $\mathbf{R} = \frac{1}{6}[11\bar{2}]$ stacking fault in the $[1\bar{1}0]$ projection of the face-centered cubic structure. The fault changes the stacking sequence of close-packed planes to ... ABC **ABCBC**ABC ...; the four highlighted planes correspond to a hexagonal close-packed structure.

Consider next the schematic in Fig. 8.11(a): the crystal is divided into two regions. Region 2 is shifted with respect to region 1 by the displacement vector **R**, which may have a component parallel to the fault plane (hkl), $\mathbf{R}_{\|}$, and a component normal to the fault plane, $\mathbf{R}_{\perp}$. If the displacement vector is not a lattice translation vector, i.e. if $\mathbf{R} \notin \mathcal{T}$, then the resulting fault is known as a *stacking fault* or SF. An example of a stacking fault in the face-centered cubic lattice is shown in the $[1\bar{1}0]$ projection of Fig. 8.11(b): the top half of the crystal is displaced by a vector $\mathbf{R} = \frac{1}{6}[11\bar{2}]$ with respect to the bottom half, and the fault plane is the (111) close-packed plane. This type of stacking fault is commonly caused by glide of a Schockley partial dislocation with Burgers vector $\mathbf{b} = \mathbf{R}$ [HB84]. Stacking faults are nearly always planar faults. If there are several possible glide planes, then the plane of the fault can change abruptly when the partial dislocation cross-slips onto another glide plane.

We already know that the contrast of defects is determined by the factor $\alpha = 2\pi\mathbf{g}\cdot\mathbf{R}$; if **R** is a lattice translation vector, then the dot product $\mathbf{g}\cdot\mathbf{R}$ is always equal to an integer, and therefore $\alpha = 2\pi n$ and there is no defect contrast. For a stacking fault $\mathbf{R}$ is not a lattice translation vector, which means that α will be a non-integer multiple of π. We will see in Section 8.4.3.3 that the bright field and dark field contrast of stacking faults is entirely determined by the value of α. The typical fringe contrast associated with stacking faults is commonly known as α-fringe contrast.

Next, assume that, upon lowering the temperature, the disordered structure of Fig. 8.10(a) can undergo two phase transformations: the first, at temperature T_1, orders the structure into the $L1_2$ or Cu_3Au structure (shown in Fig. 8.10b). The second, at temperature T_2, distorts the unit cell into a tetragonal cell with $c/a > 1$ (Fig. 8.10c).

The $L1_2$ structure has the cP Bravais lattice;† the corresponding translation group is the infinite group

$$\mathcal{T}_o = \{\mathbf{a}, \mathbf{b}, \mathbf{c}, \ldots\},$$

which is a subgroup of $\mathcal{T}$, i.e. $\mathcal{T}_o \subset \mathcal{T}$. When the ordering process begins, a particular B atom has four possible sites that it can move onto. If the B atoms occupy the sites $(0, 0, 0) + \mathcal{T}_o$ in region 1 of the crystal, and sites $\mathbf{A} + \mathcal{T}_o$ in region 2, then at the contact plane the two ordered regions will be out-of-phase. This type of fault is known as an *anti-phase boundary* or APB. The displacement vector across an APB is a translation vector of the disordered structure, but not of the ordered structure, i.e. $\mathbf{R} \in \mathcal{T}$ but $\mathbf{R} \notin \mathcal{T}_o$ [VTA74]. In the [100] projection of Fig. 8.12 the displacement vector is the centering vector $\mathbf{A}$ (upper left) or $\mathbf{B}$ (lower right), which clearly belong to $\mathcal{T}$ and not to $\mathcal{T}_o$. The two anti-phase domains are known as *translation variants*.

The total number of independent translation variants of an ordered structure is equal to the ratio of the volume of the primitive unit cell of $\mathcal{T}_o$ to the volume of the primitive unit cell of $\mathcal{T}$. In the example above, the volume of the primitive unit cell of $\mathcal{T}_o$ is equal to a^3, and for $\mathcal{T}$ we have a volume of $a^3/4$. The total number of translation variants is therefore equal to 4; there are, then, three different APBs possible between those translation variants (one less than the number of translation variants) [VTA74]. The APBs can be planar or curved. If the APB energy is isotropic, then the faults will in general be curved, since there is no preferential orientation of the fault plane. If, on the other hand, the APB energy is strongly anisotropic, then the APBs will be formed on particular lattice planes. The isotropic or anisotropic nature of the APB energy can depend strongly on the details of the crystal structure.

While stacking faults do not change the stoichiometry of the material, anti-phase boundaries may change the local composition, depending on the direction of the

† $L1_2$ has space group $\mathbf{Pm\bar{3}m}$, with atom A at the face-centers and B at the origin.

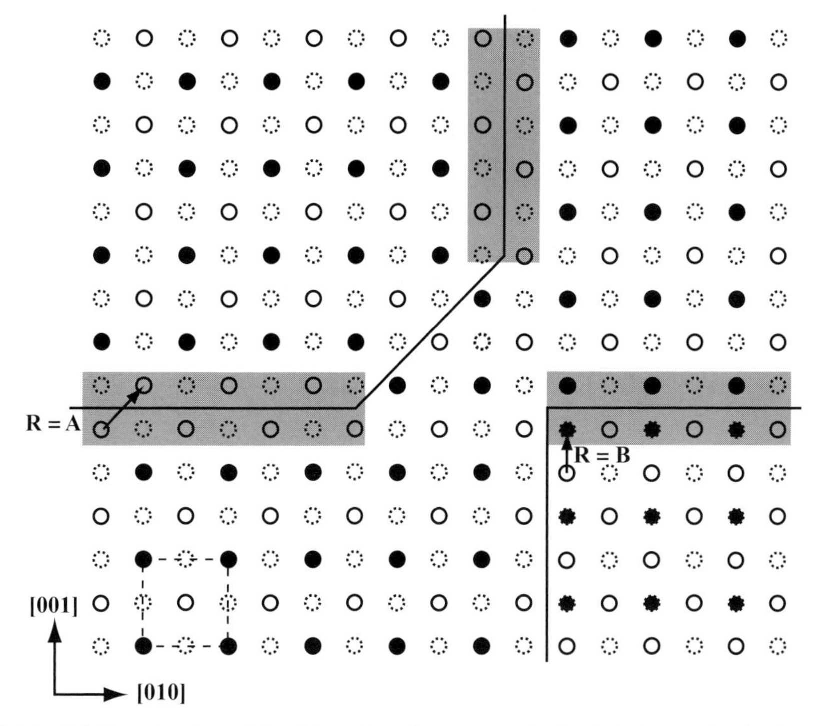

Fig. 8.12. [100] projection of the $L1_2$ ordered structure; dashed circles are located at $z = \frac{1}{2}$. The first APB (upper left) has displacement vector $\mathbf{R} = \mathbf{A}$, with two non-conservative segments (gray rectangles). The second APB has its displacement vector $\mathbf{R} = \mathbf{B}$ with a component out of the plane of the drawing. One segment of the defect interface is conservative, the other is not.

APB vector with respect to the fault plane normal **n**. If the vector is parallel to the fault plane, i.e. $\mathbf{R} \cdot \mathbf{n} = 0$, then the fault is known as a *conservative fault*, meaning that the local chemistry is conserved. If, on the other hand, $\mathbf{R} \cdot \mathbf{n} \neq 0$, then material must be inserted or removed to regain a periodic underlying lattice across the fault plane, and the fault is known as a *non-conservative fault*. The APBs in Fig. 8.12 have both conservative and non-conservative segments.

The image contrast of APBs is also determined by the number α. Since we have an ordered structure, there will be two types of reflections **g**: fundamental reflections and superlattice reflections. For all fundamental reflections we have $\alpha = 2\pi n$ and, hence, the APBs will be invisible. For superlattice reflections we have $\alpha = \pm\pi$, and the resulting fringe pattern is known as π-fringe contrast (more details in Section 8.4.3.4).

Finally, consider the transformation which takes place when the ordered phase is cooled down below the temperature T_2. For simplicity, the structure only expands along the c-axis and contracts along the normal directions, so that $c/a > 1$. The

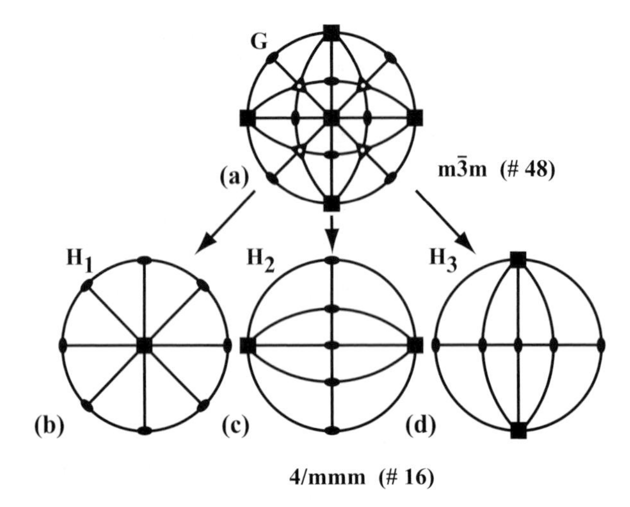

Fig. 8.13. Point group drawings for the cubic face-centered structure $G = \mathbf{m\bar{3}m}$, and the three orientation variants of the $H_i = \mathbf{4/mmm}$ point group derived from it. The groups H_i are related to each other by the three-fold rotation operator of the parent group G.

Bravais lattice changes from cP to tP, and the space group of the resulting phase is **P4/mmm**. The corresponding point groups are shown as stereographic projections in Fig. 8.13: the cubic point group $\mathbf{m\bar{3}m}$ is of order $p = 48$, while the tetragonal point group **4/mmm** has order $q = 16$. It is easy to see that the tetragonal unit cell can be formed from the cubic one in three different ways: the expansion can take place along either **a**, **b**, or **c**. The stereographic projections of the three *orientation variants* of the tetragonal phase are shown in Fig. 8.13(b)–(d), in the proper orientation with respect to the parent phase. It is easy to see that the three-fold rotation axis **3** of the cubic symmetry can take one of the orientation variants and transform its symmetry group H_i into either of the other two groups. This is quite a general observation and it can be shown that the orientation variants of the product phase are related to each other by a symmetry operation of the parent phase that is not a symmetry operation of the product phase, in this case the three-fold rotation axis. The number of orientation variants is given by the ratio of the point group orders of the parent phase to the product phase, i.e. $p/q = 48/16 = 3$ for the example structures. The statements in this paragraph can be derived rigorously from group theory, and we refer the interested reader to [VTA74] and [TT87] for a detailed description.

The boundaries between different orientation variants are, in general, planar interfaces and therefore fall in the category of planar defects. Figure 8.14 shows a possible configuration for the orientation variants H_2 and H_3 of the example structure. The [001] projection shows the H_3 variant sandwiched between two H_2 regions, separated by $(10\bar{1})$-type interfaces. It is clear that the structure fits nicely across the interface, and that the original unit cell directions are rotated slightly

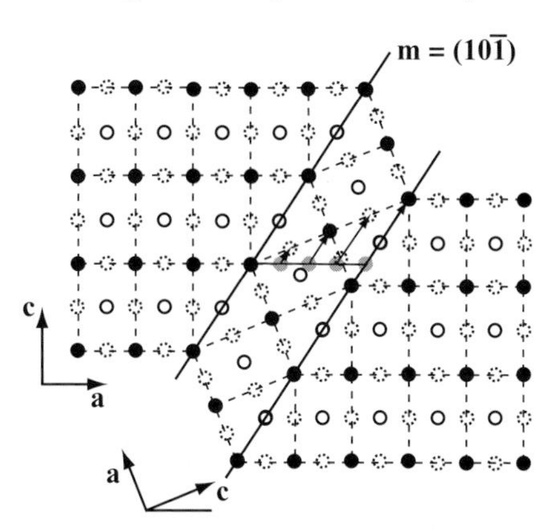

Fig. 8.14. Possible configuration of the H_2 and H_3 orientation variants of the tetragonally distorted structure. The $(10\bar{1})$ plane is the interface plane across which the structure is mirrored. The arrows in the central variant indicate how the displacement vector is a linear function of distance to the interface plane.

around the [001] direction to make everything fit (i.e. the c-axis of the H_3 variant is not parallel to the a-axis of the H_2 variant). The Miller indices of the interface plane and the rotation angle can be derived from crystallographic theories, and the reader is referred to [WLR53] and [PDGT98] for more details. The $(10\bar{1})$ interface is a mirror plane that relates the two structures to each other, and is therefore known as a *reflection twin*. Another possible interface would involve a rotation of the structure around the normal to the interface, which results in a *rotation twin*. For an extensive review of deformation twinning we refer to [CM95]. The displacement vector associated with an orientation variant boundary is in general a linear function of the distance to the interface plane, as illustrated by the arrows in the central H_3 variant. The magnitude of the displacement depends also on the lattice parameters of the tetragonally distorted structure.

In non-centrosymmetric materials, another type of orientation variant boundary can occur: the *inversion boundary*. Consider the $BaTiO_3$ structure in Fig. 8.15(a): on the left-hand side the displacement of the Ti atom is along the positive [001] direction, whereas on the right the Ti atom is shifted in the opposite direction. The two variants are related to each other by an inversion symmetry operation. In terms of the ferroelectric properties of this material we can say that the electrostatic dipole is in opposite directions across the inversion boundary, which is, therefore, also known as a 180° domain wall. In addition, the orientation variant boundary could separate two regions in which the displacements of the Ti atom are at right angles to each other, as shown in Fig. 8.15(b). This is known as a 90° domain wall. Across this boundary two of the three unit cell axes are interchanged, and for this reason the boundary is also known as a *permutation boundary*. In general, there

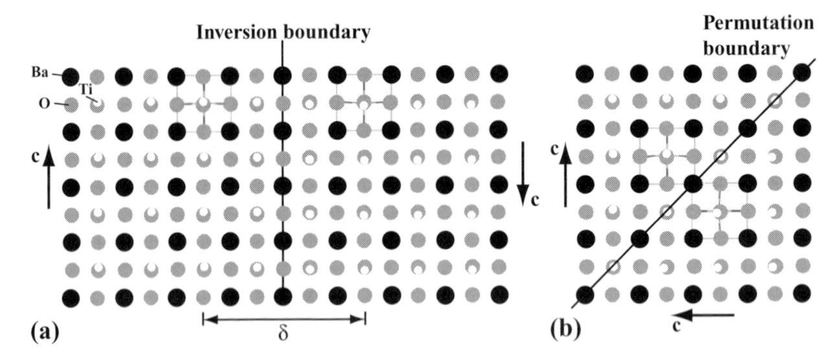

Fig. 8.15. (a) 180° inversion boundary in the $BaTiO_3$ structure. The Ti atom is indicated by a small white circle. (b) Permutation or 90° domain boundary in $BaTiO_3$.

will be a region of finite width across which the dipole moment changes orientation. This is illustrated in Fig. 8.15(a) by the gradual shift of the Ti atom upon crossing the inversion boundary plane. The width δ of this transition region is known as the *domain wall width.* In magnetic materials similar orientation variant boundaries exist between regions with magnetization along different orientations. While the presence of electrostatic dipole moments is invariably linked to atom positions, the presence of magnetic dipole moments does not necessarily give rise to atom displacements across the magnetic domain wall. It is for this reason that magnetic domain walls typically do not follow the image contrast rules that we will describe in later sections. Magnetic domain walls must be studied using the Fresnel and Foucault modes of Lorentz microscopy, introduced in Section 4.7. The types of domain walls that occur in various materials depend on the direction of the dipole moments (either electrostatic or magnetic) and on the underlying crystal structure. For instance, in a face-entered cubic material with magnetization along the $\langle 111\rangle$ directions, the domain wall types are 180°, 71°, and 109° walls. These angles are the possible angles between directions of the $\langle 111\rangle$ family.

The example structures (fcc, $L1_2$, and tetragonally distorted $L1_2$) were used to introduce the basic planar defects. There are many more possible structures and planar defects. For each structure, the nature of the defects will be different in the details, but similar in the broad features introduced in this section. The same is true for dislocations: the crystallographic details of dislocations depend in a sensitive way on the crystal structure, while many other properties are general for all dislocations. The theory of phase transformations and the resulting crystallographic microstructures is rich and continues to be an active field of research. We refer the interested reader to the proceedings of the conferences on *solid–solid phase transformations* for a review of a wide range of phase transformations and associated defects and microstructures [ALSW84, Tsa84, Lor87, JHLS94, KOM99].

More complex defects can have characteristic features of more than one of the defect types introduced in this section. In ordered alloys, for instance, stacking faults may be formed by the dissociation of perfect dislocations into more than two partial dislocations. The stacking fault will then consist of regions that have pure SF nature, and regions where the ordered lattice on one side of the fault plane is shifted by a lattice centering vector with respect to the lattice on the other side. This means that the resulting fault is a stacking fault and at the same time an anti-phase boundary. Such a defect is commonly known as a *complex fault*.

In summary, we have defined the following planar defects.

- **Stacking faults**: the displacement vector is not a Bravais lattice translation vector.
- **Anti-phase boundaries**: the displacement vector is a lattice vector of the disordered phase, but not of the ordered phase.
- **Orientation variant interfaces**: the variants are related by a symmetry operation of the point group of the parent phase, but not of the point group of the product phase. This includes domain boundaries in ferroelectric and magnetic materials, and (transformation) twin boundaries. The displacement vector is typically a linear function of the distance to the interface.

8.4.3.2 Planar defect contrast formation (two-beam case)

In Section 6.5.2, we introduced the two-beam scattering matrix formalism, in which the general solution of the two-beam case is written by means of the scattering matrix $\mathcal{S}(s, z_0)$ (equation 6.49 on page 380):

$$\mathcal{S}(s, z_0) \equiv \mathrm{e}^{-(\pi/\xi_0')z_0} \begin{pmatrix} T & S\mathrm{e}^{-\mathrm{i}\theta_{\mathbf{g}}} \\ S\mathrm{e}^{\mathrm{i}\theta_{\mathbf{g}}} & T^{(-)} \end{pmatrix}. \tag{8.40}$$

This matrix is also known as the *response matrix* of the crystal. For the centrosymmetric case, the phase factors on the off-diagonal elements are equal to unity. For simplicity, we will ignore the normal absorption exponential, although it should be taken into account for all numerical work.

In the presence of a defect with displacement field **R**, the two beam equations (6.14) and (6.15) are modified by a phase factor:

$$\frac{\mathrm{d}T}{\mathrm{d}z} + \pi \mathrm{i} s_{\mathbf{g}} T = \frac{\mathrm{i}\pi}{q_{-\mathbf{g}}} \mathrm{e}^{\mathrm{i}\alpha_{\mathbf{g}}} S; \tag{8.41}$$

$$\frac{\mathrm{d}S}{\mathrm{d}z} - \pi \mathrm{i} s_{\mathbf{g}} S = \frac{\mathrm{i}\pi}{q_{\mathbf{g}}} \mathrm{e}^{-\mathrm{i}\alpha_{\mathbf{g}}} T. \tag{8.42}$$

Consider a crystal with a planar defect as shown schematically in Fig. 8.16(a). The top section (I) of the crystal is defect-free and is, hence, described by the scattering matrix $\mathcal{S}(s_1, z_1)$. Section II has a slightly different orientation (different excitation

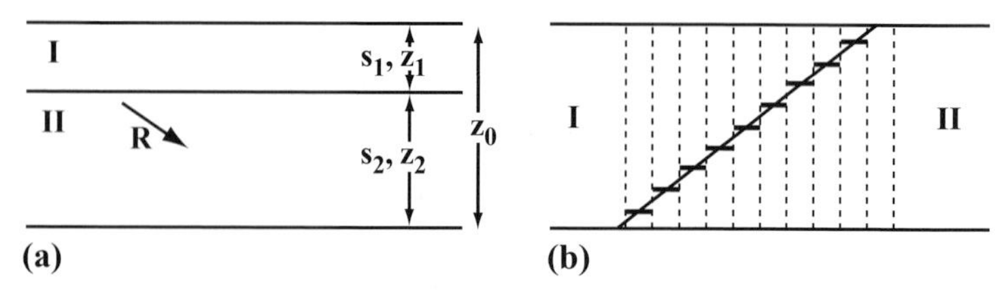

Fig. 8.16. (a) Schematic illustration of a planar defect separating crystal sections I and II; (b) inclined planar defect combined with the column approximation.

error) and may also be translated with respect to the top section. This section can also be described by a scattering matrix which we will denote by $\mathcal{M}$. We will follow Amelinckx and Van Landuyt [AVL76] and write

$$\mathcal{M}(s_2, z_2) = \begin{pmatrix} A & B \\ C & D \end{pmatrix}.$$

If we substitute $T' = T$ and $S' = \mathrm{e}^{\mathrm{i}\alpha_{\mathbf{g}}} S$, then equations (8.41) and (8.42) reduce to the equations for the perfect crystal. This means that for the first column of $\mathcal{M}(s_2, z_2)$ we have $A = T$ and $C = \mathrm{e}^{-\mathrm{i}\alpha_{\mathbf{g}}} S$. B and D can be determined in a way similar to that described in Section 6.5.2 and we find $B = \mathrm{e}^{\mathrm{i}\alpha_{\mathbf{g}}} S$ and $D = T^{(-)}$, or

$$\mathcal{M}(s_2, z_2) = \begin{pmatrix} T(s_2, z_2) & \mathrm{e}^{\mathrm{i}\alpha_{\mathbf{g}}} S(s_2, z_2) \\ \mathrm{e}^{-\mathrm{i}\alpha_{\mathbf{g}}} S(s_2, z_2) & T^{(-)}(s_2, z_2) \end{pmatrix}.$$

Writing the argument (s_2, z_2) as a subscript 2, we can split the scattering matrix into three submatrices:

$$\mathcal{M}_2 = \mathcal{P}(-\alpha_{\mathbf{g}})\mathcal{S}_2\mathcal{P}(\alpha_{\mathbf{g}}) = \begin{pmatrix} 1 & 0 \\ 0 & \mathrm{e}^{-\mathrm{i}\alpha_{\mathbf{g}}} \end{pmatrix} \begin{pmatrix} T_2 & S_2\mathrm{e}^{-\mathrm{i}\theta_{\mathbf{g}}} \\ S_2\mathrm{e}^{\mathrm{i}\theta_{\mathbf{g}}} & T_2^{(-)} \end{pmatrix} \begin{pmatrix} 1 & 0 \\ 0 & \mathrm{e}^{\mathrm{i}\alpha_{\mathbf{g}}} \end{pmatrix}. \tag{8.43}$$

The defect phase shift matrix $\mathcal{P}$ is defined by

$$\mathcal{P}(\omega) \equiv \begin{pmatrix} 1 & 0 \\ 0 & \mathrm{e}^{\mathrm{i}\omega} \end{pmatrix}. \tag{8.44}$$

The transmitted and scattered amplitudes at the exit plane of the foil are then given by

$$\begin{pmatrix} \psi_0 \\ \psi_{\mathbf{g}} \end{pmatrix} = \mathcal{P}(-\alpha_{\mathbf{g}})\mathcal{S}_2\mathcal{P}(\alpha_{\mathbf{g}})\mathcal{S}_1 \begin{pmatrix} 1 \\ 0 \end{pmatrix}. \tag{8.45}$$

If more than one planar defect overlaps along the beam direction, then we can use the additive property of the $\mathcal{P}$ matrix ($\mathcal{P}(a)\mathcal{P}(b) = \mathcal{P}(a+b)$) to write

$$\begin{pmatrix} \psi_0 \\ \psi_{\mathbf{g}} \end{pmatrix} = \left[\prod_{i=1}^{N} \mathcal{P}(\alpha_{i+1} - \alpha_i)\mathcal{S}_i \right] \begin{pmatrix} 1 \\ 0 \end{pmatrix}, \tag{8.46}$$

with $\alpha_1 = \alpha_{N+1} = 0$, α_{i+1} the phase shift introduced by the defect between crystal sections i and $i+1$, and N the number of crystal sections. Although the theory is valid for an arbitrary number of overlapping interfaces, in the remainder of this section we will restrict the discussion to a single boundary. Inclined boundaries can be described by means of the column approximation; i.e. the crystal is sliced into columns parallel to the electron beam and the above equation is used inside each column (using the appropriate thickness and excitation error for that column, see Fig. 8.16b).

For a single interface (as in Fig. 8.16a), we can explicitly work out the transmitted and diffracted amplitude:

$$\psi_{\mathbf{0}} = \mathrm{e}^{-\frac{\pi}{\xi_0'}(z_1+z_2)} \left[T_1 T_2 + S_1 S_2 \mathrm{e}^{\mathrm{i}\alpha_{\mathbf{g}}} \mathrm{e}^{\mathrm{i}(\theta_{\mathbf{g}}^I - \theta_{\mathbf{g}}^{II})} \right];$$

$$\psi_{\mathbf{g}} = \mathrm{e}^{-\frac{\pi}{\xi_0'}(z_1+z_2)} \left[S_1 T_2^{(-)} + T_1 S_2 \mathrm{e}^{-\mathrm{i}\alpha_{\mathbf{g}}} \mathrm{e}^{-\mathrm{i}(\theta_{\mathbf{g}}^I - \theta_{\mathbf{g}}^{II})} \right] \mathrm{e}^{\mathrm{i}\theta_{\mathbf{g}}^I}.$$

There are two contributions to each beam: the final transmitted beam contains the doubly transmitted term $T_1 T_2$ and the doubly scattered term $S_1 S_2 \mathrm{e}^{\mathrm{i}\alpha_{\mathbf{g}}}$. Upon crossing the interface, the diffracted amplitude suffers a phase shift, determined by the nature of the interface. These equations represent the most general solution for an arbitrary planar interface. The intensities are as before given by $I_T = |\psi_{\mathbf{0}}|^2$ and $I_S = |\psi_{\mathbf{g}}|^2$.

In the following sections, we will describe the intensity profiles for a number of different planar defects. We will use a simple computational model to represent the various fringe patterns associated with planar defect contrast. Consider the wedge-shaped crystal shown in Fig. 8.17. The wedge thickness varies from t_1 in the front to t_2 in the back. The fault plane is inclined so that it intersects the points 1, 2, and

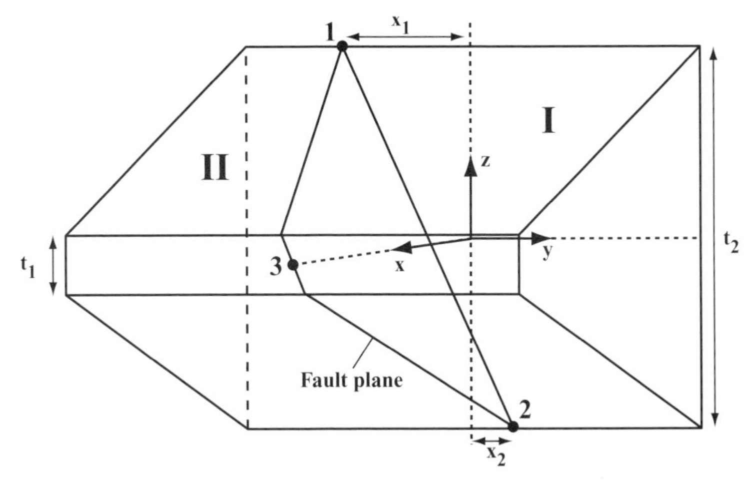

Fig. 8.17. Schematic representation of the geometry used by the pfic.pro program.

3 with coordinates $(0, -x_1, -t_2/2)$, $(0, x_2, -t_2/2)$, and $(N, 0, 0)$, respectively. N is equal to the number of image pixels times the pixel width. The equation for the plane is derived easily as

$$\frac{4x_1 + 2(x_2 - x_1)}{t_2} z = (x_2 - x_1)\left(1 - \frac{x}{N}\right) - 2y. \tag{8.47}$$

The **ION** routine pfic.pro (planar fault image contrast) can be used to compute the two-beam contrast for an arbitrary planar fault. The routine takes the following input parameters (in addition to $x_{1,2}$ and $t_{1,2}$):

- $\alpha_{\mathbf{g}}$, the fault phase shift;
- $\xi_{1,2}$, the extinction distances;
- $s_{1,2}$, the excitation errors;
- $\theta_{\mathbf{g}}^{1,2}$, the phase factors of the Fourier coefficient of the lattice potential;
- $\mathcal{A}$, the absorption ratio $\xi'_{\mathbf{g}}/\xi_{\mathbf{g}}$;
- N_x, N_y, the number of columns along x and y.

In this and the following sections, we will use two different model crystals to illustrate the contrast of various planar faults. The first crystal, called $\boxed{\text{A}}$, has a constant thickness of 350 nm. The second one, called $\boxed{\text{B}}$, is wedge-shaped as in Fig. 8.17, with thickness increasing from 200 to 500 nm. All simulated images consist of 800 × 800 columns (pixels). The diffraction conditions for a simulated image will be represented by a row vector with the following components: $[\xi_1, \xi_2, w_1, w_2, \alpha, \mathcal{A}]$. The reader is encouraged to try out different parameters and study how the defect fringe contrast varies with each parameter.

8.4.3.3 *Stacking faults and α-fringes*

Consider the simple case of a stacking fault in exact Bragg orientation; i.e. $\alpha \neq 0$ and $s_1 = s_2 = 0$. If we denote the transmitted and scattered intensities by $I_{T,S} = I_{T,S}(z_1, z_2, s, \alpha)$, then we find the following symmetry relations:

$$I_T(z_1, z_2, 0, \alpha) = I_T(z_2, z_1, 0, \alpha); \tag{8.48}$$

$$I_S(z_1, z_2, 0, \alpha) = I_S(z_2, z_1, 0, -\alpha), \tag{8.49}$$

which means that the bright field image is symmetric with respect to the center of the foil; the dark field image is asymmetric. This in turn means that the outer fringes will be the same in BF, but of opposite contrast in DF.

The detailed math is fairly involved; one can show that the total intensity can be written as the sum of three terms [GVLA65, GVLA66b]:

$$I_{T,S} = I_{T,S}^{(1)} + I_{T,S}^{(2)} + I_{T,S}^{(3)}, \tag{8.50}$$

Table 8.3. *Intensity levels of the first and last fringe in BF and DF images of a stacking fault ($\alpha = \pm 2\pi/3$). (B = bright, D = dark, centered dark field).*

	BF first	BF last	DF first	DF last
$\sin\alpha > 0$	B	B	B	D
$\sin\alpha < 0$	D	D	D	B

with

$$I^{(1)}_{T,S} = \frac{1}{2}\cos^2\frac{\alpha}{2}\,[\cosh(2\pi\sigma_i z_0) \pm \cos(2\pi\sigma_r z_0)]\,; \tag{8.51}$$

$$I^{(2)}_{T,S} = \frac{1}{2}\sin^2\frac{\alpha}{2}\,[\cosh(4\pi\sigma_i u) \pm \cos(4\pi\sigma_r u)]\,; \tag{8.52}$$

$$I^{(3)}_{T,S} = \frac{1}{2}\sin\alpha\,[\sin(2\pi\sigma_r z_1)\sinh(2\pi\sigma_i z_2) \pm \sin(2\pi\sigma_r z_2)\sinh(2\pi\sigma_i z_1)]\,. \tag{8.53}$$

The upper sign refers to the transmitted beam and the lower sign to the diffracted beam. $z_0 = z_1 + z_2$ is the total crystal thickness and $2u = z_1 - z_2$ is the distance from the foil center.

If no defect is present, $\alpha = 0$ and only $I^{(1)}_{T,S}$ is left; this means that this term represents the perfect crystal contribution, which cannot give rise to fault fringes. In sufficiently thick foils, the third term dominates: it consists of two damped sin functions, one of which vanishes at the top surface, the other at the bottom surface. The nature of the outer fringes, i.e. the fringes closest to the top and bottom surfaces, for $\alpha = \pm 2\pi/3$ is described by Table 8.3. The value $2\pi/3$ occurs for stacking faults in face-centered cubic materials. Note that we have to use the centered DF technique; for the regular DF technique, the table entries for the DF beam are opposite. Details can be found in [GVLA65], [WC96, Section 24.5.B], and [SS93].

We can derive the following general observations for α-fringes:

(i) the fringes are parallel to the closest surface;
(ii) the BF pattern is symmetric with respect to the foil center, the DF pattern is similar to the BF pattern close to the top surface, but complementary close to the back surface;
(iii) new fringes are generated at the center of the foil (for a wedge-shaped crystal).

An example of α-fringes is shown in Fig. 8.18; the simulation parameters are [95, 95, 0, 0, $-2\pi/3$, 25] for both A and B. One can observe clearly the symmetrical outer fringes for the BF image and the opposite contrast for the DF image.

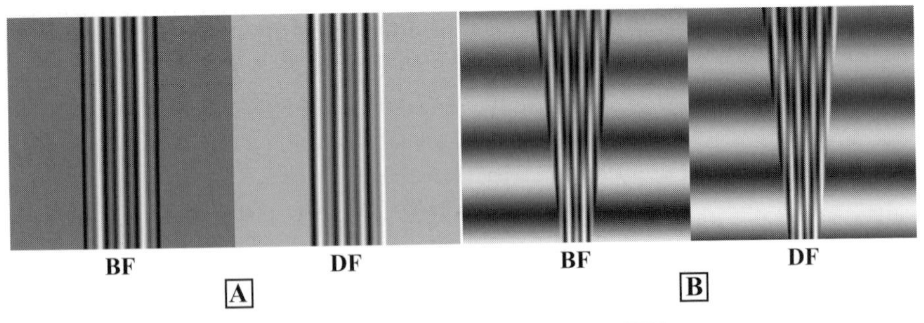

Fig. 8.18. Simulated α-fringe contrast for crystals A and B for simulation parameters $[95, 95, 0, 0, -2\pi/3, 25]$.

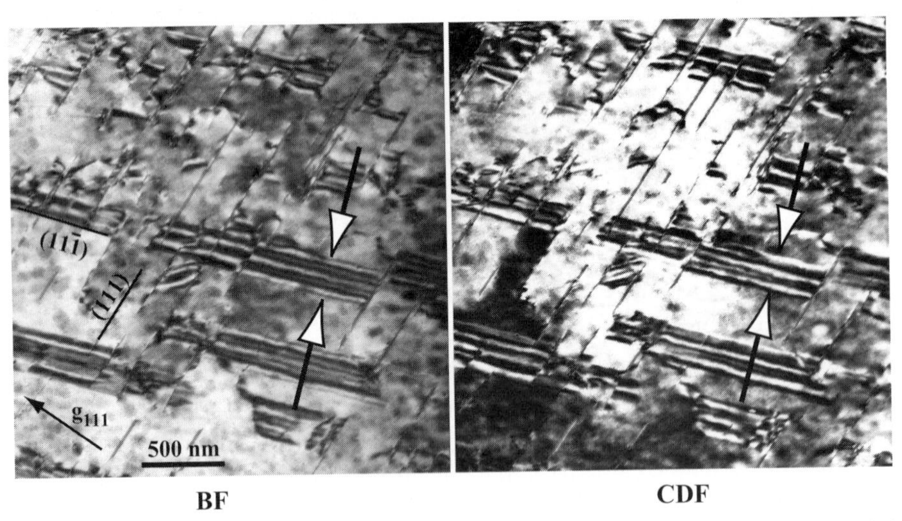

Fig. 8.19. Bright field and centered dark field images for stacking faults in Cu-15 at% Al, recorded at 200 kV. The operational reflection is $\mathbf{g}_{111}$, and the faults on the $(11\bar{1})$ planes exhibit α-fringe contrast.

Fringes run parallel to the nearest surface and new fringes are added at the center of the foil.

An experimental example of α-fringes in Cu-15 at% Al is shown in Fig. 8.19. The (111) planes are in Bragg orientation. Stacking faults on the (111) planes are viewed nearly edge-on, whereas stacking faults on the $(11\bar{1})$ planes exhibit α-fringe contrast. The BF–DF behavior is clearly visible at the arrowed fault.

In Fig. 8.20, a deviation from the Bragg condition is introduced and the simulation parameters are $[95, 95, 1, 1, -2\pi/3, 25]$. The contrast changes markedly and identification of this interface as a stacking fault becomes more difficult. Stacking faults should, therefore, always be studied in exact Bragg orientation.

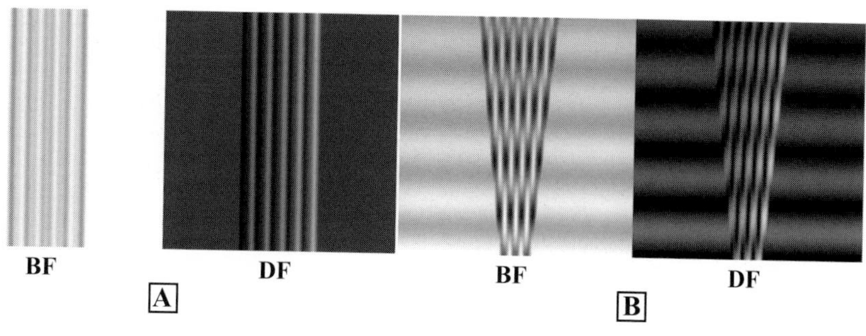

Fig. 8.20. Simulated α-fringe contrast for crystals A and B for simulation parameters [95, 95, 1, 1, $-2\pi/3$, 25].

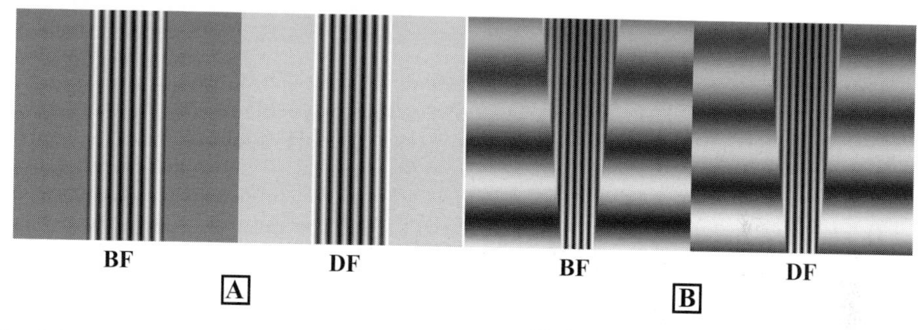

Fig. 8.21. Simulated π-fringe contrast for crystals A and B for simulation parameters [95, 95, 0, 0, π, 25].

8.4.3.4 Anti-phase boundaries and π-fringes

If $\alpha = \pi$ in the equations above, then the third term vanishes and the fringe contrast is completely described by the second term. The term $\cosh(4\pi\sigma_i u)$ describes an additional background on top of which the oscillations $\cos(4\pi\sigma_r u)$ are superimposed. Since $\sigma_r = \frac{1}{\xi}$ for $s = 0$, we can derive the following properties for π-fringes:

(i) the fringes are parallel to the foil center rather than to the closest surface;
(ii) new fringes are generated at the surfaces;
(iii) the central fringe is bright in BF, dark in DF;
(iv) BF and DF are complementary with respect to the background $\frac{1}{2}\cosh(4\pi\sigma_i u)$.

π-fringes occur for anti-phase boundaries. They are most easily analyzed in exact Bragg condition. For deviations from the Bragg condition, the images become complicated and confusion with other types of planar faults can easily arise. Figure 8.21 shows the crystals A and B for simulation parameters [95, 95, 0, 0, π, 25]; all of the above image characteristics can be observed easily in these simulated images.

Figure 8.22 shows a bright field–dark field pair of APBs in ordered Ni_4Mo. This material has a centered tetragonal $D1_a$ structure, with lattice parameters

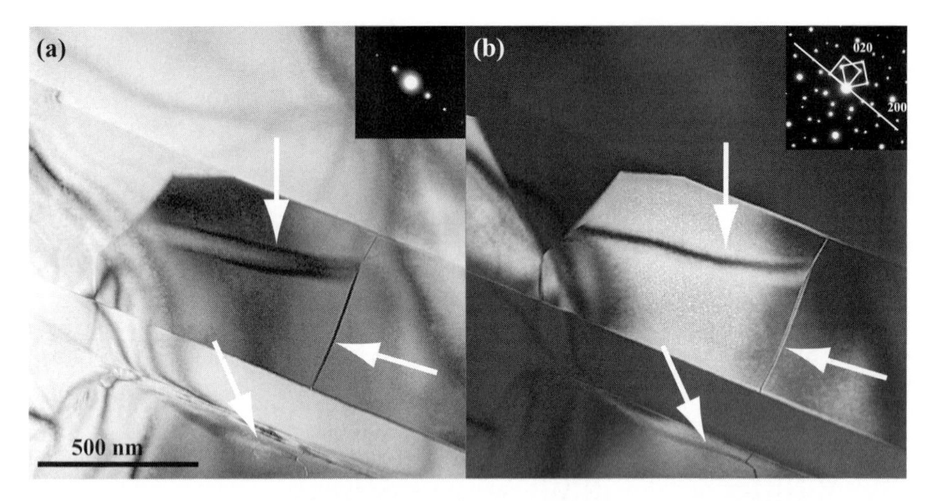

Fig. 8.22. π-fringes in ordered Ni_4Mo, recorded at 200 kV. The superlattice reflections of the type $\frac{1}{5}(\bar{4}20)$ form a systematic row. The APB contrast is visible in one of the orientation variants of the ordered structure (thin foil courtesy of G. VanTendeloo).

$a = 0.573$ nm and $c = 0.357$ nm [LHSH88]. The structure is derived from a face-centered cubic lattice, and is characterized by superlattice reflections on rows along the (420) and $(2\bar{4}0)$ reciprocal rows. The satellite spacing is $\frac{1}{5}(420)$. The [001] zone axis diffraction pattern of two orientation variants is shown as an inset in Fig. 8.22(b). The BF–DF image pair was recorded after the sample was tilted around the $\mathbf{g}_{\bar{4}20}$ reciprocal lattice vector. The APB fringe contrast is clearly visible at three arrowed locations in one of the orientation variants. APBs in this material are typically curved, and meet in junctions where particular topological conditions must be satisfied. Additional examples of APB contrast can be found in [GP59] (CuAu), [VLGA66] (TiO_2), and [DGDB88] (CuAl, CuAlZn).

It is relatively straightforward to distinguish between α- and π-fringes: π-fringes are always associated with APBs and, hence, they require an ordered structure. Since the displacement vector for an APB is always a fraction of the unit cell vectors, the product $\alpha = 2\pi\mathbf{g} \cdot \mathbf{R}$ will be $2\pi n$ for all fundamental reflections. Therefore, the APB is invisible for all fundamental reflections. Only superlattice reflections can give rise to π-fringes. For stacking faults, on the other hand, the fault disturbs the stacking of the basic lattice and, hence, α-fringes will be visible for selected fundamental reflections. Additional differences are the behavior of the outer fringes and the location where new fringes are added, as discussed above.

8.4.3.5 Inversion boundaries, permutation boundaries and δ-fringes

For an inversion boundary, also known as a 180° domain boundary, the orientation of the crystal on both sides of the defect plane is the same. The only difference is that the active reflection of the top section ($\mathbf{g}$) becomes $-\mathbf{g}$ in the bottom section. This means that only the parameter $\theta_\mathbf{g}$ changes to $-\theta_\mathbf{g}$. In terms of scattering matrices, we have for the amplitudes:

$$\begin{pmatrix} \psi_\mathbf{0} \\ \psi_\mathbf{g} \end{pmatrix} = \mathrm{e}^{-\frac{\pi}{\xi_0'}(z_1+z_2)} \begin{pmatrix} T_2 & S_2 \mathrm{e}^{\mathrm{i}\theta_\mathbf{g}} \\ S_2 \mathrm{e}^{-\mathrm{i}\theta_\mathbf{g}} & T_2^{(-)} \end{pmatrix} \begin{pmatrix} T_1 & S_1 \mathrm{e}^{-\mathrm{i}\theta_\mathbf{g}} \\ S_1 \mathrm{e}^{\mathrm{i}\theta_\mathbf{g}} & T_1^{(-)} \end{pmatrix} \begin{pmatrix} 1 \\ 0 \end{pmatrix}. \tag{8.54}$$

This can be rewritten as

$$\begin{pmatrix} \psi_\mathbf{0} \\ \psi_\mathbf{g} \end{pmatrix} = \mathrm{e}^{-\frac{\pi}{\xi_0'}(z_1+z_2)} \mathcal{P}(-\theta_\mathbf{g}) \begin{pmatrix} T_2 & S_2 \\ S_2 & T_2^{(-)} \end{pmatrix} \mathcal{P}(2\theta_\mathbf{g}) \begin{pmatrix} T_1 & S_1 \\ S_1 & T_1^{(-)} \end{pmatrix} \mathcal{P}(-\theta_\mathbf{g}) \begin{pmatrix} 1 \\ 0 \end{pmatrix}. \tag{8.55}$$

This means that an inversion boundary behaves as if it were a translation boundary with $\alpha_\mathbf{g} = 2\theta_\mathbf{g}$; i.e. the fringes are α-fringes. The α parameter is not necessarily a simple fraction of π, as it is for stacking faults and anti-phase boundaries. Figure 8.23 shows inversion boundary contrast when the only phase shifting factor is the difference between the phase factors $\theta_\mathbf{g} = -\theta_{-\mathbf{g}} = 15^\circ$. The smaller the phase factor $\theta_\mathbf{g}$, the weaker the fringe contrast.

Across a permutation boundary in a tetragonal material such as $BaTiO_3$, the electric dipole moment rotates by 90°. The reciprocal lattice points corresponding to the two domains are, therefore, not coincident. This leads to a different excitation error for the two domains, and, in general, also a different extinction distance. The mathematical details of the fringe contrast are complicated. We refer to the original paper by Gevers and coworkers [GDBA64] who proposed the name δ-fringes for contrast that does not involve a displacement vector ($\alpha = 0$). It turns out that the factor $s_1\xi_1 - s_2\xi_2 = w_1 - w_2 \equiv \delta$ determines the fringe contrast.

We will only deal with the simplest case, the symmetric case, for which $s_1 = -s_2 = s$. One can show that the transmitted and diffracted intensities again consist of three terms. The third one is the most important one for the symmetric case and

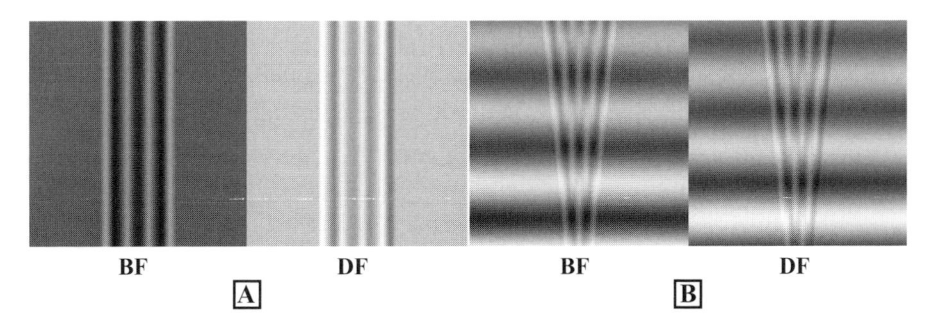

Fig. 8.23. α-fringe contrast for crystals A and B for diffraction conditions [95, 95, 0, 0, 0, 25] with $\theta_\mathbf{g} = -\theta_{-\mathbf{g}} = 15^\circ$.

Table 8.4. *Intensity levels of the first and last fringe in BF and DF images of a δ-fault at* $s_1 = -s_2 = s$. *B = bright, D = dark.*

	BF first	BF last	DF first	DF last
$\delta > 0$	B	D	B	B
$\delta < 0$	D	B	D	D

is given by [GDBA64]:

$$I^{(3)}_{T,S} \approx \frac{-\delta}{(1+w^2)^4} \left\{\cos(2\pi\sigma_{1,r}z_1)\sinh\left[2\left(\pi\sigma_{2,i}z_2 \pm \phi_2\right)\right] \right.$$
$$\left. \pm \cos\left(2\pi\sigma_{2,r}z_2\right)\sinh\left[\left(2\left(\pi\sigma_{1,i}z_1 + \phi_1\right)\right]\right\}, \qquad (8.56)$$

with $2\phi_j = \text{arcsinh}(s_j\xi_j)$. The upper sign applies to I_T, and the lower one to I_S. From this expression, we can derive that close to the front surface BF and DF images are similar, whereas they are complementary close to the back surface. This is a general feature of two-beam defect images in foils and is caused by anomalous absorption. It can be used to determine which side of the defect is close to the top surface. δ-fringes have the following properties:

(i) the fringe spacing may be different near the top and bottom surfaces;
(ii) if $\xi_1 = \xi_2$, then the DF image is symmetric;
(iii) fringes are parallel to the surface and added in the center for a wedge-shaped foil;
(iv) the nature of the outer fringes depends on the sign of δ, as shown in Table 8.4: they are of opposite nature in BF and have the same nature in DF;
(v) extinction occurs when $\delta = 0$;
(vi) the two crystals have different background contrast in BF and possibly the same in DF (for the symmetric case).

An example image simulation for the diffraction conditions $[95, 95, \frac{1}{2}, -\frac{1}{2}, 0, 25]$ is shown in Fig. 8.24. The fringe characteristics described above can be observed readily in these simulated images, as well as on the experimental images of a permutation boundary in $BaTiO_3$, shown in Fig. 8.25 (compare the contrast between the points labeled 1 and 2). In the non-symmetric case, the situation becomes more complicated and one has to resort to image simulations to determine the relevant contrast features of the interface. The example simulations shown in Fig. 8.26 for diffraction conditions $[95, 120, \frac{1}{2}, 0, 0, 25]$ (i.e. different excitation errors and extinction distances) indicate that the fringe spacing is not constant and the background intensity in both bright field and dark field images is different. The thickness fringe periodicity is also obviously different, as shown on the right in Fig. 8.26.

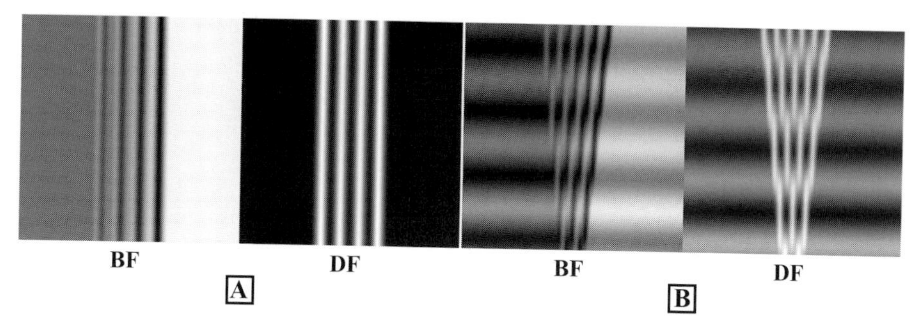

Fig. 8.24. δ-fringe contrast for crystals [A] and [B] for diffracting conditions [95, 95, $\frac{1}{2}$, $-\frac{1}{2}$, 0, 25].

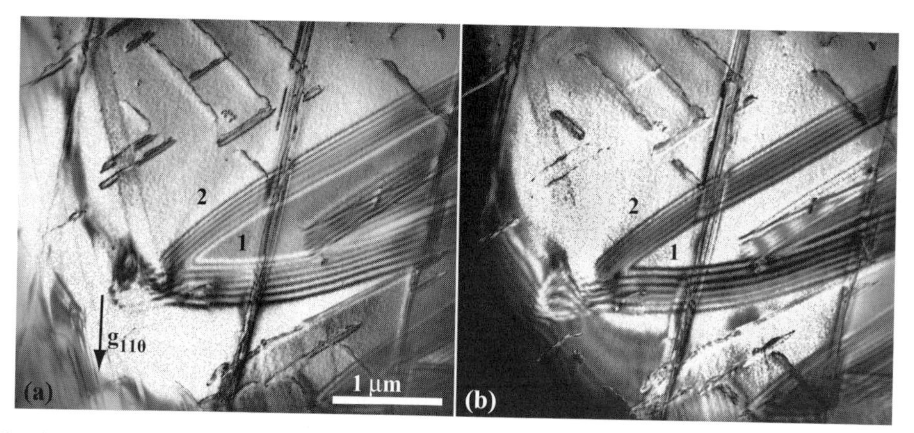

Fig. 8.25. Bright field and dark field images for a permutation boundary in $BaTiO_3$. The operative reflection is $\mathbf{g}_{110}$ at 200 kV.

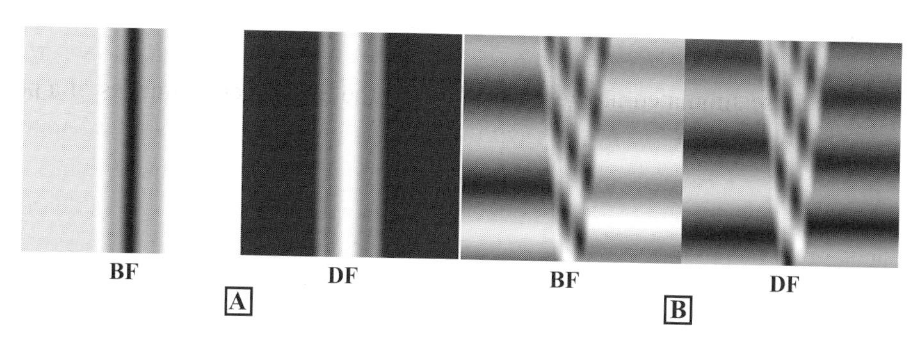

Fig. 8.26. δ-fringe contrast for crystals [A] and [B] for diffracting conditions [95, 120, $\frac{1}{2}$, 0, 0, 25].

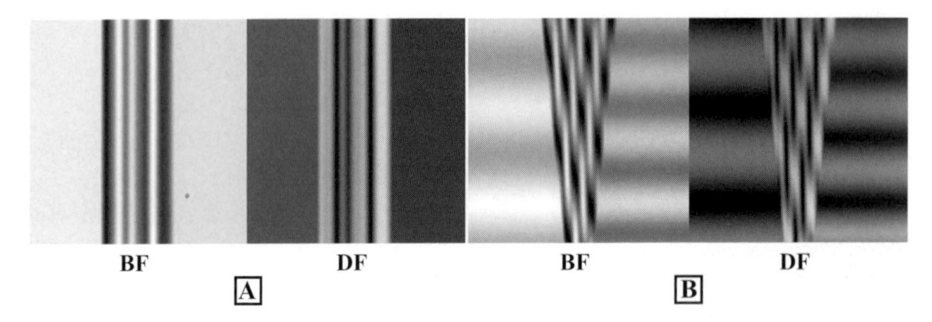

Fig. 8.27. Mixed α–δ-fringe contrast for crystals A and B for diffracting conditions $[95, 180, \frac{1}{2}, 1, -2\pi/3, 25]$.

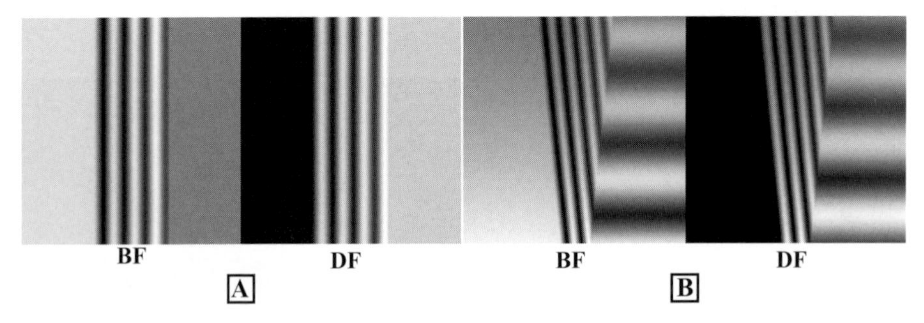

Fig. 8.28. BF–DF pairs for a high angle grain boundary, computed for diffracting conditions [95, 95, 0, 100, 0, 25].

In some cases, the fringes may have characteristics intermediate to the pure α and δ cases. This situation is rare but does occur in some materials. Needless to say, in that case one needs to use image simulations to determine the nature of the interface. An example of mixed α–δ-fringes for diffracting conditions $[95, 180, \frac{1}{2}, 1, 2\pi/3, 25]$ is shown in Fig. 8.27.

8.4.3.6 Twin boundaries and grain boundaries

For a regular high-angle grain boundary that is inclined with respect to the electron beam, we can use the same formalism to compute the two-beam contrast. Assuming that the crystal I is in Bragg orientation, we can simulate the image using the following diffraction condition: [95, 95, 0, 100, 0, 25], where the large value of s_2 indicates that this grain is far from the Bragg orientation. The simulated images are shown in Fig. 8.28. Note that this type of image is equivalent to a δ-fringe with a large value of δ. The inclined grain boundary is in effect a wedge-shaped crystal, hence the thickness contours will change direction at the boundary and turn into the crystal. This type of fringe contrast is, therefore, known as *wedge fringes*. In some

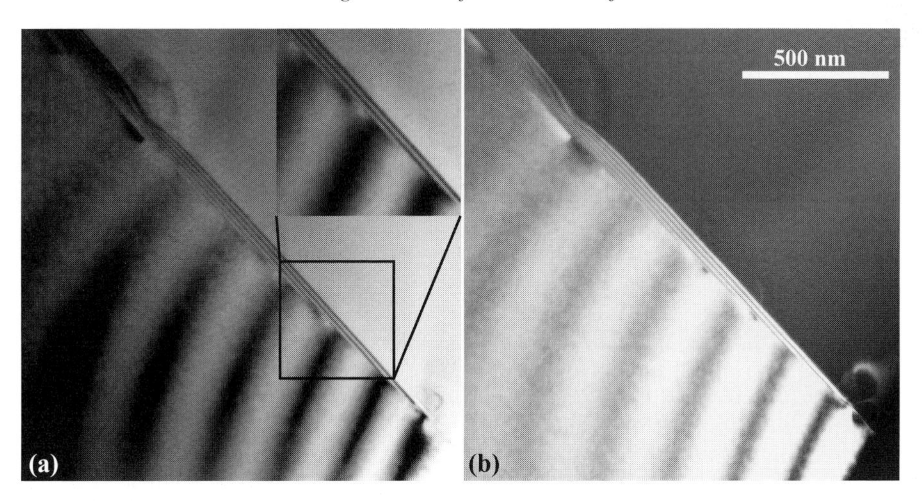

Fig. 8.29. Bright field and dark field images (200 kV) of an inclined grain boundary in Ti. The grain at the top is far from the Bragg orientation, whereas the grain at the bottom is in exact Bragg orientation. The thickness fringes change direction abruptly at the boundary.

cases the wedge fringes will have finer fringes superimposed on them. These are known as Moiré fringes and we will discuss them in more detail in Section 9.2.4. We refer to [GVLA66b] for a detailed computation of wedge and Moiré fringes for low-angle grain boundaries.

Grain boundary fringe contrast from a boundary in Ti is shown in Fig. 8.29. The boundary is inclined at a small angle with respect to the incident beam direction. The thickness fringes change direction at the boundary and run parallel to the thin edge of the diffracting grain, as shown in the inset of the bright field image.

Twin boundaries can be regarded as a special case of a grain boundary. If one twin variant is in the Bragg orientation and the other is far from the Bragg orientation, then the resulting fringe contrast for an inclined boundary will be wedge fringes. If, on the other hand, the twin is oriented so that both variants are in strongly diffracting condition with $s_1 = -s_2$, then the twin boundary will display δ-fringes.

8.4.4 Planar defects and the systematic row

The main image characteristics of planar defects do not change much when multi-beam systematic row conditions are used. The SRdefect.f90 program can be used to compute systematic row images for an arbitrary number of planar defects in the field of view. An example for a stacking fault in Cu is shown in Fig. 8.30. The fault has a displacement vector $\mathbf{R} = \frac{1}{6}[112]$, and the systematic row is along the $\mathbf{g}_{200}$ vector; seven beams were used at 200 kV for a foil thickness of 300 nm. The excitation error of all reflections changes linearly from top to bottom in the

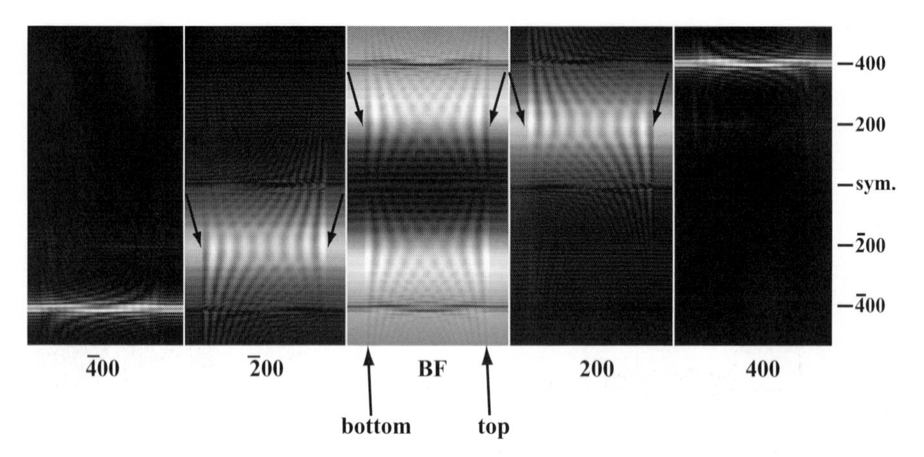

Fig. 8.30. $\mathbf{g}_{200}$-systematic row images for a stacking fault with displacement vector $\mathbf{R} = \frac{1}{6}[112]$ in a 200 nm Cu foil (200 kV). The excitation error changes linearly from top to bottom of the images, so that the bright field image shows a bend contour. Exact Bragg conditions for several reflections are indicated on the right.

images. The intersections of the stacking fault with the top and bottom foil surfaces are indicated.

At the Bragg condition for the $\pm\mathbf{g}_{200}$ reflections, indicated to the right of the figure, the fringe profile of the bright field image is symmetric (arrowed), whereas the dark field images are asymmetric, as expected. The outer fringes have opposite contrast on opposite sides of the central bend contour. The reader should be careful when comparing these images with experimental images. As stated in Chapter 4, we should use the centered dark field method to acquire the dark field images. It is important to realize that the CDF method brings the reflection $-\mathbf{g}$ into Bragg orientation by tilting the incident beam from the orientation for which $+\mathbf{g}$ was in Bragg orientation. This means that the CDF image corresponding to the $\mathbf{g}_{200}$ reflection is the $\mathbf{g}_{\bar{2}00}$ dark field image! The contrast for the outer fringes, as described in Table 8.3, is only valid for CDF conditions. If the aperture is moved to the $\mathbf{g}_{200}$ reflection, so that the dark field image is acquired with a non-centered aperture, then all contrast entries in the table must be reversed!

That the two-beam approximation is a rather good approximation for the case of copper is shown in Fig. 8.31(a). The bright field and dark field curves corresponding to Fig. 8.30 are shown for $s_g = 0$, along with the sum of all other intensity profiles; i.e. the sum of intensities for $-3\mathbf{g}$, $-2\mathbf{g}$, $-\mathbf{g}$, $2\mathbf{g}$, and $3\mathbf{g}$. The bright and dark field profiles clearly contain most of the intensity.

If we add the intensities for all images in Fig. 8.30 together on a pixel-by-pixel basis, then we obtain the image shown in Fig. 8.31(b). If there were no anomalous absorption, but only normal absorption (which is the same for each

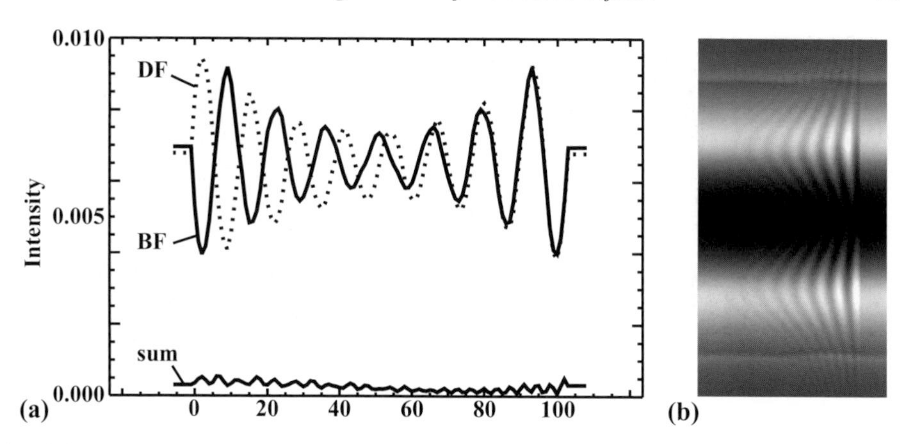

Fig. 8.31. Stacking fault intensity profile for $s_g = 0$, corresponding to the parameters of Fig. 8.30. The bottom curve represents the sum of all intensities except for the bright field and dark field intensities, indicating that the two-beam approximation is valid.

pixel since the foil thickness is constant), then this sum image would have no contrast variations. In the presence of anomalous absorption we find that the bend contour itself has significant anomalous absorption, as we have seen in Chapter 7. In addition, we see interference fringes at the location of the stacking fault. The fringes are rather similar to the thickness fringes of a wedge-shaped foil (see Fig. 7.5). We conclude that the presence of stacking fault fringes is entirely due to anomalous absorption.

We leave it as an exercise for the reader to determine the systematic row contrast of anti-phase boundaries in the Cu_3Au structure, introduced in Fig. 8.10(b), and a permutation boundary in $BaTiO_3$ for the $\mathbf{g}_{110}$ systematic row at 200 kV. For each case the reader should compare the contrast with the two-beam image characteristics derived in Sections 8.4.3.4 and 8.4.3.5.

It is a simple matter to use the SRdefect.f90 program to simulate the image contrast for overlapping planar defects. Figure 8.32(a) shows an experimental bright field image of the $\mathbf{g}_{210}$ bend contour in a Cu-21 at% Al alloy, which was quenched from 800 K to room temperature. The structure is a six-layered monoclinic martensite, which is typically indexed as an 18-layered orthorhombic phase [DGDB88]. Many stacking faults are seen nearly edge on, and where they intersect the bend contour, the contrast of the contour is interrupted. After tilting the foil, while maintaining the Bragg orientation for the (210) planes, a dark field image of the overlapping faults was recorded (Fig. 8.32b). The fringe contrast along the bend contour is complex, and the "normal" bend contour contrast is only visible in a few regions (arrowed) which do not contain any overlapping faults. We leave it as an exercise for the reader to simulate an image of overlapping stacking faults in an fcc material.

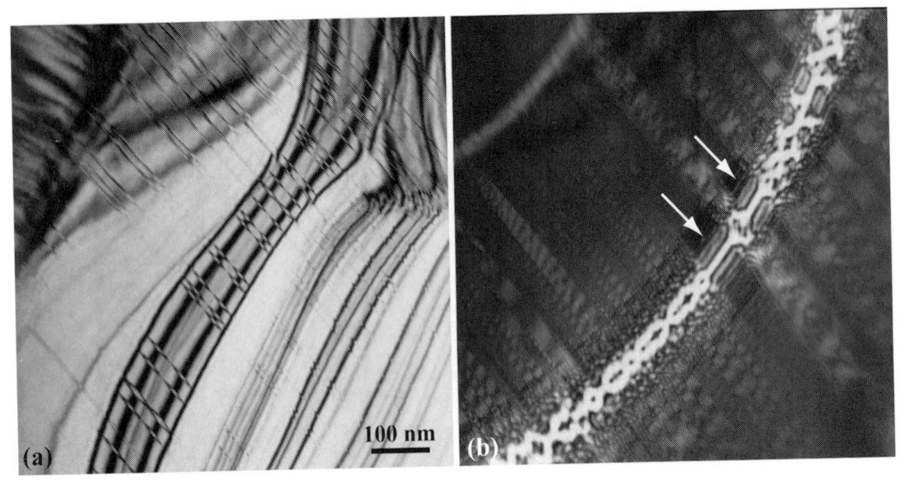

Fig. 8.32. Overlapping stacking faults on the basal plane of an orthorhombic Cu-21 at% Al martensite: (a) nearly edge-on orientation and (b) tilted along the (210) bend contour.

8.4.5 Other displacement fields

Sections 8.4.1–8.4.4 provide a first introduction to a variety of defects which are commonly observed in crystalline materials. The source code available on the website should provide the reader with a basic understanding of how defect contrast image simulations can be performed. In principle, the programs can be modified readily to incorporate defects not considered in this book. In fact, only one single routine should need to be modified: the displacement field subroutine. All the necessary crystallography and microscope geometry routines were written in a general fashion, so that the algorithm is essentially valid for triclinic crystals, and, hence, for all higher symmetries.

It is also possible to use displacement field data which is not obtainable in analytical form. For instance, displacement components derived from a finite element simulation of a particular defect configuration can be incorporated into the standard formalism. This requires a good interpolation algorithm, since finite element data are typically only available on a discrete grid. Another method would be to first approximate the finite element data with smooth spline functions for which the derivatives are known analytically. It would then be most expedient if the columns corresponding to the image pixels were to coincide with "columns" of nodes in the finite element computation. An example of such a computation can be found in [JVdBV+95]. The reader will find the source code of the SIMCON simulation package [JVDGVdB92] on the website.† This program solves the dynamical

† SIMCON is included on the website with the permission of its author, K.G.F. Janssens.

equations for the two-beam case for a variety of defects, including user-definable defects. Results from finite element computations using the ANSYS simulation package [ans] can also be imported.

The main conclusion from the defects discussed in this chapter should be that, if the displacement field can be computed in one way or another, then the image contrast for both two-beam and multi-beam situations can be computed. Such computations may be time-consuming, depending on the methods used to calculate the displacement field and the number of diffracted beams taken into account. Before embarking on extensive image simulations for a particular defect type, one should always make sure that good experimental images are available against which the simulations can be verified. Many experimental parameters are also needed to perform realistic image simulations:

- the crystal structure of the perfect crystal;
- the microscope acceleration voltage (this determines the extinction distances and, if a good model is available for the inelastic scattering events, the absorption lengths);
- the foil normal **F** and the beam direction $\mathbf{k}_0$ (this fixes both the sample orientation with respect to the microscope and the orientation of the "columns" for the contrast computation. It also fixes the value of all excitation errors);
- the foil thickness z_0 (this determines the integration limits for the dynamical equations);
- finally, a good starting model for the defect displacement field.

8.5 Concluding remarks and recommended reading

In this chapter, we have applied the dynamical theory to defect images. The intensity variations near dislocations and stacking faults were among the first contrast types to receive theoretical attention, and the reader is encouraged to consult the following papers: Whelan and Hirsch (stacking faults, 1957, [WHHB57]); Hirsch, Howie, and Whelan (dislocations, 1960, [HHW60]); Hashimoto, Howie, and Whelan (dislocations, 1962, [HHW62]); Gevers (Moiré fringe patterns, 1962, [Gev62]; planar fault fringe patterns, 1963, [Gev63]); Gevers, Delavignette, Blank and Amelinckx (coherent domain boundaries, 1964, [GDBA64]); Gevers, Van Landuyt, and Amelinckx (planar interfaces, 1965, [GVLA65]; microtwins, 1965, [VLGA65]; subgrain boundaries, 1966, [GVLA66b]; fine structure of diffraction spots due to presence of defects, 1966, [GVLA66a, VLGA66, GVLA68]); Rühle (contrast from irradiation-induced vacancy disks, 1967, [Rüh67a, Rüh67b]); Thölén (defect image simulation algorithm, 1970, [Thö70]); Stobbs (review of weak beam technique, 1970, [Sto70]); Goringe (review of defect contrast, 1970, [Gor70]); Wilkens and Rühle (small dislocation loops, 1972, [WR72]); Skalicky (simulation of defect images, 1973, [Ska73]); Mader (large inclusions, 1987, [Mad87]); etc. Many other examples of defect image contrast are scattered throughout the literature. The journals

Philosophical Magazine and *physica status solidi* provide good starting points to look for papers on defect contrast.

While the theory of dislocations was well developed by the mid 1950s, there were no direct observations of dislocations until the 1956 papers by Hirsch, Horne, and Whelan [HHW56] and Bollman [Bol56], followed in 1957 by a joint paper by the two groups [WHHB57]. The interested reader should consult Chapter 5 in the book *Out of the Crystal Maze*, edited by Hoddeson et al. [HBTW92], for a detailed history of the discovery of the dislocation and how the TEM played a crucial role in these events.

The analysis of two-beam and systematic row defect image contrast by means of the simulations described in this chapter, and the corresponding experimental methods used to acquire the images, are important tools for the study of the crystallography of a material. Some aspects of defect characterization can be carried out using high-resolution TEM methods, described in Chapter 10. This is true, in particular, for defects with a column structure; i.e. defects which can be viewed edge-on or along a single line (for instance, along a dislocation line). In many situations, however, the defect structure is three dimensional and there is no projection direction that can be used for high-resolution imaging. In such cases, the conventional bright field–dark field imaging techniques and simulations described in this chapter must be used to identify the defect.

Exercises

8.1 Use your foil from study material Cu-15 at% Al to obtain BF–DF image pairs for a stacking fault. Verify that the fringe contrast has the characteristics described in Section 8.4.3.3. Determine the crystallographic parameters of the fault (displacement vector, fault plane, see [HHC+73] for information on trace analysis). Tilt the foil slightly to obtain different excitation errors and compare the resulting images with simulated fringe profiles.

8.2 Use your foil from study material Ti to obtain BF–DF images of a fairly straight dislocation. Use $\mathbf{g} \cdot \mathbf{b}$ criteria to determine the Burgers vector of this dislocation. You may want to read about the peculiarities of indexing and dislocation studies in hexagonal systems in [Eis87].

8.3 Use your foil from study material Cu-15 at% Al to obtain a series of BF–DF images and associated diffraction patterns for a straight dislocation. Record the goniometer tilt angles, and transfer all your diffraction information onto a stereographic projection. Index all reflections consistently. Determine the dislocation line direction; then, use the hh.f program (along with xtalinfo.f90 to get the extinction distances) to simulate the images and compare them with your observations. You can use two-beam CBED to determine an approximate foil thickness at the position of the dislocation.

Table 8.5. *Material parameters for the orthorhombic form of $BaTiO_3$ (taken from [Wyc63]); [barium titanate; Pearson Symbol oA5].*

Space group	Amm2	# 38
Lattice parameters	$a = 0.399$ nm	
	$b = 0.5669$ nm	
	$c = 0.5682$ nm	
Asymmetric unit	Ba at 2a	(0, 0, 0)
	Ti at 2b	$(\frac{1}{2}, 0, 0.51)$
	O at 2a	(0, 0, 0.49)
	O at 4e	$(\frac{1}{2}, 0.253, 0.237)$

8.4 Make a Cu-2 wt% Co alloy. Solution treat it for 4 h at 1000 °C, quench to room temperature, and age for 10 h at 600 °C [AB63a]. Then, prepare a thin foil in the same way as you did for the Cu-15 at% Al material. Co will form small spherical precipitates which can be used to study strain contrast for coherently misfitting inclusions. Determine the characteristics of the strain contrast, and compare with simulated images using the SRdefect.f90 program and the Ashby–Brown displacement model.

8.5 Obtain BF–DF image pairs for both an inversion boundary and a permutation boundary in $BaTiO_3$, and verify the fringe contrast characteristics for the symmetric orientation. Before you go on the microscope, determine which reflections would give rise to the fringe contrast.

8.6 If you have access to a cooling holder for your microscope, then the following might be an interesting experiment. Use the $BaTiO_3$ sample and cool it down to below room temperature. At around +5 °C the crystal structure transforms from the tetragonal structure to an orthorhombic structure described in Table 8.5.
   (a) Determine the transformation matrix that translates crystallographic quantities from the tetragonal cell to the orthorhombic cell.
   (b) Obtain experimental evidence for this transformation by means of selected area electron diffraction observations.
   (c) Determine using group theoretical arguments how many orientation variants of the orthorhombic structure can be present in a single grain of the tetragonal structure.
   (d) Identify the character of the interface between two such orientation variants, using bright field–dark field image pairs of fringe contrast.

8.7 Carry out a systematic row image simulation, using the SRdefect.f90 program, for a series of overlapping stacking faults in an fcc material (say, copper). The faults are on (111) planes with displacement vector $\mathbf{R} = \frac{1}{6}[\bar{1}\bar{1}2]$, and are randomly spaced. Create the proper geometry, and determine which reflection you would use to image the faults.

9

Electron diffraction patterns

9.1 Introduction

In this chapter, we will take a closer look at the geometry of electron diffraction patterns. First we will discuss spot patterns and how to index them. Superpositions of spot patterns and double diffraction are explained in detail, as well as Moiré patterns and the corresponding diffraction effects. After a brief discussion of ring patterns we introduce linear features, such as streaks, Kikuchi lines, and HOLZ lines. Convergent beam electron diffraction and symmetry determination (both point group and space group) form the topic of Section 9.5. Diffraction effects in modulated structures are introduced; displacive and compositional modulations as well as interface modulations are analyzed, and the difference between commensurate and incommensurate structures is highlighted. Various types of diffuse intensity distributions are introduced next, and we conclude this chapter with a discussion of the shape function of polyhedral particles.

9.2 Spot patterns

9.2.1 Indexing of simple spot patterns

Indexing a spot pattern is perhaps one of the most important tasks of an electron microscopist and it is hence useful to outline a possible procedure. The procedure below can be used if the crystal structure of the material is known a priori. Needless to say, it is important to obtain diffraction patterns in the correct experimental conditions (eucentric height, focused SAD aperture and image, focused spots), so that the calibrated value for the camera constant can be used.

(i) Obtain a zone axis pattern with a high density of spots (i.e., short distances between the spots). Document the sample orientation (tilt angles) and the area from which you took the pattern (bright field image).
(ii) Measure the distance to the origin (in mm) for three reflections closest to the origin.

(iii) Measure the angles between these three reflections.
(iv) Convert the distances into d-spacings, using the calibrated camera length L and the wavelength λ (equation 3.71 on page 206).
(v) Compute a table of d-spacings for the known crystal structure and compare those values with the experimentally measured ones. Assign Miller indices to the reflections (in most cases you will only be able to assign a family $\{hkl\}$ to each reflection).
(vi) Compute the angles between all the members of the different families and compare them with the measured angles. Assign the correct Miller indices to the measured reflections, making sure that all three angles (1–2, 1–3, and 2–3) are correct.
(vii) Take the cross product between two reflections (counterclockwise) to obtain the zone axis of the diffraction pattern.

As an example, consider the Ti zone axis pattern shown in Fig. 4.18(b) on page 279. The pattern was obtained on a JEOL 2000EX microscope operated at 200 kV, with a calibrated camera length of $L = 1490$ mm. To identify this zone axis, we measure the distance between the origin and three different reflections. It is typically more accurate to measure the distance between $-n\mathbf{g}$ and $+n\mathbf{g}$ and then divide this distance by $2n$. The measured distances are shown in Fig. 9.1(a) (distances shown were measured on the original negative). We also measure the angles between the three reflections. It is usually best to select the shortest independent reciprocal lattice vectors.

Using the measured lattice parameters (Table 4.2 on page 241) we can calculate a table of angles between reciprocal lattice vectors. The calculated angles between

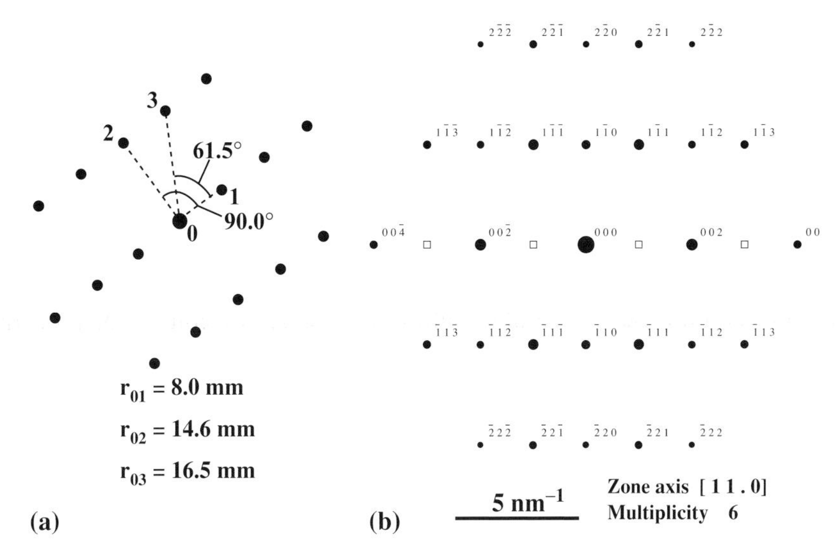

Fig. 9.1. Example of a manual indexing procedure, using the pattern for titanium shown in Fig. 4.18(b). Distances were measured on the original micrograph and are slightly magnified in this schematic.

Table 9.1. *Calculated angles between low-index plane normals for the structure of titanium. The lengths of the reciprocal space vectors were computed for a* 200 *kV microscope with a camera length of* 1490 *mm and are ranked in increasing order. All angles are in degrees; the values* 0° *and* 180° *are not shown.*

	{10.0}	{10.1}	{10.2}	{11.0}
{00.1} 7.98 mm	90.0	61.4, 118.6	42.5, 137.5	90.0
{10.0} 14.63 mm	60.0, 120.0	28.6, 64.0, 116.0, 151.4	47.5, 70.3, 109.7, 132.5	30.0, 90.0, 150.0
{10.1} 16.66 mm		52.1, 57.2, 81.0, 99.0, 103.9, 130.5	18.9, 49.5, 76.1, 86.7, 93.2, 103.9, 130.5, 161.1	40.5, 90.0, 139.5
{10.2} 21.65 mm			39.5, 71.6, 85.0, 95.0, 108.4, 140.5	54.2, 90.0, 125.8
{11.0} 25.33 mm				60.0, 120.0

plane normals are shown in Table 9.1 for the lowest-order reflections; the data for this table were generated by the Fortran program tabangle.f90. The algorithm is nearly identical to that used for the program indexZAP.f90, which will be discussed in the following paragraphs. For one member of each independent family of reflections we calculate the angles between the plane normal for that member and all other plane normals. The number of possible angles can become quite large, especially when higher-order reflections and/or low symmetry crystal structures are considered.

First, we find that the top three rows in the table correspond to the three measured distances r_{0i} in Fig. 9.1(a). The angles between the families of plane normals are also consistent with the measured values: the angles are highlighted in the first row of the table. We select reflection 1 ({00.1}) to be the (00.1) reflection; reflection 2 ({10.0}) is then equal to, say, the $(1\bar{1}.0)$ reflection. The last reflection 3 must be equal to the vector sum of the other two, i.e. $(1\bar{1}.1)$. The zone index is then computed from the cross product between (00.1) and $(1\bar{1}.0)$ and is equal to $[110] = [11\bar{2}0]$ (see Table 1.1 on page 25). The calculated pattern is shown in Fig. 9.1(b); note that the reflections of the type (00.l) with l odd are forbidden due to the non-symmorphic nature of the space group **P6$_3$/mmc**. In the experimental pattern those reflections are present because of double diffraction (see Section 9.2.3).

While the manual procedure works in principle, in practice it is found to be useful only for high-symmetry crystal structures, such as cubic and hexagonal. In lower symmetry crystals, many zone axis patterns look very similar, and an accurate measurement of the positions of the reflections may not always be sufficient for an unambiguous manual indexing of the pattern. In addition, tables such as the one shown in Table 9.1 become very large and unwieldy for low-symmetry crystal structures. In multi-phase materials, indexing is complicated by the fact that it may not always be easy to tell from which phase the pattern was obtained. Precomputed patterns, such as those produced by the xtalinfo.f90 program, may be useful if they are printed at the correct scale, so that negatives may be placed directly on the prints for visual comparison. It is also possible to use the measured values of d-spacings and angles to determine a "best-fit" pattern; it is not too difficult to write a program that will compare computed d-spacings and angles with experimental ones and identify a best fit, based on a least-squares criterion or another appropriate statistical measure. Such a program would use a look-up table of all possible angles and d-spacings, along with an evaluation criterion to select the most likely solution. It is obvious that symmetry arguments should be used to reduce the number of possibilities to the smallest number possible.

The program indexZAP.f90 implements a simple algorithm for indexing zone axis diffraction patterns from an arbitrary (but known) crystal structure. The program outline is described in pseudo code PC-20. The user must provide a crystal structure, the microscope accelerating voltage and camera length, a truncation radius in reciprocal space, and five measured quantities for each zone axis pattern. Those five quantities are the lengths of three reciprocal lattice vectors $\mathbf{g}_1$, $\mathbf{g}_2$, and $\mathbf{g}_1 + \mathbf{g}_2$, the angle α_{12} between $\mathbf{g}_1$ and $\mathbf{g}_2$, and the angle α_{13} between $\mathbf{g}_1$ and $\mathbf{g}_1 + \mathbf{g}_2$, as shown schematically in Fig. 9.2. Lengths should be measured in millimeters, angles in degrees. All angles should be less than or equal to 90°, and distances should be

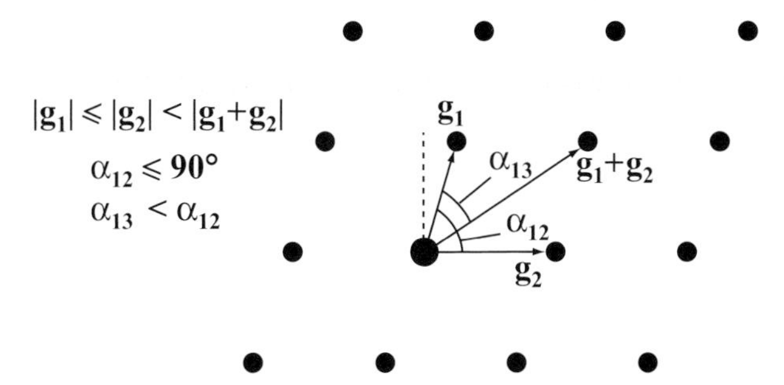

Fig. 9.2. Schematic illustration of the input parameters needed for the indexZAP.f90 program. The parameter restrictions are shown on the left of the figure.

Pseudo Code PC-20 Indexing of a zone axis diffraction pattern.

Input: PC-2, PC-3
Output: best-fit zone axis pattern identification
  ask for a truncation radius in reciprocal space
  compute all independent reflections $\mathbf{g}_{hkl}$ within this radius
  rank them in order of increasing length
  for every independent reflection **do**
    compute all **g**-vectors that enclose less than 90° with selected reflection
    eliminate doubles from this list
    compute lengths of vectors and sum vector, along with angles α_{12} and α_{13}
    store data in linked list
  end for
  repeat
    ask user for experimental data
    scale measured distances using the microscope camera constant λL
    compute Δ_i for every combination in the linked list {9.1}
    determine smallest Δ_i
    list solution
  until no further patterns to be indexed

ranked from smallest to largest. It is always possible to find three reflections that satisfy these conditions.

The algorithm first determines one reflection from each of the families that have their reciprocal d-spacing within the truncation radius. The algorithm then constructs a list of all possible combinations of two independent reflections satisfying the conditions described in the previous paragraph. Each of these combinations represents a possible zone axis orientation. A linked list is used to store the computed lengths of the three vectors $\mathbf{g}_1$, $\mathbf{g}_2$, and $\mathbf{g}_1 + \mathbf{g}_2$ and the two angles α_{12} and α_{13}. If these five numbers are represented by the symbol d_{ij}, with $j = 1, \ldots, 5$, and $i = 1, \ldots, N$, where N is the total number of possible zone axis orientations, then we can define a difference parameter Δ_i by

$$\Delta_i = \sqrt{\sum_{j=1}^{5} \left(e_j - d_{ij}\right)^2}. \qquad (9.1)$$

The parameters e_j, $j = 1, \ldots, 5$, represent the experimentally measured quantities. The zone axis orientation for which Δ_i is smallest is the most likely pattern matching the experimental data.

One complication arises when the experimental diffraction pattern contains so-called "double-diffraction" reflections. We will discuss the origin of these reflections in Section 9.2.3. The algorithm in the program indexZAP.f90 recognizes potential double-diffraction reflections and is capable of indexing experimental patterns that contain them.

The indexZAP.f90 program should function well if the microscope camera length has been accurately determined, and if the experimental parameters are accurately measured. Accuracies of a few percent are required. It is possible to extend the algorithm to take into account weighting factors to compensate for inaccuracies in distance and angle measurements. It is left to the reader to improve the algorithm for the case where the microscope camera length is not exactly known; one could even consider the camera length as a fitting parameter, to be determined by the algorithm. It is also left as an exercise for the reader to expand the algorithm to include multiple crystal structures. The algorithm should then be able to select the most likely crystal structure and zone axis out of a set of possible crystal structures.

The indexing problem in electron diffraction is similar to the indexing problem encountered in electron back-scattered diffraction or EBSD. EBSD is used to obtain orientational information from a crystalline material; the technique is known as *orientation imaging microscopy*, and is usually carried out in a scanning electron microscope (SEM) equipped with a dedicated camera system. From the EBSD pattern, the spatial orientation of several (typically around six or seven) plane normals is determined. From the orientational information, the Miller indices of all planes must then be determined. There is extensive literature on this problem and the interested reader may find more information in [SKA00].

9.2.2 Zone axis patterns and orientation relations

Obtaining low-index zone axis diffraction patterns and correctly indexing them is an important skill for the microscope user. Often, however, the most interesting diffraction patterns consist of multiple individual patterns. Let us consider a simple example: a twin in a face-centered cubic lattice. The twin geometry is shown in the [110] projection of Fig. 9.3. This projection clearly shows the *ABC* stacking sequence. The twin plane is a mirror plane (labeled **m**) parallel to the close-packed planes. Taking an *A*-plane as the twin plane, the stacking sequence on the bottom crystal is the mirror image of the sequence on the top. The twin plane is common to both twin variants, and therefore we expect the $\mathbf{g}_{111}$ reciprocal lattice vector to be common to both lattices.

We have already described in Section 1.9.5 how we can express the relative orientation of the two lattices by means of an *orientation relation*. For the fcc twin

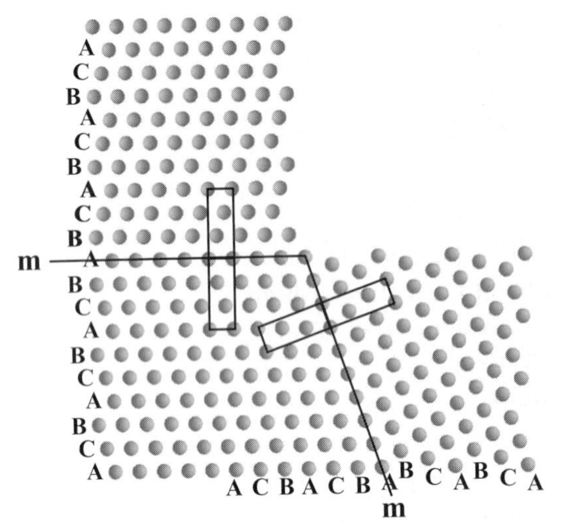

Fig. 9.3. Schematic illustration of two mirror twins in a face-centered cubic lattice in a [110] projection. The twin planes mirror the *ABC* stacking sequence into the sequence *CBA*.

this OR is given by

$$(111)_A \parallel (111)_B;$$
$$[1\bar{1}0]_A \parallel [\bar{1}10]_B.$$

While this relation was used in Chapter 1 to compute the combined stereographic projection of the two crystals, one can also use the same mathematical expressions to compute the electron diffraction pattern for the twin. After all, a zone axis diffraction pattern contains those reciprocal lattice points that lie along the equatorial circle of the stereographic projection.

Figure 9.4(a) shows experimental zone axis patterns for the Cu grain of Fig. 4.21 on page 284. The zone axis pattern for location 1 (top left in Fig. 9.4a) is a [110] pattern. When the selected area aperture is placed across variants 1 and 2, the pattern on the top right of Fig. 9.4(a) is obtained. The central nearly vertical row of reflections $\mathbf{g}_{1\bar{1}1}$ is common to both variants; the twin plane is normal to this direction and indicated by a dotted line in Fig. 9.4(b). All reflections of variant 1 are mirrored across this plane and are shown as open circles. Variants 2 and 3 result in the pattern shown on the lower left-hand side of Fig. 9.4(a). The common reflections now lie along the $\mathbf{g}_{1\bar{1}\bar{1}}$ vector, and the mirror plane is again at 90° from the common row of reflections. When the selected area aperture is placed across the three variants, the resulting pattern is a superposition of the three diffraction patterns. The individual twin mirror planes are now no longer prominent, and instead the pattern shows two new mirror planes, indicated by dashed lines in Fig. 9.4(b). In order to index all

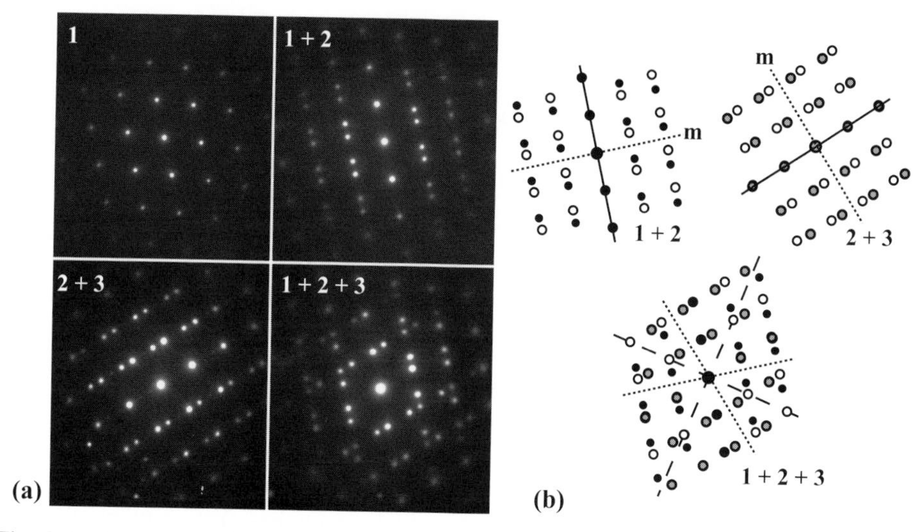

Fig. 9.4. Zone axis diffraction patterns of twins in the Cu grains shown in Fig. 4.21. The experimental patterns across several twin boundaries (a) are shown as simulated patterns in (b).

reflections in this overlap pattern and assign them to the correct twin variant, we must record the individual variant patterns as well as the overlap pattern. This may require the use of a small selected area aperture, or, in the case of a fine-grained material, *microdiffraction* techniques. The geometry of the three twin variants is shown in Fig. 9.3. The angle between the two twin planes is equal to the angle between two $\langle 111 \rangle$ directions, i.e. 70.5° or 109.5°.

The Fortran program zapOR.f90 can be used to simulate zone axis diffraction patterns for a given orientation relation. The second phase can either be identical to or different from the first phase. The output of the program is similar to that of the zap.f90 program, shown in Fig. 4.5(c).

9.2.3 Double diffraction

Consider the diamond crystal structure, shown in Fig. 9.5(a). The structure can be regarded as two interpenetrating face-centered cubic lattices, displaced by a vector $(\frac{1}{4}\frac{1}{4}\frac{1}{4})$. Consequently, the structure factor squared for this cell may be written as

$$|F_{hkl}|^2 = 4f^2\left[1 + (-1)^{h+k} + (-1)^{h+l} + (-1)^{k+l}\right]^2 \cos^2\left[\frac{\pi}{4}(h+k+l)\right]. \quad (9.2)$$

This structure factor vanishes for mixed indices h, k and l, as in the fcc case, and for $h + k + l = 4n + 2$, with n an integer (i.e. $h + k + l = 2, 6, 10, 14, \ldots$ are forbidden). Figure 9.5(b) shows the [110] zone axis pattern for the diamond

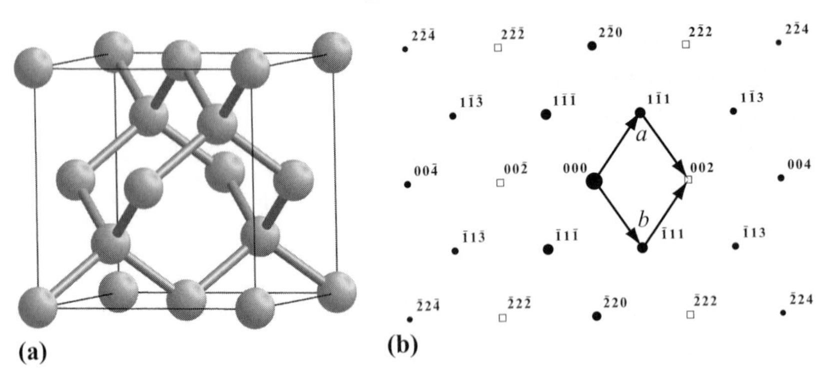

Fig. 9.5. (a) Diamond crystal structure and (b) [110] zone axis pattern.

structure. The reflections of the type {002}, {222}, etc. (indicated by squares in the figure) are forbidden because their structure factor vanishes. Yet, in many electron diffraction patterns taken along this zone axis, weak reflections are observed at these positions. How can there be intensity at a reciprocal lattice point with a vanishing structure factor, i.e. a reciprocal lattice point for which the kinematical theory predicts zero intensity?

The full dynamical scattering equation for the (002) reflection in diamond oriented along the [110] zone axis is given by

$$\frac{\mathrm{d}\psi_{002}}{\mathrm{d}z} - 2\pi \mathrm{i} s_{002}\psi_{002} = \mathrm{i}\pi \left[\cdots + \frac{\mathrm{e}^{\mathrm{i}\theta_{002}}}{q_{002}}\psi_{000} + \cdots + \frac{\mathrm{e}^{\mathrm{i}\theta_{1\bar{1}1}}}{q_{1\bar{1}1}}\psi_{\bar{1}11} + \cdots \right.$$
$$\left. \cdots + \frac{\mathrm{e}^{\mathrm{i}\theta_{\bar{1}11}}}{q_{\bar{1}11}}\psi_{1\bar{1}1} + \cdots \right]. \tag{9.3}$$

The first term on the right-hand side is zero, since $q_{002} = \infty$. The next two terms correspond to the $(1\bar{1}1)$ and $(\bar{1}11)$ reflections, both allowed for the diamond structure, so *their contributions will be non-zero*! It is easy to understand how this happens: when we introduced the coefficients q_{hkl} we stated that they are inversely proportional to the Fourier coefficient V_{hkl} of the electrostatic lattice potential. The second term on the right of the equation above contains the factor $q_{1\bar{1}1}$ along with the diffracted amplitude $\psi_{\bar{1}11}$. Electrons in this beam have a non-zero probability, described by $q_{1\bar{1}1}$ (or $V_{1\bar{1}1}$), to be scattered dynamically by the $(1\bar{1}1)$ planes. The resulting electrons will have been scattered by two planes $(\bar{1}11)$ and $(1\bar{1}1)$, and they will therefore travel in the direction given $\mathbf{k} + (\bar{1}11) + (1\bar{1}1) = \mathbf{k} + (002)$. We thus conclude that, although no electrons are directly scattered into the 002 diffracted beam, electrons which are scattered twice, once by the $(\bar{1}11)$ planes and subsequently by the $(1\bar{1}1)$ planes, do end up traveling in the direction of the 200 beam. This phenomenon is therefore known as *double diffraction*.

The Fourier coefficients $V_{1\bar{1}1}$ and $V_{\bar{1}11}$ are identical so that the two doubly diffracted paths a and b in Fig. 9.5(b) will reinforce each other. This is not always the case. Consider the titanium crystal structure, shown in Fig. 4.3. It is easy to show (exercise) that the structure factor squared for this cell may be written as

$$|F_{hkl}|^2 = 2f_{Ti}^2 \left\{1 + \cos\pi\left[\frac{2}{3}(k-h) + l\right]\right\}. \tag{9.4}$$

This structure factor vanishes whenever the cosine term is equal to -1, or, equivalently, when

$$2(k-h) = 6n + 3(1-l), \qquad n \text{ integer.}$$

Figures 9.1 and 4.18(b) show the [11.0] zone axis pattern for the titanium structure. The reflections of the type $\{00l\}$ and $(3\bar{3}l)$ with $l = 2n+1$, etc. (indicated by squares in Fig. 9.1 and encircled in Fig. 4.18b) are forbidden because their structure factor vanishes. There are now several doubly diffracted paths that will contribute to the (00.1) reflection; four possible paths are indicated by the letters a, b, c, and d in Fig. 9.6. The four reflections participating in those paths are $(\bar{1}1.0)$, $(1\bar{1}.0)$,

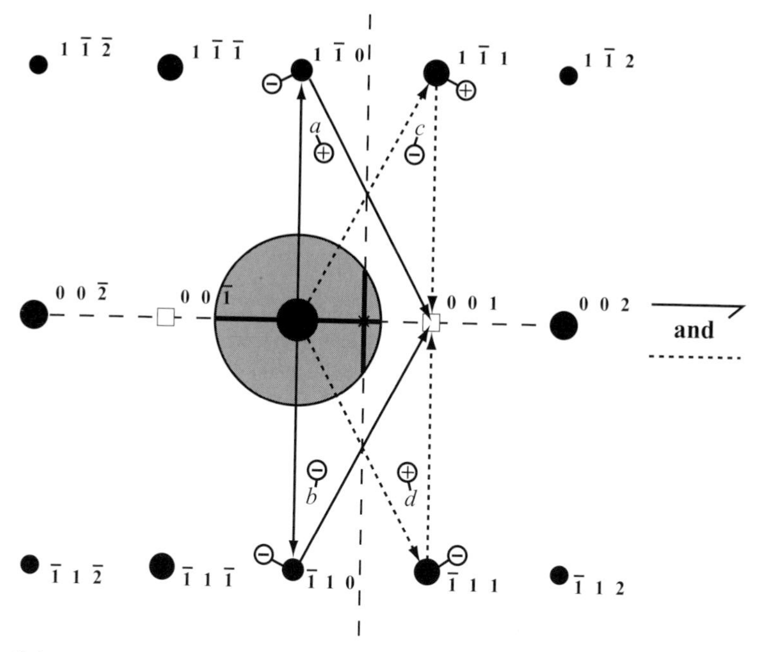

Fig. 9.6. Schematic illustration of four double diffraction paths contributing to the (00.1) reflection ([11.0] zone axis orientation). The paths a and b cancel each other for the exact zone axis orientation and for all beam directions that have their Laue center on the $\mathbf{g}_{00.1}$ row. For all incident beam directions with Laue center along the vertical dashed line the contributions of a and c cancel each other and there will be no double-diffracted intensity in the kinematically forbidden (00.1) reflection.

$(1\bar{1}.1)$, and $(\bar{1}1.1)$. The space group $\mathbf{P6_3/mmc}$ has a 6_3 screw axis along the c-direction, and also a c-glide plane. In the [11.0] zone axis orientation, the screw axis is normal to the electron beam, whereas the glide plane contains the beam direction. This is indicated to the right of Fig. 9.6. In the projected structure, the 6_3 axis effectively becomes equivalent to a 2_1 screw axis, hence the symbol for the 2_1 screw.

The presence of the translational symmetry elements enforces certain relations on the structure factors of different reflections. It is not difficult to show† that for all reflections in the [11.0] zone we have

$$F_{h\bar{h}.l} = F_{\bar{h}h.l} \qquad l = 2n;$$
$$F_{h\bar{h}.l} = -F_{\bar{h}h.l} \qquad l = 2n + 1.$$

The structure factors for the $\pm(1\bar{1}.0)$ reflections are identical and negative, whereas the $(1\bar{1}.1)$ and $(\bar{1}1.1)$ reflections have structure factors of opposite sign. This is indicated in Fig. 9.6 by means of encircled minus and plus signs attached to the reflections. In exact zone axis orientation, electrons which are doubly diffracted along the path a will experience two negative structure factors, whereas the path b would have one positive and one negative structure factor. The resulting amplitudes contributing to the (00.1) reflections are therefore equal in magnitude but have opposite signs so that the contributions of the paths a and b will cancel each other. The same happens for the paths c and d, and, indeed, for every other possible path. This means that for the exact zone axis orientation the dynamical intensity of the (00.1) reflection will remain zero *because of the presence of the translational symmetry elements*. This remains true when the incident beam is tilted so that the Laue center lies along the $\mathbf{g}_{00.1}$ row.

If we tilt the incident beam so that the (00.1) reflection is in Bragg orientation, then the Laue center will fall on the dashed vertical line (the perpendicular bisector of the vector $\mathbf{g}_{00.1}$). Contributions from a and c will cancel each other, and again there will be no intensity in the (00.1) reflection. For all other beam orientations, the cancellations will only be partial so that the (00.1) reflection will have some intensity due to one or more double-diffraction paths. Since most thin foils are not perfect foils with uniform thickness and no bending, a typical zone axis diffraction pattern will contain contributions from not just one incident beam orientation, but a range of orientations. The gray circle in Fig. 9.6 indicates such a range. Only the two line segments inside the circle would not contribute to the doubly diffracted beam because of symmetry-induced cancellations. All other beam directions would have incomplete cancellations, so that the (00.1) reflection would be visible, as it is in Fig. 4.18(b). The complete cancellations for the two line segments are

† This is related to the symmetry remarks on page 131.

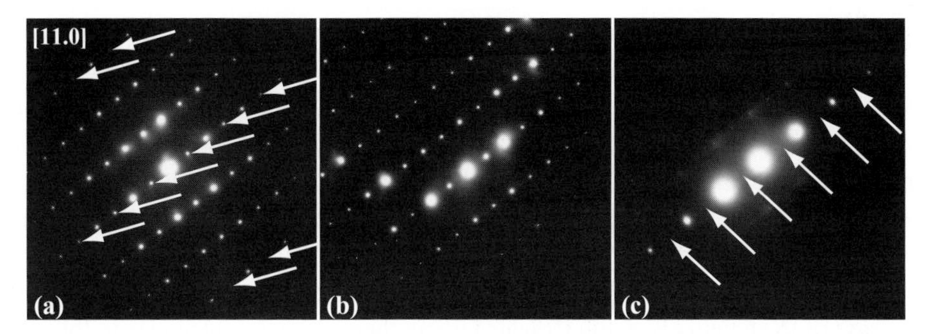

Fig. 9.7. (a) [11.0] zone axis pattern for titanium (200 kV). The arrowed reflections are caused by double diffraction; (b) and (c) the same pattern but progressively tilted around the $\mathbf{g}_{00.1}$ axis. The intensity in the (00.1) reflection is completely absent in (c).

known as *dynamical extinctions*. Dynamical extinctions can be derived from group theoretical considerations and we refer the interested reader to [PG81] for more details.

It is easy to determine experimentally whether or not a diffraction spot in a ZADP is caused by double diffraction. If we tilt the crystal, such that the excitation errors for all diffracted beams, except for reflections of the type $n\mathbf{g}_{00.1}$, become large, then the probability for the first scattering step of the double-diffraction process will be small. The second step will therefore be even less probable, and the double-diffraction contribution to the forbidden reflection will decrease. If we tilt the crystal sufficiently far from the zone axis orientation, using the plane normal $\mathbf{g}_{00.1}$ as the tilt axis, then the 00.1 reflection will vanish altogether, a clear sign that it was caused by double diffraction. This is illustrated in Fig. 9.7 for the (00.1) reflection of titanium; the [11.0] zone axis pattern is shown on the left. The intensity of the (00.1) reflection decreases significantly when the sample is tilted around the $\mathbf{g}_{00.1}$ axis, indicating that its intensity is due to double diffraction.

As a rule of thumb we can say that whenever two allowed reciprocal lattice vectors sum to a point corresponding to a forbidden reflection, double diffraction may occur if both of those reciprocal lattice points are close to the Ewald sphere. Reciprocal lattice points from higher-order Laue zones may also contribute to a forbidden reflection. We will return to this situation when we discuss G-M lines in Section 9.5.

It is important to note that double diffraction can only occur in crystal structures which show systematic absences *in addition to those caused by lattice centering operations* (in other words, only in the *non-symmorphic* space groups). The face-centered cubic lattice by itself can never have double diffraction for *any* of its forbidden reflections; when additional extinction conditions arise, usually because of the presence of glide planes or screw axes, then double diffraction may become a possibility.[†] For the titanium example above, the space group contains a 6_3 screw axis and a c glide plane, which are directly responsible for the additional systematic absences. One must therefore first consider the space group of the crystal being studied, before attributing the presence of certain reflections in an experimental pattern to double diffraction. This has obvious consequences for multi-beam computations, as it is not sufficient to include only those reciprocal lattice vectors $\mathbf{g}$ for which $F_\mathbf{g} \neq 0$! Computationally, it is easy to determine whether or not a reciprocal lattice point $\mathbf{g}$ should be taken into account: if $F_\mathbf{g} = 0$ because of the centering of the Bravais lattice, then $\mathbf{g}$ *never* contributes to the scattering process since there are no two allowed vectors $\mathbf{g}_1$ and $\mathbf{g}_2$ that sum up to $\mathbf{g}$. If $F_\mathbf{g} = 0$ due to the presence of a symmetry operation other than the Bravais lattice centering, then $\mathbf{g}$ *must* be taken into account in the dynamical computation, *despite its vanishing structure factor*. The multi-beam programs discussed in Chapter 7 exclude reciprocal lattice vectors with zero structure factor due to lattice centering, but allow all other scattered beams to have a non-zero amplitude. The xtalinfo.f90 program introduced in Chapter 4 can be used to produce zone axis diffraction patterns with labeled double diffraction reflections, as shown in Fig. 9.1(b).

9.2.4 Overlapping crystals and Moiré patterns

Double diffraction as described in the previous paragraphs can occur within a single grain of a material. The general principle is that a diffracted beam becomes an incident beam for the remainder of the crystal and can hence give rise to otherwise forbidden reflections. Another type of double diffraction commonly encountered in materials occurs when two *different* crystals or grains overlap along the beam direction; in this case each diffracted beam of the first crystal becomes an incident beam for the second crystal. This second crystal could be a precipitate embedded in a matrix or a thin film on a substrate or a thin, lamellar twin, or even a thin foil that is heavily bent and curves back on top of itself.

There are two ways to describe this type of double diffraction: (1) we write down the general dynamical diffraction equations for both crystals, e.g. using the matrix theory derived in previous chapters, and proceed from there or (2) we can employ a more intuitive approach. The first method involves quite a bit of sophisticated mathematics (for a good introduction we refer the reader to [Gev62]) and the second method is quite simple. The disadvantage of the more intuitive method is that we cannot derive quantitative results for the diffracted intensities, only geometrical characteristics can be derived.

[†] See equation (E2.2) in Chapter 2.

Let us assume that crystal A is on top of crystal B (i.e. the electron beam "sees" crystal A first). If A is in zone axis orientation for the zone axis $[uvw]_A$, and B is also in zone axis orientation for some other axis $[uvw]_B$, then each diffracted beam $\mathbf{k}_0 + \mathbf{g}_A$ of crystal A will become an incident beam for B. This essentially means that the diffraction pattern of B will be repeated around each reflection of A. Since the incident direction $\mathbf{k}_0 + \mathbf{g}_A$ is slightly off zone axis for crystal B, we will observe a Laue circle in the diffraction pattern of crystal B. For each beam $\mathbf{g}_A$ this circle is different and this gives rise to a rather complicated intensity distribution of the diffraction spots for crystal B.

The diffraction pattern of crystal A can be represented by the following summation:

$$I_A(\mathbf{q}) = \sum_{\mathbf{g}} I_{\mathbf{g}}^A \delta(\mathbf{q} - \mathbf{g}) \qquad (9.5)$$

where $\mathbf{g}$ is a reciprocal lattice vector belonging to the zone $[uvw]_A$ and $\mathbf{q}$ is an arbitrary reciprocal space vector. Figure 9.8(a) shows a square zone axis pattern for crystal A. The intensities $I_{\mathbf{g}}^A$ can be derived from the structure factor or from more extensive dynamical simulations.

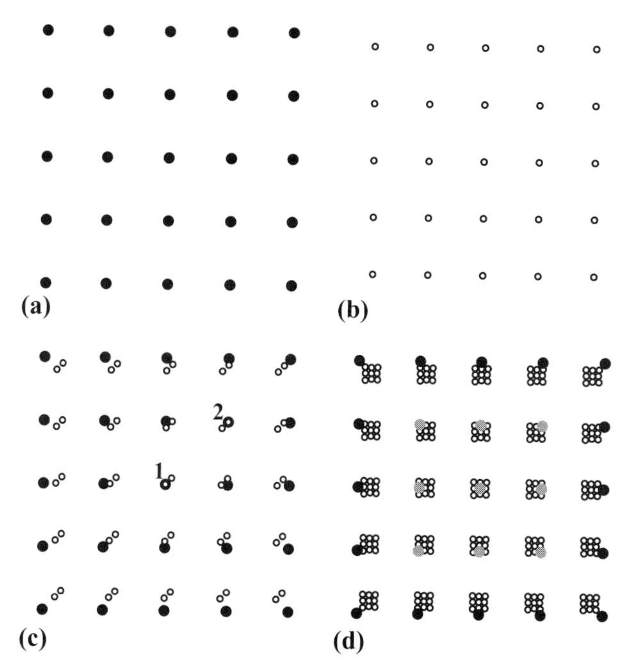

Fig. 9.8. (a) and (b) Geometrical representations of two zone axis patterns; the lattice parameter for B is 10% smaller than for crystal A. (c) Overlap of pattern (b) on two reflections of pattern (a); (d) overlap on the central nine reflections.

Similarly, the diffraction pattern of B can be represented by

$$I_B(\mathbf{q}) = \sum_{\mathbf{h}} I_{\mathbf{h}}^B \delta(\mathbf{q} - \mathbf{h}), \tag{9.6}$$

where **h** is a reciprocal lattice vector belonging to the zone $[uvw]_B$. In Fig. 9.8(b) a square pattern with lattice parameter 10% smaller than in (a) is shown. The geometry of the total intensity distribution when A and B overlap is given by the convolution product†

$$I(\mathbf{q}) = \sum_{\mathbf{g},\mathbf{h}} I_{\mathbf{g}}^A I_{\mathbf{h}}^B \delta(\mathbf{q} - \mathbf{g}) \otimes \delta(\mathbf{q} - \mathbf{h}). \tag{9.7}$$

Some simple mathematical manipulations convert this equation into

$$I(\mathbf{q}) = \sum_{\mathbf{g},\mathbf{h}} I_{\mathbf{g}}^A I_{\mathbf{h}}^B \delta(\mathbf{q} - (\mathbf{g} + \mathbf{h})). \tag{9.8}$$

We will hence observe a diffracted beam at every point $\mathbf{q} = \mathbf{g} + \mathbf{h}$. The intensities in this pattern are of course incorrect, since we did not take into account the proper boundary conditions between the two crystals, and we did not explicitly solve the dynamical scattering equations. Nevertheless, the pattern described by equation (9.8) is a reasonable representation of the actual pattern. A more realistic intensity distribution can be obtained by taking into account the location of the Laue center for each of the many incident beams from crystal A. It is easy to see that the Laue center for the incident beams $\mathbf{k}_0 + \mathbf{g}_A$ coincides with the center of the pattern, regardless of the selected beam $\mathbf{g}_A$. The intensity of the double-diffraction reflection at $\mathbf{g}_A + \mathbf{h}$ decreases with increasing distance from this point to the Laue circle, and this could be incorporated as an additional intensity scaling factor in equation (9.8). We leave it to the interested reader to implement such an intensity scaling.

The **ION**-routine moire.pro implements equation (9.8) for the overlap between two crystals; the user enters values for the lengths of two reciprocal lattice vectors for each structure, and the angle between the two patterns. The routine returns the computed diffraction pattern and a drawing of the overlap between the corresponding real-space lattices. An example of the output of the moire.pro routine is shown in Fig. 9.9.

We note in Fig. 9.9(c) that a number of reflections are centered close to the transmitted beam. If the difference in lattice parameters, i.e. the *lattice misfit*, between the two crystals is small, then the double-diffraction reflections will be close to the central beam. If we insert an aperture around the central beam to obtain a bright field image, the aperture may be too large to exclude the doubly diffracted beams

† The convolution of the two patterns repeats the pattern from crystal B around each reflection of crystal A.

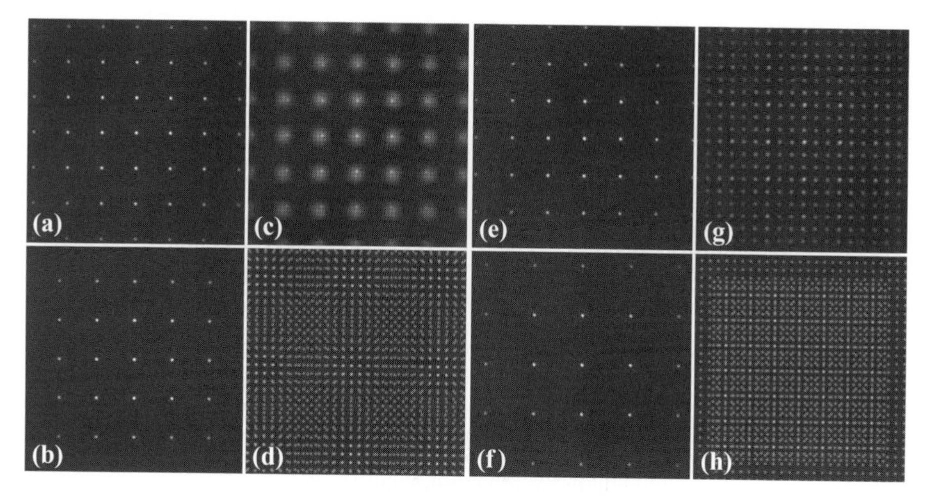

Fig. 9.9. (a) and (e) Square crystal A; (b) crystal B, 10% larger lattice parameter, (f) 33.33% larger; (c) and (g) are computed overlap patterns, and (d) and (h) show the corresponding two-dimensional real-space lattices.

close to the origin and the image will contain *fringes* because of the interference between the transmitted and doubly diffracted beam(s). In the remainder of this section we will discuss the geometrical characteristics of these fringes; they are known as *Moiré fringes*.

Crystal A is characterized by a set of lattice planes with interplanar distance d_1 and reciprocal lattice vector $\mathbf{g}_1$; the corresponding planes in crystal B have interplanar spacing d_2 with reciprocal lattice vector $\mathbf{g}_2$. Three different situations can now arise.

(i) The vectors $\mathbf{g}_1$ and $\mathbf{g}_2$ are parallel but have different lengths. In this case we can draw the difference vector (see Fig. 9.10a) $\Delta\mathbf{g}$:

$$\mathbf{g}_1 = \mathbf{g}_2 + \Delta\mathbf{g}.$$

The length of this difference vector is given by

$$|\Delta\mathbf{g}| = |\mathbf{g}_1 - \mathbf{g}_2| = \frac{d_2 - d_1}{d_1 d_2} \tag{9.9}$$

and hence we will observe fringes in the image, with fringe spacing D given by

$$D = \frac{d_1 d_2}{|d_2 - d_1|}. \tag{9.10}$$

The fringes (called *parallel* or *compression* fringes) are parallel to both sets of planes and hence they are perpendicular to the difference vector $\Delta\mathbf{g}$.

(ii) The vectors $\mathbf{g}_1$ and $\mathbf{g}_2$ have the same length but are slightly rotated with respect to each other by an angle β (Fig. 9.10b). The length of the vector $\Delta\mathbf{g}$ again determines the

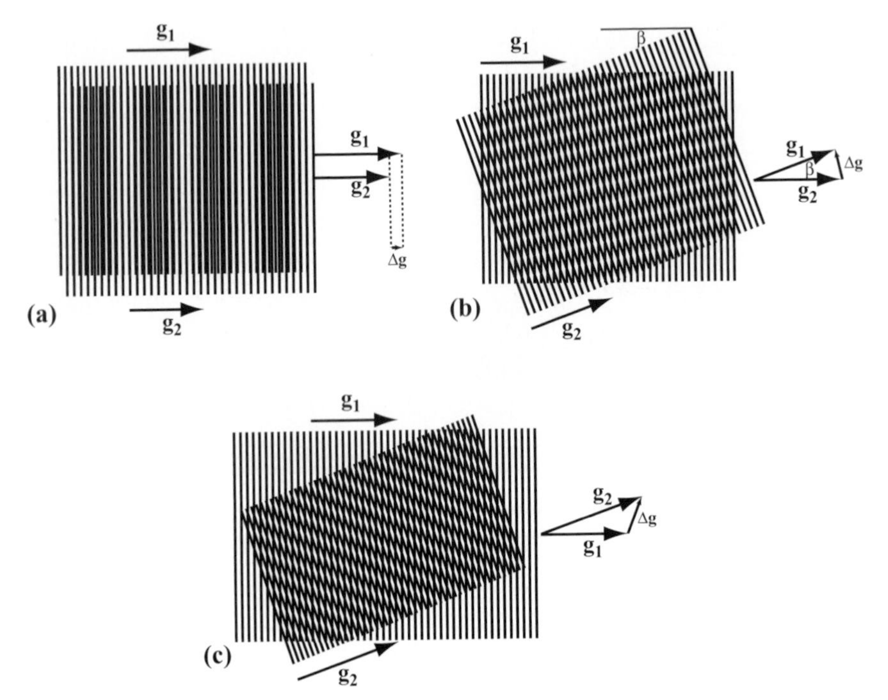

Fig. 9.10. (a) Parallel Moiré fringes, (b) rotation Moiré fringes and (c) general Moiré fringes and their relation to the difference vector $\Delta\mathbf{g}$.

spacing between the fringes, which is now given by

$$D \approx \frac{d}{\beta}, \tag{9.11}$$

with $d = d_1 = d_2$ and β is measured in radians. The fringes are perpendicular to the difference vector.

(iii) The most general case involves different lengths and a rotation β (Fig. 9.10c). Simple trigonometry shows that, for small angles β, the fringe spacing is given by

$$D \approx \frac{d_1 d_2}{\sqrt{(d_2 - d_1)^2 + d_1 d_2 \beta^2}}. \tag{9.12}$$

The fringes are again perpendicular to the difference vector $\Delta\mathbf{g}$.

The fringe spacing is always larger than either d_1 or d_2 and we can define a magnification factor M for the average spacing $\bar{d} = \frac{1}{2}(d_1 + d_2)$:

$$D = M\bar{d}. \tag{9.13}$$

For parallel fringes the magnification is given by $M = \bar{d}/|d_1 - d_2|$, for rotation fringes $M = 1/\beta$. Defects in one of the two lattices will also be magnified.

It is possible to derive a dynamical two-beam theory for the case of Moiré fringes. We refer the interested reader to [Gev62] for extensive details. Before the advent of dedicated high-resolution instruments, this type of Moiré imaging was actually quite popular because it allowed the study of lattice defects at magnifications larger than those attainable in the microscope. One would intentionally deposit a material with a selected lattice parameter on the material to be studied to generate Moiré fringes. More recently, Moiré imaging has lost much of its usefulness because with modern microscopes one can directly image the crystal lattice and most common defects. It is still useful to understand the origin of Moiré fringes, because they can occur readily in materials with complex three-dimensional microstructures, where the chances of one phase overlapping with another phase are significant. Moiré fringes are commonly observed in plan-view samples of layered materials. The reader can experiment with Moiré fringe images by evaporating a thin gold film onto a thin foil of the study material Cu-15 at% Al; the lattice parameters of gold and the Cu-alloy are sufficiently different so that double-diffraction reflections and Moiré fringes should be observable in the overlap region.

9.3 Ring patterns

We have used a ring pattern in Chapter 4 to calibrate the camera length of the microscope. Ring patterns can arise in a number of situations.

(i) If the grains in a polycrystalline material are randomly oriented or weakly textured, then the normal **g** to each diffracting plane will be oriented in all possible directions. Since the length of a particular **g** vector is a constant, the endpoint of all vectors with the same length will describe a sphere with radius $|\mathbf{g}|$. The intersection of such a sphere with the Ewald sphere is a circle, and, therefore, the diffraction pattern will consist of concentric rings, as shown schematically for GaAs in Figs 9.11(a)–(c). If the grains are sufficiently large, individual reflections can be seen in the rings (Fig. 9.11b). For

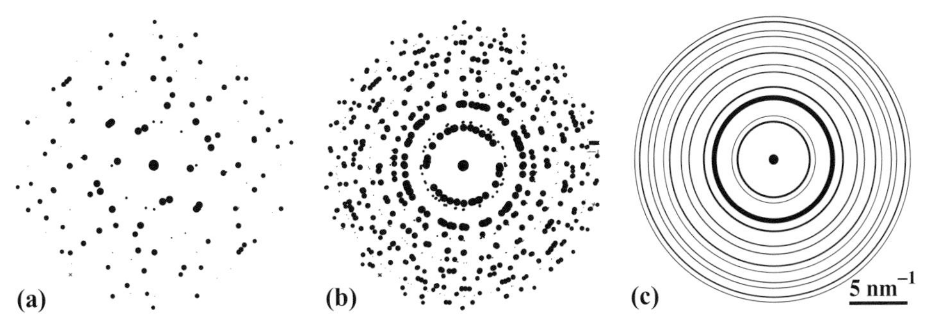

Fig. 9.11. (a) Six GaAs patterns of Fig. 4.5 superimposed with random rotations; (b) 36 randomly rotated patterns; (c) ring pattern for fine-grained polycrystalline GaAs.

nanoscale grains the diffraction pattern would look more like that shown in Fig. 9.11(c). The xtalinfo.f90 and zap.f90 programs can be used to simulate the ring pattern for an arbitrary crystal structure. The program assumes that the material is non-textured, meaning that all orientations have the same probability. If texture is present, then one or more rings may be absent, or the intensity in any particular ring may vary along the ring. From an analysis of these intensity distributions one can, in principle, derive information about the thin-foil texture (e.g. [TL96]).

(ii) Purely amorphous materials do not exhibit long-range translational or orientational order and, hence, there are no Bragg reflections. One can, however, define a statistical distribution of interatomic spacings (the so-called *radial distribution function* or RDF; e.g. [Jan92, Section 1.3]). The Fourier transform of this function gives rise to broad diffuse rings. It is not always easy to draw the line between a fine polycrystalline material and a truly amorphous material. Figure 9.12 shows a diffraction pattern of an amorphous Ge film, with small Au particles deposited on top. A rotationally averaged and a rotationally integrated profile are shown in the upper right-hand corner; in the

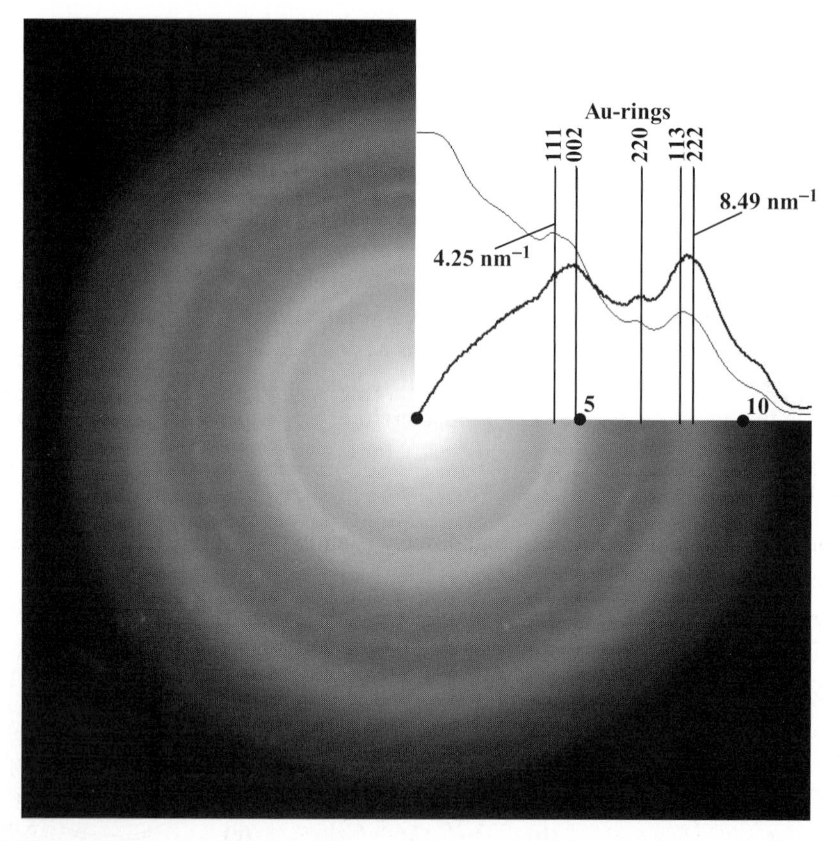

Fig. 9.12. Experimental electron diffraction pattern of an amorphous Ge film with Au particles (200 kV); rotationally integrated (thick line) and rotationally averaged (thin line) profiles along with five Au lines are shown in the upper right-hand corner.

averaged profile, each value is divided by the circumference of the circle. It is clear that amorphous Ge has significant scattered intensity well beyond 10 nm^{-1}, a fact that is important for the calibration measurements discussed in Chapter 10.

(iii) Structures that are inherently circular, such as carbon nanotubes for instance, can also give rise to diffraction patterns with partial or complete circular features. A nice example can be found in [BA96]. Several minerals of the *chrysotile* family exhibit curved lattices and the corresponding diffraction patterns show circular or polygonized diffraction features [DB95].

9.4 Linear features in electron diffraction patterns

9.4.1 Streaks

We have seen in Chapter 2, Section 2.6.3, that the shape of a reciprocal lattice point is determined by the Fourier transform of the shape function of the crystal. For a thin foil, which is thin along the beam direction, this gives rise to reciprocal lattice rods, or relrods. Relrods are "attached" to the crystal, in the same way as the excess and defect cones giving rise to Kikuchi lines (see Section 9.4.3). The main consequence of the presence of relrods is that electrons are scattered into diffracted beams for which the corresponding reciprocal lattice points do not lie precisely on the Ewald sphere. We have used the *excitation error* or *deviation parameter* s_g to take this deviation into account in the dynamical scattering equations in Chapters 5–7.

Because of the orientation of the relrods with respect to the crystal, it is unlikely that the relrods themselves can be observed directly in a diffraction pattern. This would require that the sample be tilted nearly 90° to bring the relrods tangent to the Ewald sphere. There are other situations in which relrod-like features can be observed in electron diffraction patterns. Consider a cubic structure viewed along the [001] axis. Let us assume that there exists a planar fault with the (100) plane as fault plane and a displacement vector of $\frac{1}{2}[010]$. The exact nature of the fault is not important for this discussion. The geometry is shown in Fig. 9.13a, along with the expected diffraction pattern. Some of the reflections in this pattern are elongated along the direction normal to the fault plane.

If the planar faults are randomly spaced, then the entire crystal can be subdivided into blocks of width w_i, measured along $\mathbf{g}_{100}$. The reciprocal lattice for each individual block will have the same geometry as the pattern for the perfect crystal. Each block now has *two* small dimensions; one along the foil normal, giving rise to relrods along the [001] direction, and a second small dimension along the fault plane normal, *giving rise to a second relrod in the* [100] *direction.* This second relrod is oriented normal to the incident electron beam, and is therefore tangent to the Ewald sphere. This means that the reciprocal lattice points corresponding to

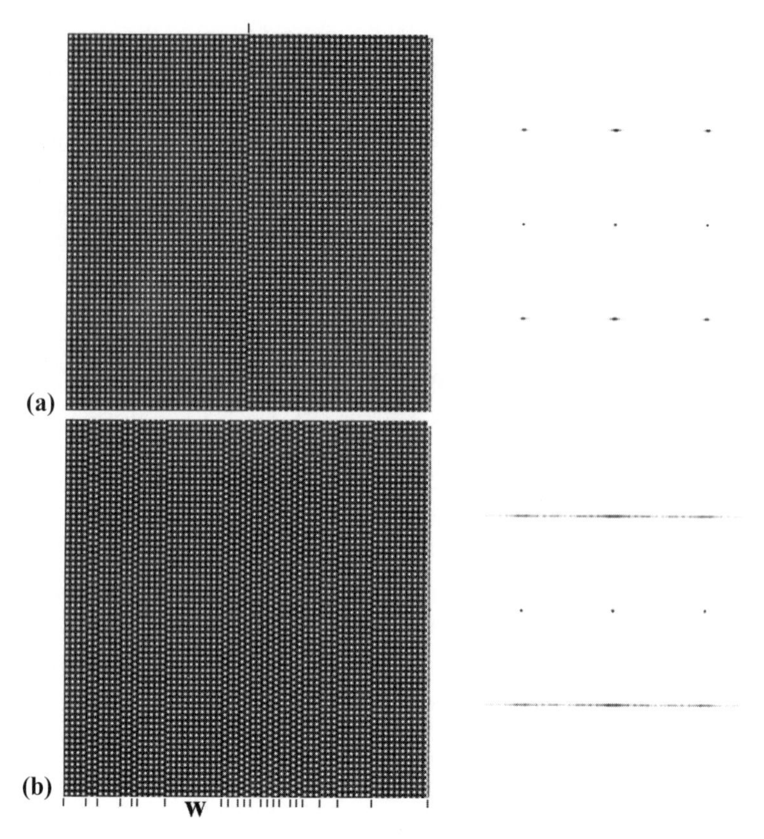

Fig. 9.13. (a) Perfect square structure with a single (100) fault plane indicated, along with the [001] ZADP (inverted contrast); (b) multiple randomly spaced blocks of width w_i, and the resulting diffraction pattern.

block i have an associated relrod of length proportional to $1/w_i$ along the [100] direction, as shown in Fig. 9.13(a).†

If the crystal consists of many blocks with a range of block widths w_i, then the corresponding diffraction pattern will have the same geometry as that of the perfect crystal, with a set of relrods of varying lengths through each reciprocal lattice point. The resulting pattern is shown in Fig. 9.13(b). Nearly continuous lines of diffuse intensity are observed in between the Bragg reflections; the direction of the lines is parallel to the fault plane normal $\mathbf{g}_{100}$. Such lines are commonly known as *streaks*.

An experimental example is shown in Fig. 9.14. Co_2NiGa is a ferromagnetic shape memory alloy with a cubic room-temperature structure [WJLC01]. When it is cooled down the structure transforms martensitically to a finely twinned structure, shown in Fig. 9.14(a). The corresponding diffraction pattern, with the

† The absence of a streak along the central row of reflections in Fig. 9.13 is due to the particular choice of fault. In general, the central row will also show a streak.

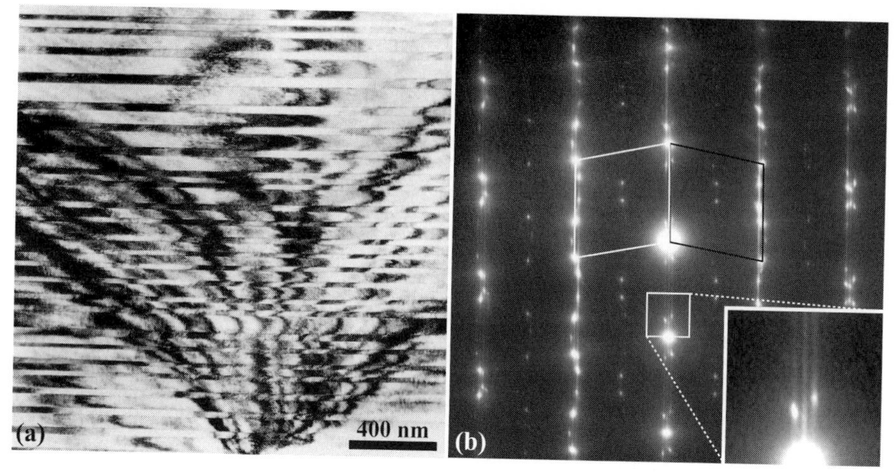

Fig. 9.14. (a) Finely twinned martensitic Co_2NiGa foil and (b) the corresponding zone axis diffraction pattern. The reciprocal unit cells of the two twin variants are outlined. The streaks are normal to the twin planes. Each reciprocal lattice point has two pairs of modulation satellites which also give rise to streaks (thin foil courtesy of Y. Kichin).

twin boundaries oriented parallel to the incident beam, is shown in Fig. 9.14(b). The 2D reciprocal unit cells corresponding to the two twin variants are outlined in white and black. Continuous streaks are present normal to the twin plane. Each reciprocal lattice point is surrounded by two pairs of satellite reflections caused by a modulation of the martensitic phase;† the streaks are also present for the satellites, giving rise to groups of three parallel streaks (see inset).

In general, any planar feature which divides crystal space into separate regions of varying widths will give rise to some sort of streak in the diffraction pattern. The direction of the streak is, of course, perpendicular to the planar feature. Planar Guinier–Preston zones in, for example, Al–Cu alloys give rise to streaks along the principal cubic directions. Streaks can give rise to misleading features in diffraction patterns; if the crystal of Fig. 9.13 were oriented away from the [001] zone axis, then each streak in the reciprocal lattice would still intersect the Ewald sphere at some point. This can easily be recognized experimentally: if the diffraction spots change location while tilting the sample, then it is likely that there are streaks and therefore disordered faults present in the crystal.

We have seen that a high density of randomly placed planar faults gives rise to streaks in reciprocal space. If those planar faults were separated by the same distance, i.e. w_i takes on only one value, then we can define a new unit cell, with a c-axis of length w_i, and the diffraction pattern will not show streaks. Instead, additional reflections will appear, as we will discuss in Section 9.6.2.3.

† See Section 9.6 for a discussion of modulated structures and their effects on diffraction patterns.

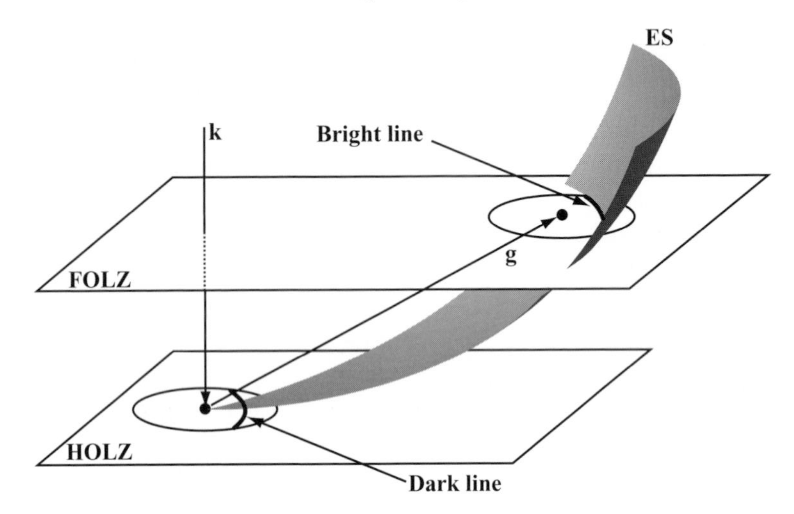

Fig. 9.15. Schematic illustration of the origin of dark HOLZ lines in the central disk: the intersection of the Ewald sphere, shown here with exaggerated curvature, with a diffraction disk in the FOLZ layer is a curved line segment that will show up as a bright line in the diffraction pattern. The corresponding part of the central disk will become dark.

9.4.2 HOLZ lines

In Section 3.9.5, we introduced the concept of higher-order Laue zones. HOLZ reflections can typically be observed in high-index zone axis diffraction patterns obtained with a small camera length (less than about 500 mm). If we increase the beam convergence angle θ_c to obtain a convergent beam pattern, then each reciprocal lattice point becomes a disk, with each point in the disk corresponding to a different incident beam direction. The HOLZ reflections also become disks that are parallel to the HOLZ layers (see Fig. 9.15). The intersection of these disks with the Ewald sphere, which is inclined with respect to the HOLZ layer, is a (curved) line segment across the disk. For the beam orientations corresponding to this line segment, electrons will be dynamically scattered out of the transmitted beam and into the HOLZ beam. This leaves a local deficit of electrons in the transmitted beam at the corresponding beam orientations. For realistic acceleration voltages, the curvature of the intersection between the Ewald sphere and the HOLZ diffraction disk is very small, and the resulting line segments will be nearly straight. The dark lines across the central disk are known as *deficient HOLZ lines*. They are caused by dynamical elastic scattering from HOLZ reflections under convergent beam conditions. Their bright counter parts are known as *excess HOLZ lines*.

Because the HOLZ lines correspond to HOLZ reflections that are described by long **g** vectors, the position of the lines will be very sensitive to the lattice

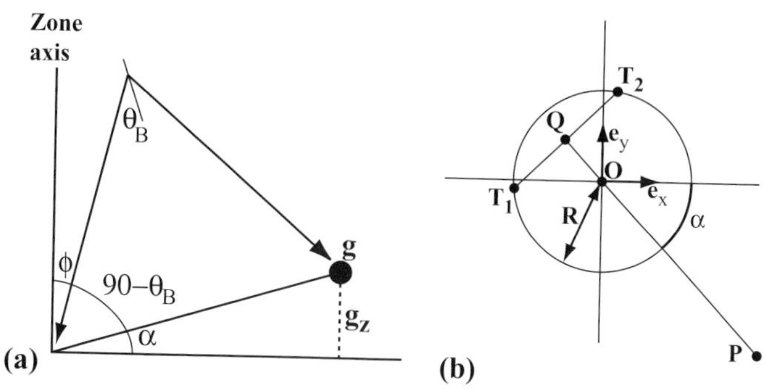

Fig. 9.16. (a) Schematic representation of a HOLZ reflection in Bragg orientation; (b) reference frame used to find the intersections T_1T_2 of the HOLZ line with the diffraction disk perimeter.

parameters. Small strains in the lattice, due to deformations around defects, will shift the lines from where they would be located normally. For the zone axis case, the location of a HOLZ line with respect to the center of the transmitted disk can be computed as follows: Fig. 9.16(a) shows a HOLZ reflection denoted by **g**. This reflection is located close to the Ewald sphere. In order to bring this reflection onto the Ewald sphere, the incident beam (which initially is parallel to the vertical zone axis direction) must be tilted over an angle ϕ. The magnitude of this angle can be computed from

$$\phi + (90^\circ - \theta_B) + \alpha = 90^\circ,$$

where θ_B is the Bragg angle associated with the reflection **g**. Using the Bragg equation we find[†]

$$\begin{aligned}\phi &= \theta_B - \alpha \\ &= \sin^{-1}\left(\frac{\lambda g}{2}\right) - \sin^{-1}\left(\frac{g_z}{g}\right),\end{aligned}$$

where g_z is the component of **g** parallel to the zone axis (i.e. the separation NH between the planes of the reciprocal lattice normal to the beam direction). If the beam is oriented along the zone axis direction, then the deficient HOLZ line corresponding to the HOLZ reflection **g** must be located at an angular distance ϕ from the origin; converting this angular distance into a real distance in reciprocal space we find the location of the deficient HOLZ line. If the radius of the central disk is given by R (in millimeters, see Fig. 9.16b), and the convergence angle is θ_c (in

[†] Other expressions can be derived for the angle ϕ. We refer the reader to [EMPH93] for a comparison of three different approximations.

radians), then the distance between the HOLZ line and the center of the diffraction disk is given by

$$|OQ| = R\frac{\phi}{\theta_c} \qquad \text{(in mm)}.$$

If the Cartesian coordinates of the HOLZ reciprocal lattice point projection (based on equation 3.77 on page 215) are given by (P_x, P_y), then it is left to the reader to show that the coordinates of the intersections of the HOLZ line T_1T_2 with the diffraction disk perimeter are given by

$$T_{x\pm} = \frac{(Q_x + Q_y \tan\alpha)\left(1 \pm \sqrt{D}\right)}{1 + \tan^2\alpha}; \tag{9.14}$$

$$T_{y\pm} = Q_y - \frac{T_{x\pm} - Q_x}{\tan\alpha}, \tag{9.15}$$

with

$$D = 1 - \left(1 + \tan^2\alpha\right)\left[1 - \left(\frac{R\tan\alpha}{Q_x + Q_y\tan\alpha}\right)^2\right],$$

$(Q_x, Q_y) = |OQ|(\cos\alpha, \sin\alpha)$, and $\tan\alpha = P_y/P_x$. The special cases of $P_x = 0$ and $P_y = 0$ are also left as an exercise. These equations have been implemented in the holz.f90 program, introduced in Chapter 3.

Figure 9.17 shows an experimental example of a zone axis pattern for the [10.4] zone axis orientation of Ti, recorded at 200 kV with a small camera length. The first and second order HOLZ rings are clearly visible. The bright line segments across the HOLZ disks are the excess HOLZ lines. The lines in between the diffraction disks are Kikuchi lines and will be discussed in the next section. The CBED pattern in Fig. 9.17(b) represents the central region of the pattern in (a), and shows faint deficit HOLZ lines inside the transmitted disk. The beam convergence angle is 5.1 mrad, measured from the radius of the disk and the distance between disks. The disks are not completely circular, a sign that the shape of the second condensor aperture was not perfectly circular.

The computations shown in Figs 9.17(c) and (d) were carried out with the holz.f90 program. On the left, the diffraction pattern including the first two HOLZ layers is shown. It is clear that for every HOLZ reflection in the computed pattern, there is an excess HOLZ line in the experimental pattern. The excess lines are all located along the intersection of the Ewald sphere with the HOLZ plane, so that they line up along the HOLZ ring. In the computed pattern, the positions of the HOLZ reflections do not necessarily coincide with the HOLZ ring.

The computed HOLZ line pattern is shown in Fig. 9.17(d). Only the lines corresponding to the experimental lines are shown, along with the indices of the

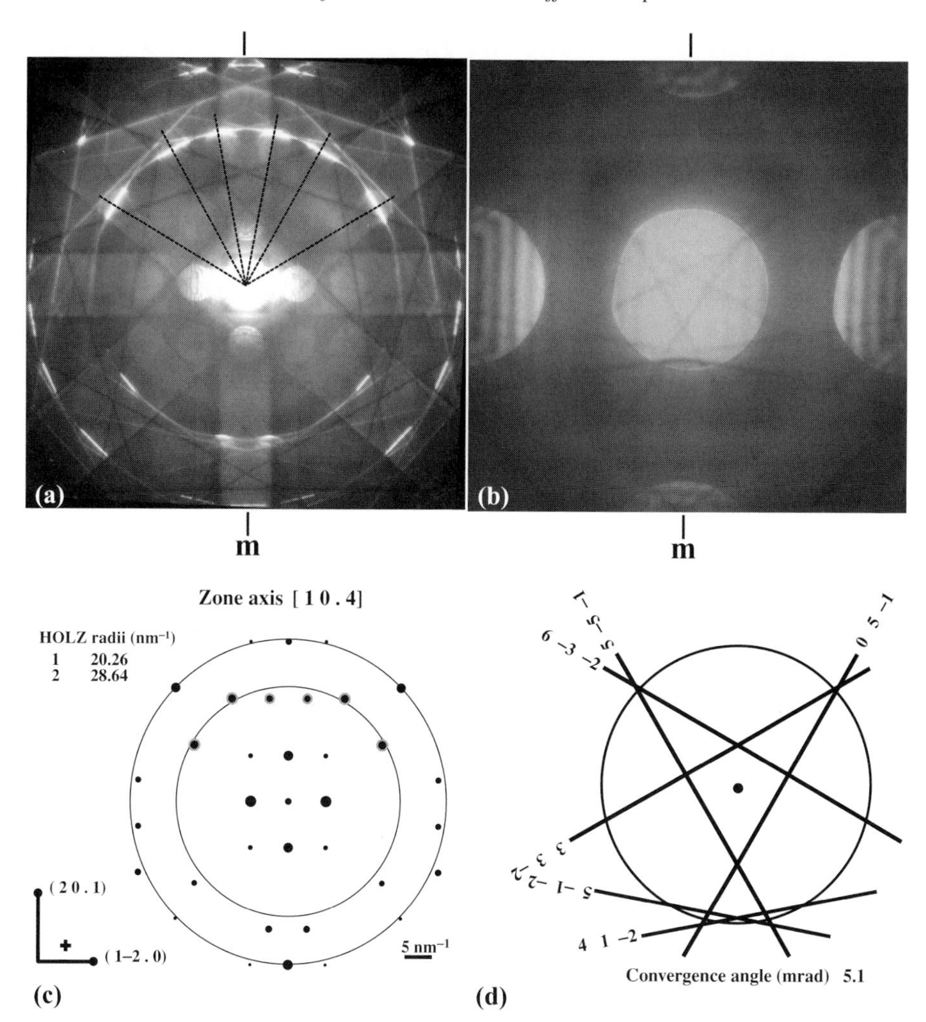

Fig. 9.17. (a) Experimental zone axis pattern for the [10.4] zone axis of Ti (200 kV). (b) Central region of (a), showing HOLZ lines inside the transmitted disk. The beam convergence angle is 5.1 mrad. (c) Simulated HOLZ diffraction pattern showing the most intense reflections. The reference frame to the left shows the two basis vectors and the location of the FOLZ with respect to the HOLZ layer. (d) Simulated 000 disk; only the lines corresponding to the experimental pattern in (b) are shown. The normals to the HOLZ lines are shown as dotted lines in (a), indicating the corresponding HOLZ reflection.

corresponding HOLZ reflection. The normals to these six lines are shown as dotted lines in Fig. 9.17(a); the HOLZ reflections giving rise to each HOLZ line can then be identified easily. The holz.f90 program predicts many more HOLZ lines than are visible in the experimental pattern. This is probably due to the fact that the foil is not necessarily strain-free. The weaker lines are often only visible when the sample is cooled down to low temperature, and when zero-loss energy filtering is employed.

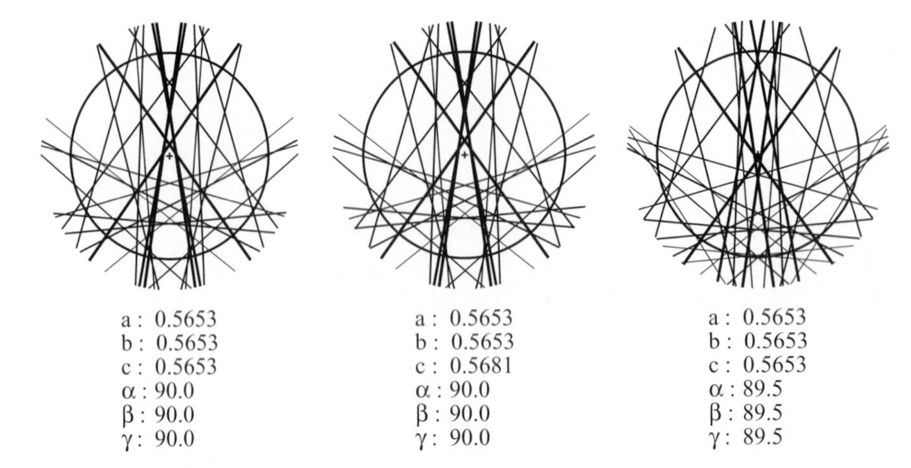

Fig. 9.18. [112] HOLZ line patterns for the first two HOLZ layers in GaAs at 100 kV: (a) nominal lattice parameters, (b) tetragonal elongation, (c) rhombohedral distortion.

HOLZ line positions can be used to determine small changes of the lattice parameters around defects. Figure 9.18 shows three HOLZ line patterns for the [112] zone axis orientation of GaAs, for the nominal lattice parameters, and for two small distortions: a 0.5% tetragonal elongation along [001], and a 0.5° rhombohedral distortion along [111]. Additionally, if the lattice parameters are accurately known from other diffraction experiments, then the position of the HOLZ lines can be used to determine the electron wavelength and hence the accelerating voltage.

The symmetry of the bright field disk, including the pattern of HOLZ lines, is determined by the diffraction group, as explained in Chapter 5. The symmetry of the [10.4] zone axis pattern is described by the diffraction group symbol $\mathbf{2}_R\mathbf{mm}_R$, as can be seen from Table 7.2 on page 438. Table 5.1 on page 340 then shows that both the bright field and the whole pattern symmetry are equal to **m**. The single mirror plane is clearly present in both Figs 9.17(a) and (b). We will say more about symmetry and convergent beam electron diffraction patterns in Section 9.5.

9.4.3 Kikuchi lines

Figure 9.19 shows a sequence of electron diffraction patterns for the [10.2] zone axis orientation of titanium. The thickness of the foil increases from left to right. The leftmost pattern was taken close to the edge of the thin foil, and the rightmost pattern corresponds to a rather thick region of the foil. The foil was tilted away from the exact zone axis orientation so that the Laue center lies at the position $\frac{1}{2}(\bar{3}0.3)$ (indicated by a small cross). As the foil thickness increases, so does the probability of *inelastic scattering*. This results in an increased intensity in between

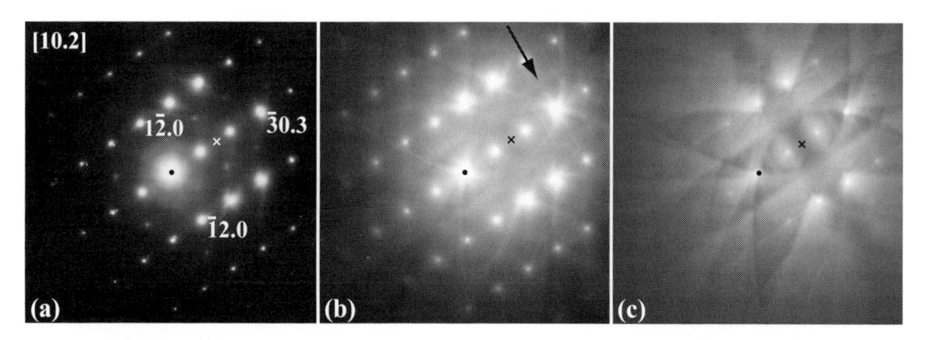

Fig. 9.19. Sequence of titanium [10.2] zone axis patterns for increasing foil thickness from (a) to (c) (200 kV).

the Bragg reflections. In addition, straight lines appear, both brighter and darker than the background intensity. These lines are known as *Kikuchi lines* and in this section we will explore their origin and geometry.

9.4.3.1 Origin and geometry of Kikuchi lines

From Fig. 2.2(b) in Chapter 2 we know that, for a given set of planes, the Bragg condition is satisfied for all electrons traveling along a conical surface with opening angle $\pi/2 - \theta$. In particular, all *inelastically* scattered electrons traveling in directions on this conical surface may be scattered *elastically*† by an angle 2θ. Let us now consider a particular set of planes **g**, and all inelastically scattered electrons which travel towards a point P on this plane, as shown schematically in Fig. 9.20(a). Those inelastically scattered electrons which travel on the conical surface along the line **OPQ** are in the proper orientation to be scattered *elastically* by the planes **g**. They will, therefore, be scattered onto the same conical surface, moving away along the direction **PR**. We will therefore have an *excess* of electrons in the direction **PR**, and a *deficit* of electrons along the direction **PQ**. Since the direction **OP** could lie anywhere on the conical surface, the top cone will have an excess of electrons and is therefore known as the *excess cone*; the bottom cone is then known as the *deficit cone*. The two cones are also known as *Kossel cones*.

The intersection of the Ewald sphere with the excess and defect cones gives rise to a pair of (curved) lines, one brighter than the background and one darker than the background intensity. The curvature of the lines is rather small and may not always be observable with commonly used camera lengths. The curvature decreases with increasing microscope accelerating voltage. The lines are known as *Kikuchi lines*, after S. Kikuchi who discovered them in 1928 [Kik28] (before the invention of

† Inelastically scattered electrons have an energy which is different from the primary beam energy. This corresponds to a different wavelength and therefore a different Bragg angle. It is a good first-order approximation to ignore the difference between these "elastic" and "inelastic" Bragg angles.

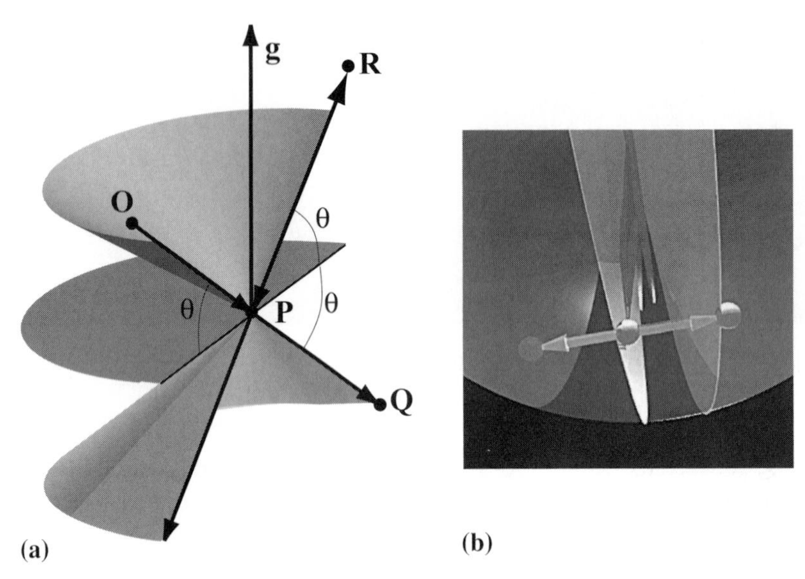

Fig. 9.20. (a) Geometry of excess and defect cones; (b) single frame from a QuickTime movie available from the website, illustrating the geometry of the defect and excess cones.

the TEM!). The distance between the two lines corresponds to the Bragg angle for elastic scattering from the planes **g**, as shown in the rendered drawing of Fig. 9.20(b). This figure is a single frame from a 200-frame QuickTime movie available from the website. The movie illustrates the geometry of the Kossel cones in relation to the reciprocal lattice points and the Ewald sphere.

It is important to realize that the excess and defect cones are "attached" to the diffracting planes **g**; when we change the orientation of the crystal, both cones move along with the sample, and hence the positions of the bright and dark lines change with respect to the reciprocal lattice point **g**. Kikuchi lines, therefore, provide us with a very sensitive way to determine the orientation of the crystal with respect to the incident beam. For each pair of reciprocal lattice points **g** and −**g**, there are two Kikuchi lines, corresponding to both sides of the plane. Which of those lines is bright depends on the relative excitation errors of the two reflections. Figure 9.21 shows the relation between crystal orientation and position of the Kikuchi lines for the reflection **g**. The defect (D) and excess (E) cones are shown as thick, short lines attached to the lattice plane **g**. In exact Bragg orientation, we have $s_{\mathbf{g}} = 0$, and the bright Kikuchi line intersects the reciprocal lattice point (Fig. 9.21a). When the crystal is tilted by a small angle $+\alpha$, such that the excitation error $s_{\mathbf{g}}$ becomes negative, then the bright Kikuchi line moves towards the origin of reciprocal space (Fig. 9.21c). A specimen tilt in the opposite direction $-\alpha$ results in a positive excitation error $s_{\mathbf{g}}$ and a bright Kikuchi line on the opposite side of the reciprocal lattice point (Fig. 9.21e).

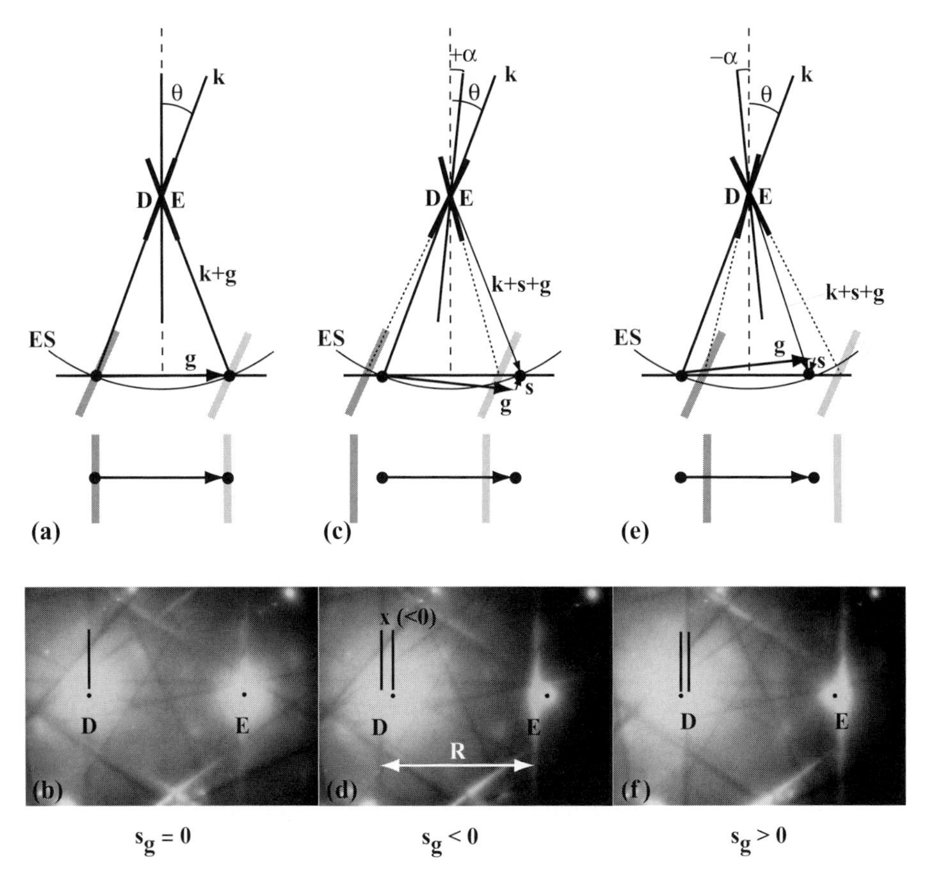

Fig. 9.21. Location of Kikuchi lines with respect to reciprocal lattice points for (a) $s_{\mathbf{g}} = 0$, (c) $s_{\mathbf{g}} < 0$, and (e) $s_{\mathbf{g}} > 0$. (b), (d) and (f) are experimental images for the (10.3) reflection of Ti, recorded at 200 kV.

Figure 9.21(b), (d) and (f) show experimental Kikuchi patterns for the (10.3) reflection of Ti, recorded at 200 kV. From the position of the Kikuchi line with respect to the reciprocal lattice point and the origin, we can derive a numerical value for the excitation error. It is easy to show (exercise) that the relation between the excitation error and the position of the bright Kikuchi line is given by

$$s_{\mathbf{g}} = \frac{x}{R}\lambda|\mathbf{g}|^2, \tag{9.16}$$

where x and R are indicated in Fig. 9.21(d). Since $|\mathbf{g}_{10.3}| = 7.5097\ \text{nm}^{-1}$ and $x/R = -1.568/21.244 = -0.073\,81$ we have for the excitation error $s = -0.01\ \text{nm}^{-1}$. If we change the position of the bright line by, say, one-tenth of the length of the **g**-vector, then this corresponds to a specimen tilt of $2\theta/10$, which is a small angle.

When the excitation errors for **g** and −**g** are equal (and therefore negative), both lines are located halfway between the origin and the reciprocal lattice point. The intensities of both lines are brighter than the surrounding background; in fact, the entire region between the two lines has a higher-than-background intensity, and one can clearly identify a *Kikuchi band*. Kikuchi bands can be used to navigate in reciprocal space. On the stereographic projection sphere, each Kikuchi band corresponds to the region enclosed by the defect and excess cones. If we tilt the sample, so that one particular band remains in the same location in the diffraction pattern, then we are effectively rotating the sample around the plane normal **g**.

The intersection of multiple Kikuchi bands corresponds to a zone axis orientation; this is similar to the intersection of bend contours in a bright field image, which also indicates a zone axis orientation. One can use the Kikuchi bands as a roadmap of reciprocal space. They are particularly useful to tilt from one zone axis orientation to another one, while observing the diffraction pattern. On older microscopes with manual specimen attitude controls, this may require considerable manual skill, as both specimen translation controls must be used to keep the sample location constant, and both specimen tilt axes to keep the Kikuchi band at a constant location. On modern microscopes with computer-controlled eucentric goniometer stages, this type of reciprocal space navigation may be easier, as the computer takes care of maintaining the eucentric sample position.

9.4.3.2 Kinematical simulation of Kikuchi lines

The location of a Kikuchi line or line pair can easily be computed using the same mathematical formalism as that used for the HOLZ line computations. The only difference is that two lines must be considered instead of just one. The mathematical formalism derived in Sections 9.4.2 and 3.9.5 provides the excitation error of a reflection for a given beam direction, foil normal, and beam tilt (Laue center). From the excitation error we can determine the location of the defect line via equations (9.14) and (9.15). The excess line is then located at a distance $|\mathbf{g}|$ away from the defect line. The program kikuchi.f90 can be used to compute diffraction patterns with superimposed Kikuchi lines. The program structure is similar to that of the holz.f90 program described in Chapter 3. The mathematical equations derived for Fig. 9.16(b) can be applied in this case as well.

Figure 9.22(a) shows a [110] zone axis pattern for aluminum. In the zone axis orientation the defect cone for **g** coincides with the excess cone for −**g**, and both Kikuchi lines have the same intensity, typically bright. This is indicated by bright lines against a gray background. If the crystal is tilted so that the Laue center is located at $(\overline{0.8}\,0.8\,1)$, as shown in Fig. 9.22(b), then the entire pattern of Kikuchi lines shifts rigidly while the intensities of the underlying Bragg reflections change. The excess lines (in this case the lines corresponding to the smallest excitation

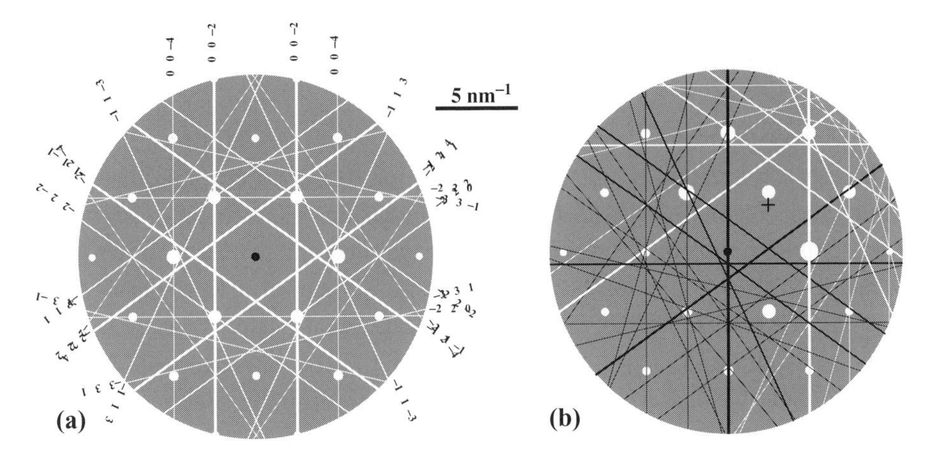

Fig. 9.22. [110] zone axis pattern for Al, with Kikuchi lines for (a) the exact zone axis orientation and (b) for an orientation with the Laue center (+ sign) at the location $(\overline{0.8}\,0.8\,1)$.

error) are indicated in white, the defect lines in black. The line width is indicative of the magnitude of the Fourier coefficient $V_{\mathbf{g}}$.†

When we introduced higher-order Laue zones (HOLZ) in Section 3.9.5, we concluded that dynamical electron diffraction is a three-dimensional process, with reflections from all HOLZ planes participating in the scattering event. When studying thick crystals, we may observe the effects of the HOLZ reflections in the form of extra Kikuchi lines which cannot be explained in terms of the regular reflections of the zone axis pattern. We know that, in zone axis orientation, the bright and dark Kikuchi line pair is replaced by a Kikuchi band with an above-background intensity, centered on the central beam. The edge of every Kikuchi band either lies halfway between the central beam and one of the diffracted beams, or it crosses a diffracted beam (which simply means that the band lies halfway between **0** and 2**g**). Often one can observe additional dark lines, without their bright partners, at locations which do not correspond to any of the Bragg reflections in the ZOLZ. These lines can be explained by taking the HOLZ reflections into account.

HOLZ lines, as defined in Chapter 3, are caused by elastic scattering of electrons from ZOLZ reflections into HOLZ reflections. Because of the slope of the Ewald sphere where it intersects the HOLZ layers (see Fig. 9.15), the result of this elastic scattering process is the appearance of dark lines in the ZOLZ diffraction disks. The corresponding bright lines can be observed along the HOLZ rings. Quite often, the dark and bright HOLZ lines continue outside of the diffraction disks. Here, they

† The kinematic theory can only predict the location of the Kikuchi lines, not their intensities. The full dynamical theory including inelastic scattering must be used to predict the intensity profile of a Kikuchi band. This involves solving the Yoshioka equations (Section 2.6.5), which is beyond the scope of this book. The interested reader may consult [Wan95].

can no longer be caused by elastic scattering processes, and they are hence Kikuchi lines. The continuation of bright and dark HOLZ lines outside the diffraction disks is a consequence of inelastic scattering processes. Since the dynamical simulation of HOLZ lines is already more complicated than a normal multi-beam simulation using the projection approximation, the inclusion of inelastic scattering effects to compute the intensity distribution of Kikuchi lines due to HOLZ reflections poses a very complex problem. We refer the interested reader to [Wan95] for a more detailed introduction to inelastic scattering theory. It is possible to compute dynamical Kikuchi patterns, starting from the Yoshioka equations of inelastic scattering (page 124). Omoto, Tsuda, and Tanaka [OTT02] have derived a full dynamical theory for inelastic scattering from a perfect crystal using the Bloch wave theory.

Using the kinematic theory, we can predict the locations of the HOLZ Kikuchi lines; in fact, these lines are simply the extensions of the HOLZ lines that we have already implemented in the program holz.f90. The kikuchi.f90 program uses the HOLZ geometry introduced in Chapter 3 for most of its computations. This means that only reflections belonging to a particular zone and its related HOLZ layers are considered. When Kikuchi bands are used for navigation, a representation of all major Kikuchi bands drawn on the asymmetric part of a stereographic projection can be useful. Such a map can be constructed from a set of experimental Kikuchi maps, or the entire pattern can be calculated and displayed on a stereogram. Such a computation only generates the geometry of the Kikuchi bands, not their intensities. The program kikmap.f90 can be used to generate sections of the asymmetric unit of the stereographic projection for an arbitrary crystal structure. The program uses purely geometrical information to determine the location of all Kikuchi bands within a conical region of opening angle α around the incident beam. This angle can be changed experimentally, by varying the diffraction camera length. For each set of planes, $\mathbf{g}$, we compute the angle β between the plane normal and the incident beam direction (see Fig. 9.23). If one or both of the angles $\beta \pm \theta_B$ is smaller than α, then a portion or the entire Kikuchi band will be visible.

The angles are converted readily into distances from the origin of reciprocal space, and the equations derived for Fig. 9.16(b) can be used to compute the locations of both edges of the Kikuchi band. Typical output of the kikmap.f90 program is shown in Fig. 9.24 for the [001] zone axis orientation of GaAs. For each band the edges of the band and the center line are indicated. Zone axis orientations occur where multiple bands intersect.

9.5 Convergent beam electron diffraction

We have seen in Chapter 5 that the diffraction process is highly sensitive to the crystal symmetry. Convergent beam electron diffraction patterns provide a direct

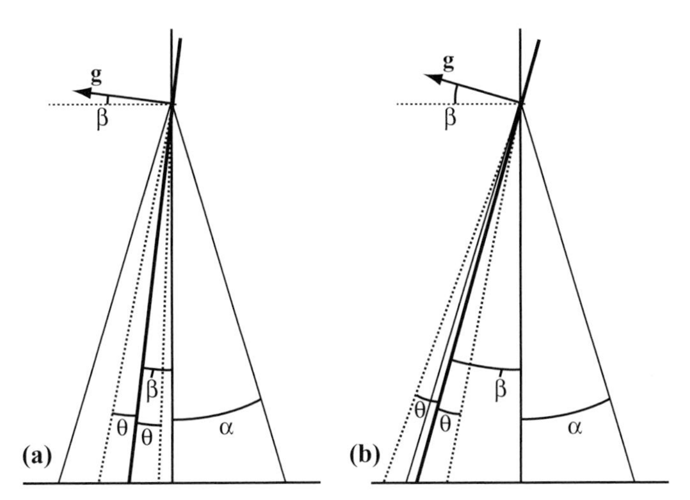

Fig. 9.23. Schematic representation of a Kikuchi band caused by the defect and excess cones intersecting the viewing screen. The band in (a) falls entirely within the viewing area corresponding to the cone with opening angle α. The band in (b) is only partially visible.

visualization of the various symmetries present in the crystal. By carefully examining the symmetry of multiple CBED patterns of a given crystal structure, it is possible to determine both the point group and the space group of that structure. In the following subsections, we will describe briefly how one can determine the symmetry groups for the study materials Cu-15 at% Al and GaAs. It is left as an exercise for the reader to determine the symmetry groups of study materials Ti and $BaTiO_3$ (see exercises).

9.5.1 Point group determination

The combined use of Tables 5.1 (page 340) and 7.2 (page 438), along with Fig. 5.11 (page 341), allows one to determine nearly all the point groups. Let us first consider the Cu-15 at% Al study material. We know that the space group is **Fm$\bar{3}$m**. Figures 9.25 and 9.26 show [001] and [111] CBED patterns acquired in a Philips CM20, operated at 80 kV. The patterns are digital combinations of individual exposures; the range of intensities in CBED patterns is often too large to capture in a single exposure. The bright field (BF) symmetry of the [001] pattern in Fig. 9.25(a) is given by **4mm**, which is also equal to the whole pattern (WP) symmetry. According to Table 5.1, this means that the diffraction group is either **4mm** or $\mathbf{4mm1}_R$. The 200 and 220 reflections are both special reflections, because they lie on the mirror planes of the group **4mm**. Accordingly, the special dark field symmetry should be equal to **2mm**. This is verified easily by tilting the incident beam so that the Bragg condition is satisfied for any one of these reflections. This is shown for the (200) reflection in Fig. 9.25(b), and for (220) in Fig. 9.25(c). In both cases, the

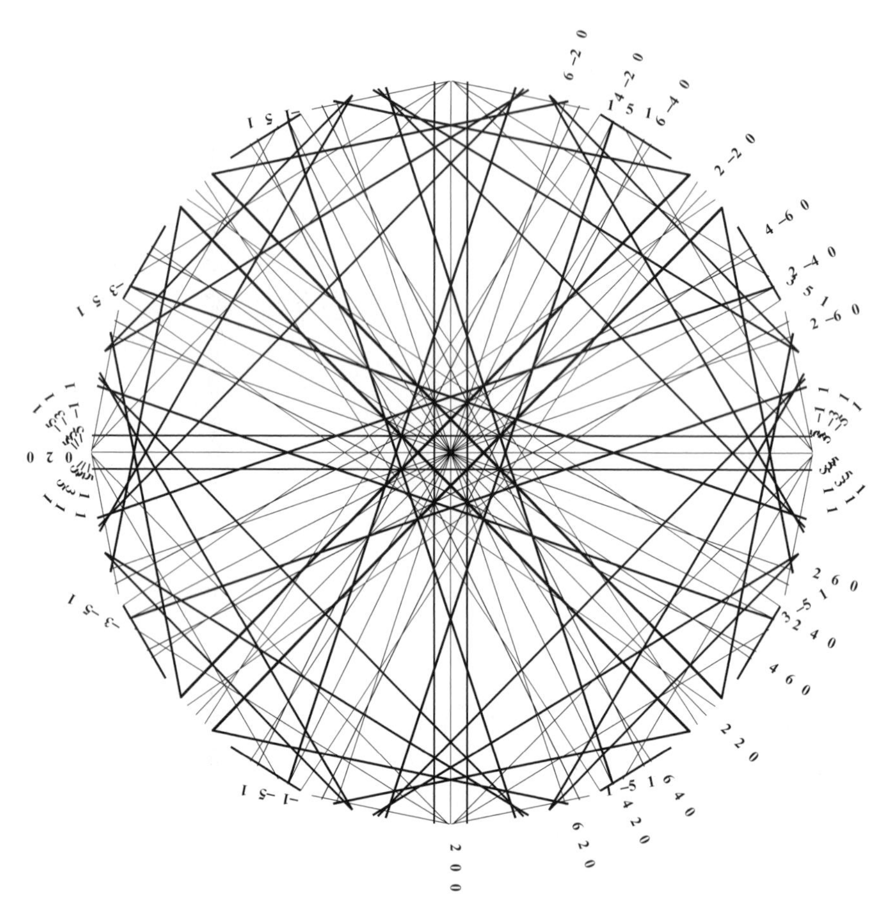

Fig. 9.24. Kikuchi band pattern for the [001] orientation of GaAs, computed with the kikmap.f90 program. Each band has a thin central line; intersections of multiple thin lines indicate the locations of zone axes.

orthogonal mirror planes are clearly visible. General reflections, such as the 240 reflection, have symmetry **2**. The dark field symmetries allow us to choose between the two possible diffraction groups, and it is clear from Table 5.1 that the diffraction group of the [001] zone axis is $\mathbf{4mm1}_R$.

According to Fig. 5.11, there are only two possible point groups that can give rise to the diffraction group $\mathbf{4mm1}_R$: $\mathbf{4/mmm}$ and $\mathbf{m\bar{3}m}$. Therefore, we need another zone axis CBED pattern to distinguish between these two point groups. Consider the [111] CBED pattern, shown in Fig. 9.26. The BF symmetry is **3m**, which is also equal to the WP symmetry (note the fine HOLZ lines in the central disk). This is consistent with only two diffraction groups: **3m** and $\mathbf{6}_R\mathbf{mm}_R$. Only $\mathbf{6}_R\mathbf{mm}_R$ is consistent with one of the two possibilities for the point group, so that the point group of Cu-15 at% Al is equal to $\mathbf{m\bar{3}m}$.

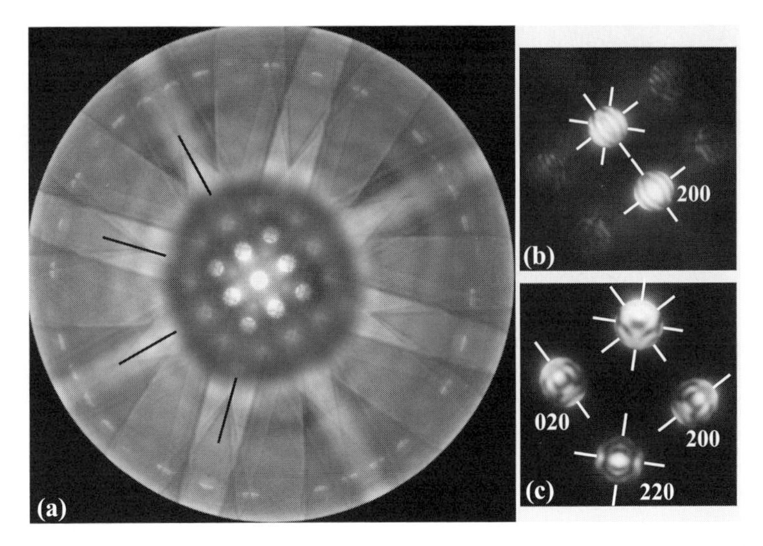

Fig. 9.25. [100] zone axis CBED patterns for Cu-15 at% Al: (a) whole pattern, (b) 200 and (c) 220 in Bragg orientation, with mirror planes indicated as short line segments (patterns recorded at 80 kV).

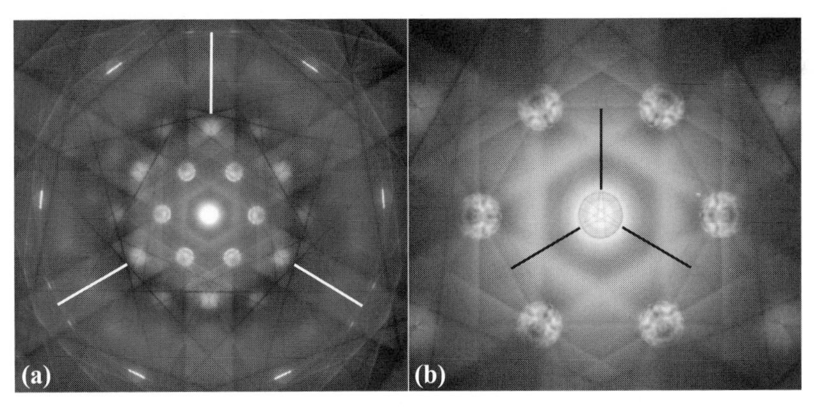

Fig. 9.26. [111] zone axis CBED pattern for Cu-15 at% Al, recorded at 80 kV. Both the whole pattern (a) and bright field (b) symmetries are equal to **3m**.

Next, let us consider GaAs as a second example. We have already determined the BF and WP symmetry of the [110] zone axis from the patterns shown in Fig. 4.25. The bright field symmetry is **2mm**, and the whole pattern symmetry is **m**. From Table 5.1, we see that there is only one possibility for the diffraction group: $\mathbf{m1}_R$. There are six point groups consistent with $\mathbf{m1}_R$: **mm2**, **4mm**, $\bar{\mathbf{4}}$**2m**, **6mm**, $\bar{\mathbf{6}}$**m2**, and $\bar{\mathbf{4}}$**3m**. A second zone axis CBED pattern is needed; the [111] CBED pattern, acquired at 80 kV, is shown in Fig. 9.27. Both the BF and WP symmetries are equal to **3m**, as is derived easily from the HOLZ lines in the central reflection. The

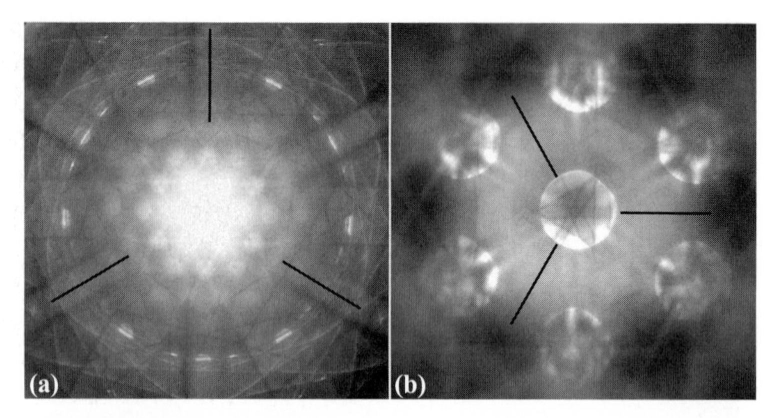

Fig. 9.27. [111] CBED patterns of GaAs, recorded at 80 kV in a Philips CM20 microscope. Both the whole pattern (a) and bright field (b) symmetries are equal to **3m** (sample courtesy of J. Neethling).

corresponding diffraction groups are then **3m** (with point groups $\bar{\mathbf{3}}\mathbf{m}$ and $\mathbf{m\bar{3}m}$) and $\mathbf{6}_R\mathbf{mm}_R$ (with point groups **3m** and $\bar{\mathbf{4}}\mathbf{3m}$). The only point group that is common between the possible groups for both zone axes is the non-centrosymmetric cubic point group $\bar{\mathbf{4}}\mathbf{3m}$.

The determination of the point group of an unknown crystal structure requires, in general, several CBED patterns, for all the highest symmetry zone axis orientations. For each zone axis you will need to record:

- the central bright field disk;
- the entire pattern, including HOLZ reflections (this typically requires a small camera length);
- individual diffraction disks, with the crystal (or beam) tilted so that each reflection in turn is in Bragg orientation;
- pairs of $\pm\mathbf{g}$ diffraction disks.

For a detailed description of the procedure, we refer to the paper by Steeds and Vincent [SV83]. Examples can be found in [WC96, Chapter 21], [FH01, Section 6.5.4] and [Man84]. The three volumes on CBED by Tanaka and coworkers [TT85, TTK88, TTT94] are mandatory reading for anyone who plans to make extensive use of CBED techniques.

9.5.2 Space group determination

We have seen in Chapter 1 that space groups are constructed by combining a Bravais lattice with a crystal point group. For both study materials considered in the previous section, the point groups belong to the cubic crystal system, which means that there are three possible Bravais lattices, cP, cI, and cF. We can determine which centering we have in two different ways.

- Index several different zone axis patterns for a given grain or location in your material. If you do this correctly, then you will find that all reflections of a particular kind are missing. For the examples above, only reflections for which all Miller indices have the same parity are allowed, which means that the lattices are face-centered.
- Obtain zone axis patterns of high symmetry with a small camera length (and possibly a lower acceleration voltage) to determine the location of the HOLZ layer reflections with respect to the ZOLZ reflections (see e.g. [WC96, Section 21.2]). This also provides information about the lattice centering.

For Cu-15 at% Al, we have the cF Bravais lattice combined with the point group $\mathbf{m\bar{3}m}$, which leads to the following possible space groups (see Tables A4.2 and A4.3): $\mathbf{Fm\bar{3}m}$, $\mathbf{Fm\bar{3}c}$, $\mathbf{Fd\bar{3}m}$, and $\mathbf{Fd\bar{3}c}$. For GaAs, we end up with the possible space groups $\mathbf{F\bar{4}3m}$ and $\mathbf{F\bar{4}3c}$. In each case, we have to choose between a symmorphic and one or more non-symmorphic space groups.

We know from structure factor considerations (e.g. [MM86] and the examples in Section 9.2.3) that the presence of glide planes and/or screw axes causes additional forbidden reflections. Such reflections may appear in zone axis diffraction patterns because of double-diffraction effects. In CBED patterns, the phases of the various structure factors contributing to a zone axis pattern may be such that the diffraction disks corresponding to kinematically forbidden reflections show Gjönnes–Moodie (G-M) lines, or dynamical extinctions [GM65], as discussed in Section 9.2.3. The presence of G-M lines (or crosses) is usually quite obvious, since their presence is independent of foil thickness and acceleration voltage, and can be used to determine whether or not glide planes or screw axes are present. G-M lines, originating from dynamical extinctions due to ZOLZ reflections, are typically broad lines, which are known as A_2 and B_2 lines. The subscript 2 denotes 2D diffraction (i.e. the projection approximation); A lines are parallel to the $\mathbf{g}$ vector of the reflection, whereas B lines are normal to the $\mathbf{g}$ vector. Double diffraction interactions with HOLZ reflections result in narrower G-M lines, labeled A_3 and B_3 lines (3, because of the 3D nature of the diffraction process).

Tanaka and coworkers have determined the occurrence of G-M lines for all high symmetry zone axis orientations of all 230 space groups. The corresponding tables can be found in [TSN83, TT85]. The tables list, for each space group, and for each zone axis orientation, which reflections are expected to have G-M lines, and which type of lines ($A_{2,3}$ and/or $B_{2,3}$). We refer the reader to the many examples in the volumes [TT85, TTK88, TTT94] for more details on the determination of glide planes and screw axes. The G-M lines allow for the determination of 185 out of the 230 different space groups.

Examination of CBED patterns of the high-symmetry zone axis orientations of Cu-15 at% Al and GaAs reveals that there are no G-M lines, and that, therefore, both space groups are symmorphic: $\mathbf{Fm\bar{3}m}$ for Cu-15 at% Al, and $\mathbf{F\bar{4}3m}$ for GaAs. The reader is encouraged to repeat the procedures described in this and the previous section to study materials Ti and $BaTiO_3$. A worked-out example for Ti can be found in [FH01, Sections 6.5.4 and 6.5.5]. Further examples of the structure determination of σ, χ, and R intermetallic phases using CBED methods can be found in [RM94].

9.6 Diffraction effects in modulated crystals

9.6.1 Modulation types

In previous chapters, we have seen that there are two different approaches to electron diffraction: the kinematical model, in which $|F_{hkl}|^2$ describes the intensity of a diffracted beam, and the dynamical theory, which permits electrons to scatter multiple times. While the kinematical theory has only limited validity when it comes to the prediction of images, the theory is quite useful for the description of the geometry and approximate intensity distributions in diffraction experiments. We have defined the structure factor F_{hkl} as a summation over all atoms in the unit cell of the electron scattering factors, appropriately phase shifted (see Chapter 2). The unit cell is repeated throughout the volume of the crystal, which determines the shape of the reciprocal lattice points. In this section, we will no longer consider the unit cell as the essential building block of the crystal. Instead, we will redefine the structure factor as a summation over *all atoms in the entire crystal*. In other words,

$$F_{\mathbf{g}}(t) = \sum_{j=1}^{N} f(\mathbf{r}_j, t) e^{2\pi i \mathbf{g} \cdot \mathbf{R}(\mathbf{r}_j, t)}. \tag{9.17}$$

The electron scattering factor $f(\mathbf{r}_j, t)$ depends on the position of the atom; the vector function $\mathbf{R}(\mathbf{r}_j, t)$ returns the position of atom j at time t, and N is the total number of atoms in the crystal. This is a very general expression which will allow us to introduce a large variety of modulated and disordered structures. Time t is a possible variable; for instance, by averaging the atom displacements due to thermal vibrations we can derive an expression for the intensity distribution due to thermal diffuse scattering (TDS), as done explicitly in Section 6.5.1. Diffusion processes that occur when a thin foil is heated inside the microscope column may also give rise to a time dependent intensity distribution. For the remainder of this section, however, we will ignore time as a variable.

The functions f and $\mathbf{R}$ completely define the state of the crystal, provided they are known for each lattice site. We can use the behavior of these functions to define different types of modulated crystals.

- **Compositional modulations.** All atoms are located on their equilibrium lattice sites ($\mathbf{R}(\mathbf{r_j}) = \mathbf{r}_j$), and the electron scattering factors $f(\mathbf{r}_j)$ vary periodically in one or more directions. In such a case, we may expand the scattering factor into a Fourier series with respect to the reciprocal lattice vectors $\mathbf{q}$ of the modulated structure:

$$f(\mathbf{r}_j) = \sum_{\mathbf{q}} a_{\mathbf{q}} e^{2\pi i \mathbf{q}\cdot\mathbf{r}_j}. \tag{9.18}$$

  It is, of course, possible for the modulated structure to have the same basis vectors as the underlying lattice (or an integer multiple of those basis vectors), in which case we end up with an *ordered structure*. The vectors $\mathbf{q}$ needed to describe this ordered structure are commonly referred to as the *ordering wave vectors*, and the corresponding complex exponentials are known as *concentration waves* [Kha83].

  Alternatively, the function f may vary slowly but periodically, so that the structure displays compositional modulations on a length scale larger than the underlying unit cell. Spinodal decomposition results in structures that display compositional modulations with well-defined periodicities.
- **Disordered and partially ordered structures.** All atoms are located at their normal lattice sites as before, but now the scattering factor f is not a periodic function of position. This does not necessarily mean that there is no order at all. There could be small regions with a particular kind of ordering, but there is no long-range correlation with other such regions. Such structures are known as *short-range ordered structures*, and we will define the mathematical behavior of f and the resulting diffraction patterns in Section 9.7.
- **Displacive modulations.** All atom sites have a well-defined composition (i.e. atom type), but the atoms are not necessarily located at their normal lattice sites. If the function $\mathbf{R}(\mathbf{r}_j)$ varies periodically in one or more directions, we end up with a *displacively modulated* structure. Such a structure may arise, for instance, during cooling if a particular phonon mode becomes "frozen in". In essence, this means that the displacement pattern of the phonon becomes locked into the crystal structure. Such a process is thought to be of importance for martensitic phase transformations. We will describe displacively modulated structures and their effects on diffraction patterns in Section 9.6.2.1.
- **Interface modulations.** In Chapter 8, we have introduced planar defects with displacement vector $\mathbf{R}$. In certain crystal systems, it may be energetically favorable for such defects to organize themselves in a periodic fashion. The crystal is, therefore, divided into perfect regions, called *blocks*, which are shifted with respect to each other at regular intervals. The function $\mathbf{R}(\mathbf{r}_j)$ is then a periodic function that is constant inside the blocks, and changes nearly discontinuously at each block boundary. Crystallographic shear planes and anti-phase boundaries can give rise to such interface modulated structures. If a crystal structure is built from a regular stacking of two or more building blocks, the resulting

structure can often also be regarded as an interface modulated structure. We will describe such structures in Section 9.6.2.3.

In many of the cases just introduced it is possible to define one or more modulation wave vectors $\mathbf{q}^{(n)}$. In general, those vectors are described in reciprocal space, which means that they can be written as a linear combination of the reciprocal basis vectors $\mathbf{a}_i^*$:

$$\mathbf{q}^{(n)} = q_i^{(n)} \mathbf{a}_i^*. \tag{9.19}$$

If all of the numbers $q_i^{(n)}$ are rational numbers, then it can be shown that there exists an integer p such that the vector $p\mathbf{q}^{(n)}$ is again a reciprocal lattice vector. This means that the modulated structure contains an integer number of unit cells of the underlying lattice. In such a case, we say that the modulation is *commensurate* with the underlying lattice. A modulation with wave vector $\mathbf{q} = \frac{1}{5}[310]$ is a commensurate modulation since $5\mathbf{q}$ is a reciprocal lattice vector. If one or more of the components $q_i^{(n)}$ are irrational, then such an integer p does not exist, and the resulting structure is *incommensurate*, i.e. the modulation wavelength is not an integer multiple of any lattice spacing.† In the following subsections, we will describe several possible commensurate and incommensurate modulations in more detail.

9.6.2 Commensurate modulations

Before dealing with general cases, let us first consider a simple example of a displacively modulated structure. For simplicity, we will work with a 1D modulation with wave vector $\mathbf{q} = \frac{1}{\Lambda}\mathbf{a}^*$. The atom positions will then be shifted by a sinusoidal function:

$$\mathbf{R}\left(\mathbf{r}_j\right) = \mathbf{r}_j + \mathbf{A}\sin(2\pi\mathbf{q}\cdot\mathbf{r}_j),$$

where $\mathbf{A}$ indicates the direction and amplitude of the modulation. We will take $\mathbf{A} = A\mathbf{a}$, i.e., a longitudinal modulation shown in Fig. 9.28(a). If the position vectors of the atoms are given by $\mathbf{r}_j = j\mathbf{a}$, then we have for the argument of the structure factor exponential:

$$\mathbf{g}\cdot\mathbf{R}\left(\mathbf{r}_j\right) = hj + hA\sin\left(2\pi\frac{j}{\Lambda}\right).$$

† The distinction between commensurate and incommensurate modulations is an academic one, since it is not possible to measure a modulation wavelength and prove that it is an irrational number. However, in most materials with incommensurate modulations, the modulation wavelength depends on an external parameter, often the temperature. If the modulation wavelength varies continuously with temperature, then the modulation is considered to be incommensurate.

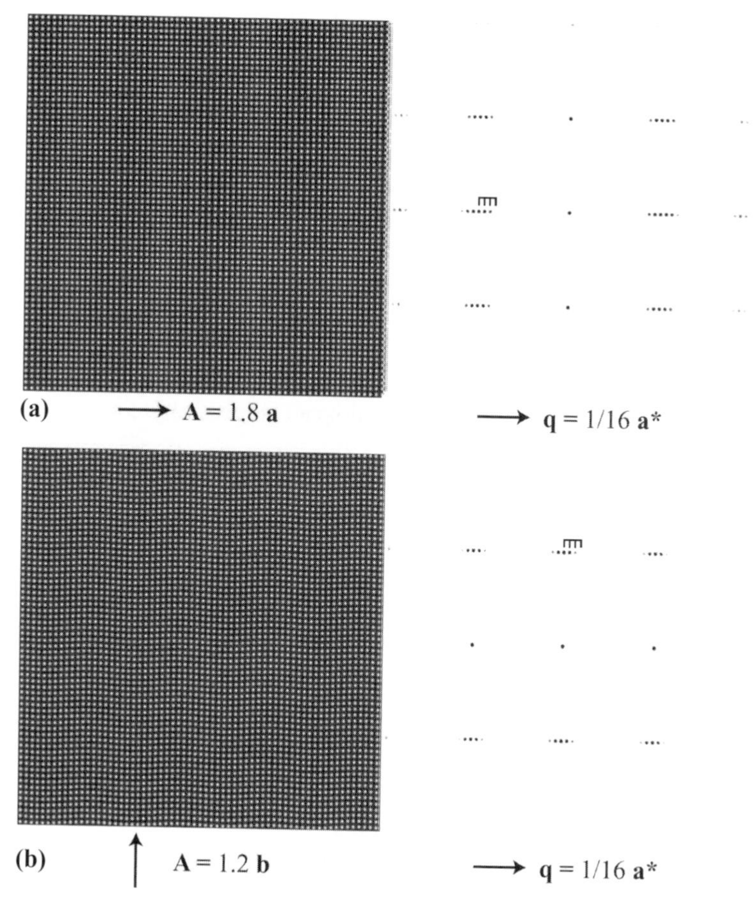

Fig. 9.28. Two examples of displacive modulations of a 2D square lattice: (a) $\mathbf{q} = \frac{1}{16}\mathbf{a}^*$, $\mathbf{A} = 1.8\mathbf{a}$ (longitudinal); (b) $\mathbf{q} = \frac{1}{16}\mathbf{a}^*$, $\mathbf{A} = 1.2\mathbf{b}$ (transverse).

If A is small, then we can expand the exponential in a Taylor series and only keep the first two terms:

$$
\begin{aligned}
F'_{\mathbf{g}} &= \sum_j f_j \mathrm{e}^{2\pi \mathrm{i} hj} \mathrm{e}^{hA\mathrm{i}\sin(2\pi j/\Lambda)}, \\
&\approx \sum_j f_j \mathrm{e}^{2\pi \mathrm{i} hj} \left[1 + 2\pi \mathrm{i} hA \sin\left(2\pi \frac{j}{\Lambda} \right) \right], \\
&= F_g + \sum_j f_j \pi hA \left[\mathrm{e}^{2\pi \mathrm{i}(h+j/\Lambda)} - \mathrm{e}^{-2\pi \mathrm{i}(h-j/\Lambda)} \right];
\end{aligned}
$$

i.e. in addition to the normal diffraction amplitude described by the structure factor $F_{\mathbf{g}}$ of the undistorted lattice, there will be satellite spots flanking each fundamental spot at a distance of $\pm j/\Lambda$. The transmitted or zero order beam does not have satellite reflections because the second term of the modulated structure factor is pre-multiplied by h. The diffraction pattern, therefore, consists of the regular diffraction spots at positions $\mathbf{g}$ and satellite reflections at the positions $\mathbf{g} \pm \mathbf{q}$.

9.6.2.1 Displacive modulations

Consider a sinusoidal displacive modulation with amplitude and direction $\mathbf{A}$ and wave vector $\mathbf{q}$, so that [AVD93]

$$\mathbf{R}(\mathbf{r}_j) = \mathbf{r}_j + \mathbf{A}\sin(2\pi\mathbf{q}\cdot\mathbf{r}_j).$$

This modulation could be longitudinal ($\mathbf{A} \parallel \mathbf{q}$) or transverse ($\mathbf{A} \perp \mathbf{q}$). The structure factor evaluated at a location $\mathbf{h}$ in reciprocal space is given by

$$\begin{aligned} F_{\mathbf{h}} &= \sum_j f_j \mathrm{e}^{2\pi i\mathbf{h}\cdot[\mathbf{r}_j+\mathbf{A}\sin(2\pi\mathbf{q}\cdot\mathbf{r}_j)]}; \\ &= \sum_j f_j \mathrm{e}^{2\pi i\mathbf{h}\cdot\mathbf{r}_j}\mathrm{e}^{2\pi i\mathbf{h}\cdot\mathbf{A}\sin(2\pi\mathbf{q}\cdot\mathbf{r}_j)}. \end{aligned}$$

Using the Jacobi–Anger relation† and changing the order of the summations we can rewrite this expression as

$$F_{\mathbf{h}} = \sum_{n=-\infty}^{+\infty} J_n(2\pi\mathbf{h}\cdot\mathbf{A})\left[\sum_j f_j \mathrm{e}^{2\pi i(\mathbf{h}+n\mathbf{q})\cdot\mathbf{r}_j}\right].$$

The second summation can only be different from zero if $\mathbf{h}+n\mathbf{q}$ equals a reciprocal lattice vector $\mathbf{g}$, or $\mathbf{h} = \mathbf{g} - n\mathbf{q}$. From this expression we can derive the following observations:

(i) reciprocal lattice points for which $\mathbf{h}\cdot\mathbf{A} \neq 0$ will have satellite reflections along the direction of the wave vector $\mathbf{q}$;
(ii) for a given value of the argument $x = 2\pi\mathbf{h}\cdot\mathbf{A}$, the Bessel function $J_n(x)$ rapidly decreases with increasing n, indicating that the amplitude of the satellites decreases rapidly with increasing n – typically, only the first three or four satellites will be observable;
(iii) for a longitudinal modulation reciprocal lattice points normal to $\mathbf{q}$ will have no satellites, since the argument of the Bessel function vanishes. For a transverse modulation reflections parallel to $\mathbf{q}$ will have no satellites. Because of double-diffraction effects satellites may be present around all reciprocal lattice points.

† If a and b are real numbers, then the following equality holds:

$$\mathrm{e}^{ia\sin b} = \sum_{n=-\infty}^{+\infty} \mathrm{e}^{inb} J_n(a),$$

where J_n are the Bessel functions of the first kind [AS77].

Figure 9.28 shows two examples of structures with 1D displacive modulations; both examples use a 2D square undistorted lattice as starting point. Fig. 9.28(a) is a longitudinal modulation along the x-direction, along with its computed diffraction pattern. Satellites can be found around all reciprocal lattice points, except for the row normal to **q**. Figure 9.28(b) shows a transverse modulation with direction along **b** and the same wave vector as in (a). In this case, the central horizontal row of reflections does not have any satellites.

The two-dimensional modulation can be regarded as two 1D modulations applied in turn. The first modulation produces sets of satellites, and then the second modulation produces satellites on all reflections present after the first modulation. This is illustrated in Figs 9.29(a) and (b), for the indicated modulation amplitude

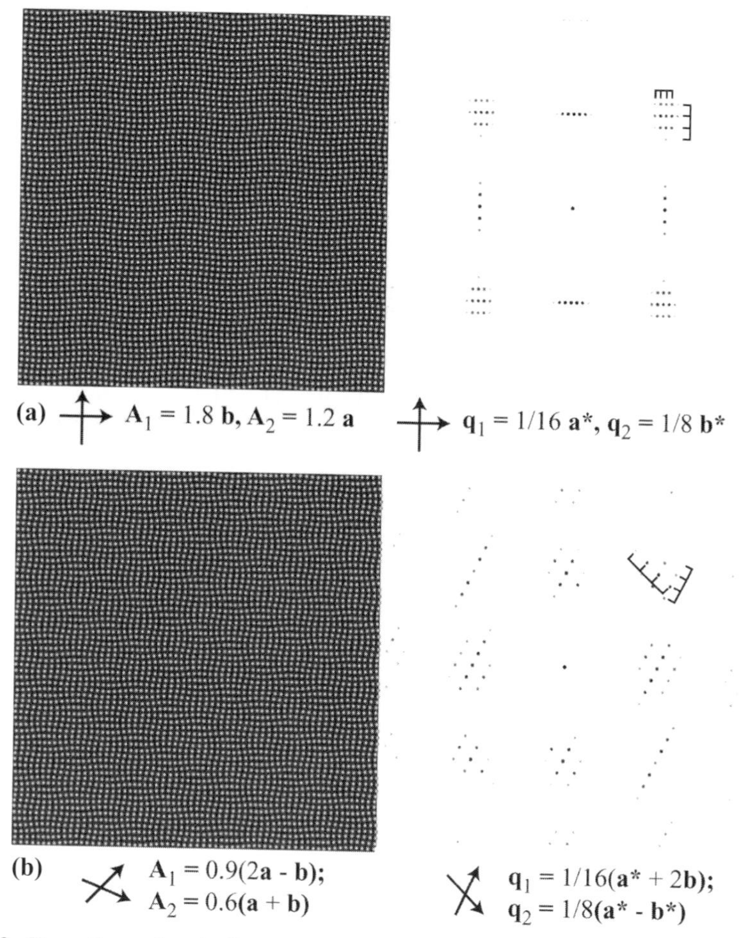

Fig. 9.29. Two-dimensional displacive modulations of a 2D square lattice: (a) $\mathbf{q}_1 = \frac{1}{16}\mathbf{a}^*$, $\mathbf{A}_1 = 1.8\mathbf{b}$, $\mathbf{q}_2 = \frac{1}{8}\mathbf{b}^*$, $\mathbf{A}_2 = 1.2\mathbf{a}$; (b) $\mathbf{q}_1 = \frac{1}{16}(\mathbf{a}^* + 2^*)$, $\mathbf{A}_1 = 0.9(2\mathbf{a} - \mathbf{b})$, $\mathbf{q}_2 = \frac{1}{8}(\mathbf{a}^* - \mathbf{b}^*)$, $\mathbf{A} = 0.6(\mathbf{a} + \mathbf{b})$.

and wave vectors. The amplitudes are somewhat exaggerated to emphasize the 1D and 2D arrays of satellite reflections.

If an experimental diffraction pattern shows 1D or 2D arrays of satellite reflections, then a careful measurement of the positions of all satellites with respect to the fundamental reciprocal lattice sites along with the directions of the modulations provides sufficient information to deduce the character of the displacive modulations. If the pattern of satellites is different for different fundamental reflections, i.e. if the satellite positions are different with respect to each of the reciprocal lattice points, then this may point to the presence of interface modulations, which we will discuss in Section 9.6.2.3. Displacive modulations occur in several of the high-T_c superconductor materials. The reader may find examples of electron diffraction studies of commensurate and incommensurate displacive modulations in the following references: [VTZA88, HII+91, ZWT99].

The **ION** routine modulation.pro on the website can be used to produce images similar to those of Figs 9.28 and 9.29 for 1D and 2D displacive modulations of a square lattice. The reader can experiment with different parameters, to see how the resulting diffraction patterns reflect the presence of the modulated structure.

9.6.2.2 Compositional modulations

For simplicity, we will continue to use the 2D square lattice to illustrate the effect of compositional modulations on the diffraction pattern. Consider a binary alloy AB, with a local concentration c_A of atoms of the type A $(0 \leq c_A \leq 1)$. The average concentration of A atoms is represented by $\bar{c}_A$. The electron scattering factor at position $\mathbf{r}_j$ can then be written as

$$f_j = c_{A,j} f_A + c_{B,j} f_B = c_{A,j} f_A + (1 - c_{A,j}) f_B.$$

In a compositionally modulated structure the concentration of A atoms varies from site to site. For a sinusoidal modulation we have

$$c_{A,j} = \bar{c}_A + \Delta c_A \sin(2\pi \mathbf{q} \cdot \mathbf{r}_j), \tag{9.20}$$

with Δc_A the maximum deviation from the average value, and $\mathbf{q}$ the modulation wave vector. If we convert the sin-function to complex exponential notation, it is easy to show that the structure factor is given by

$$F_{\mathbf{h}} = [\bar{c}_A f_A + (1 - \bar{c}_A) f_B] \sum_j e^{2\pi i \mathbf{h}\cdot\mathbf{r}_j} + \frac{1}{2i} (f_A - f_B) \Delta c_A \sum_j \left[e^{2\pi i(\mathbf{h}+\mathbf{q})\cdot\mathbf{r}_j} - e^{2\pi i(\mathbf{h}-\mathbf{q})\cdot\mathbf{r}_j} \right].$$

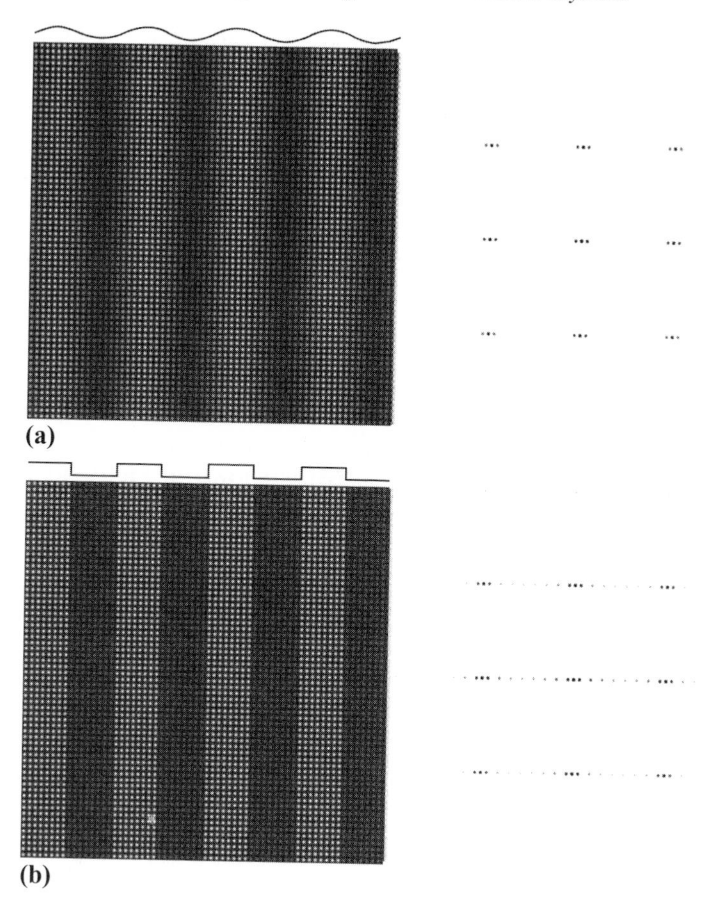

Fig. 9.30. (a) Sinusoidal and (b) square wave compositional modulations on a 2D square lattice. Simulation parameters are $\bar{c}_A = 0.6$, $\Delta c_A = 0.4$ and $\mathbf{q} = \frac{1}{16}\mathbf{a}^*$.

This expression is only different from zero if one of the vectors $\mathbf{h}$, $\mathbf{h}+\mathbf{q}$, or $\mathbf{h}-\mathbf{q}$ is equal to a reciprocal lattice vector. In other words, every reflection $\mathbf{g}$, including the transmitted beam, is surrounded by two satellite reflections at $\mathbf{g}+\mathbf{q}$ and $\mathbf{g}-\mathbf{q}$. The intensities of those reflections depend on the difference between the scattering factors for the particular elements A and B, and on the magnitude Δc_A of the compositional modulation. Figure 9.30(a) shows an example of such a modulation for $\bar{c}_A = 0.6$, $\Delta c_A = 0.4$, and $\mathbf{q} = \frac{1}{16}\mathbf{a}^*$.

If the compositional profile deviates from a purely sinusoidal profile, then additional satellite reflections will be present, depending on the actual profile. It is easy to see that the concentration profile in equation (9.20) is a Fourier series with only one term. For any other profile the Fourier series would contain additional terms, which would then correspond to additional satellite reflections at multiples of $\mathbf{q}$. For a square wave modulation, the Fourier series would contain only odd multiples

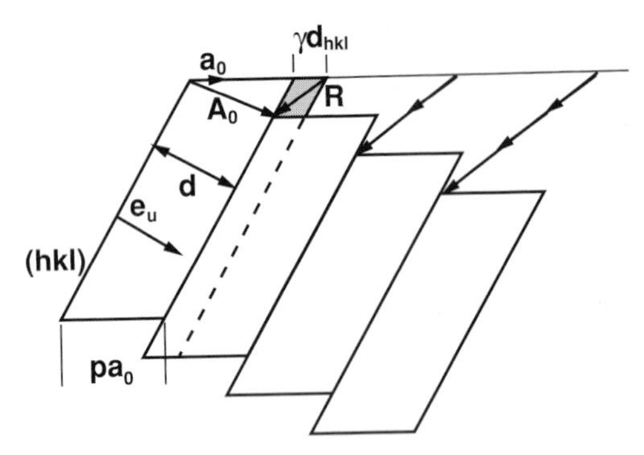

Fig. 9.31. Schematic drawing of a crystallographic shear plane defect.

of **q**, and the corresponding diffraction pattern is shown in Fig. 9.30(b) for the same parameters as in (a). The presence or absence of certain satellites provides information on the Fourier coefficients of the concentration profile, provided this profile has a single basic periodicity described by **q**. If more than one wavevector is needed, the satellite patterns become more complicated and analysis is more difficult. Compositional modulations can be found in many material systems. The most notable type of compositional modulation is perhaps *spinodal decomposition*. We refer the interested reader to [HGWA84] for details and examples.

9.6.2.3 Interface modulations: crystallographic shear planes

Consider the crystal shown in Fig. 9.31. We slice the crystal in plane-parallel blocks of thickness pd_{hkl}, where p is an integer and d_{hkl} the interplanar spacing of the planes parallel to the cut. At this cut, we remove a thin layer of material with thickness $t = \gamma d_{hkl}$, with $0 < \gamma < 1$. This is the component of the displacement vector **R** perpendicular to the cut plane. For a conservative fault, $\gamma = 0$. We then displace the second block by the vector **R** and close the gap. We have now created a new unit cell with lattice parameter $\mathbf{A}_0 = p\mathbf{a}_0 + \mathbf{R}$. This is a so-called *crystallographic shear fault*.

In many oxides, the crystal structure can be described by geometric coordination shapes, such as octahedra and tetrahedra. Often, the structure consists of a three-dimensional arrangement of corner, edge, or face-sharing polyhedra, with oxygen at the corners and metal ions at the centers of the polyhedra. In response to local chemical variations, say, an oxygen deficiency, many oxides adapt by the introduction of *crystallographic shear planes*. These defects are to oxide structures what anti-phase boundaries are for ordered intermetallics.

Since this defect can occur in a periodic fashion, we ask the question: what will we observe in the diffraction pattern? Based on the discussion of streaks in Section 9.4.1, we anticipate that the diffraction information will be located along the normal to the shear planes; we will denote the unit vector along this normal by $\mathbf{e}_u$. We expect a pattern of satellite reflections, with a spacing related to the inverse of the thickness of the individual blocks of the shear structure. In the following paragraphs, we will derive an expression for the location and intensity of the satellite reflections. The method used is very general and can be applied to any type of periodic planar defect.

We take the origin to be in the center of one of the blocks and the x-axis in the direction of the normal to the shear planes. The thickness of the blocks is thus described by $d = (p - \gamma)d_{hkl}$. We define the function $U(x)$ to be the shape function of the block, i.e.

$$U(x) = 1 \quad \text{for} -\frac{d}{2} < x < \frac{d}{2},$$

and zero elsewhere. Let the potential of the basic structure be $V(\mathbf{r})$. We will assume that the lattice potential of the sheared structure, $V_s(\mathbf{r})$, is at least approximately described by the perfect crystal potential; this is a good approximation inside the blocks but may become questionable close to the shear planes [VDBMA87]. The shear potential at position $\mathbf{r}$ is thus given by

$$V_s(\mathbf{r}) = \sum_n V_x(\mathbf{r} - n(\mathbf{a} + \mathbf{R})),$$

where we have defined $\mathbf{a} = p\mathbf{a}_0$. The block potential V_x is defined by

$$V_x(\mathbf{r}) = V(\mathbf{r})U(x).$$

The diffraction pattern associated with this potential can be computed by means of standard Fourier transformation techniques. The Fourier coefficients of the shear potential are given by

$$V_s(\mathbf{h}) = \iiint_{crystal} V_s(\mathbf{r})\mathrm{e}^{2\pi i\mathbf{h}\cdot\mathbf{r}}\,\mathrm{d}\mathbf{r},$$

with $\mathbf{h}$ (for now) an arbitrary vector in reciprocal space. Redefining the vector $\mathbf{r}$ as $\mathbf{r} = \mathbf{r}' + n(\mathbf{a} + \mathbf{R})$, we find

$$\begin{aligned}
V_s(\mathbf{h}) &= \sum_n \iiint_{crystal} V(\mathbf{r}')U(x)\mathrm{e}^{2\pi i\mathbf{h}\cdot[\mathbf{r}'+n(\mathbf{a}+\mathbf{R})]}\,\mathrm{d}\mathbf{r}',\\
&= \sum_n \mathrm{e}^{2\pi i n\mathbf{h}\cdot(\mathbf{a}+\mathbf{R})} \iiint_{block} V(\mathbf{r}')\mathrm{e}^{2\pi i\mathbf{h}\cdot\mathbf{r}'}\,\mathrm{d}\mathbf{r}',\\
&= \mathcal{I}(\mathbf{g}) \sum_{\mathbf{h}'} V_{\mathbf{h}'} \iiint_{block} \mathrm{e}^{2\pi i(\mathbf{h}-\mathbf{h}')\cdot\mathbf{r}'}\,\mathrm{d}\mathbf{r}',
\end{aligned}$$

where we have introduced the Fourier expansion of the perfect crystal potential. Since we expect the satellite reflections to occur on rows parallel to $\mathbf{e}_u$, we define a new vector in reciprocal space:

$$\mathbf{h} = \mathbf{h}' + u\mathbf{e}_u = \mathbf{h}' + \mathbf{s}.$$

The magnitude of $\mathbf{s}$ remains to be determined. We can then rewrite the equation as

$$\begin{aligned} V_s(\mathbf{h}) &= \mathcal{I}(\mathbf{h}) \sum_{\mathbf{h}'} V_{\mathbf{h}'} \iiint_{block} \mathrm{e}^{2\pi \mathrm{i}\mathbf{s}\cdot\mathbf{r}'} \,\mathrm{d}\mathbf{r}', \\ &= \mathcal{I}(\mathbf{h}) \sum_{\mathbf{h}'} V_{\mathbf{h}'} \frac{\sin \pi u d}{\pi u d}. \end{aligned}$$

The prefactor $\mathcal{I}(\mathbf{h})$ can be computed as follows:

$$\begin{aligned} \mathcal{I}(\mathbf{h}) &= \sum_n \mathrm{e}^{2\pi \mathrm{i} n\mathbf{h}\cdot(\mathbf{a}+\mathbf{R})}, \\ &= \sum_n \mathrm{e}^{2\pi \mathrm{i} n(\mathbf{h}'+\mathbf{s})\cdot(\mathbf{a}+\mathbf{R})}, \\ &= \sum_n \mathrm{e}^{2\pi \mathrm{i} n(\mathbf{R}\cdot\mathbf{h}'+ud)}, \end{aligned}$$

where we have used the fact that $\mathbf{a}_0 \cdot \mathbf{h}'$ is always an integer number and $(\mathbf{a}+\mathbf{R}) \cdot \mathbf{e}_u = d$. In the limit $n \to \infty$ this can be rewritten as

$$\mathcal{I}(\mathbf{h}) = \sum_k \delta\left[u - \frac{1}{d}(k - \mathbf{R}\cdot\mathbf{h}')\right].$$

In other words, the satellite reflections will occur at the positions

$$\mathbf{h}_k = \mathbf{h}' + \frac{1}{d}(k - \mathbf{R}\cdot\mathbf{h}')\mathbf{e}_u, \tag{9.21}$$

with k taking on all integer values. The Fourier coefficients of the shear potential at these locations are given by

$$V_{\mathbf{h}_k} = V_{\mathbf{h}'} \frac{\sin \pi u d}{\pi u d} \quad \text{with} \quad u = \frac{1}{d}(k - \mathbf{R}\cdot\mathbf{h}'), \tag{9.22}$$

as is easily verified.

Let us consider a 2D example. The upper drawing in Fig. 9.32 shows a square crystal with lattice parameter a. A periodic crystallographic shear plane is introduced with fault plane (10) (in two dimensions) and displacement vector $\mathbf{R} = \frac{1}{4}\mathbf{b}$. The block thickness is equal to $d = 8a$. The modulus squared of the Fourier transform

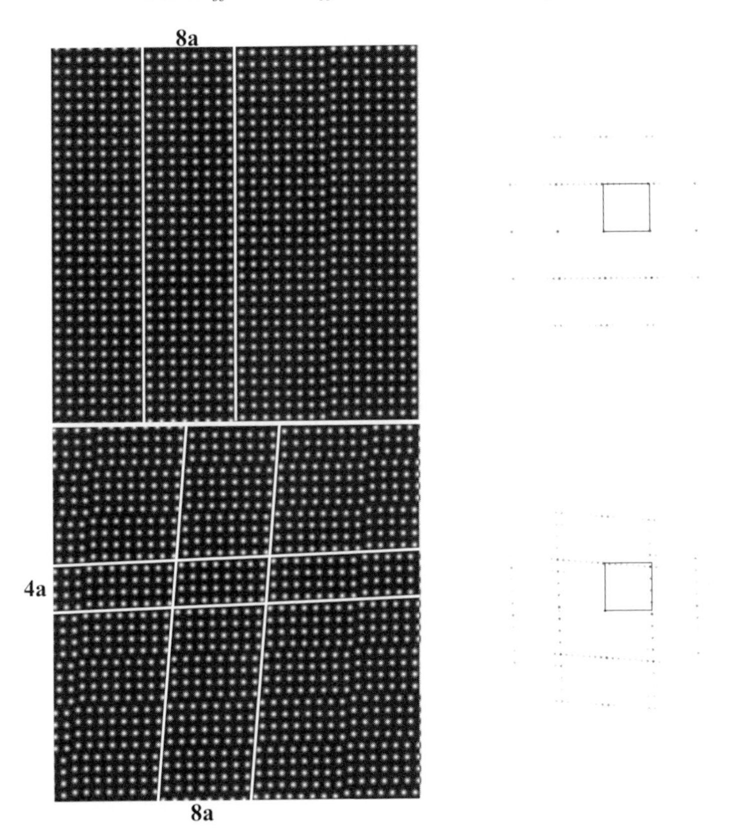

Fig. 9.32. One-dimensional (a) and two-dimensional (b) crystallographic shear plane structures for the parameters described in the text, along with their kinematical diffraction patterns.

of this pattern (i.e. the kinematic diffraction pattern) is shown on the right. The unit shear plane normal is $\mathbf{e}_u = \mathbf{a}/a = a\mathbf{a}^* = a(10)$. From the equations above, we expect satellite reflections at the positions

$$\mathbf{h}_k = \mathbf{h}' + \frac{1}{8a}\left(k - \mathbf{h}' \cdot \frac{\mathbf{b}}{4}\right)\mathbf{e}_u.$$

Consider as an example the reflection $\mathbf{h}' = (0\,1)$; for this reflection we calculate that

$$\mathbf{g}_k = (0\,1) + \frac{4k-1}{32}(1\,0).$$

In Fig. 9.32(a), we indeed find the satellite reflections at distances $-\frac{1}{32}$, $\frac{3}{32}$, and $\frac{7}{32}$ from the initial position of the (01) reflection for the unfaulted structure. For other reflections we find different patterns of satellites; in particular, for the central row of fundamental reflections $\mathbf{h}' = (n\,0)$ there are no satellites since the dot product

$\mathbf{h}' \cdot \mathbf{b}/4 = 0$. In reality, there will be satellites on this row for two reasons: first, the shear lattice potential is inadequately described near the shear planes. A more accurate description does introduce satellites along the central row [VDBMA87]. Secondly, there will be *double-diffraction* effects, as discussed in Section 9.2.3.

Crystallographic shear planes can also occur in more than one direction, as shown in the example of Fig. 9.32(b). The sheared structure has a first set of shear planes with the same parameters as in Fig. 9.32(a). A second set with parameters $\mathbf{R} = \mathbf{a}/3$, $d = 4a$, and $\mathbf{e}_u = b\mathbf{b}^*$ is superimposed on the faulted structure. As a consequence of the second series of faults, both fault planes are rotated away from the original orientation, and the rows of satellite reflections remain normal to the rotated fault planes, as is obvious from the computed diffraction pattern. For the reflection (11), the satellite reflections occur at locations of the type

$$\mathbf{g}_{k,l} = (11) + \frac{3k-1}{12}(01) + \frac{4l-1}{32}(10),$$

which generates a two-dimensional pattern of satellites around the original reflection (11).

The kinematical diffraction patterns of Figs 9.28, 9.30, and 9.32 can be computed using the interactive **ION** routine modulation.pro, available on the website. The routine creates a structure on a 1024×1024 grid, multiplies it with a Hanning window to eliminate edge effects in the subsequent Fourier transform, and then displays the central 512×512 pixels of the kinematical diffraction pattern using a logarithmic intensity scale. A Fortran-90 version of the same program is also available as modulation.f90.

9.6.2.4 Interface modulations: long period anti-phase boundary structures

A material with a completely ordered lattice, say A_3B, can undergo an additional *ordering* transformation by the introduction of a periodic array of defects in the ordering. The simplest ordering defect is the anti-phase boundary (APB), introduced in the previous chapter. Let us consider an example: the $L1_2$-structure (Fig. 8.10b). Since the B atom can be located on any of the four sublattices of the fcc structure, there will be four translation variants in the crystal (assuming heterogeneous nucleation of the ordered state). These domains are separated by APBs. Across an APB, the underlying fcc lattice is continuous, but the ordered superlattice is not. Often, the APB energy is strongly anisotropic, meaning that the APB will prefer a certain plane above all others. The simplest APB in this type of structure is then the *conservative* APB on the (001) plane, with displacement vector $\mathbf{R} = \frac{1}{2}[110]$.

This type of defect has been observed in many alloys, e.g. Cu_3Pd [BVTVL+88], Cu_3Pt [MMC+96], CuAu [ST61], Cu_3Al [BDGVT+87], etc. The spacing between the APBs determines the size of the tetragonal unit cell; the smallest spacing would

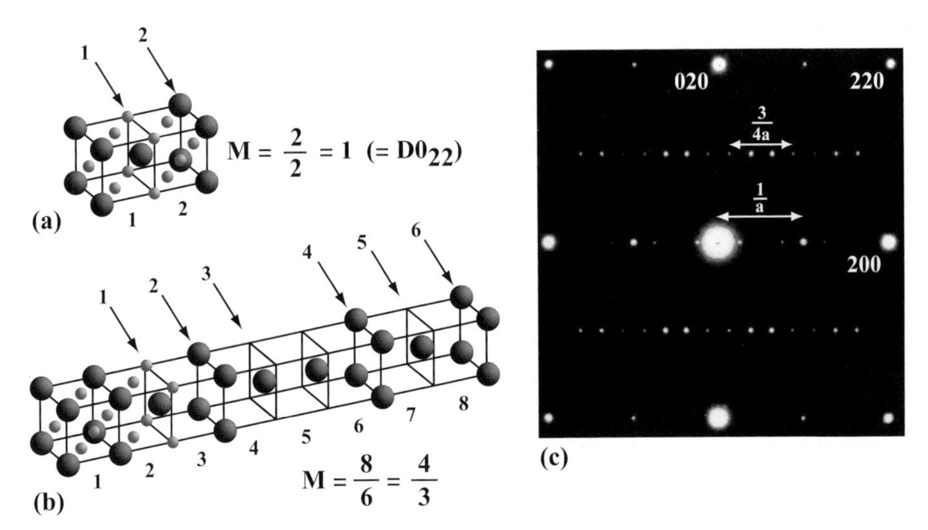

Fig. 9.33. (a) $M = 1$ or DO_{22} structure, (b) $M = \frac{4}{3}$ structure, (c) experimental [010] zone axis pattern for an $M = \frac{4}{3}$ modulated structure in a Cu-23 at% Al alloy.

correspond to the distance a. Fig. 9.33(a) illustrates the resulting structure, which is known as DO_{22}. This type of structure is generally known as a *long-period anti-phase boundary* lattice or LPAPB and can be characterized by the ratio of two integers: $M = p/q$, where p is the number of unit cells in a repeat distance and q the number of APBs within that same repeat distance. The $L1_2$ structure has $M = \infty$ and for the DO_{22} structure $M = \frac{1}{1} = 1$. Fractional values for M are also possible.

These periodic defect structures give rise to superlattice reflections along the [001]-direction; it can be shown [VDBMA87] that the satellite reflections will arise at positions of the type $(2h + 1, 0, l \pm m/2M)$, where m is an integer (this involves computation of the structure factor for an arbitrary value of M). An experimental diffraction pattern ([010] zone axis pattern) for $M = \frac{4}{3}$ is shown in Fig. 9.33(c), along with the corresponding structure in Fig. 9.33(b). The value of $1/M$ can be determined from the spacing between the superlattice reflections, as indicated.

9.6.3 Incommensurate modulations and quasicrystals

All modulation types introduced so far have in common that all reflections (fundamental and modulation satellites) can be described by the following expression [Jan86]:

$$\mathbf{g} = \sum_{i=1}^{3} g_i \mathbf{a}_i^* + \sum_{j=1}^{d} g_{3+j} \mathbf{a}_{3+j}^*, \tag{9.23}$$

where g_i and g_{3+j} are integers, and the vectors $\mathbf{a}^*_{3+j}$ describe the modulation periods and directions; i.e. they are the modulation wave vectors. For a 1D modulation, we have $d = 1$, while for a 2D modulation $d = 2$, and so on. Since the crystal is a 3D object, the additional basis vectors must be linear combinations of the original three:

$$\mathbf{a}^*_{3+j} = \sum_{i=1}^{3} \kappa_i^j \mathbf{a}^*_i. \tag{9.24}$$

For the modulation types discussed so far, the coefficients κ_i^j are rational numbers, so that the modulation period is commensurate with the underlying lattice.†

If one or more of the coefficients κ_i^j is irrational, then the modulation period becomes incommensurate with the underlying lattice. A general reflection position for a 1D modulated structure is then given by

$$\mathbf{g} = \sum_{i=1}^{3} (g_i + p\kappa_i)\mathbf{a}^*_i, \tag{9.25}$$

with p an integer labeling the satellite reflections. Strictly speaking, such an incommensurate structure no longer satisfies the Bravais lattice translation vectors, and therefore cannot be described by any of the 230 3D space groups. It is possible to consider such a structure in a $(3 + d)$-dimensional space, and there is a significant amount of literature on the concept of *superspace groups* [De 74, JJ77, JJ79]. Experimental examples of incommensurate phases studied by electron diffraction include $NbTe_4$ and $TaTe_4$ [vBM86, Bvd+87], $K_{0.3-x}MoO_3$ [Col89], and $TiSe_2$ [Fun85].

For many incommensurately modulated structures, it is straightforward to select the first three basis vectors, corresponding to the unmodulated structure. This structure can always be described by one of the 14 Bravais lattices, and, therefore, one of the 3D space groups. If the vectors $\mathbf{a}^*_i$, with $i = 1, \ldots, 3$, do not describe one of the Bravais lattices, then the resulting structures are known as *quasi-crystals*. The first experimental evidence for the existence of quasi-crystals was found in 1984 in an Al–Mn alloy by Shechtmann, Blech, Gratias, and Cahn [SBGC84]. This discovery was made by means of electron diffraction observations, and a careful analysis of the angles between two-, three-, and five-fold zone axes. Soon after this discovery, many more materials were found to exhibit either the icosahedral orientational symmetry, or octagonal, decagonal, and more exotic symmetries. The field of "quasi-crystallography" was born.

† Note that in the hexagonal system we can choose $\kappa_i^1 = (-1, -1, 0)$ to describe the fourth reciprocal basis vector (which we called $\mathbf{A}^*_3$ in equation 1.34). The hexagonal lattice can, hence, be regarded as a 1D commensurately modulated lattice.

For an excellent introduction to quasi-crystals, we refer the interested reader to the book by Janot [Jan92]. A detailed description of the indexing of diffraction patterns from quasi-crystals can be found in [CSG86]. The icosahedral quasi-crystals can be described in terms of a 6D space, in which a primitive, body-centered, or face-centered cubic lattice is decorated with atoms. When this 6D structure is projected onto 3D space along an irrational projection axis, a 3D structure is created which has no long-range translational order, but only long-range orientational order. The reader should consult the journals *Philosphical Magazine* and *Philosophical Magazine Letters* for papers on many different quasi-crystalline systems.

9.7 Diffuse intensity due to short range ordering

We know from structure factor computations that an ordered structure gives rise to *superlattice reflections*, i.e. reflections on positions that are not allowed in the disordered structure. Often, the structure factor associated with these reflections is proportional to the difference of two (for a binary alloy) atomic scattering factors. In this section, we will consider what happens when the ordering state of the crystal is not perfect, i.e. if the material is only *partially* ordered.

Let us consider a binary alloy with a primitive disordered lattice. The atoms of types A and B have atomic fractions c_A and c_B, respectively. We define a *site occupation operator*, σ_j^A, such that

$$\sigma_j^A = \begin{cases} 1 & \text{if position } j \text{ occupied by an A atom;} \\ 0 & \text{if position } j \text{ occupied by a B atom;} \end{cases} \tag{9.26}$$

σ_j^B can be defined in the same way, such that

$$\sigma_j^A + \sigma_j^B = c_A + c_B = 1 \tag{9.27}$$

for each lattice site j. Using these operators, the atomic scattering factor of a site j may be rewritten as

$$f_j = f_A\sigma_j^A + f_B\sigma_j^B. \tag{9.28}$$

As we have seen in Chapter 2, in the kinematical diffraction theory the amplitude of a diffracted beam is proportional to the structure factor for that beam, i.e.

$$A(\mathbf{q}) = \sum_j f_j\, \mathrm{e}^{-2\pi i \mathbf{q}\cdot\mathbf{r}_j} = \sum_j \left(f_A\sigma_j^A + f_B\sigma_j^B\right) \mathrm{e}^{-2\pi i \mathbf{q}\cdot\mathbf{r}_j}, \tag{9.29}$$

with $\mathbf{q}$ a vector in reciprocal space. It is customary to introduce a new operator, the scalar *Flynn operator*:

$$\bar{\sigma}_j = c_A - \sigma_j^A = \sigma_j^B - c_B \qquad (c_A > c_B). \tag{9.30}$$

The Flynn operator is simply a different way of writing the site occupation parameters. Using this operator, the scattered amplitude may be split into two contributions:

$$A(\mathbf{q}) = A_B(\mathbf{q}) + A_D(\mathbf{q}), \tag{9.31}$$

with

$$A_B(\mathbf{q}) = (c_A f_A + c_B f_B) \sum_j \mathrm{e}^{-2\pi \mathrm{i}\mathbf{q}\cdot\mathbf{r}_j}; \tag{9.32}$$

$$A_D(\mathbf{q}) = (f_B - f_A) \sum_j \bar{\sigma}_j \, \mathrm{e}^{-2\pi \mathrm{i}\mathbf{q}\cdot\mathbf{r}_j}. \tag{9.33}$$

The first amplitude is only different from zero at the Bragg positions $\mathbf{q} = \mathbf{g}$; it represents the *average disordered structure*. The second term is zero at the Bragg positions (because the average value of the Flynn operator over the whole crystal is zero) but different from zero in between the Bragg reflections. We thus find that

$$A_B(\mathbf{q}) A_D(\mathbf{q}) = 0 \qquad \text{everywhere in reciprocal space.} \tag{9.34}$$

We conclude from the above equations that any deviation from perfect disorder will give rise to diffuse intensity *in between the Bragg reflections*. This diffuse intensity is written as

$$I_D(\mathbf{q}) = |f_B - f_A|^2 \sum_{k,l} \bar{\sigma}_k \bar{\sigma}_l \, \mathrm{e}^{-2\pi \mathrm{i}\mathbf{q}\cdot(\mathbf{r}_k - \mathbf{r}_l)}. \tag{9.35}$$

For a primitive unit cell, the difference between the vectors $\mathbf{r}_k$ and $\mathbf{r}_l$ is again a lattice vector, say, $\mathbf{r}_j$, and the intensity for a crystal with N unit cells is then written as

$$I_D(\mathbf{q}) = N c_A c_B |f_B - f_A|^2 \sum_j \alpha_{0,j} \, \mathrm{e}^{-2\pi \mathrm{i}\mathbf{q}\cdot\mathbf{r}_j}. \tag{9.36}$$

The coefficients $\alpha_{0,j}$ are the *Warren–Cowley short-range order parameters* [Cow67], defined by

$$\alpha_{0,j} = \frac{\langle \bar{\sigma}_l \bar{\sigma}_{l+j} \rangle}{c_A c_B}. \tag{9.37}$$

These parameters represent the probability that, if position l is occupied by an A atom, position $l + j$ is also occupied by an A atom. The $\langle \cdots \rangle$ sign indicates a spatial average over the whole crystal.

Table 9.2. *Atom coordinates* ($\times 2$) *and corresponding SRO parameters* $\alpha_{0,j}$ *derived from diffuse neutron scattering measurements (see the second footnote on this page).*

Position	$\alpha_{0,j}$	Position	$\alpha_{0,j}$	Position	$\alpha_{0,j}$
(0, 1, 1)	−0.172	(0, 2, 4)	0.003	(0, 0, 6)	0.003
(0, 0, 2)	0.101	(2, 3, 3)	0.007	(1, 1, 6)	0.002
(1, 1, 2)	0.036	(2, 2, 4)	−0.005	(2, 3, 5)	0.003
(0, 2, 2)	0.010	(1, 3, 4)	0.005	(0, 2, 6)	0.000
(0, 1, 3)	−0.027	(0, 1, 5)	−0.007	(1, 4, 5)	0.002
(2, 2, 2)	−0.029	(1, 2, 5)	−0.004	(2, 2, 6)	−0.001
(1, 2, 3)	0.003	(0, 4, 4)	0.002	(1, 3, 6)	−0.003
(0, 0, 4)	0.021	(0, 3, 5)	−0.007	(4, 4, 4)	−0.007
(0, 3, 3)	−0.007	(3, 3, 4)	−0.002	(0, 1, 7)	0.004
(1, 1, 4)	0.012	(2, 4, 4)	−0.004	(3, 4, 5)	0.001

For perfect disorder, all parameters are zero,† except $\alpha_{0,0} = 1$. The diffuse scattered intensity away from the reciprocal lattice points $\mathbf{g}$ is then given by

$$I_D(\mathbf{q}) = N c_A c_B |f_B - f_A|^2 \quad \text{for } \mathbf{q} \neq \mathbf{g}. \tag{9.38}$$

This is a monotonically varying background intensity, known as *monotonic Laue scattering*. The variation is caused by the $|\mathbf{q}|$ dependence of the atomic scattering factors, as illustrated in Fig. 2.9 on page 113.

As an example, Table 9.2 lists the Warren–Cowley SRO parameters for Cu-15 at% Al, measured by neutron diffraction [DG89].‡ The parameters were determined on cylindrical single crystals after a heat treatment of 4 days at 505 K. Note that the value of $\alpha_{0,1}$ nearly reaches the limit value of −0.18, indicating that there are no first nearest-neighbor Al–Al pairs.

Experimental electron diffraction patterns, showing the SRO diffuse intensity in Cu-15 at% Al, can be obtained by annealing the material for a few days at 500 K, slowly cooling to room temperature (furnace cool), and preparing the thin foil as described in Chapter 4. The patterns for the [001], [110], and [112] zone axis orientations, recorded at 200 kV, are shown in Fig. 9.34(a). The patterns were obtained with long exposure times (60 s); none of the diffuse intensity was visible

† Actually, a detailed calculation shows that for perfect disorder $\alpha_{0,l} = -1/(N-1)$ for $l \neq 0$. This number approaches zero for a sufficiently large number of unit cells.

‡ Diffuse neutron scattering experiments on [001]- and [110]-oriented single crystals of Cu-15 at% Al were carried out at the "Centre d'Etudes Nucléaires, Laboratoire Léon Brillouin" at Saclay, near Paris, in the Spring of 1988. The neutrons were generated by the cold-source Orphée reactor, and a wavelength of 0.259 nm was used. More details can be found in [DG89] and [CF83].

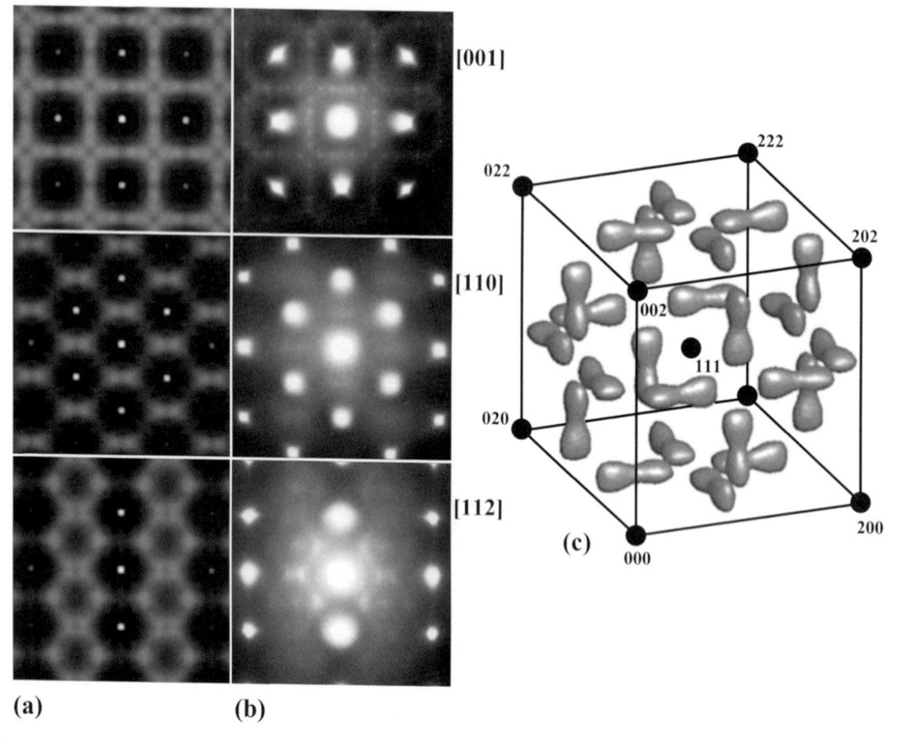

Fig. 9.34. (a) Experimental (200 kV) and (b) simulated diffuse intensity zone axis patterns for Cu-15 at% Al, annealed for 4 days at 505K. Part (c) is a 3D rendering of the iso-intensity surface of the diffuse component I_D. The simulation uses the 30 shells listed in Table 9.2.

to the naked eye on the viewing screen.† Simulated diffuse intensity patterns, using the SRO parameters of Table 9.2, are shown in Fig. 9.34(b) for the same zone axis orientations. The computations were carried out with the **ION** routine sro.pro, available from the website. The qualitative agreement between experimental and simulated patterns is rather good. It is possible to use experimental diffuse intensity patterns to derive the SRO parameters via a least-squares fitting routine.‡ The diffuse intensity in the Cu-15 at% Al system is an indicator of the various ordered alloy phases that are found for higher Al contents [BDGVT+87, BVTVL+89]. The trend of no first nearest-neighbor Al–Al pairs is also found at higher Al concentrations, and gives rise to a rich set of long-period anti-phase boundary (LPAPB) structures.

In many alloy systems, the diffuse intensity is located near or on a geometric surface in reciprocal space. This surface, with equation $\chi(\mathbf{q}) = 0$, is periodic with

† It is possible to obtain a stronger diffuse intensity by annealing the alloy at a higher temperature, around 800 K, and quenching to room temperature. This, however, does not produce an equilibrium SRO state.

‡ The reader should be warned that any combination of SRO parameters can be entered into the sro.pro program, but that not every combination is physically realizable. Conversely, there are many different atomic configurations which give rise to the same set of SRO parameters.

the same periodicity as the underlying reciprocal lattice. The diffuse amplitude $\psi_D(\mathbf{q})$ (equation 9.33) is only different from zero on the surface, so that we have

$$A_D(\mathbf{q})\chi(\mathbf{q}) = 0. \tag{9.39}$$

If we write χ as a Fourier sum,

$$\chi(\mathbf{q}) = \sum_k \omega_k e^{2\pi i \mathbf{q}\cdot\mathbf{r}_k}, \tag{9.40}$$

then we can rewrite equation (9.39) (using equation 9.33) as

$$\sum_j \sum_k \omega_k \bar{\sigma}_j \, e^{-2\pi i \mathbf{q}\cdot(\mathbf{r}_j - \mathbf{r}_k)} = \sum_j \left(\sum_k \omega_k \bar{\sigma}_{j+k} \right) e^{-2\pi i \mathbf{q}\cdot\mathbf{r}_j} = 0. \tag{9.41}$$

This is only possible if, for every j, we have

$$\sum_k \omega_k \bar{\sigma}_{j+k} = 0, \tag{9.42}$$

which is a linear relation between the Flynn operators. If we can determine the coefficients ω_k from the experimental diffuse intensity patterns (and typically there are only a few coefficients needed to represent the surface), then the linear relation leads to the identification of a cluster of lattice sites with identical composition. For examples of the application of this cluster model we refer the interested reader to [DRVTVDA76, DRVDVTA77, VDDRVTA77].

There is a very large amount of literature dealing with short-range order, order–disorder transformations, and long-range ordered structures. Theoretical models range from the pair-correlation Gorsky–Bragg–Williams model [CDGD+89], to complex multi-cluster models in the cluster variation theory [kik51, Fin87], and the statistical "axial next-nearest-neighbor interaction" model, known as the ANNNI model [FS81, Sel88]. It is possible, in principle, to use diffuse scattering information to determine the long-range ordering potential. This is often based on the Krivoglaz–Clapp–Moss theory, which relates the diffuse scattering surface to the Fourier transform of the ordering potential [CM66, CDGD+89]. The reader is referred to the proceedings of the various conferences on solid–solid phase transformations for many examples of short-range order and order–disorder transformations [ALSW84, Tsa84, Lor87, JHLS94, KOM99].

9.8 Diffraction effects from polyhedral particles

In the final section of this chapter, we will consider diffraction features caused by polyhedral inclusions in a matrix. We have seen in Chapter 2 (p. 119) that, for a finite crystal, each reciprocal lattice point assumes a shape described by the Fourier transform $D(\mathbf{q})$ of the shape function $D(\mathbf{r})$. We shall call this reciprocal shape function the *shape amplitude*:

$$D(\mathbf{q}) = \iiint D(\mathbf{r})\mathrm{e}^{-2\pi\mathrm{i}\mathbf{g}\cdot\mathbf{r}}\,\mathrm{d}\mathbf{r} = \iiint_{\mathcal{V}} \mathrm{e}^{-2\pi\mathrm{i}\mathbf{g}\cdot\mathbf{r}}\,\mathrm{d}\mathbf{r},$$

with $\mathcal{V}$ the (finite) volume of the crystal. Consider a particle with a regular polyhedral shape function, consisting of F faces, V vertices, and E edges. One can show that in 3D space the following relation is valid:

$$V + F - 2 = E; \tag{9.43}$$

this is known as *Euler's theorem*. We will now explicitly work out the Fourier integral above for an arbitrary polyhedral shape.

Following von Laue [vLR36] and Komrska [Kom87] (but with a slightly different notation) we will use the so-called *Abbe transform* to convert the volume integral into a sum of surface integrals over the faces of the polyhedron. We start from the Helmholtz equation (2.36) in Chapter 2:

$$\Delta\Psi + k^2\Psi = 0,$$

or

$$\Psi = -\frac{1}{k^2}\Delta\Psi.$$

Consider then the integral of Ψ over a finite region $\mathcal{V}$ in an N-dimensional space:

$$\int_{\mathcal{V}}^{(N)} \Psi\,\mathrm{d}^N\mathbf{r} = -\frac{1}{k^2}\int_{\mathcal{V}}^{(N)} \Delta\Psi\,\mathrm{d}^N\mathbf{r}.$$

The second integral can be rewritten using the theorem of Gauss (e.g. [Wre72, pp. 297–298]), which reduces the dimension of the integration space by 1 to $N - 1$:

$$\int_{\mathcal{V}}^{(N)} \nabla\cdot\mathbf{A}\,\mathrm{d}^N\mathbf{r} = \int_{S}^{(N-1)} \mathbf{A}\cdot\mathbf{n}\,\mathrm{d}S,$$

where S is the surface bounding the region $\mathcal{V}$, and $\mathbf{n}$ the unit outward normal on the surface S. If we select $\mathbf{A} = \nabla\Psi$, then we arrive at the *Abbe transform*:

$$\int_{\mathcal{V}}^{(N)} \Psi\,\mathrm{d}^N\mathbf{r} = -\frac{1}{k^2}\int_{S}^{(N-1)} \nabla\Psi\cdot\mathbf{n}\,\mathrm{d}S. \tag{9.44}$$

This relation is valid for every "well-behaved" function Ψ that satisfies the Helmholtz equation, and in particular it is valid for the phase factor in the Fourier transform integral. It is easy to show that the function

$$\Psi = \mathrm{e}^{-2\pi\mathrm{i}\mathbf{q}\cdot\mathbf{r}}$$

satisfies the Helmholtz equation with $k^2 = (2\pi q)^2$, with q the length of the vector $\mathbf{q}$. The Abbe transform of this function is then readily obtained as

$$\int_{\mathcal{V}}^{(N)} \mathrm{e}^{-2\pi \mathrm{i}\mathbf{q}\cdot\mathbf{r}}\, \mathrm{d}^N\mathbf{r} = \frac{\mathrm{i}}{2\pi q^2} \int_{S}^{(N-1)} \mathrm{e}^{-2\pi \mathrm{i}\mathbf{q}\cdot\mathbf{r}}\mathbf{q}\cdot\mathbf{n}\, \mathrm{d}S. \tag{9.45}$$

If the bounding surface S consists of planar sections, as is the case for a polyhedral particle, then the surface integral may be replaced by a sum of surface integrals, one over each face of the polyhedron. In such a case, the normal vector $\mathbf{n}$ is constant across a face and the dot product $\mathbf{q}\cdot\mathbf{n}$ may be taken out of the integral. The Abbe transform may then be carried out once again, this time to reduce the dimension from $N = 2$ (a plane) to a $N = 1$ (a line). It is the repeated use of this transform which permits explicit algebraic computation of the shape amplitude for a polyhedral inclusion.

An example computation for a 2D polygon will illustrate the general procedure. Consider the three-sided polygon shown in Fig. 9.35. There are $E = 3$ edges, and $V = 3$ vertices, and Euler's relation in this case reduces to $E = V$. We number the vertices from $e = 1, \ldots, E$, and the vertex position vectors are denoted by $\boldsymbol{\xi}_e$ with respect to an arbitrary origin O. It is convenient, but not necessary, to select O at the center of the polygon. The vectors $\mathbf{n}_e$ denote the unit outward normals to the edges, and the edge length is denoted by L_e. We also define the unit vector $\mathbf{t}_e$ *along the edge* e, pointing in counterclockwise direction. An arbitrary position along the edge e can then be written as

$$\mathbf{r} = \boldsymbol{\xi}_e + \ell\mathbf{t}_e,$$

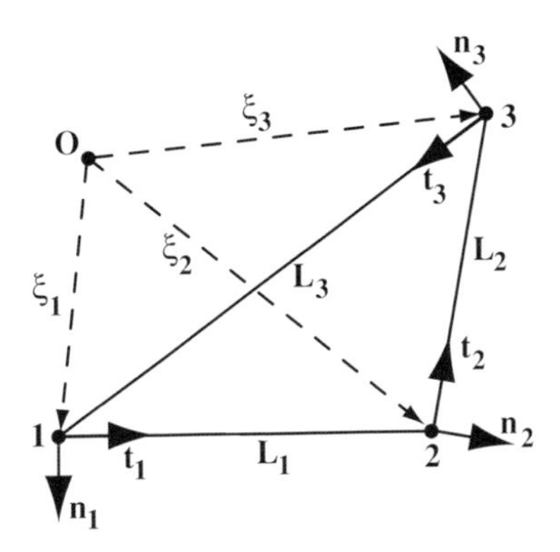

Fig. 9.35. Illustration of the notation for edges and unit outward normals for a two-dimensional three-sided polygon. The location of the origin is arbitrary.

where $\ell = 0, \ldots, L_e$ parameterizes the edge. The contribution of the edge e to the shape amplitude is now given by the integral

$$D_e(\mathbf{q}) = \frac{\mathrm{i}}{2\pi q^2}\mathbf{q}\cdot\mathbf{n}_e \int_e \mathrm{e}^{-2\pi\mathrm{i}\mathbf{q}\cdot\mathbf{r}}\,\mathrm{d}\ell.$$

Replacing $\mathbf{r}$ by the parameterization of the edge we find

$$D_e(\mathbf{q}) = \frac{\mathrm{i}}{2\pi q^2}\mathbf{q}\cdot\mathbf{n}_e \mathrm{e}^{-2\pi\mathrm{i}\mathbf{q}\cdot\boldsymbol{\xi}_e}\int_0^{L_e} \mathrm{e}^{-2\pi\mathrm{i}\mathbf{q}\cdot\mathbf{t}_e\ell}\,\mathrm{d}\ell,$$

and this integral can be solved readily. If we express the edge with respect to the center point $\boldsymbol{\xi}_e^C = \boldsymbol{\xi}_e + L_e\mathbf{t}_e/2$, then we find for the contribution of the edge e to the shape amplitude of the polygon:

$$D_e(\mathbf{q}) = \frac{\mathrm{i}}{2\pi q^2}L_e\mathbf{q}\cdot\mathbf{n}_e\frac{\sin(\pi\mathbf{q}\cdot\mathbf{t}_e L_e)}{\pi\mathbf{q}\cdot\mathbf{t}_e L_e}\mathrm{e}^{-2\pi\mathrm{i}\mathbf{q}\cdot\boldsymbol{\xi}_e^C}. \tag{9.46}$$

The complete shape amplitude is then given by

$$D(\mathbf{q}) = \sum_{e=1}^{E} D_e(\mathbf{q}).$$

For the case of the three-dimensional polyhedral particle the computations are carried out in very much the same way as for the 2D polygon. For a complete derivation of the equations we refer to the original paper by Komrska [Kom87]. The shape amplitude for a 3D polyhedron can be expressed in terms of a summation over edges and faces, edges and vertices, or faces and vertices. For numerical computations, the expressions for faces and edges are most easily implemented. While the equations appear to be quite complicated, they are really rather straightforward, and the reader is encouraged to work through the derivation in [Kom87]. The shape amplitude of a polyhedral particle with E edges and F faces is given by

$$D(\mathbf{q}) = -\frac{1}{(2\pi q)^2}\sum_{f=1}^{F}\frac{\mathbf{q}\cdot\mathbf{n}_f}{q^2-(\mathbf{q}\cdot\mathbf{n}_f)^2}\sum_{e=1}^{E_f}L_{fe}\mathbf{q}\cdot\mathbf{n}_{fe}\frac{\sin(\pi\mathbf{q}\cdot\mathbf{t}_{fe}L_{fe})}{\pi\mathbf{q}\cdot\mathbf{t}_{fe}L_{fe}}\mathrm{e}^{-2\pi\mathrm{i}\mathbf{q}\cdot\boldsymbol{\xi}_{fe}^C}. \tag{9.47}$$

This equation is only valid if the first denominator is non-zero. If $\mathbf{q} = \pm q\mathbf{n}_f$ (in other words, if $\mathbf{q}$ is parallel to any one of the face normals), then the contribution of that particular face (or faces) must be replaced by

$$D_f(\mathbf{q}) = \frac{\mathrm{i}}{2\pi}\frac{\mathbf{q}\cdot\mathbf{n}_f}{q^2}P_f\mathrm{e}^{-2\pi\mathrm{i}\mathbf{q}\cdot\mathbf{n}_f d_f}, \tag{9.48}$$

where P_f is the surface area of the face f, and d_f the distance between the origin and the face f. The symbols in equation (9.47) are defined as:

- $\mathbf{n}_f$, unit outward normal to face f;
- L_{fe}, length of the eth edge of the fth face;
- $\mathbf{t}_{fe}$, unit vector along the eth edge of the fth face, defined by

$$\mathbf{t}_{fe} = \frac{\mathbf{n}_f \times \mathbf{N}_{fe}}{|\mathbf{n}_f \times \mathbf{N}_{fe}|},$$

  where $\mathbf{N}_{fe}$ is the unit outward normal *on* the face which has the edge e in common with the face f;†
- $\mathbf{n}_{fe}$, unit outward normal *in* the face f on the edge e defined by $\mathbf{n}_{fe} = \mathbf{t}_{fe} \times \mathbf{n}_f$.

The input parameters needed to complete this computation for an arbitrary polyhedron are the V vertex coordinates $\boldsymbol{\xi}_v$ and a list of which vertices make up each face (counterclockwise when looking towards the polyhedron center). All other quantities can be computed from these parameters. In particular, the perpendicular distance d_f of a face to the origin is equal to the dot product of the position vector $\boldsymbol{\xi}$ of any one of the vertices of that face with the plane normal $\mathbf{n}_f$, i.e. $d_f = \boldsymbol{\xi} \cdot \mathbf{n}_f$. The unit face normals can be computed from the cross product of any two edge vectors for a given face.

The surface area P_f of a face can be computed as follows: consider the polygon shown in Fig. 9.36. There are E_f vertices, numbered counterclockwise, with (Cartesian) coordinates $\boldsymbol{\xi}_e$, $e = 1, \ldots, E_f$. The first step involves transforming the vertex coordinates to a two-dimensional reference frame, with origin located at the first vertex. The $\mathbf{e}_x$ vector of this reference frame is equal to the unit edge vector $\mathbf{t}_{f1}$ along the first edge of the polygon, as defined above. The $\mathbf{e}_y$ vector is equal to the opposite of the unit outward normal $\mathbf{n}_{f1}$. It is then straightforward to compute the 2D coordinates (x_e, y_e), $e = 1, \ldots, E_f$ of all vertices:

$$(x_e, y_e) = ((\boldsymbol{\xi}_e - \boldsymbol{\xi}_1) \cdot \mathbf{t}_{f1}, -(\boldsymbol{\xi}_e - \boldsymbol{\xi}_1) \cdot \mathbf{n}_{f1}). \tag{9.49}$$

The surface area of the polygon can then be computed by adding the areas of the individual triangles (shaded in Fig. 9.36) which make up the polygon. It is easy to verify that the area is then given by

$$P_f = \frac{1}{2} \sum_{e=2}^{E_f} (x_e y_{e+1} - y_e x_{e+1}). \tag{9.50}$$

† An alternative definition of the unit edge vectors could be given in terms of the vertex vectors $\boldsymbol{\xi}$:

$$\mathbf{t}_{fe} = \frac{\boldsymbol{\xi}_{e+1} - \boldsymbol{\xi}_e}{|\boldsymbol{\xi}_{e+1} - \boldsymbol{\xi}_e|}.$$

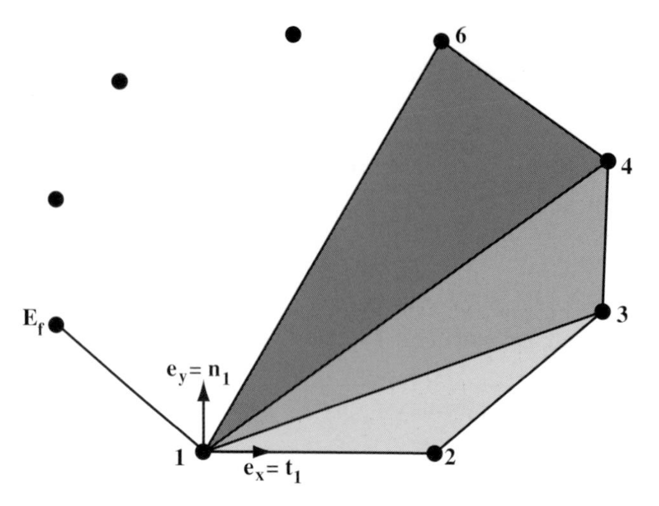

Fig. 9.36. Illustration of the computation of the surface area of an arbitrary 2D polygon. The origin of a Cartesian reference frame is placed at the first vertex, and the entire polygon is subdivided into triangular regions.

The $e = 1$ term does not contribute because the origin of the 2D reference frame is located at the first vertex. The first terms in this summation correspond to the areas of the individual triangles, and the second terms rotate the coordinate system to the base of each subsequent triangle. This relation is easily implemented in a computer program.

The Fortran routine shape.f90 implements the above formalism for the shape amplitude of an arbitrary polyhedral particle. The pseudo code PC-21 describes the computational steps involved. The equations can be simplified for the case of a centrosymmetric polyhedron, in which case the shape amplitude is real. Explicit equations for the centrosymmetric case are given in [Kom87].

As an example we will compute the shape amplitudes for a tetrahedron, an octahedron, and a rhombic dodecahedron, using the input data presented in Table 9.3. The computed images for $|D(\mathbf{q})|^2$ are shown in Fig. 9.37 for the viewing directions [100], [110], and [111]. The intensities in the images are displayed on a logarithmic scale, since the secondary maxima are about an order of magnitude weaker than the center maximum (which is arbitrarily scaled to intensity 1). The computations were carried out for a particle with unit dimensions; the size of the reciprocal space volume will change with changing particle size, but the shape of the volume will remain the same. A wire frame sketch of the three polyhedra is shown to the left of the shape amplitudes; numbers refer to the vertex coordinates in Table 9.3.

An interactive web-based implementation of the shape amplitude program is available on the website in the form of an **ION** routine called shape.pro. The user

Pseudo Code **PC-21** Compute the shape amplitude of a polyhedral particle.

```
Input: input data describing polyhedron
Output: shape amplitude in planar section
  read polyhedron data from file
  compute vertex centers and edge lengths
  compute edge and outward unit normals for each edge and face
  compute face distances and surface areas
  ask for viewing direction
  ask for offset along viewing direction
  for each vector q normal to viewing direction do
    for each face f do
      compute denominator q^2 − (q · n_f)^2                {9.47}
      if denominator equals zero then
        compute D_f(q)                                      {9.48}
      else
        for each vertex of face f do
          compute contribution to shape amplitude           {9.47}
        end for
      end if
    end for
    add all contributions together
  end for
  write real and imaginary part to file
```

can select one of a number of polyhedral shapes,[†] the viewing direction, and at which point along the viewing direction a cross-section through the shape amplitude is computed (the default cross-section contains the origin). The program then returns the calculated intensity, using a logarithmic intensity scale. The user will notice that with increasing number of polyhedral facets, the shape amplitude increasingly resembles that for a spherical particle (a set of concentric intensity shells).

In addition to the interactive shape amplitude computation, an animated GIF movie is made available for each of the polyhedra (click on the polyhedron's name), which shows a rendered image of the iso-intensity surface for intensity values between 0.01 and 0.25. Each movie is created for the same viewing direction, and

[†] The following polyhedra are available: tetrahedron, truncated tetrahedron, octahedron, truncated octahedron, cube, snub cube, truncated cube, cuboctahedron, rhombicuboctahedron, rhombitruncated cuboctahedron, dodecahedron, snub dodecahedron, truncated dodecahedron, rhombic dodecahedron, icosahedron, truncated icosahedron, icosidodecahedron, rhombicosidodecahedron, rhombitruncated icosidodecahedron, and a flat plate.

Table 9.3. *Vertex coordinates and face edges for the tetrahedron, octahedron, and rhombic dodecahedron. The vertex list is in a counterclockwise direction when viewed along the face normal to the center of the polyhedron. Coordinates refer to a standard Cartesian reference frame (see Fig. 9.37).*

N	Vertices	Face edges
	Tetrahedron	
1	$(1, \bar{1}, 1)$	1,3,2
2	$(\bar{1}, 1, 1)$	4,2,3
3	$(1, 1, \bar{1})$	3,1,4
4	$(\bar{1}, \bar{1}, \bar{1})$	4,1,2
	Octahedron	
1	$(1, 0, 0)$	1,2,5
2	$(0, 1, 0)$	2,3,5
3	$(\bar{1}, 0, 0)$	3,4,5
4	$(0, \bar{1}, 0)$	4,1,5
5	$(0, 0, 1)$	2,1,6
6	$(0, 0, \bar{1})$	3,2,6
7	–	4,3,6
8	–	1,4,6

N	Vertices	Face edges
	Rhombic dodecahedron	
1	$(1, 0, 0)$	1,4,3,7
2	$(0, 1, 0)$	2,5,3,4
3	$(0, 0, 1)$	6,3,5,8
4	$(1/2, 1/2, 1/2)$	9,7,3,6
5	$(-1/2, 1/2, 1/2)$	2,4,1,11
6	$(-1/2, -1/2, 1/2)$	8,5,2,14
7	$(1/2, -1/2, 1/2)$	9,6,8,13
8	$(\bar{1}, 0, 0)$	1,7,9,12
9	$(0, \bar{1}, 0)$	1,12,10,11
10	$(0, 0, \bar{1})$	14,2,11,10
11	$(1/2, 1/2, -1/2)$	13,8,14,10
12	$(1/2, -1/2, -1/2)$	12,9,13,10
13	$(-1/2, -1/2, -1/2)$	–
14	$(-1/2, 1/2, -1/2)$	–

the computational volume contains $128 \times 128 \times 128$ pixels. The computation of $D(\mathbf{q})$ over a 3D volume is carried out using the program shape3D.f90, also provided on the website. This program takes quite a bit more time to run than the 2D section program shape.f90; the only difference between the two programs is the range of the vector $\mathbf{q}$. An example calculation for the tetrahedron, octahedron, and rhombic dodecahedron is shown in Fig. 9.38; these images are snapshots from the movies available on the website.

The shape amplitude arguments are valid for all diffraction experiments, regardless of the nature of the radiation used. For instance, Gragg and Cohen [GC71] have analyzed the structure of Guinier–Preston zones in Al-5 at% Ag by means of x-rays, and deduced from the shape of the reciprocal lattice points that the zones have an octahedral shape.

The effect of the shape amplitude on a diffraction pattern is not necessarily restricted to the shape of the reciprocal lattice points. The shape amplitude can also become important in molecular crystals (i.e. crystals which are formed by the regular stacking of large molecules, in which the molecules do not lose their identity), and a particularly interesting example of such an effect is present in solid C_{60}. The *Buckminster-fullerene* molecule crystallizes in a face-centered cubic structure with

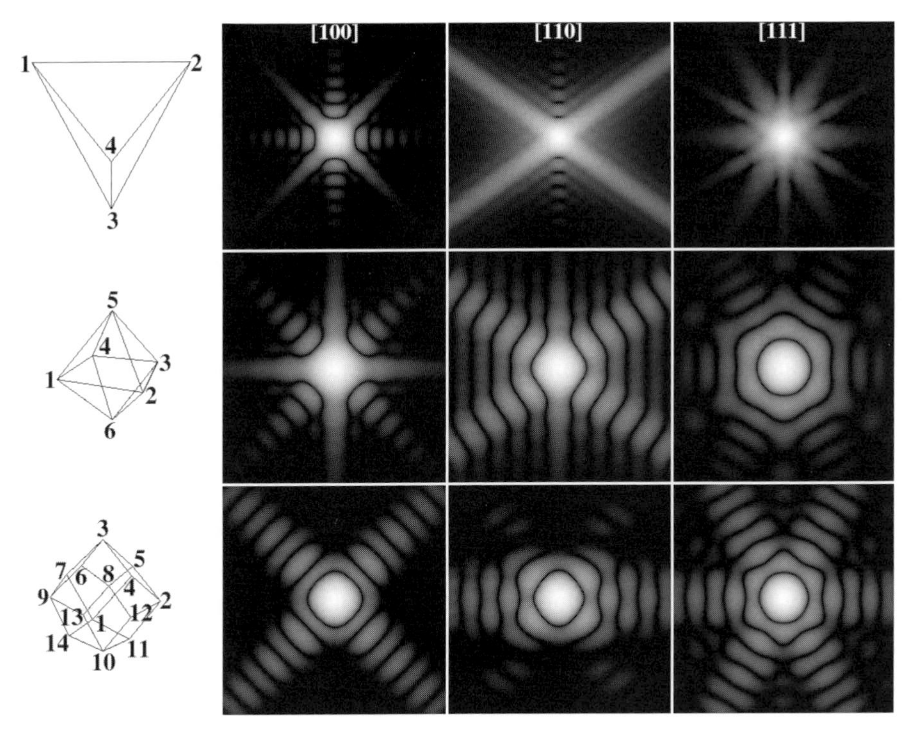

Fig. 9.37. Modulus squared of the shape amplitudes for the polyhedra shown in the left-hand column, for viewing directions [100], [110], and [111] (Cartesian reference frame). The intensities are shown on a logarithmic scale, to bring out the secondary maxima.

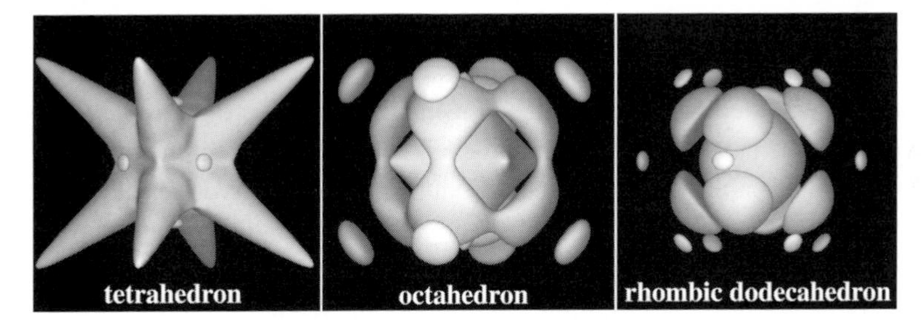

Fig. 9.38. Iso-intensity surfaces of the shape amplitudes for the polyhedra shown in the left-hand column of Fig. 9.37. The intensities are logarithmically scaled. These images are single frames taken from animations, which can be found on the website.

a lattice parameter of 1.4 nm. The radius of the "buckyball" molecule is 0.354 nm, or very nearly 1/4 of the fcc lattice parameter. If one approximates the molecule (which at room temperature spins rapidly) by a spherical shell with a finite thickness and a uniform or Gaussian charge density, then one can show [AVHVDVT92] that the shape amplitude (Fourier transform of the charge density) has zero crossings at distances corresponding to the $h00$ (h even) reciprocal lattice spacings of the fcc lattice. Therefore, the $h00$ (h even) reflections have a significantly reduced intensity from that calculated from the standard structure factor. This has been observed by means of x-ray diffraction, and also by electron diffraction (although care must be taken to properly interpret double-diffraction effects [AVHVDVT92], see Section 9.2.3).

One could incorporate the shape amplitude algorithm into a diffraction simulation program, or combine the point symmetry routines described in the first chapter of this book with the algorithm and use them to automatically generate the vertex coordinates and faces of arbitrary regular polyhedra. Neither of these algorithms are difficult to implement and the computer-oriented reader is encouraged to experiment with different algorithms.

Exercises

9.1 Consider the algorithm for the indexZAP.f90 program. Adapt the algorithm shown in pseudo code PC-20 to include the microscope camera length as a fitting parameter.

9.2 Modify the indexZAP.f90 program to include multiple crystal structures. The new algorithm should be able to distinguish between several crystal structures and decide which one is most likely, along with the identification of the correct zone axis.

9.3 Determine the possible double diffraction reflections for the diamond structure in the [011] zone axis orientation. Write down three pairs of reflections that will contribute to the forbidden 200 reflection.

9.4 Derive equations (9.14) and (9.15).

9.5 Show that the excitation error is related to the position of the bright Kikuchi line described in equation (9.16).

9.6 Obtain a zone axis orientation for a thick region in one of the study materials. From your precomputed diffraction patterns using the zap.f90 program you should be able to identify the pattern. Select one of the Kikuchi bands and tilt the sample so that this band remains on the viewing screen. Continue tilting until you reach a new zone axis orientation. Select a second Kikuchi band nearly perpendicular to the first and tilt to a third zone axis. For each orientation record the diffraction patterns and the tilt angles. Then, consistently index all three patterns and the Kikuchi bands.

9.7 Show experimentally (i.e. by recording CBED patterns) that the symmetry of the {220} diffraction disk in Cu-15 at% Al is equal to **1**. Show also that the (220) and $(\bar{2}\bar{2}0)$ diffraction disks are related to each other by the $\mathbf{2}_R$ symmetry operation.

9.8 Determine the point group of study material Ti. You should work at low acceleration voltage (80–100 kV), to have a more strongly curved Ewald sphere. Since you already know the point group, first determine which zone axis CBED pattern(s) you must acquire to make an unambiguous determination. Then record CBED patterns and analyze their symmetries.

9.9 Repeat the previous exercise for the study material $BaTiO_3$.

10

Phase contrast microscopy

10.1 Introduction

In the preceding chapters, we have studied how fast electrons interact with a crystalline thin foil. We have introduced the multi-beam dynamical diffraction equations in Chapter 5, solved the two-beam equations in Chapter 6, provided numerical methods to compute the multi-beam amplitudes for systematic row and near zone axis orientations in Chapter 7, and explained various defect contrast features and diffraction effects in Chapters 8 and 9. What we have not done yet is to describe how the microscope affects (i.e. changes) the electron wave function as it travels down the column. Actually, we did describe part of the influence of the microscope when we talked about two-beam microscopy; in a typical bright field–dark field observation, we introduce an aperture in the OL back focal plane, and this determines which component of the wave function reaches the observation screen. Moving the aperture around (or, more practically, tilting the incident electron beam) changes the image contrast, and this is indeed an example of how the microscope settings affect the image contrast. Another example is the excitation current of the objective lens: when we change the current, the image goes from under-focus to over-focus, and at the in-focus condition we observe an interpretable image. It is important to realize that *the same information is present in all images, regardless of the focus condition*; the human brain is just better at interpreting the contrast of the in-focus image, and we do not usually care too much about the out-of-focus condition. We will see in this chapter that there are observation conditions for which the in-focus image is not necessarily the easiest image to interpret, and, indeed, does no longer hold a special position in a through-focus series of images.

We will describe how the microscope affects the electron wave function as it travels through the objective lens. We will see how various lens aberrations influence both the amplitude and phase of the exit wave function. Then we will describe a method to determine the magnitude of these aberrations. In the remainder of this chapter we will describe the simulation of images acquired with two important observation modes: high-resolution mode and Lorentz mode. We conclude this final chapter with a few general remarks about currently active areas of research in the TEM field.

10.1.1 A simple experimental example

Figure 10.1(d) shows an example of a high-resolution electron micrograph for the [100] zone axis orientation of $BaTiO_3$. The image was recorded on a JEOL 4000EX,

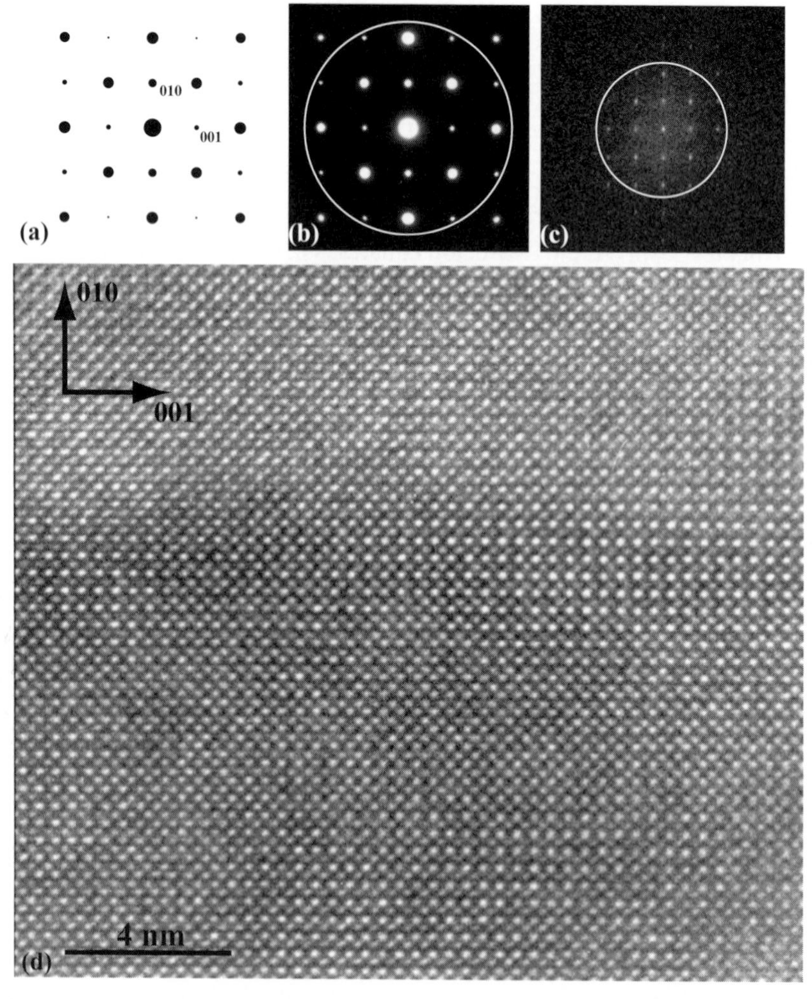

Fig. 10.1. (a) Schematic [100] zone axis pattern for $BaTiO_3$; (b) experimental pattern with an outline of the diffraction aperture; (c) computed diffractogram of the high-resolution image in (d). The aperture is represented as a white circle, and several diffraction spots can be seen outside the circle (e.g. 022 reflection), indicating non-linear image contributions. Images acquired at 400 kV in a JEOL 4000EX.

operated at 400 kV. The experimental diffraction pattern is shown in Fig. 10.1(b), along with a circle indicating the diffraction aperture (radius about 6 nm^{-1}). All diffracted beams inside the circle contribute to the image. The interference between individual beams gives rise to an interference pattern, Fig. 10.1(d). The microscope operator has some control over the details of this pattern through, primarily, the microscope defocus. The image is very sensitive to several external parameters, including: the precise zone axis orientation, the alignment of the incident beam along the coma-free axis, the foil thickness, the presence of astigmatism, sample drift, vibrations, and so on. The main challenge of experimental phase contrast observations is, then, to create an environment in which all residual aberrations (to be defined in Section 10.2.5) can be kept within reasonable bounds, so that the image can be interpreted, using the theory and modeling described in this chapter.

The computed diffractogram in Fig. 10.1(c) is essentially the modulus-squared of the Fourier transform of the image. This pattern shows which spatial frequencies are present in the image. It is clear that there are higher-order frequencies, larger than the aperture radius, in this image. This is due to non-linear image contributions, which we will discuss in Section 10.2.2. The diffractogram is an important tool in HREM observations, as it can be used to analyze the conditions under which the micrograph was recorded. Several aberrations and related artifacts can be visualized using diffractograms, as discussed in Sections 10.2.4 and 10.2.5.2.

One of the basic questions which arises in the interpretation of Fig. 10.1(d) is: where are the atoms with respect to the bright and dark contrast features in this image? In other words, do the brightest dots correspond to the locations of the Ba atoms? Or do they coincide with the Ti positions? Can we resolve the oxygen positions? What does "resolve" mean in the context of wave interference? What is the smallest distance that we can observe with a given microscope? Can we obtain an image in which all projected atom positions are visible? We will attempt to answer all of these questions, and several more, in Sections 10.2.1.4 and 10.2.2.

There are many excellent texts covering the basics of high-resolution electron microscopy: [Spe88], [BCE88], [Hor94], [WC96], and [FH01]. We refer the reader to these citations for more details on the experimental aspects of HREM observations. We will be concerned mostly with the underlying image formation theory, and the simulation of phase contrast images.

10.2 The microscope as an information channel

The experimental high-resolution (or multi-beam) image shown in the introduction cannot be interpreted simply by direct visual comparison with a crystal structure model. In many cases, the crystal or defect structure may not be known *a priori*, and we must resort to *image simulations* to confirm a particular model structure. We have already seen in Chapters 5–7 how we can compute the exit plane wave for any given crystal structure or atom arrangement. The multi-beam dynamical scattering formalism allows us to take a configuration of atoms and an incident beam with a given energy and orientation, and calculate how the amplitude and phase of this beam are modified by the sample. Ultimately, we are interested in finding out where all the atoms are, and what types of atoms are present at particular positions. In a sense, we are fortunate that there are only 92 possibilities for the atom species; in addition, for perfect crystal structures we can make use of the powerful machinery of symmetry and group theory. The total number of independent variables may, therefore, be very small. For instance, for pure copper we would need the lattice parameter a, the atomic number of Cu (29), the space group number, and the position of one atom (0, 0, 0), to unambiguously define the entire structure.

When it comes down to determining a structure (i.e. all atom positions and types in the asymmetric unit), we must make sure that there is sufficient information available to determine unambiguously each of the independent parameters. The following question must therefore be answered: *how much information can be transferred by the microscope?* Or, equivalently, how many independent degrees of freedom can be determined? To answer this question we will follow Van Dyck and de Jong [VDdJ92].

Since the microscope generates 2D images, we need to determine how many independent degrees of freedom are transferred per unit area of the image. We introduce the parameter ρ as the smallest image dimension that can contain meaningful information; we will see later on that ρ is the so-called *information limit* of the microscope. For now, it is sufficient to simply assume that this distance can be calculated in terms of the electron optical parameters of the microscope. The smallest meaningful area in the image is then equal to ρ^2. The distance ρ corresponds to a distance $q = 1/\rho$ in reciprocal space; q represents the largest spatial frequency that can meaningfully contribute to the image. All higher frequencies should be excluded from the image by means of the diffraction aperture. The area covered by this aperture is $\pi q^2 = \pi \frac{1}{\rho^2}$. Inside the aperture, we have several reciprocal lattice points. Each reciprocal lattice point is characterized by the real and imaginary parts of a complex number, which corresponds to two degrees of freedom for each point. Because of Friedel's law the diffraction pattern is centrosymmetric, which means that there is really only one degree of freedom per reciprocal lattice point. For a crystal structure with lattice parameter a the area of the reciprocal unit cell is equal to $1/a^2$. The total number of degrees of freedom (i.e. reciprocal lattice points) inside the diffraction aperture is then given by

$$n_f = \frac{\pi a^2}{\rho^2}.$$

The number of degrees of freedom per unit area of the image $\bar{n}_f$ is given by n_f/a^2 or

$$\bar{n}_f = \frac{\pi}{\rho^2},$$

which means that there are about three degrees of freedom per area ρ^2. These degrees of freedom could represent atom coordinates or atom types.

We will see later on that typically $\rho = 0.05$–0.15 nm, which is in the range of typical interatomic distances. If we have a crystal structure which is oriented such that atom columns line up along the beam direction (i.e. a low-index zone axis orientation), then the projected crystal structure will contain a small number of projected atom positions per unit area, and if our microscope has the proper ρ, then we can, in principle, determine those atom positions. If the crystal is tilted with respect to the electron beam, so that atoms overlap each other in the projected image, then the number of atoms per unit area will be too large, and it will be impossible to determine the coordinates of all the atoms. This means that there is virtually no chance of determining the atom coordinates of a thin amorphous foil using transmission electron microscopy methods. For high-symmetry zone axis orientations, on the other hand, the microscope may transfer sufficient information for the determination of all independent atom coordinates. It is, therefore, quite important that we learn how the microscope affects the structural information as the electron wave travels down the column. In the following sections, we will describe in great detail how lens aberrations affect the phase and amplitude of the electron exit wave.

10.2.1 The microscope point spread and transfer functions

In the theoretical Chapters 5–7, we described how the exit wave function can be calculated for a variety of observation conditions: two-beam, systematic row, and zone axis. In this section, we will analyze how the exit wave propagates to the observation plane. We already know that for an ideal microscope the intensity distribution $I(\mathbf{r})$ in the image plane is given by $|\psi(\mathbf{r})|^2$, where $\psi(\mathbf{r})$ is the exit plane wave function.† This means that there is a one-to-one correspondence between the exit wave and the image plane wave; all electrons that leave a single exit plane point converge to a single image plane point, regardless of their trajectory. For a real lens the electrons which leave a single exit plane point will not, in general, arrive at the same image plane point, but instead they will arrive in a neighborhood around this point. The resulting image will be a "blurred" version of the ideal image. The illuminated region on the sample typically has an area in the range of several tens

† Throughout this chapter we will ignore both image magnification and rotation, since these are simple linear transformations which do not distort the wave function.

of square nanometers to several hundreds of square micrometers, which is small compared with the physical dimensions of the lens. We do not expect the imaging properties of the lens to vary much across this region of interest, and this allows us to introduce the so-called *isoplanatic approximation*: *the imaging properties of the objective lens are the same across the entire region of interest.*

When we discussed the double-slit experiment in Chapter 2, only two beams originating from a single source were considered. In the present case, every exit plane point is effectively a point source from which spherical waves originate. The amplitude and phase of these waves is described by the amplitude and phase of $\psi(\mathbf{r})$. The electrons which leave a given exit plane point will travel in discrete directions described by the Bragg equation, and, similar to the double-slit experiment, electrons in different scattered beams can interfere with each other. This means that the mathematical formalism describing the image formation process must allow for such an interference between *all* pairs of beams. We will postpone a more detailed discussion of this formalism until Section 10.2.2. For now, we will assume that the transmitted beam is significantly stronger than any of the scattered beams, so that *only interference effects between a scattered beam and the transmitted beam are taken into account*. This is known as *linear image formation*.

The two approximations in the preceding paragraphs allow us to write the imaging properties of the lens as a simple convolution product of the exit plane wave function with the function $T(\mathbf{r})$, the *point spread function*. The image plane wave function is then given by

$$\psi_i(\mathbf{r}) = \psi(\mathbf{r}) \otimes T(\mathbf{r}). \tag{10.1}$$

For an ideal lens, the point spread function reduces to the Dirac delta-function, and $\psi_i = \psi$. For a real lens, the point spread function has a finite width, and it should be obvious to the reader that the narrower this width, the closer the image wave will approximate the exit plane wave. However, intuition is not going to take us very far because the point spread function is a complex function which will affect the amplitude and phase of the exit wave in different ways. In particular, the amplitude and phase of the exit plane wave will become *mixed* in the image plane, which means that image interpretation will not be straightforward.

The Fourier transform of equation (10.1) is given by

$$\psi_i(\mathbf{q}) = \mathcal{F}[\psi(\mathbf{r}) \otimes T(\mathbf{r})] = \psi(\mathbf{q})T(\mathbf{q}), \tag{10.2}$$

where $\mathbf{q}$ is a vector in the lens back focal plane. We see that the effect of the objective lens under isoplanatic, linear imaging conditions is simply a multiplication by the function $T(\mathbf{q})$ in the back focal plane of the lens. $T(\mathbf{q})$ is known as the *coherent transfer function* of the lens, or, more commonly, as the microscope transfer function. For the ideal lens, we have $T(\mathbf{q}) = 1$, since the Fourier transform of a delta-function is a constant. In what follows, we will describe the components of the transfer function in detail.

10.2.1.1 The aperture function

We can eliminate all but one or a few scattered beams by means of a physical aperture in the back focal plane (BFP) of the objective lens, so we will need to have a mathematical description in the form of an *aperture function*. We introduce the *diffraction aperture function* $A(\mathbf{q})$, which is defined by

$$A(\mathbf{q}-\mathbf{q}_0) = \begin{cases} 1 & |\mathbf{q}-\mathbf{q}_0| < q_a \\ 0 & |\mathbf{q}-\mathbf{q}_0| \geq q_a, \end{cases} \tag{10.3}$$

where q_a is the aperture radius in reciprocal length units, and $\mathbf{q}$ is a 2D vector in the BFP (i.e. a reciprocal space vector). The center of the aperture is determined by the position vector $\mathbf{q}_0$; for most observations the aperture is centered around the optical axis and therefore $\mathbf{q}_0 = \mathbf{0}$. We will see in Section 10.4 on Lorentz microscopy methods that the aperture need not always be centered on the optical axis. In mathematical terms, the aperture provides a truncation point for the reciprocal space Fourier expansion of the lattice potential. Only lattice periodicities with spatial frequencies smaller than the reciprocal aperture radius can contribute to the image.[†]

It is a trivial matter to measure the reciprocal radius of the diffraction aperture. The "true" (i.e. direct space) radius of the aperture is usually known, since the microscope manufacturer specifies the diameters of all the apertures in the column. Even when custom aperture strips are used, it is still a trivial matter to calibrate the reciprocal aperture radius. All that is needed is a material with a known lattice parameter, say Ti with $a = 0.255$ and $c = 0.468$ nm. Insert the sample in the column and obtain a zone axis orientation, say [11.0]. Since the lengths of the reciprocal lattice vectors are known ($1/d_{1\bar{1}.0} = 3.914$ nm^{-1} and $1/d_{00.2} = 4.272$ nm^{-1}), a simple multiple exposure of the zone axis pattern with and without the aperture provides an immediate value of the aperture radius, as shown schematically in Fig. 10.2. The aperture radii are $R_1 = 7.94$ nm^{-1} = 19.9 mrad, $R_2 = 4.45$ nm^{-1} = 11.2 mrad, and $R_3 = 1.83$ nm^{-1} = 4.58 mrad. The location of the aperture center can also be measured easily and is typically expressed in the same way as the Laue center; i.e. as a fractional coordinate with respect to two $\mathbf{g}$-vectors.

10.2.1.2 The phase transfer function

We have already seen one way in which the objective lens can change the phase of the exit plane wave. In Section 5.6.2, we derived that the Fresnel propagator

[†] Assuming linear imaging conditions hold, non-linear contributions may have higher spatial frequencies than those admitted by the aperture.

Fig. 10.2. Calibration of the reciprocal diffraction aperture radius using the [11.0] zone axis pattern of Ti as an internal standard.

expresses how a wave travels over a distance in free space. When we change the current through the objective lens, we effectively change the location of the image plane along the optical axis by an amount Δf, which we will call the image *defocus*. In the back focal plane, the Fresnel propagator can be written as

$$\mathcal{P}_{\Delta f}(\mathbf{q}) = \mathrm{e}^{-\pi \mathrm{i}\lambda\Delta f q^2}, \tag{10.4}$$

and this phase factor expresses how the phase of the exit plane wave is affected by a deviation from the perfect (Gaussian) focus. We have for the phase shift

$$\Delta\chi_{\Delta f}(q) = -\pi\lambda\Delta f q^2. \tag{10.5}$$

A magnetic lens will impart other phase changes to the back focal plane wave. We have seen in Chapter 3 that the main aberration of the objective lens is *spherical aberration*, expressed by the coefficient C_s. Spherical aberration results in a shorter effective focal length for electrons which travel at an angle β with respect to the optical axis. Following Reimer [Rei93, page 82], we will compute the path length difference between a trajectory at an angle β with respect to the optical axis and a trajectory along the optical axis.

Consider the trajectories in Fig. 10.3. The trajectory at an angle β would, in the absence of spherical aberration, intersect the optical axis in the image plane. Spherical aberration causes an additional bending towards the optical axis by an angle ω. The trajectory intersects the image plane at a distance $C_s\beta^3 M$, with M the lens magnification. This causes a path length difference given by the distance ds,

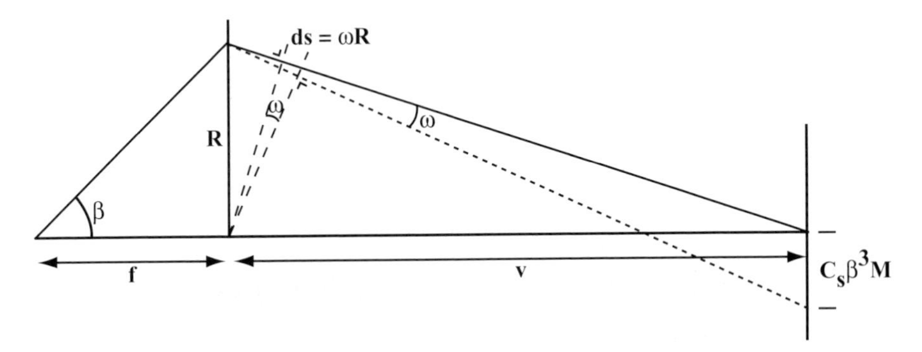

Fig. 10.3. Schematic of the path length difference computation for spherical aberration.

which is equal to ωR. The path length difference increases with increasing distance R, so the total path length difference Δs is given by the integral

$$\Delta s = \int_0^R \mathrm{d}s = \int_0^R \omega \, \mathrm{d}R.$$

The angle ω is given by

$$\omega = \frac{C_s \beta^3 M}{v} = \frac{C_s \beta^3}{f} = \frac{C_s R^3}{f^4},$$

where we have used the fact that $v = fM$ and $R = f\beta$. This is a trivial integral and we find

$$\Delta s = \int_0^R \omega \, \mathrm{d}R = \frac{1}{4} C_s \beta^4. \tag{10.6}$$

If we convert from the angle β to the back focal plane variable q, using $\beta = \lambda q$, we find for the path length difference

$$\Delta s = \frac{1}{4} C_s \lambda^4 q^4. \tag{10.7}$$

This can be converted into a phase difference by multiplication with the factor $2\pi/\lambda$ to result in

$$\Delta\chi_{C_s}(q) = \frac{\pi}{2} C_s \lambda^3 q^4. \tag{10.8}$$

Combining equations (10.5) and (10.8), we find for the total phase shift due to defocus and spherical aberration:

$$\chi(q) = \pi \left(\frac{C_s}{2} \lambda^3 q^4 - \lambda \Delta f q^2 \right). \tag{10.9}$$

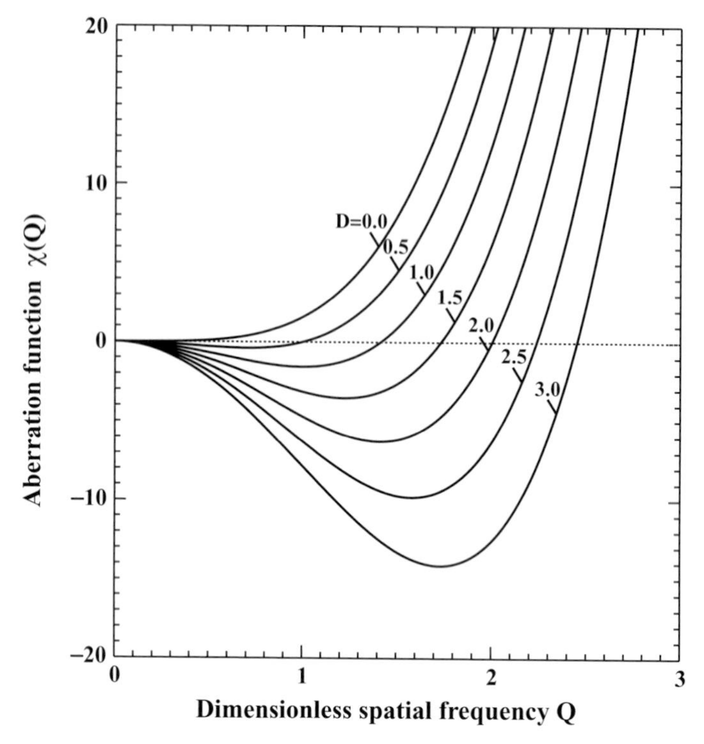

Fig. 10.4. Aberration function (10.12) for several values of the dimensionless defocus value D.

Positive values of Δf are known as *underfocus* values, whereas negative values are *overfocus* values.† It is convenient to use dimensionless variables in equation (10.9), so that the equation becomes independent of the electron wavelength. Following Kirkland [Kir98, page 14], we define the following variables:

$$Q = q(C_s\lambda^3)^{1/4}; \tag{10.10}$$

$$D = \frac{\Delta f}{\sqrt{C_s\lambda}}. \tag{10.11}$$

The quantity $\sqrt{C_s\lambda}$ is known as 1 *Scherzer* or 1 Sch; $(C_s\lambda^3)^{1/4}$ is 1 *Glaser* or 1 Gl. It is easy to show that the total phase shift $\chi(Q)$ becomes

$$\chi(Q) = \pi\left(\frac{1}{2}Q^4 - DQ^2\right). \tag{10.12}$$

This function is known as the *aberration function*‡ and is represented in Fig. 10.4 for several values of the dimensionless defocus D. This function is valid for

† It is possible to use an alternate sign convention for which underfocus is negative. In that case the second term in equation (10.9) would be preceded by a + sign. The reader should always check which sign convention is being used, in particular when entering parameters into a simulation program.

‡ Some authors do not include the factor 2π in the definition of the aberration function.

every microscope, regardless of electron wavelength and spherical aberration constant.

Example 10.1 *Consider a microscope operated at 300 kV, with a spherical aberration of $C_s = 1.2$ mm and a defocus value of $\Delta f = 45$ nm. What are the dimensionless parameters D and Q for the (111) planes of aluminum? What would be the phase shift for this situation?*

Answer: *Aluminum has a lattice parameter of 0.4049 nm, so the q value for the (111) planes is equal to $q = 1/d_{111} = 4.277$ nm^{-1}. The electron wavelength $\lambda = 1.969$ pm, so we find*

$$Q = q\left(C_s\lambda^3\right)^{1/4} = 4.277[1.2 \times 10^6 \times (1.969 \times 10^{-3})^3]^{1/4} = 1.323.$$

The dimensionless defocus is given by

$$D = \frac{\Delta f}{\sqrt{C_s\lambda}} = \frac{45}{\sqrt{1.2 \times 10^6 \times 1.969 \times 10^{-3}}} = 0.926.$$

The phase shift is given by equation (10.12):

$$\chi(Q) = \pi\left(\frac{1}{2}Q^4 - DQ^2\right) = 3.1416 \times \left(0.5 \times 1.323^4 - 0.926 \times 1.323^2\right)$$
$$= -0.28 \text{ rad}.$$

Example 10.2 *For the parameters of the previous example, what would be the spherical aberration and the defocus if a 400 kV microscope were used with the calculated values of Q and D?*

Answer: *The electron wavelength is $\lambda = 1.644$ pm, so the spherical aberration is given by*

$$C_s = \frac{Q^4}{\lambda^3 q^4} = \frac{1.323^4}{\left(1.644 \times 10^{-3}\right)^3 \times 4.277^4} = 2.06 \text{ mm}.$$

The defocus is then given by

$$\Delta f = D\sqrt{C_s\lambda} = 0.926\sqrt{2.06 \times 10^6 \times 1.644 \times 10^{-3}} = 53.89 \text{ nm}.$$

In Chapter 3, we introduced the five primary or Seidel aberrations: spherical aberration, coma, distortion, field curvature, and astigmatism. Each of these aberrations can be expressed as an equivalent phase shift. It is easy to see that astigmatism is essentially equivalent to an azimuthally varying defocus. We leave it to the reader to verify that the phase shift due to astigmatism with strength C_a and phase angle

ϕ_a is given by

$$\Delta\chi_{C_a}(q) = -\pi C_a \lambda q^2 \cos[2(\phi - \phi_a)], \quad (10.13)$$

where ϕ_a is a reference angle in the back focal plane. The wavelength and q dependence of this phase shift are identical to those of the defocus phase shift, so that we can combine the two:

$$\Delta\chi(q) = -\pi\lambda q^2[\Delta f + C_a \cos 2(\phi - \phi_a)].$$

The dimensionless form of C_a scales in the same way as Δf. The two-fold astigmatism is illustrated in Fig. 10.5(a); the height of the solid line above or below the dashed circle is equal to the astigmatism contribution. While spherical aberration is usually in the range 0.3–3.0 mm, astigmatism is measured in micrometers; if the stigmator coils are turned off completely, then the *intrinsic astigmatism* is usually in the range $C_a = 1$–2 μm. Two-fold astigmatism can be corrected completely using the objective lens stigmator coils.

Astigmatism is caused primarily by imperfections in the lens pole piece and there is no real reason why these imperfections should have two-fold symmetry with respect to the lens axis. In fact, the defocus variation around the unit circle in Fig. 10.5(a) would probably look more like that shown in Fig. 10.5(d).

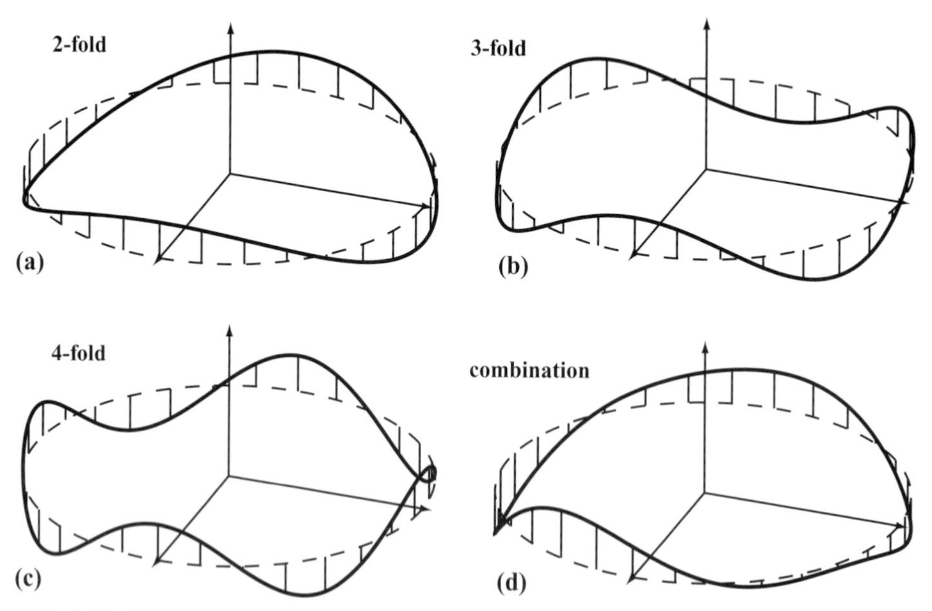

Fig. 10.5. Illustration of two-fold, three-fold, and four-fold astigmatism as a defocus change around a unit circle. A more general case consisting of random contributions of all three astigmatism orders is shown in (d).

Mathematically, this can be described by an expansion in terms of two-fold, three-fold, four-fold terms, and so on. The phase shift corresponding to the three-fold astigmatism is given by [Kri94]:

$$\Delta\chi_G = -\frac{2\pi}{3} G\lambda^2 q^3 \cos 3(\phi - \phi_G), \tag{10.14}$$

where G is the magnitude and ϕ_G is the phase angle of the three-fold astigmatism. This is a third-order contribution in q and the typical magnitude of G is around 1 μm for modern high-resolution instruments. Another third-order contribution is given by the *axial coma* K:

$$\Delta\chi_K = -2\pi K\lambda^2 q^3 \cos(\phi - \phi_K), \tag{10.15}$$

where K is the magnitude and ϕ_K is the phase angle of the axial coma.

We can, of course, continue on with higher-order aberrations, such as fifth-order spherical aberration C_5. Four-fold astigmatism contributes a fourth-order term in q, which is the same order as the spherical aberration term C_s. For a long time, equation (10.9) has been used as *the* phase shift needed to understand image formation in a high-resolution microscope. As the quality of the lenses improved, the importance of higher-order terms in the phase shift has increased. For microscopes with spherical aberration correctors, it is important to have a clear understanding of the contributions of *all* third-order aberrations and several of the fifth-order aberrations. The correction of these aberrations requires progressively more complex optical elements: e.g. three-fold astigmatism can be corrected with two sextupole coils rotated 30° with respect to each other. The more aberrations one wishes to correct, the more complex the microscope becomes.

10.2.1.3 The weak phase object approximation

We have seen in Section 5.6.1 that the exit wave function may be calculated by repeated operation of the phase grating function and the Fresnel propagator, either in the multi-slice formalism or the equivalent real-space formalism. If the object is very thin, then it is a good approximation to represent the exit wave function by the first phase grating factor, and ignore the Fresnel propagator altogether; i.e. the entire foil is considered as a single slice. In that case the exit wave function is given by

$$\phi(\mathbf{r}) = e^{i\sigma V_p(\mathbf{r})}. \tag{10.16}$$

The object is therefore a pure phase object, and this approximation is known as the *strong phase object approximation* or *SPOA*.

If the thin foil consists of low atomic number elements, then the projected potential will be small and the argument of the phase grating exponential will be small (recall that the interaction constant σ is in the range 0.005–0.009 $\mathrm{V}^{-1}\ \mathrm{nm}^{-1}$). This means that we can expand the wave function in powers of the exponential argument, and only keep terms of first order in V_p:

$$\phi(\mathbf{r}) = 1 + \mathrm{i}\sigma V_p(\mathbf{r}) + \cdots \approx 1 + \mathrm{i}\sigma V_p(\mathbf{r}). \tag{10.17}$$

This is known as the *weak phase object approximation* or *WPOA*.

In the back focal plane of the objective lens, the wave function becomes

$$\phi(\mathbf{q}) = \mathcal{F}[1 + \mathrm{i}\sigma V_p(\mathbf{r})] = \delta(\mathbf{q}) + \mathrm{i}\sigma V_p(\mathbf{q}). \tag{10.18}$$

In the absence of an aperture, the microscope transfer function consists only of the phase shift factor, so that the aberrated wave function is given by

$$\phi_a(\mathbf{q}) = [\delta(\mathbf{q}) + \mathrm{i}\sigma V_p(\mathbf{q})]\mathrm{e}^{-\mathrm{i}\chi(q)}. \tag{10.19}$$

Working out the product we find:

$$\phi_a(\mathbf{q}) = [\cos\chi(q) - \mathrm{i}\sin\chi(q)]\delta(\mathbf{q}) + \mathrm{i}\sigma V_p(\mathbf{q})\cos\chi(q) + \sigma V_p(\mathbf{q})\sin\chi(q). \tag{10.20}$$

The first term only contributes for $\mathbf{q} = \mathbf{0}$ and is equal to unity. For the remaining terms we must first consider the behavior of the aberration function. Fig. 10.6 shows

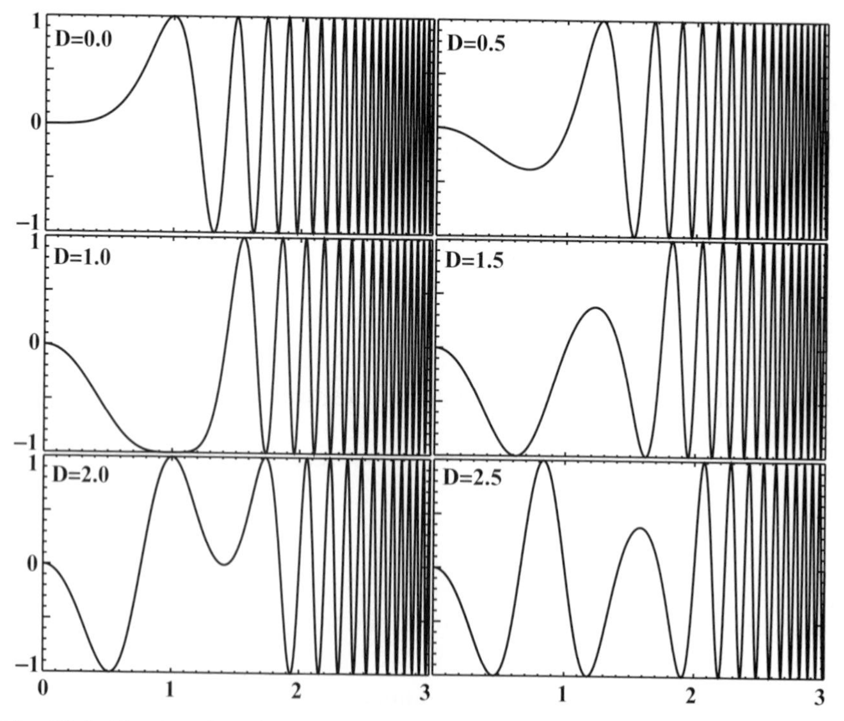

Fig. 10.6. The function $\sin\chi(Q)$ for several of the defocus values D of Fig. 10.4.

the function $\sin \chi(q)$ for the same dimensionless defocus values as Fig. 10.4. The function oscillates wildly for large values of q, but there is a range of defocus values D for which the function changes relatively slowly for smaller values of q. In particular, around $D = 1$ the function $\sin \chi(q)$ is close to -1 for a reasonably large range of q. This range of q is known as a *passband*, since the value of $\sin \chi(q)$ is nearly constant and equal to -1 in this range. The value of $\cos \chi(q)$ is then nearly zero. Let us now assume that we can tune the microscope defocus value so that $\sin \chi(q) = -1$ *for all values of* q; this is obviously another approximation, since the q-range for which this can be done is finite, not infinite. The aberrated wave function under this assumption becomes

$$\phi_a(\mathbf{q}) \approx 1 - \sigma V_p(\mathbf{q}). \tag{10.21}$$

In the image plane we find

$$\phi_a(\mathbf{r}) = \mathcal{F}^{-1}[\phi_a(\mathbf{q})] = 1 - \sigma V_p(\mathbf{r}). \tag{10.22}$$

The image intensity is then given by

$$I(\mathbf{r}) = |\phi_a(\mathbf{r})|^2 = 1 - 2\sigma V_p(\mathbf{r}) + \sigma^2 V_p^2(\mathbf{r}) \approx 1 - 2\sigma V_p(\mathbf{r}), \tag{10.23}$$

where the quadratic term in σV_p is neglected, since we only keep terms linear in V_p.

Equation (10.23) can be interpreted as follows: we know from Section 2.6.7 that the interatomic potential $V(\mathbf{r})$ has sharp peaks at the atom positions (see also Fig. 2.12). The projected potential along a given zone axis direction will also show large positive peaks at the projected atom positions. Therefore, the intensity $I(\mathbf{r})$ will be less than unity at the projected atom positions. In other words, if we can find a microscope setting for which $\sin \chi(q) = -1$ over a significant range of q, then the corresponding image will show dark regions at the projected atom positions. This is known as *dark atom contrast*. In the following section, we will explore how closely we can actually approximate this idealized microscope setting in a real microscope.

10.2.1.4 The microscope point resolution

The function $\sin \chi(Q)$ is shown as a grayscale plot in Fig. 10.7. White (black) areas correspond to positive (negative) values. It is clear that this function oscillates rapidly for large spatial frequencies Q, as is also obvious from Fig. 10.6; for small spatial frequencies, the behavior for under-focus and over-focus conditions is different. For under-focus ($D > 0$) conditions, the first minimum (dark band) appears for $D \approx 1$. At this point the function $\sin \chi(Q)$ reaches its minimum of -1, which

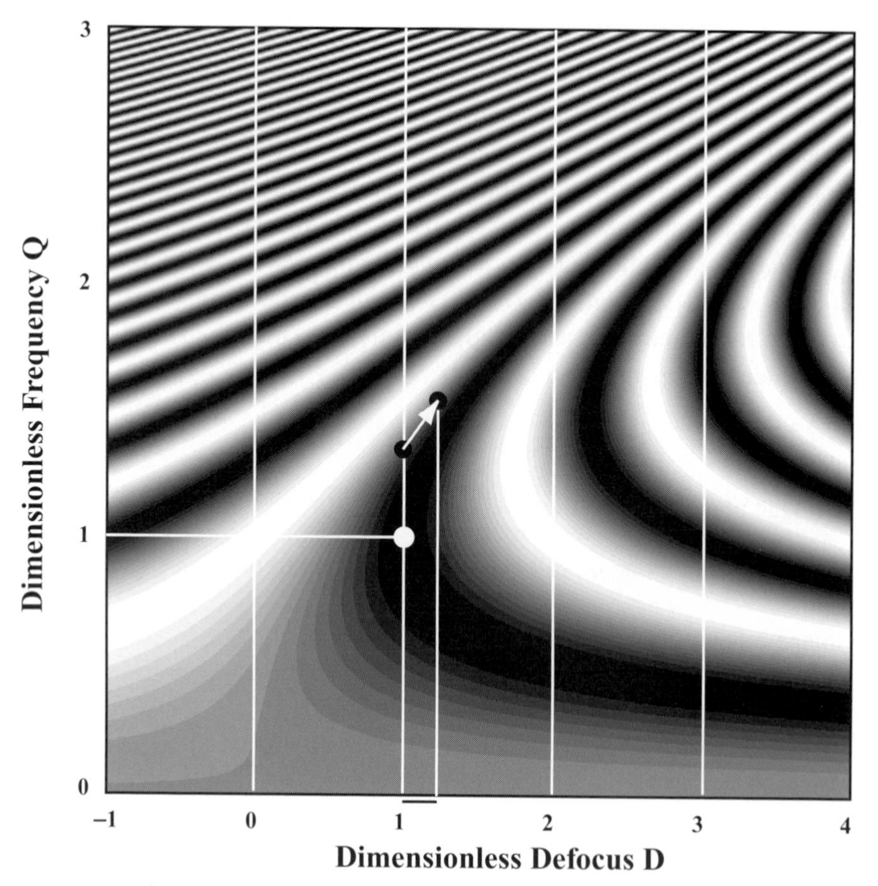

Fig. 10.7. The function $\sin \chi(Q)$ displayed as a grayscale plot; vertical lines indicate integer values of the dimensionless defocus D and correspond to the left-hand column of Fig. 10.6.

can be expressed by the following two conditions:

$$\pi\left(\frac{1}{2}Q^4 - DQ^2\right) = -\frac{\pi}{2}(4n+1) \qquad (n \geq 0); \tag{10.24}$$

$$\frac{\partial \chi}{\partial Q} = 2\pi Q(Q^2 - D) = 0. \tag{10.25}$$

The first relation expresses that $\sin \chi(Q) = -1$, the second that $\chi(Q)$ reaches a minimum. It is easy to see that minima are located at $Q = 0$ and $Q = \sqrt{D}$. Substituting this in the first relation results in

$$D^2 = \frac{8n+2}{2}, \tag{10.26}$$

which, using equation (10.11), translates to

$$\Delta f_n = \sqrt{C_s\lambda\left(\frac{8n+2}{2}\right)}. \tag{10.27}$$

The first minimum lies at the position $(D, Q) = (1, 1)$, as indicated by the filled white circle in Fig. 10.7. If we move to a slightly larger defocus D, then the minimum splits into two minima with a small bump in between. It is customary to allow this bump to increase to a value of $\sin\chi \approx -0.7$, which can be accomplished by replacing the factor $8n+2$ in equation (10.27) by $8n+3$. As a consequence of this change, the first zero crossing of $\sin\chi$ (black filled circle) moves towards a larger value of Q, as indicated by the white arrow in Fig. 10.7. The defocus value D moves from 1 to $\sqrt{\frac{3}{2}} = 1.2247$.

For values of n larger than 0, the same situation occurs: the function $\sin\chi$ shows two minima separated by a bump at which $\sin\chi \approx -0.7$. The general expression for the microscope defocus values at which this bump occurs is given by

$$\Delta f_n = \sqrt{C_s\lambda\left(\frac{8n+3}{2}\right)}. \tag{10.28}$$

Of particular importance is the value for $n = 0$ ($D = \sqrt{1.5}$):

$$\Delta f_0 = \sqrt{\frac{3}{2}C_s\lambda}, \tag{10.29}$$

which is known as the *Scherzer defocus*. The corresponding $\sin\chi$ function is shown in Fig. 10.8. The first zero crossing of $\sin\chi$ at Scherzer defocus is determined from

$$\pi\left(\frac{1}{2}Q^4 - \sqrt{\frac{3}{2}}Q^2\right) = 0, \tag{10.30}$$

which is solved easily to result in $Q = 6^{1/4}$. Using equation (10.10) we find

$$q_S \equiv 6^{1/4}\left(C_s\lambda^3\right)^{-1/4}. \tag{10.31}$$

In real space, the frequency q_S corresponds to the distance

$$\rho_S \equiv 6^{-1/4}\left(C_s\lambda^3\right)^{1/4} \approx 0.64\left(C_s\lambda^3\right)^{1/4} = 0.64\ \mathrm{Gl}. \tag{10.32}$$

The number ρ_S is known as the *point resolution* of the microscope[†] and it is the number stated by microscope manufacturers as one of the microscope parameters.

[†] It is also possible to take the high-frequency edge of the first passband and define it to be the Scherzer frequency; the point resolution then becomes $\rho_S = 0.67$ Gl.

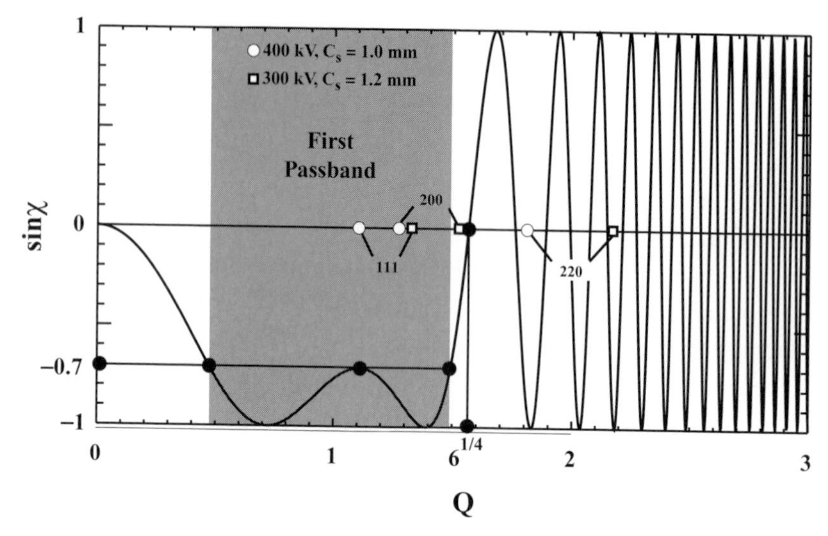

Fig. 10.8. $\sin\chi(Q)$ at the Scherzer defocus $D = \sqrt{1.5}$. The {111}, {200}, and {220} reciprocal lattice spacings of aluminum are superimposed on the drawing for 300 and 400 kV microscopes with $C_s = 1.2$ and 1.0 mm, respectively.

Example 10.3 *Consider a 400 kV microscope with a spherical aberration of 1.1 mm. What is the Scherzer defocus Δf_0? What is the defocus value for the third passband?*

Answer: *From Table 2.2 we have for the electron wavelength $\lambda = 0.001\,644$ nm. The Scherzer defocus is given by $D = \sqrt{1.5}$ which translates to*

$$\Delta f_0 = \sqrt{1.5 C_s \lambda} = 52.08 \text{ nm}.$$

For the third passband we have $n = 2$ or $D = \sqrt{9.5}$ which translates to $\Delta f_2 =$ 131.07 nm.

Example 10.4 *Consider a 300 kV microscope with $C_s = 1.2$ mm, and a 400 kV microscope with $C_s = 1.0$ mm. Determine for each microscope which reflections of the [110] zone axis pattern of aluminum will fall within the first passband.*

Answer: *The spatial frequencies corresponding to the first three reflections in the [110] zone are given by: $q_{111} = 4.2777$, $q_{200} = 4.9395$, and $q_{220} = 6.9855$ nm^{-1}. The conversion factors from q to Q are 0.309 37 Gl for the 300 kV machine, and 0.258 18 Gl for the 400 kV machine. Therefore, the reflections at 300 kV correspond to $Q_{111} = 1.323$, $Q_{200} = 1.528$ and $Q_{220} = 2.161$; at 400 kV we find $Q_{111} = 1.104$, $Q_{200} = 1.275$, and $Q_{220} = 1.803$. Since the Scherzer frequency is equal to $6^{1/4} =$ 1.565, the {111} reflections fall within the first passband for both microscopes, the {200} reflections are near the edge of the passband at 300 kV, and the {220}* *reflections are well beyond the first passband, as indicated by the open circles and open squares in Fig. 10.8.*

The preceding example shows that the microscope parameters determine which spatial frequencies of a crystalline sample (in this case aluminum) will be transmitted with the same phase shift and therefore similar contrast. For the 400 kV microscope, the 111 and 200 reflections are transmitted with nearly identical phase shifts, whereas the 300 kV microscope would have nearly zero transmission for the 200 reflections. Reflections with $Q > Q_S$ are also transmitted (for coherent illumination), but the phase shifts vary rapidly with defocus and spatial frequency, resulting in images which are difficult to interpret.

The definition of *microscope resolution* is not a straightforward one in the case of transmission electron microscopy. For an *optical* microscope, resolution is defined by the *Rayleigh criterion* [Str99, BW75], which states that two intensity distributions (originally described by $(\sin x/x)^2$-type functions) can be distinguished from each other if the minimum intensity at the center is smaller than 0.81 times the maximum intensity. As shown in Fig. 10.9, this corresponds to the second function maximum being located at the first zero of the first function. If the two functions are closer together then they can no longer be resolved. This criterion can be reformulated for more general intensity distributions, in particular for distributions which do not have zeros anywhere [Pap68].

The attentive reader may notice that, if we know a priori the analytical shape of the functions, then a simple curve fit will allow separation of the two functions *even when they are spaced more closely than shown in Fig. 10.9* [van92]. This type of curve fitting is commonly used in many branches of science. Since careful measurements of intensity distributions along with curve fitting (taking into account

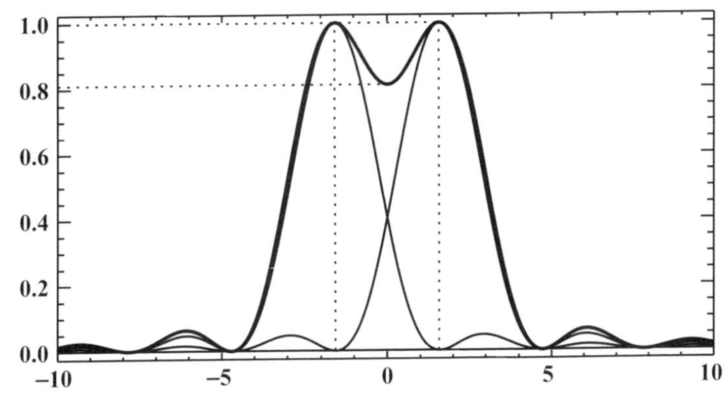

Fig. 10.9. Illustration of the Rayleigh resolution criterion for optical observations (based on Fig. 7.62 in [BW75]).

all measurement errors) are currently possible, the Rayleigh resolution criterion is not a useful criterion for electron optical observations.

To conclude these comments on resolution it is useful to consider the influence of other aberrations, such as two-fold, three-fold, and four-fold astigmatism. Following Krivanek [Kri94] we write the aberration function $\chi(q)$ as follows:

$$\begin{aligned}\frac{\chi(q)}{2\pi} &= -\frac{1}{2}\lambda q^2\left[\Delta f + C_a\cos 2(\phi-\phi_a)\right]\\ &\quad -\frac{1}{3}\lambda^2 q^3\left[G\cos 3(\phi-\phi_G) + 3K\cos(\phi-\phi_K)\right]\\ &\quad -\frac{1}{4}\lambda^3 q^4\left[H\cos 4(\phi-\phi_H) - C_s\right],\end{aligned} \tag{10.33}$$

where (H, ϕ_H) represent the four-fold astigmatism. The maximum possible value for this function would occur when all cosines are simultaneously equal to $+1$, assuming that the phase angles are all equal to each other. The maximum phase difference between a perfect microscope (with zero phase shift) and the phase shift of equation (10.33) is then given by

$$\max\left(\frac{\chi(q)}{2\pi}\right) \equiv \delta = \frac{1}{2}\lambda q^2(\Delta f + C_a) + \frac{1}{3}\lambda^2 q^3(G+3K) + \frac{1}{4}\lambda^3 q^4(H - C_s). \tag{10.34}$$

Let us consider each aberration separately. If only intrinsic two-fold astigmatism is present, i.e. the stigmator coils are turned off, then the maximum phase shift δ is equal to

$$\delta = \frac{1}{2}\lambda q^2 C_a.$$

This function is shown as a log–log plot in Fig. 10.10 for a typical value of $C_a = 1\,\mu\text{m}$ and $\lambda = 2.508$ pm (200 kV). The function appears as a straight line with slope 2. If we consider a phase shift of $\pi/2$ to be the maximum tolerable phase shift due to aberrations (indicated by the horizontal dashed line at 0.25 in Fig. 10.10), then only spatial frequencies smaller than about 0.45 nm^{-1} will experience small phase shifts. This will limit the point resolution of the microscope to about 2.2 nm. Before the invention of quadrupole stigmators, this was indeed the best attainable resolution in any TEM. The quadrupole stigmator allows us to correct the two-fold astigmatism so that it is no longer the resolution-limiting aberration.

The next aberration in equation (10.34) is spherical aberration. The maximum associated phase shift is given by

$$\delta = \frac{1}{4}\lambda^3 q^4 C_s,$$

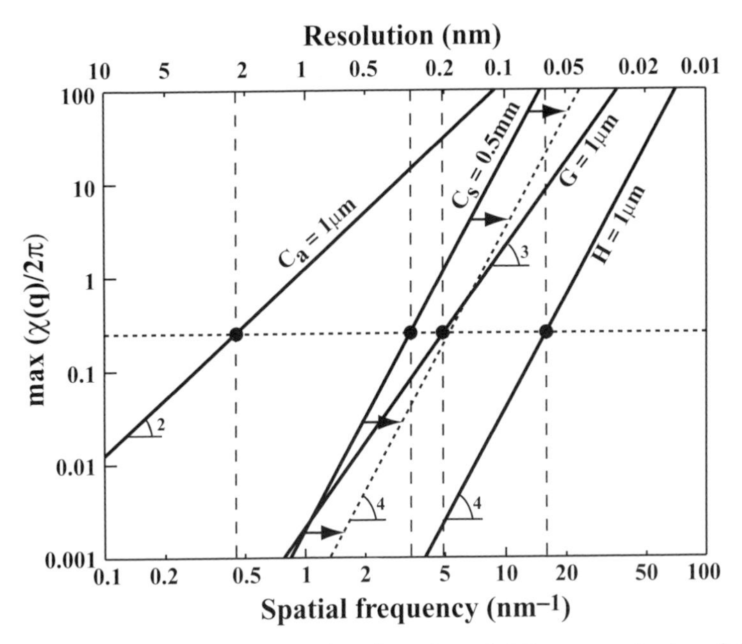

Fig. 10.10. Maximum phase shifts as a function of spatial frequency for two-fold, three-fold, and four-fold astigmatism, and for spherical aberration. This figure is based on Fig. 8 in [Kri94].

which, for $C_s = 0.5$ mm, results in the leftmost straight line with slope 4 in Fig. 10.10. This line can be moved sideways by selecting the Scherzer defocus, for which the effect of spherical aberration is partially canceled. This shift is indicated by arrows pointing to the dashed line with slope 4. The resulting point resolution is slightly better than 0.2 nm, which is what many current microscopes are capable of.

The next aberration is three-fold astigmatism G, along with axial coma K which has the same functional dependence. The maximum associated phase shift is

$$\delta = \frac{1}{3}\lambda^2 q^3 (G + 3K);$$

taking $G + 3K = 1\,\mu\text{m}$, we have the third line with slope 3 in Fig. 10.10. This line intersects the horizontal line at a point corresponding to a point resolution of about 0.2 nm, *slightly larger than the Scherzer point resolution*! This means that the point resolution is limited by three-fold astigmatism and/or axial coma, and that these aberrations must be corrected along with the spherical aberration.

If three-fold astigmatism and axial coma are corrected, and the spherical aberration is reduced using a C_s-corrector, then the next resolution-limiting aberration will be four-fold astigmatism, which has a maximum phase shift of

$$\delta = \frac{1}{4}\lambda^3 q^4 H,$$

shown as the rightmost line with slope 4 in Fig. 10.10. Four-fold astigmatism limits the resolution to about 0.06 nm. The precise positions of these lines depend of course on the actual values of the aberration coefficients, including the phase angles of the aberrations. The main point of this comment is that resolution improvements are only possible if the next level of aberrations can be corrected. To bring the resolution from 2–3 nm down to 0.2 nm required the removal of two-fold astigmatism by means of quadrupole stigmators. Improvements down to 0.1 nm require correction of C_s, three-fold astigmatism and axial coma, whereas 0.05 nm resolution requires correction of (among others) four-fold astigmatism.

10.2.2 The influence of beam coherence

In Chapter 3, we introduced the concept of *coherence* and we distinguished between temporal and spatial coherence. Perfect coherence is only possible if the electron gun produces perfectly monochromatic electrons which all travel precisely along the same direction, in other words, the electron source produces the ideal plane wave. In reality, each electron gun has an associated energy spread which is small compared with the beam energy but is not zero. The electrons travel with a range of wave vectors that is sharply peaked along the optical axis, but we cannot ignore the fact that the beam has a finite (non-zero) divergence angle θ_c. Both of these factors will contribute to the fact that the electron beam is not perfectly coherent.

The double-slit thought experiment described in Chapter 2 can be used to introduce the concept of *partial coherence*, which is crucial to a realistic description of the information transfer in the TEM. In the *incoherent* case, electrons which travel through either slit will not interfere with each other and the intensity pattern at a point Q on the screen is simply the sum of the intensity patterns due to each of the slits:

$$I_{inc}(Q) = |\Psi_1(Q)|^2 + |\Psi_2(Q)|^2 = I_1(Q) + I_2(Q). \qquad (10.35)$$

In the coherent case, *amplitudes* are superimposed instead of intensities, and we have

$$I_{coh}(Q) = |\Psi_1(Q) + \Psi_2(Q)|^2 = I_1(Q) + I_2(Q) + 2\sqrt{I_1(Q)}\sqrt{I_2(Q)}\gamma_{12}\cos 2\pi\beta, \qquad (10.36)$$

where the factor γ_{12} is a slowly varying function describing the amplitude of the intensity oscillations (fringes) and β is a function which depends on the geometry and determines the frequency and position of the fringes (see Fig. 2.1). If there is no interference (incoherent case), then we have $\gamma_{12} = 0$. Hence, we can combine the coherent and incoherent cases into a single equation:

$$I(Q) = \gamma_{12}\left[I_1(Q) + I_2(Q) + 2\sqrt{I_1(Q)}\sqrt{I_2(Q)}\cos 2\pi\beta\right] + (1-\gamma_{12})\,[I_1(Q) + I_2(Q)]. \tag{10.37}$$

The factor γ_{12} is known as the *degree of coherence*.† Partial coherence simply means that $0 < \gamma_{12} < 1$. The degree of coherence refers to the amplitude of the interference fringes and it is easy to see that γ_{12} is related to the interference term $\Psi_1\Psi_2^* + \Psi_1^*\Psi_2$.

Consider an exit wave $\psi(\mathbf{r})$ which, in the perfectly coherent case, is modified by the point spread function $T(\mathbf{r})$. The image intensity is then given by

$$\begin{aligned} I(\mathbf{r}) &= |\psi(\mathbf{r}) \otimes T(\mathbf{r})|^2; \\ &= [\psi(\mathbf{r}) \otimes T(\mathbf{r})]\,[\psi^*(\mathbf{r}) \otimes T^*(\mathbf{r})]. \end{aligned}$$

Taking the Fourier transform of this expression and using the convolution and multiplication theorems we find [Kir98]:

$$\begin{aligned} I(\mathbf{q}) &= [\psi(\mathbf{q})T(\mathbf{q})] \otimes [\psi^*(\mathbf{q})T^*(\mathbf{q})]; \\ &= \int \psi(\mathbf{q}')T(\mathbf{q}')\psi^*(\mathbf{q}'+\mathbf{q})T^*(\mathbf{q}'+\mathbf{q})\,\mathrm{d}\mathbf{q}'; \\ &\equiv \int T_{cc}(\mathbf{q}', \mathbf{q}'+\mathbf{q})\psi(\mathbf{q}')\psi^*(\mathbf{q}'+\mathbf{q})\,\mathrm{d}\mathbf{q}'. \end{aligned} \tag{10.38}$$

The new function T_{cc} is known as the *transmission cross coefficient* and it depends on two spatial frequency vectors $\mathbf{q}'$ and $\mathbf{q}' + \mathbf{q}$.

The transmission cross coefficient (or TCC for short) has a simple interpretation: equation (10.38) expresses the fact that each spatial frequency $\mathbf{q}'$ in the exit wave function will interfere with every other spatial frequency $\mathbf{q}' + \mathbf{q}$ to produce the final image intensity. The TCC describes the *strength* of this pairwise interference. For the case of the *linear image theory* discussed in Section 10.2.1, only interference effects between the forward scattered beam and the diffracted beams are considered. The more general case is described by equation (10.38). In the following section, we will take a closer look at the nature of the TCC and its consequences on image formation under partial coherence conditions.

† A more complete and rigorous analysis reveals that γ_{12} is actually the modulus of a complex number, which is known as the *complex degree of coherence*. We refer the reader to Chapter 10 in [BW75] for more details.

10.2.2.1 Derivation of the damping envelope

The coherent TCC in equation (10.38) is defined by the product

$$T_{cc}^{coh}(\mathbf{q}, \mathbf{q}') \equiv T(\mathbf{q})T^*(\mathbf{q}') = A(\mathbf{q})A(\mathbf{q}')\mathrm{e}^{-\mathrm{i}[\chi(\mathbf{q})-\chi(\mathbf{q}')]}, \qquad (10.39)$$

where we have used the explicit form of the transfer function in terms of the aperture function and the aberration function. In the case of partial coherence, there is a range of incident beam directions and also a range of beam energies which must be taken into account. We will denote by $s(\mathbf{q})$ the normalized *source spread function* which for a Gaussian distribution is given by

$$s(\mathbf{q}) \equiv \frac{1}{\pi q_0^2}\mathrm{e}^{-(q/q_0)^2}. \qquad (10.40)$$

Similarly, we can define the *defocus spread function* $f(\Delta f)$ by the Gaussian distribution

$$f(\Delta f) \equiv \frac{1}{\sqrt{\pi}\,\Delta}\mathrm{e}^{-(\Delta f/\Delta)^2}. \qquad (10.41)$$

The quantities q_0 and Δ are the e^{-1} half-width values of the Gaussian distributions, as indicated in the figure. The full width at half maximum (FWHM) value δ is related to Δ as follows:

$$\Delta = \frac{\delta}{2\sqrt{\ln 2}} \approx 0.60\delta. \qquad (10.42)$$

For partial coherence, the TCC in equation (10.39) must be replaced by an integration over both spread functions:

$$T_{cc}^{pc}(\mathbf{q}, \mathbf{q}'; \Delta f) = \iint s(\mathbf{q}'')f(\Delta f')T_{cc}^{coh}(\mathbf{q}'' + \mathbf{q}, \mathbf{q}'' + \mathbf{q}'; \Delta f + \Delta f')\,\mathrm{d}\mathbf{q}''\,\mathrm{d}\Delta f', \qquad (10.43)$$

where we have explicitly written the defocus dependence of the transfer functions. The image intensity can then be computed by substituting this integral into equation (10.38). This integral is not trivial and the reader is referred to the following sources for a more detailed computation: Frank [Fra73], Ishizuka [Ish80], and Kirkland [Kir98]. We will compute an approximate expression for the integral by considering the spread functions one at a time.

We will assume that the exit wave function $\psi(\mathbf{r})$ can be written in the following form [Fra73]:

$$\psi(\mathbf{r}) = 1 + O(\mathbf{r}),$$

such that

$$\psi(\mathbf{q}) = \delta(\mathbf{q}) + O(\mathbf{q}).$$

Substituting ψ into equation (10.38) and working out the integrations leads to

$$\begin{aligned} I(\mathbf{q}) &= \int T_{cc}(\mathbf{q}', \mathbf{q}' + \mathbf{q})[\delta(\mathbf{q}') + O(\mathbf{q}')][\delta(\mathbf{q}' + \mathbf{q}) + O^*(\mathbf{q}' + \mathbf{q})]\,\mathrm{d}\mathbf{q}'; \\ &= T_{cc}(\mathbf{0}, \mathbf{q})\delta(\mathbf{q}) \\ &\quad + [T_{cc}(\mathbf{0}, \mathbf{q})O^*(\mathbf{q}) + T_{cc}(-\mathbf{q}, \mathbf{0})O(-\mathbf{q})] \\ &\quad + \int T_{cc}(\mathbf{q}', \mathbf{q}' + \mathbf{q})O(\mathbf{q}')O^*(\mathbf{q}' + \mathbf{q})\,\mathrm{d}\mathbf{q}'. \end{aligned} \tag{10.44}$$

The first term describes the background intensity; the second term is the linear term since it only depends on a single spatial frequency vector $\mathbf{q}$. The last term is the non-linear component of the image. We will first consider the linear term and determine how it is affected by the presence of the defocus spread function $f(\Delta f)$.

From the definition of the partially coherent TCC we have

$$T_{cc}(-\mathbf{q}, \mathbf{0}) = \int f(\Delta f')T_{cc}^{coh}(-\mathbf{q}, \mathbf{0}; \Delta f + \Delta f')\,\mathrm{d}\Delta f';$$

$$T_{cc}(\mathbf{0}, \mathbf{q}) = \int f(\Delta f')T_{cc}^{coh}(\mathbf{0}, \mathbf{q}; \Delta f + \Delta f')\,\mathrm{d}\Delta f' = T_{cc}^*(-\mathbf{q}, \mathbf{0}).$$

The explicit computation of $T_{cc}(-\mathbf{q}, \mathbf{0})$ for the defocus spread proceeds as follows: after substitution of the transfer function and replacing $-\mathbf{q}$ by $\mathbf{q}$ we have

$$T_{cc}(\mathbf{q}, \mathbf{0}) = \int f(\Delta f')A(\mathbf{q})\mathrm{e}^{-\mathrm{i}\chi(\mathbf{q};\Delta f+\Delta f')}\,\mathrm{d}\Delta f'. \tag{10.45}$$

Since the defocus spread function is strongly peaked around its central value, we can expand the function in the exponential in a Taylor expansion with respect to Δf and keep only the linear term:

$$\chi(\mathbf{q}; \Delta f + \Delta f') \approx \chi(\mathbf{q}; \Delta f) + \Delta f'\frac{\partial\chi}{\partial\Delta f} + \cdots. \tag{10.46}$$

For all vectors $\mathbf{q}$ inside the aperture we have

$$T_{cc}(\mathbf{q}, \mathbf{0}) = \frac{\mathrm{e}^{-\mathrm{i}\chi(\mathbf{q};\Delta f)}}{\Delta\sqrt{\pi}} \int \mathrm{e}^{-(\Delta f'/\Delta)^2}\mathrm{e}^{-\mathrm{i}\Delta f'\partial\chi/\partial\Delta f}\,\mathrm{d}\Delta f'. \tag{10.47}$$

The derivative is calculated easily as

$$\frac{\partial\chi}{\partial\Delta f}=\frac{\partial}{\partial\Delta f}\left[\pi\left(\frac{C_s}{2}\lambda^3q^4-\lambda\Delta f q^2\right)\right]=-\pi\lambda q^2.$$

Substituting in equation (10.47) and using the following standard integral $\mathcal{F}[\mathrm{e}^{-x^2}]=\sqrt{\pi}\mathrm{e}^{-\pi^2q^2}$ we arrive at

$$T_{cc}(\mathbf{q},\mathbf{0})=\mathrm{e}^{-\mathrm{i}\chi(\mathbf{q})}\mathrm{e}^{-\pi^2(\Delta\lambda q^2)^2/2}=\mathrm{e}^{-\mathrm{i}\chi(\mathbf{q})}\mathrm{e}^{-\pi^2(\delta\lambda q^2)^2/16\ln 2}\equiv\mathrm{e}^{-\mathrm{i}\chi(\mathbf{q})}E_t(\mathbf{q}).\quad(10.48)$$

The function $E_t(\mathbf{q})$ is known as the *temporal incoherence envelope function.* It is an exponential damping function which depends on the magnitude of the defocus spread Δ, but is independent of the actual defocus value.

By assuming that the defocus spread function is sharply peaked, we have been able to write the cross coefficient as a simple multiplication by an exponential damping factor. We can repeat this exercise for the source spread function (10.40), and work out the following integral:

$$T_{cc}(\mathbf{q},\mathbf{0})=\int s(\mathbf{q}')A(\mathbf{q}'+\mathbf{q})A(\mathbf{q}')\mathrm{e}^{-\mathrm{i}[\chi(\mathbf{q}'+\mathbf{q})-\chi(\mathbf{q}')]}\,\mathrm{d}\Delta f'.\quad(10.49)$$

Using a Taylor expansion for small $\mathbf{q}'$ we find:

$$\chi(\mathbf{q}+\mathbf{q}')\approx\chi(\mathbf{q})+\mathbf{q}'\cdot\nabla\chi(\mathbf{q})+\cdots.$$

Working through the integration as before we find for the TCC:

$$T_{cc}(\mathbf{q},\mathbf{0})=\mathrm{e}^{-\mathrm{i}\chi(\mathbf{q})}\mathrm{e}^{-\frac{1}{4}q_0^2|\nabla\chi|^2}\equiv\mathrm{e}^{-\mathrm{i}\chi(\mathbf{q})}E_s(\mathbf{q}).\quad(10.50)$$

The function $E_s(\mathbf{q})$ is known as the *spatial incoherence envelope function.* It is again an exponential damping function which depends on the magnitude of the gradient of the aberration function $|\nabla\chi|$. This gradient is given explicitly by

$$\nabla\chi(\mathbf{q})=2\pi\left(C_s\lambda^3q^2-\lambda\Delta f\right)\mathbf{q}.\quad(10.51)$$

The gradient becomes large for the higher spatial frequencies, as is obvious from Fig. 10.4, and depends on the defocus. The explicit expression for E_s is then given by

$$E_s(\mathbf{q})=\mathrm{e}^{-\pi^2q_0^2(C_s\lambda^3q^3-\lambda\Delta f q)^2}.\quad(10.52)$$

The evaluation of the integral (10.43) for both defocus spread and source spread functions simultaneously is rather lengthy and we refer to Ishizuka [Ish80] and Kirkland [Kir98] for a detailed derivation. The primary complication stems from the fact that the Taylor expansion of the aberration function must now be carried out with respect to both $\mathbf{q}'$ and $\Delta f'$, and *all* terms linear in either of these quantities must be kept, resulting in the following expansion:

$$\chi(\mathbf{q}+\mathbf{q}', \Delta f + \Delta f') \approx \chi(\mathbf{q}, \Delta f) + \Delta f' \frac{\partial \chi}{\partial \Delta f} + \mathbf{q}' \cdot \nabla \chi(\mathbf{q}) + \Delta f' \mathbf{q}' \cdot \frac{\partial^2 \chi}{\partial \mathbf{q}' \partial \Delta f} + \cdots .$$

The last terms couples the effects of both spread functions. For completeness we provide the partially coherent microscope transfer function in the presence of defocus and source spread:

$$\begin{aligned} T(\mathbf{q}) &= A(\mathbf{q}) \mathrm{e}^{-\mathrm{i}\chi(\mathbf{q})} E_s(\mathbf{q}) E_t(\mathbf{q}); \\ &= A(\mathbf{q}) \mathrm{e}^{-\mathrm{i}\chi(\mathbf{q})} \exp\left[-\frac{(\pi\lambda\Delta)^2}{2u} q^4 \right] \exp\left[-\frac{\pi^2 \theta_c^2}{\lambda^2 u} \left(C_s \lambda^3 q^3 - \Delta f \lambda q \right)^2 \right], \end{aligned} \tag{10.53}$$

where

$$u = 1 + 2(\pi \theta_c \Delta)^2 q^2,$$

and the beam divergence angle θ_c is related to q_0 by

$$\theta_c = \lambda q_0.$$

Equation (10.53) partially accounts for the effects of incoherence. While it is not exact, it is a useful expression since the temporal and spatial envelope functions appear in the form of multiplicative factors. For the exact case, we would need to numerically integrate equation (10.43) without applying the Taylor expansions for the aberration function.

To wrap things up, we substitute the function $T(\mathbf{q})$ into equation (10.44) to determine the linear part of the image intensity. Using $O(\mathbf{q}) = O_r(\mathbf{q}) + \mathrm{i}O_i(\mathbf{q})$ and the fact that $O_{r,i}(\mathbf{q}) = O^*_{r,i}(-\mathbf{q})$ (Friedel's law) we have [Fra73]:

$$\begin{aligned} I^{linear}(\mathbf{q}) &= T(\mathbf{q}) O^*(\mathbf{q}) + T(-\mathbf{q}) O(-\mathbf{q}); \\ &= O_r(\mathbf{q})[T(\mathbf{q}) + T(-\mathbf{q})] + \mathrm{i}O_i[T(-\mathbf{q}) - T(\mathbf{q})]; \\ &= 2A(\mathbf{q}) E_s(\mathbf{q}) E_t(\mathbf{q}) \left[O_r(\mathbf{q}) \cos\chi(\mathbf{q}) + O_i(\mathbf{q}) \sin\chi(\mathbf{q}) \right]. \end{aligned} \tag{10.54}$$

This expression clearly shows that the real and imaginary parts of the exit wave function (or, equivalently, the phase and amplitude) are mixed by the coherent microscope transfer function and attenuated by the incoherent envelope functions. The non-linear component of equation (10.44) is more complicated to compute. For crystalline samples the integration can be replaced by a summation over reciprocal lattice vectors.

10.2.2.2 Significance of the damping envelope

After the lengthy derivations of the previous section, we are now ready to determine how the partially coherent transfer function is different from the coherent transfer function. As before, we can introduce dimensionless quantities based on equations (10.10) and (10.11) and the envelope function becomes (taking $u = 1$):

$$E_s(Q)E_t(Q) = \exp\left\{-\frac{\pi^2}{2}\left[\bar{\Delta}^2 Q^4 + 2\bar{q}_0^2(Q^3 - DQ)^2\right]\right\}, \tag{10.55}$$

with

$$\bar{\Delta} = \frac{\Delta}{\sqrt{C_s\lambda}},$$

and

$$\bar{q}_0 = q_0(C_s\lambda^3)^{1/4}.$$

Figure 10.11 shows the total envelope function $E_s E_t$ for several combinations of dimensionless parameters. In the top half of the figure $\bar{q}_0 = 0.05$, and the solid lines indicate the defocus values $D = 1, 2, 3,$ and 4. In the bottom half $\bar{q}_0 = 0.1$. The dash-dotted line is the temporal incoherence envelope E_t for $\bar{\Delta} = 0.1$ and $\bar{q}_0 = 0.0$. The curves in this figure reveal that the temporal incoherence envelope

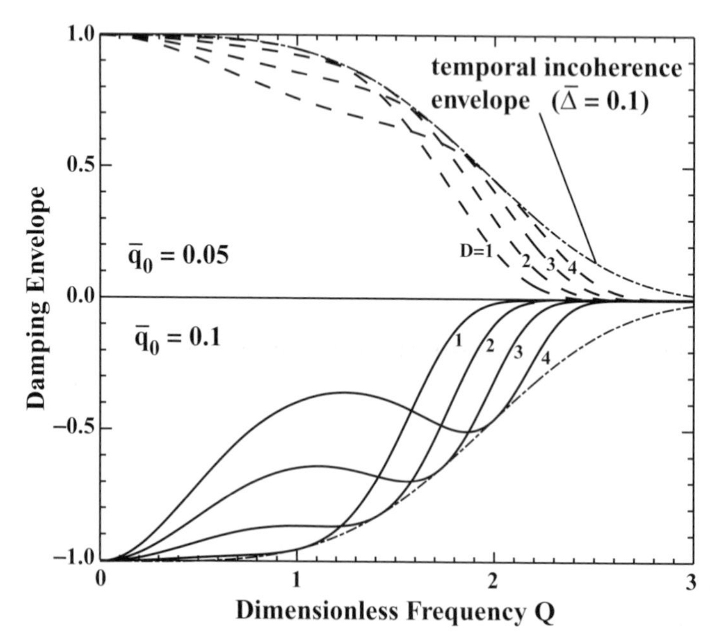

Fig. 10.11. Incoherent damping envelope for a few selected values of the dimensionless parameters $\bar{\Delta}$ and $\bar{q}_0$ and several values of D.

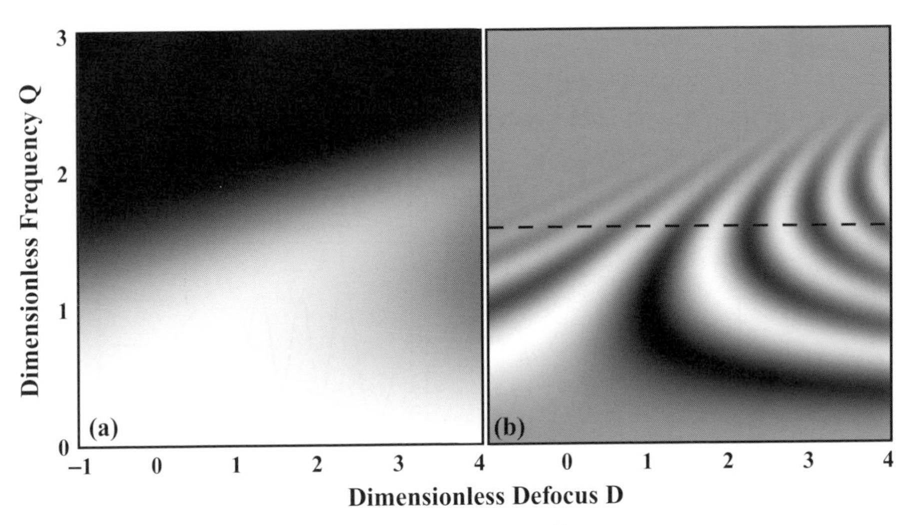

Fig. 10.12. Incoherent damping envelope $E_s E_t$ for $\bar{\Delta} = 0.1$ and $\bar{q}_0 = 0.1$ shown as a grayscale plot (left) and multiplied by $\sin \chi(q)$ (right).

is a smoothly varying function whereas the spatial incoherence envelope shows a single "dip" around $Q = 1$.

Alternatively, we can represent the damping envelope by a grayscale plot, similar to that of Fig. 10.7. This is shown in Fig. 10.12(a) for $\bar{\Delta} = 0.1$ and $\bar{q}_0 = 0.1$. The product of $\sin \chi(Q)$ and $E_s(Q)E_t(Q)$ is shown in Fig. 10.12(b); comparison with Fig. 10.7 reveals that the higher-frequency oscillations are strongly damped, and that the point where damping becomes severe moves to larger Q with increasing defocus D. This is a consequence of the defocus dependence of the spatial incoherence envelope. The dashed line in Fig. 10.12(b) represents the Scherzer frequency $Q = 6^{1/4}$.

Vertical sections through Fig. 10.12 at the defocus values corresponding to the first four passbands D_n are shown in Fig. 10.13. Several trends can be observed from this figure:

- the width of the passband decreases with increasing order n;
- the number of oscillations of $\sin \chi(Q)$ before the passband is equal to the order of the passband;
- the cutoff frequency of the incoherent envelope function shifts to larger spatial frequencies with increasing defocus D.

Beyond a certain spatial frequency Q, the incoherent damping envelope becomes so small that virtually no signal can be passed through. It is therefore useful to define a parameter which indicates the frequency beyond which no further signal information can be passed through the lens. This point is known as the *information limit*.

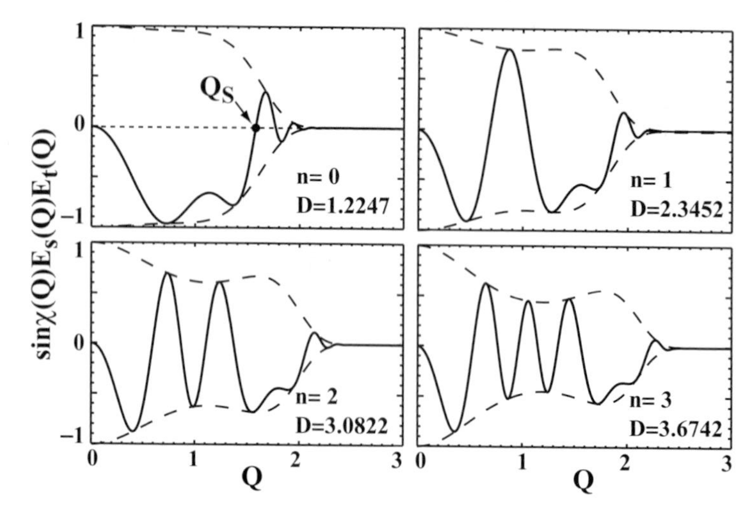

Fig. 10.13. Vertical sections through the function $\sin \chi E_s E_t$ for the first four passbands $D = \sqrt{(8n+3)/2}$. The Scherzer spatial frequency Q_S is indicated for $n = 0$.

We should point out that the above derivation assumes that defocus and spherical aberration are the only relevant lens aberrations, and that two-fold astigmatism has been properly corrected. If additional aberrations are to be taken into account (say, three-fold astigmatism and axial coma), then the expression for the aberration function becomes more complicated and so will the spatial incoherence envelope, since E_s depends on the gradient $\nabla\chi$. The spatial incoherence envelope must therefore be recomputed every time a new aberration is taken into account [KS95].

The damping envelope attenuates the signal frequencies and at some point the product of the signal and the damping envelope will fall below the noise level of the signal. If we define the signal-to-noise ratio as $s \equiv S/N$, then we can introduce the information limit due to the damping envelope E by determining at which spatial frequency $SE(q) = N$, or $E(Q) = 1/s$. Each damping envelope will have a characteristic frequency Q_i for which this equality is true.

- *Temporal incoherence envelope*. The function $E_t(Q)$ reaches the value $1/s$ at

$$Q_i = \left(\frac{\sqrt{2\ln s}}{\pi\bar{\Delta}}\right)^{1/2}.$$

Converting the dimensionless quantities we find

$$q_i = \left(\frac{\sqrt{2\ln s}}{\pi\lambda\Delta}\right)^{1/2}.$$

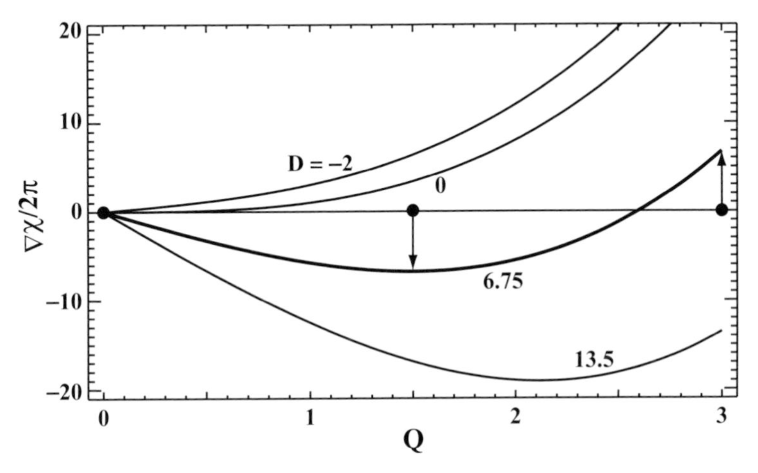

Fig. 10.14. Illustration of the defocus dependence of the gradient of the aberration function $\chi(Q)$.

The spatial distance corresponding to q_i is commonly denoted by ρ_c and is known as the *chromatic information limit*:†

$$\rho_c \equiv \left(\frac{\pi \lambda \Delta}{\sqrt{2 \ln s}} \right)^{1/2}. \tag{10.56}$$

- *Spatial incoherence envelope*. The function $E_s(Q)$ is a bit more complicated than $E_t(Q)$, since its argument also depends on the defocus value D. The spatial incoherence envelope will have a minimal effect if the argument of the exponential varies as little as possible across the frequency range of interest. This means that we must look for the defocus value for which $\nabla\chi$ has a minimum variation over the interval $[0, Q_{max}]$. Figure 10.14 shows the function $\nabla\chi/2\pi = Q^3 - DQ$ for the frequency range $Q = [0, 3]$ and for several values of D. For over-focus conditions ($D < 0$) the gradient varies substantially across the interval $[0, 3]$ and the only minimum occurs at $Q = 0$. For under-focus conditions ($D > 0$) the gradient has a minimum located at the value $Q_{min} = \sqrt{D/3}$ (exercise).

  We will now determine the defocus value for which the value of $\nabla\chi$ at the minimum is equal to the negative of the value at Q_{max}, i.e. [Lic91]:

$$\nabla\chi(Q_{min}) = -\nabla\chi(Q_{max}),$$

  which translates into

$$\left(\frac{D}{3}\right)^{3/2} - D\left(\frac{D}{3}\right)^{1/2} = -\left(Q^3_{max} - DQ_{max}\right).$$

† "Chromatic" because the temporal incoherence envelope deals with the energy spread of the electron beam.

This is a third-order polynomial equation in D which has three solutions (e.g. [AS77]): $D_1 = \frac{3}{4}Q_{max}^2$ and $D_{2,3} = 3Q_{max}^2$. Only D_1 corresponds to a minimum within the range $[0, Q_{max}]$; the other value corresponds to the minimum at $Q = Q_{max}$. The defocus value D for which the function $\nabla\chi(Q)$ is minimized over the interval $[0, Q_{max}]$ is known as the *optimum focus* or the *Lichte focus*, after H. Lichte who first introduced it in the context of electron holography [Lic91].

Converting from dimensionless quantities we find for the optimum defocus:

$$\Delta f_{opt} = \frac{3}{4}C_s\lambda^2 q_{max}^2. \tag{10.57}$$

At the optimum defocus and near the maximum spatial frequency the spatial incoherence envelope can be rewritten as

$$E_s(Q_{max}) = \mathrm{e}^{-\pi^2\bar{q}_0^2\frac{1}{16}Q_{max}^6} = \mathrm{e}^{-\pi^2\bar{q}_0^2a^2Q_{max}^6},$$

where we have introduced the variable $a^2 = \frac{1}{16}$. At the optimum defocus $a = \frac{1}{4}$; it is easy to show that at Gaussian defocus ($D = 0$) we have $a = 1$. The effect of the spatial incoherence envelope at the optimum focus is therefore smaller than the effect at the Gaussian defocus.

Figure 10.15 shows the spatial incoherence envelope at optimum defocus for two different electron sources: a thermionic LaB_6 source with $\bar{q}_0 = 0.1$, and a field emission source with $\bar{q}_0 = 0.01$. Since the field emission source has a much smaller beam divergence angle θ_c (which is essentially the meaning of $\bar{q}_0$), the envelope extends to larger

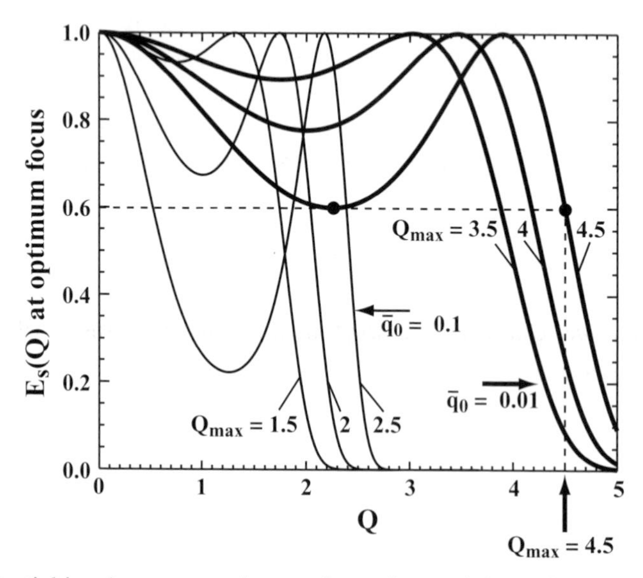

Fig. 10.15. Spatial incoherence envelope at the optimum defocus for two different source spread parameters $\bar{q}_0$ typical for a LaB_6 source (0.1) and a field emission source (0.01). For each source spread three different values of Q_{max} are shown. The dashed line indicates that the value of the envelope at Q_{max} is equal to the value at the lowest point of the dip.

spatial frequencies than that of the thermionic source. For each source three different values of Q_{max} are shown. The dashed lines on the $Q_{max} = 4.5$ curve indicate that the value of $E_s(Q)$ at Q_{max} is equal to the value at the minimum of the dip, which is located at $Q_{max}/2$.

The spatial incoherence envelope is equal to the inverse of the signal-to-noise ratio s when the following relation is satisfied:

$$\ln s = \pi^2 \bar{q}_0^2 a^2 Q^6 = \left(\frac{\pi a \theta_c}{\lambda}\right)^2 C_s^2 \lambda^6 q^6,$$

where we have replaced all dimensionless quantities. From expression (10.32) we derive that $C_s^2 \lambda^6 = 36 \rho_S^8$. Rearranging terms, we find the following expression for the *spatial incoherence information limit*:

$$\rho_{\theta_c} \equiv \left(\frac{6\pi a \theta_c}{\lambda \sqrt{\ln s}} \rho_S^4\right)^{1/3}. \tag{10.58}$$

Before we analyze representative values of the two information limits introduced in this section we will first discuss three additional envelope functions which may have an effect on the information limit.

10.2.2.3 *Additional envelope functions*

There are three additional envelope functions which can suppress the higher spatial frequencies (following de Jong and Van Dyck [dJVD93]).

(i) **Sample drift**. If the sample moves with a constant velocity $\mathbf{v}_d$ in a given direction,† then the total sample displacement during the exposure time t_e is given by the vector $\mathbf{d} = \mathbf{v}_d t_e$. Only lattice planes that contain the drift direction are unaffected, which can be expressed by the dot product between the reciprocal lattice vector (the plane normal) and the drift vector. It can be shown (see Frank [Fra69]) that the resulting damping envelope is described by the function

$$E_d(\mathbf{q}) = \frac{\sin(\pi \mathbf{q} \cdot \mathbf{d})}{\pi \mathbf{q} \cdot \mathbf{d}} \approx \mathrm{e}^{-\frac{1}{6}(\pi \mathbf{q} \cdot \mathbf{d})}, \tag{10.59}$$

from which we can derive a *drift information limit* (using the exponential approximation):

$$\rho_d \equiv \frac{\pi d}{\sqrt{6 \ln s}}. \tag{10.60}$$

(ii) **Sample vibration**. If the sample vibrates with an amplitude u and a frequency which is significantly higher than the inverse of the exposure time t_e, then it can be shown [Fra69]

† Sample drift is most probably due to thermal instabilities in the microscope or microscope room. Small temperature gradients in the room or column may cause the sample holder to drift at several hundredths of a nanometer per second. Thermal equilibrium is essential to minimize the effects of sample drift.

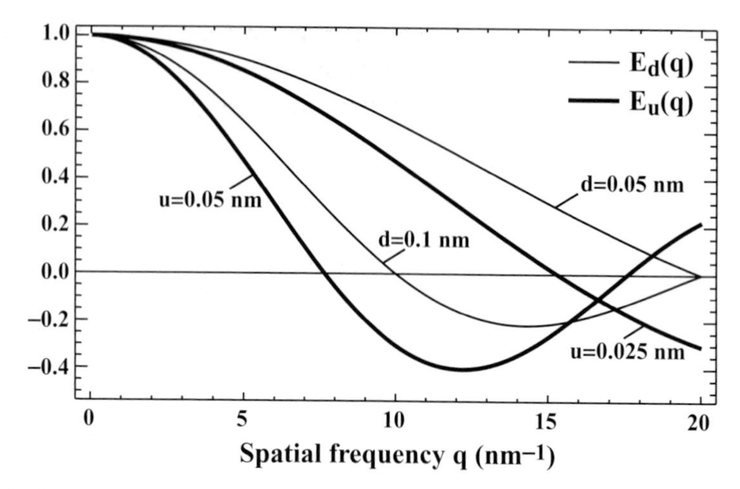

Fig. 10.16. The drift envelope (thin line) for $d = 0.1$ and 0.05 nm and the vibration envelope (thick line) for $u = 0.05$ and 0.025 nm.

that the corresponding damping envelope is given by

$$E_u(\mathbf{q}) = J_0(2\pi\mathbf{q}\cdot\mathbf{u}) \approx \mathrm{e}^{-(\pi\mathbf{q}\cdot\mathbf{u})^2}, \tag{10.61}$$

where $J_0(x)$ is a Bessel function of the first kind. The *vibration information limit* can be derived from the exponential approximation as

$$\rho_u \equiv \frac{\pi u}{\sqrt{\ln s}}. \tag{10.62}$$

The drift and vibration envelope functions have a similar functional shape, as shown in Fig. 10.16 for $d = 0.1$ and 0.05 nm and $u = 0.05$ and 0.025 nm. The spatial frequencies range from 0 to 20 nm^{-1}, which corresponds to real space distances of 0.05 nm. The meaning of these functions is obvious: if a high-resolution image is to contain information at a length scale of 0.1 nm, then the experimental conditions must be such that both the drift envelope $E_d(q)$ and the vibration envelope $E_u(q)$ are significantly larger than zero at $q = 10$ nm^{-1}. If $d = 0.1$ nm, then the drift envelope reaches zero at $q = 10$ nm^{-1}, according to Fig. 10.16, and the highest spatial frequencies will be absent from the image.

The drift and vibration envelopes are partly determined by a proper design of the microscope and the microscope room. The highest spatial frequencies (say around $q = 10$ nm^{-1} or higher) can only contribute to the image if they are not attenuated by vibrations or drift. In practice, this means that the sample must be in thermal equilibrium with its surroundings, which requires that the sample holder and the objective lens cooling circuits are properly designed. The temperature in the room must also be controlled to better than 1 °C, and there should be no air drafts caused by temperature gradients and/or air conditioning systems. Vibrations can be kept to a minimum by a proper design of the column suspension system (possibly with active vibration control) and the floor on which the microscope is mounted. Often, the entire column sits on an acoustically isolated concrete block. Finally, the microscope operator represents both a heat source and a source of acoustic vibrations, so ideally the operator should control the microscope from a remote location. While this is possible in the latest microscope models, in most situations the operator must be in the same room as the instrument. Recording images which contain high spatial frequency information therefore requires some self-discipline from the operator: sounds must be minimized, and preferably the operator should not move around at all since this will set up air currents and thermal gradients. For more details on the environmental considerations of high-resolution work we refer the reader to Chapters 9 and 10 in [Spe88]. A nice example of the importance of drift and vibration considerations in the design of a microscopy facility can be found in [Lic01].

(iii) **Detector envelope function**. The point spread function $T(\mathbf{r})$ (PSF) of the microscope is a sharply peaked (complex) function with oscillating tails which depend strongly on the objective lens defocus value. An example PSF for a 400 kV microscope with $C_s = 1$ mm at Scherzer defocus is shown in Fig. 10.17(a). The real and imaginary parts of both PSF and CTF are shown for a beam divergence angle of $\theta_c = 0.8$ mrad, and a defocus spread Δ of 6 nm. Fig. 10.17(b) shows similar curves for a 200 kV instrument with identical spherical aberration, but with $\theta_c = 0.4$ mrad and $\Delta = 3$ nm. This latter machine has a more coherent electron source which is reflected in a wider PSF. If the incoherent envelope functions extend to higher spatial frequencies, then the corresponding PSF will be wider, and this has consequences for digitally recorded images.

Consider a CCD camera with $M \times M$ pixels. The camera captures an image for a particular microscope setting, i.e. a particular PSF. The information originating from a single point in the exit plane of the sample will be spread over an area determined by the lateral extent of the PSF. The more coherent the source, the larger this area becomes, i.e. the more *delocalized* the information becomes. This means that image pixels which are close to the edge of the CCD camera will detect electrons which originated from exit plane points that are not part of the acquired image. Figure 10.18 shows schematically that there is a central area on the CCD camera that detects no electrons from outside the region of interest. The total width of the "incomplete" edge region is N pixels. The finite size of the detector causes a decrease in the information in the captured image, and hence the detector itself has an associated attenuation envelope. Since the width of the PSF is determined by the spatial incoherent envelope function $E_s(q)$, it should come as no surprise that the information limit imposed by the detector depends on the same factors as the spatial incoherence information limit. For a detailed derivation we refer the interested reader to [dJVD93]. The detector envelope function is given by

$$E_D(q) = \exp\left[-2\pi^2 \frac{\left(aC_s\lambda^3 q^4\right)^2}{N^2}\right], \qquad (10.63)$$

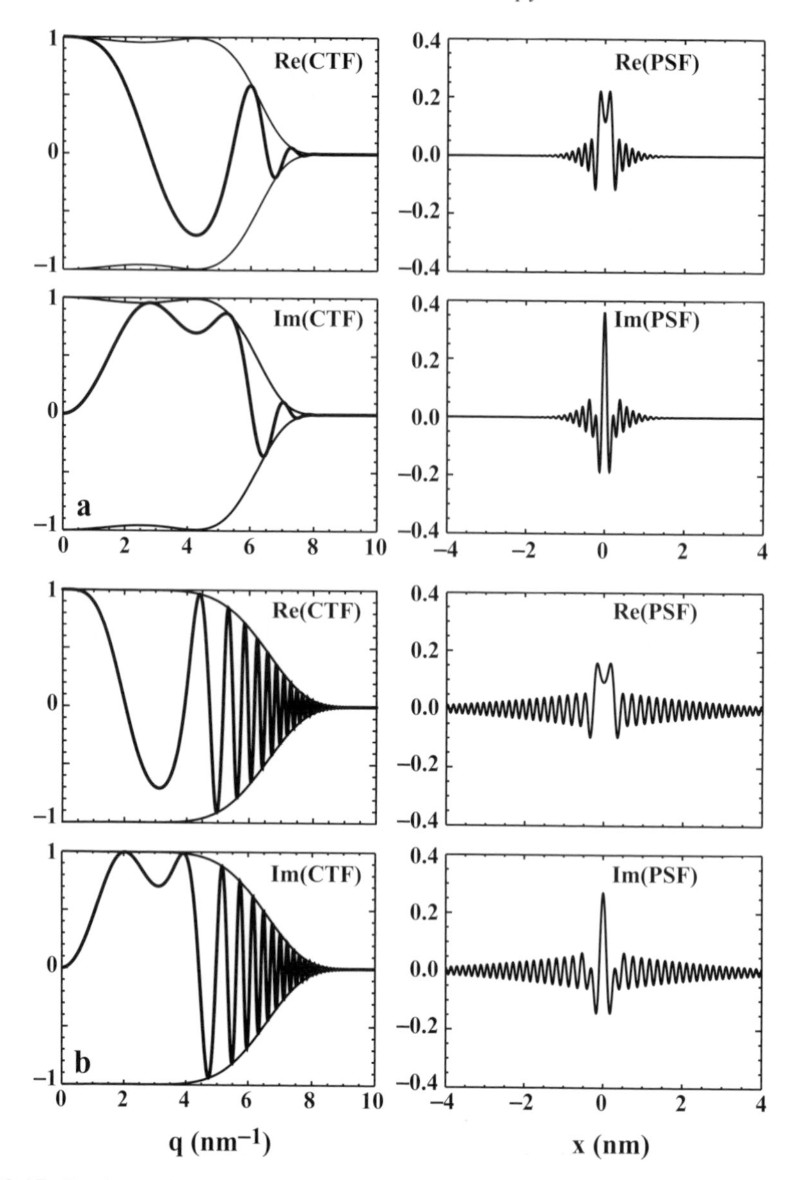

Fig. 10.17. Real and imaginary parts of the CTF and PSF functions for (a) a 400 kV instrument with $C_s = 1$ mm, $\theta_c = 0.8$ mrad, and $\Delta = 6$ nm, and (b) a 200 kV instrument with $C_s = 1$ mm, $\theta_c = 0.4$ mrad and $\Delta = 3$ nm.

where N is the number of detector pixels in the unusable border region. The information limit corresponding to the detector envelope is then given by

$$\rho_D \equiv \left(\frac{12\sqrt{2}\pi a}{N\sqrt{\ln s}} \rho_S^4 \right)^{1/4}. \tag{10.64}$$

Typical values of N are in the range of 200–400 pixels.

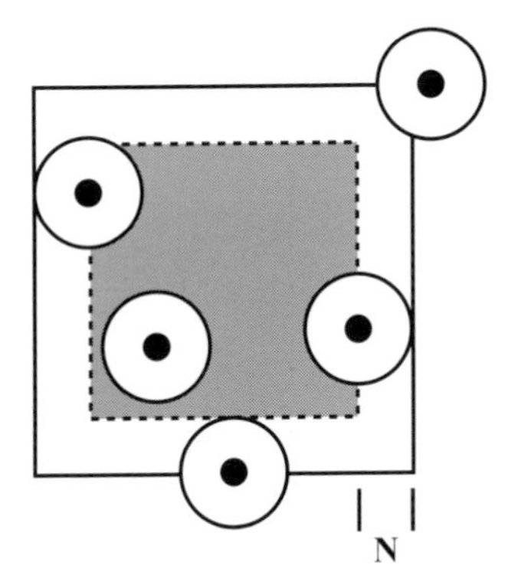

Fig. 10.18. Illustration of the effect of delocalization on the reliable area of an image acquired by a CCD camera. The circles represent the extent of the PSF; only for points inside the central gray square is the delocalized information completely captured in the image.

10.2.3 Plug in the numbers

Most of the derivations in the preceding sections were expressed in terms of dimensionless quantities, which makes the equations essentially independent of any particular microscope model. Now it is time to actually plug some realistic numbers into the equations. We will consider three 200 kV instruments: I_1 is equipped with a LaB_6 source, I_2 with a field emission source, and I_3 has a field emission source and a spherical aberration corrector.† The electron optical parameters of all three machines along with several derived and dimensionless quantities are shown in Table 10.1.

Figures 10.19 and 10.20 show the function $E_s(q)E_t(q)\sin\chi(q)$ for the frequency range $q = [0, 12]$ nm^{-1} and the defocus range $\Delta f = [-100, 400]$ nm for all three microscopes I_1, I_2, and I_3. The horizontal white lines indicate the location of the Scherzer frequency q_S and the chromatic information limit ρ_c. The vertical lines indicate the Scherzer defocus Δf_0 and the optimum defocus Δf_{opt} (for $q_{max} =$ 12 nm^{-1}). The dash-dotted line corresponds to the Gaussian defocus $\Delta f = 0$.

The following observations can be made regarding the transfer functions in Figs. 10.19 and 10.20.

- The higher coherence of a field emission source increases the frequency range transferred by the microscope (compare I_1 with I_2). The defocus spread, determined by the chromatic aberration of the objective lens and the energy spread of the source according to equation (3.61), sets the chromatic aberration limit, while the beam divergence angle determines the spatial incoherence envelope. The source energy spread and hence the

† Commercial spherical aberration correctors have only recently become available, and it is anticipated that many instruments will be equipped with a C_s-corrector in coming years.

Table 10.1. *Electron optical and operational parameters for three hypothetical $E = 200$ kV microscopes.*

Parameter	I_1	I_2	I_3
C_s [mm]	1.0	0.5	Variable
C_c [mm]	1.6	1.0	1.0
$\sigma(E)$ [eV]	1.0	0.4	0.4
$\sigma(V)/V$	10^{-6}	10^{-6}	10^{-6}
$\sigma(I)/I$	10^{-6}	10^{-6}	10^{-6}
θ_c [mrad]	0.6	0.2	0.2
$\ln s$	2.0	3.5	3.5
Derived quantities			
$\sigma(E)/E$	5×10^{-6}	2×10^{-6}	2×10^{-6}
Δ [nm]	8.8	3.0	3.0
q_0 [nm^{-1}]	0.239	0.079	0.079
ρ_S [nm]	0.227	0.191	Variable
Δf_0 [nm]	61.33	43.37	Variable
ρ_c [nm]	0.186	0.094	0.094
Δf_{opt} [nm]	136.36	266.95	266.95
Dimensionless quantities			
$\bar{\Delta}$	0.175	0.085	0.085
$\bar{q}_0$	0.085	0.023	0.023

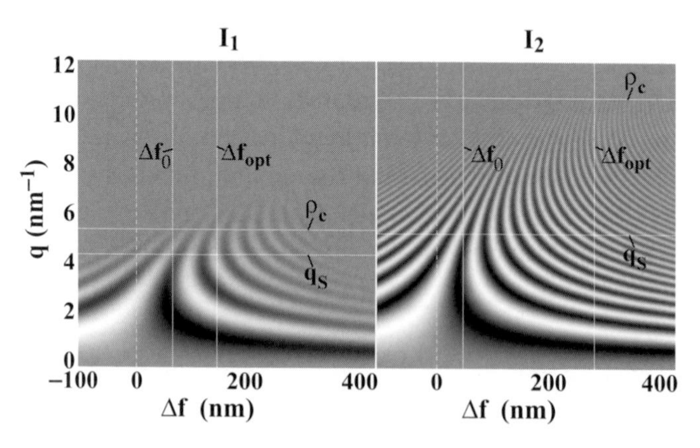

Fig. 10.19. Attenuated phase contrast transfer function $E_s(q)E_t(q)\sin\chi(q)$ for the microscopes I_1 and I_2 from Table 10.1. See the text for more details.

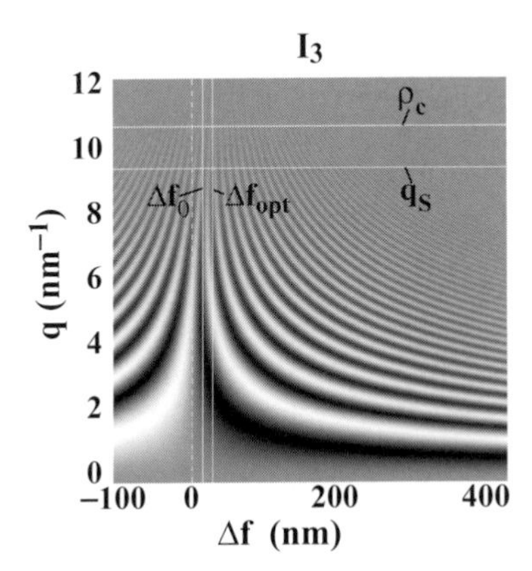

Fig. 10.20. Same as Fig. 10.19 for microscope I_3 with $C_s = 50$ μm.

defocus spread can be reduced by means of a post-source *monochromator*. There are several candidate monochromator designs available [Ros90, Tie99, MK00] and only time will tell which monochromator(s) will become commercially available. The main effect of a monochromator will be an improved chromatic information limit, in addition to a better energy resolution for electron energy loss observations. The reader is referred to the literature cited for more details.

- A decrease of the spherical aberration improves the point resolution ρ_S but the improvement is not as dramatic as that of the information limit when the beam divergence angle is changed. This is due to the fact that the point resolution depends only on the $\frac{1}{4}$-power of C_s. Alternatively, the point resolution could be improved by increasing the accelerating voltage which in turn decreases the electron wavelength. This is the main reason for the existence of intermediate-voltage and high-voltage transmission electron microscopes. The improved resolution of such machines may come at the expense of higher irradiation damage to the sample.
- A reduced C_s also "straightens out" the curved contours of constant $\sin\chi$, as can be seen by comparison of Fig. 10.19 (I_2) with Fig. 10.20 for which $C_s = 50$ μm. This corresponds to the fact that the transfer function oscillations are pushed to higher frequencies and would disappear altogether for $C_s = 0$ and $D = 0$. Spherical aberration correctors can be used to select a given C_s value (either positive or negative) so that the point resolution can be made equal to the information limit. We refer the interested reader to the following papers for more details on C_s correctors: [HU97, KDL99].

The moduli of the point spread functions corresponding to the microscopes described in Table 10.1 are shown in Fig. 10.21. The delocalization effect associated with the field emission source (I_2) is clearly visible. C_s-correction removes part of the delocalization and sharpens the PSF. The two curves on the right are for the third set of microscope parameters, with $C_s = 50$ μm and $C_s = 5$ μm, respectively. It is clear that the PSF becomes more sharply peaked and approaches the ideal δ-function shape.

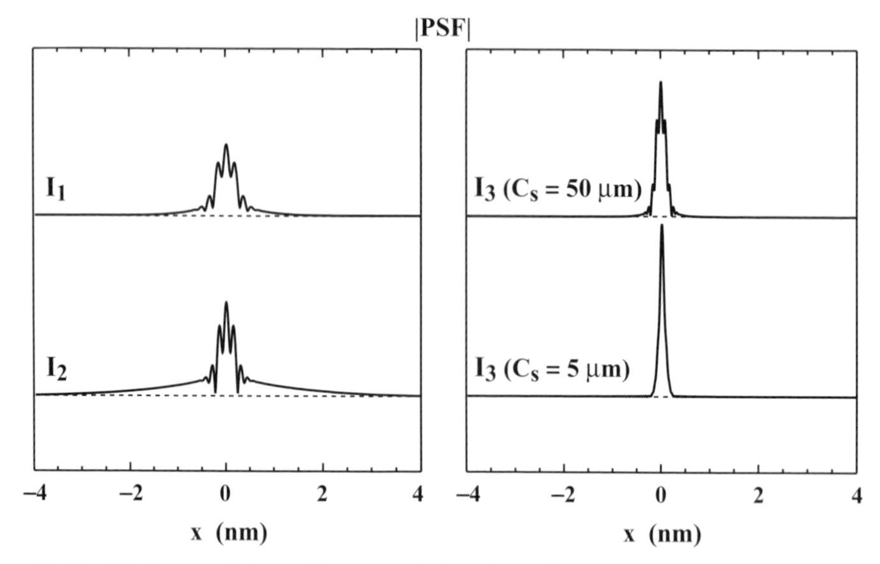

Fig. 10.21. Modulus of the point spread functions for the microscopes described in Table 10.1 (arbitrary vertical scale, Scherzer defocus); the microscope I_3 PSF is shown for two different values of the spherical aberration constant.

Example 10.5 *Consider microscope I_2 of Table 10.1. What is the maximum acceptable vibration amplitude u for this instrument to be used out to its chromatic information limit?*

Answer: *The chromatic information limit equals $\rho_c = 0.094\,nm$. If sample vibration is to be small enough so that the vibration information limit ρ_u is smaller than ρ_c, then we have*

$$u < \frac{\rho_c\sqrt{\ln s}}{\pi} \approx 0.6\rho_c = 0.056\,nm.$$

The vibration amplitude must therefore be kept below about $0.5\,\text{Å}$ to keep the vibration information limit smaller than ρ_c!

The contrast transfer function can be displayed using the ctf1.pro routine, available on the website. The routine displays the real and imaginary parts of the attenuated CTF, along with a 2D representation similar to Fig. 10.12, and the modulus of the point spread function. A second routine, ctf2.pro, can be used to display grayscale images of $E(q)\sin\chi(q)$, where $E(q)$ is the complete damping envelope. An example of the output of ctf2.pro is shown in Fig. 10.22: the effect of two-fold, three-fold, and four-fold astigmatism and axial coma is illustrated for a 200 kV instrument operated at Scherzer defocus ($C_s = 1$ mm, $C_c = 1$ mm, $\Delta E = 1$ eV,

Table 10.2. *Total number of lattice planes with d-spacings larger than a given point resolution ρ_S for a few technologically important cubic materials.*

Material	a [nm]	S.G.	0.20 nm	0.15 nm	0.10 nm
Fe	0.2866	Im$\bar{3}$m	1	1	4
GaAs	0.5653	F$\bar{4}$3m	2	5	10
Ni_3Al	0.3572	Pm$\bar{3}$m	3	5	12
CuZn	0.2959	Pm$\bar{3}$m	2	3	7
$BaTiO_3$	0.4012	Pm$\bar{3}$m	4	6	15

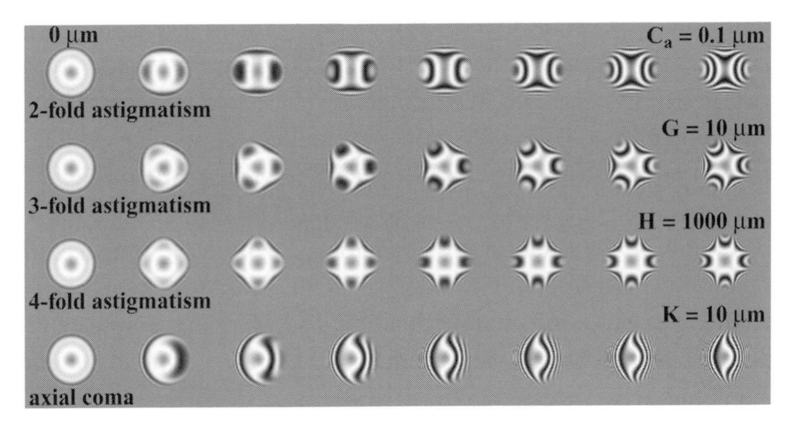

Fig. 10.22. Illustration of the effect of two-fold, three-fold, four-fold astigmatism, and axial coma on the imaginary part of the attenuated transfer function for a 200 kV LaB_6 instrument operated at Scherzer defocus (other parameters in text).

$\theta_c = 0.5$ mrad). The maximum values of the aberrations are $C_a = 0.1$ μm, $G =$ 10 μm, $H = 1$ mm, and $K = 10$ μm.

The reader may ask the question: why do we need a microscope with an information limit of better than 0.1 nm? The answer to this question is rather simple: consider the technologically important cubic materials listed in Table 10.2. The table lists for each material how many families of lattice planes have a d-spacing larger than a given point resolution. For pure Fe, a rather important material, only one single d-spacing can be resolved on a microscope with $\rho_S = 0.2$ nm ($d_{110} = 0.203$ nm and $d_{200} = 0.143$ nm). This means that the only zone axis orientation for which a high-resolution image can be obtained on a machine with an information limit of $\rho_c = 0.15$ nm is the [111] zone axis, which contains three independent reflections of the {110} family. Similarly, for GaAs there are only three families of planes for which the d-spacing is larger than 0.2 nm: $\{11\bar{1}\}$, {111}, and {200}, which corresponds to only two different d-spacings. Microscopes with a point resolution close to or better than 0.1 nm allow us to obtain high-resolution images of many different zone axis orientations, which is useful for the study of the structure of defects and interfaces. One must always bear in mind that the maximum number of degrees of freedom that can be transferred by the microscope is of the order of three per area ρ_c^2; this means that it may not be possible to determine the atomic structure of a defect if the density of projected atom positions is too high.

10.2.4 Image formation for an amorphous thin foil

To illustrate the effect of the transfer function on the electron wave function exiting a given thin foil, it is useful to start from a wave function which has all spatial frequencies present with about the same weight. Such a wave function would result from scattering from a random 3D arrangement of identical atoms; in a 2D projection (*any* projection) all possible projected interatomic spacings will be present, and therefore the frequency spectrum of such a projected structure will be rather flat. To get as many electrons as possible to scatter at large angles, we must make an amorphous foil with the heaviest possible atom, ideally U. In practice, it is easier to make amorphous foils out of Ge, or, occasionally, W. Amorphous carbon is very easy to obtain by evaporation methods, but it has a low atomic number and does not scatter strongly enough at the spatial frequencies corresponding to the information limit of modern microscopes.

Every microscopy laboratory will probably have a few amorphous calibration foils available for measurement of various microscope parameters. The basic numerical tool used for the analysis of images of amorphous objects is the fast Fourier transform. The FFT of a high-resolution image of an amorphous thin foil will allow us to determine many microscope parameters, as described in more detail in Section 10.2.5. The modulus squared of the FFT of a high-resolution image is known as a *diffractogram*. In the early days of high-resolution observations, diffractograms were created by optical means, using a laser and an optical bench [GMWW86]. Nowadays, digital image recording makes it possible to observe diffractograms in real time.

To create a physically realistic structure model for the amorphous film, we would need to use, in principle, a real radial distribution function. Fan and Cowley [FC87] have shown that an amorphous object function can be constructed using a simple uniform random number generator: for each image pixel assign a unit amplitude and a random phase in the interval $[-\pi, +\pi]$, i.e. create a weak phase object with randomly varying phase,

$$\psi(\mathbf{r}) = \mathrm{e}^{\mathrm{i}\phi(\mathbf{r})}.$$

The Fourier transform of this weak phase object is then multiplied by the microscope transfer function, and after an inverse Fourier transform the image intensity is given by

$$I(\mathbf{r}) = \left|\mathcal{F}^{-1}\left[\mathcal{F}[\mathrm{e}^{\mathrm{i}\phi(\mathbf{r})}]E(q)\mathrm{e}^{-\mathrm{i}\chi(q)}\right]\right|^2, \tag{10.65}$$

where $E(q)$ represents the complete damping envelope. The modulus-squared of the Fourier transform of this image intensity is then known as the *diffractogram*† I_d:

$$I_d(\mathbf{q}) \equiv |\mathcal{F}[I(\mathbf{r})]|^2. \tag{10.66}$$

To obtain a more realistic version of the diffractogram, the Fourier transform $\mathcal{F}[\psi]$ could be multiplied by the electron scattering factor $f^e(q)$ for the particular element that makes up the amorphous film.

Figure 10.23 shows several computed diffractograms for an amorphous thin foil, created using the random phase algorithm. Images for two different microscopes are represented: the first microscope is equivalent to I_1 of Table 10.1 (with $\theta_c = 1$ mrad), and the second is identical to I_2. Row (a) of Fig. 10.23 shows four diffractograms and corresponding images for the first four passbands Δf_n for machine I_1. The passband is visible for the first two defocus values but rapidly disappears below the noise level for higher-order passbands. This attenuation is mostly due to the large beam divergence angle θ_c. For machine I_2 (row c), the higher-order passbands are still clearly visible in the diffractogram. The half or quarter circles indicate 6, 8 and 10 nm^{-1}, respectively.

Rows (b) and (d) in Fig. 10.23 reveal the effect of two-fold astigmatism on the diffractogram. The astigmatism values are indicated as magnitude–angle pairs (magnitude in nm) in the upper right-hand corner of each diffractogram. The astigmatism clearly affects the shape of the diffractogram rings, whereas it is not always clear in the corresponding images. For the images with higher beam coherence (row d), astigmatism becomes harder to recognize visually.

The bottom row of Fig. 10.23 shows the effect of limiting the bandwidth of the transmitted signal. The diffractograms are computed for machine I_1 with decreasing aperture radii indicated in nm^{-1} in the upper right-hand corner of each diffractogram. The effect of frequency truncation on the image contrast and detail is dramatic.

Experimental diffractograms of amorphous thin foils can be used to align the microscope and to determine various electron optical parameters. In the next section, we will discuss in detail how microscope alignment can be performed and also how electron optical parameters can be measured.

† A diffractogram is also known as a *Thon diagram*, after F. Thon who first employed them in the area of electron optics [Tho66].

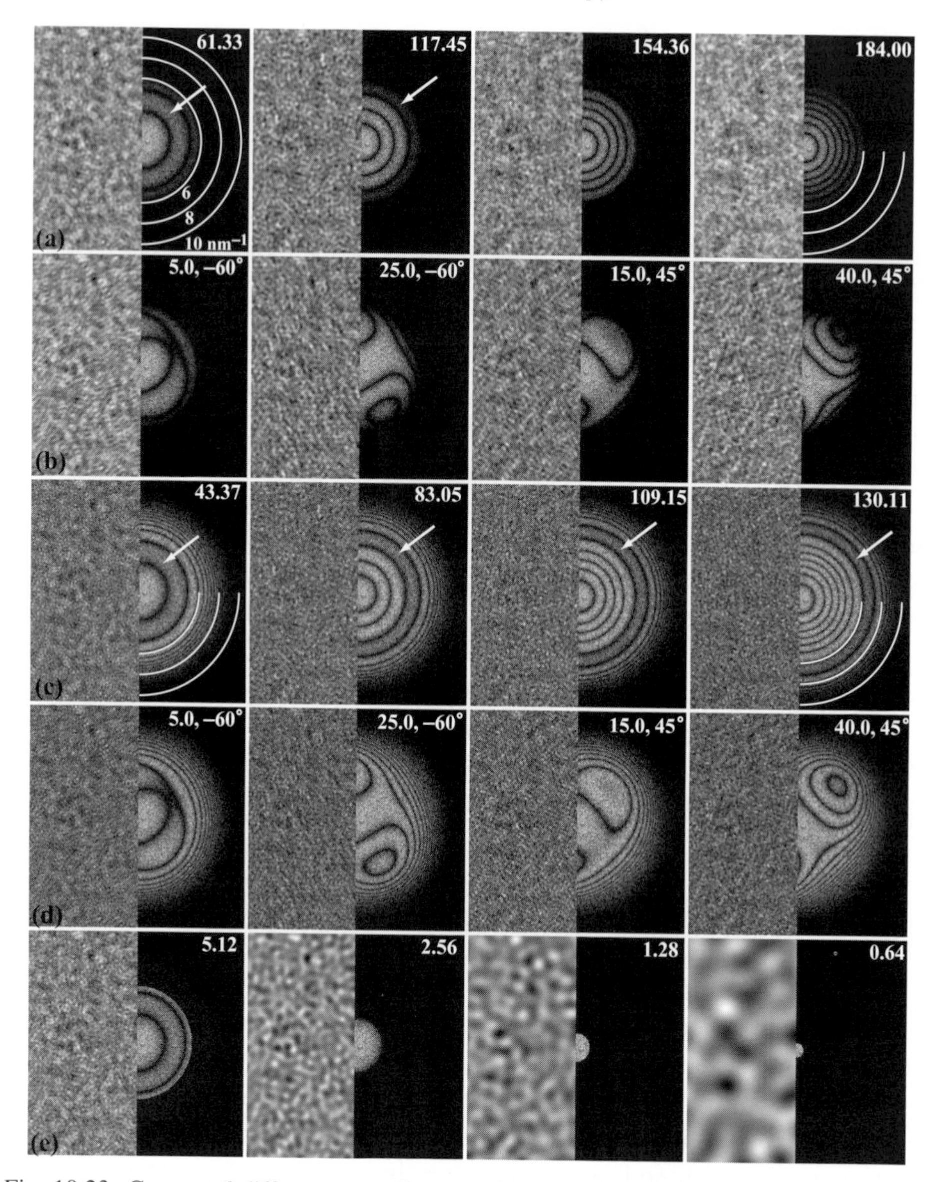

Fig. 10.23. Computed diffractograms for two different 200 kV microscopes: (a), (b) and (e) conventional LaB_6 instrument, (c) and (d) FEG instrument. (a) and (c) represent the first four passband images for each machine (defocus indicated in upper right-hand corner). The passbands are indicated by white arrows. (b) and (d) represent images with varying amounts of astigmatism, indicated as a magnitude-angle pair in the upper right corner of each diffractogram. The bottom row shows the effect of limiting the bandwidth of the image with an aperture. The aperture diameter is indicated in nm^{-1}.

10.2.5 Alignment and measurement of various imaging parameters

10.2.5.1 The importance of beam tilt

In the previous sections, we have implicitly assumed that the microscope column is perfectly aligned, in other words, that the electron beam is incident along a direction for which aberrations such as coma are minimized or, ideally, absent. In reality, this is not necessarily the case. When you begin observations on a sample, the microscope alignment will nearly always be less than perfect and it is important to understand how an imperfect alignment affects the aberration function $\chi(q)$. Starting from equation (10.33) we will first rewrite the aberration function in Cartesian coordinates, and then introduce a beam tilt. We will ignore four-fold astigmatism, described by the term in H in equation (10.33), but include all other terms (defocus, two-fold and three-fold astigmatism, coma, and spherical aberration).

The derivation below is based on [TOCL96] with a slightly different notation and a different sign convention. Starting from equation (10.33), we introduce Cartesian coordinates $(u, v) \equiv q(\cos\phi, \sin\phi)$ and $q^2 = u^2 + v^2$. Making use of standard trigonometric identities and defining $C_{a,x} \equiv C_a \cos 2\phi_a$, $C_{a,y} \equiv C_a \sin 2\phi_a$, $G_x \equiv G \cos 3\phi_G$, $G_y \equiv G \sin 3\phi_G$, $K_x \equiv K \cos\phi_K$, and $K_y \equiv K \sin\phi_K$, it is easy to show that

$$\begin{aligned}\frac{\chi(q)}{2\pi} = &-\frac{1}{2}\lambda(u^2+v^2)\Delta f - \frac{1}{2}\lambda[C_{a,x}(u^2-v^2) + 2C_{a,y}uv]\\ &-\frac{1}{3}\lambda^2[G_x(u^3-3uv^2) + G_y(3u^2v - v^3)] - \lambda^2(u^2+v^2)[uK_x + vK_y]\\ &+\frac{1}{4}\lambda^3(u^2+v^2)^2 C_s. \end{aligned} \qquad (10.67)$$

Next, we introduce a beam tilt. Since a beam tilt in direct space is equivalent to a translation in the diffraction pattern, a beam tilt is simply described by a translation vector $\mathbf{t} = (t_x, t_y)$; the aberration function for tilted illumination is derived from equation (10.67) by the following substitution: $u \rightarrow u + t_x$; $v \rightarrow v + t_y$. The mathematical operations are elementary but somewhat lengthy and we obtain the following expression:

$$\begin{aligned}\frac{\chi^t(q)}{2\pi} = &-\frac{1}{2}\lambda(u^2+v^2)\overline{\Delta f} - \frac{1}{2}\lambda\left[\overline{C_{a,x}}(u^2-v^2) + 2\overline{C_{a,y}}uv\right]\\ &-\frac{1}{3}\lambda^2\left[G_x(u^3-3uv^2) + G_y(3u^2v - v^3)\right] - \lambda^2(u^2+v^2)\\ &\times\left[u\overline{K_x} + v\overline{K_y}\right] + \frac{1}{4}\lambda^3(u^2+v^2)^2 C_s + uD_x + vD_y, \end{aligned} \qquad (10.68)$$

with

$$\begin{aligned}
\overline{\Delta f} &\equiv \Delta f + 4\lambda(K_x t_x + K_y t_y) - 2\lambda^2 C_s\left(t_x^2 + t_y^2\right);\\
\overline{C_{a,x}} &\equiv C_{a,x} + 2\lambda\left[t_x(G_x + K_x) + t_y(G_y - K_y)\right] - \lambda^2 C_s\left(t_x^2 - t_y^2\right);\\
\overline{C_{a,y}} &\equiv C_{a,y} + 2\lambda\left[t_x(G_y + K_y) + t_y(K_x - G_x)\right] - 2\lambda^2 C_s t_x t_y;\\
\overline{K_x} &\equiv K_x - \lambda C_s t_x;\\
\overline{K_y} &\equiv K_y - \lambda C_s t_y;\\
uD_x + vD_y &\equiv -\lambda\Delta f[ut_x + vt_y] - \lambda[C_{a,x}(ut_x - vt_y) + C_{a,y}(ut_y + vt_x)]\\
&\quad - \lambda^2\left[G_x\left(u\left(t_x^2 - t_y^2\right) - 2vt_xt_y\right) + G_y\left(v\left(t_x^2 - t_y^2\right) + 2ut_xt_y\right)\right]\\
&\quad - \lambda^2\left[K_x\left(u\left(3t_x^2 + t_y^2\right) + 2vt_xt_y\right) + K_y\left(v\left(3t_y^2 + t_x^2\right) + 2ut_xt_y\right)\right]\\
&\quad + \lambda^3 C_s(ut_x + vt_y)\left(t_x^2 + t_y^2\right).
\end{aligned}$$

The final two terms in equation (10.68) represent the image displacement that is induced by the beam tilt. If the microscope is perfectly aligned (i.e. no coma, no two-fold astigmatism, and corrected three-fold astigmatism) then the image displacement depends only on the spherical aberration and the defocus, as we have already seen in Section 4.4.2.4. The relation between image displacement, defocus and C_s can be used to determine the value of C_s with very high precision, as described in [KdJ91]; the method relies on the fact that the displacement is described by a third-order polynomial in the beam tilt.

While the relations derived above are somewhat complicated, we can learn a lot from them.

- Coma can be removed by the appropriate beam tilt. If we tilt the beam so that $t_x = K_x/\lambda C_s$ and $t_y = K_y/\lambda C_s$, then the effective coma coefficients $\overline{K_x}$ and $\overline{K_y}$ vanish and the beam is aligned along the so-called *coma-free axis* (see Section 10.2.5.2).
- The effective defocus of the image is not only determined by the objective lens current. A beam tilt changes the defocus by an amount proportional to the spherical aberration constant (in the absence of coma). For instance, for a beam tilt $\theta_t = 15$ mrad and a spherical aberration of $C_s = 0.65$ mm, the induced defocus change is equal to $-2C_s\theta_t^2 = -292$ nm. In a properly aligned, C_s-corrected instrument, there would be no difference between the effective defocus $\overline{\Delta f}$ and Δf, even for a large beam tilt (assuming that higher-order aberrations can be ignored).
- It is always possible to use the objective stigmator coils to correct the effective two-fold astigmatism, *even in the presence of coma and beam tilt*. Therefore, it is not sufficient to correct astigmatism of the image and then assume that the microscope is well aligned.
- The three-fold astigmatism coefficients are not affected by beam tilt (assuming that higher-order aberrations can be ignored). This means that it is possible to correct three-fold astigmatism once and for all using special correction coils.

- If coma and three-fold astigmatism are properly corrected, then the effective two-fold astigmatism depends only on beam tilt and spherical aberration. If the astigmatism at zero beam tilt is properly corrected, then the beam-tilt-induced astigmatism is given by

$$\left.\begin{aligned}\overline{C_{a,x}} &= -\lambda^2 C_s t^2 \cos 2\phi_t;\\ \overline{C_{a,y}} &= -\lambda^2 C_s t^2 \sin 2\phi_t,\end{aligned}\right\} \quad (10.69)$$

  where the beam tilt is written as $(t_x, t_y) = t(\cos\phi_t, \sin\phi_t)$. This means that one of the major axes of the astigmatism diffractogram will always lie in the plane formed by the tilted beam direction and the coma-free axis.
- The aberrations in equation (10.68) fall into two categories: those that are even in the coordinates u and v (defocus, two-fold astigmatism, spherical aberration), and those that are odd (three-fold astigmatism and coma). The even aberrations can be visualized using diffractograms from an amorphous region in the foil; odd aberrations do not affect the diffractogram, and one must resort to the acquisition of a so-called *Zemlin tableau*, a diffractogram series for a given beam tilt and varying beam azimuth (see Section 10.2.5.2).

It is obvious that a correct alignment of the microscope requires a detailed understanding of what it is we are trying to correct. The ideal aberration function $\chi(q)$ contains only contributions from the defocus and spherical aberration terms. All other terms are *residual aberrations* and their contributions must be kept as small as possible through a proper alignment.

10.2.5.2 Coma-free alignment

When we introduced the Seidel aberrations in Chapter 2, we have seen that *coma* is an aberration that "smears out" the image of a point object. This aberration can be corrected by aligning the incident beam onto the so-called *coma-free axis*. Since HREM images are often difficult to interpret, even when we know what the crystal structure looks like, it is nearly impossible to visually recognize the presence of coma in an experimental image. A simple procedure for the correction of coma was suggested by Zemlin and coworkers [ZWS+78] and involves a series of HREM images of an amorphous object, recorded for a given beam tilt. The procedure for determining whether or not coma is present works as follows.

- Place an amorphous region in eucentric position, correct the astigmatism, and record an image at an under-focus condition ($\Delta f \approx 100$ nm or so). The corresponding diffractogram should show multiple concentric rings, as shown in the central diffractogram in Fig. 10.24. The fact that the rings are concentric indicates that astigmatism was properly corrected for this particular incident beam direction. It does not indicate that the microscope was properly aligned.
- Switch to dark field mode and tilt the beam by an angle of 10–15 mrad. This requires that the tilt controls have been calibrated. Record an HREM micrograph. Tilt the beam by the same amount to the opposite direction and record another HREM micrograph.

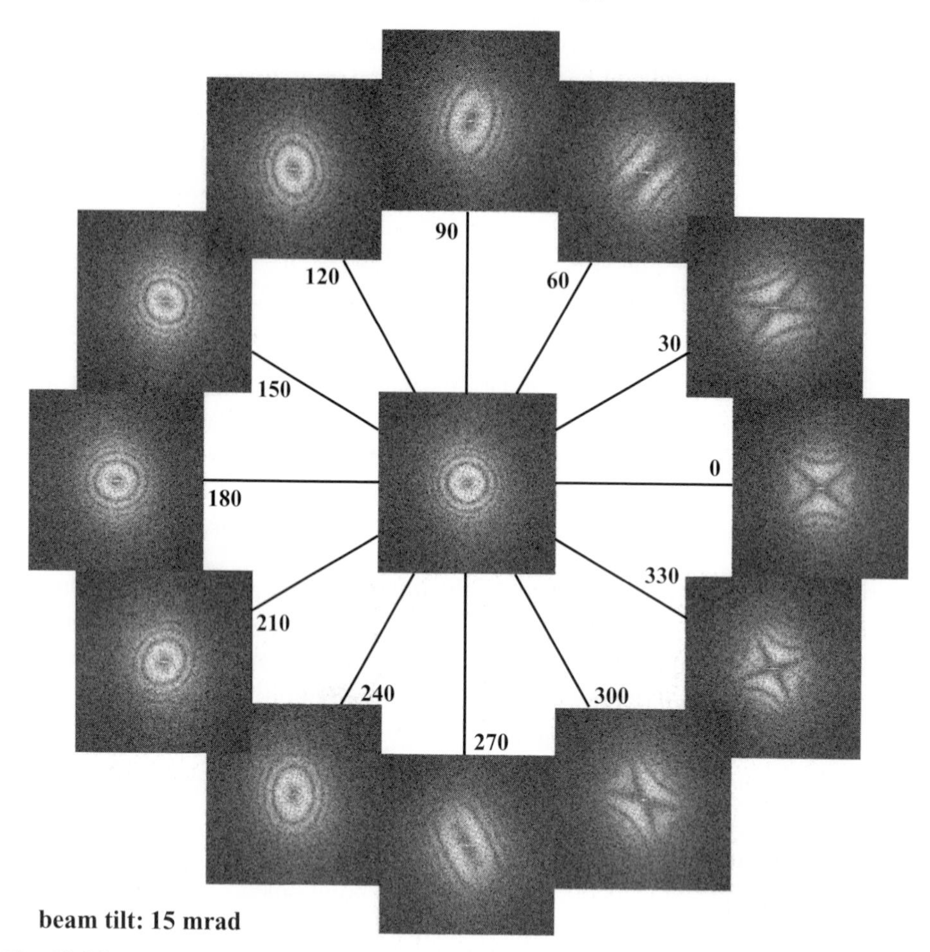

Fig. 10.24. Zemlin tableau recorded using a 3 nm thick amorphous carbon film on a Philips CM30 UltraTwin field emission microscope operated at 300 kV. The central diffractogram shows concentric rings; the six tilt pairs were acquired in dark field mode for a beam tilt of 15 mrad. The asymmetry of the diagonal running from top left to bottom right indicates that the central beam orientation does not coincide with the coma-free axis.

The diffractograms will appear to be astigmatic, but should have the same defocus value. In other words, the pattern of distorted rings must be the same for opposite tilt angles.

- Repeat this procedure for several additional pairs of beam tilts, with different azimuthal angles for the tilt. If you record half a dozen or so different tilt directions, you can then assemble all the diffractograms into a so-called *Zemlin tableau*, such as the one shown in Fig. 10.24.

A coma-free alignment essentially requires that the pairs of diffractograms across the Zemlin tableau be identical with respect to the defocus. If they are not identical, as is the case for the tableau shown in Fig. 10.24, then the central beam orientation with respect to which the tableau was recorded does not coincide with the coma-free axis. The coma-free alignment then involves changing the beam tilt of the central image while maintaining zero astigmatism until the entire tableau is symmetric with respect to the defocus. Many modern microscopes have special alignment controls for this type of alignment, and we refer to the standard operating procedures for more details. The resulting Zemlin tableau should look similar to that shown in Fig. 10.25: the ring pattern appears to be astigmatic, and the main axis of the elliptical distortion points towards the central diffractogram. If three-fold astigmatism is present then the elliptical patterns will all look alike, but will be oriented at different angles with respect to the central diffractogram. For the particular microscope used to obtain Figs 10.24 and 10.25, three-fold astigmatism

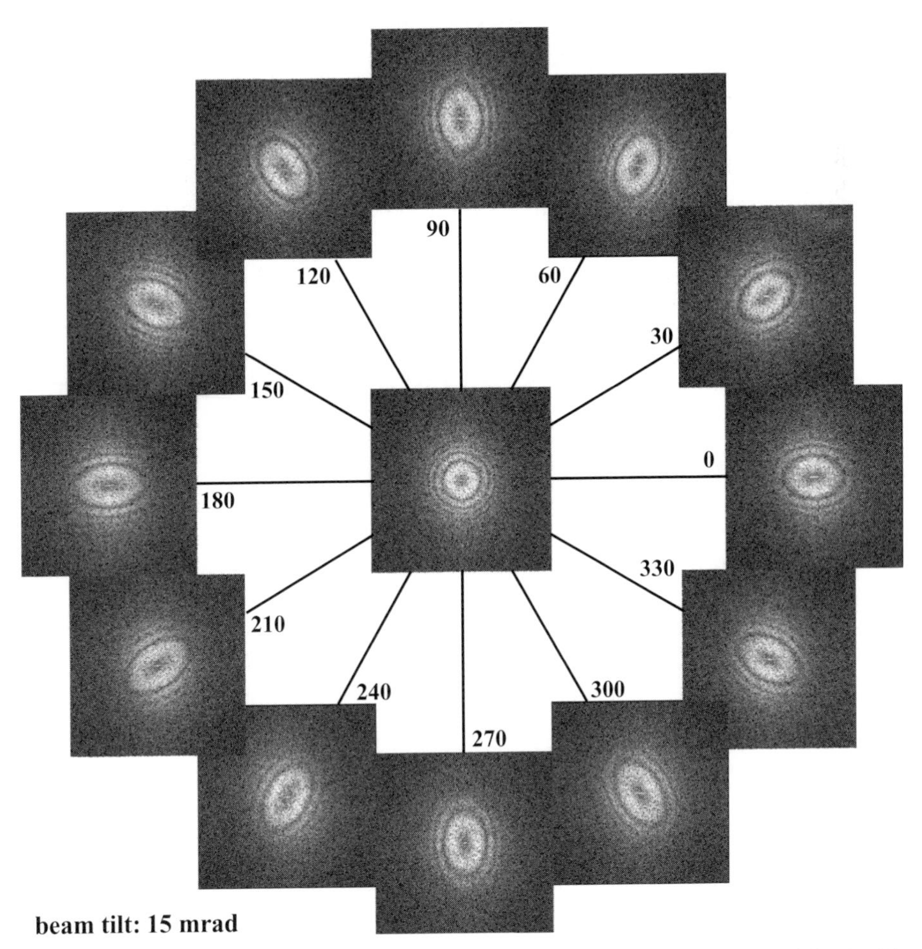

Fig. 10.25. Same as Fig. 10.24, but with the correct coma-free orientation of the incident beam.

was corrected by the insertion of a properly excited correction coil, and all ellipses point towards the center diffractogram. The apparent astigmatism present in the tilted diffractograms is due to spherical aberration; in a C_s-corrected microscope the Zemlin tableau would show circular patterns for all tilt angles.

The manual acquisition of a Zemlin tableau is somewhat tedious. If the microscope can be controlled by a computer, then a script can be used to automatically acquire the diffractograms, determine the location of the coma-free axis, correct the beam tilt and acquire a final tableau. Additional procedures for various alignments can be found in [SSE82] and [Spe88].

10.2.5.3 Measurement of electron optical parameters

When the residual aberrations have been corrected, the aberration function $\chi(Q)$ only contains the defocus (Δf) and spherical aberration (C_s) contributions. Assuming that drift and vibration can be kept reasonably small, the only other parameters needed for a full characterization of the image formation process are the beam convergence angle θ_c and the defocus spread Δ. The spherical aberration and defocus can be determined using ring diffractograms from an amorphous area. If the magnifications and camera lengths of the microscope have been accurately calibrated, then the location of the zeros of the ring pattern can be measured. The zeros occur whenever the total phase shift of equation (10.9) is equal to a multiple of π, or

$$n = -\lambda \Delta f q^2 + \frac{1}{2} C_s \lambda^3 q^4; \tag{10.70}$$

if we introduce the new variable $p \equiv \lambda q^2$, then we have the following relation:

$$\frac{n}{p} = -\Delta f + \frac{1}{2} C_s \lambda p. \tag{10.71}$$

The left-hand side is represented graphically by a hyperbola, the right-hand side is a straight line. Figure 10.26 shows the hyperbolic curves for values of $n = -1, \ldots, -20$. Superimposed are nine straight lines for defocus values $\Delta f_i = 100 \times i$ and $C_s = 1$ mm. The electron wavelength is $\lambda = 2.508$ pm (i.e. 200 kV). The straight lines have slope $\frac{1}{2} C_s \lambda$ and intercepts $-\Delta f$. At each intersection of a straight line with a hyperbola the corresponding value of p indicates the location $q = \sqrt{p/\lambda}$ of the zero of the diffractogram.

Krivanek [Kri76] proposed in 1976 to use this relation in a least-squares fit to determine C_s and Δf, given a series of zeros for a single or multiple defocus values. Coene and Denteneer [CD91] improved the method and used several diffractograms and a modified least-squares procedure to obtain a better fit for the

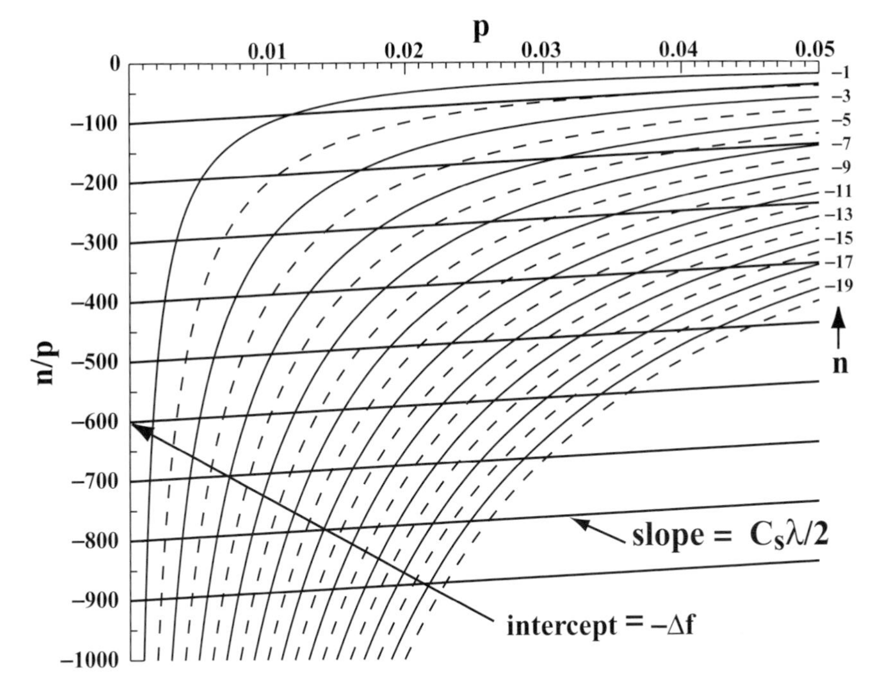

Fig. 10.26. Graphical representation of Krivanek's method to determine the spherical aberration coefficient and defocus from a series of ring diffractograms.

spherical aberration coefficient. The improved method uses the parameter $r = q^2$ and the relation

$$n = -\lambda \Delta f r + \frac{1}{2} C_s \lambda^3 r^2 \tag{10.72}$$

to determine the intersections of a series of straight lines (left-hand side) and a parabola (right-hand side). The explicit least-squares relations can be found in [CD91]. The method can result in a standard deviation of less than 1% for the spherical aberration coefficient.

To improve the accuracy of these fitting methods the diffractograms must be obtained for the coma-free incident beam orientation. The accuracy of the zero locations is improved if the diffractograms are rotationally averaged to increase the signal-to-noise ratio. It is not difficult to write a program to carry out the least-squares fit; we refer the reader to the "Numerical Recipes" book for details on least-squares algorithms [PFTV89].

Buddinger and Glaeser [BG76] proposed a method based on image displacements between bright field and dark field images to measure C_s; an accuracy of 3% was reported. The image displacement method was subsequently improved by Koster and de Jong [KdJ91] to a precision of better than 1%.

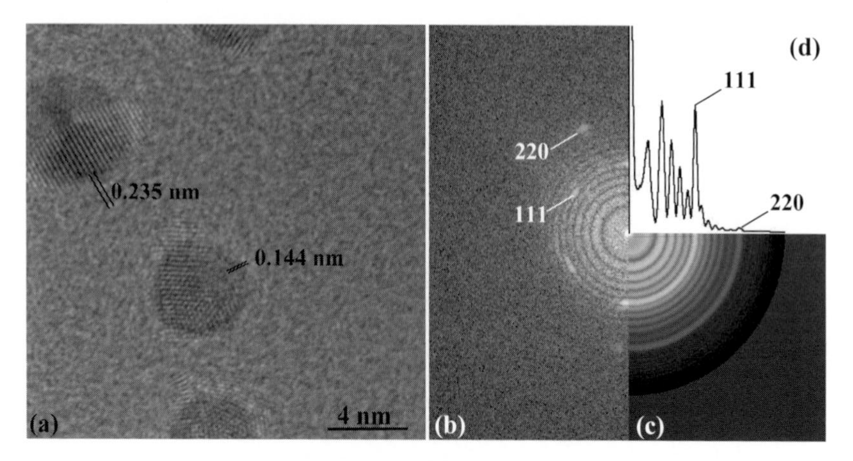

Fig. 10.27. (a) High-resolution micrograph of Au islands on an amorphous Ge film (Philips CM30 Ultra-Twin, operated at 300 kV); (b) diffractogram of (a); (c) rotationally averaged diffractogram and (d) diffractogram profile, with Au 111 and 220 spacings indicated.

As an example, we determined the spherical aberration constant of a Philips CM30 FEG microscope with an Ultra-Twin objective lens, operated at 300 kV. The manufacturer-provided value for C_s is 0.65 mm for the eucentric sample position. We used an amorphous Ge sample with Au islands to record a through-focus series of 50 images with a defocus step size of around 4 nm. A single image of the series is shown in Fig. 10.27(a). Several Au particles are visible, and two lattice spacings are indicated. The diffractogram is shown in Fig. 10.27(b). The rotational average is shown in both (c) and (d); from measurements of the 111 and 220 Au peaks on several diffractograms, we can deduce that the diffractogram has a scale of 0.0491 ± 0.0007 nm^{-1} per pixel.

Using the weighted least-squares method proposed by Coene and Denteneer [CD91], we used 189 zero crossings in 50 diffractograms to refine the initial estimates of $C_s = 0.65$ mm, the first defocus value of -10 nm, and the defocus step size of 4 nm. The resulting values are: $C_s = 0.663 \pm 0.038$ mm, initial defocus -5.86 nm, and defocus step size 3.47 nm. The error bars on C_s are entirely due to the standard deviation on the diffractogram scale. A better magnification calibration would result in a narrower range for C_s. The fitted zero-crossing curves are shown in Fig. 10.28, along with the experimental data points for the zeros of the diffractograms. The inset in the upper left-hand corner is a combination of all 50 diffractogram profiles. The Scherzer defocus (44.2 nm), and Scherzer spatial frequency ($q_S = 5.86$ nm^{-1}) are also indicated.

Measurement of the defocus spread and the beam divergence angle are a bit more complicated. The chromatic aberration coefficient C_c for the CM30 microscope is 1.14 mm. For an energy spread of 0.9 eV, and voltage and current stabilities of 10^{-6},

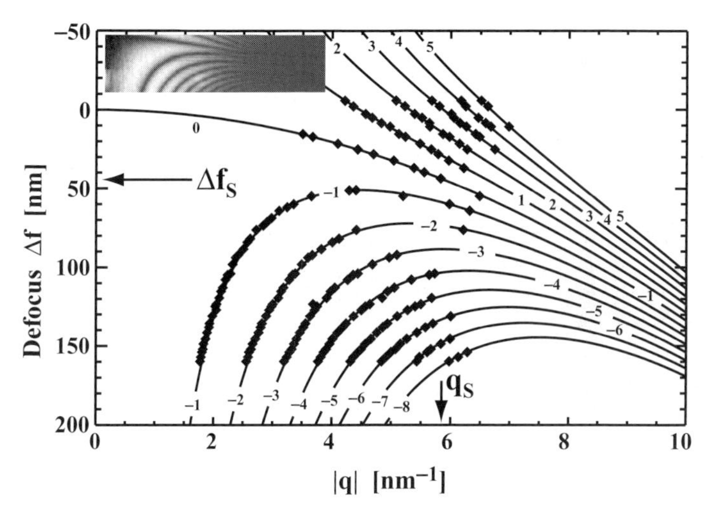

Fig. 10.28. Weighted least-squares fit of the zero-crossings of 50 experimental diffractograms (Philips CM30, 300 kV). The diamonds indicate the experimental measurements, the solid lines are the results of the fitting procedure.

the defocus spread (equation 3.61) amounts to 4.26 nm. The chromatic information limit ρ_c (equation 10.56) is then approximately equal to 0.1 nm (at 300 kV). A direct measurement of ρ_c is possible in principle, if we use an amorphous film which has a sufficiently large electron scattering factor out to $q = 10$ nm^{-1}. An amorphous germanium film may do the trick, but it is better to use a heavier element, such as tungsten.

The information limit may be estimated using the Young's fringes technique, first introduced by J. Frank and coworkers in 1970 [FBLH70, Fra75, Fra76]. The method uses two electron micrographs of the same region of the sample, with one translated intentionally by a small amount with respect to the first one. The Fourier transform of the sum of the two images then contains a sinusoidal modulation, which can be used to estimate which part of the diffractogram contains meaningful information. An example is shown in Fig. 10.29 for a Ge amorphous film with Au particles. The images were obtained at 300 kV in a Philips CM30 microscope. The region of the diffractogram where the fringes are clearly visible is indicated by an ellipse in Figs 10.29(b) and (c). In the absence of sample drift and vibrations, the ellipse should become a circle, so that the information limit is isotropic. The information limits corresponding to the major and minor axes of the ellipse are $1/8.4 = 0.119$ nm and $1/6.4 = 0.156$ nm, respectively; both limits are larger than the chromatic limit derived above.

The beam divergence angle can in principle be determined by analyzing the defocus dependence of the extent of the Young's fringes. This analysis is based on

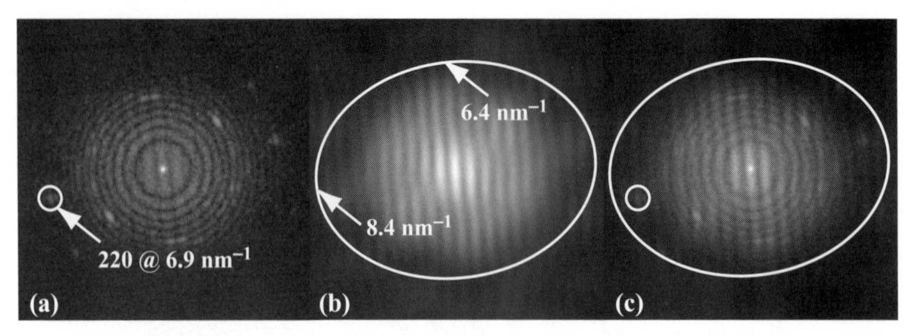

Fig. 10.29. (a) Diffractogram of the sum of two micrographs with a small relative translation, showing Young fringes; (b) Fourier filtered image using only the Young fringe periodicity; (c) same as (a), but with enhanced fringe contrast. The ellipse indicates the point at which the Young fringes are no longer detectable.

equation (10.58) on page 617 and requires a series of image pairs for increasing defocus. An alternative method for the determination of the effective source size, which is related to the damping envelope, can be found in [HSI91]. A full and accurate characterization of all electron optical parameters is essential if the information content of an image or series of images is to be pushed towards the information limit. The reader is referred to volume 64 of the journal *Ultramicroscopy* for a series of papers related to the acquisition of images with an information limit of 0.1 nm. That it is possible to push the information limit well below 0.1 nm has been shown by O'Keefe and coworkers [OHW$^+$01,ONWA01] in the *One Ångstrom Microscope* project (OAM) at the Lawrence Berkeley Laboratory.

10.3 High-resolution image simulations

We have all the ingredients necessary for a HREM image simulation: the exit wave (from the BWEW.f90 program), and the microscope contrast transfer function (computed via the ctf.f90 module). The computation of a single high-resolution image, using the linear image formation model, is now straightforward and consists of the following steps:

- for a given exit wave $\phi(\mathbf{r})$, compute its Fourier transform $\phi(\mathbf{q})$;
- compute the transfer function $T(\mathbf{q})$ for a given set of microscope parameters;
- multiply $\phi(\mathbf{q})$ and $T(\mathbf{q})$, and apply the aperture mask;
- inverse Fourier transform $\phi(\mathbf{q})T(\mathbf{q})$ to the image plane and take the modulus-squared to convert to intensities.

There are obviously many parameters which can be varied: defocus, defocus spread, beam divergence, foil thickness, acceleration voltage, and so on. In addition, the residual aberrations discussed in Section 10.2.5 can also be included

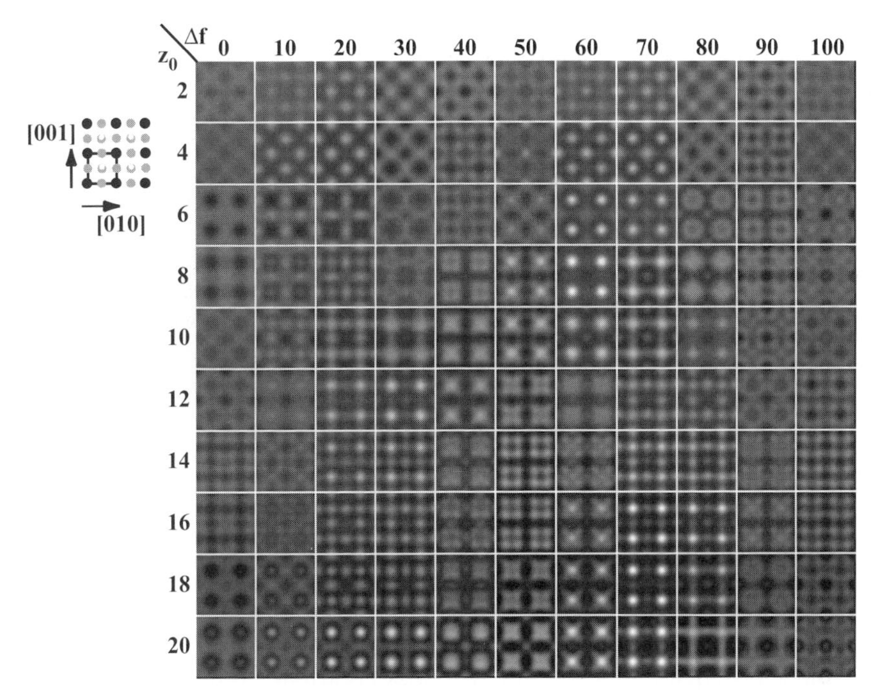

Fig. 10.30. HREM image simulation for the [100] zone axis of $BaTiO_3$, using the electron optical parameters stated in the text. The exit wave computation was carried out with the BWEW.f90 program, and the images were simulated with HREMExample.f90. All images use a common intensity scale.

in the simulations. Typically, however, only two parameters are changed: the foil thickness and the microscope defocus. The resulting images are assembled in a matrix of through-focus series, similar to the one shown in Fig. 10.30. The simulations were carried out for the [100] zone axis orientation of $BaTiO_3$, using 400 kV electrons and a sampling grid of 32×32 pixels. The maximum $|\mathbf{g}|$ included in the exit wave computation is 35 nm^{-1}, resulting in a 615-beam Bloch wave computation (BWEW.f90). The exit waves were computed for 10 foil thicknesses from 2 to 20 nm. The HREMExample.f90 program implements the steps outlined above. The program uses a *namelist* file to store the electron optical parameters for a given microscope. The following parameters were used for Fig. 10.30: $C_s = 1$ mm, $\theta_c = 0.8$ mrad, $\Delta = 8$ nm, and the aperture radius is 6 nm^{-1}. Underfocus values from $\Delta f = 0$ to 100 nm were used.

Comparison of the simulated images with the projected structure model to the left of Fig. 10.30 shows that for most thickness–defocus combinations, the image consists of bright or dark contrast at the metal atom positions (Ba or Ti). In some cases, the O positions also show bright or dark contrast. The Scherzer point resolution (at $\Delta f_S = 49.7$ nm) is 0.165 nm, so that the (020) and (002) interplanar

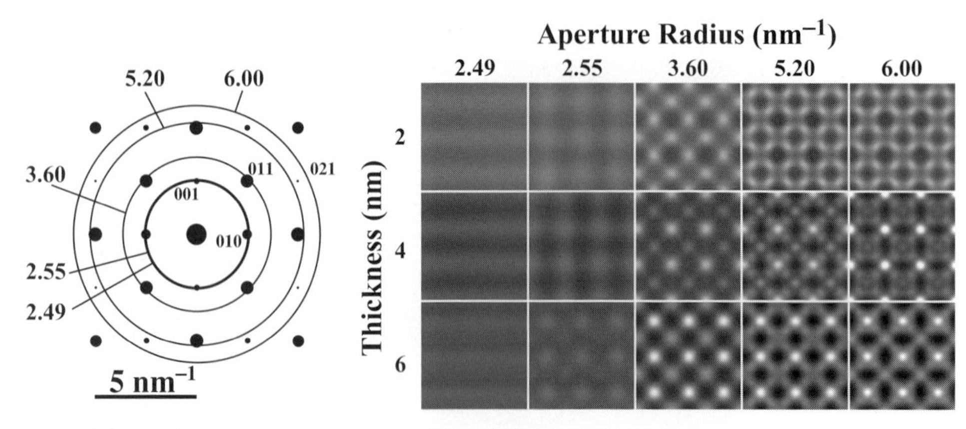

Fig. 10.31. Influence of the aperture size on simulated HREM images, for five different aperture sizes indicated on the $BaTiO_3$ [100] zone axis pattern. See the text for more details.

spacings can be resolved directly. It is also clear from the images that the bright or dark features do not always coincide with the projected atom positions.

Figure 10.31 illustrates the effect of the diffraction aperture. The aperture radii were selected so that only selected reflections contribute to the image. The smallest aperture, with a radius of 2.49 nm^{-1}, only admits the {001} reflections, and the resulting image consists of a sinusoidal modulation with the d_{001} spacing as wavelength. A slightly larger aperture allows the {010} reflections to contribute as well, and the fringe pattern becomes two dimensional. The strong reflections of the {011} family are admitted with the 3.6 nm^{-1} aperture, and the image contrast increases significantly. Larger apertures then admit the {021} and the {020} and {002} families. Since the chromatic information limit for the selected set of parameters is $\rho_c = 0.144$ nm, there is no point in using an aperture larger than the inverse of this limit. No further image details would be introduced by such a larger aperture.

The HREMExample.f90 program is provided on the website as an example of how high-resolution image simulations can be performed. The reader is encouraged to adapt the program, for instance to include non-rectangular projected unit cells. There are several other HREM simulation programs available, including:

- NCEMSS: developed at Lawrence Berkeley Laboratories, runs on various UNIX platforms, `ncem.lbl.gov/frames/software.htm`;
- EMS: developed at the Ecole Polytechnique Fédérale de Lausanne, runs on most platforms, `cimesg1.epfl.ch/CIOL/ems.html`, [Sta87];
- ACEM: programs developed by E.J. Kirkland as part of the book on *Advanced Computing in Electron Microscopy*, [Kir98];
- MacTempas: developed by R. Kilaas, runs on the Macintosh platform, `www.totalresolution.com`.

There are many other packages available and the reader is encouraged to search on the web for commercial, shareware and free software. Details on simulation methods can be found in [Kir98, KO89, BCE88, Spe88].

10.4 Lorentz image simulations

In Section 7.5.3, we have seen that a magnetic thin foil affects only the phase of the electron wave, and not its amplitude. Lorentz microscopy is, therefore, a phase contrast method, and the magnetic object is a *strong phase object*. The phase shift can easily be several radians large so that the weak phase object approximation of Section 10.2.1.3 is of little or no use for Lorentz microscopy. It is possible, however, to simplify the microscope transfer function derived earlier in this chapter. The Lorentz deflection angle θ_L is of the order of a few µrad, and hence the relevant spatial frequency vectors $\mathbf{q}$ are short vectors. This, in turn, means that the dominant terms in the phase transfer function are the lowest-order terms in q, which are the quadratic and fourth-order terms. The contrast transfer function (10.53) can be rewritten by grouping terms of order q^2 and q^4, putting $u = 1$, and ignoring all higher-order terms:

$$\mathcal{T}_L(\mathbf{q}) = A(\mathbf{q} - \mathbf{q}_0)\mathrm{e}^{z_2 q^2}\mathrm{e}^{z_4 q^4} \tag{10.73}$$

with

$$z_2 = -(\pi\theta_c\Delta f)^2 + \mathrm{i}\pi\lambda\,[\Delta f + C_a\cos 2(\phi - \phi_a)];$$

$$z_4 = -\frac{(\pi\lambda\Delta)^2}{2} + 2(\pi\theta_c\lambda)^2\,\Delta f C_s - \mathrm{i}\frac{\pi}{2}C_s\lambda^3.$$

We have included two-fold astigmatism as a component of the transfer function. From the aberrated wave function in the back focal plane, the image intensity can be derived by an inverse Fourier transform:

$$I(\mathbf{r}) = |\mathcal{F}^{-1}[\psi(\mathbf{q})]|^2 = |\psi(\mathbf{r}) \otimes \mathcal{T}_L(\mathbf{r})|^2, \tag{10.74}$$

where $\otimes$ indicates the convolution product and the function $\mathcal{T}_L(\mathbf{r})$ is the *point spread function* appropriate for Lorentz microscopy.

One might wonder whether or not spherical aberration, the most important lens aberration for most observation modes, is still important. It turns out that we can safely ignore the phase shifts caused by spherical aberration, as illustrated by the following example: at 400 kV, a sample with saturation induction of $B_0 = 1$ tesla, and thickness $t = 100$ nm will give rise to a Lorentz deflection angle of $eB_0\lambda t/h =$ 40 µrad. This angle corresponds to a spatial frequency of $q = \lambda\theta_L = 0.024\ \mathrm{nm}^{-1}$. The phase shift caused by spherical aberration (C_s in meters) at this spatial frequency is then equal to $\Delta\phi = 2.44 \times 10^{-6}C_s$. This phase shift remains smaller than 1 mrad for spherical aberration constants of up to 409 m. Since the spherical aberration of a dedicated Lorentz pole piece is still significantly smaller than several hundred meters, it is safe to simply ignore phase shifts caused by C_s in all Lorentz image simulations. This means that z_4 is effectively a real number.

For an in-focus image ($\Delta f = 0$) the first term of z_4 becomes the leading component of the damping envelope, so we must take it into account for all Lorentz image simulations. The defocus spread Δ is significantly larger for Lorentz microscopy when a dedicated Lorentz pole piece is used. The chromatic aberration constant C_c of such a lens is in the range of meters rather than millimeters, similar to the spherical aberration constant. We cannot ignore this damping factor, since for $\Delta f = 0$, ignoring Δ would amount to ignoring the complete damping envelope.

A Lorentz image simulation proceeds along the same lines as a high-resolution simulation: first we must compute the phase of the electron wave function, using either an analytical approach to solving the Aharonov–Bohm integral (7.35), or a numerical approach based on the Mansuripur algorithm introduced in equation (7.42) on page 456. An example of an analytical approach to the computation of the phase shift for a uniformly magnetized spherical particle can be found in [DGNM99].

Pseudo code PC-22 outlines a typical Lorentz image computation, as implemented in the **ION** routine lorentz.pro. The routine asks the user for several microscope parameters (accelerating voltage, defocus, defocus spread, astigmatism, aperture radius and position) and computes either a through-focus series (Fresnel mode) or an aperture shift series (Foucault mode). The actual computation is carried out by a Fortran program LorentzExample.f90; this program computes the magnetization configuration, the resulting phase shift, and then applies the Lorentz contrast transfer function T_L to the reciprocal wave function. The resulting images are stored in a TIFF file or sent back to the remote browser in the case of the **ION** version. Next, we will discuss two example magnetization configurations: two parallel 180° domain walls and the 2D configuration of Fig. 7.35(a).

10.4.1 Example Lorentz image simulations for periodic magnetization patterns

The simulations in this section were carried out using the **ION** implementation of pseudo code PC-22. The reader may experiment with this program and vary all parameters to study their influence on the image features.

Figure 10.32 shows four Fresnel series for a pair of 180° domain walls (a and b), and for the intersecting 71° and 180° domain walls of Fig. 7.35(a). The microscope parameters are listed in the figure caption; the image sequences (c) and (d) have the same sequence of defocus values as the labeled values in Fig. 10.32(a). The first Fresnel series (a) shows that domain walls show up as bright or dark lines for

Pseudo Code PC-22 Outline of a Lorentz image simulation.

```
Input: 2D periodic magnetization configuration
Output: exit wave phase and images
  DFFT of magnetization components → M_mn
  compute phase φ for given beam direction, foil thickness        {7.42}
  DFFT of exit wave e^{iφ}
  ask user for microscope data
  if Fresnel mode then
    for each defocus value do
      compute contrast transfer function
      multiply with reciprocal wave function
      compute modulus squared of inverse DFFT
    end for
  end if
  if Foucault mode then
    for each aperture position do
      compute contrast transfer function
      multiply with reciprocal wave function
      compute modulus squared of inverse DFFT
    end for
  end if
  output to TIFF file
```

out-of-focus images. We will show below that the Fresnel image contrast occurs wherever the phase profile has non-zero curvature.

The through-focus series in Fig. 10.32(b) differs from that in (a) only in the value of the beam divergence angle θ_c; smaller values indicate a more coherent beam, and the wall images consist of parallel fringes, as predicted in Section 7.5.2. Figure 10.33 shows how the fringe contrast at a 180° domain wall changes with image defocus for a beam divergence angle of $\theta_c = 10^{-5}$ and 10^{-6} rad and a defocus range of ± 0.5 mm. The fringe contrast is strong for the convergent wall image (right-hand side) but nearly vanishes for the divergent wall image. The fringe spacing is determined by the product $B_0 t$, as described in Section 7.5.2. A smaller beam divergence angle extends the defocus range for which Fresnel fringes are visible.

The through-focus series in Fig. 10.32(c) shows that the contrast of 180° domain walls is significantly stronger than that of 71° walls, implying that the larger the change in magnetization (either magnitude or direction) across the domain wall, the more pronounced the image contrast. The through-focus series in Fig. 10.32(d) is

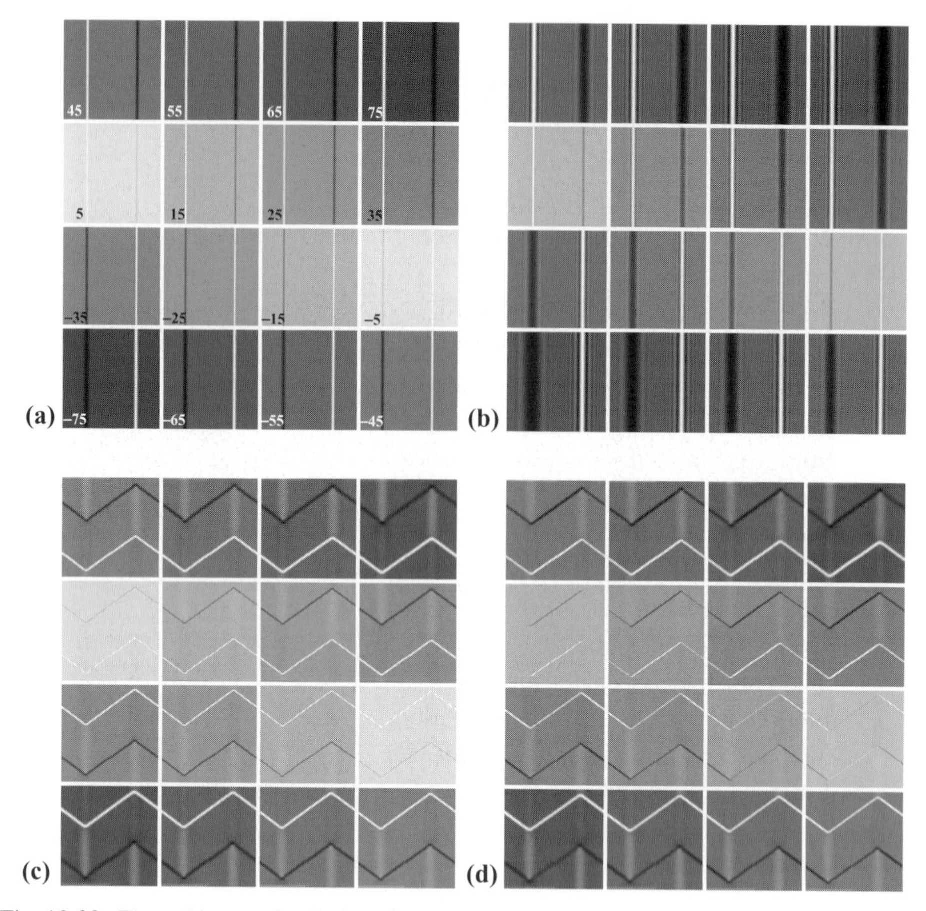

Fig. 10.32. Fresnel image simulations for two parallel 180° domain walls (a) and (b), and for the domain configuration in Fig. 7.35(a), (c) and (d). The microscope accelerating voltage is 200 kV, $B_0t = 100$ nm T, $\theta_c = 0.1$ mrad, aperture radius $q_a = 0.2\,\text{nm}^{-1}$. (a) and (c) show calculated images for a defocus range from -75 to $+75$ μm, (b) has $\theta_c = 0.001$ mrad and a defocus range of ± 750 μm, and (d) is similar to (c), except that astigmatism is included in the computation ($C_a = 8$ μm, $\phi_a = 60°$). The intensity of each individual image was scaled between 0 and 255.

identical to that in (c), except for the presence of two-fold astigmatism ($C_a = 8$ μm, $\phi_a = 60°$). Comparing the image for a defocus of $+5$ μm to the corresponding image without astigmatism shows that astigmatism can completely remove the contrast for domain walls in certain orientations, while enhancing the contrast for other orientations. It is hence imperative that astigmatism be corrected properly, so that for the in-focus condition no domain wall contrast is observed.

We will now show that Fresnel contrast occurs wherever the phase profile has a non-zero curvature. Consider a magnetic thin foil which gives rise to a phase shift $\phi(\mathbf{r})$, where $\mathbf{r}$ is a vector in the plane normal to the electron beam. The phase shift

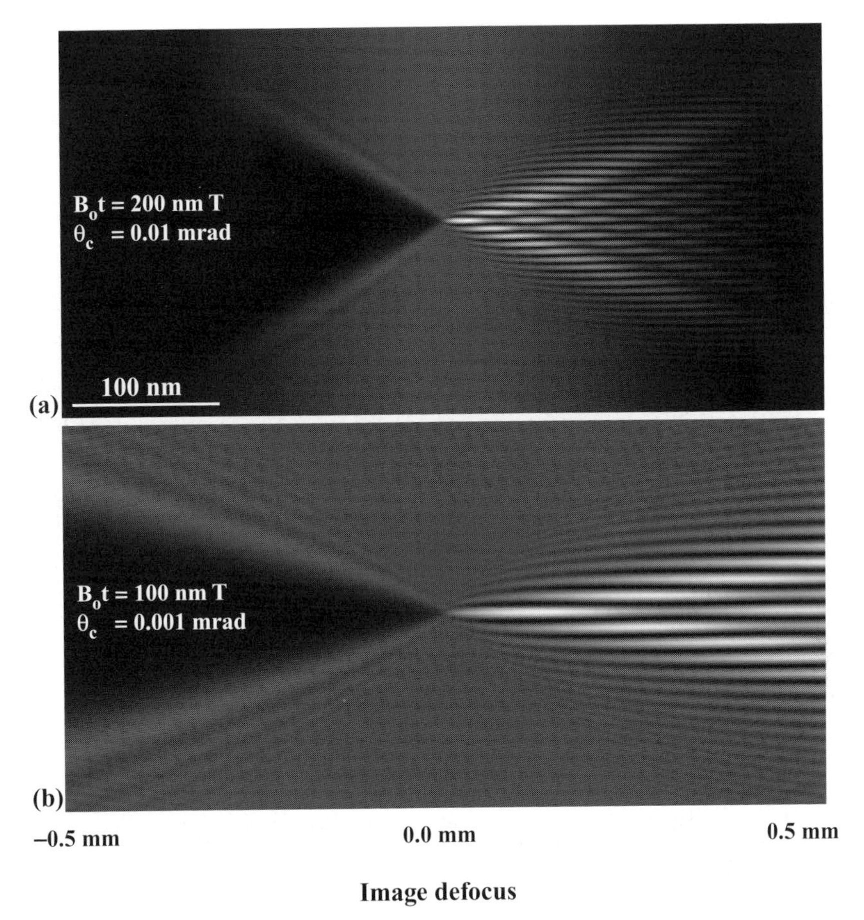

Fig. 10.33. Fresnel through-focus montage for a 180° domain wall of width $\delta = 60$ nm. The beam divergence angle is $\theta_c = 10^{-4}$ rad for the upper image and $\theta_c = 10^{-6}$ rad for the bottom one.

includes both magnetic and electrostatic contributions, as described in Chapter 7 (equation 7.35). The exit wave function can be written as

$$\psi(\mathbf{r}) = a(\mathbf{r})e^{i\phi(\mathbf{r})}.$$

The sample thickness may be non-uniform. We start from the Lorentz transfer function (10.73). For small deflection angles (i.e. small $|\mathbf{q}|$), and small defocus values Δf, we can ignore the fourth-order terms in z_4 and we write for the paraxial wave function in the back focal plane:

$$\psi(\mathbf{q}) = \mathcal{F}[\psi(\mathbf{r})] \left(1 - z_2 q^2\right), \tag{10.75}$$

where $z_2 = z_{2r} + iz_{2i}$ (see equation 10.73). We have conveniently ignored the aperture function $A(\mathbf{q})$; we will assume that the aperture has a small enough radius to exclude all Bragg reflections. This means that the exit wave $\psi(\mathbf{r})$ is essentially the bright field wave, which may be split into multiple components by the magnetic domain structure of the sample, as illustrated in Fig. 7.35(c). The wave function in the image plane is given by the inverse Fourier transform of equation (10.75), which is (dropping the argument $\mathbf{r}$):

$$\psi = a\mathrm{e}^{\mathrm{i}\phi} - z_2\mathcal{F}^{-1}\left[\mathcal{F}\left[a\mathrm{e}^{\mathrm{i}\phi}\right]q^2\right]. \tag{10.76}$$

Writing $\mathcal{F}[a\mathrm{e}^{\mathrm{i}\phi}] = f(\mathbf{q})$ we have

$$\begin{aligned}\mathcal{F}^{-1}\left[\mathcal{F}\left[a\mathrm{e}^{\mathrm{i}\phi}\right]q^2\right] &= \iint q^2 f(\mathbf{q})\mathrm{e}^{2\pi\mathrm{i}\mathbf{q}\cdot\mathbf{r}}\,\mathrm{d}\mathbf{q}\\ &= \frac{-1}{4\pi^2}\iint f(\mathbf{q})\nabla^2\mathrm{e}^{2\pi\mathrm{i}\mathbf{q}\cdot\mathbf{r}}\,\mathrm{d}\mathbf{q}\\ &= \frac{-1}{4\pi^2}\nabla^2\left[\mathcal{F}^{-1}\left[f(\mathbf{q})\right]\right]\\ &= \frac{-1}{4\pi^2}\nabla^2\left[a\mathrm{e}^{\mathrm{i}\phi}\right]\end{aligned}$$

and therefore

$$\psi = a\mathrm{e}^{\mathrm{i}\phi} + \frac{z_2}{4\pi^2}\nabla^2\left[a\mathrm{e}^{\mathrm{i}\phi}\right]. \tag{10.77}$$

The image intensity is then given by the modulus squared of this expression, and we leave it to the reader to show that (ignoring terms in λ^2 and in the absence of astigmatism):

$$I = a^2 - \frac{\lambda\Delta f}{2\pi}\nabla\cdot\left(a^2\nabla\phi\right) + \frac{(\theta_c\Delta f)^2}{2}\left[a\nabla^2 a - a^2(\nabla\phi)^2\right]. \tag{10.78}$$

For a uniform background intensity we have $a^2 = 1$, and therefore

$$I = 1 - \frac{\lambda\Delta f}{2\pi}\nabla^2\phi - \frac{(\theta_c\Delta f)^2}{2}(\nabla\phi)^2. \tag{10.79}$$

For realistic conditions, the last term will be small compared to the other two, and we find that the image intensity is unity everywhere except when $\nabla^2\phi$ is non-zero, or, equivalently, wherever the curvature of the phase is non-zero. For a constant foil thickness, magnetic contrast will be visible only in those regions where the gradient of the phase $\nabla\phi$ is not constant, i.e. at the domain walls. The image contrast is also linear in the defocus Δf, which explains the contrast reversal between under-focus and over-focus images.

Foucault domain contrast can also be simulated using the LorentzExample.f90 program by simply moving the aperture to an off-axis location. Figure 10.34 shows three sets of simulated Foucault images for a pair of 180° domain walls (a), and for

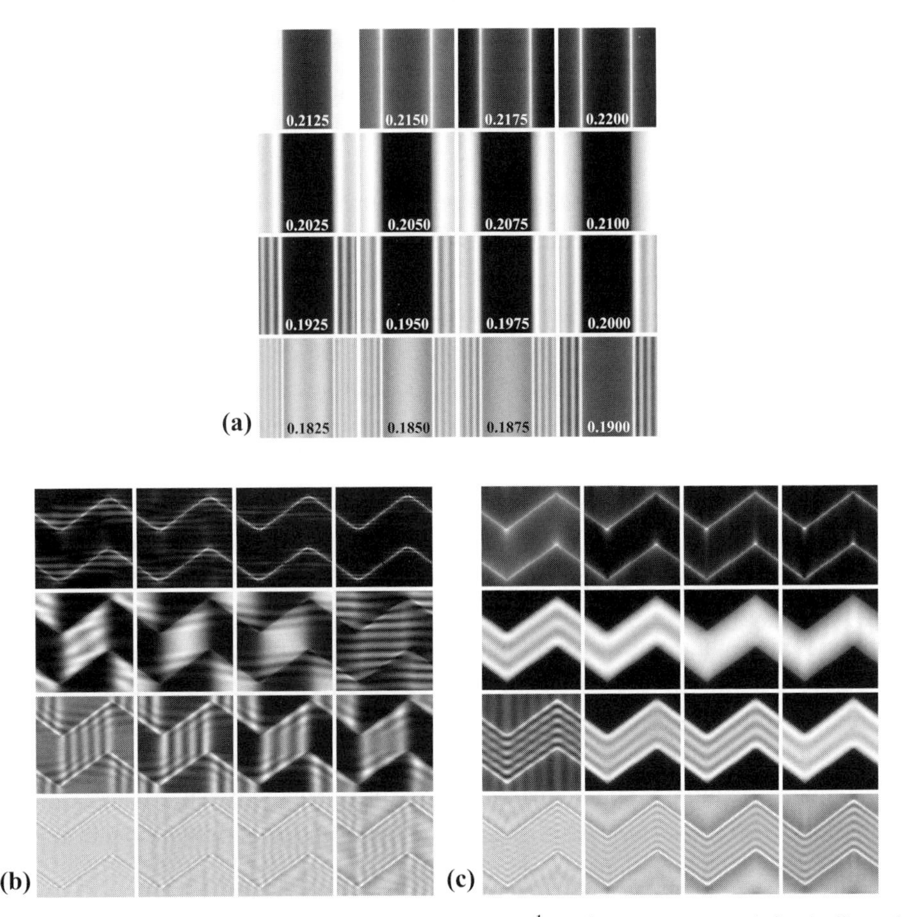

Fig. 10.34. Simulated Foucault series for a 0.2 nm^{-1} radius aperture and the indicated aperture shifts. Imaging parameters are identical to those used for Fig. 10.32. The aperture shift is in the horizontal direction for (a) and (b), and along the vertical direction for (c).

the intersecting domains of Fig. 7.35 (b and c). The material parameters are identical to those used in Fig. 10.32. The aperture radius is $q_a = 0.2$ nm^{-1}; the aperture shifts are indicated by the numbers in Fig. 10.34(a). Foucault contrast appears when the aperture begins to intersect the beams deflected by the various magnetic domains. The easiest way to visualize this is with respect to Fig. 7.35(c): imagine an aperture that just encloses all four deflected beams. When this aperture is moved in any direction, one or more of the beams will be cut off and the corresponding magnetic domain will become dark. If the aperture shift directions are calibrated with respect to the image orientation, then it is in principle possible to derive the local magnetization direction from a pair of Foucault images with aperture shifts in independent directions.

10.4.2 Fresnel fringes for non-magnetic objects

In the previous section, we saw that Fresnel (fringe) contrast occurs wherever the phase of the electron wave has a non-vanishing second derivative (curvature). This is valid for *any* phase, not just for phase of a magnetic origin, so in this section we will explore briefly the nature of Fresnel fringes for non-magnetic objects. From the Aharonov–Bohm equation we see that a non-magnetic object will cause a phase shift which is entirely due to the electrostatic lattice potential. At low magnifications, significantly lower than those used for high-resolution observations, the only important component of the electrostatic lattice potential is the mean inner potential V_0, so that the phase shift due to a foil of thickness t is given by $\phi_e = \sigma V_0 t$, where σ is the interaction constant. From here on we will drop the subscript 0 on the mean inner potential and simply write V.

The **ION** routine fresnel.pro can be used to display Fresnel images for objects with a spatially varying mean inner potential. The routine provides the following pre-defined sample "shapes": a strip of material, a foil with a circular hole, a disk-shaped island, a square island, and a thin strip of material with a slightly lower mean inner potential than the surrounding matrix. The routine calls a Fortran program that is very similar to the LorentzExample.f90 program discussed earlier. The only difference is that the phase computation is much simpler than for a magnetic object: if the geometry of the object is known, then the local thickness can be computed for a given orientation, and hence the phase shift will be known.

Figure 10.35 shows four pairs of simulated images for a circular hole in a matrix. The matrix is 20 nm thick and has a mean inner potential of 30 V; the normal absorption length is taken to be 50 nm. The exit wave function is then equal to

$$\phi(x, y) = \mathrm{e}^{-t/\xi'} \mathrm{e}^{\mathrm{i}\sigma V t} \qquad \text{for } x^2 + y^2 > r^2,$$

with r the radius of the hole. The origin is taken at the center of the hole. The exit wave is unity inside the hole. The image pair a–b of Fig. 10.35 shows Fresnel fringes for the defocus values $\pm 150\,\mu\text{m}$ (same values for all images in this figure). The beam divergence angle is $\theta_c = 0.05$ mrad, and several Fresnel fringes are visible near the edge of the hole, which is indicated by a thin white circle in all simulated images. The first fringe inside the hole is bright for positive defocus and dark for negative defocus. In the presence of astigmatism ($C_a = 100\,\mu\text{m}$ and $C_a = 60°$), the distance between the first fringe and the edge of the hole is no longer constant and varies around the hole, as shown in image pair c–d. Conversely, if the first fringe is not at the same distance from the edge for all points along the hole, then astigmatism must be present. Correcting the astigmatism is then simply reduced to making the Fresnel fringe continuous around the hole at a uniform distance from the edge.

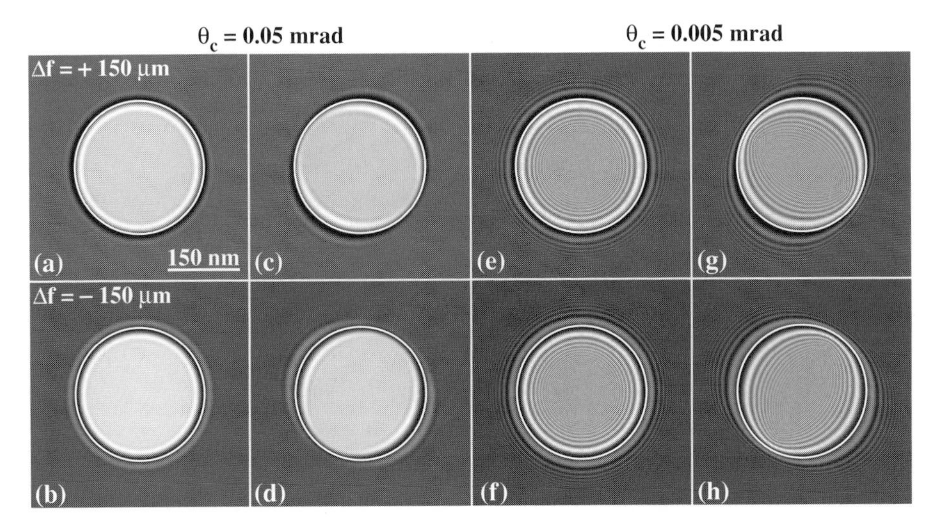

Fig. 10.35. Fresnel fringe simulations for a 300 nm diameter circular hole in a 20 nm thick matrix with a mean inner potential of 30 V. The defocus is $+150$ μm for the top row, and -150 μm for the bottom row. The four images on the left have $\theta_c = 0.05$ mrad, while the others have $\theta_c = 0.005$ mrad. Image pairs c–d and g–h contain astigmatism ($C_a = 100$ μm and $C_a = 60°$). The images were computed using the **ION** routine fresnel, available from the website.

If the beam divergence angle is decreased to $\theta_c = 0.005$ mrad (this is possible for a field emission microscope), then a much larger number of Fresnel fringes can be observed, as illustrated by image pair e–f in Fig. 10.35. The presence of astigmatism (g–h) can be recognized easily from the asymmetry of the fringes. The number of fringes and the defocus range for which they are visible are illustrated in Fig. 10.36. The top image represents a through-focus series (Δf from -150 to $+150$ μm) for a solid–vacuum interface; the foil thickness is 20 nm, with a mean inner potential of 30 V. The interface is indicated by a dashed line. The beam divergence angle is $\theta_c = 0.1$ mrad, which represents a low beam coherence. The Fresnel fringes are visible for small defocus magnitudes but disappear for larger values. For a beam divergence of 0.001 mrad, many fringes are visible (bottom of Fig. 10.36) even for large defocus values. The fringes inside the foil may not always be visible, depending on the thickness and absorption length of the foil.

10.5 Exit wave reconstruction

10.5.1 What are we looking for?

In virtually every TEM experiment we are ultimately interested in the structure and/or chemistry of a particular material or defect. This means that we would

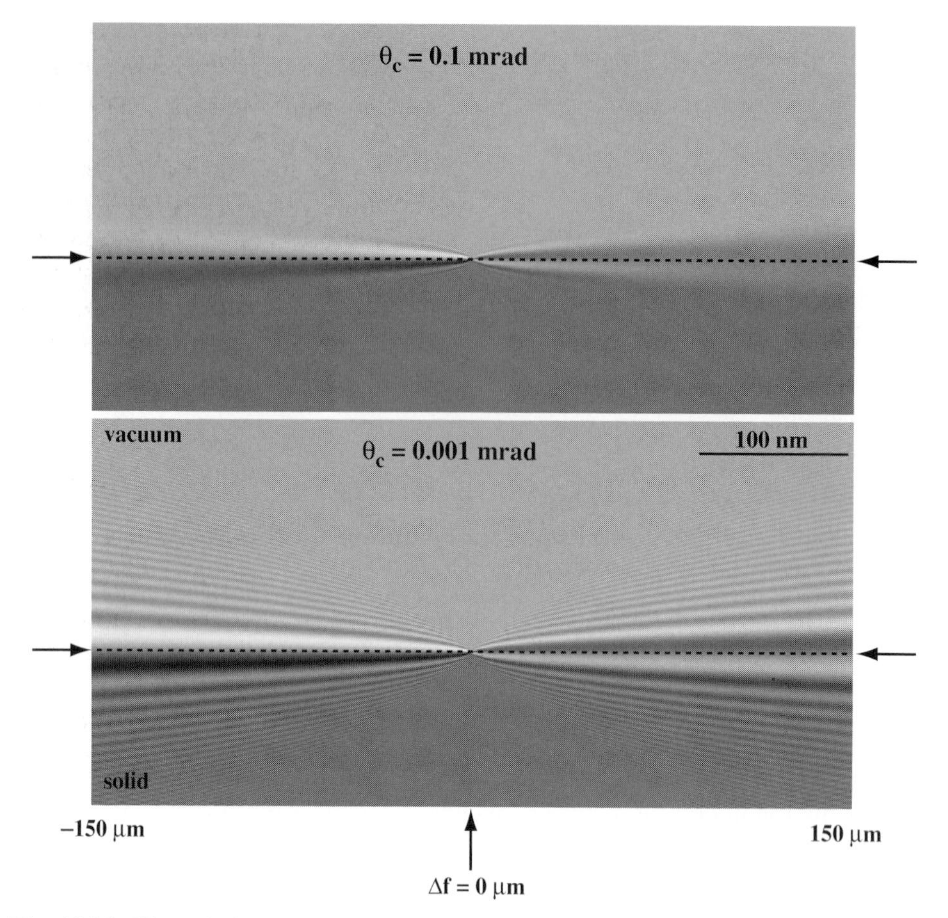

Fig. 10.36. Through-focus series of Fresnel fringes near a straight edge. The microscope accelerating voltage is 200 kV, foil thickness 20 nm, mean inner potential 30 V, and two different beam divergence angles are shown.

like to find out where the individual atoms are located, or how the magnetization is distributed throughout the thin foil, or where charges accumulate near an electrically active boundary, and so on. For non-periodic features, such as most defects, we may have a structure model that we can use to simulate images which can then be compared with the observations. This is the standard approach which we will call the *forward approach*:

$$\text{atom positions} \rightarrow \text{lattice potential} \rightarrow \text{exit wave function} \rightarrow \text{simulated images.} \tag{10.80}$$

From the differences between observed and simulated images we can often derive information to refine the model, and we may be able to iterate the model parameters until the simulated images converge to the observed ones.

In the forward approach, we typically do not modify the observed intensities, other than perhaps some noise filtering and intensity and/or magnification scaling. We use the observed images (they could also be diffraction patterns) as the *reference state* against which to validate the model. We could also think about an opposite approach, in which we attempt to derive the structure *directly* from the observed images, with as few adjustable parameters as possible. This is known as the *inverse approach* or, more commonly, as the *inverse problem*. While it may seem obvious that this would be a possible approach, the inverse problem is actually a highly non-trivial problem, since it requires that we reconstruct both the amplitude and the phase of the electron wave at the exit plane of the sample. Phase information is present in the images, but generally not in a form that can be extracted easily. Ultimately, we want to find out what the electrostatic lattice potential looks like, since it contains information on both the atom types and locations. This means that we must invert the Schrödinger equation, in addition to removing the transfer function of the microscope. The inverse problem can then be stated schematically as follows:

$$\text{experimental images} \rightarrow \text{exit wave function} \rightarrow \text{lattice potential} \rightarrow \text{atom positions.} \tag{10.81}$$

The first step, from image to wave function, is known as *exit wave restoration* or *reconstruction*. It is a highly non-linear problem, since it essentially involves the deconvolution of the microscope transfer function according to equation (2.56) on page 107. We know from the first section of this chapter that the microscope transfer function has multiple zero-crossings, so that a simple deconvolution is not possible. The second step requires inversion of the Schrödinger equation, and this is again a non-trivial step. Both problems have received considerable attention, even since the early days of TEM research. There are several solutions available in the literature, but none of them is entirely satisfactory in the sense that no single method can be applied to all possible situations. Exit wave reconstruction remains a topic of significant interest, in particular now that spherical aberration can be removed from the image formation process using an aberration corrector.

There is a significant body of literature on wave function restoration. The main ideas go back to the work by Gerchberg and Saxton [GS72], Schiske [Sch68], and ultimately to Cowley and Moodie [CM57]. The phase of the image wave can be found from combinations of multiple images acquired at different microscope defocus values, sometimes combined with diffraction information. Schiske's original work introduces so-called *filter functions*, which are essentially weighting functions used to combine images taken at multiple defocus values in such a way that a least-squares best estimate of the exit wave function is obtained.

This procedure is now known as a *generalized Wiener filter*, as described in [HK94, Chapter 73].

Kirkland and coworkers [KSUF80, Kir82, KSUF82] have successfully implemented a reconstruction scheme based on a minimum mean-square formalism and applied it to what has effectively become a "standard sample", chlorinated Cu-phthalocyanine. Adaptive least-squares filters were introduced by Hawkes [Haw74], and a review of various filter methods can be found in [Sch84]. The *Gerchberg–Saxton* algorithm makes use of an image and a diffraction pattern taken from the same region [GS72, GS73]; a *multiple image* algorithm, proposed by Misell [Mis73], and elaborated by many others, can in principle be used to recover the phase from two images at different microscope defocus, but no experimental verification of the method has been reported. Kirkland's *multiple input maximum a posteriori* method (MIMAP) [Kir84, KSUF85] again works in principle, but does not function satisfactorily for realistic experimental conditions. Non-linear reconstruction schemes based on the earlier work have also been proposed [CJOdBVD92] and have been shown to work in selected experimental situations. For an in-depth review of other reconstruction schemes we refer to Chapters 73 and 74 in [HK94]. A review of exit wave reconstruction using electron holography methods can be found in [VAJ99].

In the next section, we will illustrate the process of exit wave reconstruction for Lorentz imaging. The reconstruction process for Lorentz images is much more straightforward than for conventional high-resolution images, since we can make good use of the fact that the Lorentz deflection angle is small compared with the Bragg angle. As we will see in the next section, this makes it possible to derive the phase of the exit wave from just three experimental images.

10.5.2 Exit wave reconstruction for Lorentz microscopy

We start from equation (10.78) and write the in-focus intensity a^2 as $I(\mathbf{r}, 0)$; the out-of-focus image intensity is then written as $I(\mathbf{r}, \Delta f)$. Next, we derive an equation for the phase ϕ; consider an under-focus and over-focus image, for the same defocus magnitude $|\Delta f|$:

$$\begin{aligned} I(\mathbf{r}, |\Delta f|) = I(\mathbf{r}, 0) &- \frac{\lambda|\Delta f|}{2\pi}\nabla \cdot (I(\mathbf{r}, 0)\nabla\phi) \\ &+ \frac{(\theta_c|\Delta f|)^2}{2\ln 2}\left[\sqrt{I(\mathbf{r}, 0)}\nabla^2\sqrt{I(\mathbf{r}, 0)} - I(\mathbf{r}, 0)(\nabla\phi)^2\right]; \\ I(\mathbf{r}, -|\Delta f|) = I(\mathbf{r}, 0) &+ \frac{\lambda|\Delta f|}{2\pi}\nabla \cdot (I(\mathbf{r}, 0)\nabla\phi) \\ &+ \frac{(\theta_c|\Delta f|)^2}{2\ln 2}\left[\sqrt{I(\mathbf{r}, 0)}\nabla^2\sqrt{I(\mathbf{r}, 0)} - I(\mathbf{r}, 0)(\nabla\phi)^2\right]. \end{aligned}$$

Subtracting the second equation from the first, and rearranging terms we have:

$$-\frac{2\pi}{\lambda}\frac{I(\mathbf{r},|\Delta f|)-I(\mathbf{r},-|\Delta f|)}{2|\Delta f|}=\nabla\cdot[I(\mathbf{r},0)\nabla\phi]. \tag{10.82}$$

In the limit of vanishingly small defocus the left-hand side becomes a derivative and we arrive at the so-called *transport-of-intensity equation* (TIE)† [Tea83, GRN95b, GRN95a, GN96, PN98]:

$$\boxed{\nabla\cdot[I(\mathbf{r},0)\nabla\phi]=-\frac{2\pi}{\lambda}\frac{\partial I(\mathbf{r},0)}{\partial z}.} \tag{10.83}$$

The TIE equation was also derived by Van Dyck and Coene in 1987 starting from equation (5.35) with $\bar{V}=0$ (i.e. the free-space propagation equation) [VDW87]. Note that the beam divergence term, which is even in the defocus, cancels out in the difference between under-focus and over-focus images. This means that phase reconstruction for Lorentz microscopy images does not require a highly coherent electron beam.

10.5.2.1 Solving the TIE equation

Equation (10.83) can be solved numerically in a number of different ways. Perhaps the most straightforward method relies on the use of fast Fourier transforms. The Fortran source code for the algorithm can be found on the website in the routine TIE.f90. A formal solution to the TIE equation can be derived by introducing a new variable Ψ, such that

$$\nabla\Psi(\mathbf{r},0)\equiv I(\mathbf{r},0)\nabla\phi. \tag{10.84}$$

If the phase is entirely determined by the magnetic structure of the thin foil, then its gradient can be used to compute the in-plane magnetic induction configuration:

$$\nabla\phi=\frac{\nabla\Psi}{I(\mathbf{r},0)}=-\frac{e}{\hbar}[\mathbf{B}(\mathbf{r})\times\mathbf{n}]\,t(\mathbf{r}), \tag{10.85}$$

where $\mathbf{n}$ is the unit normal in the beam direction and t is the local sample thickness.

Using the function Ψ defined in equation (10.84), the TIE becomes a Poisson-type equation [PN98]:

$$\nabla^2\Psi(\mathbf{r},0)=-k\frac{\partial I(\mathbf{r},0)}{\partial z}. \tag{10.86}$$

† The name of this equation was first coined by M.R.Teague [Tea83], who showed that this formalism could be applied to phase retrieval.

While a direct analytical solution is difficult to obtain, it is possible to write down a *symbolic solution* as follows:

$$\Psi(\mathbf{r},0) = -k\nabla^{-2}\frac{\partial I(\mathbf{r},z)}{\partial z}. \tag{10.87}$$

The symbol ∇^{-2} refers to the two-dimensional *inverse Laplacian operator*, and we will describe shortly a simple numerical implementation. Taking the two-dimensional gradient of this equation we obtain:

$$\nabla\phi(\mathbf{r},0) = -\frac{k}{I(\mathbf{r},0)}\nabla\nabla^{-2}\frac{\partial I(\mathbf{r},0)}{\partial z}, \tag{10.88}$$

where we have used equation (10.84). After applying the two-dimensional divergence operator $\nabla\cdot$, and bringing the Laplacian operator to the right-hand side we find the symbolic solution for the phase [PN98]:

$$\phi(\mathbf{r},0) = -k\nabla^{-2}\left[\nabla\cdot\left(\frac{1}{I(\mathbf{r},0)}\nabla\left\{\nabla^{-2}\left[\frac{\partial I(\mathbf{r},0)}{\partial z}\right]\right\}\right)\right]. \tag{10.89}$$

It is straightforward to show that the same equation can be used to recover the phase for an out-of-focus image; the argument $(\mathbf{r},0)$ can then be replaced by $(\mathbf{r},\Delta f)$.

While equation (10.89) looks rather menacing, it is actually not very difficult to implement an efficient numerical procedure to solve for the phase. In fact, we can take the formal solution one step further and note that (at least formally) we have

$$\nabla^{-2}\nabla = \nabla^{-1} = \nabla\nabla^{-2}, \tag{10.90}$$

and we arrive at the formal solution

$$\boxed{\phi(\mathbf{r},0) = -k\nabla^{-1}\cdot\left(\frac{1}{I(\mathbf{r},0)}\left\{\nabla^{-1}\left[\frac{\partial I(\mathbf{r},0)}{\partial z}\right]\right\}\right).} \tag{10.91}$$

From equation (10.82) we see that we need to acquire two out-of-focus images, $I(\mathbf{r},-\Delta f)$ and $I(\mathbf{r},+\Delta f)$. In addition, according to equation (10.89), we need the in-focus image $I(\mathbf{r},0)$. The longitudinal derivative $\partial I/\partial z$ can be computed readily, provided both images are perfectly aligned and have precisely the same magnification.† From equations (10.76) and (10.77) we can derive the following relation:

$$\nabla^2[\cdots] = -4\pi^2\mathcal{F}^{-1}\left[\mathcal{F}[\cdots]|\mathbf{q}|^2\right], \tag{10.92}$$

† Recall that the magnification changes with defocus, in particular when a low-field lens with a long focal length is used.

where $[\cdots]$ indicates a function of the coordinate variable $\mathbf{r}$. We leave it to the reader to show that

$$\nabla^{-2}[\cdots] = -\frac{1}{4\pi^2}\mathcal{F}^{-1}\left[\frac{\mathcal{F}[\cdots]}{|\mathbf{q}|^2}\right]. \tag{10.93}$$

If we combine the inverse Laplacian with the gradient operator ∇, then we can show that with respect to the Cartesian reference frame $\mathbf{e}_i$ we have for every function $\alpha(\mathbf{r})$

$$\begin{aligned}
\nabla^{-1}\alpha(\mathbf{r}) &= \nabla\nabla^{-2}\alpha(\mathbf{r}) = \nabla^{-2}\nabla\alpha(\mathbf{r});\\
&= -\frac{1}{4\pi^2}\frac{\partial}{\partial x_j}\left[\mathcal{F}_{\mathbf{q}}^{-1}\left[\frac{\mathcal{F}_{\mathbf{r}}[\alpha(\mathbf{r})]}{|\mathbf{q}|^2}\right]\right]\mathbf{e}_j;\\
&= -\frac{\mathrm{i}}{2\pi}\iint \alpha(\mathbf{r}')\left[\iint \frac{\mathbf{q}}{|\mathbf{q}|^2}\mathrm{e}^{2\pi\mathrm{i}\mathbf{q}\cdot(\mathbf{r}-\mathbf{r}')}\mathrm{d}\mathbf{q}\right]\mathrm{d}\mathbf{r}';\\
&= \frac{\mathrm{i}}{2\pi}\iint \alpha(\mathbf{r}-\mathbf{R}_\perp)\left[\iint \frac{\mathbf{q}}{|\mathbf{q}|^2}\mathrm{e}^{2\pi\mathrm{i}\mathbf{q}\cdot\mathbf{R}_\perp}\mathrm{d}\mathbf{q}\right]\mathrm{d}\mathbf{R}_\perp;\\
&\equiv \frac{\mathrm{i}}{2\pi}\iint \alpha(\mathbf{r}-\mathbf{R}_\perp)\mathbf{G}(\mathbf{R}_\perp)\mathrm{d}\mathbf{R}_\perp;\\
&= \frac{\mathrm{i}}{2\pi}\mathbf{G}(\mathbf{r})\otimes\alpha(\mathbf{r}).
\end{aligned}$$

The vector function $\mathbf{G}(\mathbf{r})$ is equal to the inverse Fourier transform of the vector function $\mathbf{q}/|\mathbf{q}|^2$, and is not defined for $\mathbf{q} = \mathbf{0}$. The formal solution (10.91) to the TIE equation can then be expressed as a double convolution product:

$$\phi(\mathbf{r}, \Delta f_j) = \frac{\delta^2}{64\pi^3\lambda\Delta f}(\mathbf{G}(\mathbf{r})\otimes)\cdot\left(\frac{\mathbf{G}(\mathbf{r})\otimes\partial I(\mathbf{r}, \Delta f_j)/\partial z}{I(\mathbf{r}, \Delta f_j)}\right). \tag{10.94}$$

It is understood that the integrations are carried out on a grid with unit grid spacing. In component notation we have

$$\mathbf{G}(\mathbf{r})\otimes[\cdots] = \mathbf{e}_i\gamma_i\otimes[\cdots] = \mathbf{e}_i\left[\iint \frac{q_i}{|\mathbf{q}|^2}\mathrm{e}^{2\pi\mathrm{i}\mathbf{q}\cdot\mathbf{r}}\mathrm{d}\mathbf{q}\right]\otimes[\cdots],$$

and for numerical implementation we can use

$$\phi = -\frac{\delta^2}{64\pi^3\lambda\Delta f}\sum_{i=1}^{2}\gamma_i\otimes\left(\frac{\gamma_i\otimes\partial I/\partial z}{I}\right), \tag{10.95}$$

where γ_i are the components of the vector function $\mathbf{G}$. The prefactor follows from a simple dimensional analysis of the formal solution (10.91). This particular form of the symbolic solution shows clearly that the accuracy of the magnitude of the reconstructed phase depends directly on a careful calibration of the microscope

---

Pseudo Code PC-23 Reconstruct exit wave phase from Lorentz–Fresnel images.

Input: Three or five images, δ, Δf, λ {Microscope calibration!}
Output: Reconstructed exit wave phase
  compute array of $\frac{\mathbf{q}}{|\mathbf{q}|^2}$ vectors
  compute longitudinal derivative $\hat{\nabla}_z I$
  verify conservation criterion
  convolve with $\mathbf{G}(\mathbf{r})$ (using FFT operations)
  divide by the in-focus image
  convolve with $\mathbf{G}(\mathbf{r})$ (using FFT operations)
  normalize with the prefactor $\frac{\delta^2}{64\pi^3\lambda\Delta f}$

---

magnification (through δ) and the defocus step size (Δf). Pseudo code for the implementation of equation (10.95) can be found in PC-23. Since part of the reconstruction involves a convolution product, one must take care of aliasing effects and truncate the Fourier transforms at $\frac{2}{3}q_c$. Since the inverse Laplacian is effectively a low-pass filter, this does not cause any loss of information.

In addition to the anti-aliasing measures, we must also *symmetrize* the input images, to ensure that the solution to the TIE equation is unique. The reasons for this symmetrization are beyond the scope of this textbook and we refer the interested reader to [VZDG02] for the details. In practice, we double the size of the images by application of two orthogonal mirror planes, as shown in Fig. 10.37. The algorithm is then applied to the symmetrized images. The main reason for this symmetrization is the use of the fast Fourier transform, which requires a periodic input image. An example computation is shown in the following section.

Finally, the right-hand side derivative in the TIE must average to zero, indicating that no electrons are lost from the image with varying defocus. This conservation criterion guarantees that the solution to the TIE is unique [GRN95a].

While we have used the small-angle approximation to the transfer function, it can be shown (e.g. [BPN00, Chapter 5] and [VDW87]) that the TIE equation is valid for a much more general class of experimental configurations. It is possible to apply the same formalism to high-resolution phase contrast images, and one can then reconstruct the phase of the *image* wave from a through focus series consisting of at least three images. When the phase of the image wave is known, a Fourier transform propagates the wave back to the back focal plane of the objective lens, where one can then deconvolve the transfer function. This phase reconstruction method is rather promising when used in combination with a spherical aberration

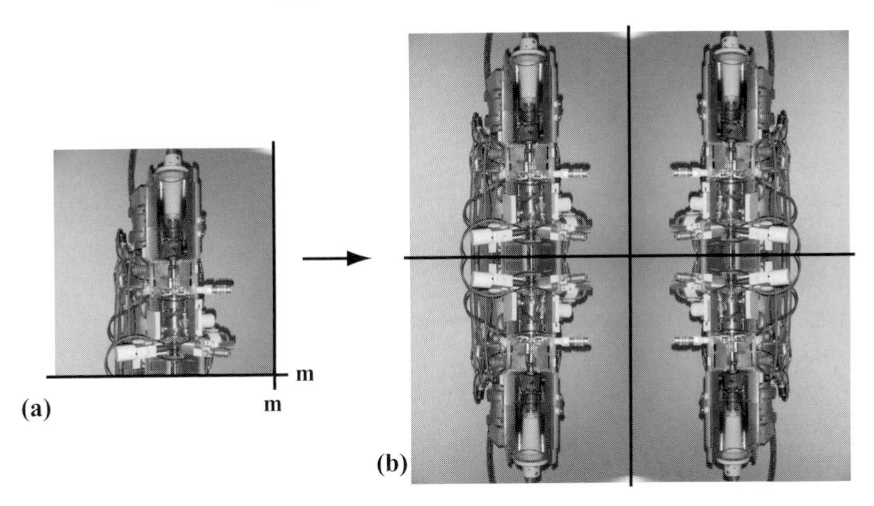

Fig. 10.37. Symmetrization of the images needed to solve the transport-of-intensity equation using a fast Fourier transform algorithm.

corrector, since the zero-crossings of the transfer function can then be avoided entirely.

10.5.2.2 Example application of the TIE formalism

The TIE solution algorithm has been applied to the Fresnel images of Fig. 4.28. The area indicated by a square was taken as the input image (160×160 pixels). After aligning the three images using a cross-correlation algorithm, the algorithm in pseudo code PC-23 was applied. The resulting reconstructed phase is shown as a grayscale image in Fig. 10.38(a) and a contour plot in (b). The x and y components of the gradient of ϕ are proportional to the y and x components of the integrated magnetic induction and are shown in Figs 10.38(c) and (d); bright areas indicate positive components. A more familiar representation of the magnetic induction in the foil is shown in Fig. 10.39: each vector represents the average induction over a 4×4 pixel area. Vector lengths are scaled with respect to the longest vector which has unit magnitude. The vortex and cross-tie wall are clearly visible in the reconstructed map.

As a final example we consider a 2×1 μm rectangular Permalloy island deposited on a Si_3N_4 membrane, observed in Fresnel mode on a JEOL 4000EX microscope equipped with a Gatan post-column energy filter (GIF). The under- and over-focus images are shown in Figs 10.40(a) and (b), along with the numerically computed derivative $\partial I/\partial z$, shown in (c). The image consists of 300×300 pixels. The reconstructed phase is shown as a grayscale plot (d) and as $\cos\phi$ (e). The square region outlined in (c) is shown in (f) as a vector plot, obtained from the x and y components of $\nabla_\perp\phi$.

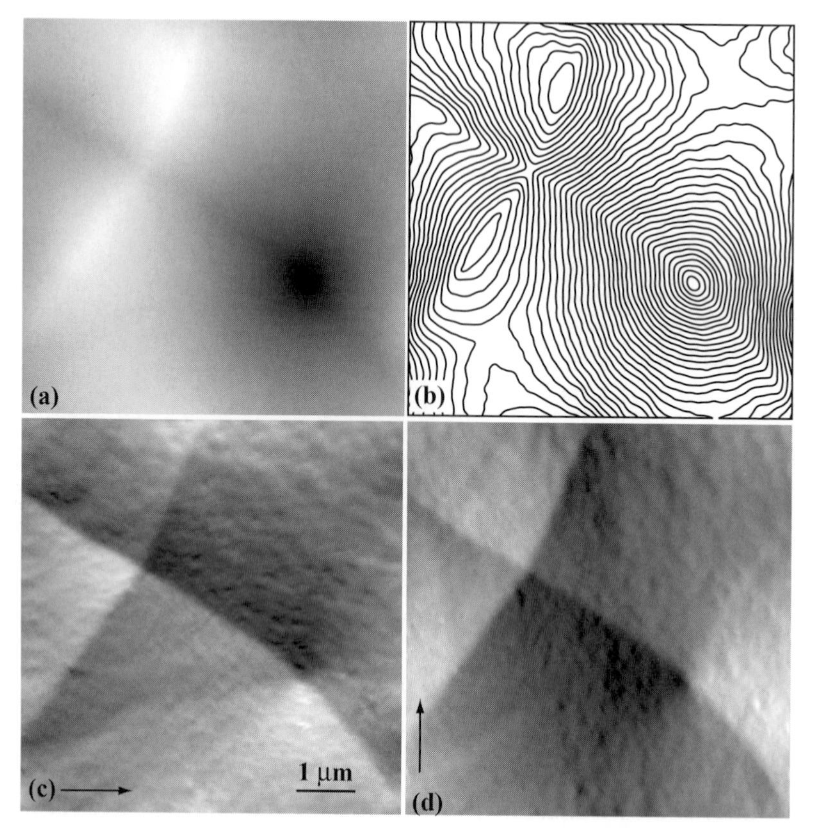

Fig. 10.38. (a) Reconstructed phase for the square area outlined in Fig. 4.28(c). (b) Contour plot of the phase $\phi(\mathbf{r})$; the vortex is clearly visible to the lower right. (c) B_x and (d) B_y maps, obtained by numerical differentiation of the phase. The maps are only qualitatively correct, since the microscope defocus was not calibrated.

Exercises

10.1 Show that the paraxial image intensity in Lorentz mode is given by equation (10.78).

10.2 Derive equation (10.93) for the inverse Laplacian operator.

10.3 Show that the gradient of the aberration function $\chi(Q)$ reaches a minimum at $Q = \sqrt{D/3}$.

10.4 Consider the microscope I_2 from Table 10.1. If this machine is equipped with a 1K×1K CCD camera with a pixel size of $25 \times 25\ \mu\text{m}^2$, then what microscope magnification should be used to record images that contain spatial frequencies out to the chromatic information limit (assuming the sample contains such spatial frequencies)? How many pixels of the CCD camera cannot be used because of the detector envelope function?

10.5 Adapt the HREMExample.f90 program to use any pair of parameters as variables for the output matrix. As an example, you may want to study the source code of the ctf2.pro **ION** program on the website.

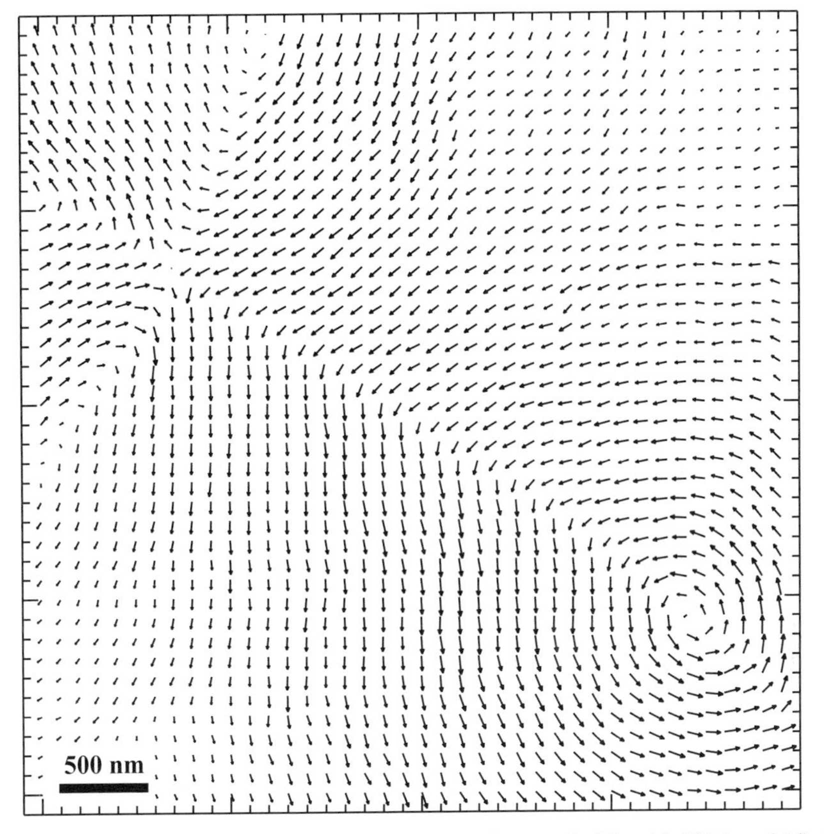

Fig. 10.39. Vector map derived from the component images in Figs 10.38(c) and (d). Each vector is an average over a 4 × 4 pixel square region (or 100 × 100 nm^2).

10.6 Use an amorphous Ge film (or, if you do not have one, a C film) to determine the spherical aberration of the microscope you use for your observations. How accurately can you determine C_s?

10.7 Use your thin foil of GaAs to obtain a [110] high-resolution through-focus series for two different diffraction aperture diameters. Simulate the images, and compare simulations and experiments.

10.8 The transport-of-intensity equation can be used to reconstruct the phase of the electron wave for Lorentz-type observations. Can you think of a way to adapt the formalism to high-resolution images? What would happen if the formalism were applied to images from a C_s-corrected microscope?

10.6 Final remarks

Well, you have reached the end of this book. Hopefully, you found it to be a useful book. If you were new to microscopy, it is hoped that this text has provided you

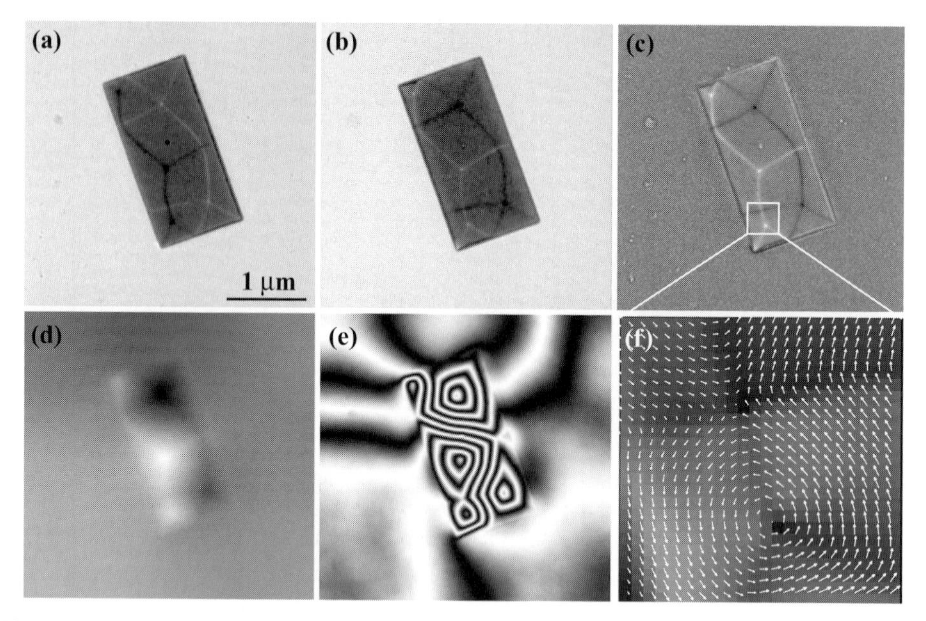

Fig. 10.40. (a) and (b) under- and over-focus Fresnel images (400 kV, zero-loss filtered) of a $1 \times 2\,\mu$m Permalloy island on an Si_3N_4 support membrane. The derivative $\partial I/\partial z$ is shown in (c), along with the reconstructed phase ϕ in (d), and $\cos\phi$ in (e). (f) is a vector representation of the square area outlined in (c). (Sample courtesy of J. Chapman, Univ. Glasgow.)

with a solid introduction to the physical principles underlying the interaction of electrons with the specimen, and the subsequent propagation of those electrons down the microscope column. If you like computers and simulations, then perhaps the source code accompanying this book was just what you needed to get started. If you are an experienced microscopist, it is hoped that you could find something new in this text, perhaps a derivation that suddenly makes sense, or an illustration that provoked what the German language so efficiently describes as an "Aha-Erlebnis".†

The field of electron optical methods is vast, and what you have learned in this book is truly just an introduction to the operation and use of one the most important instruments of materials science. The material covered in this text should give you a basic knowledge of elastic scattering in crystalline matter. You can build on this knowledge when you expand your studies to include inelastic scattering and analytical observation methods.

† An *Aha-Erlebnis* is "Ein eigenartiges im Denkverlauf auftretendes-lustbetontes Erlebnis, das sich bei plötzlicher Einsicht in einen zuerst undurchsichten Zusammenhang einstellt." (K. Bühler, speech psychologist.)

Appendix A1

Explicit crystallographic equations

In Chapter 1, we have derived the general equations for angles and distances in both direct and reciprocal space. Sometimes it is useful to have the explicit versions of those equations for a particular crystal system. For instance, the d-spacing for the planes (hkl) in a cubic crystal is given by

$$d_{hkl} = \frac{a}{\sqrt{h^2 + k^2 + l^2}},$$

and for manual calculations this is indeed easier to use than the more formal tensor relations. For implementations in a computer program, it is much easier to use the tensor formalism, because then one need not distinguish between the seven crystal systems (but, beware of the hexagonal system!). The equations for the hexagonal reference frame are only valid for the 3-index system.

In this appendix, we list the explicit equations for the direct and reciprocal metric tensors, the length of a vector $\mathbf{t}_{uvw}$ (Table A1.1), the angle between two vectors $\mathbf{t}_{u_1v_1w_1}$ and $\mathbf{t}_{u_2v_2w_2}$ (Table A1.2), the length of a reciprocal lattice vector $\mathbf{g}_{hkl}$ (Table A1.3), and the angle between two reciprocal lattice vectors $\mathbf{g}_{h_1k_1l_1}$ and $\mathbf{g}_{h_2k_2l_2}$ (Table A1.4). All equations in the tables were directly derived from the expressions for the direct and reciprocal metric tensors below.

(i) **Direct metric tensors:**

$$g^c = \begin{bmatrix} a^2 & 0 & 0 \\ 0 & a^2 & 0 \\ 0 & 0 & a^2 \end{bmatrix}, \quad g^t = \begin{bmatrix} a^2 & 0 & 0 \\ 0 & a^2 & 0 \\ 0 & 0 & c^2 \end{bmatrix},$$

$$g^o = \begin{bmatrix} a^2 & 0 & 0 \\ 0 & b^2 & 0 \\ 0 & 0 & c^2 \end{bmatrix}, \quad g^h = \begin{bmatrix} a^2 & -a^2/2 & 0 \\ -a^2/2 & a^2 & 0 \\ 0 & 0 & c^2 \end{bmatrix},$$

$$g^r = \begin{bmatrix} a^2 & a^2\cos\alpha & a^2\cos\alpha \\ a^2\cos\alpha & a^2 & a^2\cos\alpha \\ a^2\cos\alpha & a^2\cos\alpha & a^2 \end{bmatrix}, \quad g^m = \begin{bmatrix} a^2 & 0 & ac\cos\beta \\ 0 & b^2 & 0 \\ ac\cos\beta & 0 & c^2 \end{bmatrix},$$

$$g^a = \begin{bmatrix} a^2 & ab\cos\gamma & ac\cos\beta \\ ba\cos\gamma & b^2 & bc\cos\alpha \\ ca\cos\beta & cb\cos\alpha & c^2 \end{bmatrix}.$$

(ii) **Reciprocal metric tensors :**

$$g_c^* = \begin{bmatrix} 1/a^2 & 0 & 0 \\ 0 & 1/a^2 & 0 \\ 0 & 0 & 1/a^2 \end{bmatrix}, \quad g_t^* = \begin{bmatrix} 1/a^2 & 0 & 0 \\ 0 & 1/a^2 & 0 \\ 0 & 0 & 1/c^2 \end{bmatrix},$$

$$g_o^* = \begin{bmatrix} 1/a^2 & 0 & 0 \\ 0 & 1/b^2 & 0 \\ 0 & 0 & 1/c^2 \end{bmatrix}, \quad g_h^* = \begin{bmatrix} 4/3a^2 & 2/3a^2 & 0 \\ 2/3a^2 & 4/3a^2 & 0 \\ 0 & 0 & 1/c^2 \end{bmatrix},$$

$$g_r^* = \frac{1}{W^2}\begin{bmatrix} 1+\cos\alpha & -\frac{1-\tan^2\alpha/2}{2} & -\frac{1-\tan^2\alpha/2}{2} \\ -\frac{1-\tan^2\alpha/2}{2} & 1+\cos\alpha & -\frac{1-\tan^2\alpha/2}{2} \\ -\frac{1-\tan^2\alpha/2}{2} & -\frac{1-\tan^2\alpha/2}{2} & 1+\cos\alpha \end{bmatrix},$$

with

$$W^2 = a^2(1+\cos\alpha - 2\cos^2\alpha),$$

$$g_m^* = \begin{bmatrix} \frac{1}{a^2\sin^2\beta} & 0 & -\frac{\cos\beta}{ac\sin^2\beta} \\ 0 & \frac{1}{b^2} & 0 \\ -\frac{\cos\beta}{ac\sin^2\beta} & 0 & \frac{1}{c^2\sin^2\beta} \end{bmatrix}$$

$$g_a^* = \frac{1}{\Omega^2}\begin{bmatrix} b^2c^2\sin^2\alpha & abc^2\mathcal{F}(\alpha,\beta,\gamma) & ab^2c\mathcal{F}(\gamma,\alpha,\beta) \\ abc^2\mathcal{F}(\alpha,\beta,\gamma) & a^2c^2\sin^2\beta & a^2bc\mathcal{F}(\beta,\gamma,\alpha) \\ ab^2c\mathcal{F}(\gamma,\alpha,\beta) & a^2bc\mathcal{F}(\beta,\gamma,\alpha) & a^2b^2\sin^2\gamma \end{bmatrix}$$

where

$$\mathcal{F}(\alpha,\beta,\gamma) = \cos\alpha\cos\beta - \cos\gamma$$

and

$$\Omega^2 = a^2b^2c^2(1-\cos^2\alpha - \cos^2\beta - \cos^2\gamma + 2\cos\alpha\cos\beta\cos\gamma).$$

Table A1.1. *Expressions for the length t of a vector $\mathbf{t}_{uvw}$ in the seven crystal systems. The primes on the hexagonal components indicate that the 3-index notation must be used.*

System	l	Expression
Cubic	${}^{c}t$	$a\{u^2 + v^2 + w^2\}^{1/2}$
Tetragonal	${}^{t}t$	$\{a^2(u^2 + v^2) + c^2w^2\}^{1/2}$
Orthorhombic	${}^{o}t$	$\{a^2u^2 + b^2v^2 + c^2w^2\}^{1/2}$
Hexagonal	${}^{h}t$	$\{a^2(u'^2 + v'^2 - u'v') + c^2w'^2\}^{1/2}$
Rhombohedral	${}^{r}t$	$a\{u^2 + v^2 + w^2 + 2\cos\alpha[uv + uw + vw]\}^{1/2}$
Monoclinic	${}^{m}t$	$\{a^2u^2 + b^2v^2 + c^2w^2 + 2acuw\cos\beta\}^{1/2}$
Triclinic	${}^{a}t$	$\{a^2u^2 + b^2v^2 + c^2w^2 + 2bcvw\cos\alpha$ $+2acuw\cos\beta + 2abuv\cos\gamma\}^{1/2}$

Table A1.2. *Expressions for the cosine of the angle α between two vectors $\mathbf{t}_{u_1v_1w_1}$ and $\mathbf{t}_{u_2v_2w_2}$ in the seven crystal systems. The primes on the hexagonal components indicate that the 3-index notation must be used.*

System	$\cos\alpha$
Cubic	$\dfrac{a^2(u_1u_2+v_1v_2+w_1w_2)}{{}^{c}t_1\times{}^{c}t_2}$
Tetragonal	$\dfrac{a^2(u_1u_2+v_1v_2)+c^2w_1w_2}{{}^{t}t_1\times{}^{t}t_2}$
Orthorhombic	$\dfrac{a^2u_1u_2+b^2v_1v_2+c^2w_1w_2}{{}^{o}t_1\times{}^{o}t_2}$
Hexagonal	$\dfrac{a^2(u'_1u'_2+v'_1v'_2-\frac{1}{2}(u'_1v'_2+v'_1u'_2))+c^2w'_1w'_2}{{}^{h}t_1\times{}^{h}t_2}$
Rhombohedral	$\dfrac{a^2(u_1u_2+v_1v_2+w_1w_2+\cos\alpha[u_1(v_2+w_2)+v_1(u_2+w_2)+w_1(u_2+v_2)])}{{}^{r}t_1\times{}^{r}t_2}$
Monoclinic	$\dfrac{a^2u_1u_2+b^2v_1v_2+c^2w_1w_2+ac(w_1u_2+u_1w_2)\cos\beta}{{}^{m}t_1\times{}^{m}t_2}$
Triclinic	$\dfrac{a^2u_1u_2+b^2v_1v_2+c^2w_1w_2+bc(v_1w_2+v_2w_1)\cos\alpha+ac(u_1w_2+u_2w_1)\cos\beta+ab(u_1v_2+u_2v_1)\cos\gamma}{{}^{a}t_1\times{}^{a}t_2}$

Table A1.3. *Expressions for the length $|\mathbf{g}| = 1/d_{hkl}$ of a reciprocal lattice vector $\mathbf{g}_{hkl}$ in the seven crystal systems.*

System	g	Expression
Cubic	${}^c g$	$\frac{1}{a}\left\{h^2+k^2+l^2\right\}^{1/2}$
Tetragonal	${}^t g$	$\left\{\frac{1}{a^2}(h^2+k^2)+\frac{1}{c^2}l^2\right\}^{1/2}$
Orthorhombic	${}^o g$	$\left\{\frac{1}{a^2}h^2+\frac{1}{b^2}k^2+\frac{1}{c^2}l^2\right\}^{1/2}$
Hexagonal	${}^h g$	$\left\{\frac{4}{3a^2}(h^2+k^2+hk)+\frac{1}{c^2}l^2\right\}^{1/2}$
Rhombohedral	${}^r g$	$\frac{1}{a}\left\{\frac{(1+\cos^2\alpha)(h^2+k^2+l^2)-(1-\tan^2\alpha/2)(hk+kl+lh)}{1+\cos\alpha-2\cos^2\alpha}\right\}^{1/2}$
Monoclinic	${}^m g$	$\left\{\frac{1}{a^2}\frac{h^2}{\sin^2\beta}+\frac{1}{b^2}k^2+\frac{1}{c^2}\frac{l^2}{\sin^2\beta}-\frac{2hl\cos\beta}{ac\sin^2\beta}\right\}^{1/2}$
Triclinic	${}^a g$	$\frac{1}{\Omega}\{h^2b^2c^2\sin^2\alpha+k^2a^2c^2\sin^2\beta+l^2a^2b^2\sin^2\gamma$ $+ 2hkabc^2\mathcal{F}(\alpha,\beta,\gamma)+2kla^2bc\mathcal{F}(\beta,\gamma,\alpha)$ $+ 2lhab^2c\mathcal{F}(\gamma,\alpha,\beta)\}^{1/2}$ with $\Omega=\{a^2b^2c^2(1-\cos^2\alpha-\cos^2\beta-\cos^2\gamma$ $+2\cos\alpha\cos\beta\cos\gamma)\}^{1/2}$

Table A1.4. *Expressions for the cosine of the angle α between two vectors $\mathbf{g}_{h_1k_1l_1}$ and $\mathbf{g}_{h_2k_2l_2}$ in the seven crystal systems.*

System	$\cos\alpha$
Cubic	$\frac{1}{a^2}\frac{h_1h_2+k_1k_2+l_1l_2}{{}^c g_1\times{}^c g_2}$
Tetragonal	$\frac{\frac{1}{a^2}(h_1h_2+k_1k_2)+\frac{1}{c^2}l_1l_2}{{}^t g_1\times{}^t g_2}$
Orthorhombic	$\frac{\frac{1}{a^2}h_1h_2+\frac{1}{b^2}k_1k_2+\frac{1}{c^2}l_1l_2}{{}^o g_1\times{}^o g_2}$
Hexagonal	$\frac{\frac{4}{3a^2}(h_1h_2+k_1k_2+\frac{1}{2}(h_1k_2+k_1h_2))+\frac{1}{c^2}l_1l_2}{{}^h g_1\times{}^h g_2}$
Rhombohedral	$\frac{(1+\cos\alpha)(h_1h_2+k_1k_2+l_1l_2)-\frac{1}{2}(1-\tan^2\alpha/2)(h_1(k_2+l_2)+k_1(h_2+l_2)+l_1(h_2+k_2))}{a^2(1+\cos\alpha-2\cos^2\alpha)\times{}^r g_1\times{}^r g_2}$
Monoclinic	$\frac{\frac{1}{a^2\sin^2\beta}h_1h_2+\frac{1}{b^2}k_1k_2+\frac{1}{c^2\sin^2\beta}l_1l_2+(l_1h_2+h_1l_2)\frac{\cos\beta}{ac\sin^2\beta}}{{}^m g_1\times{}^m g_2}$
Triclinic	$\frac{1}{\Omega^2\times{}^a g_1\times{}^a g_2}\{h_1h_2b^2c^2\sin^2\alpha+k_1k_2a^2c^2\sin^2\beta+l_1l_2a^2b^2\sin^2\gamma$ $+abc^2(k_1h_2+k_2h_1)\mathcal{F}(\alpha,\beta,\gamma)$ $+ab^2c(h_1l_2+h_2l_1)\mathcal{F}(\beta,\gamma,\alpha)$ $+a^2bc(k_1l_2+k_2l_1)\mathcal{F}(\gamma,\alpha,\beta)\}$

Appendix A2

Physical constants

Table A2.1. *Some physical constants and unit conversions often used in electron microscopy.*

Name	Symbol	Value	Units
Speed of light in vacuum	c	299 792 458	m s^{-1}
Planck's constant	h	$6.626\,075 \times 10^{-34}$	J s
	h/e	$4.135\,669 \times 10^{-15}$	eV s
Boltzmann's constant	k	$1.380\,658 \times 10^{-23}$	J K^{-1}
	k/e	$8.617\,385 \times 10^{-5}$	eV K^{-1}
Permeability of vacuum	μ_0	$4\pi \times 10^{-7}$	–
Permittivity of vacuum	ϵ_0	$8.854\,187\,817 \times 10^{-12}$	F m^{-1}
Magnetic flux quantum	$\Phi_0 = h/2e$	$2.067\,834 \times 10^{-15}$	Wb
Elementary charge	e	$1.602\,177 \times 10^{-19}$	C
	e/h	$2.417\,988 \times 10^{14}$	A J^{-1}
Electron rest mass	m_0	$9.109\,389 \times 10^{-31}$	kg
Electron rest energy	m_0c^2	$\times 10$	J
		$5.109\,991 \times 10^{8}$	eV
Magnetic moment	μ_e	$9.284\,770 \times 10^{-24}$	J T^{-1}
Electronvolt	eV	$1.602\,177 \times 10^{-19}$	J

Appendix A3

Space group encoding and other software

This appendix contains the conversion table for the space group encoding scheme. Table A3.1 lists the point symmetry parts of the 14 basic matrices, and Table A3.2 lists the conversions for the components of translation vectors.

Due to space limitations, it is not possible to provide a detailed description of all the Fortran-90 code provided with this book. The reader will find such a description in PDF format on the website.

Table A3.1. *Explicit point symmetry matrices for the 14 matrices used to encode the space group generators.*

$a = \begin{pmatrix} 1 & 0 & 0 \\ 0 & 1 & 0 \\ 0 & 0 & 1 \end{pmatrix}$	$b = \begin{pmatrix} -1 & 0 & 0 \\ 0 & -1 & 0 \\ 0 & 0 & 1 \end{pmatrix}$	$c = \begin{pmatrix} -1 & 0 & 0 \\ 0 & 1 & 0 \\ 0 & 0 & -1 \end{pmatrix}$
$d = \begin{pmatrix} 0 & 0 & 1 \\ 1 & 0 & 0 \\ 0 & 1 & 0 \end{pmatrix}$	$e = \begin{pmatrix} 0 & 1 & 0 \\ 1 & 0 & 0 \\ 0 & 0 & -1 \end{pmatrix}$	$f = \begin{pmatrix} 0 & -1 & 0 \\ -1 & 0 & 0 \\ 0 & 0 & -1 \end{pmatrix}$
$g = \begin{pmatrix} 0 & -1 & 0 \\ 1 & 0 & 0 \\ 0 & 0 & 1 \end{pmatrix}$	$h = \begin{pmatrix} -1 & 0 & 0 \\ 0 & -1 & 0 \\ 0 & 0 & -1 \end{pmatrix}$	$i = \begin{pmatrix} 1 & 0 & 0 \\ 0 & 1 & 0 \\ 0 & 0 & -1 \end{pmatrix}$
$j = \begin{pmatrix} 1 & 0 & 0 \\ 0 & -1 & 0 \\ 0 & 0 & 1 \end{pmatrix}$	$k = \begin{pmatrix} 0 & -1 & 0 \\ -1 & 0 & 0 \\ 0 & 0 & 1 \end{pmatrix}$	$l = \begin{pmatrix} 0 & 1 & 0 \\ 1 & 0 & 0 \\ 0 & 0 & 1 \end{pmatrix}$
$m = \begin{pmatrix} 0 & 1 & 0 \\ -1 & 0 & 0 \\ 0 & 0 & -1 \end{pmatrix}$	$n = \begin{pmatrix} 0 & -1 & 0 \\ 1 & -1 & 0 \\ 0 & 0 & 1 \end{pmatrix}$	

Table A3.2. *Conversions for the components of translation vectors in the space group encoding scheme.*

$A = \frac{1}{6}$	$B = \frac{1}{4}$	$C = \frac{1}{3}$	$D = \frac{1}{2}$	$E = \frac{2}{3}$	$F = \frac{3}{4}$
$G = \frac{5}{6}$	$O = 0$	$X = -\frac{3}{8}$	$Y = -\frac{1}{4}$	$Z = -\frac{1}{8}$	

Appendix A4

Point groups and space groups

This appendix contains detailed information on the 32 crystallographic point groups. For each point group we list: the group symbol (both International and Schœnflies notations), the group order, the crystal class, the Laue class, the diffraction groups (see Chapter 5), the number of space groups related to the point group, the stereographic projection, and a rendered 3D view [DG98]. We also list the space group symbols, separated into the symmorphic and non-symmorphic groups.

Table A4.1. *The 32 point groups, with International–[Schœnflies] symbols, stereographic projection (S.P.), rendered 3D view, and important properties.*

Point group information	Stereographic projection	Rendered 3D view
Order: 1 **Crystal Class:** a **Laue Class:** $\bar{1}$ **Diffraction Groups:** **1** **# Space Groups:** 1	$\mathbf{1}-[C_1]$	
Order: 2 **Crystal Class:** a **Laue Class:** $\bar{1}$ **Diffraction Groups:** $\mathbf{2}_R$ **# Space Groups:** 1	$\bar{\mathbf{1}}-[C_i]$	

Table A4.1. *(cont.)*.

Point group information	Stereographic projection	Rendered 3D view
Order: 2 **Crystal Class:** m **Laue Class: 2/m** **Diffraction Groups:** **1**, **2**, $\mathbf{m}_R$ **# Space Groups:** 3	**2** – $[C_2]$ 	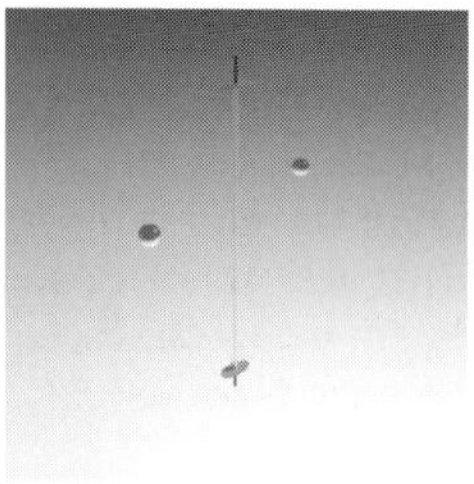
Order: 2 **Crystal Class:** m **Laue Class: 2/m** **Diffraction Groups:** **1**, $\mathbf{1}_R$, **m** **# Space Groups:** 4	**m** – $[C_s]$ 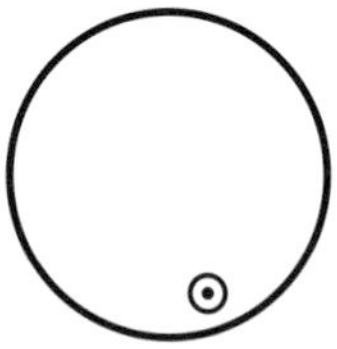	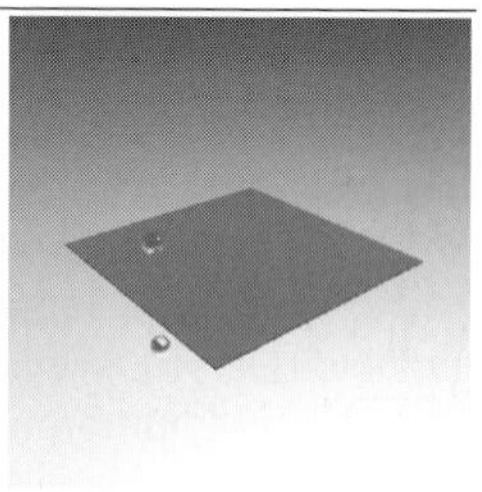
Order: 4 **Crystal Class:** m **Laue Class: 2/m** **Diffraction Groups:** $\mathbf{2}_R$, $\mathbf{21}_R$, $\mathbf{2}_R\mathbf{mm}_R$ **# Space Groups:** 6	**2/m** – $[C_{2h}]$ 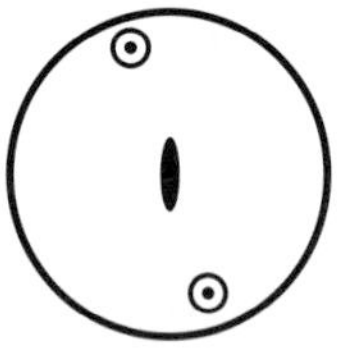	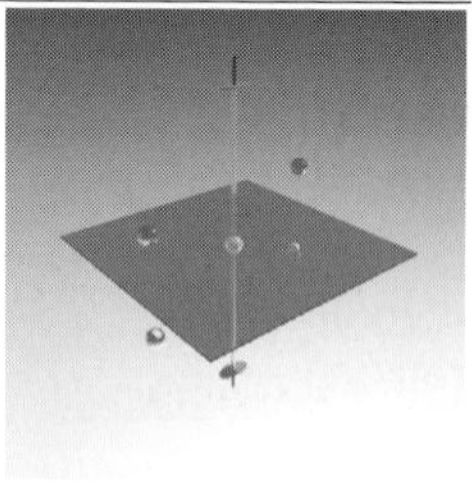
Order: 4 **Crystal Class:** o **Laue Class: mmm** **Diffraction Groups:** **1**, $\mathbf{m}_R$, $\mathbf{2m}_R\mathbf{m}_R$ **# Space Groups:** 9	**222** – $[D_2]$ 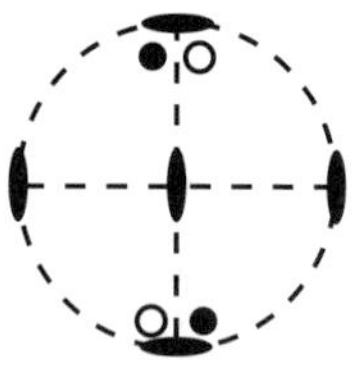	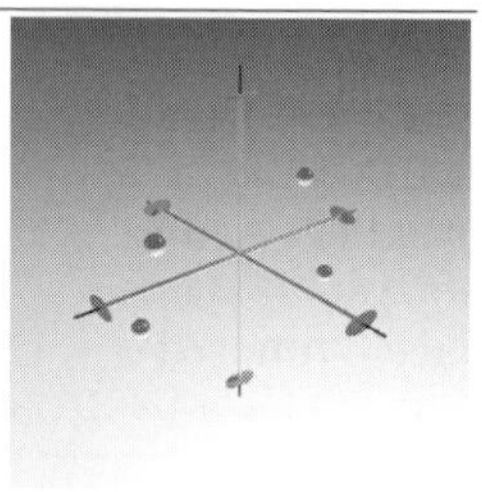
Order: 4 **Crystal Class:** o **Laue Class: mmm** **Diffraction Groups:** **1**, $\mathbf{m}_R$, **m**, $\mathbf{m1}_R$, **2mm** **# Space Groups:** 22	**mm2** – $[C_{2v}]$ 	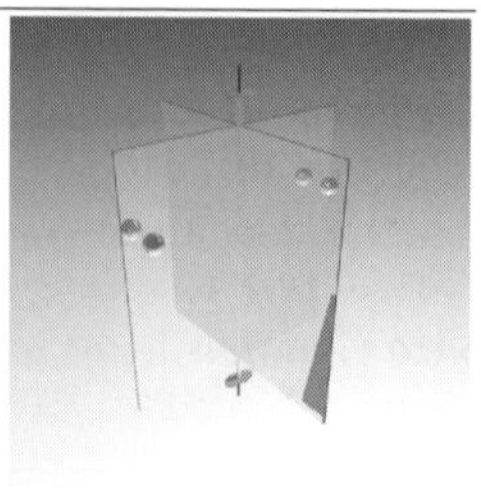

Table A4.1. *(cont.)*.

Point group information	Stereographic projection	Rendered 3D view
Order: 8 **Crystal Class:** o **Laue Class: mmm** **Diffraction Groups:** $\mathbf{2}_R$, $\mathbf{2}_R\mathbf{mm}_R$, $\mathbf{2mm1}_R$ **# Space Groups:** 28	**mmm** – $[C_{2v}]$ 	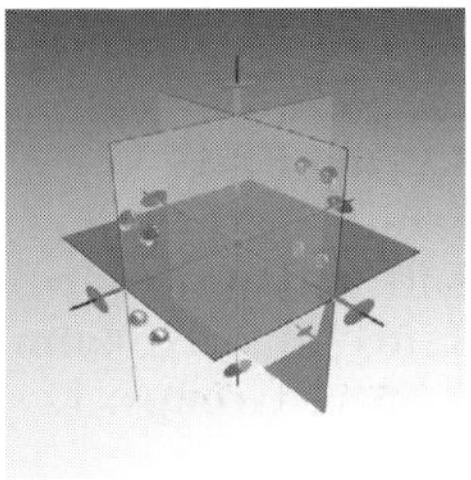
Order: 4 **Crystal Class:** t **Laue Class: 4/m** **Diffraction Groups:** **1**, $\mathbf{m}_R$, **4** **# Space Groups:** 6	**4** – $[C_4]$ 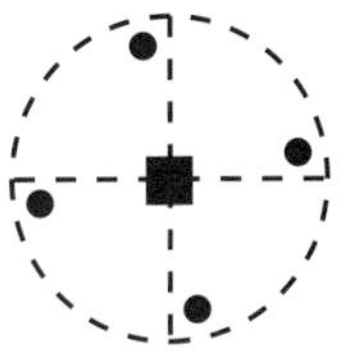	
Order: 4 **Crystal Class:** t **Laue Class: 4/m** **Diffraction Groups:** **1**, $\mathbf{m}_R$, $\mathbf{4}_R$ **# Space Groups:** 2	$\bar{\mathbf{4}}$ – $[S_4]$ 	
Order: 8 **Crystal Class:** t **Laue Class: 4/m** **Diffraction Groups:** $\mathbf{2}_R$, $\mathbf{2}_R\mathbf{mm}_R$, $\mathbf{41}_R$ **# Space Groups:** 6	**4/m** – $[C_{4m}]$ 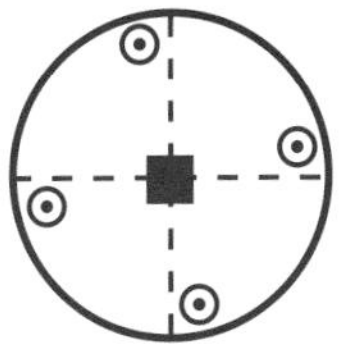	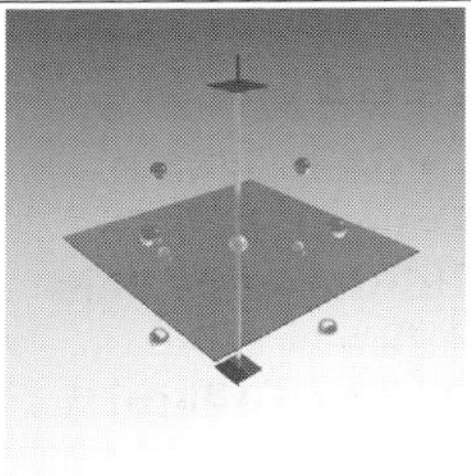
Order: 8 **Crystal Class:** t **Laue Class: 4/mmm** **Diffraction Groups:** **1**, $\mathbf{m}_R$, $\mathbf{2m}_R\mathbf{m}_R$, $\mathbf{4m}_R\mathbf{m}_R$ **# Space Groups:** 10	**422** – $[D_4]$ 	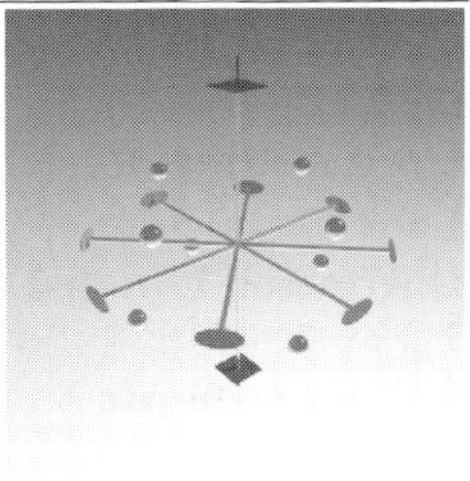

Table A4.1. *(cont.)*.

Point group information	Stereographic projection	Rendered 3D view
Order: 8 **Crystal Class:** t **Laue Class: 4/mmm** **Diffraction Groups:** **1**, $\mathbf{m}_R$, **m**, $\mathbf{m1}_R$, **4mm** **# Space Groups:** 12	**4mm** – $[C_{4v}]$ 	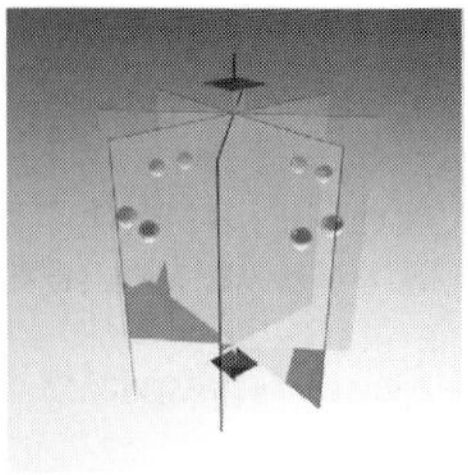
Order: 8 **Crystal Class:** t **Laue Class: 4/mmm** **Diffraction Groups:** **1**, $\mathbf{m}_R$, **m**, $\mathbf{m1}_R$, $\mathbf{2m}_R\mathbf{m}_R$, $\mathbf{4}_R\mathbf{mm}_R$ **# Space Groups:** 12	$\bar{\mathbf{4}}$**2m** – $[D_{2d}]$ 	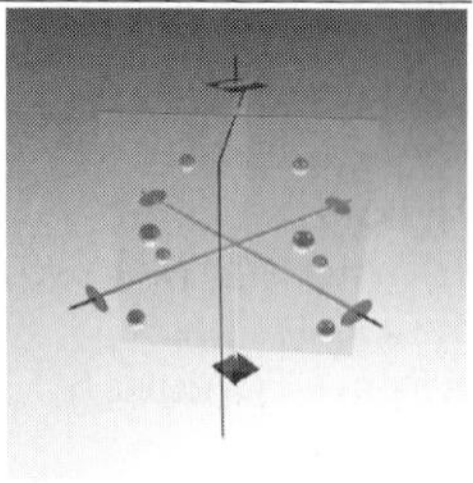
Order: 16 **Crystal Class:** t **Laue Class: 4/mmm** **Diffraction Groups:** $\mathbf{2}_R$, $\mathbf{2}_R\mathbf{mm}_R$, $\mathbf{2mm1}_R$, $\mathbf{4mm1}_R$ **# Space Groups:** 20	**4/mmm** – $[D_{4h}]$ 	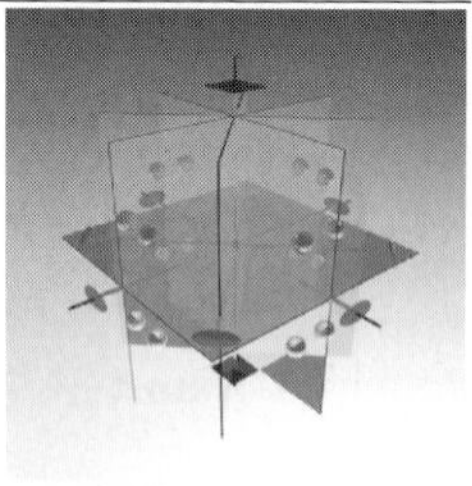
Order: 3 **Crystal Class:** R **Laue Class:** $\bar{\mathbf{3}}$ **Diffraction Groups:** **1**, **3** **# Space Groups:** 4	**3** – $[C_3]$ 	
Order: 6 **Crystal Class:** R **Laue Class:** $\bar{\mathbf{3}}$ **Diffraction Groups:** $\mathbf{2}_R$, $\mathbf{6}_R$ **# Space Groups:** 2	$\bar{\mathbf{3}}$ – $[C_{3i}]$ 	

Table A4.1. *(cont.)*.

Point group information	Stereographic projection	Rendered 3D view
Order: 6 **Crystal Class:** R **Laue Class:** $\bar{\mathbf{3}}\mathbf{m}$ **Diffraction Groups:** **1**, **2**, $\mathbf{m}_R$, $\mathbf{3m}_R$ **# Space Groups:** 7	**32** – [D_3] 	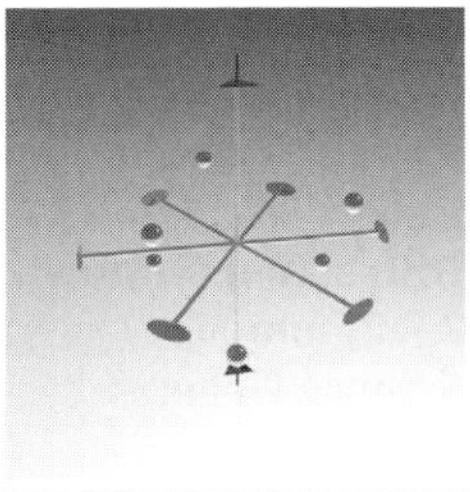
Order: 6 **Crystal Class:** R **Laue Class:** $\bar{\mathbf{3}}\mathbf{m}$ **Diffraction Groups:** **1**, $\mathbf{1}_R$, **m**, **3m** **# Space Groups:** 6	**3m** – [C_{3v}] 	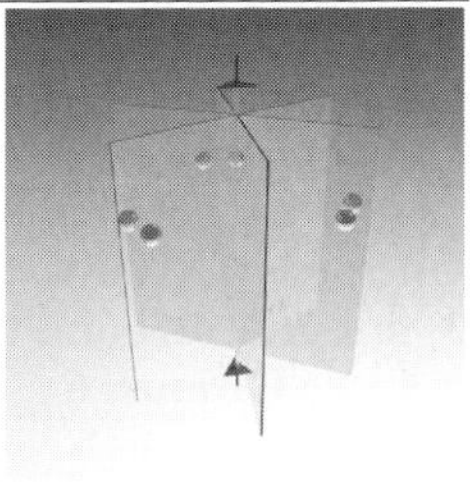
Order: 12 **Crystal Class:** R **Laue Class:** $\bar{\mathbf{3}}\mathbf{m}$ **Diffraction Groups:** **2**, $\mathbf{21}_R$, $\mathbf{2}_R\mathbf{mm}_R$ $\mathbf{6}_R\mathbf{mm}_R$ **# Space Groups:** 6	$\bar{\mathbf{3}}\mathbf{m}$ – [D_{3d}] 	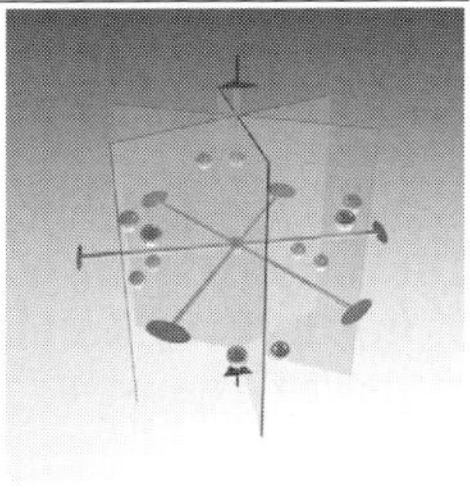
Order: 6 **Crystal Class:** h **Laue Class: 6/m** **Diffraction Groups:** **1**, $\mathbf{m}_R$, **6** **# Space Groups:** 6	**6** – [C_6] 	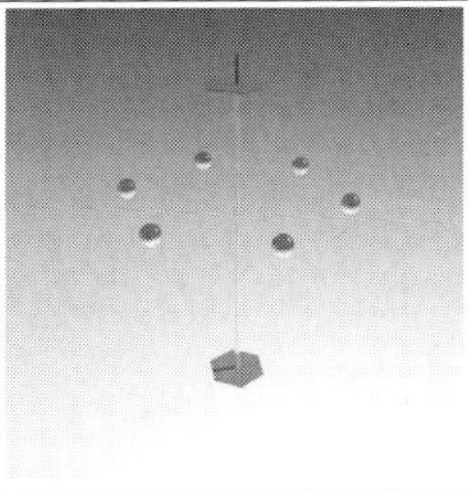
Order: 6 **Crystal Class:** h **Laue Class: 6/m** **Diffraction Groups:** **1**, **m**, $\mathbf{31}_R$ **# Space Groups:** 6	$\bar{\mathbf{6}}$ – [C_{3h}] 	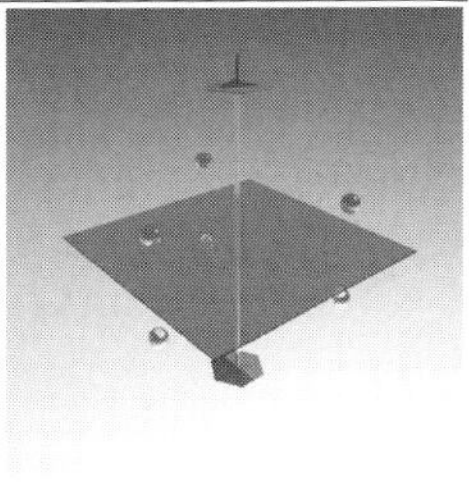

Table A4.1. *(cont.)*.

Point group information	Stereographic projection	Rendered 3D view
Order: 12 **Crystal Class:** h **Laue Class: 6/m** **Diffraction Groups:** $\mathbf{2}_R$, $\mathbf{2}_R\mathbf{mm}_R$, $\mathbf{61}_R$ **# Space Groups:** 2	**6/m** – $[C_{6h}]$ 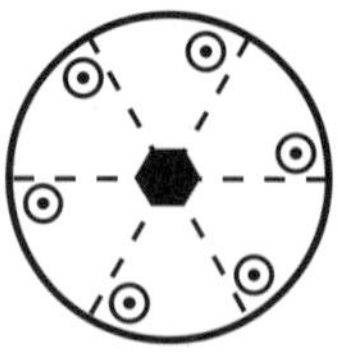	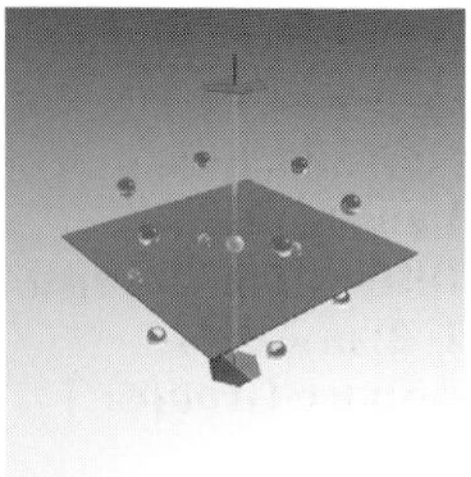
Order: 12 **Crystal Class:** h **Laue Class: 6/mmm** **Diffraction Groups:** **1**, $\mathbf{m}_R$, $\mathbf{2m}_R\mathbf{m}_R$, $\mathbf{6m}_R\mathbf{m}_R$ **# Space Groups:** 6	**622** – $[D_6]$ 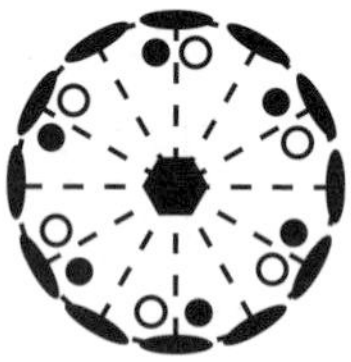	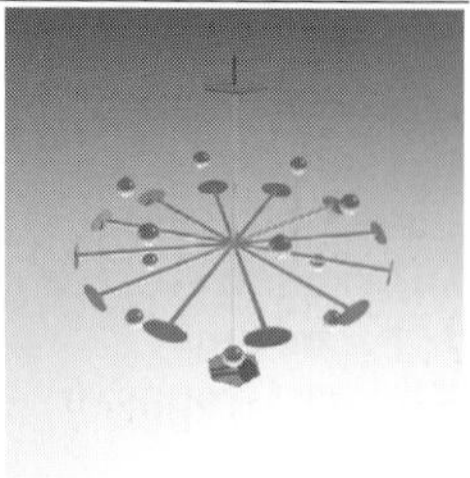
Order: 12 **Crystal Class:** h **Laue Class: 6/mmm** **Diffraction Groups:** **1**, $\mathbf{m}_R$, **m**, $\mathbf{m1}_R$, **6mm** **# Space Groups:** 4	**6mm** – $[C_{6v}]$ 	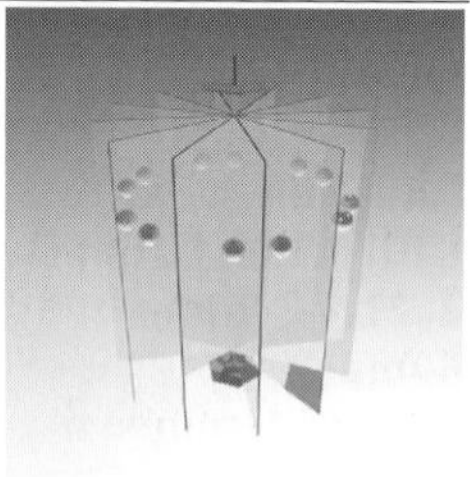
Order: 12 **Crystal Class:** h **Laue Class: 6/mmm** **Diffraction Groups:** **1**, $\mathbf{m}_R$, **m**, $\mathbf{m1}_R$, **2mm**, $\mathbf{3m1}_R$ **# Space Groups:** 4	$\bar{\mathbf{6}}$**m2** – $[D_{3h}]$ 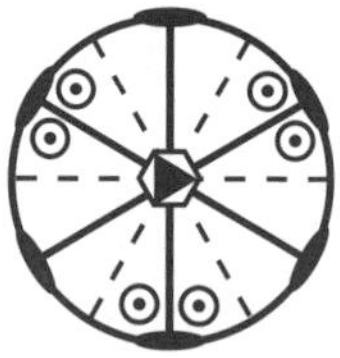	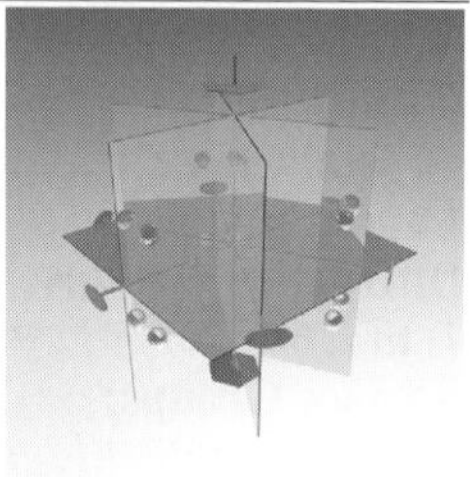
Order: 24 **Crystal Class:** h **Laue Class: 6/mmm** **Diffraction Groups:** $\mathbf{2}_R$, $\mathbf{2}_R\mathbf{mm}_R$, $\mathbf{2mm1}_R$, $\mathbf{6mm1}_R$ **# Space Groups:** 4	**6/mmm** – $[D_{6h}]$ 	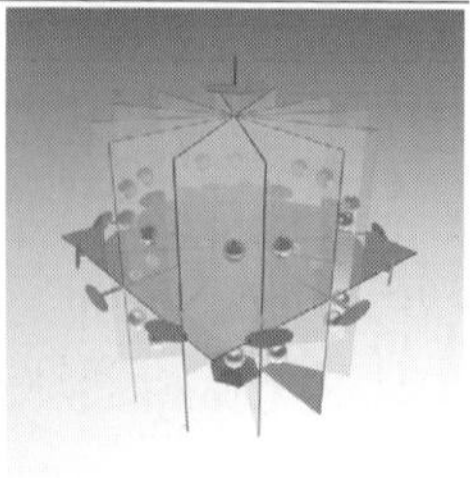

Table A4.1. *(cont.)*.

Point group information	Stereographic projection	Rendered 3D view
Order: 12 **Crystal Class:** c **Laue Class:** $\mathbf{m\bar{3}}$ **Diffraction Groups:** **1**, $\mathbf{m}_R$, $\mathbf{2m}_R\mathbf{m}_R$, **3**, **# Space Groups:** 5	**23** – [T] 	
Order: 24 **Crystal Class:** c **Laue Class:** $\mathbf{m\bar{3}}$ **Diffraction Groups:** $\mathbf{2}_R$, $\mathbf{2}_R\mathbf{mm}_R$, $\mathbf{2mm1}_R$, $\mathbf{6}_R$ **# Space Groups:** 7	$\mathbf{m\bar{3}}$ – [T_h] 	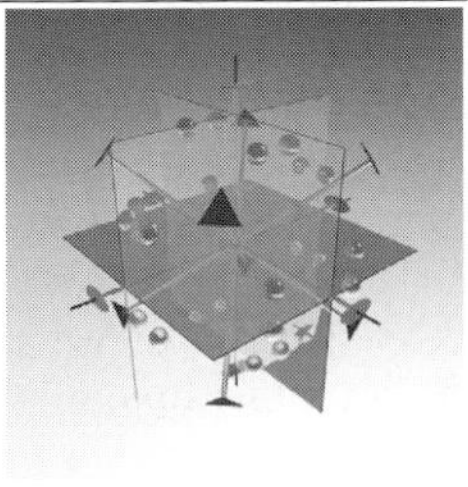
Order: 24 **Crystal Class:** c **Laue Class:** $\mathbf{m\bar{3}m}$ **Diffraction Groups:** **1**, $\mathbf{m}_R$, $\mathbf{2m}_R\mathbf{m}_R$, $\mathbf{4m}_R\mathbf{m}_R$, $\mathbf{3m}_R$ **# Space Groups:** 8	**432** – [O] 	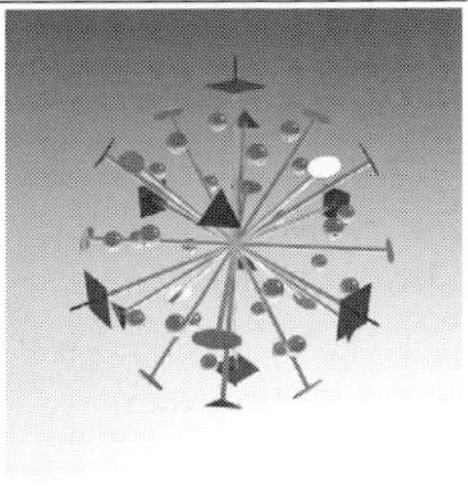
Order: 24 **Crystal Class:** c **Laue Class:** $\mathbf{m\bar{3}m}$ **Diffraction Groups:** **1**, $\mathbf{m}_R$, **m**, $\mathbf{m1}_R$, $\mathbf{4}_R\mathbf{mm}_R$, **3m** **# Space Groups:** 6	$\mathbf{\bar{4}3m}$ – [T_d] 	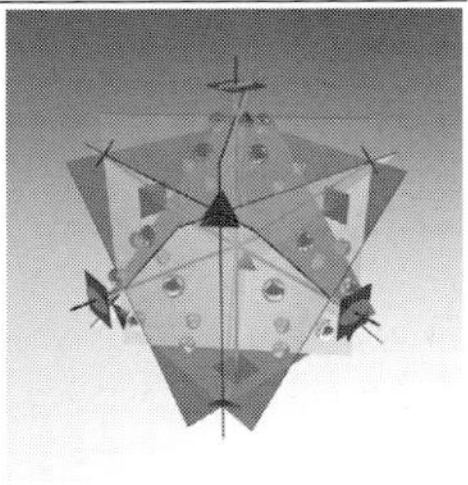
Order: 48 **Crystal Class:** c **Laue Class:** $\mathbf{m\bar{3}m}$ **Diffraction Groups:** $\mathbf{2}_R$, $\mathbf{2}_R\mathbf{mm}_R$, $\mathbf{2mm1}_R$, $\mathbf{4mm1}_R$, $\mathbf{6}_R\mathbf{mm}_R$ **# Space Groups:** 10	$\mathbf{m\bar{3}m}$ – [O_h] 	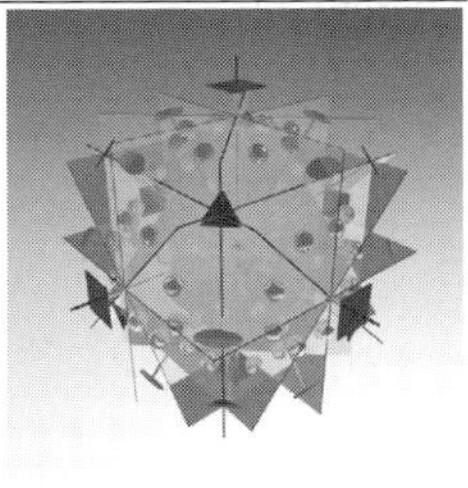

Table A4.2. *The 73 symmorphic space groups, with sequential number and corresponding point group.*

S.G.#	PG	Symbol	S.G.#	PG	Symbol
1	1	**P1**	139		**I4/mmm**
2	$\bar{1}$	**P$\bar{1}$**	143	3	**P3**
3	2	**P2**	146		**R3**
5		**C2**	147	$\bar{3}$	**P$\bar{3}$**
6	m	**Pm**	148		**R$\bar{3}$**
8		**Cm**	149	32	**P312**
10	$2/m$	**P2/m**	150		**P321**
12		**C2/m**	155		**R32**
16	222	**P222**	156	$3m$	**P3m1**
21		**C222**	157		**P31m**
22		**F222**	160		**R3m**
23		**I222**	162	$\bar{3}m$	**P$\bar{3}$1m**
25	$mm2$	**Pmm2**	164		**P$\bar{3}$m1**
35		**Cmm2**	165		**R$\bar{3}$m**
38		**Amm2**	168	6	**P6**
42		**Fmm2**	174	$\bar{6}$	**P$\bar{6}$**
44		**Imm2**	175	$6/m$	**P6/m**
47	mmm	**Pmmm**	177	622	**P622**
65		**Cmmm**	183	$6mm$	**P6mm**
69		**Fmmm**	187	$\bar{6}m2$	**P$\bar{6}$m2**
71		**Immm**	189		**P$\bar{6}$2m**
75	4	**P4**	191	$6/mmm$	**P6/mmm**
79		**I4**	195	23	**P23**
81	$\bar{4}$	**P$\bar{4}$**	196		**F23**
82		**I$\bar{4}$**	197		**I23**
83	$4/m$	**P4/m**	200	$m\bar{3}$	**Pm$\bar{3}$**
87		**I4/m**	202		**Fm$\bar{3}$**
89	422	**P422**	204		**Pm$\bar{3}$**
97		**I422**	207	432	**P432**
99	$4mm$	**P4mm**	209		**F432**
107		**I4mm**	211		**I432**
111	$\bar{4}2m$	**P$\bar{4}$2m**	215	$\bar{4}3m$	**P$\bar{4}$3m**
115	$\bar{4}m2$	**P$\bar{4}$m2**	216		**F$\bar{4}$3m**
119	$\bar{4}m2$	**I$\bar{4}$m2**	217		**I$\bar{4}$3m**
121	$\bar{4}2m$	**I$\bar{4}$2m**	221	$m\bar{3}m$	**Pm$\bar{3}$m**
123	$4/mmm$	**P4/mmm**	225		**Fm$\bar{3}$m**
			229		**Im$\bar{3}$m**

Table A4.3. *The non-symmorphic space groups, with sequential number and corresponding point group.*

S.G.#	PG	Symbol	S.G.#	PG	Symbol
4	2	$\mathbf{P2_1}$	57		**Pbcm**
7	*m*	**Pc**	58		**Pnnm**
9		**Cc**	59		**Pmmn***
11	2/*m*	$\mathbf{P2_1/m}$	60		**Pbcn**
13		**P2/c**	61		**Pbca**
14		$\mathbf{P2_1/c}$	62		**Pnma**
15		**C2/c**	63		**Cmcm**
17	222	$\mathbf{P222_1}$	64		**Cmca**
18		$\mathbf{P2_12_12}$	66		**Cccm**
19		$\mathbf{P2_12_12_1}$	67		**Cmma**
20		$\mathbf{C222_1}$	68		**Ccca***
24		$\mathbf{I2_12_12_1}$	70		**Fddd***
26	*mm*2	$\mathbf{Pmc2_1}$	72		**Ibam**
27		**Pcc2**	73		**Ibca**
28		**Pma2**	74		**Imma**
29		$\mathbf{Pca2_1}$	76	4	$\mathbf{P4_1}$
30		**Pnc2**	77		$\mathbf{P4_2}$
31		$\mathbf{Pmn2_1}$	78		$\mathbf{P4_3}$
32		**Pba2**	80		$\mathbf{I4_1}$
33		$\mathbf{Pna2_1}$	84	4/*m*	$\mathbf{P4_2/m}$
34		**Pnn2**	85		**P4/n***
36		$\mathbf{Cmc2_1}$	86		$\mathbf{P4_2/n}^*$
37		**Ccc2**	88		$\mathbf{I4_1/a}^*$
39		**Abm2**	90	422	$\mathbf{P41_12}$
40		**Ama2**	91		$\mathbf{P4_122}$
41		**Aba2**	92		$\mathbf{P4_12_12}$
43		**Fdd2**	93		$\mathbf{P4_222}$
45		**Iba2**	94		$\mathbf{P4_22_12}$
46		**Ima2**	95		$\mathbf{P4_322}$
48	*mmm*	**Pnnn***	96		$\mathbf{P4_32_12}$
49		**Pccm**	98		$\mathbf{I4_122}$
50		**Pban***	100	4*mm*	**P4bm**
51		**Pmma**	101		$\mathbf{P4_2cm}$
52		**Pnna**	102		$\mathbf{P4_2nm}$
53		**Pmna**	103		**P4cc**
54		**Pcca**	104		**P4nc**
55		**Pbam**	105		$\mathbf{P4_2mc}$
56		**Pccn**	106		$\mathbf{P4_2bc}$

*indicates multiple origin settings.

Table A4.3. *(cont.)*.

S.G.#	PG	Symbol	S.G.#	PG	Symbol
108	$4mm$	**I4cc**	170		**P6$_5$**
109		**I4$_1$md**	171		**P6$_2$**
110		**I4$_1$cd**	172		**P6$_4$**
112	$\bar{4}2m$	**P$\bar{4}$2c**	173		**P6$_3$**
113		**P$\bar{4}$2$_1$m**	176	$6/m$	**P6$_3$/m**
114		**P$\bar{4}$2$_1$c**	178	622	**P6$_1$22**
116		**C$\bar{4}$c2**	179		**P6$_5$22**
117		**P$\bar{4}$b2**	180		**P6$_2$22**
118		**P$\bar{4}$n2**	181		**P6$_4$22**
120		**I$\bar{4}$c2**	182		**P6$_3$22**
122		**I$\bar{4}$2d**	184	$6mm$	**P6cc**
124	$4/mmm$	**P4/mcc**	185		**P6$_3$cm**
125		**P4/nbm***	186		**P6$_3$mc**
126		**P4/nnc***	188	$\bar{6}m2$	**P$\bar{6}$c2**
127		**P4/mbm**	190		**P$\bar{6}$2c**
128		**P4/mnc**	192	$6/mmm$	**P6/mcc**
129		**P4/nmm***	193		**P6$_3$/mcm**
130		**P4/ncc***	194		**P6$_3$/mmc**
131		**P4$_2$/mmc**	198	23	**P2$_1$3**
132		**P4$_2$/mcm**	199		**I2$_1$3**
133		**P4$_2$/nbc***	201	$m\bar{3}$	**Pn$\bar{3}$***
134		**P4$_2$/nnm***	203		**Fd$\bar{3}$***
135		**P4$_2$/mbc**	205		**Pa$\bar{3}$**
136		**P4$_2$/mnm**	206		**Ia$\bar{3}$**
137		**P4$_2$/nmc***	208	432	**P4$_2$32**
138		**P4$_2$/ncm***	210		**F4$_1$32**
140		**I4/mcm**	212		**P4$_3$32**
141		**I4$_1$/amd***	213		**P4$_1$32**
142		**I4$_1$/acd***	214		**I4$_1$32**
144	3	**P3$_1$**	218	$\bar{4}3m$	**P$\bar{4}$3n**
145		**P3$_2$**	219		**F$\bar{4}$3c**
151	32	**P3$_1$12**	220		**I$\bar{4}$3d**
152		**P3$_1$21**	222	$m\bar{3}m$	**Pn$\bar{3}$n***
153		**P3$_2$12**	223		**Pm$\bar{3}$n**
154		**P3$_2$21**	224		**Pn$\bar{3}$m***
158	$3m$	**P3c1**	226		**Fm$\bar{3}$c**
159		**P31c**	227		**Fd$\bar{3}$m***
161		**R3c**	228		**Fd$\bar{3}$c***
163	$\bar{3}m$	**P$\bar{3}$1c**	230		**Ia$\bar{3}$d**
165		**P$\bar{3}$c1**			
167		**R$\bar{3}$c**			
169	6	**P6$_1$**			

List of symbols

Latin letters

(B_r, B_φ, B_z) Magnetic field components in cylindrical coordinates (p. 145)

A Elastic anisotropy factor (p. 488)

$A(\mathbf{q})$ Diffraction aperture function (p. 591)

a, b, c Lengths of Bravais basis vectors (p. 3)

a^*, b^*, c^* Length of reciprocal basis vectors (p. 14)

a_{ij} Direct structure matrix (p. 56)

B Debye–Waller factor (p. 127)

$B(z)$ Axial magnetic flux density (p. 145)

b_{ij} Reciprocal structure matrix (p. 59)

C Normalization constant (p. 84)

$C(\mathbf{r})$ Bloch wave function (p. 321)

$C, K, k, A,$ Seidel aberration

a, F, D, d Coefficients for a round lens (p. 168)

C_c Chromatic aberration coefficient (p. 192)

C_s Spherical aberration coefficient (p. 169)

$C_{a,x}, C_{a,y}$ Two-fold astigmatism (Cartesian) (p. 629)

c_{ijkl} Elastic moduli tensor (p. 482)

$C_\mathbf{g}, C_\mathbf{g}^{(j)}$ Bloch wave coefficient (p. 321)

D Distance between two points (p. 5)

D Normalized defocus (p. 594)

D Photographic density (p. 228)

$D(i, j)$ Background image intensity (p. 224)

$D(U_n)$ Transmission coefficient (p. 181)

$D(\mathbf{r})$ Shape function for crystal (p. 119)

$D^{(k)}$ Point symmetry matrix (p. 45)

$D^{*(k)}$ Reciprocal point symmetry matrix (p. 45)

d_{hkl} Interplanar spacing for (hkl) planes (p. 13)

D_{ij} Point symmetry matrix (p. 40)

D_i Depth of focus (p. 161)

D_o Depth of field (p. 161)

D_x, D_y Image displacement (Cartesian) (p. 630)

E Electron exposure (p. 228)

E Electrostatic acceleration potential (p. 92)

e Electron charge (p. 87)

$E_s(\mathbf{q})$ Spatial incoherence envelope function (p. 610)

$E_t(\mathbf{q})$ Temporal incoherence envelope function (p. 610)

$E_d(q)$ Drift envelope function (p. 617)
E_F Fermi energy level (p. 180)
e_{ijk} Normalized permutation symbol (p. 19)
E_t Total energy of a system (p. 84)
F Electric field applied to filament (p. 180)
f Focal length (p. 151)
f physical observable (p. 81)
$f(E, T)$ Fermi–Dirac distribution (p. 180)
$f(\Delta f)$ Defocus spread function (p. 608)
$f^e(\mathbf{s})$ Atomic scattering factor for electrons (p. 111)
$f^X(\mathbf{s})$ Atomic scattering factor for x-rays (p. 112)
f_i Asymptotic image focal length (p. 157)
f_n Eigenvalue for eigenstate $|\Psi_n\rangle$ (p. 82)
f_o Asymptotic object focal length (p. 157)
$g(E)$ Normalized energy distribution for electron emission (p. 190)
$G(i, j)$ Gain reference image intensity (p. 224)
G^*_{ij} Reciprocal 4-index metric tensor (p. 27)
g^*_{ij} Reciprocal metric tensor (p. 14)
G_{ij} 4-index metric tensor (hexagonal) (p. 25)
g_{ij} Direct metric tensor (p. 6)
g_n Normal component of reciprocal lattice vector (p. 325)
G_x, G_y Three-fold astigmatism (Cartesian) (p. 629)
g_{max} Maximum spatial frequency (p. 441)
H Hamiltonian (classical) (p. 84)
h Planck's constant (p. 83)
H_{mn} Matrix elements of beam–crystal interaction (p. 124)
$I_D(\mathbf{q})$ Diffuse intensity (p. 572)
I_S Dark field intensity (p. 350)
I_T Bright field intensity (p. 350)
I_d Intensity of a diffractogram (p. 627)
j_S Schottky current density (p. 183)
j_{TF} Thermal field emission current density (p. 184)
j_T Thermionic current density (p. 183)
k Wave number (p. 85)
k_B Boltzmann constant (p. 127)
K_0 General wave number (p. 94)
k_0 Wave number corrected for refraction (p. 305)
k_n Normal component of wave vector (p. 325)
K_x, K_y Coma coefficients (Cartesian) (p. 629)
L Camera length (p. 206)
L Column width (p. 466)
L_x, L_y Dimensions of sample (p. 347)
L_c Column width (p. 347)
$L_{\mathbf{gh}}$ Bloch wave matrix for ALCHEMI-like computations (p. 415)
M Transverse magnification of round lens (p. 158)
m Electron mass (relativistic) (p. 90)
m_0 Electron rest mass (p. 83)
M_α Angular magnification of round lens (p. 159)
n Diffraction order (in Bragg equation) (p. 97)
n Electron-optical refractive index (p. 95)
n_e Linear electron density (p. 195)
N_x, N_y Number of pixels in image (p. 347)
p Momentum (classical or relativistic) (p. 83)
Q Normalized spatial frequency (p. 594)
$q^{(j)}$ Imaginary part of Bloch wave vector $\mathbf{k}^{(j)}$ (p. 328)

q_c Nyquist critical frequency (p. 222)
q_S Frequency corresponding to point resolution (p. 601)
$q_{\mathbf{g}}$ Complex combination of extinction distance and absorption length (p. 309)
R 180° symmetry operator of a diffraction disk (p. 336)
r_1, r_2 General solutions of paraxial ray equation (p. 156)
s Signal-to-noise ratio (p. 614)
$S(z)$ Amplitude of the diffracted beam (p. 353)
$s(\mathbf{q})$ Source spread function (p. 608)
S_n Amplitude of scattered beam (p. 311)
$S_{\mathbf{gh}}$ Structure matrix for ALCHEMI-like computations (p. 415)
$s_{\mathbf{g}}$ Excitation error/deviation parameter (p. 121)
T Absolute temperature (p. 127)
T Kinetic energy (p. 83)
$T(z)$ Amplitude of the transmitted beam (p. 353)
$T(\mathbf{r})$ Point spread function (p. 223)
t, t' Time variable (p. 90)
T_{cc} Transmission cross-coefficient (p. 607)
$U'(\mathbf{r})$ Scaled electrostatic lattice potential (imaginary part) (p. 305)
$U(\mathbf{r})$ Scaled electrostatic lattice potential (real part) (p. 305)
$U(\mathbf{r})$ Scaled electrostatic lattice potential (p. 95)
U_n, U_t Normal and tangential energy component of emitted electron (p. 181)
u_o, u_a, u_i Complex coordinates in the object, lens, and image planes (p. 168)
V Potential energy (p. 84)
v Velocity (p. 83)
$V(z)$ Barrier potential (p. 180)
$V(\mathbf{r})$ Electrostatic lattice potential (p. 95)
$V^a(\mathbf{r})$ Electrostatic atom potential (p. 112)
$V_f(\mathbf{r})$ Finite crystal lattice potential (p. 119)
V_0 Mean inner potential (p. 304)
$V_c(\mathbf{r})$ Complex electrostatic lattice potential (p. 125)
V_p Projected potential (p. 317)
V_s Potential of sheared structure (p. 565)
V_x Block potential (p. 565)
$V_{\mathbf{g}}$ Fourier coefficients of electrostatic lattice potential (p. 116)
W Work function (p. 180)
w Dimensionless excitation error (p. 350)
$W(\mathbf{r})$ Optical potential (p. 125)
Z Atomic number (p. 112)
z_0 Crystal thickness (p. 312)
z_b Crystal thickness along beam direction (p. 466)
z_{Fi}, z_{Fo} Asymptotic object and image focal points (p. 156)
z_{No}, z_{Ni} Asymptotic object and image nodal points (p. 159)
z_{Pi}, z_{Po} Asymptotic object and image principal points (p. 157)
z_ξ Foil thickness in units of extinction distance (p. 350)
$\chi(q)$ Aberration function (p. 594)
δ Lattice misfit parameter (p. 479)
$\hat{T}$ Kinetic energy operator (p. 84)
$\hat{V}$ Potential energy operator (p. 84)
$\bar{I}$ Average image intensity (p. 223)
$\bar{U}_{\mathbf{g}}$ Dynamical or Bethe potential (p. 436)
$\bar{V}$ Scaled complex electrostatic lattice potential (p. 316)

$\bar{\mathbf{r}}$	Vector in normal coordinates (p. 41)
$\bar{\mathcal{A}}$	Bloch wave structure matrix (p. 325)
$\hat{B}_j$	Slice operator (p. 440)
$\mathbf{A}$	Magnetic vector potential (p. 146)
$\mathbf{a}, \mathbf{b}, \mathbf{c}$	Bravais lattice basis vectors (p. 2)
$\mathbf{A}, \mathbf{B}, \mathbf{C}, \mathbf{I}$	Lattice centering vectors (p. 3)
$\mathbf{a}^*, \mathbf{b}^*, \mathbf{c}^*$	Reciprocal basis vectors (p. 11)
$\mathbf{A}_i^*$	Reciprocal hexagonal basis vectors (4-index) (p. 27)
$\mathbf{a}_j^*$	Reciprocal basis vectors (p. 11)
$\mathbf{a}_i$	Bravais lattice basis vectors (p. 2)
$\mathbf{B}$	Magnetic field vector (p. 144)
$\mathbf{b}$	Burgers vector (p. 481)
$\mathbf{B}_\perp$	In-plane magnetic induction (p. 291)
$\mathbf{C}$	Center of Ewald sphere (p. 98)
$\mathbf{C}^{(j)}$	Column vector of Bloch wave coefficients (p. 325)
$\mathbf{E}$	Electric field vector (p. 146)
$\mathbf{e}_i^*$	Reciprocal Cartesian basis vectors (p. 19)
$\mathbf{e}_i$	Cartesian basis vectors (p. 19)
$\mathbf{e}_r, \mathbf{e}_\varphi, \mathbf{e}_z$	Basis vectors in cylindrical coordinates (p. 147)
$\mathbf{F}$	Foil normal (p. 122)
$\mathbf{F}$	Force vector (p. 90)
$\mathbf{F}_L$	Lorentz force (p. 291)
$\mathbf{G}$	Shortest reciprocal lattice vector of the FOLZ (p. 213)
$\mathbf{g}$	Reciprocal lattice vector (p. 11)
$\mathbf{g}_{hkl}$	Normal to the plane (hkl) (p. 13)
$\mathbf{J}$	Electron flux (p. 125)
$\mathbf{k}, \mathbf{k}', \ldots$	Wave vectors (p. 98)
$\mathbf{k}^{(j)}$	Wave vector of (j)th Bloch wave (p. 322)
$\mathbf{k}_0$	Wave vector corrected for refraction (p. 305)
$\mathbf{k}_t$	Tangential component of incident wave vector (p. 334)
$\mathbf{k}_{xy}$	2D wave vector (p. 316)
$\mathbf{k}_\mathbf{g}$	Tangential component of diffracted wave vector (p. 334)
$\mathbf{n}_\mathbf{m}$	Screw axis (p. 37)
$\mathbf{n}$	Unit normal (p. 13)
$\mathbf{O}$	Origin of reciprocal space (p. 98)
$\mathbf{p}$	Momentum vector (p. 83)
$\mathbf{p}, \mathbf{q}$	General position vectors (p. 5)
$\mathbf{q}$	Position vector in reciprocal space (p. 101)
$\mathbf{q}^{(k)}$	Star of $\mathbf{q}$ (p. 46)
$\mathbf{r}$	Position vector (p. 5)
$\mathbf{R}(\mathbf{r})$	Displacement field (p. 461)
$\mathbf{r}^{(k)}$	Orbit of $\mathbf{r}$ (p. 46)
$\mathbf{R}_1, \mathbf{R}_2$	Rhombohedral centering vectors (p. 54)
$\mathbf{S}$	Column vector of diffracted amplitudes (p. 312)
$\mathbf{S}$	Scattering vector (p. 373)
$\mathbf{s}$	Scattering vector (p. 111)
$\mathbf{T}$	Pitch vector of a screw axis (p. 37)
$\mathbf{t}$	Translation vector (p. 2)
$\mathbf{u}$	Line direction (dislocation) (p. 481)
$\mathbf{u}(t)$	Instantaneous atomic vibration amplitude (p. 126)
$\mathbf{U}(z)$	Paraxial ray vector (p. 157)
$\mathbf{u}_i(t)$	Instantaneous displacement of atom i (p. 374)
$\mathbf{v}$	Velocity vector (pp. 90, 146)
$\mathcal{A}$	Crystal transfer matrix (p. 311)
$\mathcal{C}$	Matrix of Bloch wave eigenvectors (p. 325)
$\mathcal{E}(z)$	Diagonal matrix with thickness dependent part of $\mathcal{S}$ (p. 327)
$\mathcal{G}$	General group (p. 42)

$\mathcal{I}_{jk}$ Matrix used for thickness integrated intensity computations (p. 415)
$\mathcal{L}_f$ Lens phase factor (p. 164)
$\mathcal{P}$ Phase matrix for planar defect (p. 500)
$\mathcal{P}_u$ Fresnel propagator (p. 163)
$\mathcal{R}, \mathcal{R}', \ldots$ Signal functions (p. 107)
$\mathcal{S}$ Scattering matrix (p. 312)
$\mathcal{T}$ Bravais lattice (p. 2)
$\mathcal{T}$ Diagonal part of $\mathcal{A}$ (p. 314)
$\mathcal{T}$ Instrument point spread function (p. 107)
$\mathcal{T}$ Lens transfer matrix (p. 158)
$\mathcal{T}^*$ Reciprocal lattice (p. 12)
$\mathcal{V}$ Off-diagonal part of $\mathcal{A}$ (p. 314)
$\hbar$ Planck's constant divided by 2π (p. 83)
$\hat{H}$ Hamiltonian operator (p. 84)
$\hat{H}_{cr}$ Crystal Hamiltonian (p. 123)
$\hat{H}_{int}$ Interaction Hamiltonian (p. 123)
$\hat{p}$ Momentum operator (p. 83)
$\hat{f}$ Quantum mechanical operator (p. 82)
$\{a, c\}$ Hexagonal lattice parameters (p. 54)
c Velocity of light (p. 90)
DQE Detective quantum efficiency (p. 223)
$var(I)$ Variance of image intensity (p. 223)

Greek letters

α Angle between $\mathbf{k}+\mathbf{g}$ and foil normal (p. 122)
α, β, γ Angles between Bravais basis vectors (p. 3)
$\alpha^{(j)}$ Bloch wave excitation amplitude (p. 322)
$\alpha^*, \beta^*, \gamma^*$ Angles between reciprocal basis vectors (p. 14)
$\alpha_{0,j}$ Warren–Cowley short-range order parameters (p. 572)
α_{ij} General coordinate transformation matrix (p. 48)
α_{jk} Difference between Bloch wave absorption coefficients (p. 415)
$\alpha_{\mathbf{g}}(\mathbf{r})$ Defect phase factor (p. 462)
β Beam brightness (p. 185)
β Velocity in units of velocity of light (p. 90)
β_{jk} Difference between Bloch wave eigenvalues (p. 415)
$\beta_{\mathbf{g}}$ Phase difference between real and imaginary part of lattice potential (p. 310)
Δ Path length difference (p. 162)
δ Magnetic domain wall width (p. 296)
δ sampling interval (p. 222)
$\delta(\mathbf{p}-\mathbf{q})$ Dirac delta-function (p. 84)
$\Delta\mathbf{p}$ Momentum transfer vector (p. 99)
δ_{ij} Kronecker delta (unit matrix) (p. 9)
ϵ $0.978\,45 \times 10^{-6}\ \mathrm{V}^{-1}$ (p. 93)
ϵ Slice thickness (p. 313)
ϵ_0 Permittivity of vacuum (p. 112)
ϵ_{ij} Strain tensor (p. 482)
$\eta_{\mathbf{g}}^{(j)}$ Shorthand notation for Bethe potential computation (p. 435)
γ Relativistic correction factor (p. 90)
$\Gamma^{(j)}$ Normal part of Bloch wave vector $\mathbf{k}^{(j)}$ with absorption (p. 328)
$\gamma^{(j)}$ Normal component of Bloch wave vector (p. 323)
$\hat{\Psi}$ Relativistic acceleration potential (p. 93)
$\hat{\Psi}_c$ Relativistic acceleration potential inside crystal (p. 95)
Λ Diagonal matrix with Bloch wave eigenvalues (p. 326)

λ Wavelength (p. 85)
λ_C Compton wavelength (p. 94)
λ_c Relativistic wavelength inside crystal (p. 95)
λ_{nr} Non-relativistic electron wavelength (p. 87)
$\lambda_r \equiv \lambda$ Relativistic electron wavelength (p. 92)
$\lambda_{sc,a}$ Spatial coherence width at condensor aperture plane (p. 194)
λ_{sc} Spatial coherence width (p. 193)
λ_{tc} Temporal coherence length (p. 193)
$\bar{\Delta}$ Two-dimensional complex differential operator (p. 316)
$\bar{\sigma}_j$ Flynn operator (p. 572)
$\hat{\Psi}_c$ Refraction corrected relativistic acceleration potential (p. 305)
μ Magnetic permeability (p. 145)
ν Frequency of a wave (p. 85)
ν Poisson's ratio (p. 482)
Ω Unit cell volume (p. 11)
ω Angular frequency of a wave (p. 85)
ω Scalar magnetic potential (p. 144)
$\Omega(S)$ Solid angle (p. 185)
Ω^* Volume reciprocal unit cell (p. 21)
ω_N FFT twiddle factors (p. 108)
Φ Electrostatic potential (p. 146)
$\Phi, \Psi, \ldots$ Wave function (p. 81)
$\phi_a(\mathbf{q})$ Aberrated wave function (p. 599)
$\psi, \phi, \ldots$ General wave functions (p. 88)
$\psi_{\mathbf{g}}$ Fourier component of wave function ψ (p. 306)
ρ Total charge density (p. 111)
ρ_c Chromatic information limit (p. 615)
ρ_D Detector information limit (p. 620)
ρ_d Drift information limit (p. 617)
ρ_e Electron charge density (p. 111)
ρ_n Nuclear charge density (p. 111)
ρ_S Microscope point resolution (p. 601)
ρ_{θ_c} Spatial incoherence information limit (p. 617)
σ Interaction constant (p. 93)
σ Two-beam parameter (p. 354)
$\sigma(I)$ Standard deviation of image intensity (p. 223)
σ_j^A Site occupation operator (p. 571)
σ_{ij} Stress tensor (p. 482)
τ Exposure time (p. 228)
θ Angle between two vectors (p. 7)
θ Bragg angle (p. 96)
θ_c Beam divergence angle (p. 198)
θ_{ij} Angle between basis vectors i and j (p. 6)
θ_L Lorentz deflection angle (p. 292)
$\theta_{\mathbf{g}}, \theta'_{\mathbf{g}}$ Phase of $U_{\mathbf{g}}$ and $U'_{\mathbf{g}}$ (p. 309)
$\xi'_{\mathbf{0}}$ Normal absorption length (p. 310)
$\xi'_{\mathbf{g}}$ Absorption length (p. 309)
$\xi_{\mathbf{g}}$ Extinction distance (p. 309)

Special symbols

(hkl) Miller indices of a plane (p. 10)
(r, φ, z) Cylindrical coordinates (p. 144)
$(x, y, z) = r_i$ Fractional atom coordinates (p. 5)
$(\mathbf{D}|\mathbf{t})$ Seitz symbol for symmetry operator (p. 40)
$(\odot, \mathbf{k}_t)$ decomposition of wave vector incident from below crystal (p. 334)

$(\otimes, \mathbf{k}_t)$	decomposition of wave vector incident from above crystal (p. 334)
$f \otimes g$	Convolution product (p. 106)
$\|a_n\rangle$	Crystal eigenstates (p. 123)
$\|\mathbf{q}\|$	Length or norm of vector **q** (p. 3)
$\|\Psi_n\rangle$	Eigenstate of quantum mechanical system (p. 82)
$\langle f \rangle$	Quantum mechanical expectation value (p. 81)
$\langle uvw \rangle$	Family of directions (p. 45)
$\langle \Phi\| f \|\Psi\rangle$	Scalar product of functions (p. 82)
Δ_{xy}	2D Laplacian operator (p. 316)
$\mathbf{1}$	Identity matrix (p. 312)
$\mathbf{p} \cdot \mathbf{q}$	Vector dot product (p. 6)
$\mathbf{p} \times \mathbf{q}$	Vector cross product (p. 18)
$\mathcal{F}^{-1}[\,]$	Inverse Fourier transform operator (p. 104)
$\mathcal{F}[\,]$	Direct Fourier transform operator (p. 104)
$\mathcal{W}$	Symmetry matrix in normal coordinates (p. 41)
∇	Gradient differential operator (p. 83)
∇^2, Δ	Laplacian operator (p. 84)
∇_{xy}	2D gradient operator (p. 316)
$\mathcal{O}$	General symmetry operator (p. 39)
$\{a_r, \alpha\}$	Rhombohedral lattice parameters (p. 54)
$\{hkl\}$	Family of planes (p. 45)
$[uvtw]$	Miller–Bravais direction indices (p. 23)
$[uvw]$	Crystallographic direction (p. 5)

Bibliography

[AB59] Y. Aharonov and D. Bohm. Significance of Electromagnetic Potentials in the Quantum Theory. *Phys. Rev.*, 115:485–491, 1959.

[AB63a] M.F. Ashby and L.M. Brown. Diffraction Contrast from Spherically Symmetrical Coherency Strains. *Phil. Mag.*, 8:1083–1103, 1963.

[AB63b] M.F. Ashby and L.M. Brown. On Diffraction Contrast from Inclusions. *Phil. Mag.*, 8:1649–1676, 1963.

[ABB$^+$99] E. Anderson, Z. Bai, C. Bischof, S. Blackford, J. Demmel, J. Dongarra, J. Du Croz, A. Greenbaum, S. Hammarling, A. McKenney, and D. Sorensen. *LAPACK Users' Guide, Third Edition.* Society for Industrial and Applied Mathematics, Philadelphia, 1999.

[AGRVL70] S. Amelinckx, R. Gevers, G. Renault, and J. Van Landuyt, editors. *Modern Diffraction and Imaging Techniques in Material Science.* North-Holland/American Elsevier, Amsterdam, 1970.

[AH82] S.M. Allen and E.L. Hall. Foil Thickness Measurements from Convergent-Beam Diffraction Patterns. An Experimental Assessment of Errors. *Phil. Mag. A*, 46:243–253, 1982.

[All81] S.M. Allen. Foil Thickness Measurements from Convergent-Beam Diffraction Patterns. *Phil. Mag. A*, 43:325–335, 1981.

[ALSW84] H.I. Aaronson, D.E. Lsughlin, R.F. Sekerka, and C.M. Wayman, editors. *Proceedings of International Conference on Solid→Solid Phase Transformations.* The Metallurgical Society of AIME, 1984.

[And90] R.M. Anderson, editor. *Specimen Preparation for Transmission Electron Microscopy of Materials II*, volume 199 of *Mat. Res. Soc. Symp. Proc.* Materials Research Society, Pittsburgh, PA, 1990.

[ans] `http://www.ansys.com`.

[AR93] L.J. Allen and C.J. Rossouw. Delocalization in Electron-Impact Ionization in a Crystalline Environment. *Phys. Rev. A*, 47:2446–2452, 1993.

[AS77] S. Abramovitz and I. Stegun. *Tables of Mathematical Functions.* Dover Publications, Inc., New York, 1977.

[ATB92] R.M. Anderson, B. Tracy, and J.C. Bravman, editors. *Specimen Preparation for Transmission Electron Microscopy of Materials III*, volume 254 of *Mat. Res. Soc. Symp. Proc.* Materials Research Society, Pittsburgh, PA, 1992.

[AVD93] S. Amelinckx and D. Van Dyck. Electron Diffraction Effects due to Modulated Structures. In J.M. Cowley, editor, *Electron Diffaction Techniques*, volume 2, pages 309–372. Oxford Science Publications, International Union of Crystallography, 1993.

[AVHVDVT92] S. Amelinckx, C. Van Heurck, D. Van Dyck, and G. Van Tendeloo. A Peculiar Diffraction Effect in FCC Crystals of C_{60}. *phys. stat. sol. (a)*, 131:589–604, 1992.

[AVL76] S. Amelinckx and J. Van Landuyt. Contrast Effects at Planar Interfaces. In H.R. Wenk, editor, *Electron Microscopy in Mineralogy*, pages 68–112. Springer-Verlag, Berlin, 1976.

[AW97] R.M. Anderson and S.D. Walck, editors. *Specimen Preparation for Transmission Electron Microscopy of Materials IV*, volume 480 of *Mat. Res. Soc. Symp. Proc.* Materials Research Society, Pittsburgh, PA, 1997.

[BA96] D. Bernaerts and S. Amelinckx. Different Forms of Carbon. In S. Amelinckx, D. Van Dyck, J. Van Landuyt, and G. Van Tendeloo, editors, *Handbook of Microscopy, vol. III*, pages 437–482. VCH, 1996.

[BAM88] J.C. Bravman, R.M. Anderson, and M.L. McDonald, editors. *Specimen Preparation for Transmission Electron Microscopy of Materials*, volume 115 of *Mat. Res. Soc. Symp. Proc.* Materials Research Society, Pittsburgh, PA, 1988.

[Bar94] W. Barlow. Über die Geometrischen Eigenschaften homogener starrer Strukturen. *Z. Kristallogr. Mineral.*, 23:1–63, 1894.

[BCE88] P. Buseck, J. Cowley, and L. Eyring, editors. *High-Resolution Transmission Electron Microscopy and Associated Techniques.* Oxford University Press, Oxford, 1988.

[BD95] C. Barreteau and F. Ducastelle. The Orthogonalized Plane-Wave Method Applied to the Calculation of Dynamical Effects in Electron Diffraction. *Ultramicroscopy*, 57:11–15, 1995.

[BDGVT$^+$87] D. Broddin, M. De Graef, G. Van Tendeloo, J. Van Landuyt, L. Delaey, and S. Amelinckx. Long Period Superstructures in Cu-Al. *Micr. Microsc. Acta*, 18:239–240, 1987.

[BESR76] B.F. Buxton, J.A. Eades, J.W. Steeds, and G.M. Rackham. The Symmetry of Electron Diffraction Zone Axis Patterns. *Phil. Trans. R. Soc.*, 281:171–194, 1976.

[Bet28] H.A. Bethe. Theorie der Beugung von Elektronen an Kristallen. *Ann. d. Physik*, 87:55–129, 1928.

[BG76] T.F. Buddinger and R.M. Glaeser. Measurement of Focus and Spherical Aberration of an Electron Microscope Objective Lens. *Ultramicroscopy*, 2:31–41, 1976.

[BG90] G. Burns and A.M. Glazer. *Space Groups for Solid State Scientists.* Academic Press, Inc., New York, 2nd edition, 1990.

[BG94] R.W. Brankin and I. Gladwell. *RKSUITE90, Release 1.0*, 1994.

[Bir84] J.L. Birman. *Theory of Crystal Space Groups and Lattice Dynamics.* Springer-Verlag, Berlin, 1984.

[Bir89] D.M. Bird. Theory of Zone Axis Electron Diffraction. *J. Electron Microsc. Techn.*, 13:77–97, 1989.

[BK90] D.M. Bird and Q.A. King. Absorptive Form Factors for High Energy Electron Diffraction. *Acta Crystall. A*, 46:202–208, 1990.

[BL80] J.D.E. Beynon and D.R. Lamb. *Charge-coupled Devices and their Applications.* McGraw-Hill Book Company, London, 1980.

[Blo29] F. Bloch. Bemerkung zur Elektronentheorie des Ferromagnetismus und der elektrischen Leitfähigkeit. *Z. Phys.*, 57:545–555, 1929.

[BLS78] B.F. Buxton, J.E. Loveluck, and J.W. Steeds. Bloch Waves and their Corresponding Atomic and Molecular Orbitals in High Energy Electron Diffraction. *Phil. Mag. A*, 38:259–278, 1978.

[Boe54] H. Boersch. Experimentelle Bestimmung der Energieverteilung in thermisch ausgelösten Elektronenstrahlen. *Z. Phys.*, 139:115–146, 1954.

[Boh13a] N.H. Bohr. Constitution of Atoms and Molecules. *Phil. Mag.*, 26:1–25, 1913.

[Boh13b] N.H. Bohr. Constitution of Atoms and Molecules II. *Phil. Mag.*, 26:476–502, 1913.

[Boh13c] N.H. Bohr. Constitution of Atoms and Molecules III. *Phil. Mag.*, 26:857–875, 1913.

[Bol56] W. Bollman. Interference Effects in the Electron Microscopy of Thin Crystal Foils. *Phys. Rev.*, 103:1588–1589, 1956.

[BPN00] A. Barty, D. Paganin, and K. Nugent. Phase Retrieval in Lorentz Microscopy. In *Magnetic Microscopy and its Applications to Magnetic Materials*, Chapter 5. Academic Press, New York, 2000.

[Bra50] A. Bravais. Memoire sur les Systèmes Formé par des Points Distribués Regulierement sur un Plan ou dans l'Espace. *J. Ec. Polytech.*, 19:1–128, 1850.

[Bra15] W.H. Bragg. IX-th Bakerian Lecture: X-rays and Crystal Structure. *Phil. Mag.*, 215:253–274, 1915.

[Bra86] R.N. Bracewell. *The Fourier Transform and its Applications*. McGraw-Hill International Editions, New York, 2nd edition, 1986.

[BS70] W.S. Boyle and G.E. Smith. Charge-coupled Semiconductor Devices. *Bell Syst. Tech, Journ.*, 49:587–593, 1970.

[BT79] A.I. Borisenko and I.E. Tarapov. *Vector and Tensor Analysis with Applications* (translated from Russian). Dover Publications, Inc., New York, 1979.

[Bus26] H. Busch. Calculation of the Paths of Kathode Rays in Axial Symmetric Electromagnetic Fields. *Ann. Phys.*, 81:974–993, 1926.

[Bvd$^+$87] K.D. Bronsema, S. van Smaalen, J.L. de Boer, G.A. Wiegers, F. Jellinek, and J. Mahy. The Determination of the Commensurately Modulated Structure of Tantalum Tetratelluride. *Acta Crystall. B*, 43:305–313, 1987.

[BVTVL$^+$88] D. Broddin, G. Van Tendeloo, J. Van Landuyt, S. Amelinckx, and A. Loiseau. Chaotic and Uniform Regimes in Incommensurate Anti-Phase Boundary Modulated Cu_3Pd. *Phil. Mag. B*, 57:31–48, 1988.

[BVTVL$^+$89] D. Broddin, G. Van Tendeloo, J. Van Landuyt, S. Amelinckx, and M. De Graef. The Long Period Antiphase Boundary Modulated Structures in $Cu_{3+x}Al_{1-x}$ Alloys. *Phil. Mag. A*, 59:979–998, 1989.

[BW75] M. Born and E. Wolf. *Principles of Optics*. Pergamon, Oxford, 1975.

[CBWF78] J.N. Chapman, P.E. Batson, E.M. Waddell, and R.P. Ferrier. The Direct Determination of Magnetic Domain Wall Profiles by Differential Phase Contrast Electron Microscopy. *Ultramicroscopy*, 3:203–214, 1978.

[CCK97] G.H. Campbell, D. Cohen, and W.E. King. Data Preparation for Quantitative High-Resolution Electron Microscopy. *Micros. Microanal.*, 3:299–310, 1997.

[CD91] W. Coene and T.J.J. Denteneer. Improved Methods for the Determination of the Spherical Aberration Coefficient in High-Resolution Electron Microscopy from Micrographs of an Amorphous Object. *Ultramicroscopy*, 38:225–233, 1991.

[CDGD$^+$89] G. Ceder, M. De Graef, L. Delaey, J. Kulik, and D. de Fontaine. Gorsky–Bragg–Williams Approach to the Study of Long-Period Superlattice Phases in Binary Alloys. *Phys. Rev. B*, 39:381–385, 1989.

[CF83] R. Caudron and A. Finel. Application de la diffusion diffuse des neutrons aux études de thermodynamique des alliages, rapport technique n° 11/1221 m. Technical report, ONERA, 1983.

[CG00] E. Chu and A. George. *Inside the FFT Black Box*. CRC Press, Boca Raton, 2000.

[Chi78] S. Chikazumi. *Physics of Magnetism*. R.E. Krieger Publishing Co., Malabar, Florida, 1978.

[CJOdBVD92] W. Coene, G. Janssen, M. Op de Beeck, and D. Van Dyck. Phase Retrieval through Focus Variation for Ultra-Resolution in Field Emission Transmission Electron Microscopy. *Phys. Rev. Lett.*, 69:3743–3746, 1992.

[CK80] A.V. Crewe and D. Kopf. A Sextupole System for the Correction of Spherical Aberration. *Optik*, 55:1–10, 1980.

[CM57] J.M. Cowley and A.F. Moodie. The Scattering of Electrons by Atoms and Crystals. I. A New Theoretical Approach. *Acta Crystall.*, 10:609–619, 1957.

[CM66] P.C. Clapp and S.C. Moss. Correlation Functions of Disordered Binary Alloys. *Phys. Rev.*, 142:418–427, 1966.

[CM95] J.W. Christian and S. Mahajan. Deformation Twinning. *Prog. Mater. Sci.*, 39:1–57, 1995.

[Col89] D. Colaitis. Modulated Incommensurable Structure in Blue Potassium Molybdenum Bronze $K_{0.3-x}MoO_3$. *J. Sol. Stat. Chem.*, 83:158–169, 1989.

[Com73] ASM Handbook Committee. *Metals Handbook, Volume 8: Metallography, Structures and Phase Diagrams*. American Society for Metals, 8th edition, 1973.

[Cow67] J.M. Cowley. Kinematical Diffraction from Solid Solutions with Short Range Order and Size Effect. *Acta Crystall. A*, 24:557–563, 1967.

[Cow81] J.M. Cowley. *Diffraction Physics*. North-Holland, Amsterdam, second revised edition, 1981.

[Cre77] A.V. Crewe. Ideal Lenses and the Scherzer Theorem. *Ultramicroscopy*, 2:281–284, 1977.

[CSG86] J. Cahn, D. Shechtman, and D. Gratias. Indexing of Icosahedral Quasiperiodic Crystals. *J. Mater. Res.*, 1:13–26, 1986.

[CT65] J.W. Cooley and J.W. Tukey. An Algorithm for the Machine Calculation of Complex Fourier Series. *Math. Comp.*, 19:297–301, 1965.

[CT95] E.R. Cohen and B.N. Taylor. The Fundamental Physical Constants. *Physics Today*, 48:BG9–BG13, 1995.

[CVD84a] W. Coene and D. Van Dyck. The Real Space Method for Dynamical Electron Diffraction Calculations in High Resolution Electron Microscopy. II. Critical Analysis of the Dependency on the Input Parameters. *Ultramicroscopy*, 15:41–50, 1984.

[CVD84b] W. Coene and D. Van Dyck. The Real Space Method for Dynamical Electron Diffraction Calculations in High Resolution Electron Microscopy. III. A Computational Algorithm for the Electron Propagation with its Practical Applications. *Ultramicroscopy*, 15:287–300, 1984.

[CVDOdBVL97] J.H. Chen, D. Van Dyck, M. Op de Beeck, and J. Van Landuyt. Computational Comparisons between the Conventional Multislice Method and the Third-Order Multislice Method for Calculating High-Energy Electron Diffraction and Imaging. *Ultramicroscopy*, 69:219–240, 1997.

[CVDVTVL85] W. Coene, D. Van Dyck, G. Van Tendeloo, and J. Van Landuyt. Computer Simulation of High-Energy Electron Scattering by Non-Periodic Objects. The Real Space Patching Method as an Alternative to the Periodic Continuation Technique. *Phil. Mag. A*, 52:127–143, 1985.

[DB95] B. Devouard and A. Baronnet. Axial Diffraction of Curved Lattices: Geometrical and Numerical Modeling. Application to Chrysotile. *European J. Mineralogy*, 7:835–847, 1995.

[DBMS79] J.J. Dongarra, J.R. Bunch, C.B. Moler, and G.W. Stewart. *LINPACK Users' Guide*. Society for Industrial and Applied Mathematics, Philadelphia, 1979.

[DDCHH88] J. Dongarra, J. Du Croz, S. Hammarling, and R.J. Hanson. An Extended Set of FORTRAN Basic Linear Algebra Subprograms. *ACM Trans. Math. Soft.*, 14:1–17, 1988.

[DDG97] J. Dooley and M. De Graef. Energy Filtered Lorentz Microscopy. *Ultramicroscopy*, 67:113–132, 1997.

[de 23] L. de Broglie. Ondes et quanta. *Comptes Rendues Hebd. Acad. Sci. Paris*, 177:507–510, 1923.

[De 74] P.M. De Wolff. The Pseudo-Symmetry of Modulated Crystal Structures. *Acta Crystall. A*, 30:777–785, 1974.

[DG27] C.J. Davisson and L.H. Germer. Diffraction of Electrons by a Nickel Crystal. *Phys. Rev.*, 30:705–740, 1927.

[DG89] M. De Graef. *Bijdrage tot de studie van fazestabiliteit voor en na de martensitische transformatie in β-Cu-Zn-Al legeringen*. PhD thesis, Catholic University of Leuven, Belgium, 1989.

[DG98] M. De Graef. A Novel Way to Represent the 32 Crystallographic Point Groups. *Journal of Materials Education*, 20:31–42, 1998.

[DGDB88] M. De Graef, D. Delaey, and D. Broddin. High Resolution Electron Microscopic Study of the X-Phase in Cu-Al and Cu-Al-Zn Alloys. *phys. stat. sol. (a)*, 107:597–609, 1988.

[DGNM99] M. De Graef, N.T. Nuhfer, and M.R. McCartney. Phase Contrast of Spherical Magnetic Particles. *J. Microscopy*, 194:84–94, 1999.

[Dir28a] P.A.M. Dirac. The Quantum Theory of the Electron. *Proc. Roy. Soc. A*, 117:610–624, 1928.

[Dir28b] P.A.M. Dirac. The Quantum Theory of the Electron, Part II. *Proc. Roy. Soc. A*, 118:351–361, 1928.

[Dir47] P.A.M. Dirac. *The Principles of Quantum Mechanics*. Clarendon Press, Oxford, third edition, 1947.

[dJCVD87] A.F. de Jong, W. Coene, and D. Van Dyck. On the Phase-Object Function, Used in Dynamical Electron Diffraction Calculations. *Ultramicroscopy*, 23:3–16, 1987.

[dJVD93] A.F. de Jong and D. Van Dyck. Ultimate Resolution and Information in Electron Microscopy: II. The Information Limit of Transmission Electron Microscopes. *Ultramicroscopy*, 49:66–80, 1993.

[Don69] W.F. Donoghue Jr. *Distributions and Fourier Transforms*. Academic Press, New York, 1969.

[DPVC01] D. Delille, R. Pantel, and E. Van Cappellen. Crystal Thickness and Extinction Distance Determination using Energy Filtered CBED Pattern Intensity Measurement and Dynamical Diffraction Theory Fitting. *Ultramicroscopy*, 87:5–18, 2001.

[dR95] W.J. de Ruijter. Imaging Properties and Applications of Slow-Scan Charge-Coupled Device Cameras Suitable for Electron Microscopy. *Micron*, 26:247–275, 1995.

[dRMK93] W.J. de Ruijter, P.E. Mooney, and O.L. Krivanek. Signal Transfer Efficiency of Slow-scan CCD Cameras. In G.W. Bailey and C.L. Rieder, editors, *Proc. 51st Ann. Meet. MSA*, pages 1062–1063. Springer-Verlag, New York, 1993.

[DRVDVTA77] R. De Ridder, D. Van Dyck, G. Van Tendeloo, and S. Amelinckx. A Cluster Model for the Transition State and Its Study by Means of Electron Diffraction. ii. Application to some Particular Systems. *phys. stat. sol. (a)*, 40:669–683, 1977.

[DRVTVDA76] R. De Ridder, G. Van Tendeloo, D. Van Dyck, and S. Amelinckx. A Cluster Model for the Transition State and Its Study by Means of Electron Diffraction. i. Theoretical Model. *phys. stat. sol. (a)*, 38:663–674, 1976.

[DT68] P.A. Doyle and P.S. Turner. Relativistic Hartree–Fock X-ray and Electron Scattering Factors. *Acta Crystall. A*, 24:390–397, 1968.

[DVvV91] J.L.C. Daams, P. Villars, and J.H.N. van Vucht. *Atlas of Crystal Structure Types for Intermetallic Phases*. ASM International, Materials Park, OH, 1991.

[Ead80] J.A. Eades. Zone-Axis Patterns Formed by a New Double-Rocking Technique. *Ultramicroscopy*, 5:71–74, 1980.

[Ead90] J.A. Eades. Laue Zones: a Clarification of Nomenclature. *Ultramicroscopy*, 32:183, 1990.

[Ead92] J.A. Eades. Convergent beam diffraction. In J.M. Cowley, editor, *Electron Diffraction Techniques, Volume I*, pages 313–359. Oxford Science Publications, Oxford, 1992.

[Ead96] J.A. Eades. Some Reflections on how much Information is there on a Piece of Film, on how Film compares with CCD Cameras and what Features a Scanner would need to Digitize TEM Negatives. *Microscopy Today*, 96-1:24–26, 1996.

[Edi76] J.W. Edington. *Practical Electron Microscopy in Materials Science*. Reprinted from original series by TechBooks, Herndon, VA, 1976.

[Ein05] A. Einstein. Zur Elektrodynamik bewegter Körper. *Ann. Phys.*, 17:891–921, 1905.

[Eis87] H.M. Eissa. Burgers Vector Determination in Anisotropic Hexagonal Single Crystals (Zn). *Prakt. Met.*, 24:280–288, 1987.

[EMPH93] J.A. Eades, S. Moore, T. Pfullmann, and J. Hangas. Discrepancies in Kinematic Calculations of HOLZ Lines. *Microsc. Res. Techn.*, 24:509–513, 1993.

[Far39] M. Faraday. *Experimental Researches in Electricity*. Quaritch, London, 1839.

[FBLH70] J. Frank, P. Bussler, R. Langer, and W. Hoppe. Einige Erfahrungen mit der rechnerischen Analyse und Synthese von elektronenmikroskopischen Bildern hoher Auflösung. *Ber. Bunsenges. Phys. Chem.*, 74:1105–1115, 1970.

[FC87] G.Y. Fan and J.M. Cowley. The Simulation of High Resolution Images of Amorphous Carbon Films. *Ultramicroscopy*, 21:125–130, 1987.

[FCRT$^+$99] M.J. Fransen, Th.L. Can Rooy, P.C. Tiemeijer, M.H.F. Overwijk, J.S. Faber, and P. Kruit. On the Electron-Optical Properties of the ZrO/W Schottky Electron Emitter. In P.W. Hawkes, editor, *Advances in Imaging and Electron Physics*, volume 111, pages 92–167. Academic Press, New York, 1999.

[Fed90] E.S. Fedorov. *The Symmetry of Regular Systems of Figures*. A. Yakob, St Petersburg, 1890.

[FH01] B. Fultz and J.M. Howe, editors. *Transmission Electron Microscopy and Diffractometry of Materials*. Springer-Verlag, Berlin, 2001.

[FHS86] H.A. Ferwerda, B.J. Hoenders, and C.H. Slump. Fully Relativistic Treatment of Electron-Optical Image Formation based on the Dirac Equation. *Optica Acta*, 33:145–157, 1986.

[Fin87] A. Finel. *Contribution á l'étude des effects d'ordre dans le cadre du modèle d'Ising: états de base et diagrammes de phase*. PhD thesis, Université Pierre et Marie Curie, 1987.

[FKZ59] H. Frieser, E. Klein, and E. Zeitler. Das Verhalten photographischer Schichten bei Elektronenbestrahlung (II). *Z. Angew. Phys.*, 11:190–199, 1959.

[FLS63] R.P. Feynman, R.B. Leighton, and M. Sands. *The Feynman Lectures on Physics, Vol. I*. Addison-Wesley Publishing Company, Reading, MA, 1963.

[Fra69] J. Frank. Nachweis von Objektbewegungen im lichtoptischen Diffraktogramm von elektronenmikroskopischen Aufnahmen. *Optik*, 30:171–180, 1969.

[Fra73] J. Frank. The Envelope of Electron Microscopic Transfer Functions for Partially Coherent Illumination. *Optik*, 38:519–536, 1973.

[Fra75] J. Frank. A Practical Resolution Criterion in Optics and Electron Microscopy. *Optik*, 43:25–34, 1975.

[Fra76] J. Frank. Determination of Source Size and Energy Spread from Electron Micrographs using the Method of Young's Fringes. *Optik*, 44:379–391, 1976.

[FS81] M.E. Fisher and W. Selke. Low Temperature Analysis of the Axial Next-Nearest Neighbour Ising Model near its Multiphase Point. *Phil. Trans. Roy. Soc.*, 302:1–44, 1981.

[fTM98] American Society for Testing and Materials. *Annual Book of ASTM Standards*, volume 03.01, standard E112-96, pages 229–251. ASTM, Philadelphia, PA, 1998.

[Fuj61] K. Fujiwara. Relativistic Dynamical Theory of Electron Diffraction. *J. Phys. Soc. Japan*, 16:2226–2238, 1961.

[Fuj86] H. Fujita, editor. *History of Electron Microscopes*. Komiyama Printing Co., Ltd., Tokyo, 1986.

[Fuk66] A. Fukuhara. Many-Ray Approximation in the Dynamical Theory of Electron Diffraction. *J. Phys. Soc. Japan*, 21:2645–2662, 1966.

[Fun85] K.K. Fung. Direct Observation of Antiphase Discommensurations in $TiSe_2$ by Transmission Electron Microscopy. *J. Physics C*, 18:L489–491, 1985.

[GBA66] R. Gevers, H. Blank, and S. Amelinckx. Extension of the Howie–Whelan Equations for Electron Diffraction to Non-Centro Symmetrical Crystals. *phys. stat. sol.*, 13:449–465, 1966.

[GBDM77] B.S. Garbow, J.M. Boyle, J.J. Dongarra, and C.B. Moler. *Matrix Eigensystem Routines – EISPACK Guide Extension*, volume 51 of *Lecture Notes in Computer Science*. Springer-Verlag, Berlin, 1977.

[GC71] J.E. Gragg and J.B Cohen. The Structure of Guinier–Preston Zones in Al–5 at% Ag. *Acta Metall.*, 19:507–519, 1971.

[GC87] H. Gong and J.N. Chapman. On the Use of Divergent Wall Images in the Fresnel Mode of Lorentz Microscopy for the Measurement of the Widths of Very Narrow Domain Walls. *J. Magn. Magn. Mater.*, 67:4–8, 1987.

[GDBA64] R. Gevers, P. Delavignette, H. Blank, and S. Amelinckx. Electron Microscope Transmission Images of Coherent Domain Boundaries. I. Dynamical Theory. *phys. stat. sol.*, 4:383–410, 1964.

[Gev62] R. Gevers. Dynamical Theory of Moiré Fringe Patterns. *Phil. Mag.*, 7:1681–1720, 1962.

[Gev63] R. Gevers. On the Dynamical Theory of Different Types of Electron Microscopic Fringe Patterns. *phys. stat. sol.*, 3:1672–1683, 1963.

[Gev70] R. Gevers. Kinematical Theory of Electron Diffraction. In S. Amelinckx, R. Gevers, G. Remaut, and J. Van Landuyt, editors, *Modern Diffraction and Imaging Techniques in Materials Science*, pages 1–34. North-Holland, Amsterdam, 1970.

[GM65] J. Gjönnes and A.F. Moodie. Extinction Conditions in the Dynamical Theory of Electron Diffraction. *Acta Crystall.*, 19:65–67, 1965.

[GM74] P. Goodman and A.F. Moodie. Numerical Evaluation of N-Beam Wave Functions in Electron Scattering by the Multi-Slice Method. *Acta Crystall. A*, 30:280–290, 1974.

[GMWW86] S.R. Glanvill, A.F. Moodie, H.J. Whitfield, and I.J. Wilson. Experimental Determination of the Imaging Properties of a 200 kV Electron Microscope. *Aust. J. Phys.*, 39:71–92, 1986.

[GN96] T.E. Gureyev and K.A. Nugent. Phase Retrieval with the Transport-of-Intensity Equation. II. Orthogonal Series Solution for Nonuniform Illumination. *J. Opt. Soc. Am. A*, 13:1670–1682, 1996.

[Gol76] E. Goldstein. Vorläufige Mittheilungen über elektrische Entladungen in verdünnten Gasen. *Monatsber. Ak. der Wiss. Berlin*, pages 279–296, 1876.

[Gol78] H. Goldstein. *Classical Mechanics*. Addison-Wesley Publishing Company, Reading, MA, 1978.

[Goo68] J.W. Goodman. *Introduction to Fourier Optics*. McGraw-Hill, San Francisco, 1968.

[Goo84] P.J. Goodhew. *Specimen Preparation for Transmission Electron Microscopy of Materials*. Oxford University Press, Oxford, 1984.

[Gor70] M.J. Goringe. Diffraction Contrast from Imperfect Crystals. In S. Amelinckx, R. Gevers, G. Remaut, and J. Van Landuyt, editors, *Modern Diffraction and Imaging Techniques in Materials Science*, pages 553–589. North-Holland, Amsterdam, 1970.

[GP59] A.B. Glossop and D.W. Pashley. The Direct Observation of Anti-Phase Domain Boundaries in Ordered Copper-Gold (CuAu) Alloy. *Proc. Roy. Soc. A*, pages 132–146, 1959.

[GP99] H.X. Gao and L.M. Peng. Parametrization of the Temperature Dependence of the Debye–Waller Factors. *Acta Crystall. A*, 55:926–932, 1999.

[Gre90] W. Greiner. *Relativistic Quantum Mechanics: Wave Equations*. Springer-Verlag, Berlin, 1990.

[GRN95a] T.E. Gureyev, A. Roberts, and K.A. Nugent. Partially Coherent Fields, the Transport-of-Intensity Equation, and Phase Uniqueness. *J. Opt. Soc. Am. A*, 12:1942–1946, 1995.

[GRN95b] T.E. Gureyev, A. Roberts, and K.A. Nugent. Phase Retrieval with the Transport-of-Intensity Equation: Matrix Solution with Use of Zernike Polynomials. *J. Opt. Soc. Am. A*, 12:1932–1941, 1995.

[GS72] R.W. Gerchberg and W.O. Saxton. A Practical Algorithm for the Determination of Phase from Image and Diffraction Plane Pictures. *Optik*, pages 237–246, 1972.

[GS73] R.W. Gerchberg and W.O. Saxton. Wave Phase from Image and Diffraction Plane Pictures. In P.W. Hawkes, editor, *Image Processing and Computer-Aided Design in Electron Optics*, pages 66–81. Academic Press, London, 1973.

[GVLA65] R. Gevers, J. Van Landuyt, and S. Amelinckx. Intensity Profiles for Fringe Patterns due to Planar Interfaces as Observed by Electron Microscopy. *phys. stat. sol.*, 11:689–709, 1965.

[GVLA66a] R. Gevers, J. Van Landuyt, and S. Amelinckx. The Fine Structure of Spots in Electron Diffraction Resulting from the Presence of Planar Interfaces and Dislocations. I. General Theory and its Application to Stacking Faults and Anti-Phase Boundaries. *phys. stat. sol.*, 18:343–361, 1966.

[GVLA66b] R. Gevers, J. Van Landuyt, and S. Amelinckx. On a Simple Derivation of the Amplitudes of the Electron Beams Transmitted and Scattered by a Crystal Containing Planar Interfaces – Images of Subgrain Boundaries. *phys. stat. sol.*, 18:325–342, 1966.

[GVLA68] R. Gevers, J. Van Landuyt, and S. Amelinckx. The Fine Structure of Spots in Electron Diffraction Resulting from the Presence of Planar Interfaces and Dislocations. III. Domain Boundaries. *phys. stat. sol.*, 26:577–590, 1968.

[GW87] R.C. Gonzalez and P. Wintz. *Digital Image Processing*. Addison-Wesley, Reading, MA, 1987.

[Hah96] Theo Hahn, editor. *International Tables for Crystallography, Volume A: Space-Group Symmetry (4th revised edition)*. Kluwer, Dordrecht, 1996.

[Hal53] C.E. Hall. *Introduction to Electron Microscopy*. McGraw-Hill, San Francisco, 1953.

[Has64] H. Hashimoto. Energy Dependence of Extinction Distance and Transmissive Power for Electron Waves. *J. Appl. Phys.*, 35:277–290, 1964.

[Haü84] René-Juste Haüy. *Essai d'une Theorie sur la Structure des Cristaux*. Cogu & Né de la Rochelle, Paris, 1784.

[Haw74] P.W. Hawkes. Transforms for Discrete Electron Image Processing and Filtering. *Optik*, 40:539–556, 1974.

[Haw85] P.W. Hawkes. The Beginnings of Electron Microscopy. In *Advances in Electronics and Electron Physics*, volume 16 (supplement). Academic Press, Orlando, 1985.

[HB84] D. Hull and D.J. Bacon. *Introduction to Dislocations (3rd Edition)*. Pergamon Press, Oxford, 1984.

[HBTW92] L. Hoddeson, E. Braun, J. Teichmann, and S. Weart, editors. *Out of the Crystal Maze. Chapters from the History of Solid State Physics*. Oxford University Press, New York, 1992.

[Hei25] W.C. Heisenberg. Über den anschaulichen Inhalt der quantumtheoretischen Kinematik und Mechanik. *Zeits. f. Phys.*, 33:879–893, 1925.

[Hei49] R.D. Heidenreich. Electron Microscope and Diffraction Study of Metal Crystal Textures by Means of Thin Sections. *J. Appl. Phys.*, 20:993–1010, 1949.

[HGWA84] P. Haasen, V. Gerold, R. Wagner, and M.F. Ashby, editors. *Decomposition of Alloys: the Early Stages*. Pergamon Press, Oxford, 1984.

[HH65] C.H. Hall and P.B. Hirsch. The Effect of Thermal Diffuse Scattering on Propagation of High Energy Electrons through Crystals. *Proc. Roy. Soc.*, A286:158–177, 1965.

[HHC$^+$73] A.K. Head, P. Humble, L.M. Clarebrough, A.J. Morton, and C.T. Forwood. *Computed Electron Micrographs and Defect Identification*, volume 7 of *Defects in Crystalline Solids, S. Amelinckx, R. Gevers, and J. Nihoul, editors*. North-Holland, Amsterdam, 1973.

[HHN$^+$77] P.B. Hirsch, A. Howie, R.B. Nicholson, D.W. Pashley, and M.J. Whelan. *Electron Microscopy of Thin Crystals (2nd revised edition)*. Krieger, Malabar, FL, 1977.

[HHW56] P.B. Hirsch, R.W. Horne, and M.J. Whelan. Direct Observations of the Arrangement and Motion of Dislocations in Aluminium. *Phil. Mag.*, 1:677–684, 1956.

[HHW60] P.B. Hirsch, A. Howie, and M.J. Whelan. A Kinematical Theory of Diffraction Contrast of Electron Transmission Microscope Images of Dislocations and Other Defects. *Phil. Trans. Roy. Soc. Lond.*, A 252:499–529, 1960.

[HHW62] H. Hashimoto, A. Howie, and M.J. Whelan. Anomalous Electron Absorption Effects in Metal Foils: Theory and Comparison with Experiment. *Proc. Roy. Soc. A*, 269:80–103, 1962.

[HII$^+$91] Y. Hirotsu, Y. Ikeda, Y. Ichinose, S. Nagakura, T. Komatsu, and K. Matsushita. Modulated Structure of the High-T_c Superconductor $Bi_{2-x}Pb_xSr_2CaCu_2O_y$. *J. Electr. Microsc.*, 40:147–156, 1991.

[HK89a] P.W. Hawkes and E. Kasper. *Principles of Electron Optics: Applied Geometrical Optics*, volume 1. Academic Press, New York, 1989.

[HK89b] P.W. Hawkes and E. Kasper. *Principles of Electron Optics: Basic Geometrical Optics*, volume 2. Academic Press, New York, 1989.

[HK94] P.W. Hawkes and E. Kasper. *Principles of Electron Optics: Wave Optics*, volume 3. Academic Press, New York, 1994.

[HL52] N.F.M. Henry and D. Londsdale, editors. *International Tables for Crystallography*. Kynoch Press, Birmingham, 1952.

[HL68] J.P. Hirth and J. Lothe. *Theory of Dislocations*. McGraw-Hill, New York, 1968.

[HM54] G. Honjo and K. Mihama. Fine Structure due to Refraction Effect in Electron Diffraction Pattern of Powder Sample. *J. Phys. Soc. Japan*, 9:184–198, 1954.

[HM79] M.J. Howes and D.V. Morgan, editors. *Charge-coupled Devices and Systems*. Wiley, Chichester, 1979.

[Hor94] S. Horiuchi. *Fundamentals of High-Resolution Transmission Electron Microscopy*. North-Holland, Amsterdam, 1994.

[HS80] F. Herman and S. Skillman. *Atomic Structure Calculations*. Prentice Hall, Englewood Cliffs, NJ, 1980.

[HSI91] F. Hosokawa, M. Suzuki, and K. Ibe. Determination of the Effective Source Size from its Image in the Backfocal Plane of the Objective Lens. *Ultramicroscopy*, 36:367–373, 1991.

[HU97] M. Haider and S. Uhlemann. Seeing is not always Believing: Reduction of Artifacts by an Improved Point Resolution with a Spherical Aberration Corrected 200 kV Transmission Electron Microscope. *Microscopy and Microanalysis*, 3:1179–1180, 1997.

[Hum79] C.J. Humphreys. The Scattering of Fast Electrons by Crystals. *Rep. Prog. Phys.*, 42:1825–1887, 1979.

[HW61] A. Howie and M.J. Whelan. Diffraction Contrast of Electron Microscope Images of Crystal Lattice Defects. II The Development of a Dynamical Theory. *Proc. R. Soc. London*, A263:217–237, 1961.

[IFO$^+$99] T.C. Isabell, P.E. Fischione, C. O'Keefe, M.U. Guruz, and V.P. Dravid. Plasma Cleaning and its Applications for Electron Microscopy. *Microscopy and Microanalysis*, 5:126–135, 1999.

[Ish80] K. Ishizuka. Contrast Transfer of Crystal Images in TEM. *Ultramicroscopy*, 5:55–65, 1980.

[IU77] K. Ishizuka and N. Uyeda. A New Theoretical and Practical Approach to the Multislice Method. *Acta Crystall. A*, 33:740–749, 1977.

[Jac75] J.D. Jackson. *Classical Electrodynamics, Second Edition*. Wiley, New York, 1975.

[Jac87] A.G. Jackson. Prediction of HOLZ Pattern Shifts in Convergent Beam Diffraction. *J. Electron Microscopy Techn.*, 5:373–377, 1987.

[Jan86] T. Janssen. Crystallography of Quasi-Crystals. *Acta Crystall. A*, 42:261–271, 1986.

[Jan92] C. Janot. *Quasicrystals. A Primer.* Clarendon Press, Oxford, 1992.

[JH68] T.H. James and G.C. Higgins. *Fundamentals of Photographic Theory*. Morgan and Morgan, New York, 1968.

[JHJF99] N. Jiang, D.H. Hou, I.P. Jones, and H.L. Fraser. Optimizing the ALCHEMI Technique. *Phil. Mag. A*, 79:2525–2538, 1999.

[JHLS94] W.C. Johnson, J.M. Howe, D.E. Laughlin, and W.A. Sofa, editors. *Proceedings of International Conference on Solid→Solid Phase Transformations*. The Minerals, Metals, and Materials Society, 1994.

[JJ77] A. Janner and T. Janssen. Symmetry of Periodically Distorted Crystals. *Phys. Rev. B*, 15:643–658, 1977.

[JJ79] A. Janner and T. Janssen. Superspace Groups. *Physica A*, 99:47–76, 1979.

[JK01] M.L. Jenkins and M.A. Kirk. *Characterization of Radiation Damage by Transmission Electron Microscopy*. Institute of Physics Publishing, Bristol, 2001.

[Jon97] T.E. Jones. `http://www.utmem.edu/personal/thjones/hist/hist_mic.htm`, 1997.

[JVdBV$^+$95] K.G.F. Janssens, O. Van der Biest, J. Vanhellemont, H.E. Maes, R. Hull, and J.C. Bean. Localized Strain Characterization in Semiconductor Structures using Electron Diffraction Contrast Imaging. *Mater. Sci. Techn.*, 11:66–71, 1995.

[JVDGVdB92] K.G.F. Janssens, J. Vanhellemont, M. De Graef, and O. Van der Biest. SIMCON: a Versatile Software Package for the Simulation of Electron Diffraction Contrast Images of Arbitrary Displacement Fields. *Ultramicroscopy*, 45:323–335, 1992.

[Kai55] Y. Kainuma. The Theory of Kikuchi Patterns. *Acta Crystall.*, 8:247–257, 1955.

[Kas82] E. Kasper. Field Electron Emission Systems. In P.W. Hawkes, editor, *Advances in Imaging and Electron Physics*, volume 8, pages 207–260. Academic Press, New York, 1982.

[Kau97] W. Kaufmann. Magnetischen Ablenkbarkeit der Kathodenstrahlen und ihre Abhängigkeit von Enladungspotential. *Ann. der Phys. und Chem.*, 61:544–568, 1897.

[KdJ91] A.J. Koster and A.F. de Jong. Measurement of the Spherical Aberration Coefficient of Transmission Electron Microscopes by Beam-tilt-induced Image Displacements. *Ultramicroscopy*, 38:235–240, 1991.

[KDL99] O.L. Krivanek, N. Dellby, and A.R. Lupini. STEM without Spherical Aberration. *Microscopy and Microanalysis*, 5:670–671, 1999.

[Kha83] A.G. Khachaturyan. *Theory of Structural Transformations in Solids*. Wiley, New York, 1983.

[Kik28] S. Kikuchi. Diffraction of Cathode Rays by Mica. *Jap. J. Phys.*, 5:83–96, 1928.

[Kik51] R. Kikuchi. A Theory of Cooperative Phenomena. *Phys. Rev.*, 81:988–1003, 1951.

[Kil87] R. Kilaas. Interactive Software for Simulation of High Resolution TEM Images. In *22nd. Annual Conference of the Microbeam Analysis Society*, pages 293–300, 1987.

[Kir82] E.J. Kirkland. Nonlinear High Resolution Image Processing of Conventional Transmission Electron Micrographs: I. Theory. *Ultramicroscopy*, 9:45–64, 1982.

[Kir84] E.J. Kirkland. Improved High Resolution Image Processing of Bright Field Electron Micrographs: I. Theory. *Ultramicroscopy*, 15:151–172, 1984.

[Kir98] E.J. Kirkland. *Advanced Computing in Electron Microscopy*. Plenum Press, New York, 1998.

[KJBN75] P.M. Kelly, A. Jostsons, R.G. Blake, and J.G. Napier. The Determination of Foil Thickness by Scanning Transmission Electron Microscopy. *phys. stat. sol. (a)*, 31:771–780, 1975.

[KLF74] K. Kambe, G. Lehmpfuhl, and F. Fujimoto. Interpretation of Electron Channeling by the Dynamical Theory of Electron Diffraction. *Z. Naturforsch.*, 29a:1034–1044, 1974.

[KO89] W. Krakow and M.A. O'Keefe, editors. *Computer Simulation of Electron Microscope Diffraction and Images*. TMS, Warrendale, 1989.

[KOK87] R. Kilaas, M.A. O'Keefe, and K.M. Krishnan. On the Inclusion of Upper Laue Layers in Computational Methods in High Resolution Transmission Electron Microscopy. *Ultramicroscopy*, 21:47–62, 1987.

[Kom87] J. Komrska. Algebraic Expressions of Shape Amplitudes of Polygons and Polyhedra. *Optik*, 80:171–183, 1987.

[KOM99] M. Koiwa, K. Otsuka, and T. Miyazaki, editors. *Proceedings of International Conference on Solid→Solid Phase Transformations*. The Japan Institute of Metals, 1999.

[KR32a] M. Knoll and E. Ruska. The Electron Microscope. *Z. Physik*, 78:318–339, 1932.

[KR32b] M. Knoll and E. Ruska. Geometric Electron Optics. *Ann. Physik*, 12:607–640, 1932.

[Kri76] O.L. Krivanek. A Method for Determining the Coefficient of Spherical Aberration from a Single Electron Micrograph. *Optik*, 45:97–101, 1976.

[Kri88] K.M. Krishnan. Atomic Site and Species Determinations using Channeling and Related Effects in Analytical Electron Microscopy. *Ultramicroscopy*, 24:125–142, 1988.

[Kri94] O.L. Krivanek. Three-fold Astigmatism in High-Resolution Transmission Electron Microscopy. *Ultramicroscopy*, 55:419–433, 1994.

[KS82] H.S. Kim and S.S. Sheinin. An Assessment of the High Energy Approximation in the Dynamical Theory of Electron Diffraction. *phys. stat. sol. (b)*, 109:807–816, 1982.

[KS95] O.L. Krivanek and P.A. Stadelmann. Effect of Three-fold Astigmatism on High Resolution Electron Micrographs. *Ultramicroscopy*, 60:103–114, 1995.

[KSUF80] E.J. Kirkland, B.M. Siegel, N. Uyeda, and Y. Fujiyoshi. Digital Reconstruction of Bright Field Phase Contrast Images from High Resolution Electron Micrographs. *Ultramicroscopy*, 5:479–503, 1980.

[KSUF82] E.J. Kirkland, B.M. Siegel, N. Uyeda, and Y. Fujiyoshi. Nonlinear High Resolution Image Processing of Conventional Transmission Electron Micrographs: II. Experiment. *Ultramicroscopy*, 9:65–74, 1982.

[KSUF85] E.J. Kirkland, B.M. Siegel, N. Uyeda, and Y. Fujiyoshi. Improved High Resolution Image Processing of Bright Field Electron Micrographs: I. Experiment. *Ultramicroscopy*, 17:87–104, 1985.

[LHMF72] J.S. Lally, C.J. Humphreys, A.J.F. Metherell, and R.M. Fisher. The Critical Voltage Effect in High Voltage Electron Microscopy. *Phil. Mag.*, 25:321–343, 1972.

[LHSH88] K.H. Lee, K. Hiraga, D. Shindo, and M. Hirabayashi. High Resolution Electron Microscopic Study of the Ordering Processes in Ni_4Mo Alloy. *Acta Metall.*, 36:641–649, 1988.

[Lic91] H. Lichte. Optimum Focus for Taking Electron Holograms. *Ultramicroscopy*, 38:12–22, 1991.

[Lic01] H. Lichte, et al. The Triebenberg Laboratory – Designed for Highest Resolution Electron Microscopy and Holography. In G.W. Bailey, R.L. Price, E. Völkl, and I.H. Musselman, editors, *Microscopy and Microanalysis*, volume 7, Supp. 2, pages 894–895. Springer-Verlag, Berlin, 2001.

[LL74] L.D. Landau and E.M. Lifshitz. *Quantum Mechanics*. Pergamon Press, Oxford, 1974.

[Lor04] H.A. Lorentz. Electromagnetic Phenomena in Systems moving with Velocity less than that of Light. *Proc. K. Ak. Amsterdam*, 12:986–1009, 1904.

[Lor87] G.W. Lorimer, editor. *Proceedings of International Conference on Solid→Solid Phase Transformations*. The Institute of Metals, 1987.

[Lyn74] D.F. Lynch. An Alternative Method for the Calculation of a Phase-Grating Function for Use in Dynamic Electron Diffraction Calculations. *Acta Crystall. A*, 30:101–102, 1974.

[Mac40] C.H. MacGillavry. Zur Prüfung der Dynamischen Theorie der Elektronenbeugung am Kristallgitter. *Physica*, 7:329–343, 1940.

[Mad87] W. Mader. On the Electron Diffraction Contrast caused by Large Inclusions. *Phil. Mag. A*, 55:59–83, 1987.

[Man62] M. Mannami. Electron Microscopic Images of Lattice Imperfections. *J. Phys. Soc. Japan*, 17:1423–1433, 1962.

[Man84] J.F. Mansfield. *Convergent Beam Electron Diffraction of Alloy Phases*. Adam Hilger, Bristol, 1984.

[Man91] M. Mansuripur. Computation of Electron Diffraction Patterns in Lorentz Electron Microscopy of Thin Magnetic Films. *J. Appl. Phys.*, 69:2455–2464, 1991.

[Met75] A.J.F. Metherell. Diffraction of Electrons by Perfect Crystals. In U. Valdrè and E. Ruedl, editors, *Electron Microscopy in Materials Science*, pages 401–552. Commission of European Communities, Luxembourg, 1975.

[Mil39] W.H. Miller. *A Treatise on Crystallography*. Deighton, Cambridge, 1839.

[Mis73] D.L. Misell. An Examination of an Iterative Method for the Solution of the Phase Problem in Optics and Electron Optics. I. Test Calculations; II. Sources of Error. *J. Phys. D: Appl. Phys.*, 6:2200–2216 and 2217–2225, 1973.

[MK00] H.W. Mook and P. Kruit. Optimization of the Short Field Monochromator Configuration for a High Brightness Electron Source. *Optik*, 111:339–346, 2000.

[MM86] D. McKie and C. McKie. *Essentials of Crystallography*. Blackwell Scientific Publications, Oxford, 1986.

[MMC+96] A. Matsumoto, K. Matsuno, N. Chiwata, N. Kuwano, and K. Oki. *In-situ* TEM Observation of Phase Transformation Processes in Cu_3Pt with a Long-Period Superstructure. *J. Elec. Microsc.*, 45:442–447, 1996.

[Mol39] K. Molière. Uber den Einfluß der Absorption auf den Brechungseffekt der Elektronenstrahlen. *Ann. d. Phys. Lpz.*, 34:461–472, 1939.

[MR84] V.W. Maslen and C.J. Rossouw. Implications of (e, 2e) Scattering for Inelastic Electron Diffraction in Crystals, I. Theoretical. *Phil. Mag. A*, 49:735–742, 1984.

[MTW73] C.W. Misner, K.S. Thorne, and J.A. Wheeler. *Gravitation*. Freeman, New York, 1973.

[Mul96] T.. Mulvey, editor. *The Growth of Electron Microscopy*, volume 96. Academic Press, 1996.

[NDW93] J. Neider, T. Davis, and M. Woo. *OpenGL Programming Guide: The Official Guide to Learning OpenGL, Release 1*. Addison-Wesley, Reading, MA, 1993.

[OBI74] M.A. O'Keefe, P.R. Buseck, and S. Iijima. Computed Crystal Structure Images for High Resolution Electron Microscopy. *Nature*, 274:322–324, 1974.

[Ode79] J.T. Oden. *Applied Functional Analysis, A First Course for Students of Mechanics and Engineering Science*. Prentice-Hall, Englewood Cliffs, NJ, 1979.

[OHW+01] M.A. O'Keefe, C.J.D. Hetherington, Y.C. Wang, E.C. Nelson, J.H. Turner, C. Kisielowski, J.-O. Malm, R. Mueller, J. Ringnalda, M. Pan, and A. Thust Sub-Ångstrom High-Resolution Transmission Electron Microscopy at 300 keV. *Ultramicroscopy*, 89:215–241, 2001.

[OK89] M.A. O'Keefe and R. Kilaas. *NCEMSS Users Guide, LBL Publication 658*, 1989.

[ONWA01] M.A. O'Keefe, E.C. Nelson, Y.C. Wang, and A. Thust Sub-Ångstrom Resolution of Atomistic Structures below 0.8 Å. *Phil. Mag. B*, 81:1861–1878, 2001.

[OP85] S. Olariu and S.I. Popescu. The Quantum Effects of Electromagnetic Fluxes. *Rev. Mod. Phys.*, 57:339–436, 1985.

[OT68] P.R. Okamoto and G. Thomas. On the Four-Axis Hexagonal Reciprocal Lattice and its Use in Indexing of Transmission Electron Diffraction Patterns. *phys. stat. sol.*, 25:81–91, 1968.

[OTT02] K. Omoto, K. Tsuda, and M. Tanaka. Simulations of Kikuchi Patterns due to Thermal Diffuse Scattering on MgO Crystals. *J. Electron Microsc.*, 51:67–78, 2002.

[Pai86] A. Pais. *Inward Bound: of Matter and Forces in the Physical World*. Clarendon Press, Oxford, 1986.

[Pap68] A. Papoulis. *Systems and Transforms with Applications in Optics*. McGraw-Hill, New York, 1968.

[Pau25] W. Pauli. Relation between the Closing in of Electron Groups in the Atom and the Structure of Complexes in the Spectrum. *Zeits. f. Phys.*, 31:765–783, 1925.

[PDGT98] X. Pierron, M. De Graef, and A.W. Thompson. On the Effect of Hydrogen on the Microstructure of α_2-Ti_3Al+Nb Alloys. *Phil. Mag. A*, 77:1399–1421, 1998.

[Pen88] S.J. Pennycook. Delocalization Corrections for Electron Channeling Analysis. *Ultramicroscopy*, 26:239–248, 1988.

[PFTV89] W.H. Press, B.P. Flannery, S.A. Teukoslky, and W.T. Vetterling. *Numerical Recipes: The Art of Scientific Computing (Fortran Version)*. Cambridge University Press, Cambridge, 1989.

[PG81] R. Portier and D. Gratias. Diffraction Symmetries for Elastic Scattering. In *Proc. EMAG 81*, volume 61, pages 275–278. The Institute of Physics, Bristol, 1981.

[Pie95] X. Pierron. *A Study of the Hydrogen Effects on the Microstructure and Tensile Properties of* $(\alpha_2 + B2)$ *Ti-25Al-10Nb-3V-1Mo Titanium Aluminide Alloy*. PhD thesis, Carnegie Mellon University, Pittsburgh, PA, 1995.

[PN98] D. Paganin and K.A. Nugent. Noninterferometric Phase Imaging with Partially Coherent Light. *Phys. Rev. Lett.*, 80:2586–2589, 1998.

[PRDW96] L.-M. Peng, G. Ren, S.L. Dudarev, and M.J. Whelan. Debye–Waller Factors and Absorptive Scattering Factors of Elemental Crystals. *Acta Crystall. A*, 52:456–470, 1996.

[PT68] A.P. Pogany and P.S. Turner. Reciprocity in Electron Diffraction and Microscopy. *Acta Crystall. A*, 24:103–109, 1968.

[Rei93] L. Reimer. *Transmission Electron Microscopy: Physics of Image Formation and Microanalysis*, 3rd Edition. Springer-Verlag, Berlin, 1993.

[Res98] Research Systems Inc., Boulder, CO. *The Interactive Data Language (Version 5)*, 1998.

[Rez78] P. Rez. Virtual Inelastic Scattering in High-Energy Electron Diffraction. *Acta Crystall. A*, 34:48–51, 1978.

[RM84] C.J. Rossouw and V.M. Maslen. Implications of (e, 2e) Scattering for Inelastic Electron Diffraction in Crystals, II. Application of the Theory. *Phil. Mag. A*, 49:743–757, 1984.

[RM94] A. Redjaïmia and J.P. Morniroli. Application of Microdiffraction to Crystal Structure Identification. *Ultramicroscopy*, 53:305–317, 1994.

[Roh01] G.S. Rohrer. *Structure and Bonding in Crystalline Materials*. Cambridge University Press, Cambridge, 2001.

[Ros46] A. Rose. Unified Approach to Performance of Photographic Film, Television Pickup Tubes and Human Eye. *J. Soc. Motion Pict. Eng.*, 47:273–294, 1946.

[Ros71] H. Rose. Properties of Spherically Corrected Achromatic Electron-lenses. *Optik*, 33:1–24, 1971.

[Ros90] H. Rose. Electrostatic Energy Filter as Monochromator of a Highly Coherent Electron Source. *Optik*, 86:95–98, 1990.

[RRG94] D. Rez, P. Rez, and I. Grant. Dirac–Fock Calculations of X-Ray Scattering Factors and Contributions to the Mean Inner Potential for Electron Scattering. *Acta Crystall. A*, 50:481–497, 1994.

[RTW88] C.J. Rossouw, P.S. Turner, and T.J. White. Axial Electron-Channeling Analysis of Perovskite. I. Theory and Experiment for $CaTiO_3$. *Phil. Mag. B*, 57:209–225, 1988.

[Rüh67a] M. Rühle. Elektronenmikroskopie kleiner Fehlstellenagglomerate in bestrahlten Metallen. i. Theorie des Kontrastes und experimetalle Methoden zur Ermittlung des Defekttyps. *phys. stat. sol.*, 19:263–278, 1967.

[Rüh67b] M. Rühle. Elektronenmikroskopie kleiner Fehlstellenagglomerate in bestrahlten Metallen. ii. Untersuchungen an neutronen- und ionenbestrahlten Kupfer sowie neutronenbestrahlten Nickel. *phys. stat. sol.*, 19:279–295, 1967.

[Rus34a] E. Ruska. Advances in Building and Performance of the Magnetic Electron Microscope. *Z. Physik*, 87:580–602, 1934.

[Rus34b] E. Ruska. Magnetic Objective for the Electron Microscope. *Z. Physik*, 89:90–128, 1934.

[Rus92] J.C. Russ. *The Image Processing Handbook*. CRC Press, Boca Raton, FL, 1992.

[SB62] G.H. Smith and R.E. Burge. The Analytical Representation of Atomic Scattering Amplitudes for Electrons. *Acta Crystall.*, A15:182–186, 1962.

[SBGC84] D. Shechtmann, I. Blech, D. Gratias, and J. Cahn. Metallic Phase with Long-Range Orientational Order and no Translational Symmetry. *Phys. Rev. Lett.*, 53:1951–1954, 1984.

[SBH95] H.V. Sorensen, C.S. Burrus, and M.T. Heideman. *Fast Fourier Transform Database*. PWS Publishing Co., Boston, 1995.

[Sch91] A.M. Schœnflies. *Kristallsysteme und Kristallstruktur*. Teubner, Leipzig, 1891.

[Sch26] E. Schrödinger. Quantisierung als Eigenwertproblem. *Ann. Phys.*, 79:361–376, 1926.

[Sch68] P. Schiske. Zur Frage der Bildrekonstruktion durch Fokusreihen. In *Proc. Fourth European Regional Conference on Electron Microscopy*, volume I, pages 145–146, 1968.

[Sch84] P. Schiske. Linear Filters for Image Series in Matrix Notation. *Optik*, 69:13–16, 1984.

[Sel88] W. Selke. The ANNNI Model – Theoretical Analysis and Experimental Application. *Physics Reports*, 170:213–264, 1988.

[Shm96] U. Shmueli, editor. *International Tables for Crystallography, Volume B: Reciprocal Space*. Kluwer, Dordrecht, 1996.

[Sin67] R.C. Singleton. A Method for Computing the Fast Fourier Transform with Auxiliary Memory and Limited High-speed Storage. *IEEE Trans. Audio and Electroacoustics*, 15:91–98, 1967.

[Ska73] P. Skalicky. Computer-Simulated Electron Micrographs of Crystal Defects. *phys. stat. sol. (a)*, 20:11–52, 1973.

[SKA00] A.J. Schwartz, M. Kumar, and B.L. Adams, editors. *Electron Backscatter Diffraction in Materials Science*. Kluwer, Dordrecht, 2000.

[SM84] H. Shimoyama and S. Maruse. Theoretical Considerations on Electron Optical Brightness for Thermionic, Field and T-F Emissions. *Ultramicroscopy*, 15:239–254, 1984.

[SOBS83] P.G. Self, M.A. O'Keefe, P.R. Buseck, and A.E.C. Spargo. Practical Computation of Amplitudes and Phases in Electron Diffraction. *Ultramicroscopy*, 11:35–52, 1983.

[Spe88] J.C.H. Spence. *Experimental High-Resolution Electron Microscopy*. Oxford University Press, Oxford, 1988.

[Spe93] J.C.H. Spence. On the Accurate Measurement of Structure-Factor Amplitudes and Phases by Electron Diffraction. *Acta Crystall. A*, 49:231–260, 1993.

[Spi68] M.R. Spiegel. *Mathematical Handbook of Formulas and Tables (Schaum's Outline Series in Mathematics)*. McGraw-Hill, New York, 1968.

[Spr97] M. Springford, editor. *Electron, a Centenary Volume*. Cambridge University Press, Cambridge, 1997.

[SQM93] J.C.H. Spence, W. Qian, and A.J. Melmed. Experimental Low-voltage Point-projection Microscopy and its Possibilities. *Ultramicroscopy*, 52:437–477, 1993.

[SS87] R. Salmon and M. Slater. *Computer Graphics, Systems and Concepts*. Addison-Wesley, Wokingham, 1987.

[SS93] M. Schäublin and P. Stadelmann. A Method for Simulating Electron Micrscope Dislocation Images. *Mater. Sci. Eng. A*, 164:373–378, 1993.

[SSE82] W.O. Saxton, D.J. Smith, and S.J. Erasmus. Procedures for Focusing, Stigmating and Alignment in High Resolution Microscopy. *J. Microsc.*, 130:187–201, 1982.

[ST61] H. Sato and S. Toth. Effect of Additional Elements on the Period of CuAu II and the Origin of the Long Period Superlattice. *Phys. Rev.*, 124:1833–1847, 1961.

[ST83] J.C.H. Spence and J. Taftö. ALCHEMI: a New Technique for Locating Atoms in Small Crystals. *J. Microscopy*, 130:147–154, 1983.

[Sta87] P.A. Stadelmann. EMS – A Software Package for Electron Diffraction Analysis and HREM Image Simulation in Materials Science. *Ultramicroscopy*, 21:131–146, 1987.

[Sto91] G.J. Stoney. Cause of Double Lines and of Equidistant Sattelites in the Spectra of Gases. *Transactions of the Royal Dublin Society*, 4:563–569, 1891.

[Sto70] W.M. Stobbs. The Weak Beam Technique. In S. Amelinckx, R. Gevers, G. Remaut, and J. Van Landuyt, editors, *Modern Diffraction and Imaging Techniques in Materials Science*, pages 591–645. North-Holland, Amsterdam, 1970.

[Str99] J.W. Baron Rayleigh Strutt. In *Scientific Papers by John William Strutt, Baron Rayleigh*, volume 1, pages 1869–1881. Cambridge University Press, Cambridge, 1899.

[Str58] A.N. Stroh. Dislocations and Cracks in Anisotropic Elasticity,. *Phil. Mag.*, 3:625–646, 1958.

[Stu62] L. Sturkey. The Calculation of Electron Diffraction Intensities. *Proc. Phys. Soc.*, 80:321–354, 1962.

[Suz77] M. Suzuki. On the Convergence of Exponential Operators – the Zassenhaus Formula, BCH Formula and Systematic Approximants. *Commun. Math. Phys.*, 57:193–200, 1977.

[SV83] J.W. Steeds and R. Vincent. Use of High-Symmetry Zone Axes in Electron Diffraction in Determining Crystal Point and Space Groups. *J. Appl. Cryst.*, 16:317–324, 1983.

[SW69] T. Suzuki and C.H. Wilts. Domain Wall Width Measurements in Cobalt Films by Lorentz Microscopy. *J. Appl. Phys.*, 40:1216–1217, 1969.

[SWP68] T. Suzuki, C.H. Wilts, and C.E. Patton. Lorentz Microscopy Determination of Domain-Wall Width in Thick Ferromagnetic Films. *J. Appl. Phys.*, 39:1983–1986, 1968.

[SZ92] J.C.H. Spence and J.M. Zuo. *Electron Microdiffraction*. Plenum Press, New York, 1992.

[Szi88] M. Szilagyi. *Electron and Ion Optics*. Plenum Press, New York, 1988.

[Tea83] M.R. Teague. Deterministic Phase Retrieval: a Green's Function Solution. *J. Opt. Soc. Am.*, 73:1434–1441, 1983.

[Tha92] B. Thaller. *The Dirac Equation*. Springer-Verlag, Berlin, 1992.

[Tho97] J.J. Thomson. Cathode Rays. *Phil. Mag.*, 44:311–335, 1897.

[Tho99] J.J. Thomson. The Masses of the Ions in Gases at Low Pressures. *Phil. Mag.*, 48:547–567, 1899.

[Tho66] F. Thon. Zur Defokussierungsabhängigkeit des Phasenkontrastes bei der elektronenmikroskopischen Abbildung. *Z. Naturf.*, 21:476–478, 1966.

[Thö70] A.R. Thölén. A Rapid Method for Obtaining Electron Microscope Contrast Maps of Various Lattice Defects. *Phil. Mag.*, 22:175–182, 1970.

[Tie99] P.C. Tiemeijer. Measurement of Coulomb Interactions in an Electron Beam Monochromator. *Ultramicroscopy*, 78:53–62, 1999.

[TL82] J. Taftö and Z. Liliental. Studies of the Cation Atom Distribution in $ZnCr_xFe_{2-x}O_4$ Spinels using the Channeling Effect in Electron-induced X-ray Emission. *J. Appl. Cryst.*, 15:260–265, 1982.

[TL96] L. Tang and D.E. Laughlin. Electron Diffraction Patterns of Fibrous and Lamellar Textured Polycrystalline Thin Films. I. Theory. *J. Appl. Cryst.*, 29:411–418, 1996.

[TOCL96] A. Thust, M.H.F. Overwijk, W.M.J. Coene, and M. Lentzen. Numerical Correction of Lens Aberrations in Phase-Retrieval HRTEM. *Ultramicroscopy*, 64:249–264, 1996.

[TOM+86] A. Tonomura, N. Okasabe, T. Matsuda, T. Kawasaki, J. Endo, S. Yano, and H. Yamada. Evidence for Aharonov–Bohm Effect with Magnetic Field Completely Shielded from Electron Wave. *Phys. Rev. Lett.*, 56:792–795, 1986.

[Tou61] M. Tournarie. Théorie Dynamique Rigoureuse de la Propagation Cohérente des

Electrons à travers une Lame Cristalline Absorbante. *Comptes Rend. Hebd. S. l'Acad. des Sciences*, 252:2862–2864, 1961.

[Tsa84] T. Tsakalakos, editor. *Phase Transformations in Solids*. North-Holland, New York, 1984.

[TSN83] M. Tanaka, H. Sekii, and T. Nagasawa. Space-Group Determination by Dynamic Extinction in Convergent-Beam Electron Diffraction. *Acta Crystall. A*, 39:825–837, 1983.

[TT85] M. Tanaka and M. Terauchi. *Convergent Beam Electron Diffraction*. JEOL Ltd., 1985.

[TT87] J.-C. Tolédano and P. Tolédano. *The Landau Theory of Phase Transitions: Application to Structural, Incommensurate, Magnetic, and Liquid Crystal Systems*. World Scientific, Singapore, 1987.

[TTK88] M. Tanaka, M. Terauchi, and T. Kaneyama. *Convergent Beam Electron Diffraction II*. JEOL Ltd., 1988.

[TTT94] M. Tanaka, M. Terauchi, and K. Tsuda. *Convergent Beam Electron Diffraction III*. JEOL Ltd., 1994.

[TW93] S.B. Tochilin and M.J. Whelan. Combined Basis Algorithm for HREM Wave Field Calculations. *Ultramicroscopy*, 50:313–320, 1993.

[VAJ99] E. Völkl, L.F. Allard, and D.C. Joy, editors. *Introduction to Electron Holography*. Kluwer, Dordrecht/Plenum, New York, 1999.

[Val79] U. Valdre. Electron Microscope Stage Design and Applications. *J. Microsc.*, 117:55–75, 1979.

[van92] A. van den Bos. Ultimate Resolution: a Mathematical Framework. *Ultramicroscopy*, 47:298–306, 1992.

[Var71] C.F. Varley. Experiment on the Discharge of Electricity through Rarefied Media and the Atmosphere. *Proc. Roy. Soc.*, 19:236–242, 1871.

[vBM86] S. van Smaalen, K.D. Bronsema, and J. Mahy. The Determination of the Incommensurately Modulated Structure of Niobium Tetratelluride. *Acta Crystall. B*, 42:43–50, 1986.

[VC85] P. Villars and L.D. Calvert. *Handbook of Crystallographic Data for Intermetallic Phases*. American Society for Metals, 1985.

[VD75] D. Van Dyck. The Path Integral Formalism as a New Description for the Diffraction of High-Energy Electrons in Crystals. *phys. stat. sol. (b)*, 72:312–336, 1975.

[VD76] D. Van Dyck. The Importance of Backscattering in High-Energy Electron Diffraction Calculations. *phys. stat. sol. (b)*, 71:301–308, 1976.

[VDBMA87] D. Van Dyck, D. Broddin, J. Mahy, and S. Amelinckx. Electron Diffraction of Translation Interface Modulated Structures. *phys. stat. sol. (a)*, 103:357–373, 1987.

[VDC84] D. Van Dyck and W. Coene. The Real Space Method for Dynamical Electron Diffraction Calculations in High Resolution Electron Microscopy. I. Principles of the Method. *Ultramicroscopy*, 15:29–40, 1984.

[VDC99] D. Van Dyck and J.H. Chen. Towards an Exit Wave in Closed Analytical Form. *Acta Crystall. A*, 55:212–215, 1999.

[VDdJ92] D. Van Dyck and A.F. de Jong. Ultimate Resolution and Information in Electron Microscopy: I. General Principles. *Ultramicroscopy*, 47:266–281, 1992.

[VDDRVTA77] D. Van Dyck, R. De Ridder, G. Van Tendeloo, and S. Amelinckx. A Cluster Model for the Transition State and its Study by Means of Electron Diffraction. iii. Generalisations of the Theory and Relation to the SRO Parameters. *phys. stat. sol. (a)*, 43:541–552, 1977.

[VDOdB96] D. Van Dyck and M. Op de Beeck. A Simple Intuitive Theory for Electron Diffraction. *Ultramicroscopy*, 64:99–108, 1996.

[VDW87] D. Van Dyck and Coene W. A New Procedure for Wave Function Restoration in High Resolution Electron Microscopy. *Optik*, 77:125–128, 1987.

[Vil97] P. Villars. *Pearson's Handbook (Desk Edition) Crystallographic Data for Intermetallic Phases*. American Society for Metals, 1997.

[vL35] M. von Laue. Die Fluoreszenzröntgenstrahlung von Einkristallen (mit einem Anhang über Elektronenbeugung). *Z. Phys.*, 23:705–746, 1935.

[VLGA65] J. Van Landuyt, R. Gevers, and S. Amelinckx. Dynamical Theory of the Images of Microtwins as Observed in the Electron Microscope. *phys. stat. sol.*, 9:135–155, 1965.

[VLGA66] J. Van Landuyt, R. Gevers, and S. Amelinckx. The Fine Structure of Spots in Electron Diffraction Resulting from the Presence of Planar Interfaces and Dislocations. II. Observations on Crystals Containing Stacking Faults. *phys. stat. sol.*, 18:363–378, 1966.

[vLR36] M. von Laue and K.-H. Riewe. Der Kristallformfaktor für das Oktaeder. *Zeitschrift für Kristallographie*, 55:408–420, 1936.

[Voi87] W. Voigt. Über das Doppler'sche Princip. *Goett. Nachr.*, 2:41–52, 1887.

[VTA74] G. Van Tendeloo and S. Amelinckx. Group-Theoretical Considerations Concerning Domain Formation in Ordered Alloys. *Acta Crystall. A*, 30:431–439, 1974.

[VTZA88] G. Van Tendeloo, H.W. Zandbergen, and S. Amelinckx. Electron Diffraction and High Resolution Electron Microscopic Study of the 20 K Superconducting Phase in the Bi-Sr-Cu-O System. *Sol. St. Comm.*, 66:927–930, 1988.

[VZDG02] V.V. Volkov, Y. Zhu, and M. De Graef. A New Symmetrized Solution for Phase Retrieval Using the Transport of Intensity Equation. *Micron*, 33:411–416, 2002.

[Wan79] R.K. Wangsness. *Electromagnetic Fields*. Wiley, New York, 1979.

[Wan95] Z.L. Wang. *Elastic and Inelastic Scattering in Electron Diffraction and Imaging*. Plenum Press, New York, 1995.

[WC96] D.B. Williams and C.B. Carter. *Transmission Electron Microscopy, a Textbook for Materials Science*. Plenum Press, New York, 1996.

[Wey22] Hermann Weyl. *Space–Time–Matter*. Dover, New York, reprint of English translation of Fourth Edition published in 1922.

[WHHB57] M.J. Whelan, P.B. Hirsch, R.W. Horne, and W. Bollman. Dislocations and Stacking Faults in Stainless Steel. *Proc. Roy. Soc.*, 240:524–538, 1957.

[Wie97] J.E. Wiechert. Wesen der Elektric. und Exper. über Kathodenstrahlen. *Schriften der Phys.-Ökonomischen Ges. zu Königsberg (Abh)*, 38:3–19, 1897.

[WJLC01] M. Wuttig, L. Jian-Li, and C. Craciunescu. A New Ferromagnetic Shape Memory Alloy System. *Scripta Mater.*, 44:2393–2397, 2001.

[WK91] A. Weickenmeier and H. Kohl. Computation of Absorptive Form Factors for High-Energy Electron Diffraction. *Acta Crystall. A*, 47:590–597, 1991.

[WLR53] M.S. Wechsler, D.S. Lieberman, and T.A. Read. On the Theory of the Formation of Martensite. *Trans. AIME*, 197:1503–1515, 1953.

[Woh71] D. Wohlleben. Magnetic Phase Contrast. In U. Valdrè, editor, *Electron Microscopy in Material Science*, pages 713–757. Academic Press, New York, 1971.

[WP99] A.J.C. Wilson and E. Prince, editors. *International Tables for Crystallography, Volume C: Mathematical, Physical and Chemical Tables (2nd edition)*. Kluwer, Dordrecht, 1999.

[WR72] M. Wilkens and M. Rühle. Black-White Contrast Figures from Small Dislocation Loops. *phys. stat. sol. (b)*, 49:749–760, 1972.

[Wre72] R.C. Wrede. *Introduction to Vector and Tensor Analysis*. Dover, New York, 1972.

[Wyc63] R.W.G. Wyckoff. *Crystal Structures (Volumes 1–6)*. Interscience Publishers, New York, 2nd edition, 1963.

[Yos57] H. Yoshioka. Effect of Inelastic Waves on Electron Diffraction. *J. Phys. Soc. Japan*, 12:618–628, 1957.

[Zei92] E. Zeitler. The Photographic Emulsion as Analog Recorder for Electrons. *Ultramicroscopy*, 46:405–416, 1992.

[Zuo96] J.M. Zuo. Electron Detection Characteristics of Slow-scan CCD Camera. *Ultramicroscopy*, 66:21–33, 1996.

[ZWS+78] F. Zemlin, K. Weiss, P. Schiske, W. Kunath, and K.-H. Herrmann. Coma-free Alignment of High Resolution Electron Microscopes with the Aid of Optical Diffractograms. *Ultramicroscopy*, 3:49–60, 1978.

[ZWT99] Y. Zhu, L. Wu, and J. Taftö. A Study of Charge Distribution in $Bi_2Sr_2CaCu_2O_{8+\delta}$ Superconductors using Novel Electron-Diffraction and Imaging Techniques. *J. Microscopy*, 194:21–29, 1999.

Index